Reference Series in Phytochemistry

Series Editors

Jean-Michel Mérillon, Faculty of Pharmaceutical Sciences, Institute of Vine and Wine Sciences, University of Bordeaux, Villenave d'Ornon, France

Kishan Gopal Ramawat, Department of Botany, University College of Science, M. L. Sukhadia University, Udaipur, Rajasthan, India

This series provides a platform for essential information on plant metabolites and phytochemicals, their chemistry, properties, applications, and methods.

By the strictest definition, phytochemicals are chemicals derived from plants. However, the term is often also used to describe the large number of secondary metabolic compounds found in and derived from plants. These metabolites exhibit a number of nutritional and protective functions for human wellbeing and are used e.g. as colorants, fragrances and flavorings, amino acids, pharmaceuticals, hormones, vitamins and agrochemicals.

The series offers extensive information on various topics and aspects of phytochemicals, including their potential use in natural medicine, their ecological role, role as chemo-preventers and, in the context of plant defense, their importance for pathogen adaptation and disease resistance. The respective volumes also provide information on methods, e.g. for metabolomics, genetic engineering of pathways, molecular farming, and obtaining metabolites from lower organisms and marine organisms besides higher plants. Accordingly, they will be of great interest to readers in various fields, from chemistry, biology and biotechnology, to pharmacognosy, pharmacology, botany and medicine.

The Reference Series in Phytochemistry is indexed in Scopus.

More information about this series at http://link.springer.com/series/13872

Jean-Michel Mérillon • Hippolyte Kodja
Editors

Orchids Phytochemistry, Biology and Horticulture

Fundamentals and Applications

With 122 Figures and 50 Tables

Springer

Editors
Jean-Michel Mérillon
Faculty of Pharmaceutical Sciences
Institute of Vine and Wine Sciences
University of Bordeaux
Villenave d'Ornon, France

Hippolyte Kodja
Qualisud, Université de La Réunion
Université Montpellier
CIRAD, Institut Agro, Avignon Université
Montpellier, France

ISSN 2511-834X ISSN 2511-8358 (electronic)
ISBN 978-3-030-38391-6 ISBN 978-3-030-38392-3 (eBook)
ISBN 978-3-030-38393-0 (print and electronic bundle)
https://doi.org/10.1007/978-3-030-38392-3

This Springer imprint is published by the registered company Springer Nature Switzerland AG.
The registered company address is: Gewerbestrasse 11, 6330 Cham, Switzerland

Preface

Orchidaceae is one of the largest flowering plant families along with Asteraceae, comprising of 28000 species in about 1000 genera. Orchids have attained significant economic importance as potted plants and cut flowers due to very attractive flowers in many species, and are finding their way in cosmetic, food (as fragrance, e.g., vanilla), and medicine industry. This book encompasses all aspect of biodiversity, biology, biotechnology, and varied applications of orchids from horticulture to medicine.

The book is a timely compilation of recent developments on orchids and is divided into six parts, covering the entire gamut of orchid research and applications:

Part I: Biogeography, Biodiversity, and Environmental Factors
Part II: Biology
Part III: Horticulture
Part IV: Agri-food Applications
Part V: Phytochemistry and Medicinal Properties
Part VI: Cosmetic Applications

The 26 well-illustrated chapters are prepared by experts working in this field, who have been selected from all over the world. This book is envisioned as a reference work providing comprehensive information on orchids. It is intended to serve the needs of graduate students, scholars, and researchers in the field of botany, horticulture, pharmacy, cosmetology, biotechnology, and phytochemistry; industrial scientists; and those involved in marketing flowers and phytochemicals, plants, and plant extracts.

We would like to acknowledge the cooperation, patience, and support of our contributors who have put their serious efforts to ensure the high scientific quality of this book with up-to-date information. We are thankful to the staff at Springer, namely Dr. S. Blago and J. Klute, for their professional support in this project.

Villenave d'Ornon, France
Saint Denis, La Réunion, France
January 2022

Jean-Michel Mérillon
Hippolyte Kodja
Editors

Contents

About the Editors

Professor Dr. Jean-Michel Mérillon received his MPharma (1979) and PhD (1984) from the University of Tours, France. He joined the University of Tours as assistant professor in 1981, became associate professor in 1987. In 1993, he moved to the Faculty of Pharmacy, University of Bordeaux, France, accepting a position as full professor. He has been the director of the research group on biologically active plant substances for over 15 years, at the Institute of Vine and Wine Sciences (University of Bordeaux, France), which comprises 25 scientists and research students. The group has been working on phenolic compounds from vine and wine for many years, mainly complex stilbenes and their involvement in health. He is involved in developing courses on plant biology, natural bioactive compounds, and biotechnology. Prof. Mérillon has published more than 180 research papers in internationally recognized journals, and has co-edited books and reference works on secondary metabolites and biotechnology. In 2004, he founded the technology transfer unit "Polyphenols Biotech," providing support for R&D programs for SMEs and major groups in the cosmetic, pharmaceutical, agricultural, and health-nutrition sectors. He is currently the manager of this unit.

Professor Dr. Hippolyte Kodja received his PhD in plant cell biotechnology from the Faculty of Pharmacy, University of Tours, France, in 1988, and later, he joined the University of La Reunion as Professor of Plant Physiology. With a vast teaching and research experience, Prof. Kodja has been exploring the contribution of fungal and bacterial endophytes of vanilla on biosynthesis of vanilla flavor and aroma metabolites, and he is also interested in the characterization of diversity in vanilla flavor production from Madagascar, Maurice, and La Réunion by phytochemical, microbiology, and sensory analyses, and in the comparative inventories of mycorrhizal fungi of vanilla. Prof. Kodja has published several research papers in internationally recognized journals on plant biology, physiology, and biochemistry, and has co-authored a chapter in the volume *Fungal Metabolites* from the Reference Series in Phytochemistry.

Contributors

Anjana Adhikari-Devkota Graduate School of Pharmaceutical Sciences, Kumamoto University, Kumamoto, Japan

Thoungamba Amom Department of Biotechnology, Manipur University, Canchipur, Manipur, India

Adilson Anacleto State University of Paraná, Paranaguá, Brazil

Nandeibam Apana Department of Biotechnology, Manipur University, Canchipur, Manipur, India

Carlos Fernando Araujo-Lima Laboratory of Environmental Mutagenesis, Department of Biophysics and Biometry, Rio de Janeiro State University, Rio de Janeiro, Brazil

Roberto Alcantara Gomes Institute of Biology, Universidade do Estado do Rio de Janeiro, UERJ, Rio de Janeiro, Brazil

Cesar Arriagada Laboratorio de Biorremediación, Facultad de Ciencias Agropecuarias y Forestales, Departamento de Ciencias Forestales, Universidad de La Frontera, Temuco, Chile

Cláudia Baider The Mauritius Herbarium, Agricultural Services, Ministry of Agro-Industry and Food Security, Réduit, Mauritius

Adama Bakayoko UFR des Sciences de la Nature (SN), Université NANGUI ABROGOUA, Abidjan, Ivory Coast

Centre Suisse de Recherches Scientifiques en Côte d'Ivoire, Abidjan, Ivory Coast

Tarun Belwal College of Biosystems Engineering and Food Science, Key Laboratory for Agro-Products Postharvest Handling of Ministry of Agriculture and Rural Affairs, Zhejiang Key Laboratory for Agro-Food Processing, Zhejiang University, Hangzhou, China

Indra D. Bhatt G.B. Pant National Institute of Himalayan Environment, Almora, Uttarakhand, India

Paromik Bhattacharyya Research Centre for Plant Growth and Development, School of Life Sciences, University of KwaZulu-Natal Pietermaritzburg, Scottsville, South Africa

Hari Datta Bhattarai Central Department of Botany, Tribhuvan University, Kathmandu, Nepal

Ravish Choudhary DSST, Indian Agricultural Research Institute, New Delhi, India

Bertrand Côme La Vanilleraie, Sainte-Suzanne, La Réunion, France

Maurizio D'Auria Dipartimento di Scienze, Università della Basilicata, Potenza, Italy

M. Dayamma ICAR-National Research Centre for Orchids, Darjeeling Campus, Darjeeling, West Bengal, India

Hari Prasad Devkota Graduate School of Pharmaceutical Sciences, Kumamoto University, Kumamoto, Japan

Abhijit Dey Department of Life Sciences, Presidency University, Kolkata, India

Vladan Djordjević Faculty of Biology, Institute of Botany and Botanical Garden, University of Belgrade, Belgrade, Serbia

Simonetta Fascetti School of Agricultural Forestry, Food, and Environmental Science, Università della Basilicata, Potenza, Italy

Israel Felzenszwalb Laboratory of Environmental Mutagenesis, Department of Biophysics and Biometry, Rio de Janeiro State University, Rio de Janeiro, Brazil

Roberto Alcantara Gomes Institute of Biology, Universidade do Estado do Rio de Janeiro, UERJ, Rio de Janeiro, Brazil

F. B. Vincent Florens Tropical Island Biodiversity, Ecology and Conservation Pole of Research, Department of Biosciences and Ocean Studies, University of Mauritius, Réduit, Mauritius

Alejandra Fuentes Laboratorio de Biorremediación, Facultad de Ciencias Agropecuarias y Forestales, Departamento de Ciencias Forestales, Universidad de La Frontera, Temuco, Chile

Seeja G Department of Plant Breeding and Genetics, College of Agriculture, Ambalavayal, India

Shubhpriya Gupta Research Centre for Plant Growth and Development, School of Life Sciences, University of KwaZulu-Natal Pietermaritzburg, Scottsville, South Africa

Héctor Herrera Laboratorio de Biorremediación, Facultad de Ciencias Agropecuarias y Forestales, Departamento de Ciencias Forestales, Universidad de La Frontera, Temuco, Chile

Nipawan Jitsopakul Department of Plant Science, Textile and Design, Faculty of Agriculture and Technology, Rajamangala University of Technology Isan, Surin, Thailand

Mayuree Kanlayavattanakul School of Cosmetic Science, Mae Fah Luang University, Chiang Rai, Thailand

Phytocosmetics and Cosmeceuticals Research Group, Mae Fah Luang University, Chiang Rai, Thailand

Saranjeet Kaur Department of Chemistry, University Institute of Sciences, Chandigarh University, Mohali, Punjab, India

Shahnoo Khoyratty Natural Products Laboratory, Institute of Biology, Leiden University, Leiden, The Netherlands

Qualisud, Université de La Réunion, Université Montpellier, CIRAD, Institut Agro, Avignon Université, Montpellier, France

Pavel Kindlmann Global Change Research Institute, Academy of Sciences of the Czech Republic, České Budějovice, Czech Republic

Institute for Environmental Studies, Faculty of Science, Charles University, Benátská 2/Prague 2, Czech Republic

Hippolyte Kodja Qualisud, Université de La Réunion, Université Montpellier, CIRAD, Institut Agro, Avignon Université, Montpellier, France

Tanoh Hilaire Kouakou UFR des Sciences de la Nature (SN), Université NANGUI ABROGOUA, Abidjan, Ivory Coast

Vinay Kumar Department of Biotechnology, Modern College of Arts, Science and Commerce, Savitribai Phule Pune University, Pune, India

Rajan Logesh TIFAC-CORE in Herbal Drugs, Department of Pharmacognosy and Phytopharmacy, JSS College of Pharmacy (JSS Academy of Higher Education and Research), Udhagamandalam, Tamil Nadu, India

Nattaya Lourith School of Cosmetic Science, Mae Fah Luang University, Chiang Rai, Thailand

Phytocosmetics and Cosmeceuticals Research Group, Mae Fah Luang University, Chiang Rai, Thailand

Andrea Furtado Macedo Integrated Laboratory of Plant Biology, Department of Botany, Institute of Biosciences, Federal University of Rio de Janeiro State, UNIRIO, Rio de Janeiro, Brazil

Keshika Mahadeo Qualisud, Université de La Réunion, Université Montpellier, CIRAD, Institut Agro, Avignon Université, Montpellier, France

Laboratoire de Chimie des Produits Naturels, Université de la Réunion, Faculté des Sciences et Technologies, 15 Avenue René Cassin, CS 92 003, 97 744 St Denis Cedex 9, La Réunion, France

Djah François Malan UFR des Sciences de la Nature (SN), Université NANGUI ABROGOUA, Abidjan, Ivory Coast

N. K. Meena ICAR-National Research Center on Seed and Spices, Tabiji, Ajmer, India

Danho Fursy-Rodelec Neuba UFR des Sciences de la Nature (SN), Université NANGUI ABROGOUA, Abidjan, Ivory Coast

Potshangbam Nongdam Department of Biotechnology, Manipur University, Canchipur, Manipur, India

Noufou Doudjo Ouattara UFR des Sciences de la Nature (SN), Université NANGUI ABROGOUA, Abidjan, Ivory Coast

Centre Suisse de Recherches Scientifiques en Côte d'Ivoire, Abidjan, Ivory Coast

Ram Pal ICAR-National Research Centre for Orchids, Darjeeling Campus, Darjeeling, West Bengal, India

Tony L. Palama Université Sorbonne Paris Nord, Laboratoire de Chimie, Structures, Propriétés de Biomatériaux et d'Agents Thérapeutiques, CSPBAT, CNRS, UMR 7244, Villetaneuse, France

Bijaya Pant Central Department of Botany, Tribhuvan University, Kathmandu, Nepal

R. P. Pant ICAR- Indian Agricultural Research Institute, Pusa Campus, New Delhi, India

Mukti Ram Paudel Central Department of Botany, Tribhuvan University, Kathmandu, Nepal

Angamba Meetei Potshangbam Department of Biotechnology, Manipur University, Canchipur, Manipur, India

Sasikarn Prasongsom Pathum Wan District, Bangkok, Thailand

Aziz Purwantoro Faculty of Agriculture, Universitas Gadjah Mada, Yogyakarta, Indonesia

Ika Puspita Sari Faculty of Pharmacy, Universitas Gadjah Mada, Yogyakarta, Indonesia

Rocco Racioppi Dipartimento di Scienze, Università della Basilicata, Potenza, Italy

R. S. Rawal G.B. Pant National Institute of Himalayan Environment, Almora, Uttarakhand, India

Sandeep Rawat G.B. Pant National Institute of Himalayan Environment, Sikkim Regional Centre, Gangtok, Sikkim, India

Vito Antonio Romano Dipartimento di Scienze, Università della Basilicata, Potenza, Italy

Dušan Romportl Department of Physical Geography and Geoecology, Faculty of Science, Charles University, Albertov 6, Czech Republic

Leonardo Rosati School of Agricultural Forestry, Food, and Environmental Science, Università della Basilicata, Potenza, Italy

Sreekumar S Biotechnology and Bioinformatics Division, KSCSTE- Jawaharlal Nehru Tropical Botanic Garden and Research Institute, Thiruvananthapuram, India

Rajkumari Sanayaima DDU College, University of Delhi, New Delhi, India

Luciane Scheuer State University of Paraná, Paranaguá, Brazil

Endang Semiarti Faculty of Biology, Universitas Gadjah Mada, Yogyakarta, Indonesia

Varsha Shriram Department of Botany, Prof. Ramkrishna More Arts, Commerce and Science College, Savitribai Phule Pune University, Pune, India

D. R. Singh ICAR-National Research Centre for Orchids, Pakyong, East Sikkim, Sikkim, India

Javiera Soto Laboratorio de Biorremediación, Facultad de Ciencias Agropecuarias y Forestales, Departamento de Ciencias Forestales, Universidad de La Frontera, Temuco, Chile

Zuzana Štípková Global Change Research Institute, Academy of Sciences of the Czech Republic, České Budějovice, Czech Republic

Institute for Environmental Studies, Faculty of Science, Charles University, Benátská 2/Prague 2, Czech Republic

Renu Suyal G.B. Pant National Institute of Himalayan Environment, Almora, Uttarakhand, India

Kanchit Thammasiri Department of Plant Science, Faculty of Science, Mahidol University, Bangkok, Thailand

Leimapokpam Tikendra Department of Biotechnology, Manipur University, Canchipur, Manipur, India

Bi Fézan Honora Tra UFR des Sciences de la Nature (SN), Université NANGUI ABROGOUA, Abidjan, Ivory Coast

Spyros Tsiftsis Department of Forestry and Natural Environment, International Hellenic University, Drama, Greece

Global Change Research Institute, Academy of Science of the Czech Republic, Brno, Czech Republic

Rafael Valadares Instituto Tecnologico Vale, Belém, PA, Brazil

Johannes Van Staden Research Centre for Plant Growth and Development, School of Life Sciences, University of KwaZulu-Natal Pietermaritzburg, Scottsville, South Africa

Andrea Vasas Department of Pharmacognosy, University of Szeged, Szeged, Hungary

Rob Verpoorte Natural Products Laboratory, Institute of Biology, Leiden University, Leiden, The Netherlands

Akoua Clémentine Yao UFR des Sciences de la Nature (SN), Université NANGUI ABROGOUA, Abidjan, Ivory Coast

Centre Suisse de Recherches Scientifiques en Côte d'Ivoire, Abidjan, Ivory Coast

Part I

Biogeography, Biodiversity, and Environmental Factors

The Role of Ecological Factors in Distribution and Abundance of Terrestrial Orchids

1

Vladan Djordjević and Spyros Tsiftsis

Contents

V. Djordjević (✉)
Faculty of Biology, Institute of Botany and Botanical Garden, University of Belgrade, Belgrade, Serbia
e-mail: vdjordjevic@bio.bg.ac.rs

S. Tsiftsis
Department of Forestry and Natural Environment, International Hellenic University, Drama, Greece

Global Change Research Institute, Academy of Science of the Czech Republic, Brno, Czech Republic
e-mail: stsiftsis@for.ihu.gr

J.-M. Mérillon, H. Kodja (eds.), *Orchids Phytochemistry, Biology and Horticulture*, Reference Series in Phytochemistry, https://doi.org/10.1007/978-3-030-38392-3_4

Abstract

Distributed throughout the continents, terrestrial orchids are known for their great species richness and specificity in relation to pollinators and mycorrhizal symbionts. Moreover, a large number of them are rare and sensitive to environmental changes. This chapter is mainly focused on the terrestrial orchids of Europe and reviews the major environmental factors affecting the patterns of their distribution, abundance, and richness (elevation, latitude, longitude, area size, climatic factors, geological substrates, soil characteristics, vegetation types, effects of disturbance), as well as the significance of mycorrhizal fungi and pollination systems. Some new data, especially regarding the responses of orchids to climate change and their occurrence on specific geological and soil substrates and vegetation types, are presented. Although the distribution and abundance of terrestrial orchids are associated with the joint effects of most of the examined factors, some factors have emerged as crucial, especially on the northern and southern borders of their distribution. Furthermore, the role of environmental factors depends largely on the belowground strategies of orchids. The chapter highlights the importance of exploring the level of specialization of orchids with respect to habitat conditions as an important basis for their conservation.

Keywords

Orchidaceae · Ecology · Distribution · Elevation · Climate · Geological substrates · Soil characteristics · Vegetation · Effects of disturbance · Specialists and generalists

1 Introduction

The family Orchidaceae is one of the most species-rich families in the plant kingdom and includes approximately 26,000 species within 749 genera [1]. Epiphytic orchids of tropical and subtropical areas have the largest number of representatives, whereas terrestrial orchids comprise either one-fourth or one-third of the total number of described orchid taxa, depending on the resolution of taxonomic issues [2, 3]. Terrestrial orchids are classified into three out of five subfamilies of the family Orchidaceae (Cypripedioideae, Orchidoideae, and Epidendroideae) [4]. Representatives of the family occur on all continents, and the most important centers of their diversity are Indochina, Southwest Australia, Europe, Northern Asia, and North America [2, 3, 5].

A wide variety of life histories have been identified among terrestrial orchids, both in terms of their development, morpho-anatomical, and physiological adaptations, life expectancy and survival of adverse periods during the year and with respect to their pollination systems and mycorrhizal associations [6, 7].

All representatives of the family Orchidaceae in Europe are terrestrial, belonging to the life form of cryptophytes, i.e., geophytes. This life form category encompasses an ecological group of plants whose aboveground parts are completely extinct during the climatically unfavorable season, whereas the buds from which new sprouts emerge the next year are found in underground organs (tubers or rhizomes) in the soil. The evolutionary development of underground organs in terrestrial orchids has produced variants ranging from rhizomes to tubers (Fig. 1), the most primitive representatives of European orchids being species of the genera *Cephalanthera*, *Epipactis*, and *Cypripedium* with short rhizomes [8]. In addition to the ones mentioned above, other known European genera that include rhizomatous species are *Corallorhiza*, *Epipogium*, *Goodyera*, *Limodorum*, and *Neottia* (Fig. 1a) [9]. An important moment in the evolution of terrestrial orchids was the development of tubers, the organ for storing reserves of matter for survival during unfavorable periods of the year [2]. Among tuberoid orchids, *Pseudorchis albida* is the most primitive species, whereas representatives of the genera *Coeloglossum*, *Dactylorhiza*, *Nigritella*, and *Gymnadenia* have palmate tubers (Fig. 1b), and species of the genus *Platanthera* are characterized by fusiform tubers (Fig. 1c) [2, 8, 10, 11]. Representatives of the genera *Anacamptis*, *Herminium*, *Himantoglossum*, *Neotinea*, *Ophrys*, *Orchis*, *Spiranthes*, and *Traunsteinera* are particularly prominent among the group of orchids with ovoid tuberoids (Fig. 1d) [2, 8, 11].

Over 300 species of orchids have been reported in Europe, North Africa, and the Middle East [5]. The estimated orchid species richness in Europe varies depending on the applied taxonomic concept. According to one author, about 250 species and subspecies from 35 genera of orchids occur in Europe [12]. However, orchids occurring in Europe are not uniformly distributed throughout the area. The most important center of orchid diversity in Europe is the eastern Mediterranean area, i.e., the area around the Aegean Sea [5, 13, 14]. Significant centers of orchid diversity in Europe also include other parts of the Balkan Peninsula, the Alps, Sicily, and the Caucasus [5]. Among countries with a high diversity of orchids in Europe, it is especially important to single out Greece, with 193 species and subspecies [13, 14]. Considerable richness of orchid species has also been reported in Italy (*c.* 175 species and subspecies) [16]; in Turkey (*c.* 170 species and subspecies) [17]; on the Iberian Peninsula (122 species and subspecies) [19]; and in Croatia (113 species and subspecies) [15]. On the island of Corsica, 77 orchid species and a density of 88 orchid species per 1000 km^2 have been recorded, making it one of the areas with the highest known value of the orchid diversity index [18].

Ophrys, *Dactylorhiza*, *Orchis*, *Anacamptis*, and *Epipactis* are the genera with the largest number of species and subspecies in Europe [9, 12, 20]. The genus *Ophrys* includes *c.* 90 species and subspecies [21], which are distributed in Europe, North Africa, and Southwest Asia [9]. The genus *Dactylorhiza* is represented by 86 species and subspecies, which are widespread in Europe, Asia, North Africa, and Alaska, whereas the genus *Orchis* is represented by 32 species and subspecies, mostly distributed in Europe, temperate Asia, and North Africa [12, 13, 20, 21]. The genus *Anacamptis* includes 11 species and 7 subspecies, which are distributed in Europe, North Africa, and Southwest and West Asia [13, 21]. Finally, the genus

Fig. 1 Root systems of terrestrial orchids. (**a**) rhizome *Neottia ovata*, (**b**) palmate tubers *Gymnadenia conopsea*, (**c**) fusiform tubers *Platanthera bifolia*, (**d**) ovoid tubers *Anacamptis coriophora* (photos V. Djordjević)

Epipactis includes a total of 79 species and subspecies, which are widespread in Europe, Asia, North Africa, and North America [21, 22].

Terrestrial orchids are known for their complex ecology, rarity, and ability to inhabit almost all terrestrial ecosystems [3, 6]. Many representatives have limited distribution due to their mycorrhizal specificity, pollinator specialization, and special needs when it comes to habitat conditions [7]. Numerous orchids are sensitive to changes in the ecosystem's balance, especially changes in the light regime, moisture content, availability of nutrients, and the level of competition. All of these changes have a significant impact on the survival and ability of seedlings to germinate and successfully reach adulthood [3]. Due to changes in ecological factors, land-use practices, and habitat loss, the ranges and abundance of many terrestrial orchids have declined [18, 23]. Consequently, many terrestrial orchids are protected by laws and included in Red Data Books [6]. In general, rarer terrestrial species and those at high risk of extinction are geographically restricted to a small area and are characterized by a small population size, have a higher level of specialization in terms of habitat conditions, and exhibit greater dependence on mycorrhizal symbionts and pollinators [3, 24]. Understanding the role of environmental conditions in determining the distribution and abundance of orchid species is therefore necessary in order to properly and adequately organize the protection of these plants [11, 25, 26].

In this chapter, we summarize knowledge about the ecological preferences of terrestrial orchids, with major emphasis on European orchids. Our main objective was to provide an overview of environmental factors that significantly affect the distribution, richness, and abundance of terrestrial orchids. The chapter highlights the role of elevation, latitude, longitude, the species-area relationship, climatic factors, geological substrates, soil factors, vegetation types, and effects of disturbance. The importance of mycorrhizal associations and pollination systems in orchid distribution and abundance is also discussed.

2 Elevation, Latitude, Longitude, and the Orchid Species-Area Relationship

2.1 Elevation

Elevation is one of the most important factors affecting the richness and composition of orchid species [11, 27]. Investigations of orchid diversity patterns along elevation gradients have been performed mainly outside Europe, more specifically in the countries of Asia [27–29], Africa [30], and America [31, 32]. However, several studies on the relationship between elevation and orchids in Europe indicated that elevation influences orchid abundance and distribution patterns and affects the separation of ecological niches of orchid species [11, 24, 25, 33, 34]. Other studies showed that orchid breeding systems and floral traits [30] and the relative proportion of food-deceptive orchids [35] also changed with elevation. It can be asserted that the most important factors responsible for variation of orchid species richness along the elevation gradient are climatic conditions [27, 28], bearing in mind that as elevation increases, the air temperature, total atmospheric pressure, and partial pressure of all atmospheric gasses decrease, whereas precipitation, relative humidity, and UV radiation usually increase [36]. Researchers found that mean annual temperature and mean annual rainfall are the most important factors influencing the abundance of orchid species along the elevational gradient in the Yunnan province of China [28] and in the central and eastern Himalayas [27]. Other important factors affecting the variation of orchid species abundance and composition along the elevational gradient are reduction of land area per bioclimatic unit, differences of nutrient availability, changes in vegetation types, evolutionary history, ecotone effects, different kinds of disturbance, and biotic processes [30, 36].

A mid-elevational bulge in orchid diversity was noted in several orchid studies [27, 37], whereas a decrease in the number of orchid species with increasing elevation was observed less frequently [30]. In the central Balkans (western Serbia), it was found that most orchid species and subspecies occur at middle elevations (1000–1100 m) [26, 34]. In Greece, the total orchid species richness increased with elevation up to 2000 m and then slightly decreased [11]. Moreover, recent research has revealed differences in the distribution of specific life forms of orchids and their root types along the elevational gradient [11, 27, 28, 34]. In the Himalayas and China (Yunnan), researchers found that species richness peaked at a higher elevation in the case of terrestrial orchids than in that of epiphytic species [28]. In Greece, tuberous orchids showed a significantly unimodal response to the elevation gradient with a peak at *c.* 1000 m, whereas the richness of rhizomatous orchids and orchids with palmate and fusiform tubers monotonously increased with elevation [11].

It is assumed that orchid species richness is highest in the mid-elevation zones of most European countries. This can be explained by several hypotheses. The concept of the mid-domain effect (MDE) predicts that orchid richness will peak in the mid-elevation zone because geometric constraints result in an increased overlap of species ranges near the midpoint of the center of the domain [38]. The climate-gradient hypothesis predicts that maximum orchid species richness occurs at a particular elevation where the combination of growing conditions proves optimal for the species [28]. Finally, the species-area relationship (SAR) hypothesis posits that the greatest number of species grow in elevation zones that cover the largest area [28]. A smaller number of orchid species in high-elevation areas can also be attributed to lower pollinator diversity and lower pollinator visitation rates [30].

Among European orchids that prefer lower elevations, several Mediterranean species with ovoid tubers from the genera *Serapias* and *Ophrys* stand out, along with certain species from the genera *Himantoglossum*, *Anacamptis*, *Neotinea*, and *Orchis* [13, 20]. On the other hand, orchid species known to grow in the highest-elevation zones are mainly orchids with palmate and fusiform tubers – *Chamorchis alpina*, *Coeloglossum viride*, *Pseudorchis albida*, *Traunsteinera globosa*, species from the genus *Nigritella*, and many *Dactylorhiza* species (Fig. 2) [9, 20].

Recent studies have emphasized that the altitudinal ranges of terrestrial orchids in Europe can vary greatly depending on geographical regions. Thus, many species primarily characteristic of Central and Northern Europe have found similar ecological conditions at higher elevations in the southern parts of their ranges [25, 26]. For instance, the altitudinal ranges in Greece for the species *Coeloglossum viride*, *Corallorhiza trifida*, *Epipactis purpurata*, *Epipogium aphyllum*, *Neottia cordata*, *Orchis militaris*, and *Pseudorchis albida* are 700–2200 m, 800–1900 m, 1200–1500 m, 1150–1800 m, 1300–1800 m, 1200–1800 m, and 1800–1900 m,

Fig. 2 Terrestrial orchids occurring mostly in high-altitude areas. (**a**) *Coeloglossum viride*, (**b**) *Nigritella rhellicani*, (**c**) *Pseudorchis albida*, (**d**) *Traunsteinera globosa*, (**e**) *Chamorchis alpina* (**a–d** photos V. Djordjević; **e** photo M. Bobocea)

respectively [13]. By way of contrast, the altitudinal ranges of these species in Europe are 0–2970 m, 0–2350 m, 50–1400 m, 80–1900 m, 10–2300 m, 0–2000 m, and 0–2700 m, respectively [20].

Interestingly, in some studies, the influence of habitat heterogeneity on orchid diversity was expressed by the difference between elevation of the highest peak and that of the lowest point in the country [32]. Researchers found that increase in the total number of orchid species with increasing altitudinal amplitude plays an important role in tropical but not in temperate areas of Latin America [32]. The authors hypothesized that this is because in temperate countries, only a few orchid species occur in high-elevation zones, so that there is no significant increase in the number of orchid species with increasing elevation [32].

2.2 Latitude and Longitude

It should be emphasized that latitude is one of the most important determinants of orchid richness, distribution, and abundance [11, 32, 39, 40]. Several hypotheses about the significance of latitude (postulation of a latitudinal diversity gradient, the latitude niche-breadth hypothesis, and Rapoport's latitude rule) were recently tested to determine the patterns of diversity of orchid taxa [11]. Latitude and elevation have a similar effect on the climatic conditions of an area: areas at high latitudes and in high-elevation areas are both characterized by lower temperatures, greater seasonality, and higher environmental stochasticity than areas at low latitudes and in low-elevation zones [11, 36]. In general, the number of terrestrial orchid species increases with decreasing latitude from the poles to tropical areas, and this pattern is among the most consistent ones in biogeography [32]. In Europe, it is expressed from the North Pole to the Mediterranean. Thus, Northern Europe has the smallest number of orchid species, which is a consequence of its climate, geology, vegetation, and history, whereas the number of orchid species gradually increases toward the southern areas of Europe, being highest in the Mediterranean area [5, 9]. However, many specific orchid species belonging to the boreal and Central European chorological groups occur mainly in Northern and Central Europe. Among boreal orchids, *Corallorhiza trifida*, *Epipogium aphyllum*, *Goodyera repens*, and *Neottia cordata* should be emphasized. Central Europe is one of the important centers of diversity and abundance of the genera *Epipactis* and *Dactylorhiza* [9, 20, 22]. On the other hand, the Mediterranean area of Southern Europe represents the center of diversity of the genera *Ophrys*, *Serapias*, *Anacamptis*, *Orchis*, *Neotinea*, and *Himantoglossum* [9, 11, 16].

In general, latitude when viewed on a large scale has a more important role than energy availability in determining patterns of orchid diversity [40]. It is important to note that a significant influence of latitude on orchid species richness was recorded for Europe and the Americas, whereas no significant influence of latitude was identified in Africa and Asia [40]. The importance of latitude and longitude in the distribution and abundance of orchids has also been demonstrated on regional scales. Thus, the spatial distribution of Greek orchids is associated with a combination of latitude, elevation, and climate [11]. With increasing latitude, orchids in Greece were more widely distributed and had greater mean niche breadths. The same study indicated that the numbers of orchid species with palmate and fusiform tubers increased with increasing latitude and slightly decreased with increasing longitude. At the same time, the numbers of tuberous orchid species and the total number of orchid species decreased with increasing latitude and showed an inverse unimodal trend with respect to longitude [11]. In northeastern Greece and western Serbia, both latitude and longitude (but especially latitude) were found to have a significant impact on the distribution and abundance of orchids, as well as on the separation of ecological niches of orchids [24–26]. Studies of orchid species richness on islands around the world showed that effects of latitude exist, but are weak compared to those of other factors, such as island size and elevation [39]. Moreover, for relatively small archipelagos, the effects of latitude can be ignored [39].

Attention should be called to a study of morphological variation in populations of some orchid species (*Cypripedium calceolus*, *Neottia ovata*, *Platanthera bifolia*, and *Gymnadenia conopsea*) along a latitudinal gradient in three European countries, viz., Russia [Murmansk (66–70° N) and Tver (55–59° N)], the Netherlands (50–54° N), and Italy (35–47° N) [41]. This study showed that the mean values of shoot structural characters in the majority of orchid species increase from the peripheral parts to the centers of their ranges and are higher on their southern than on their northern boundaries [41]. The results of this study also indicated that the number of flowers in inflorescences and the number of bracts were the most labile traits in most investigated species, whereas *Cypripedium calceolus* had the most conservative shoot structure, with little variation between different parts of its range [41].

2.3 Orchid Species-Area Relationship

The species-area relationship hypothesis postulates that the number of species increases with size of the area. Specifically, greater habitat heterogeneity and greater availability of resources in vast areas contribute to greater species diversity. The relative importance of area size has been tested in several orchid studies [32, 39, 40]. One study treated both total areas and protected areas of 67 countries from 5 continents and demonstrated that area viewed on a large scale is always very important and that size of a protected area gives a better fit than the total area of a country in most cases [40]. However, the results of this study indicated that in Europe, where the orchid flora consists solely of terrestrial species, total area size predicted the number of orchid species better than protected area size [40]. A possible explanation of this lies in the fact that many orchids in Europe grow in semi-natural habitats such as grasslands and meadows, which are maintained by regular mowing or grazing [42, 43]. In addition, numerous orchids in Europe grow in forest ecosystems, which are often not protected [40].

3 Climate

Climatic factors strongly affect the distribution, richness, abundance, and population dynamics of terrestrial orchids [11, 28, 29, 44, 45]. The patterns of diversity of orchid species on large geographic scales are influenced mainly by macroclimate, whereas on regional and local scales, meso- and macroclimate play an important role affecting the richness and abundance of orchids [44]. Among the most important climatic factors are temperature and precipitation, as well as atmospheric humidity and light.

3.1 Temperature

The influence of temperature on orchid populations varies depending on altitude and latitude, which is especially evident when comparing the centers and boundaries of species ranges [44, 46]. From a phytogeographical point of view, it is known that in temperate regions, orchid species richness is lower in areas of cold compared to warm climatic conditions, which is consistent with the general pattern observed among vascular plant species. The greatest number of terrestrial orchid species in Europe grows in the Mediterranean area, which is characterized by warm and dry summers and mild, humid winters, whereas the smallest number of species occurs in the northernmost, coldest areas [5, 8]. However, it is unclear whether temperature, precipitation, or the combination of these two factors most strongly affects orchid distributions.

Most terrestrial orchid species occurring in Europe that can tolerate colder conditions with great success belong to the group of orchids with palmate and fusiform tubers, e.g., some species from the genera *Dactylorhiza*, *Coeloglossum*, *Gymnadenia*, *Chamorchis*, *Nigritella*, and *Pseudorchis* [2, 8, 13, 20, 47]. This is understandable considering the evolutionary development of tuberoid orchids. Specifically, the formation and development of the first orchids with palmate and fusiform tubers are associated with the Alpine orogenesis and the formation of mountain habitats with low temperatures [8]. The same author claims that these orchids significantly expanded their distribution areas at the end of the Neogene and in the Pleistocene, when temperatures were low. The fact that these orchids inhabit colder areas with high precipitation is confirmed by their phytogeographical affiliation, since they are mostly representatives of the Boreal, Central European Mountain, Arctic-Alpine, and Central European chorological groups [34]. In addition, among cold-adapted orchids, there are some rhizomatous species (*Neottia cordata* and *Corallorhiza trifida*) [48, 49] and *Malaxis monophyllos*, a species that forms one basal pseudobulb [50].

On the other hand, most orchids that best tolerate high temperatures and dry conditions belong to a group of orchids with ovoid tubers, especially species of the genera *Ophrys*, *Orchis*, *Serapias*, *Neotinea*, and *Himantoglossum*, as well as some species of the genus *Anacamptis* (e.g., *A. papilionacea* and *A. pyramidalis*) (Fig. 3). Orchids with this root system represent a terminal phase in the evolution of underground organs of terrestrial orchids that enable orchids to survive in warm and dry conditions [8, 10].

It is known that most orchid species require relatively stable temperature conditions [29], whereas it has been established that extreme temperatures within the season (high temperatures during summer and low temperatures during winter) equally adversely affect orchid populations in Europe [44, 51]. Studies about the effect of temperature on the abundance of orchids at the northern boundaries of orchid distribution in Russia showed that air temperature during the previous and current growing season strongly affects orchid populations [44, 52]. At the same time, it has been proved that temperatures both at the beginning and end of the growing season strongly influence the performance of orchid populations. Among climatic factors, seasonal temperature changes and low winter temperatures especially are factors that exert significant influence on orchid species richness in China [29]. Furthermore, scientists have determined that most terrestrial orchids in China prefer a lower temperature (10.1–13.3 °C) compared to epiphytic orchids (15.0–17.4 °C) [28].

Fig. 3 Terrestrial orchids that tolerate high temperatures and dry conditions. (**a**) *Anacamptis pyramidalis*, (**b**) *Anacamptis papilionacea*, (**c**) *Orchis italica*, (**d**) *Himantoglossum robertianum*, (**e**) *Serapias orientalis*, (**f**) *Ophrys apifera*, (**g**) *Ophrys argolica*, (**h**) *Ophrys homeri*, (**i**) *Ophrys regis-ferdinandii*, (**j**) *Ophrys umbilicata* (**a**, **f** photos V. Djordjević; **b**–**e**, **g**–**j** photos S. Tsiftsis)

Air temperature affects germination and seedling establishment, photosynthetic activity, the flowering period, fruiting, abundance, and the density of orchid populations [33, 53–57]. Climatic variables, including air temperature, have a significant influence on the fruiting of orchids since they are related to weather conditions favorable for pollinating insects [53]. Studies have shown that very warm weather in spring and early summer can lead to insufficient water supply, which causes abortion of inflorescences in *Dactylorhiza sambucina* [55], *Himantoglossum hircinum* [58], and *Ophrys apifera* [59]. During our field investigations, we found that this is also the case with the *Epipactis* taxa occurring in Southern Europe. Specifically, under warm climatic conditions during late spring and early summer, flowers of these species do not open and in several cases dry up and fall on the ground without being pollinated. Contrary to the situation in spring and summer, low temperatures in winter may negatively influence orchid populations, since seed germination and flowering of orchids are sensitive to temperature [27]. Moreover, most pollinators of orchids are specific insects that are very sensitive to winter coldness [29].

A study in the north of Russia indicates that the abundance of *Dactylorhiza maculata*, *Gymnadenia conopsea*, *Neottia ovata*, and *Epipogium aphyllum* is negatively correlated with the rise in temperature in the previous few seasons, indicating that lower temperatures are favorable for these orchids [44]. Furthermore, a study from the Czech Republic shows that *Dactylorhiza fuchsii* is very likely to be present in areas with more frost days per year and that the species is almost absent in areas where there are only a few frost days per year [60]. Cool seasons may have positive effects on the performance of orchid species by exerting negative influence on their competitors [54, 61]. Furthermore, species of the genus *Dactylorhiza* and many other European orchid species require a chilling period to develop their leaves, whereas most species of the genera *Serapias* and *Ophrys* from Southern Europe do not require vernalization [62]. In Central Norway, a positive and statistically significant correlation was found between the population density of *Dactylorhiza lapponica* and temperature of the previous growing season [54].

3.2 Precipitation

The results of numerous studies indicate that a second, very important climatic factor that affects the distribution patterns of orchids, their population dynamics, flowering period, and the height of individuals is the amount of precipitation [63]. Scientists have found that reduced precipitation combined with high temperatures, especially in May and June, are the most important factors that lead to decrease in the size of individuals and decrease in the number of individuals of *Himantoglossum hircinum* at the southern boundaries of its distribution [46]. The same authors found that drought during May and June may lead to a reduction in the flowering of this species, whereas an increased amount of precipitation in May most likely leads to an increase of photosynthetic efficiency, resulting in higher heights of individuals, higher probabilities of flowering, and increased flower production. The importance of precipitation and its effects on orchid populations were also demonstrated by an investigation in England, where it was established that height of the inflorescence and the number of flowers in *Ophrys apifera* were positively correlated with the amount of precipitation in the current and previous growing season [63]. A positive correlation between annual precipitation of the previous growing season and the number of flowering specimens and flowers was also found for *Himantoglossum adriaticum*, whereas these traits are negatively correlated with the number of frost days in the previous growing season [64]. It has been shown that the probability of flowering of *Orchis simia* strongly depends mainly on weather conditions in the current year [65], whereas according to the results of a study conducted in Sweden, the flowering of *Dactylorhiza sambucina* and *Neottia ovata* is negatively correlated with the summer drought period of the previous growing season [55]. Similarly, a decline in the proportion of flowering individuals of *Neottia ovata* in a particular year as a result of drought in the previous year was reported by Brzosko [66]. A negative effect of drought in combination with high temperature on future growth and flowering was also found in *Herminium monorchis* [67] and *Gymnadenia conopsea* [54], whereas the typical wetland species *Epipactis palustris* does not tolerate drought [68]. In general, drought is linked not only with soil moisture but also with the loss of the orchid mycorrhizal fungi [61]. Changes in precipitation and temperature directly affect orchid seedlings and indirectly influence the carbon source and availability of fungi [56].

Several scientists have investigated the impact of precipitation and temperature on the morphological characteristics of individual orchids. Thus, it was determined that June precipitation is positively correlated and May precipitation negatively associated with the total leaf area of *Dactylorhiza majalis*, whereas a highly significant positive correlation was determined between May temperatures and total leaf area of this orchid species [42]. Furthermore, lip size and spur length in *Himantoglossum jankae* were found to be influenced by precipitation during the warmest quarter, larger lips being recorded in Greek areas with more precipitation during the flowering period of this species [69].

A study in the north of Russia showed that the correlation between the number of individuals and snow depth in most populations of the species *Cypripedium calceolus*, *Dactylorhiza maculata*, *Platanthera bifolia*, and *Gymnadenia conopsea* was negative but the abundance of populations of the two species *Dactylorhiza maculata* and *Coeloglossum viride* was positively correlated with snow depth [44].

3.3 Atmospheric Humidity

The relationship between atmospheric humidity and abundance of orchid populations has been insufficiently explored. However, the results indicate that different species react differently to changes in atmospheric humidity. Specifically, according to the results of an investigation conducted at the northern border of orchid populations in Russia, the number of individuals of *Dactylorhiza maculata* and *D. incarnata* was negatively correlated with atmospheric humidity recorded in the previous growing season, whereas populations of *Gymnadenia conopsea* and *Cypripedium calceolus* showed a positive correlation [44]. The same author found that high atmospheric humidity in the current growing season negatively influences the abundance of many orchid populations and that an especially negative correlation between these parameters was recorded in *Neottia ovata*. Some authors have determined favorable values of relative humidity for the cultivation of European terrestrial orchids in greenhouses. Thus, the recommended humidity for orchids is 50–60% in winter and 60–70% in summer [70].

3.4 Light

The light regime is also an important factor in defining patterns of orchid distribution and abundance [60, 71–73]. This explains the abundance of orchids mainly at the microhabitat level, i.e., in an area smaller than 4 m^2 [72]. Studies have shown that morphological and physiological characteristics of terrestrial orchids, as well as the density of orchid populations, depend on the light regime of their habitats [57, 71, 74]. It is known that sunlight is one of the major environmental factors influencing photosynthesis, growth, and reproduction of terrestrial orchids [75]. However, specific light requirements of orchid species may depend on their life form, developmental phase, and nutritive regime [57]. Furthermore, light plays a significant role in determining the population dynamics and flowering patterns of orchid species [76]. Moreover, specific morphological features were found to be related to light conditions of the studied sites. It has been discovered that differences of light conditions from site to site can cause differences in length of the inflorescence and the number of flowers in specific orchids, but this was not always the case. The hypothesis of light influence is supported by findings that individuals of *Orchis punctulata* occurring in shade and semi-shade in forest habitats had larger inflorescences with more flowers compared to those recorded under conditions of full light in grassland habitats [77]. In contrast, no significant differences were recorded in the cases of *Orchis purpurea* and *O. mascula* [77–80].

It is assumed that most terrestrial orchids grow at sites with illumination of 50–100% [49], whereas a significantly smaller number of orchid species grow in habitats with conditions of deep shade (e.g., *Cephalanthera damasonium*, *Corallorhiza trifida*, *Epipogium aphyllum*, *Epipactis helleborine*, *E. purpurata*, *Neottia cordata*, and *Goodyera repens*) (Fig. 4). Among the orchids that can tolerate shade conditions, *Neottia nidus-avis* and *Epipogium aphyllum* are characterized by special ecological and biological adaptations. They do not photosynthesize and can be said to be extremely adapted to heterotrophy in view of the fact that throughout their life cycle they depend on mycorrhizal symbionts and are associated with basidiomycete fungi which form ectomycorrhizae on tree roots [81, 82]. In general, mycoheterotrophic orchids are usually independent of the light regime [57]. However, in the case of partially mycoheterotrophic orchids, low illumination leads to strong mycoheterotrophy, whereas greater illumination stimulates orchids to autotrophy [83]. It should be noted that the photosynthetic apparatus of orchid

Fig. 4 Terrestrial orchids growing in habitats with deep shade conditions. (**a**) *Neottia nidus-avis*, (**b**) *Epipogium aphyllum*, (**c**) *Goodyera repens*, (**d**) *Corallorhiza trifida* (**a**–**c** photos V. Djordjević; **d** photo S. Tsiftsis)

species inhabiting a forest floor can acclimate to reduced illumination by changing the Chl a/b ratio to manage coordination between photosystem II (PSII) and photosystem I (PSI) [57, 75].

Some species that require full light conditions are *Anacamptis morio* [84], *Epipactis palustris* [68], and *Pseudorchis albida* [85]. On the other hand, a significant number of orchids have a wide ecological valence in relation to light (e.g., *Epipactis helleborine*, *Cephalanthera longifolia*, *Dactylorhiza fuchsii*, *Neottia ovata*, and *Platanthera bifolia*) [9, 73]. A good example is *Cypripedium guttatum*, which grows in both shady and open habitats where illumination can vary between 22 and 76% of full sunlight [75]. The successful acclimatization of this species to a wide range of light levels probably allows for its wide geographical distribution [75]. These are general trends, but in some cases, especially at the boundaries of species ranges, orchids can have specific ecological requirements. Although it is known that *Platanthera bifolia* occurs both in open habitats with full light conditions and at shady sites within forest ecosystems, a study from the Jeseníky Mountains in the Czech Republic showed that solar radiation is the most important factor associated with distribution of the given species, which occurs mainly in places with low values of this factor [60]. Furthermore, on the southernmost border of the range of *Pseudorchis albida* in Europe (northeastern Greece), this species is found only in semi-shadow and shadow conditions of dense *Pinus heldreichii* forest patches on north-facing slopes, whereas it is absent from well-lit neighboring sites [86]. In addition, the well-known light-demanding species *Epipactis palustris*, which rarely occurs at places where the illumination in summer falls to below 40% of full sunlight, in some cases was found to grow in damp forests and scrub ecosystems where light penetration is significantly lower than in open habitats [68]. Recent studies from Serbia showed that certain typical grassland species (e.g., *Anacamptis morio*, *A. pyramidalis*, *Dactylorhiza maculata* subsp. *transsilvanica*, *Gymnadenia conopsea*, *Neotinea ustulata*, and *N. tridentata*) also occur in semi-shadow and even shadow conditions in forests of *Pinus sylvestris*, *P. nigra*, *Quercus*, and *Ostrya carpinifolia* [34]. Such examples are not rare in the existing literature, and within the range of distribution of each species, differences between sites with respect to ecological requirements are often recorded.

The way that light conditions affect orchid species can also be seen in cases where habitats are altered due to specific natural or artificial changes (e.g., natural successions of vegetation and management treatments). For example, natural deforestation and a certain degree of disturbance of forest ecosystems, leading to opening of the forest structure, positively influence the development of populations of some forest orchids [73, 87, 88]. Thus, the probability of flowering and fruiting of *Cypripedium calceolus* increased at intensively harvested sites of spruce forests, whereas its survival and population density increased at moderately harvested broadleaf forest sites [89]. On the other hand, reduction of light intensity caused by closing of the forest structure leads to an extension of the dormant period and postponement of the flowering period in some orchids (*Cephalanthera rubra* and *Cypripedium calceolus*) [74]. Furthermore, it has been found that weakened illumination leads to a reduction of flower production in the species *Cephalanthera longifolia* and to a decrease in the height of individuals of *Cypripedium calceolus* [90].

3.5 Orchid Species Distribution and Its Patterns in a World Subjected to Climate Change

In the last few decades, we have been experiencing the phenomenon of climate change, which affects different regions of the world differently. The most commonly observed consequences are increasing temperature, less pronounced seasonality, and lower environmental stochasticity. Influencing different areas of the world, all these changes, regardless of their degree, are expected to strongly affect the distribution of orchids and their population dynamics [91, 92]. Recent studies explore the possibility of predicting distributional, ecological, and evolutionary consequences of climate change [91–95]. A study in southwestern China showed that climate warming in the region, along with a reduced level of soil moisture, has a negative influence on most of the investigated orchid species [93]. The same authors found that orchid populations that occur on limestone are especially highly subject to the danger of extinction due to the lack of high places to which they can migrate, whereas heavy rainfall can initiate stronger erosion, which also adversely affects orchid populations [93]. However, a study treating the pattern of distribution of Sardinian orchids under conditions of climate change has shown that a consequence of the trend of increasing temperature and decreasing precipitation is a widening of areas suitable for orchids [95]. Moreover, although prognoses regarding climate change and the abundances of orchid populations on the northern boundaries of orchid distribution in Russia (Murmansk region) show that global warming is generally favorable for many orchid species, the predicted global warming and aridization could have negative impacts on the species *Dactylorhiza incarnata*, *Hammarbya paludosa*, and *Epipogium aphyllum*, since they are sensitive to changes in moisture conditions [44].

Some authors used herbarium records to explore climatic patterns of orchid diversification [96] and clarify phenological cues that reveal the consequences of climate change on pollination and reproductive success [97, 98]. Scientists have found that orchids generally flower earlier than in the past in Hungary and that deceptive, autogamous, early flowering and long-lived terrestrial orchids with mainly Mediterranean distributions (e.g., *Orchis simia* and *Anacamptis pyramidalis*) follow global warming more closely, whereas nectar-rewarding, later-flowering, and short-lived orchids with non-Mediterranean distributions (e.g., *Coeloglossum viride*) do not respond or respond less markedly to changing climate [97]. The first indication that climate change upsets the relationships between orchid species and their pollinators was obtained, thanks to a study led by Prof. Michael Hutchings at the University of Sussex, who studied how rising temperatures since the mid-seventeenth century have disturbed the relationship between the deceptive orchid *Ophrys sphegodes* and its pollinator – male bees of the species *Andrena nigroaenea* [98]. The results of this study showed that global warming has changed the timing of phenological events that are critical to the reproductive success of the given orchid species. For optimal pollination of this orchid species, male bees have to emerge before female bees and before orchid flowering, whereas orchid flowering has to occur before female bees emerge. The increase in temperature led to earlier orchid flowering and earlier flying of male and female bees. However, the timing of these events did not change to the same extent. To be specific, female bees now fly before orchid flowering occurs and male bees consequently will mate with female bees rather than pseudocopulate with the orchid flower. The pollination of this orchid species therefore is less likely today than during the spring season in the past when temperatures were lower, suggesting that global warming will increase the period in which this orchid species experiences reproductive failure [98].

Recent studies explore the importance of climatic variables in determining the distribution of orchid species [11, 29, 45, 60, 99]. Climatic data freely available through specific databases (e.g., WorldClim database) [100, 101] constitute a very useful tool in exploring trends of orchid distribution and the effects of climatic conditions. Thus, most recent research carried out in Greece showed unimodal associations between the total number of orchid species and the maximum temperature in the warmest month and indicated that the number of rhizomatous and intermediate orchid species decreases with increasing maximum temperature in the warmest month [11]. Moreover, this study showed that the number of rhizomatous orchids and ones with palmate and fusiform tubers is greatest in the areas of Greece with the coldest winters, the highest orchid richness occurring in areas with harsh and mild winters, whereas the lowest orchid richness is in areas with moderately cold winters. Furthermore, scientists explored the probability of orchid occurrence in relation to climatic variables such as annual precipitation and annual mean temperature on Crete [99]. The results of this study showed *Anacamptis coriophora* subsp. *fragrans*, *A. papilionacea* subsp. *heroica*, *Ophrys bombyliflora*, *O. mammosa*, *Serapias bergonii*, and *S. orientalis* to be highly positively correlated with the mean annual temperature, whereas *Epipactis cretica*, *Cephalanthera cucullata*, *C. damasonium*, *Neotinea tridentata*, and *Orchis prisca* were highly negatively correlated [99]. Moreover, these latter orchids were also found to prefer habitats with high annual precipitation. Important studies have also explored the key climatic factors that affect survival of orchid species on the southernmost border of their distribution in Europe. For example, the most significant factors for the distribution of *Neottia cordata* on the southernmost border of its distribution were found to be precipitation during the warmest quarter and the seasonality of precipitation, thereby highlighting the importance of the amount of summer rainfall [45].

4 Geological Substrates and Soil Properties

The geological substrate and soil characteristics are important factors affecting the distribution and abundance of orchids primarily at the regional and local levels [11, 24–26, 99, 102–104]. In general, variation in the availability of soil resources (water and nutrients) across geological substrates significantly influences the richness and composition of orchid species [24, 26, 34].

4.1 Geological Substrates

Calcareous geological substrates and soils can be considered to be the most important substrates on which terrestrial orchids grow in Europe, which means that the greatest number of species occur on limestone, chalk, dolomite, and carbonate clastites (carbonate shales, sands, sandstones, clays, gravels, conglomerates, and marls) [11, 13, 25, 105–107]. This pattern represents one of the most consistent features of species richness, and it also applies to other vascular plants in Europe [108]. While the richness of orchids on carbonate substrates is very prominent in Central Europe, the patterns of distribution and abundance of orchids in Southern and Southeast Europe have somewhat different shapes. To be specific, recent research from the central Balkans (western Serbia) and Greece indicated that a large number of orchid species grow on non-calcareous bedrock types, i.e., on various silicate substrates, including felsic, intermediate, mafic, and even ultramafic igneous rocks, as well as on metamorphic and silicate sedimentary rocks [24–26, 107]. These studies highlighted the important role of the gradient of geological substrates in separating niches of orchid species. Differences in the chemical and physical composition of geological substrates and soils not only lead to differences in the richness and composition of species; they also affect the size of orchid populations [24].

Terrestrial orchids that grow exclusively or mainly on carbonate geological substrates include the following species: *Anacamptis papilionacea*, *Orchis militaris*, *O. pauciflora*, *O. anthropophora*, *Gymnadenia conopsea*, *G. odoratissima*, *Nigritella rhellicani*, *Neotinea ustulata*, *Dactylorhiza fuchsii*, *Epipogium aphyllum*, *Epipactis palustris*, *E. purpurata*, *E. muelleri*, and many *Ophrys* and *Himantoglossum* species (Fig. 5) [9, 12, 20, 61, 68, 109, 110]. However, most of these species were also recorded on different silicate substrates in Greece [13, 25], in the central Balkans [24, 26, 34], and in Italy [16]. Some orchid species are known to occur on a large number of bedrock types, indicating their great ecological plasticity (*Anacamptis coriophora*, *Cephalanthera longifolia*, *Dactylorhiza sambucina*, *Gymnadenia conopsea*, *Epipactis helleborine*, *Neottia nidus-avis*, *N. ovata*, *Platanthera bifolia*, etc.) [9, 12, 26, 34, 107, 111, 112].

Recent studies have highlighted the importance of ophiolitic mélanges as an important bedrock type for the survival and development of many terrestrial orchids in the Balkans [26, 34, 107]. Great orchid richness on this bedrock type can be attributed to its heterogeneous composition, bearing in mind that these volcanogenic-sedimentary formations usually contain diabase (basic igneous rock) and cherts (silicate sedimentary rocks, primarily composed of quartz, the mineral form of silicon dioxide). Although it is known that Si plays an important role in the alleviation of abiotic and biotic stress, its significance for orchid growth has been insufficiently investigated. Some orchids that have significant abundance on ophiolitic mélanges are the Balkan endemic *Himantoglossum calcaratum* subsp. *calcaratum*, two subendemics (*Dactylorhiza maculata* subsp. *transsilvanica* and *D. cordigera*), certain species of Central or North European origin (*Anacamptis morio*, *Traunsteinera globosa*, *Neotinea ustulata*, *Dactylorhiza incarnata*, *Coeloglossum viride*, *Epipactis leptochila* subsp. *neglecta*, and *E. purpurata*), as well as *Dactylorhiza saccifera* [26, 34, 107].

Fig. 5 Terrestrial orchids that grow mainly or exclusively on calcareous geological substrates. (**a**) *Ophrys oestrifera* (syn. *Ophrys scolopax* subsp. *cornuta*), (**b**) *Orchis militaris*, (**c**) *Orchis anthropophora*, (**d**) *Orchis pauciflora* (**a**, **b** photos V. Djordjević; **c**, **d** photos S. Tsiftsis)

A significant number of terrestrial orchid species in the Balkans occur on metamorphic rocks, including schists, gneisses, and phyllites [25, 26, 34, 107]. Among orchid species found on these bedrock types, due to the significant abundance and frequency of findings on them, two species (*Nigritella rhellicani* and *Gymnadenia frivaldii*) are defined as indicators of these geological substrates [26]. In addition, a study from the Czech Republic shows that *Dactylorhiza fuchsii* has a high probability of occurring on phyllites [60]. Although the species *Epipactis pontica* is known to occur mainly on calcareous substrates, in recent studies populations of this species were recorded on gneisses in Bulgaria and Greece [25, 113]. Furthermore, many terrestrial orchids also occur on intermediate igneous rocks (andesite, dacite, and porphyrite) and flysch, whereas some species were found to grow on Quaternary sediments, including proluvial and alluvial deposits, eluvial-deluvial sediments, and river terraces [26, 107]. The majority of species found on Quaternary sediments prefer habitats with high soil moisture (e.g., *Dactylorhiza incarnata*, *D. saccifera*, and *Anacamptis palustris*) [26].

Recent studies show that many terrestrial orchids grow on ultramafic rocks in the Balkans [24, 26, 34, 107], Malaysia [114, 115], and Russia [116], which at first glance is unexpected considering the stressful conditions of these substrates. To be specific, soils derived from serpentine (ultramafic) rocks are known for high concentrations of Ni, Cr, and Co; low concentrations of macronutrients (P, K, and N); low Ca/Mg ratios; and properties unfavorable for plant life, such as a wide range of diurnal temperature fluctuations and poor water-retaining capacity [24]. In the central Urals (Russia), scientists found *Epipactis atrorubens* on this bedrock type [116]. Investigations in the central Balkans (western Serbia) showed that 29 orchid species and subspecies occur on ultramafic rocks, the following orchids having notably large populations on this type of substrate: *Anacamptis morio*, *Gymnadenia conopsea*, *Dactylorhiza sambucina*, *D. maculata* subsp. *transsilvanica*, and *Platanthera bifolia* (Fig. 6) [34, 107]. Interestingly, two species (*Gymnadenia conopsea* and *Anacamptis morio*) have statistically significantly larger populations in serpentine areas than in non-serpentine areas of the Valjevo Mountains in Serbia [24].

All orchid species recorded on serpentine substrates in the central Balkans are serpentine-facultative plant species, since they are also present on other bedrock types [24]. However, due to notably large populations and high frequency of occurrences, two orchids (*Dactylorhiza maculata* subsp. *transsilvanica* and *Platanthera bifolia*) are defined as indicator species of serpentine substrates in western Serbia [26]. The explanation for the existence of a significant number of orchid species and large orchid populations on this bedrock type lies in the physical and chemical properties of serpentine soils, primarily their low content of nutrients [116], considering that most orchid species are sensitive to high levels of phosphorus and nitrogen in the soil [103]. Furthermore, it is also assumed that mycorrhizal fungi play a key role in increasing tolerance to high content of heavy metals in serpentine soils.

It should be noted that serpentine substrates allow for the development of open habitats with a generally low degree of competition, which makes possible the survival of poorly competitive orchid species that have a high demand for light [24]. Furthermore, scientists found that conditions of the open structure of forest

Fig. 6 Orchid species that grow on ultramafic substrates in the central Balkans (western Serbia). (**a**) an ultramafic landscape on Mt. Zlatibor (Serbia), (**b**) *Dactylorhiza sambucina*, (**c**) *Anacamptis morio*, (**d**) *Gymnadenia conopsea*, (**e**) *Platanthera bifolia*, (**f**) *Dactylorhiza maculata* subsp. *transsilvanica* (photos V. Djordjević)

ecosystems on serpentine substrates are an important factor influencing the richness of both terrestrial and epiphytic orchids in Kinabalu Park in Malaysia [114]. These authors assume that adaptations to the stressful chemical conditions of ultramafic (serpentine) substrates may be an important factor for the high level of endemism of terrestrial orchids in this part of the world [114]. Two terrestrial orchids found on serpentine substrates in Malaysia are *Crepidium metallicum* and *Paphiopedilum dayanum* [114]. It is interesting that Rothschild's slipper orchid (*Paphiopedilum rothschildianum*), one of the most famous orchid species in the world, now widely available thanks to its propagation in culture, in fact originated and grows on serpentine substrates in Kinabalu Park in Malaysia [115]. Detailed ecological study showed that extreme substrate conditions are reflected in the leaf chemistry of this species, while experiments involving translocation onto non-ultramafic substrates were successful, indicating that substrate chemistry is not a factor limiting the growth of this species [115].

A smaller number of orchid species are recorded to grow on acidic igneous rock [107]. Species found on quartz latite include *Anacamptis coriophora*, *Cephalanthera longifolia*, *Dactylorhiza incarnata*, *D. maculata* subsp. *maculata*, *D. sambucina*, *Nigritella rhellicani*, *Gymnadenia conopsea, Coeloglossum viride*, *Neottia cordata*, and *Traunsteinera globosa* [26, 34, 111]. Ones recorded on granodiorites are mainly forest orchids such as *Cephalanthera damasonium*, *C. longifolia*, *C. rubra*, *Epipactis helleborine*, *E. microphylla*, *Limodorum abortivum*, *Neottia nidus-avis*, *Platanthera bifolia*, and *P. chlorantha* [34, 107]. Among species growing on acidic igneous rocks in Greece, many species were found to occur on rhyolite, quartz-monzonite, granodiorite, and granite (e.g., *Epipactis pontica*, *Goodyera repens*, *Neottia cordata*, etc.) [13, 25, 45].

Overall, recent studies suggest that non-calcareous bedrock types, i.e., different siliceous substrates, are favorable for many Central European and Boreal orchid species that occur in the central Balkans and Greece [25, 26, 34, 45, 107]. This can be attributed to the higher water-retaining capacity of these substrates in comparison with calcareous ones and to the fact that they are present mainly in high-altitude areas where the humidity is greater [26]. Therefore, future research should examine whether areas with silicate substrates can represent significant orchid reserves. Further research should also explore the impact of physical and chemical characteristics of the soils of different bedrock types on certain characteristics of orchids, such as their potential for accumulation of trace elements and the phytochemical properties of these plants.

4.2 Soil Properties

Physical and chemical properties of the soil are known to have a significant impact both on the growth of terrestrial orchids and on their abundance and distribution [25, 103, 104, 117]. Among numerous physical and chemical properties of the soil, scientists have primarily studied its moisture, pH, nutrients (nitrogen, phosphorus, and potassium), content of calcium and magnesium, and organic content in soils that support terrestrial orchid species (Tables 1 and 2). These properties vary in response to differences in bedrock types, climatic conditions, and vegetation, and they affect the performances of orchid populations [102, 117]. Because many orchid species are ecologically specialized to limited ranges of soil factors, adequate ecophysiological adaptations can be expected [103].

4.2.1 Soil Moisture

Soil moisture represents an important factor influencing orchid distribution patterns mainly on a microsite scale [72, 118]. Studies have shown that soil moisture up to 10 cm depths significantly affects orchid population density and distribution, whereas the absence or presence of orchids depends on soil moisture to a lesser extent below a depth 10 cm [119, 120]. Of course, this is because of the depth where orchids have their root system, which rarely exceeds 10–15 cm below the surface of the ground [62]. Detailed data on soil moisture have been obtained for some orchid species, for example, in nine forest populations of *Neottia ovata* in Belgium, where measured soil moisture ranged from 8.8% to 28.6% [121].

An important property of the soil influencing moisture content – through its retaining capacity – is soil texture, but corresponding analyses focusing exclusively on orchids are lacking in the existing literature. Apart from other advantages of its availability, dissolvable cations and nitrogen compounds can be transported through the underground flow of water. This transport can potentially affect pH and nitrogen content in the soil, and as a result the dynamics of moisture in space and time can indirectly influence the composition of vegetation, its diversity, and species distribution [118]. Moreover, the effects of soil moisture can be seen in the growth of orchids, their germination, establishment of their seedlings, diversity of mycorrhizal fungi, and soil temperature characteristics [118, 122]. As regards orchids specifically, a recent study from the central Balkans indicated that soil moisture is one of the key factors affecting the distribution and abundance of orchid species within grasslands and herbaceous wetlands [26]. Moreover, the results of this research showed that the number of orchid species and their occurrences decrease at the extreme ends of the moisture gradient and that the majority of the orchid species occur in habitats with moderately moist soils (the data follow a unimodal distribution).

However, such a unimodal trend is not the case for every region or country, as orchid trends are strongly dependent on the orchid flora of the studied region or country. In Europe, orchids can be found from totally dry sites of the Mediterranean region to moist and wet places of the North European countries [5]. Specifically, many orchid species are totally dependent on high levels of water availability, and such orchids can become extinct when their habitats dry up or when the amount of soil moisture is below a specific threshold for some months. Some of the best-known terrestrial orchids that grow exclusively on high-moisture soils in Europe are *Anacamptis palustris*, *Epipactis palustris*, *Hammarbya paludosa*, *Liparis loeselii*, *Gymnadenia frivaldii*, and most species of the genus *Dactylorhiza* [9, 26, 47]. On the other hand, orchids that tolerate dry soil conditions are most representatives of the genera *Serapias*, *Ophrys*, *Orchis*, *Himantoglossum*, and *Anacamptis* (e.g., *Anacamptis papilionacea* and *A. pyramidalis*) (Fig. 3) [9, 13, 26].

Table 1 Soil pH at sites with some terrestrial orchid species

Orchid taxon	The soil pH range	Soil pH values/literature sources
Anacamptis boryi	7.1–8.1	7.1–7.2 [131], 7.2–8.1 [132]
Anacamptis collina	6.7–8.6	7.5–8.4 [131], 6.7–8.6 [132]
Anacamptis coriophora subsp. *coriophora*	4.74–8.3	4.74–5.72 [25], 6.0–8.3 [126], 6.3 [131], 5.5–8.0 [133]
Anacamptis coriophora subsp. *fragrans*	4.58–8.1	4.58–7.74 [25], 7.7–8 [131], 8.1 [132], 7.33–7.53 [134]
Anacamptis laxiflora	3.5–8.0	4.89–7.13 [25], 7.9–8.0 [132], 3.5–5.5 [133], 7.32–7.7 [134], 5.0–8.0 [135]
Anacamptis morio subsp. *caucasica*	4.13–8.0	4.13–7.70 [25], 5.5–8.0 [133], 4.7–7.21 [134]
Anacamptis morio subsp. *morio*	4.5–8.1	4.5–7.5 [49], 5.5–8.1 [126], 4.5–8.0 [133], 4.9–6.8 [136]
Anacamptis morio subsp. *champagneuxii*	4.5–7.5	4.5–7.5 [135]
Anacamptis palustris subsp. *elegans*	5.5–8.0	5.91 [25], 7.05–7.75 [126], 5.5–8.0 [133]
Anacamptis palustris subsp. *palustris*	6.4–8.01	6.4–8.01 [126]
Anacamptis papilionacea subsp. *papilionacea*	4.67–7.43	4.67–7.43 [25], 5.04–6.4 [134]
Anacamptis pyramidalis	4.94–8.7	4.94–7.99 [25], 7.17–8.25 [126], 7.5–8.5 [131], 8.0–8.7 [132], 5.5–8.0 [133], 5.75–7.58 [134], 5.5–7.5 [135]
Calypso bulbosa	4.5–7.5	6.9 [131], 4.5–7.5 [133]
Cephalanthera damasonium	4.3–8.0	4.45–7.65 [25], 5.76–7.98 [126], 6.0 [131], 5.5–8.0 [133], 6.0–7.5 [135], 4.3–7.4 [137]
Cephalanthera longifolia	4.0–8.6	7.85 [25], 4.59–8.0 [126], 6.5–8.6 [131], 5.5–8.0 [133], 4.34–5.07–7.03 [134], 4.0–7.0 [135], 4.3–7.4 [137]
Cephalanthera rubra	4.3–8.2	4.38–7.74 [25], 5.16–7.83 [126], 5.9–8.2 [131], 5.5–8.0 [133], 4.3–7.4 [137]
Chamorchis alpina	5.5–8.0	5.5–7.5 [131], 5.5–8.0 [133]
Coeloglossum viride	3.5–8.0	4.58–7.54 [25], 4.4–8.0 [49], 5.0–7.9 [126], 5.7–7.9 [131], 3.5–7.5 [133]
Corallorhiza trifida	3.5–7.5	4.14–7.48 [25], 5.4 [117], 6.0–7.5 [126], 6.5–7.5 [131], 3.5–7.5 [133], 4.1–4.4 [137]
Cypripedium calceolus	4.5–8.26	5.7–8.26 [126], 7.3–7.9 [131], 4.5–8.0 [133], 7.1–7.4 [136], 5.4–7.4 [138]
Dactylorhiza aristata	4.5–7.5	4.5–7.5 [133]
Dactylorhiza baltica	4.5–8.0	4.5–8.0 [133]
Dactylorhiza cordigera	4.03–7.5	4.03–6.64 [25], 4.5–7.5 [133]
Dactylorhiza cruenta	4.5–7.5	6.7 [131], 4.5–7.5 [133]
Dactylorhiza elata	5.0–7.5	5.0–7.5 [135]
Dactylorhiza euxina	4.5–8.0	4.5–8.0 [133]

(continued)

Table 1 (continued)

Orchid taxon	The soil pH range	Soil pH values/literature sources
Dactylorhiza flavescens	3.5–8.0	3.5–8.0 [133]
Dactylorhiza foliosa	4.9–5.9	4.9–5.9 [131]
Dactylorhiza fuchsii	4.5–7.5	5.8–7.5 [126], 4.5–7.5 [133]
Dactylorhiza iberica	7.3–8.2	7.3–8.2 [131]
Dactylorhiza incarnata subsp. *incarnata*	4.5–8.3	5.71–5.97 [25], 5.5–8.3 [126], 4.5–7.5 [133], 5.0 [135]
Dactylorhiza incarnata subsp. *ochroleuca*	7.59–7.7	7.59–7.7 [126]
Dactylorhiza insularis	4.5–7.0	4.5–7.0 [135]
Dactylorhiza lapponica	7.5	7.5 [126]
Dactylorhiza macedonica	5.58–5.71	5.58–5.71 [25]
Dactylorhiza maculata	3.5–7.6	3.5–7.5 [133], 5–7.6 [136]
Dactylorhiza majalis	4.5–7.8	5.1–7.8 [126], 4.5–7.5 [133]
Dactylorhiza markusii	5.0–5.5	5.0–5.5 [135]
Dactylorhiza romana	4.13–8.0	4.13–4.6 [25], 6.7 [132], 5.5–8.0 [133]
Dactylorhiza russowii	4.5–8.0	4.5–8.0 [133]
Dactylorhiza saccifera	5.26–7.6	5.26–7.6 [25]
Dactylorhiza sambucina	3.5–8.0	4.29–7.68 [25], 4.2–8.0 [111], 4.3–6.4 [126], 3.5–8.0 [133], 5.0 [135], 4.9–5.0 [136], 4.0–5.6 [139]
Dactylorhiza traunsteineri	4.5–7.5	4.5–7.5 [133]
Dactylorhiza urvilleana	4.5–8.0	4.5–8.0 [133]
Epipactis albensis	4.5–7.65	4.5–7.65 [126], 6.98 [140]
Epipactis atrorubens	5.5–9.0	6.52–7.7 [25], 6.0–8.1 [126], 7.5–9.0 [131], 5.5–8.0 [133], 5.6–7.0 [141]
Epipactis bugacensis	7.9–8.1	7.9–8.1 [126]
Epipactis exilis	4.27–7.48	4.27–7.48 [25], 4.5 [126], 7.17 [134], 4.3–7.4 [137]
Epipactis futakii	5.9–7.1	5.9–7.1 [126]
Epipactis helleborine subsp. *helleborine*	4.27–8.0	4.27–7.74 [25], 5.2–7.7 [126], 4.5–8.0 [133], 7.1–7.86 [134], 4.3–7.4 [137]
Epipactis leptochila subsp. *leptochila*	5.9–7.6	5.9–7.6 [126]
Epipactis leptochila subsp. *nauosaensis*	6.63–7.36	6.63–7.36 [25]
Epipactis leptochila subsp. *neglecta*	5.1–7.5	5.1–7.5 [126]
Epipactis mecsekensis	5.2–7.7	5.2–7.7 [126]
Epipactis microphylla	5.12–8.0	5.12–7.39 [25], 5.2–7.9 [126], 6.3–7.9 [131], 5.5–8.0 [133], 5.6 [137]
Epipactis moravica	6.6–7.7	6.6–7.7 [126]
Epipactis muelleri	6.3–7.9	6.3–7.9 [126]
Epipactis nordeniorum	4.5–7.9	4.5–7.9 [126]

(continued)

Table 1 (continued)

Orchid taxon	The soil pH range	Soil pH values/literature sources
Epipactis palustris	4.97–8.5	4.97–7.94 [25], 6.6–8.3 [126], 6.5–8.5 [131], 5.5–8.0 [133]
Epipactis placentina	5.7–6.7	5.7–6.7 [126]
Epipactis pontica	4.48–7.5	4.54–5.47 [25], 5.1–7.5 [126], 4.48–5.65 [127], 5.7–6.3 [128], 6.2–7.1 [131]
Epipactis purpurata	5.2–8.0	5.2–6.8 [126], 5.5–8.0 [133]
Epipactis tallosii	5.5–7.9	5.5–7.9 [126]
Epipactis tremolsii	7.0–7.5	7.0–7.5 [135]
Epipactis voethii	7.2–7.8	7.2–7.8 [126]
Epipogium aphyllum	4.5–8.0	4.72–6.75 [25], 4.5 [82], 4.6–7.6 [126], 6.0–7.6 [131], 4.5–8.0 [133]
Goodyera repens	4.27–7.5	4.27–7.03 [25], 4.27–7.03 [126], 4.6–7.5 [131], 4.5–7.5 [133]
Gymnadenia conopsea	4.5–8.5	4.88–7.73 [25], 5.0–8.0 [126], 4.5–8.5 [131], 5.5–8.0 [133]
Gymnadenia densiflora	7.6–7.8	7.6–7.8 [126]
Gymnadenia frivaldii	4.97	4.97 [25]
Gymnadenia odoratissima	5.4–8.0	7.5–8.0 [126], 5.4–7.7 [131], >6.5 [133]
Hammarbya paludosa	3.5–7.5	4.0–5.9 [126], 4.5–6.0 [131], 3.5–7.5 [133], 4.73–5.35 [142]
Herminium monorchis	5.5–8.1	7.4–8.1 [126], 5.7–8.1 [131], 5.5–8.0 [133]
Himantoglossum adriaticum	6.3–7.85	6.3–7.5 [64], 7.35–7.85 [126]
Himantoglossum comperianum	5.5–8.0	5.5–8.0 [133]
Himantoglossum hircinum	5.0–8.0	5.0–8.0 [135], 6.9–7 [136]
Himantoglossum jankae	4.43–8.0	4.43–7.72 [25], 7.34–8.0 [126], 5.5–8.0 [133], 7.19–7.51 [134]
Himantoglossum robertianum	5.0–8.5	7.6–8.5 [131], 8.2 [132], 5.0–8.0 [135]
Limodorum abortivum	4.19–8.5	4.19–7.99 [25], 5.38–8.01 [126], 7.1–8.5 [131], 6.9 [132], 5.5–8.0 [133], 5.9 [134], 5.0–7.5 [135]
Limodorum trabutianum	5.5	5.5 [135]
Liparis loeselii	6.77–8.5	6.77–7.72 [126], 7.3–8.5 [131]
Malaxis monophyllos	3.5–8.1	6.7–8.1 [126], 6.9–7.7 [131], 3.5–8.0 [133]
Neotinea lactea	8.4	8.4 [132]
Neotinea maculata	4.0–8.3	6.2–8.3 [131], 7.4 [132], 4.0–7.5 [135]
Neotinea tridentata	4.59–8.0	4.59–7.99 [25], 5.7–8.0 [126], 5.5–8.0 [133], 5.04–7.3 [134], 6.5 [135], 7.0–7.4 [136]
Neotinea ustulata var. *aestivalis*	5.0–7.1	5.0–7.1 [126]
Neotinea ustulata var. *ustulata*	4.32–8.5	4.32–7.6 [25], 4.9–7.6 [110], 5.46–8.0 [126], 5.3–8.5 [131], 5.5–8.0 [133], 5–7.5 [135], 5.2–7.3 [136], 6.0–8.5 [143]
Neottia cordata	2.8–6.0	2.8–5.5 [48], 4.0–6.0 [131], 3.0–4.5 [133]

(continued)

Table 1 (continued)

Orchid taxon	The soil pH range	Soil pH values/literature sources
Neottia nidus-avis	4.14–8.5	4.14–7.74 [25], 5.2–8.0 [126], 6.4–8.5 [131], 5.5–8.0 [133]
Neottia ovata	4.29–8.5	4.29–7.63 [25], 5.5–7.5 [112], 5.74–7.54 [121], 4.6–7.98 [126], 6.4–8.5 [131], 4.5–7.5 [133]
Neottianthe cucullata	3.5–8.0	3.5–8.0 [133]
Nigritella rhellicani	4.5–7.5	5.23–6.18 [25], 4.5–7.5 [144]
Ophrys apifera	4.89–8.5	4.89–7.6 [25], 7.4–7.9 [126], 7.4–8.5 [131], 5.5–8.0 [133], 7.37–7.48 [134], 5.0–8.0 [135]
Ophrys apulica	7.5	7.5 [135]
Ophrys atlantica	7.5	7.5 [135]
Ophrys attica	6.5	6.5 [135]
Ophrys argolica	7.1–8.2	7.1–8.2 [131]
Ophrys bertolonii	7.3–8.4	7.3–8.4 [131], 7.5 [135], 7.83 [145]
Ophrys bombyliflora	5.0–9.4	7.1–9.4 [131], 5.0–7.5 [135]
Ophrys cretica subsp. *cretica*	8.1–8.2	8.1–8.2 [132]
Ophrys ferrum-equinum	7.0–8.2	7.5–8.2 [131], 7.0 [135]
Ophrys fuciflora s.str.	7.0–7.7	7.2–7.7 [126], 7.0–7.3 [136]
Ophrys grammica	6.12–7.85	6.12–7.85 [25]
Ophrys heldreichii	8.0–8.6	8.0–8.6 [132]
Ophrys helenae	6.5	6.5 [135]
Ophrys insectifera	5.5–8.5	7.8–8.0 [126], 7.1–8.5 [131], 5.5–8.0 [133], 7.0–7.5 [135], 7.1 [136]
Ophrys mammosa	4.87–8.1	4.87–7.85 [25], 8.1 [132], 6.5–8.0 [135]
Ophrys oestrifera (syn. *Ophrys scolopax* subsp. *cornuta*)	5.5–8.0	5.64–7.77 [25], 7.6–8.0 [126], 5.5–8.0 [133], 7.19–7.51 [134]
Ophrys reinholdii	5.15–7.8	5.15–6.29 [25], 7.5–7.8 [131], 7.0 [135]
Ophrys speculum	7.1–8.1	7.1–8.1 [131]
Ophrys sphegodes	5.0–8.3	7.2–8.3 [126], 5.0–7.5 [135]
Ophrys spruneri	6.5–8.0	6.5–8.0 [135]
Ophrys tenthredinifera	5.5–8.4	6.4–8.4 [131], 7.3 [132], 5.5–8.0 [135]
Ophrys zeusii	5.64–5.81	5.64–5.81 [25]
Orchis anatolica	6.8–8.2	6.9–8.2 [131], 6.8–7.2 [132]
Orchis anthropophora	5.5–8.6	7.0–8.6 [131], 5.5–8.0 [133], 6.5–8.0 [135]
Orchis italica	5.1–9.1	5.1–7.6 [25], 6.8–9.1 [131], 5.26–7.6 [134], 6.5–8.0 [135]
Orchis mascula subsp. *mascula*	4.18–8.5	4.18–7.99 [25], 4.8–8.5 [131], 5.5–8.0 [133], 6.9–7.27 [134], 5.0–7.0 [135]
Orchis mascula subsp. *olbiensis*	5.0–8.0	5.0–8.0 [135]
Orchis mascula subsp. *speciosa*	5.6–7.7	5.6–7.7 [126]
Orchis militaris subsp. *militaris*	5.5–9.0	7.25–7.54 [25], >7.5 [109], 6.0–8.3 [126], 7.4–9.0 [131], 5.5–8.0 [133], 7.5 [135]

(continued)

Table 1 (continued)

Orchid taxon	The soil pH range	Soil pH values/literature sources
Orchis pallens	4.83–8.0	4.83–7.54 [25], 6.4–7.5 [126], 6.7–7.0 [131], 5.5–8.0 [133], 6.3–7.3 [136]
Orchis pauciflora	7.04–8.3	7.04–7.85 [25], 7.5–8.3 [131], 8.2 [132], 8.0 [135]
Orchis provincialis	4.59–8.0	4.59–5.87 [25], 5.5–7.8 [131], 5.5–8.0 [133], 7.0 [135]
Orchis punctulata	5.5–8.0	5.5–8.0 [133]
Orchis purpurea subsp. *purpurea*	4.67–8.7	4.67–7.84 [25], 6.2–8.0 [126], 7.5–8.7 [131], 5.5–8.0 [133], 5.5–7.0 [135], 6.9–7.3 [136]
Orchis quadripunctata	5.09–8.0	5.09–7.85 [25], 7.5–8.0 [131], 7.8 [132], 7.1–7.3 [134], 6.5–8.0 [135]
Orchis simia subsp. *simia*	4.74–8.5	4.74–7.85 [25], 7.4–7.9 [126], 7.5–8.5 [131], 7.9 [132], 5.5–8.0 [133], 7.3–7.33 [134], 5.5–7.5 [135]
Orchis spitzelii subsp. *spitzelii*	5.0–8.1	7.6–8.1 [131], 5.0–8.0 [135]
Platanthera bifolia	4.3–8.0	4.45–6.73 [25], 4.6–8.0 [126], 4.3–7.5 [131], 4.5–8.0 [133]
Platanthera chlorantha	3.98–8.4	3.98–7.85 [25], 5.3–8.0 [126], 5.4–8.4 [131], 5.5–8.0 [133], 6.9–7.1 [134], 7.4 [137]
Pseudorchis albida	3.5–8.0	4.7–6.2 [131], 3.5–8.0 [133], 4.5–7.2 [146], 4.1–4.6 [147]
Serapias bergonii	4.67–7.2	4.67 [25], 5.27–7.2 [134]
Serapias cordigera subsp. *cordigera*	4.0–5.0	4.67–4.84 [25], 4.0–5.0 [135]
Serapias lingua	4.0–9.0	7.1–9.0 [131], 4.0–7.5 [135]
Serapias orientalis subsp. *orientalis*	7.0–8.0	7.0–8.0 [135]
Serapias parviflora	4.0–8.0	5.7–7.8 [131], 4.0–8.0 [135]
Serapias perez-chiscanoi	5.0	5.0 [135]
Serapias strictiflora	7.5	7.5 [135]
Serapias vomeracea	4.5–7.5	4.99–6.44 [25], 4.5–7.5 [133], 5.27–7.37 [134], 6.5–7.0 [135]
Spiranthes aestivalis	5.5–7.9	5.5–7.9 [126], 6.8 [131]
Spiranthes spiralis	4.47–8.1	4.47–7.52 [25], 5.7–8.1 [126], 5.0–7.6 [131], 4.5–7.5 [133], 7.4 [136]
Steveniella satyrioides	5.5–8.0	6.4–7.9 [131], 5.5–8.0 [133]
Traunsteinera globosa	5.0–8.0	5.0–6.27 [126], 5.5–8.0 [133]
Traunsteinera sphaerica	5.5–8.0	5.5–8.0 [133]

The proportion of orchids belonging to each one of these categories strongly depends on both the latitudinal gradient and the altitudinal gradient [11]. Around the Mediterranean Sea, species that are totally dependent on wet meadows and other habitat types characterized by high levels of soil moisture are few and restricted to microsites of suitable land patches, whereas in the countries of Northern Europe,

Table 2 Chemical elements and organic matter in soils at sites with some terrestrial orchid species

Orchid taxa	C	N	K	P	Ca	Mg	Organic matter
Cypripedium calceolus	–	1–2 mg 100 g^{-1} [136], 5.9–55.3 mg kg^{-1} [138]	9–29 mg 100 g^{-1} [136], 30–100 mg kg^{-1} [138]	8–62 mg 100 g^{-1} [136], 4.6–10 mg kg^{-1} [138]	300–980 mg 100 g^{-1} [136], 3500–10,000 mg kg^{-1} [138]	6–12 mg 100 g^{-1} [136], 60–990 mg kg^{-1} [138]	–
Dactylorhiza maculata	–	1 mg 100 g^{-1} [136]	1.5–14 mg 100 g^{-1} [136]	7–20 mg 100 g^{-1} [136]	129–181 mg 100 g^{-1} [136]	5–34 mg 100 g^{-1} [136]	–
Dactylorhiza sambucina	–	1 mg 100 g^{-1} [136], 160.3–418.5 mg 100 g^{-1} [139]	0.6–8.9 g kg^{-1} [111], 14–23 mg 100 g^{-1} [136], 0.05–0.2 g kg^{-1} [139]	1.17–6.68 mg 100 g^{-1} [25], 2.1–14.6 mg kg^{-1} [111], 5–8 mg 100 g^{-1} [136], 1.8–14.9 mg kg^{-1} [139]	0.3–40.4 g kg^{-1} [111], 79–83 mg 100 g^{-1} [136], 0.5–1.8 g kg^{-1} [139]	11–16 mg 100 g^{-1} [136]	0.97–35.93% [25]
Epipactis atrorubens	–	28.3 ± 2.3 mg kg^{-1} [116], 59.0 ± 4.0 mg kg^{-1} [116]	–	22.0 ± 2.1 mg kg^{-1} [116], 71.3 ± 5.2 mg kg^{-1} [116], 1.69–4.39 mg 100 g^{-1} [25]	–	–	7.93–30.18% [25], 0.3–1.0% [141]
Epipactis pontica	4.10–5.30% [128]	0.38–0.49% [128]	13.07–18.34 mg 100 g^{-1} [128]	2.61–8.78 mg 100 g^{-1} [25], 1.08–1.40 mg 100 g^{-1} [128]	–	–	2.97–15.32% [25], Humus: 7.01–9.09% [128]
Hammarbya paludosa	–	–	0.2–0.37 g kg^{-1} [142]	0.32–0.76 g kg^{-1} [142]	0.6–2.89 g kg^{-1} [142]	0.37–0.89 g kg^{-1} [142]	96.5–98.3% [142]
Himantoglossum hircinum	–	1–1.5 mg 100 g^{-1} [136]	28 mg 100 g^{-1} [136]	41–46 mg 100 g^{-1} [136]	162–203 mg 100 g^{-1} [136]	10–42 mg 100 g^{-1} [136]	–
Neotinea tridentata	–	1–2 mg 100 g^{-1} [136]	24–30 mg 100 g^{-1} [136]	21–29 mg 100 g^{-1} [136]	448–484 mg 100 g^{-1} [136]	8–32 mg 100 g^{-1} [136]	–
Neotinea ustulata	–	0.261–0.516% [110], 1–2 mg 100 g^{-1} [136]	25.1–263.2 mg kg^{-1} [110], 12–44 mg 100 g^{-1} [136]	14.81–23.54 mg kg^{-1} [110], 1.69–4.66 mg 100 g^{-1} [25],	1636–6001 mg kg^{-1} [110], 357–780 mg 100 g^{-1} [136]	69.7–368.2 km kg^{-1} [110], 7–65 mg 100 g^{-1} [136]	5.4–10.76% [110], 1.68–27.99% [25]

				8–27 mg 100 g^{-1} [136]			
Neottia ovata	–	41.9–194.2 mg kg^{-1} [121]	–	3.1–31.6 mg kg^{-1} [121], 1.17–8.78 mg 100 g^{-1} [25]	–	–	8.8–28.6% [121], 0.78–28.41% [25]
Ophrys fuciflora s.str.	–	1–2 mg 100 g^{-1} [136]	26–44 mg 100 g^{-1} [136]	5–46 mg 100 g^{-1} [136]	162–620 mg 100 g^{-1} [136]	7–42 mg 100 g^{-1} [136]	–
Ophrys insectifera	–	2 mg 100 g^{-1} [136]	44 mg 100 g^{-1} [136]	27 mg 100 g^{-1} [136]	620 mg 100 g^{-1} [136]	7 mg 100 g^{-1} [136]	–
Orchis militaris	–	120–360 μg g^{-1} [109]	68–160 mg kg^{-1} [109]	0.037–0.10% [109], 3.92–4.71 mg 100 g^{-1} [25]	54–94% [109]	50–91 mg kg^{-1} [109]	20.23–21.03% [25]
Orchis pallens	–	2–3 mg 100 g^{-1} [136]	31–38 mg 100 g^{-1} [136]	12–28 mg 100 g^{-1} [136], 1.76–4.25 mg 100 g^{-1} [25]	420–665 mg 100 g^{-1} [136]	14–16 mg 100 g^{-1} [136]	1.68–24.03% [25]
Orchis purpurea	–	1–3 mg 100 g^{-1} [136]	16–30 mg 100 g^{-1} [136]	8–41 mg 100 g^{-1} [136]	203–510 mg 100 g^{-1} [136]	6–12 mg 100 g^{-1} [136]	–
Pseudorchis albida	7.2–16.3% [147]	0.4–1.1% [147]	4.0–112.0 mg 100 g^{-1} [146],	0.4–20.6 0 mg 100 g^{-1} [146]	–	–	4.3–89.4% [146]
Spiranthes spiralis	–	1 mg 100 g^{-1} [136]	24 mg 100 g^{-1} [136]	21 mg 100 g^{-1} [136], 1.5–5.33 mg 100 g^{-1} [25]	440 mg 100 g^{-1} [136]	8 mg 100 g^{-1} [136]	1.28–7.45% [25]

these orchids can prevail or be present in a proportion notably higher than in the countries of Southern Europe. For example, in the British Isles, 32.7% of the orchid species and subspecies (19 out of 58 taxa) belong to the group of orchids which require wet conditions, while the corresponding figure is 25.7% (18 out of 70 taxa) for the Czech Republic and 0.07% (14 out of 193 orchid taxa) for Greece [13, 14, 123, 124].

4.2.2 Soil pH

Soil pH is one of the most important soil factors affecting the distribution and abundance of terrestrial orchids [25, 26, 105, 117]. Since the soil reaction controls the uptake of minerals, either directly or through the mycorrhizal association, the pH gradient can be considered to be a resource gradient [108]. In general, it can be considered that most terrestrial orchids in Europe occur on soils that range from slightly acidic to slightly alkaline, whereas just a small portion of orchids can tolerate and survive in conditions of low (<4–4.5) or high (>8.5) pH values. The smaller number of orchids on strongly acidic soils can be attributed to the fact that acidic soils with pH < 4.5 contain high concentrations of harmful H^+ and Al^{3+} ions [103, 108]. Increasing richness of orchid species with increase of soil pH can be attributed to the fact that increased pH can positively affect the availability of nutrients. However, in highly alkaline soils, and also in very acidic ones, mycorrhiza cannot survive, which in turn can cause a reduction in the number of orchids occurring at a site [103].

Soil pH is often considered to be one of the most important factors separating habitats of similar and related species. Thus, *Dactylorhiza fuchsii* is generally linked to soils with higher pH values than *Dactylorhiza maculata*, which is often associated with acidic soils [125]. In addition to *D. maculata*, other well-known terrestrial orchids that occur on acidic soils in Europe are *Neottia cordata*, *Anacamptis coriophora*, *Dactylorhiza romana*, *D. cordigera*, *Hammarbya paludosa*, *Corallorhiza trifida*, *Malaxis monophyllos*, *Epipactis pontica*, *E. purpurata*, and *Serapias cordigera* (Fig. 7, Table 1), whereas the group of species that prefer alkaline soils includes *Orchis militaris*, *O. pauciflora*, *Epipactis atrorubens*, *Anacamptis pyramidalis*, *Cypripedium*, and most *Ophrys* species (Table 1) [9, 25, 48, 49, 109, 126–128]. A significant number of species of orchids that prefer acidic soils grow in high-altitude areas, which is understandable in view of the fact that soil pH is negatively correlated with altitude, since at higher altitudes the decomposition of organic matter is slower and the acidification process more intense due to higher precipitation [26, 129]. Although this is a general rule and is frequently observed in Central Europe, it is not the case in the southernmost areas of Europe, where values of soil pH in many high mountains are greater than 7.0 [130].

It should be noted that several orchids have a wide range of soil pH at which they occur, while others have a narrower pH range (Table 1). Specifically, *Anacamptis laxiflora*, *Coeloglossum viride*, *Dactylorhiza flavescens*, *D. incarnata*, *D. sambucina*, *Epipactis helleborine* subsp. *helleborine*, *Malaxis monophyllos*, *Neotinea maculata*, *Neottia nidus-avis*, *N. ovata*, *Neottianthe cucullata*, *Orchis mascula* subsp. *mascula*, *O. purpurea* subsp. *purpurea*, *Platanthera chlorantha*, *Pseudorchis albida*, and

Fig. 7 Terrestrial orchids occurring mostly on acidic soils. (**a**) *Anacamptis coriophora*, (**b**) *Dactylorhiza romana*, (**c**) *Serapias cordigera*, (**d**) *Epipactis pontica*, (**e**) *Neottia cordata* (**a** photo V. Djordjević; **b**–**e** photos S. Tsiftsis)

Serapias lingua are orchids that occur over a wide range of soil pH values, thereby indicating their considerable ecological plasticity. Furthermore, the recorded pH values can differ from one geographical area to another. Thus, the pH ranges indicated by Sundermann [131] for the species *Orchis simia*, *Corallorhiza trifida*, and *Cephalanthera longifolia* were 7.5–8.5, 6.5–7.5, and 6.5–8.6, respectively, whereas pH values measured in northeastern Greece were 4.74–7.85, 4.14–7.48, and 4.34–7.85, respectively [25].

4.2.3 Nitrogen, Phosphorus, and Potassium

According to the results of numerous studies, the content of nutrients in the soil plays an important role in the growth, development, and distribution of orchids [25, 57, 62, 103, 117, 148]. There are several ways in which the availability of nutrients can affect the growth of orchids [103]. Nutrient availability can directly affect the growth of orchids, a situation determined by ecophysiological characteristics of the orchids. In addition, the content of nutrients in the soil influences the growth of associated mycorrhizal fungi and affects the relationship between mycorrhizal fungi and orchids [103]. Furthermore, increased content of nutrients in the soil can also increase the growth of surrounding strongly competitive plant species that endanger the survival of orchids [42].

Although it is known that terrestrial orchids can absorb both nitrate nitrogen (NO_3^-) and ammonium nitrogen (NH_4^+), the absorption rate is higher in the case of NO_3^- [57]. Some studies explored the transfer of nutrients from mycorrhizal fungi to orchids [149]. While all orchids in the early growth stages are entirely dependent on mycorrhizal fungi, many autotrophic adult orchids still obtain nutrients through mycorrhizal fungi [57]. Furthermore, the transfer of nitrogen and carbon from substrates or trees to orchids through mycorrhizal fungi has been demonstrated using radiocarbon and stable isotopes [150].

The widespread distribution of some terrestrial orchids (e.g., *Platanthera bifolia* and *Neottia ovata*) can be explained by the fact that they tolerate high nitrogen content in the soil, i.e., they are euryvalent with respect to the amount of nitrogen in the soil [151]. Interesting results of research were obtained in Greece, where it was established that the content of phosphorus is the most important factor affecting the distribution of *Goodyera repens* on the southern border of its distribution [104].

Although most terrestrial orchids are sensitive to increased content of nutrients in the soil [103], species respond differently to changes of nutrient content. Thus, soil enrichment with phosphorus, nitrogen, and potassium has a negative effect on populations of *Dactylorhiza majalis* [152]. It has been concluded that *Dactylorhiza maculata*, *Platanthera bifolia*, and *Neottia ovata* can survive if the content of calcium or calcium together with nitrogen in the soil increases but that they cannot survive when the levels of calcium, nitrogen, and phosphorus increase together [151]. The negative effect of high content of nutrients is frequently reflected in disturbance of mycorrhizal relationships [103]. To be specific, studies have shown that a high concentration of nitrogen in the soil negatively influences the development of protocorms in *Dactylorhiza incarnata* [148]. Moreover, in soil with high carbon and nitrogen content, protocorms of some orchid species rejected mycorrhizal fungi [148]. A negative impact of soil fertilization was shown by research in Flanders and the Netherlands, where investigators found a significant decrease in the abundance of *Anacamptis morio* due to soil enrichment with phosphorus [153].

Several studies provided detailed insight into nitrogen, phosphorus, and potassium content in the soil on which terrestrial orchids grow (Table 2). Overall, it is considered that the majority of orchid species occur on soils that are relatively poor in nutrients [103]. Well-known terrestrial orchid species that occur on oligotrophic soils are *Goodyera repens*, *Epipactis atrorubens*, and *Spiranthes spiralis* [49]. Moreover, *Dactylorhiza maculata* and *D. praetermissa* were found to grow primarily on low-phosphorus soils [154]. Among orchid species that grow on soils with very low phosphorus content in northeastern Greece were *Anacamptis pyramidalis*, *Neotinea tridentata*, *Orchis quadripunctata*, *Goodyera repens*, *Epipogium aphyllum*, and *Serapias* species, whereas species that grow on soils with high phosphorus content were *Cephalanthera damasonium*, *C. longifolia*, *C. rubra*, *Corallorhiza trifida*, *Epipactis helleborine*, *E. microphylla*, and *Neottia nidus-avis* [25].

4.2.4 Calcium and Magnesium

Most terrestrial orchids in Europe are known to grow on calcium-rich soils [9, 11, 34]. Calcium in the soil is found primarily in the form of calcium carbonate, but also as phosphoric and silicic acid salts, as well as salts of various organic acids [155]. Terrestrial orchids mostly occur on typical calcium-rich soils that are developed on the sedimentary rock limestone, and little is known about orchids that grow on soils containing high concentrations of calcium derived from other geological substrates [34]. In addition, a significant number of species grow on soils developed on dolomite, a sedimentary rock composed mostly of the mineral dolomite, i.e., calcium magnesium carbonate – $CaMg(CO_3)_2$ [34, 107]. Recent studies suggested that numerous terrestrial orchids can also be found on serpentine soils that are rich in magnesium, with lower calcium content [24, 26, 107, 116]. The calcium and magnesium content in soils with some terrestrial orchid species has been presented in several studies (Table 2).

Calcium affects both physical and chemical properties of the soil. Most often, calcareous soils are permeable to water and therefore are warm, dry, and well aerated. Consequently, these soils are suitable for most orchids that prefer warm and dry habitat types, especially ones from the genera *Orchis*, *Ophrys*, *Anacamptis*, *Himantoglossum*, *Serapias*, and *Neotinea* [9]. Since these soils contain large amounts of Ca^{2+} and HCO_3^- ions, the negative effects of H^+ and Al^{3+} ions are suppressed. At the same time, calcareous soils are characterized by high activity of nitrogen-fixing bacteria. On the other hand, due to their high calcium concentration and high pH, some elements become less available for plants (phosphorus is converted to insoluble calcium phosphate, whereas iron and manganese are poorly soluble), which is a situation favorable to a large number of orchids that are sensitive to high levels of phosphorus and heavy metals [103, 104]. Free calcium at high concentrations may be toxic to orchid cells. Therefore, orchids often produce calcium oxalate crystals to remove excess calcium or oxalic acid [156]. The presence of calcium oxalate crystals has been reported in almost all orchid parts, especially in tubers, roots, stems, anthers, and lips. Moreover, calcium oxalate crystals have a prominent role in plant defense, osmotic regulation, and maintenance of ionic balance [156].

4.2.5 Organic Matter

Orchids at their initial growing stages are fully mycoheterotrophic, obtaining carbohydrates and mineral nutrients through the mycorrhizal fungi with which they are associated [62]. Contrary to this fully mycoheterotrophic relationship, orchids at maturity obtain carbon both through photosynthesis and from ectomycorrhizal fungi, thereby becoming only partially mycoheterotrophic. Soil organic matter contains the main source of carbon and nutrients that are transferred to orchids through their mycorrhizal fungi [62]. For example, albino specimens of *Cephalanthera longifolia* depend on their fungi for carbon (mycoheterotrophy), whereas photosynthetic green specimens recover just 33% of their carbon from fungal partners (myxotrophy) [157]. So far it has been found that the germination rate of orchids is positively correlated with the content of organic matter in the soil [158]. Moreover, the content of organic matter in soils significantly affects the abundance of orchids at the population level, i.e., in areas larger than 400 m^2 [72]. Species that grow on soil with a high percentage content of organic matter include *Orchis simia* (1.46–47.99%), *Dactylorhiza cordigera* (2.96–45.10%), *D. incarnata* (2.03–41.72%), *Cephalanthera damasonium* (0.78–38.96%), *C. rubra* (0.97–38.96%), and *Corallorhiza trifida* (1.86–38.96%), whereas other species grow on soil with low percentage content of organic matter – *Ophrys reinholdii* (1.85–2.47%), *O. zeusii* (2.07–3.57%), *O. apifera* (1.87–2.47%), and *Dactylorhiza romana* (3.35–4.09%) [25].

The humus content of the soil, i.e., specific mixtures of organic and mineral materials, mainly in the colloidal state, formed by the decomposition of organic residues in soil under the influence of microorganisms, was highlighted in a recent study treating the orchid species *Epipactis pontica* [128].

4.2.6 Orchids Growing on Anthroposols

In other studies, some terrestrial orchid species were found to grow on anthroposols, i.e., azonal soils highly modified or constructed by human activities (e.g., fly ash, mine waste deposits and soils in the vicinity of roads) [159–162]. Orchids encounter a range of stressful and growth-limiting factors on anthroposols, factors such as drought, high temperature, and light intensity; low/high pH; unfavorable mechanical composition; high concentrations of soluble salts; lack of essential nutrients (N, P, K); toxic concentrations of As, B, Cd, Cr, Cu, Hg, Mn, Mo, Ni, Zn, Pb, and Se; reduced presence of mycorrhizal fungi; and reduced abundance of nitrogen-fixing microorganisms [163]. It is assumed that some orchids grow on anthroposols primarily due to favorable light regimes and reduced competition between plants [164].

Orchids that are found to occur on anthroposols primarily belong to the genera *Epipactis* and *Dactylorhiza* [160–162, 164, 165]. Among orchids which have the ability to colonize and grow on soils in the vicinity of roads, it is especially important to emphasize species from the genus *Himantoglossum* [166, 167], as well as *Epipactis helleborine*, *Neottia ovata*, *Dactylorhiza saccifera*, *Ophrys apifera*, and *Cephalanthera longifolia* [26, 165, 167]. In southern Poland, *Dactylorhiza majalis*, *Epipactis atrorubens*, and *E. helleborine* were found to grow on zinc mine tailings consisting of post-flotation material and characterized by high concentrations of Zn, Pb, and Cd [159]. In that study, mycorrhizal fungi were found to play an important role in the detoxification of heavy metals in the case of *E. atrorubens*, in which Pb and Zn are accumulated in the roots [159]. The same authors showed that heavy metals are mostly stored in the fungal cell walls, a circumstance which is probably associated with the presence of fungal pigments such as melanin, known for the fact that it is involved in detoxification mechanisms [159]. In this way, it was demonstrated that the mycorrhizal mycelium takes heavy metals from the soil and stores them, thereby reducing their toxicity to the orchid plant host. However, heavy metals can also be transported directly through plant root cells, which may explain the high concentration of heavy metals in the shoots of *E. atrorubens* [159].

Several orchid species were found growing on mine tailings in northeastern Estonia, viz., *Epipactis atrorubens*, *Orchis militaris*, and *Dactylorhiza baltica* [160]. It has been established that these orchids and their mycorrhizal fungi are not limited to environmentally polluted sites compared to natural sites. Moreover, the same fungi associated with *E. atrorubens* and *O. militaris* were found on mine tailings and in natural habitats [160]. Furthermore, *Epipactis palustris* was found to grow in former lignite mining areas in eastern Germany [161]. The results of molecular investigation indicated that genetic diversity values for populations of this species from mining areas do not differ from those of populations from natural habitats [161]. *Epipactis palustris* was found on Pb-Zn polymetallic deposits of the Rudnik orefield in Serbia (unpublished data). To be specific, specimens of this species were discovered on flotation tailings that are in the initial phase of colonization and sparsely covered with vegetation.

Interesting findings come from Australia, where orchids were found to grow on bauxite-mined land, suggesting that mycorrhiza fungi also play an important role in orchid colonization [168, 169]. The latter authors determined that orchids occur for the first time in 2-year-old rehabilitated areas, but increase in abundance over time [169]. Especially at rehabilitated sites more than 10 years old, the density of orchids was found to have increased significantly [168]. Among orchids growing in the rehabilitated areas, the most abundant species were *Caladenia flava*, *Disa bracteata*, *Diuris longifolia*, *Microtis media*, *Pterostylis mint*, and *Pterostylis vittata* [168].

In general, there have been few studies on the ecophysiological responses of orchid species to the stressful conditions of anthroposols, and it is unclear whether habitats developed on these substrates can act as refuges for endangered and rare orchid species.

5 Vegetation Types

Terrestrial orchids occur in almost all known ecosystems with various types of vegetation from coasts to highlands, usually in forest and grassland communities, as well as in wetlands, on steppes, and in desert oases [5, 9]. However, orchids are absent or less abundant in salt marshes, in extremely dry deserts, and on cultivated agricultural lands [5]. Although certain species in Europe occur exclusively in forests and scrub vegetation, and some inhabit only herbaceous vegetation, a significant number of species grow in both forests and herbaceous communities [9, 26]. However, a large number of orchids grow in forest-grassland and scrub-grassland ecotones and on the very edges of forests and herbaceous communities, which makes it harder to determine vegetation units more precisely. In general, vegetation types significantly affect the distribution and abundance of orchids, representing one of the key gradients influencing separation of the ecological niche of orchids [24–26].

The degree of representation of orchids in plant communities is rarely greater than 30% [49]. Therefore, they rarely represent diagnostic species of plant communities.

However, some orchid species appear as dominant species in several plant communities on the Balkan Peninsula. For example, *Ophrys sphegodes*, *Anacamptis coriophora*, and *A. papilionacea* have significant abundance in the *Orchido-Chrysopogonetum* community of the thermo-Mediterranean calcicolous phrygana of northern Greece (*Dorycnio-Coridothymion capitati*) [34]. Furthermore, the *Trolio-Orchidetum bosniacae* (*Calthion palustris*) community was described in Bosnia and Herzegovina, whereas the *Orchido-Schoenetum nigricantis* community of the alliance *Caricion davallianae* is represented in Slovenia and Croatia [34]. It should be noted that the association of orchid species and surrounding species within plant communities is determined not only by abiotic factors, such as moisture, the light regime, and soil properties, but also by many complex biotic factors, as well as by age and evolutionary development of the ecosystem. For example, the association between orchid species and the surrounding forest trees is further strengthened by the fact that many orchids form associations with fungi that are ectomycorrhizal on the roots of neighboring trees [7, 81, 170], indicating that the phylogeographic dynamics of many orchid species can be linked to the phylogeographic patterns of trees [22].

In this section, we present an overview of the main vegetation types at sites with terrestrial orchids in Europe, with special reference to the Balkan Peninsula. The names of syntaxonomic units follow the phytocenological nomenclature proposed by Mucina et al. [171].

5.1 Forest and Scrub Vegetation

5.1.1 Deciduous Forests

Among the types of forest vegetation in Europe, plant communities of the class *Carpino-Fagetea sylvaticae* are particularly important owing to the great diversity of terrestrial orchids, in terms of both the number of species and their abundance [34]. These are mesic deciduous and mixed forests of temperate Europe, Anatolia, the Caucasus, and Southern Siberia, consisting primarily of different beech and mixed fir-beech forests, as well as oak-hornbeam and mesic oak forests [171]. Orchid studies indicate that the largest number of *Epipactis*, *Cephalanthera*, and *Neottia* species occurs in beech-dominated forests in Europe [9, 12, 20, 106, 126, 127]. The great richness of orchid species in these forest types can be attributed to (i) the strong presence of these forests, both horizontally and vertically; (ii) the fact that they appear on different geological substrates (calcareous and non-calcareous) and in different climatic zones; and (iii) the age of these forests. In general, the beech-dominated forests of Southern Europe, including the Iberian, Apennine, and Balkan Peninsulas, are important refugia of many orchid species [22]. To be specific, during the last glaciations, many orchid species were distributed in beech forests and restricted to these peninsulas as well as the Caucasus. However, during the climate amelioration that began c 10,000 BC, beech forests moved slowly to the north and to the west, reaching Scandinavia around AD 500 [9]. Scientists believe that this second arrival of beech forests in Central Europe made possible the process of evolutionary radiation of many orchid species, especially ones of the genus *Epipactis*. Some recent research provides an insight into the exact number of orchid taxa growing in these forest types. For example, 26 orchid species and subspecies were found to grow in beech-dominated forests in western Serbia, representing the forest communities richest in orchid taxa in this part of the central Balkans [34]. Furthermore, 22 orchid species were found to inhabit these forests in northeastern Greece [13]. It should be noted that many boreal and Central European orchids that have the southern boundaries of their distribution on the Balkan Peninsula (e.g., *Epipogium aphyllum*, *Epipactis leptochila* subsp. *neglecta*, *E. pontica*, and *E. purpurata*) were found in such beech and beech-fir forests [13, 25, 128, 172].

The most common orchid species that grow in oak-hornbeam forests (*Carpinetalia betuli*) are *Epipactis helleborine*, *E. purpurata*, *E. pontica*, *Neottia nidus-avis*, *N. ovata*, *Orchis mascula*, *O. pallens*, *O. purpurea*, and *Platanthera bifolia* [9, 34, 112, 127, 173]. Overall, these forests are inhabited by a large number of Central European orchid species that require mesophilic conditions. In the central Balkans (western Serbia), ten orchid species were found in oak-hornbeam forests, especially in the community *Querco-Carpinetum betuli* [34].

Terrestrial orchids have significant representation in open oak, mixed deciduous, and conifer forests of warm regions in the temperate zone, i.e., in communities of the class *Quercetea pubescentis*. These forests are important vegetation types for many orchid species in Central and Southern Europe and in Mediterranean regions, Asia Minor, and the Middle East [171]. Within this vegetation class, orchids are especially well represented in the order *Quercetalia pubescenti-petraeae* and the alliances *Fraxino orni-Ostryion*, *Quercion petraeo-cerridis*, *Quercion confertae*, *Quercion petraeae*, *Aceri tatarici-Quercion*, and *Carpinion orientalis* [34]. Studies from the Balkan Peninsula provide precise data on how many orchid species grow in oak forests. For example, 21 orchid species were found to occur in forest communities dominated by the oaks *Quercus frainetto* and *Q. petraea* subsp. *medwediewii* in northeastern Greece [174]. Furthermore, 21 orchid species and subspecies were registered in thermophilous montane oak forests of the alliance *Quercion petraeo-cerridis*, whereas 16 orchid species were found to inhabit thermophilous deciduous oak forests on slightly acidic deep soils (*Quercion confertae*) in western Serbia [34].

Other studies highlighted the importance of amphi-Adriatic calcareous sub-Mediterranean and inland oak and hop-hornbeam forests of the alliance *Fraxino orni-Ostryion* [24, 25, 34]. The large number of orchids in these forests is explained by the fact that they are mainly widespread on limestones, especially in gorges and canyons, where they are sheltered from extreme climatic influences. The importance of Tertiary relict *Ostrya carpinifolia* forests is reflected in the great abundance of orchid species from the sub-Mediterranean and Mediterranean chorological groups (e.g., many *Ophrys*, *Orchis*, and *Himantoglossum* species) and in the presence of certain orchids that are primarily cenobionts of grassland ecosystems [34]. Not strongly influenced by anthropogenic factors, these forests include some of the most highly specialized orchid species [24]. In western Serbia, 24 orchid species and subspecies were found to grow in these forest types [34]. Similarly, 22 orchid species were registered in communities consisting of *Ostrya carpinifolia*, *Carpinus orientalis*, and *Fraxinus ornus* in northeastern Greece [174].

Terrestrial orchids in Europe also occur in acidophilous oak and oak-birch forests on nutrient-poor soils of Europe (*Quercetea robori-petraeae*) [171]. For example, in Ukraine, *Dactylorhiza sambucina* was found to grow in *Quercion robori-petraeae* forests [111]. Within this vegetation class, orchids inhabit acidophilous chestnut-oak forests (*Castaneo-Quercion petraeae*) of Southeast Europe. In forests dominated by chestnut (*Castanea sativa*) in northeastern Greece, the following species were recorded: *Cephalanthera longifolia*, *C. rubra*, *Dactylorhiza saccifera*, *Epipactis helleborine*, *E. microphylla*, *Limodorum abortivum*, *Neottia nidus-avis*, and *Platanthera chlorantha* [174]. Orchids inhabiting similar forest communities in western Serbia include *Epipactis microphylla*, *Neottia nidus-avis*, *Ophrys scolopax* subsp. *cornuta*, and *Orchis purpurea* (authors' unpublished data).

Several terrestrial orchids were found to grow in forest communities of the class *Brachypodio pinnati-Betuletea pendulae*. These forests represent hemiboreal pine and birch-pine forests on fertile soils of the southern Urals and Southern Siberia and relict birch-poplar forests of Europe [171]. For example, *Epipactis helleborine* and *Platanthera bifolia* were found to have significant populations in birch forests of Russia and the former Soviet Union [133]. Furthermore, some orchid species inhabit relict extrazonal temperate deciduous birch-poplar forests on mineral soils (*Fragario vescae-Populion tremulae*). In western Serbia, orchid representatives include mainly species that are characteristic cenobionts of grassland communities (e.g., *Anacamptis morio*, *Dactylorhiza sambucina*, *Gymnadenia conopsea*, and *Platanthera bifolia*) [34], whereas *Dactylorhiza sambucina*, *Epipactis helleborine*, *Neottia ovata*, and *Platanthera chlorantha* were recorded in birch forests in northeastern Greece [174]. The smaller number of orchids in birch forests on the Balkan Peninsula is explained by the limited distribution of these forests and the fact that they generally represent an unstable stage in the succession of forest vegetation [34]. To be specific, birch forests in western Serbia and Greece are widespread within the belt of oak, pine, and beech forests and arise mainly after forest fires or logging.

5.1.2 Coniferous and Mixed Broadleaved-Coniferous Forests

A significant number of orchids grow in communities of the vegetation class *Vaccinio-Piceetea*, i.e., in Holarctic coniferous and boreo-subarctic birch forests of the boreal zone and in high-elevation areas of mountains in the nemoral zone of Eurasia [9, 12, 13, 19, 34, 73, 133, 175]. Within this vegetation class, many orchid species occur in European boreo-montane and subalpine spruce and pine forests on nutrient-poor soils (*Piceetalia excelsae*); in boreo-temperate pine forests on nutrient-poor and hydromorphic soils (*Pinetalia sylvestris*); in boreo-montane spruce, fir, and pine forests on nutrient-rich soils (*Athyrio filicis-feminae-Piceetalia*); and in boreal mesophilous coniferous forests developed on podzolic soils in easternmost European Russia, the Urals, and Siberia (*Piceo obovatae-Pinetalia sibiricae*) [34, 171]. These forests are inhabited by many orchids from the boreal chorological group, which tolerate shade and cooler temperatures, species such as *Epipogium aphyllum*,

Goodyera repens, *Corallorhiza trifida*, and *Neottia cordata* [9]. In Southern Europe, most of these orchid species grow in forests of spruce (*Picea abies*), pine (*Pinus sylvestris*), and mixed forests of spruce, fir, and beech at higher altitudes on the southernmost boundaries of their distribution in Europe [34, 45, 104, 175]. In the central Balkans (western Serbia), 26 orchid species and subspecies were found to grow in communities of the class *Vaccinio-Piceetea* [34]. In this region, many orchids were found to grow in stands of these communities at lower altitudes, primarily due to temperature inversion and higher rainfall [34]. Recent research from Greece treated the importance of *Picea abies* and *Pinus sylvestris* forests for the growth and survival of *Neottia cordata*, and the results indicated that *Picea abies* forests are more suitable for its conservation than those of *Pinus sylvestris* [45].

Members of the family Orchidaceae are also represented in different pine forests and related scrub communities [9, 13, 24, 34, 133]. Numerous orchid species were found in such communities, especially in pine forests on calcareous and ultramafic geological substrates in the Balkans, the Alps, the Carpathians, and the Crimea within the vegetation class *Erico-Pinetea*. Furthermore, orchids are also registered in communities of the vegetation class *Roso pendulinae-Pinetea mugo*, which include pine krummholz in subalpine belts of the nemoral mountain ranges of Europe [48, 133].

Orchids are significantly abundant in relict *Pinus sylvestris* forests on calcareous substrates of the Alps, the Hercynicum, and the Massif Central (*Erico carneae-Pinion*). Moreover, the high abundance of some orchids (e.g., *Cephalanthera rubra* and *Epipactis muelleri*) resulted in corresponding names of syntaxa, such as *Cephalanthero rubrae-Pinion sylvestris* and *Epipactido muelleri-Pinion sylvestris* [171]. On the Balkan Peninsula, orchids inhabit *Pinus sylvestris* forests on calcareous substrates of the central and eastern Dinarides (*Seslerio rigidae-Pinion*); *Pinus nigra* forests on calcareous substrates of the central and southern Balkans (*Fraxino orni-Pinion nigrae*); *Pinus nigra* forests on dolomite and ultramafic geological substrates of the Dinarides (*Erico-Fraxinion orni*); and *Pinus heldreichii* forests on calcareous and ultramafic substrates of the southern Balkans (*Pinion heldreichii*) [24, 25, 34]. Particularly high diversity of orchids in pine forests has been reported in Greece [13]. Specifically, in northeastern Greece, 19 orchid species were found in *Pinus sylvestris* forests, whereas 18 orchid species were recorded in *Pinus nigra* forests [174]. In communities of the vegetation class *Erico-Pinetea* in western Serbia, a total number of 22 orchid species was registered [34]. In this area, orchids were often found in pine forest communities developed on ultramafic substrates [34]. Due to significant illumination, many orchids characteristic of grassland ecosystems have been recorded in these pine forests, species such as *Anacamptis morio*, *Dactylorhiza sambucina*, *D. maculata* subsp. *transsilvanica*, *Gymnadenia conopsea*, *Platanthera bifolia*, and *Traunsteinera globosa* [24, 34]. In addition, *G. conopsea* has been registered in Central Europe in communities of the alliance *Erico-Pinion* and the *Molinio-Pinetum* community in eastern Switzerland [61].

In the central Balkans, some orchid species were registered in *Picea omorika* forests (*Erico carneae-Piceion omorikae*) within the vegetation class *Erico-Pinetea*. Among the 11 orchid species recorded in these forests in western Serbia, especially the boreal species *Goodyera repens* and *Neottia cordata* are significantly abundant

[34, 176]. Interestingly, the community *Goodyero-Piceetum omorikae* has been described in Bosnia and Herzegovina, which confirms the association between *Goodyera repens* and paleoendemic coniferous species (*Picea omorika*) [34, 176].

5.1.3 Mire and Swamp Forests

Among damp forest types, orchids are especially common in communities within the classes *Alnetea glutinosae* (alder swamp forests and willow scrub occurring in habitats with a permanently high water table) and *Franguletea* (willow carr of Western Europe and the sub-Atlantic regions of Central Europe) [19, 34]. Thus, *Dactylorhiza majalis* was found to inhabit *Alnetea glutinosae* communities in Central Europe [42], whereas *Neottia ovata* was registered in the *Stellario nemorum-Alnetum glutinosae* community [112]. In western Serbia, *Dactylorhiza saccifera* and *Neottia ovata* were found to occur in similar communities with *Alnus glutinosa* as a diagnostic species [34]. In Slovakia and Romania, *Epipactis albensis* was found to grow in alluvial forests and on banks with willows and poplars [140].

5.1.4 Broadleaved Evergreen Forests, Coniferous Forests of the Mediterranean, and Scrub Vegetation

Communities of the class *Quercetea ilicis* (thermo-Mediterranean pine and oak forests and associated macchia) are inhabited by a significant number of orchids in the Mediterranean region of Europe [16, 19, 174]. Furthermore, *Junipero-Pinetea sylvestris* communities (relict oro-Mediterranean and sub-Mediterranean pine forests, juniper woods, and related scrub of the Mediterranean) also support many terrestrial orchids [19]. A great number of orchids from the Mediterranean and sub-Mediterranean chorological groups are distributed in scrub vegetation of types such as *Ononido-Rosmarinetea* (Mediterranean scrub on calcareous substrates under the names garrigue, phrygana, tomillar, espleguer, romeral, and batha), *Cisto-Lavanduletea stoechadis* (Mediterranean scrub on siliceous and ultramafic substrates known as matorral, garrigue, phrygana, and jaral), and *Cytisetea scopario-striati* (Mediterranean and sub-Atlantic scrub on acidic substrates) [19].

Scrub and mantle vegetation seral or marginal to broadleaved forests in the sub-Mediterranean regions and nemoral zone of Europe (class *Crataego-Prunetea*) [171] also represent an important vegetation type for growth of many terrestrial orchid species. The communities richest in orchid species on the Balkan Peninsula belong to the vegetation order *Paliuretalia* (thermophilous mantle, pseudomaquis, and shrubbery fringing oak forests of the sub-Mediterranean regions of Southeast Europe) [171]. These vegetation types are distributed mainly in low- and middle-altitude areas, and they contain a large number of sub-Mediterranean and Mediterranean orchids of the genera *Ophrys*, *Orchis*, *Anacamptis*, *Serapias*, and *Himantoglossum* [25].

Representatives of the orchid family also inhabit communities of the vegetation class *Robinietea* [34]. This vegetation type represents seral forest clearings and anthropogenic successional scrub and thickets on nutrient-rich soils in temperate Europe [171]. Specifically, *Dactylorhiza saccifera* and *Neottia ovata* were found in communities of elder, willow, and hazel scrub on nutrient-rich soils in forest clearings (*Sambuco-Salicion capreae*) [34].

5.2 Herbaceous Vegetation

Terrestrial orchids are widely represented in various types of herbaceous vegetation, including grasslands, meadows, heaths, montane-subalpine tall-herb vegetation, and tall-grass perennial swards on mobile coastal dunes, as well as fens, bogs, and marshes [19, 26, 133, 177]. Recent studies of herbaceous communities in Europe indicated that herbaceous vegetation types significantly influence the patterns of orchid diversity, in regard to both composition and abundance of species [26, 105, 178, 179].

5.2.1 Grasslands, Meadows, and Heaths

Many terrestrial orchids grow in communities of the class *Festuco-Brometea* [24, 26, 34, 105, 133, 178, 180–182]. This is dry and semi-dry grassland and steppe vegetation in the sub-Mediterranean, nemoral, and hemiboreal zones of Europe [171]. The large number of orchids within communities of the *Festuco-Brometea* class indicates that these represent a major herbaceous vegetation type for orchids, which can be explained by the significant distribution of this vegetation from lowland to highland areas, as well as by the diversity of geological substrates and soils on which the communities are developed. Although these communities are mainly distributed on calcareous substrates [171], many orchids are also widely represented in communities on various siliceous and ultramafic substrates [24, 26]. In communities of this vegetation class, a total of 31 orchid species and subspecies were found in western Serbia, suggesting that the given vegetation class is the richest with orchid species in this region of the central Balkans [34]. Moreover, according to the European Union's Habitats Directive (92/43/EEC), it is also emphasized that "semi-natural dry grasslands and scrubland facies on calcareous substrates (*Festuco-Brometalia*) (∗ important orchid sites)" are one of the priority vegetation types [179].

Within the class *Festuco-Brometea*, orchids are especially well represented in the order *Brachypodietalia pinnati*, which encompasses meso-xerophytic basiphilous grasslands of Western Europe and sub-Atlantic Central Europe (*Bromion erecti*); meso-xerophytic basiphilous grasslands of the subcontinental regions of Central and Southeast Europe (*Cirsio-Brachypodion pinnati*); and dry and semi-dry grasslands on deep soils over siliceous bedrocks in the colline to submontane belts of the southern and central Balkans (*Chrysopogono-Danthonion calycinae*) [2, 34, 171, 182]. Well-known orchids registered on carbonate substrates in communities of the alliance *Bromion erecti* (often called *Mesobromion*) in Central Europe include *Anacamptis pyramidalis*, *A. morio*, *Gymnadenia conopsea*, *Herminium monorchis*, *Neotinea tridentata*, *N. ustulata*, *Neottia ovata*, *Ophrys apifera*, *O. sphegodes*, *O. holoserica*, *O. insectifera*, *Orchis mascula*, *O. militaris*, *O. purpurea*, *O. simia*, *Platanthera bifolia*, and *Spiranthes spiralis* [61, 112, 180–182]. However, orchids from

Chrysopogono-Danthonion calycinae communities have only recently been studied, in investigations where researchers found that orchids growing in these communities in the central Balkans are common both on ultramafic substrates and on calcareous as well as siliceous soils [24, 26]. Orchids that form large colonies in these grassland communities are *Anacamptis morio*, *A. coriophora*, *A. papilionacea*, *A. pyramidalis*, *Gymnadenia conopsea*, *Neotinea tridentata*, *N. ustulata*, *Platanthera bifolia*, and *Dactylorhiza sambucina* [24, 26, 34]. A significant number of orchids were found to inhibit the meso-xerophytic basiphilous open grassland communities of sub-Atlantic and sub-Mediterranean Europe (order *Artemisio albae-Brometalia erecti*), also known as *Xerobrometalia* [171, 178]. In Italy, common species that inhabit these communities are *Anacamptis morio*, *A. pyramidalis*, *Himantoglossum adriaticum*, *Neotinea tridentata*, *N. ustulata*, *Ophrys sphegodes*, *O. benacensis*, and *Serapias vomeracea* [178, 179].

Another important vegetation order is *Festucetalia valesiacae*, representing steppes and rocky steppe grasslands in the steppe and forest-steppe zones of Europe and northwestern Central Asia [26, 171]. Orchids were recorded especially on calcareous substrates in communities of the alliance *Festucion valesiacae* [26, 182, 183]. Some of the most common orchids found in these communities in the central Balkans (western Serbia) are *Anacamptis morio*, *A. pyramidalis*, *Gymnadenia conopsea*, *Himantoglossum calcaratum* subsp. *calcaratum*, *Neotinea tridentata*, *N. ustulata*, and *Ophrys scolopax* subsp. *cornuta* [26, 34].

Within the class *Festuco-Brometea*, a small number of orchid species have been registered in communities of the *Halacsyetalia sendtneri* order (ultramafic and siliceous xeric rocky grasslands in continental regions of the Balkan Peninsula), which can be attributed to extreme habitat conditions [34]. Specifically, *Anacamptis morio* was the only species registered in the community *Poo molinieri-Plantaginetum holostei*, of which it represents an indicator species [26]. Furthermore, a few species such as *Dactylorhiza saccifera* and *Gymnadenia conopsea* were recorded in communities of the order *Scorzoneretalia villosae* (amphi-Adriatic dry steppic sub-Mediterranean pastures of the Prealpine, Illyrian, and Dinaric regions) [34].

Terrestrial orchids in Europe that require mesophilous and hygro-mesophilous habitat conditions have especially significant abundance in communities of the vegetation class *Molinio-Arrhenatheretea*, which includes meadows, pastures, and tall-herb meadow fringes [19, 24, 26, 34, 42, 111, 112, 133, 154]. Within this vegetation class, many orchid species were found to occur in communities of the order *Arrhenatheretalia elatioris* (meadows and pastures on well-drained mineral soils) and the alliance *Arrhenatherion elatioris* [34]. Some of the orchids most frequently observed in these communities are *Anacamptis morio*, *A. coriophora*, *A. pyramidalis*, *Dactylorhiza sambucina*, *D. maculata*, *D. majalis*, *D. saccifera*, *Epipactis palustris*, *Gymnadenia conopsea*, *Neotinea ustulata*, *Neottia ovata*, *Orchis mascula*, *Platanthera bifolia*, and *Traunsteinera globosa* [26, 34, 110–112, 180, 183, 184].

The vegetation order *Molinietalia caeruleae* (mown meadows on mineral and peaty soils) also supports many terrestrial orchid species, including a significant number of *Dactylorhiza* species, which have large populations in this vegetation type [26, 34]. The most common orchids that were found to grow in communities of the alliance

Molinion caeruleae are *Dactylorhiza cordigera*, *D. incarnata*, *D. majalis*, *D. maculata* subsp. *maculata*, *Epipactis palustris*, *Gymnadenia conopsea*, *G. densiflora*, *Neotinea ustulata*, and *Neottia ovata* [9, 61, 68, 112, 184]. Recent studies indicated that stands of *Molinion caeruleae* communities on ultramafic substrates are especially suitable for *Dactylorhiza maculata* subsp. *transsilvanica* and *Platanthera bifolia*, whereas *Molinion caeruleae* communities on Quaternary sediments support significant populations of *Dactylorhiza incarnata* and *Anacamptis palustris* [26]. In addition, orchid species that frequently occur in communities of the alliance *Calthion palustris* include *Dactylorhiza incarnata*, *D. maculata*, *D. majalis*, *D. praetermissa*, *D. saccifera*, *D. cordigera*, *Epipactis palustris*, *Gymnadenia conopsea*, and *Neottia ovata*, whereas *Dactylorhiza incarnata*, *D. saccifera*, *Epipactis palustris*, *Gymnadenia conopsea*, and *Platanthera bifolia* are orchids that have significant populations in *Deschampsion cespitosae* communities [26, 34, 42, 61, 68, 112, 154, 185].

Based on recent studies, it can be stated that some orchids are significantly represented in communities of the vegetation orders *Trifolio-Hordeetalia* (wet meadows distributed on the Apennine and Balkan Peninsulas) and *Filipendulo ulmariae-Lotetalia uliginosi* (tall-herb wet meadow fringe vegetation on mineral soils) [34]. For example, *Dactylorhiza incarnata* and *Epipactis palustris* have particularly high abundance in *Mentho longifoliae-Juncion inflexi* communities within the second mentioned vegetation order [26, 34]. Furthermore, certain species have been reported to grow in high-altitude mesofilous meadows and pastures in the mountain ranges of Europe (*Poo alpinae-Trisetetalia*) [34, 171]. Within this vegetation order, several high-altitude orchid species (*Traunsteinera globosa*, *Dactylorhiza cordigera*, *D. sambucina*, *Gymnadenia conopsea*, and *Orchis mascula* subsp. *speciosa*) were found to grow in communities of the endemic alliance *Pancicion serbicae* in the central Balkans [26, 34].

Among herbaceous vegetation, tall-herb vegetation at high altitudes of Europe, Siberia, and Greenland (class *Mulgedio-Aconitetea*) supports the growth and development of some terrestrial orchids [26, 34]. Thus, recent studies showed that *Gymnadenia conopsea* inhabits communities of the alliance *Calamagrostion villosae* [34], i.e., tall-grass and herb-rich communities on dry acidic soils in the upper montane and subalpine belts of the investigated mountain ranges [171].

Secondary mat-grass swards on nutrient-poor soils of the temperate, boreal, and subarctic regions of Europe (class *Nardetea strictae*) also represent an important vegetation type for the existence of populations of numerous orchid species [34]. Some of the most frequently recorded species in these communities in Central and Western Europe are *Gymnadenia conopsea*, *Dactylorhiza maculata*, *D. majalis*, *D. sambucina*, *Neottia ovata*, *Orchis militaris*, and *Pseudorchis albida* [42, 61, 85, 109, 111, 112, 125]. In the central Balkans, 23 orchid species were found to grow in similar communities, especially within the alliance *Nardo-Agrostion* [34]. The most common species in these communities are *Dactylorhiza sambucina*, *Traunsteinera globosa*, *Anacamptis coriophora*, *A. morio*, *Gymnadenia conopsea*, *Orchis mascula* subsp. *speciosa*, and *Neotinea ustulata* [26, 34].

Terrestrial orchids requiring acidic soils have significant representation within the vegetation class *Juncetea trifidi*, which includes acidophilous grasslands in the alpine belt of Europe, in the Caucasus, and in the boreo-arctic zones of Northern Europe and Greenland [171]. The importance of this type of vegetation is reflected in the presence of significant populations of high-altitude orchid species such as *Coeloglossum viride*, *Nigritella gabasiana*, *N. rhellicani*, *Pseudorchis albida*, *Traunsteinera globosa*, and many taxa of the genus *Dactylorhiza* (*D. incarnata*, *D. maculata* subsp. *maculata*, *D. maculata* subsp. *transsilvanica*, and *D. sambucina*) [34, 85, 111, 125, 171]. Within this vegetation class, certain orchid species are cenobionts in communities of the order *Nardetalia strictae* (secondary mat-grass swards on nutrient-poor soils at low and middle altitudes of temperate, boreal, and subarctic regions of Europe), the order *Festucetalia spadiceae*, and especially the alliance *Nardion strictae* (mat-grass swards in the subalpine and alpine belts of the Alps, Carpathians, and northern Apennines) and the alliance *Potentillo ternatae-Nardion* (oligotrophic mat-grass swards of mountain ranges of the southern and central regions of the Balkan Peninsula) [34, 171]. Furthermore, a few species were found to grow in communities of the order *Seslerietalia comosae* and the alliance *Poion violaceae*, which represent alpine and subalpine silicicolous grasslands of the Balkan Peninsula [34]. According to the most recent studies, *Dactylorhiza sambucina* represents an indicator species of this vegetation type in the central Balkans (western Serbia) [26, 34].

Orchids are less prevalent in communities of the vegetation class *Calluno-Ulicetea* [34]. This vegetation represents a heath on acidic and nutrient-poor soils in the temperate and boreal zones of Europe [171]. In the central Balkans, communities of the alliance *Bruckenthalion spiculifoliae* (dwarf heath on siliceous substrates of the southern Carpathians and the Dinarides) are inhabited by *Dactylorhiza maculata* subsp. *maculata* and *Neotinea tridentata* [34].

5.2.2 Vegetation of Bogs and Fens

Fens, transitional mires, and bogs in the temperate, boreal, and Arctic zones of the Northern Hemisphere (class *Scheuchzerio palustris-Caricetea fuscae*) represent important vegetation types for many moisture-demanding orchid species [19, 25, 26, 133, 186]. Orchids are primarily registered in communities of the orders *Caricetalia fuscae* (sedge-moss vegetation of acidic fens in the boreal and temperate zones and in the supra-Mediterranean belt of mountains in Southern Europe) and *Caricetalia davallianae* (sedge-moss vegetation of calcareous and extremely mineral-rich fens in Eurasia) [34, 171]. The following orchid taxa were recognized as diagnostic taxa of this vegetation class: *Anacamptis palustris* subsp. *palustris*, *Dactylorhiza cordigera*, *D. fuchsii*, *D. incarnata* subsp. *cruenta*, *D. incarnata* subsp. *incarnata*, *D. incarnata* subsp. *ochroleuca*, *D. incarnata* subsp. *pulchella*, *D. lapponica* subsp. *angustata*, *D. lapponica* subsp. *lapponica*, *D. maculata* subsp. *maculata*, *D. maculata* subsp. *savogiensis*, *D. majalis* subsp. *majalis*, *D. majalis* subsp. *sphagnicola*, *D. russowii*, *D. traunsteineri* subsp. *curvifolia*, *D. traunsteineri* subsp. *traunsteineri*, *Epipactis palustris*, *Gymnadenia densiflora*, *G. frivaldii*, *Hammarbya paludosa*, *Herminium monorchis*, *Liparis loeselii*, *Spiranthes aestivalis*, and *S. sinensis* (Fig. 8) [13, 42, 68, 122, 126, 142, 171, 186, 187]. Other orchid taxa that are cited as cenobionts of these vegetation communities are *Dactylorhiza maculata* subsp. *transsilvanica*, *Gymnadenia conopsea*, *Neottia ovata*, *Platanthera bifolia*, *Pseudorchis albida*, and *Malaxis monophyllos* (Fig. 8h) [26, 50, 61, 85, 112, 186]. Recent research from the central Balkans indicated that this vegetation type is inhabited by 12 orchid species and subspecies [34], including 4 indicator species (*Dactylorhiza cordigera*, *D. maculata* subsp. *maculata*, *D. saccifera*, and *Gymnadenia frivaldii*) (Fig. 8) [26]. Moreover, it has been found that endemic orchids (*Dactylorhiza cordigera* subsp. *bosniaca* and *Gymnadenia frivaldii*) occur mainly in moderately mineral-rich relict oro-Mediterranean fens of the Balkans (*Narthecion scardici*), calcareous mineral-rich fens (*Caricion davallianae*), and moderately calcium-rich to low-calcium slightly acidic fens (*Caricion fuscae*) [34].

Members of the family Orchidaceae are less frequently found in communities of the class *Oxycocco-Sphagnetea*, which represent dwarf-shrub, sedge, and peat-moss vegetation of Holarctic bogs and wet heaths on extremely acidic soils [171]. This vegetation type is inhabited by a few orchid species that require strongly acidic soils. In Poland, *Hammarbya paludosa* was recorded in the community *Sphagnetum magellanici* [142], whereas in Central Europe *Neottia cordata* was found in the community *Andromedo polifoliae-Sphagnetum magellanici* [48] from the alliance *Sphagnion medii*. It should be noted that *Dactylorhiza majalis* subsp. *sphagnicola* is recognized as a diagnostic taxon of this vegetation class [171].

5.2.3 Marshland Vegetation

Some water-demanding orchid species were found to grow in communities of the vegetation class *Phragmito-Magnocaricetea* [34]. This is reedy swamp, sedge bed, and herb-land vegetation of freshwater or brackish water bodies and streams of Eurasia [171]. For example, it was found that *Epipactis palustris* inhabits communities of *Magnocaricion elatae* (marsh vegetation on oligotrophic to mesotrophic organic sediments of temperate Europe) [68]. In Germany, the Czech Republic, and Hungary, *Dactylorhiza incarnata*, *D. majalis*, *Epipactis palustris*, *Hammarbya paludosa,* and *Liparis loeselii* have been reported to grow in communities where *Phragmites australis* is highly represented [122, 185]. In the central Balkans, this vegetation type is inhabited especially by *Dactylorhiza incarnata* and *Epipactis palustris* within the vegetation orders *Magnocaricetalia* and *Phragmitetalia* [26, 34].

5.3 Anthropogenic Vegetation

Terrestrial orchid species have significant representation in various anthropogenic habitat types such as roadsides; abandoned quarries; railway embankments; forests influenced by industrial emissions, industrial terrains, and wasteland; city parks and hedges; plantations of ecologically alien and non-native trees; sandpits; and clay pits [162, 164, 166]. Of particular interest are studies showing that a large number of orchid species grow in cemeteries: 86 species in Turkey [188] and 29 species in

Fig. 8 Orchids inhabiting the vegetation of bogs and fens. (**a**) *Epipactis palustris*, (**b**) *Gymnadenia frivaldii*, (**c**) *Dactylorhiza cordigera*, (**d**) *Dactylorhiza incarnata*, (**e**) *Hammarbya paludosa*, (**f**) *Herminium monorchis*, (**g**) *Liparis loeselii*, (**h**) *Malaxis monophyllos* (**a–d** photos V. Djordjević; **e–h** photos M. Bobocea)

Albania [189]. Moreover, in research on 3 Mediterranean islands (Cyprus, Crete, and Lesbos), investigators found 77 orchid species and almost 7,000 individuals of these plants in 2 anthropogenic habitat types (cemeteries and roadside verges) [167]. The same authors found that *Serapias bergonii*, *Anacamptis sancta*, *A. coriophora* subsp. *fragrans*, and *Himantoglossum robertianum* were the most abundant species. These studies suggest that graveyards and roadside verges have a great conservation potential and can be considered as important refuges for orchids [167, 188]. Linear habitats along roadsides act as dispersion corridors that can contribute to linking of landscapes and create better connection between orchid populations [167]. Generally, the explanation for why roadside habitats are important for the survival and growth of many orchid species lies in the fact that competitively weak orchid species often inhabit newly created sites that are poorly covered by dominant plant species and that roadsides can act as ecotones, which are favorable for many orchid species [26, 167].

Several studies provide detailed insights into the performances of orchids growing in anthropogenic vegetation types, including reproductive success, the height of individuals, and inflorescences and pollinator diversity [165, 190]. For example, it has been found that *Epipactis helleborine* in ruderal vegetation in Poland had higher reproductive success than in natural vegetation types, which can be attributed to greater diversity of pollinator species and higher frequency of visits by pollinators, as well as to the larger size of plants growing in such habitats [165]. Investigations of morphology and size of the genome of this species in anthropogenic and natural habitats showed that ten biometric features of flowers differed significantly between habitats [190]. These plants were taller in anthropogenic habitats than those from natural populations, and genome size in some populations was significantly different between plants growing in natural and anthropogenic habitats [190]. In addition, it was found that reduced seed weight is characteristic of *E. helleborine* in anthropogenic habitats, which is in part associated with its inability to adapt to the lower nutrient availability in soils of these habitats [191]. Moreover, scientists found that orchids inhabiting anthropogenic vegetation most often have rapid growth and flower production [164]. In the case of *Himantoglossum* species, it has been established that their reproductive success is weaker near roads [166]. It is nevertheless considered that traditionally managed roadside verges can ensure long-term persistence of the species in question and be important refuges for them [166, 167].

Representatives of the family Orchidaceae often inhabit non-native poplar plantations. For example, in Poland, *Cephalanthera longifolia*, *Epipactis helleborine*, *Epipactis × schmalhausenii*, *Platanthera bifolia*, *Neottia ovata*, and *Dactylorhiza majalis* were registered in fast-growing hybrid poplar (*Populus × canadensis*) plantations [164]. The authors emphasized that orchid populations occupying these anthropogenous habitats often have a high population density and abundance, which is explained primarily by good light conditions, limited competition from other plants, and disturbance of the upper soil levels [164]. For example, 677 individuals of *Platanthera bifolia* per 100 m^2 and 275 individuals of *Epipactis × schmalhausenii* per 100 m^2 were recorded in poplar plantations in Poland [164]. Moreover, orchids growing under a canopy in poplar plantations reached a significant height (up to 130 cm in the case of *E.* × *schmalhausenii* and up to 92 cm in the case of *Platanthera bifolia*) [164].

Some studies provide detailed knowledge about the anthropogenic vegetation units in which orchids grow. Thus, in the central Balkans (western Serbia), orchids were reported in some ruderal communities of the class *Artemisietea vulgaris* [34]. Specifically, *Anacamptis morio* and *Neotinea ustulata* were found in communities of the order *Agropyretalia intermedio-repentis* (semiruderal grasslands and herblands and weed segetal vegetation of perennial crops in the nemoral, forest-steppe, and subboreal zones of Europe), whereas *A. morio* and *Orchis purpurea* were registered in communities of the order *Onopordetalia acanthi* (subxeric ruderal vegetation dominated by short-lived perennials) [34, 171]. Moreover, *O. purpurea* was the only species recognized as an indicator species of the ruderal vegetation type in western Serbia [26].

Disa bracteata (the South African orchid) and *Microtis media* (the common mignonette orchid) are two relatively ruderal (weedy) orchid species in Australia [192]. In the case of *M. media*, it can be asserted that mycorrhizal association with species of the fungal genus *Tulasnella* enables this orchid to inhabit anthropogenic vegetation types, bearing in mind that fungal pathways contribute significantly to its ability to exploit the minimal phosphorus reserves in the soil [193].

6 Effects of Disturbance

Based on Grime's theory of population strategies [194], orchids are classified as stress tolerators [2], whereas other authors classify them between ruderal plants and stress tolerators [5]. This position in Grime's C-S-R (competitors – stress tolerators – ruderals) triangle indicates that orchids can grow under conditions of certain stress and disturbance and are not competitively strong. Moreover, specific studies have shown that some degree of disturbance may reduce the competition between orchids and other plants and thus have a beneficial effect on the development of orchid populations [42], whereas the growth and survival of some terrestrial orchids require natural disturbances such as fires that maintain the forest openings necessary for long-term orchid survival [195].

Since many terrestrial orchids grow in grasslands and meadows, i.e., secondary forest-free areas created by humans, appropriate management of these habitats is needed for the existence of these orchid populations [42, 196]. Activities and management practices such as mowing and grazing actually help to maintain the relatively low-growing herbaceous vegetation and are favorable for most terrestrial orchids of these habitats. Thus, regular annual mowing, especially late in the season, is essential for the optimal development of many species of the genus *Dactylorhiza* [42, 43]. Scientists indicate that mowing improves the performance of *Dactylorhiza majalis* because the presence of constantly larger individuals is maintained and the rate of disappearance of individuals is reduced [42, 43]. The negative effects of unmown grasslands on orchids have to do with reduction of light reaching the lower parts of grassland vegetation, which negatively affects orchid populations and their viability [42]. Community stands with a high degree of coverage of dominant *Filipendula ulmaria* were identified as especially bad sites for growth and development of *D. majalis* [42]. This can be explained by the fact that orchid species with shorter stature are often unable to overcome the increased competition from associated tall-growing species and consequent decrease of light availability within these stands [23]. It has been shown that *Dactylorhiza incarnata* is capable of withstanding reduced light availability to a certain extent by increasing its vertical growth [185]. Since populations of this species increased exponentially during the first 10 years after the re-introduction of mowing, the authors recommended an initially higher mowing frequency at previously abandoned sites [185]. Mowing of roadside verges is suggested as a highly favorable practice from the point of view of orchid conservation because the regularly mowed 0–2 m margin of roads is the part of the roadside verge most important for the growth of orchid individuals [167]. Certain research has provided insights into when and to what extent mowing best influences the performance of orchid populations, highlighting the importance of modified mowing regimes. Thus, in Merseyside (England), it was found that delaying spring/summer mowing until 15 July each year led to an increase in the population of *Anacamptis morio*, whereas populations of adjacent habitats subjected to frequent regular mowing or early and heavy grazing showed no such increase [197]. In the Netherlands, site management with mowing in July and sheep grazing in winter is beneficial to populations of *Coeloglossum viride* [196]. Scandinavian authors suggested that *Dactylorhiza lapponica* in central Norway would benefit the most from traditional mowing performed after seed dispersal, but would also have a high probability of future survival in the absence of mowing [198].

It is therefore important to note that abandoning traditional practices such as mowing or grazing can influence the survival of many orchid species, bearing in mind that species which require open habitats are threatened by vegetation succession (development of forest and scrub vegetation). In the absence of traditional land management practices, orchids growing in fen communities are particularly endangered in view of the fact that vegetation overgrowth with woody trees occurs under such conditions. Moreover, research in the UK showed that orchids inhabiting calcareous grasslands are the most endangered ones, a circumstance associated with the loss of a very high percentage of these traditionally sheep-grazed grassland ecosystems [23]. In agreement with the results obtained in Europe, controlled grazing was found to have a positive effect on certain orchids in North America. Specifically, it was found that reduction in the level of competition from shrubs and trees has a direct beneficial effect on the development of populations of *Spiranthes* species and *Cypripedium reginae* in North America [5].

There are studies showing that some degree of natural and anthropogenic disturbance of forest communities has a positive impact on the survival and development of orchid populations [88]. For example, the authors highlighted the importance of regular coppicing of forests in maintaining viable populations of *Orchis mascula* over the long term, as it increases the probability of flowering, fruit and seed set, and seedling recruitment [87]. Furthermore, it was found that fencing in *Pinus sylvestris* forests of Scotland favors the establishment of *Goodyera repens* colonies [199]. It seems that the maintenance of a complete canopy cover is necessary to restrict competition from other plants and provides the shade conditions to which this species is adapted [104]. In Greece, however, pine forests, which constitute the main habitat of *G. repens*, are in rapid decline, primarily due to their replacement with beech and spruce forests, which is a consequence of natural succession and reduced levels of disturbance caused by grazing and fires [104]. Scientists investigating the response of the lady's slipper orchid (*Cypripedium calceolus*) to tree removal [89] demonstrated that tree removal sites had greater orchid density, higher odds of orchid survival, and better flowering and fruiting of orchids than the control sites. Although most of these effects disappeared after 3 years, when the canopy gaps closed, this study suggests that selective tree harvesting represents a suitable management practice that increases the size of populations of the given orchid species [89]. Other investigators explored the influence of clear-cutting, green-tree retention and artificial drainage on the abundance of 11 terrestrial orchid species in Estonia [73]. Although the highest number of species was found on artificially drained plots and in mature stands, most sites were inhabited by shade-tolerant orchid species that disappeared after timber harvesting. Moreover, this study showed that cut-over areas (3–7 years after harvest) did not host any species that were not present from uncut forest stands and that keeping solitary trees had no impact on orchid abundance in comparison with clear-cut areas ones [73].

Fire is an important disturbance factor, especially in areas characterized by the Mediterranean climate [2]. In Europe, the effects of fire have not been well studied, and there is little information about its effects on orchids. However, it is known that forest fires favor orchid populations, as they maintain open space in phrygana, maquis, and other scrublands. In contrast to the European continent, the effects of fire on orchid performance have been investigated in detail on some terrestrial orchids in Australia. Thus, it was established that management of the habitats of *Prasophyllum correctum* should involve frequent burning, the benefits of which for this species include reduction of competition from grasses, shortening of dormancy periods, reduction of mortality, and enhancement of flowering [195]. Moreover, there are some orchids that need fire for their growth, so that accidental fires provide conditions favorable for their flowering. *Burnettia cuneata*, *Pyrorchis nigricans*, *Leptoceras menziesii*, and *Prasophyllum australe* are such fire-dependent orchid species [200]. The same author states that there are also fire-stimulated species (many species of the genera *Caladenia*, *Diuris*, *Prasophyllum*, and *Thelymitra*), fire-neutral species (some representatives of the genus *Pterostylis*), fire-sensitive species (*Pterostylis alveata* and *Corunastylis despectans*), and fire-killed species (ones that die due to intense fire, e.g., *Thynninorchis huntianus*) [200].

7 Orchid Specialists and Generalists

Terrestrial orchids, depending on the degree of habitat specialization, can be divided into two major groups, generalists and specialists [25, 26]. However, these two categories can only be considered as the extreme ends of the continuum. It should be noted that the extent of tolerance of orchid species to different environmental factors depends on the level of competition, geographical and historical factors, the level of pollinator specialization, and the level of specialization with associated mycorrhizal fungi [3].

Generalist orchid species are those occurring across a wide range of vegetation types with widely varying conditions, such as different geological substrates, different light conditions, varying degrees of soil moisture and atmospheric humidity, and a wide range of altitudes. Specifically, the altitudinal range where a particular species occurs is directly related to breadth of the ecological niche of a given species, which means that species occurring from lowland to high-altitude areas have a greater physiological tolerance and ability to inhabit different vegetation types and thus to grow sympatrically with different species. Moreover, generalists are known to be usually widespread species, whereas specialists tend to occur in more restricted areas [201, 202]. The most significant orchids in Europe considered to be generalists are *Epipactis helleborine*, *Neottia ovata*, *Platanthera bifolia*, *Cephalanthera longifolia*, *Anacamptis morio*, *Dactylorhiza sambucina*, *D. fuchsii*, *D. saccifera*, *Gymnadenia conopsea*, and *Neotinea ustulata* [9, 12, 24–26, 73, 112, 133]. Recent studies in the central Balkans have particularly highlighted species that inhabit a wide range of vegetation units (*Gymnadenia conopsea*, *Anacamptis morio*, *Dactylorhiza saccifera*, *D. sambucina*, *Neottia ovata*, and *Platanthera bifolia*) and species growing on a large number of geological substrates (*Cephalanthera longifolia*, *Dactylorhiza sambucina*, *Epipactis helleborine*, *Gymnadenia conopsea*, *Neottia nidus-avis*, and *Platanthera bifolia*) [34, 107]. The obtained results emphasize the high ecological plasticity and adaptability of these orchids.

On the other hand, specialist terrestrial orchids are ones having a narrow range of ecological requirements, inhabiting a small number of vegetation communities, occurring in a narrow range of altitudes and represented on a small number of geological substrates. Among the most common specialist orchids in Europe are the following: *Dactylorhiza traunsteineri*, *D. lapponica*, *D. macedonica*, *D. cordigera*, *Gymnadenia frivaldii*, *Nigritella rhellicani*, *Hammarbya paludosa*, *Herminium monorchis*, *Liparis loeselii*, *Spiranthes aestivalis*, *Cephalanthera cucullata*, and *Epipactis cretica* [9, 25, 26, 99]. It is noticeable that a large number of orchid specialists are those which inhabit wet habitats, such as wet meadows or fen communities, as well as species that inhabit warm and dry habitats. This is consistent with the general statement that the majority of highly specialized species occur at the extreme ends of environmental gradients and in extreme and rare vegetation types [201, 202].

Certain authors found that orchid taxa are relatively common and have large populations close to the center of their geographic distribution, whereas they are rarer and characterized by smaller populations toward the boundaries of their ranges [46]. At the same time, the levels of specialization of particular orchid species vary depending on the geographical area, i.e., on the position of populations of different species in relation to the center/edges of their ranges. Although this is a situation perceived rather intuitively by many orchidologists, recent studies from the Balkan Peninsula provide rigorous and thoughtful arguments to substantiate it on the basis of numerical analyses [24–26, 99]. Thus, it was found that orchids belonging to the

Central European and boreal chorological groups generally have higher levels of specialization in northeastern Greece and on Crete than in the central Balkans, which is attributable primarily to differences in the climates of these areas [24, 26]. In the central Balkans (western Serbia), there is a humid and continental climate that allows the widespread distribution of wet and mesophilous habitat types, whereas in northeastern Greece, and especially on Crete, the Mediterranean climate has strong influence [26]. At the same time, most orchids of Central or North European origin that have the southern boundaries of their distribution on the Balkan Peninsula occur at high altitudes in Greece, whereas in Central and Northern Europe, they grow in habitats ranging from lowlands to highlands [25]. On the other hand, orchids belonging to the Mediterranean and sub-Mediterranean chorological groups (e.g., *Anacamptis papilionacea*, *A. pyramidalis*, *A. laxiflora*, *Orchis simia*, and *Ophrys scolopax* subsp. *cornuta*) have a higher level of specialization in the central Balkans compared to the degree of their specialization in northeastern Greece and on Crete [24–26, 99].

Particularly noticeable differences of orchid specialization are evident when orchids growing in Central Europe are compared with those occurring on the Balkan Peninsula. Thus, some moisture-demanding orchids (e.g., *Epipactis palustris*, *Dactylorhiza incarnata*) have a higher level of specialization in Southern Europe (e.g., in Greece) [25] than in Central Europe [9, 68]. *Epipactis pontica* grows exclusively in communities of *Fagion sylvaticae* and is considered a specialist species in the central Balkans [128], but it is less specialized in Slovakia, where it grows in forest communities of the vegetation alliances *Luzulo-Fagion sylvaticae*, *Fagion sylvaticae*, *Quercion confertae-cerris*, and *Carpinion betuli* [127]. In addition, *Dactylorhiza fuchsii* is considered to be one of the least specialized species in Central Europe [9], but it is very rare and specialized in the central Balkans [26, 177]. This general trend is obvious in the case of Crete, where some of the most widespread species of European shrimp (*Cephalanthera damasonium*, *C. longifolia*, and *Neottia ovata*) are categorized as specialists [99]. Overall, the results of these analyses suggest that levels of specialization of orchid species increase from the center to the edges of species ranges [24, 26].

8 Orchid Mycorrhizal Fungi

Mycorrhizal fungi play a crucial role in the growth of orchids, especially in the early stage of development that is during germination and phases of seedling establishment [62]. Mycorrhizal dependency varies depending on the stage of orchid development. While the need for specialized fungi during seed germination is well documented, the degree of dependence in adult orchid specimens is less known and varies throughout autotrophic and heterotrophic phases of the orchids [3]. As orchid plants mature, their dependency changes to partial mycoheterotrophy, in which plants utilize carbon both from their fungal partners and from photosynthesis [203]. However, recent studies showed that partial mycoheterotrophy plays a great role in rhizoctonia-associated orchids, even under full light conditions in open meadow habitats [203]. Some terrestrial orchids do not photosynthesize at all throughout their lives, so they are completely mycoheterotrophic. It should be noted that mycoheterotrophy exists in several plant families but is most common in the family Orchidaceae, in which over 100 fully mycoheterotrophic species have been recorded [7]. Since orchids have stages that depend fully or partially on fungi for carbon and other elements, the distribution and abundance of orchid populations depend on the distribution and abundance of orchid mycorrhizal fungi [204, 205]. Scientists found that seed germination and protocorm development reflect the distribution and abundance of mycorrhizal fungi and that fungal "hot spots" represent a key factor in the dynamics of orchid populations [206]. The distribution of mycorrhizal fungi is probably independent of the distribution of orchids and the ability of orchids to engage in symbiosis with fungi that otherwise live freely is their unique feature [7]. Most fungi found to be associated with terrestrial orchids are basidiomycetes (*Ceratobasidiaceae*, *Sebacinales*, and *Tulasnellaceae*) [7, 204, 207]. Among the best-known genera, *Rhizoctonia*, *Ceratobasidium*, *Sebacina*, and *Tulasnella* [208] should be particularly emphasized. Some orchids are associated with ascomycetes, primarily members of the genera *Tuber*, *Peziz*, *Wilcoxina*, *Tricharina*, and *Phialophora* [204, 207, 209]. The majority of fully mycoheterotrophic orchids and certain partially mycoheterotrophic ones mainly associate with ectomycorrhizal agaricomycetes, including many species of *Russulales*, *Thelephorales*, and *Sebacinales* [204], whereas some orchids associate with ectomycorrhizal ascomycetes [209].

Many studies showed that orchids use carbohydrates, minerals, and water of mycorrhizal fungi without adequate reciprocal benefit for the fungi [3]. However, other studies suggested that there are mutually beneficial interactions between orchids and mycorrhizal fungi: carbon produced by plant photosynthesis is obtained in exchange for access to nutrients from the soil such as nitrogen, phosphorus, vitamins, and amino acids [7]. Fungi supply orchids with carbon, vitamins, hormones, minerals, and other physiologically active substances necessary for growth by decomposing organic compounds from the substrate, especially cellulose and lignin. However, the relationship between orchids and fungi ranges from parasitism on fungi to commensalism and may approach neutralism [210]. On the other hand, mutualism was recorded in *Goodyera repens* [149, 210]. The dependence of orchids on their symbiotic fungi is obvious in the case of albino orchids and when orchids spend long periods of dormancy. Both cases demonstrate that the fungi do not rely on carbon produced by orchids through their photosynthesis and can grow well even without these orchid plants [204].

It is worth mentioning that the interactions between orchid plants and fungi show a higher degree of specialization than those recorded in other plants [3]. Thus, the rarity and long-term survival of some orchid species can be explained by the degree of mycorrhizal specificity and the fact that the mycorrhizal association is restricted to fungal species limited in distribution [7, 211]. The fact is that a high level of dependency between orchid species and mycorrhizal fungi is seen as an increased risk of extinction [3]. Moreover, this risk is linked not only to the level of specialization but also to the ability of orchid species to switch from one fungal partner to another under changed environmental conditions [3].

It has been hypothesized that orchids that can swap and share mycorrhizal fungi have a broader distribution than those species that are highly specialized for fungi with a limited distribution [3]. However, recent research indicated that the rarity and persistence of orchids are not necessarily related to fungal diversity, and it is suggested that other factors may play a more important role in determining the persistence of orchids [211]. This has been proved in the case of endemic Australian terrestrial orchids with limited ranges, whose rarity and distribution patterns are affected to a greater extent by soil characteristics and pollination systems than by the narrow distributions of fungal species [212]. In Europe, study of *Cypripedium calceolus* showed that this orchid has a fungal partner with one of the narrowest of ranges, but is nevertheless a widespread species [213]. In general, orchids with a high degree of exploitation of mycorrhizal fungi exhibit a high degree of specialization toward mycorrhizal symbionts and above all toward fungi that form ectomycorrhizal relationships with trees [3, 7]. These orchids therefore can be considered plant epiparasites because they exploit ectomycorrhizal networks between fungi and other neighboring plant species [7]. Examples include *Corallorhiza maculata* and *C. mertensiana*, which specialize in ectomycorrhizal species from the family *Russulaceae* [7], and *Neottia nidus-avis*, which specializes in fungi from the family *Sebacinaceae*, orchids that form ectomycorrhizae with trees [81]. Overall, high mycorrhizal specialization occurs primarily in species that are completely mycoheterotrophic, whereas the specificity of photosynthetic orchids can vary [7]. Thus, it has been found that 16 photosynthetic species of Mediterranean orchids from the genera *Anacamptis*, *Orchis*, *Ophrys*, and *Serapias* have a low degree of specialization in mycorrhizal fungi [214]. Mycorrhizal specialization is most likely associated with the one-sided nature of the relationship between orchids and fungi [7]. To be specific, if the fungi have little or no benefit from the symbiosis, then orchids can be considered parasites, and they often exhibit high specialization due to selection driven by evolutionary “arms-races” [7]. On the other hand, some authors assumed that the degree of specialization is correlated with the degree of heterotrophy [170].

It is important to emphasize that narrow orchid specificity toward fungi has a major impact on the ecology and distribution of orchids [7, 204, 215]. At the same time, some authors believe that there may be an impact on orchid diversity [7, 213]. Specifically, mycorrhizal specialization can encourage orchid diversification by affecting orchid distribution patterns. Fragmentary fungal distribution, together with high mycorrhizal specialization, may be responsible for extremely dispersed orchid populations [7]. The consequences of long-range orchid seed transmission can be small effective population sizes and reduced gene flow, results that create conditions favorable for the drift-selection model of orchid speciation [7, 216]. The mycorrhizal relationship between orchids and fungi thus may increase the potential for faster emergence of new species in the family Orchidaceae [7].

9 Pollination Systems

Orchid pollination has intrigued scientists from the time of Darwin, primarily because of its complexity and great diversity [3]. Orchid pollination systems are often mistakenly considered to be the outcome of co-evolutionary processes [217]. However, co-evolution between orchid species and their pollinators is most likely rare, and evolutionary changes in orchids are largely unilateral, with no evolutionary changes in pollinators [218]. Insects of the order Hymenoptera represent the most numerous pollinators of European orchids, followed by butterflies (Lepidoptera), whereas members of the order Diptera represent a smaller number of pollinators [217, 219]. Furthermore, the fertilization of some terrestrial orchids can result from autogamy or even cleistogamy.

The most common pollination system among orchids is the system of rewarding, where insects receive mostly nectar as a reward for pollination. However, a specific characteristic of the family Orchidaceae is very high representation of non-rewarding pollination systems. Specifically, as many as one-third of the total number of orchid species (between 6500 and 9000 species) deceive their pollinators [218]. There are several mechanisms by which orchids deceive their pollinators, mechanisms such as generalized food deception for example, where orchids advertise floral signals that are characteristic of rewarding plant species [7, 218]. In sexual deception, orchid flowers imitate the mating signals of female insects, especially using odors that mimic female insect pheromones and tactile or visual cues, and are then pollinated by male insects trying to copulate with the flowers [7, 218]. Other deception mechanisms include Batesian floral mimicry, brood-site imitation, shelter imitation, pseudo-antagonism, and rendezvous attraction [218]. The pollination system that an orchid species uses has a direct effect on pollination success. Specifically, it has been found that non-rewarding orchids, on average, have lower frequencies of visitation by pollinators and lower reproductive success than rewarding orchids [220]. Darwin therefore suggested that low levels of fruit set, which are typical of non-rewarding species, might be a key factor in determining orchid rarity [221]. However, results obtained in a recent study conducted in Belgium and the Netherlands showed that orchid distribution patterns are not related to nectar reward and that the relationship between nectar rewards and extinction of orchids is not significant [222]. Another recent study indicated that the relative occurrence of food-deceptive orchids decreases with increasing altitude on the territory of Switzerland and in the Vaud Mountains [35]. The authors of that study found that this may be linked to altitude-dependent climatic factors such as temperature and precipitation and factors that cause decrease in the pollinator visitation rate at high altitudes. In high-altitude areas, the growing season becomes shorter, and consequently plant species tend to flower simultaneously, which results in increasing the levels of competition among plants for access to pollinators and reduced access to pollinators that have not learned to recognize rewarding and non-rewarding orchid flowers [35].

Many authors have stressed that pollinator limitation and specialization may be an important factor affecting the distribution of orchids, especially near the margins of their distribution [3, 6, 219]. It has been found that orchids that are pollinated by a large number of diverse pollinators most often have a widespread distribution. An example of such a species, *Epipactis helleborine*, has great pollinator diversity, and among other things, this fact allows it to grow in both natural and anthropogenic habitats and helps to explain why it is one of the few orchids to have successfully colonized North America [165]. On the other hand, some species of the genus *Ophrys* are highly specialized and this causes their rarity. *Ophrys* pollinators can be divided into three groups: food generalists, food specialists, and parasitic specialists [219]. The sexual deception present in this genus imposes a high degree of specialization, as insect pheromones are species-specific in most cases. Moreover, a large number of *Ophrys* pollinators include solitary bees that have pollen specialization (oligolecty) [219]. In general, researchers found that sexually deceptive orchids are usually pollinator specialists, whereas the majority of food-deceptive orchids are pollinator generalists [7]. Some authors suggested that specialized pollination strategies influence the diversification of orchids and increase the risk of extinction, especially if environmental changes affect their long-term survival and evolutionary potential [3]. It should be noted that a great number of orchids are pollinated by specialized insects, which often require specific conditions (e.g., specific nesting sites, the presence of surrounding nectar plants and host plants for egg-laying, existence of brood cell parasites, and pollen specialization), suggesting that the relationship between orchids and pollinators is fragile and that many orchids are thereby rendered prone to extinction [3, 219].

Self-pollination occurs in a small number of representatives of the family Orchidaceae, more precisely in about 3% [217] or between 5 and 20% of the total number of orchid species [223]. The number of completely self-pollinating orchid species is small, and self-pollinating species more often are ones that have cross-pollination in addition to self-pollination [2]. Facultative autogamy occurs in many orchid species and is an appropriate strategy when the frequency of cross-pollination is low [223]. In Europe, self-pollination is present primarily in species from the genera *Epipactis*, *Cephalanthera*, and *Neottia*, followed by members of the genera *Corallorhiza* (*C. trifida*), *Limodorum* (*L. trabutianum*), *Neotinea* (*N. maculata*), *Ophrys* (*O. apifera*), and *Serapias* (*S. parviflora*) [13, 219]. It has been found that the number of self-pollinated orchids increases with increasing latitude, as well as in isolated geographical areas [216]. Moreover, most self-pollinated orchids occur in high-altitude areas and by self-pollination overcome the lack of pollinator availability there [30, 223]. The highest percentage of self-pollinating orchids (about 50%) was found in boreal regions [223] and on Réunion Island [30]. In eastern Canada, self-pollination was reported in 17% of the total number of orchid species, whereas in Europe it occurs in 27–50% of orchid species [216, 223].

Self-pollination in orchids may be advantageous in habitats where high levels of disturbance cause uncertain activity of pollinators [223]. It is important to emphasize that self-pollination reduces the rate of pollen export and the number of seed embryos. At the same time, autogamy reduces the level of genetic variation and can lead to inbreeding depression, resulting in fewer offspring and a lower survival rate [216].

10 Conclusions

Terrestrial orchids include a great variety of species that are widespread on all continents and characterized by specific life histories and varying sensitivity to changing habitat conditions. The importance of environmental factors affecting the distribution, abundance, and richness of orchids varies depending on the geographical scale. It can be concluded that latitude, area size, macroclimate, and evolutionary history have an important impact on the distribution and abundance of orchids on a large scale. On the other hand, soil moisture, the light regime, geological substrates, soil characteristics, the disturbance regime, and the specificity of mycorrhizal fungi and pollinators play an important role on a fine scale. The importance of the elevation range is recognized both on the macroscale level and at regional levels.

In general, the number of terrestrial orchid species increases with decreasing latitude, which is among the most consistent of discernible patterns. Thus, the northern countries of Europe have the smallest number of orchid species, while the number of orchid species gradually increases toward southern areas of the continent, reaching a peak in the Mediterranean area. It is assumed that orchid species richness in most European countries has a hump-shaped pattern with the highest species number at middle elevations. This unimodal pattern can be caused by climatic factors along the elevational gradient, by some spatial aspects like geometric constraints, by size of the region, and by biotic processes.

Temperature, precipitation, and the light regime play an important role in determining patterns of growth, development, flowering, population dynamics, abundance, and distribution of terrestrial orchids. The role of temperature and precipitation is particularly pronounced on the southern and northern borders of species distribution. Studies have indicated that climatic parameters, most often in the previous and current seasons, influence the number of flowering individuals. As a result of global warming and climate change, many terrestrial orchids have altered their performances. However, the degree of such changes in the performances of orchids depends on the type of their life history and especially on pollination systems. Furthermore, the importance of climate variables in predicting the distribution of terrestrial orchids should be emphasized.

Variation in availability of soil resources (water and nutrients) across geological substrates significantly affects the richness and composition of orchid species. Most orchids in Europe grow on calcareous geological substrates and soils, moderately damp soils, slightly acidic to slightly alkaline soils, and soils that are relatively poor in nutrients. However, a surprising number of orchids have recently been discovered on non-calcareous substrates, primarily felsic, intermediate, mafic, and even ultramafic igneous rocks, as well as on metamorphic and silicate sedimentary rocks in the central Balkans. Future research should therefore be focused on the study of ecophysiological characteristics, the potential for trace element accumulation, and the phytochemistry of orchids growing on these substrates.

Vegetation types significantly affect patterns of distribution and abundance of orchids, as well as separation of their ecological niches. Although terrestrial orchids inhabit almost all known vegetation types in Europe, the greatest species diversity is recorded in various deciduous forests (beech, oak, and hornbeam forests); coniferous and mixed coniferous-deciduous forests (spruce, fir, and pine forests); Mediterranean vegetation types, especially scrubs; different grassland and meadow types including heaths; montane-subalpine tall-herb vegetation; and fens, bogs, and marshes. Moreover, anthropogenic vegetation types are also inhabited by terrestrial orchids, and it is often asked whether they can play an important role as refuges for endangered species. The importance of disturbance effects should be emphasized, since traditional management practices such as mowing and grazing may reduce the competition between orchids and other plants and provide favorable light conditions, thus having a beneficial effect on the development of orchid populations.

The patterns of distribution and abundance of terrestrial orchids are largely determined by characteristics of their life histories, especially by the existing three belowground strategies determined by their root systems. Most terrestrial orchids that can tolerate colder conditions and high moisture soils in Europe are species with palmate and fusiform tubers. At the same time, these species are the ones most prevalent at higher altitudes. On the other hand, most orchids that best tolerate high temperatures and dry conditions are species with ovoid tubers. Consequently, these species are dominant in low- to mid-elevation areas.

Although the influence of environmental factors on the distribution and abundance of terrestrial orchids can be viewed individually, these factors in reality act cumulatively in nature. According to the degree of specialization in relation to habitat conditions, terrestrial orchids can roughly be divided into two large groups – specialists and generalists. It has been pointed out that the levels of specialization of orchid species increase from the center to the edges of their ranges. Moreover, the abundance and distribution of orchids largely depend on the abundance and distribution of adequate mycorrhizal fungi, pollinator limitation, and specialization. To sum up, it can be stated that awareness of the relationships between terrestrial orchids and ecological factors is a necessary prerequisite for the success of future research and conservation of this intriguing group of plants.

Acknowledgments This study was supported by the Ministry of Education, Science and Technological Development of the Republic of Serbia under Grant [number 173030]. ST was partially supported by the Ministry of Education, Youth and Sports of the Czech Republic within the National Sustainability Program I (NPU I) [LO1415]. The authors would like to thank Mihai Bobocea for providing photos of specific orchids. We are grateful to Prof. Dr. Vladimir Stevanović, Prof. Dr. Slobodan Jovanović, and Prof. Dr. Dmitar Lakušić for useful suggestions and information. We are very grateful to Mr. Raymond Dooley, native English editor for the proofreading of the manuscript.

References

1. Chase M, Christenhusz M, Mirenda T (2017) The book of orchids: a life-size guide to six hundred species from around the world. Ivy Press, London
2. Dressler RL (1981) The orchids: natural history and classification. Harvard University Press, Cambridge, MA
3. Swarts ND, Dixon KW (2009) Terrestrial orchid conservation in the age of extinction. Ann Bot 104:543–556
4. Chase MW, Cameron KM, Freudenstein JV, Pridgeon AM, Salazar G, van den Berg C, Schuiteman A (2015) An updated classification of Orchidaceae. Bot J Linn Soc 177:151–174
5. Hágsater E, Dumont V (eds) (1996) Orchids: status, survey and conservation action plan. IUCN, Gland/Cambridge, UK
6. Whigham DF, Willems JH (2003) Demographic studies and life-history strategies of temperate terrestrial orchids as a basis for conservation. In: Dixon KW, Kell SP, Barrett RL, Cribb PJ (eds) Orchid conservation. Natural History Publications, Kota Kinabalu

7. Waterman RJ, Bidartondo MI (2008) Deception above, deception below: linking pollination and mycorrhizal biology of orchids. J Exp Bot 59:1085–1096
8. Averyanov LV (1990) A review of the genus *Dactylorhiza*. In: Arditti J (ed) Orchid biology: reviews and perspectives, vol V. Timber Press, Oregon
9. Delforge P (2006) Orchids of Europe, North Africa and the Middle East. A & C Black, London
10. Tatarenko I (2007) Growth habits of temperate terrestrial orchids. In: Cameron KM, Arditti J, Kull T (eds) Orchid biology – reviews and perspectives, vol IX. The New York Botanical Garden Press, Bronx/New York
11. Tsiftsis S, Štípková Z, Kindlmann P (2019) Role of way of life, latitude, elevation and climate on the richness and distribution of orchid species. Biodivers Conserv 28:75–96
12. Buttler PK (1991) Field guide to orchids of Britain and Europe. The Crowood Press, Swindon
13. Tsiftsis S, Antonopoulos Z (2017) Atlas of the Greek orchids, vol I. Mediterraneo Editions, Rethymno
14. Antonopoulos Z, Tsiftsis S (2017) Atlas of the Greek orchids, vol II. Mediterraneo Editions, Rethymno
15. Nikolić T (2019) Flora Croatica Vol. 4: vascular flora of the Republic of Croatia. Excursion flora, Alfa d.d., Zagreb
16. GIROS (2009) Orchidee d'Italia – Guida alle orchidee spontanee. Il Castello, Cornaredo (Milano)
17. Kreutz CAJ, Çolak AH (2009) Türkiye Orkideleri. Rota Yayınlari, Istanbul
18. Vogt-Schilb H, Pradel R, Geniez P, Hugot L, Delage A, Richard F, Schatz B (2016) Responses of orchids to habitat change in Corsica over 27 years. Ann Bot 118:115–123
19. Bernardos S, García-Barriuso M, Sánchez-Anta MA, Amich F (2007) Composition, geographical affinities and endemism of the Iberian Peninsula orchid flora. Nord J Bot 25:227–237
20. Baumann H, Künkele S, Lorenz R (2006) Die Orchideen Europas. Mit angrenzenden Gebieten. Eugen Ulmer KG, Stuttgart
21. WCSP (2019) World checklist of selected plant families. Royal Botanic Gardens, Kew. http://apps.kew.org/wcsp/
22. Tranchida-Lombardo V, Cafasso D, Cristaudo A, Cozzolino S (2011) Phylogeographic patterns, genetic affinities and morphological differentiation between *Epipactis helleborine* and related lineages in a Mediterranean glacial refugium. Ann Bot 107:427–436
23. Kull T, Hutchings MJ (2006) A comparative analysis in decline in the distribution ranges of orchid species in Estonia and the United Kingdom. Biol Conserv 129:31–39
24. Djordjević V, Tsiftsis S, Lakušić D, Stevanović V (2016) Niche analysis of orchids of serpentine and non-serpentine areas: implications for conservation. Plant Biosyst 150:710–719
25. Tsiftsis S, Tsiripidis I, Karagiannakidou V, Alifragis D (2008) Niche analysis and conservation of the orchids of east Macedonia (NE Greece). Acta Oecol 33:27–35
26. Djordjević V, Tsiftsis S, Lakušić D, Jovanović S, Stevanović V (2016) Factors affecting the distribution and abundance of orchids in grasslands and herbaceous wetlands. Syst Biodivers 14:355–370
27. Acharya KP, Vetaas OR, Birks HJB (2011) Orchid species richness along Himalayan elevational gradients. J Biogeogr 38:1821–1833
28. Zhang SB, Chen WY, Huang JL, Bi YF, Yang XF (2015) Orchid species richness along elevational and environmental gradients in Yunnan, China. PLoS One 10:e0142621
29. Zhang Z, Yan Y, Tianb Y, Lib J, Hea JS, Tanga Z (2015) Distribution and conservation of orchid species richness in China. Biol Conserv 181:64–72
30. Jacquemyn H, Micheneau C, Roberts DL, Pailler T (2005) Elevational gradients of species diversity, breeding system and floral traits of orchid species on Réunion Island. J Biogeogr 32:1751–1761
31. Ackerman JD, Trejo-Torres JC, Crespo-Chuy Y (2007) Orchids of the West Indies: predictability of diversity and endemism. J Biogeogr 34:779–786
32. Štípková Z, Traxmandlová I, Kindlmann P (2016) Determinants of orchid species diversity in Latin America. Lankesteriana 16:293–297
33. Barbaro L, Dutoit T, Grossi JL (2003) Influence des facteurs agro-écologiques sur les assemblages d'orchidées dans les pelouses calcicoles du Vercors (Préalpes, France). Bot Helv 113:63–79

34. Djordjević V (2018) Spatial distribution and ecology of orchids (Orchidaceae) of western Serbia. Dissertation, University of Belgrade (in Serbian with English abstract)
35. Pellissier L, Vittoz P, Internicola AI, Gigord LDB (2010) Generalized food-deceptive orchid species flower earlier and occur at lower altitudes than rewarding ones. J Plant Ecol 3:243–250
36. Körner C (2007) The use of 'altitude' in ecological research. Trends Ecol Evol 22:569–574
37. Chen YK, Yang XB, Yang Q, Li DH, Long WX, Luo WQ (2014) Factors affecting the distribution pattern of wild plants with extremely small populations in Hainan Island, China. PLoS One 9:e97751
38. Colwell RK, Lees DC (2000) The mid-domain effect: geometric constraints on the geography of species diversity. Trends Ecol Evol 15:70–76
39. Traxmandlová I, Ackerman JD, Tremblay RL, Roberts DL, Štípková Z, Kindlmann P (2018) Determinants of orchid species diversity in world islands. New Phytol 217:12–15
40. Schödelbauerová I, Roberts DL, Kindlmann P (2009) Size of protected areas is the main determinant of species diversity in orchids. Biol Conserv 142:2329–2334
41. Blinova IV (2012) Intra-and interspecific morphological variation of some European terrestrial orchids along a latitudinal gradient. Russ J Ecol 43:111–116
42. Wotavová K, Balounová Z, Kindlmann P (2004) Factors affecting persistence of terrestrial orchids in wet meadows and implications for their conservation in a changing agricultural landscape. Biol Conserv 118:271–279
43. Janečková P, Wotavová K, Schödelbauerová I, Jersáková J, Kindlmann P (2006) Relative effects of management and environmental conditions on performance and survival of populations of a terrestrial orchid, *Dactylorhiza majalis*. Biol Conserv 129:40–49
44. Blinova IV (2008) Populations of orchids at the northern limit of their distribution (Murmansk Oblast): effect of climate. Russ J Ecol 39:26–33
45. Tsiftsis S, Djordjević V, Tsiripidis I (2019) *Neottia cordata* (Orchidaceae) at its southernmost distribution border in Europe: threat status and effectiveness of Natura 2000 network for its conservation. J Nat Conserv 48:27–35
46. Pfeifer M, Passalacqua NG, Bartram S, Schatz B, Croce A, Carey PD, Kraudelt H, Jeltsch F (2010) Conservation priorities differ at opposing species borders of a European orchid. Biol Conserv 143:2207–2220
47. Pillon Y, Fay MF, Shipunov AB, Chase MW (2006) Species diversity versus phylogenetic diversity: a practical study in the taxonomically difficult genus *Dactylorhiza* (Orchidaceae). Biol Conserv 129:4–13
48. Kotilínek M, Tatarenko I, Jersáková J (2018) Biological Flora of the British Isles: *Neottia cordata*. J Ecol 106:444–460
49. Vakhrameeva MG, Denissova LV, Nikitina SV, Samsonov SK (1991) Orchids of our country. Nauka, Moscow (in Russian)
50. Jermakowicz E, Brzosko E, Kotowicz J, Wróblewska A (2017) Genetic diversity of orchid *Malaxis monophyllos* over European range as an effect of population properties and post-glacial colonization. Pol J Ecol 65:69–86
51. Tamm CO (1991) Behaviour of some orchid populations in a changing environment. Observations on permanent plots, 1943–1990. In: Wells TCE, Willems JH (eds) Population ecology of terrestrial orchids. SPB Academic Publishers, The Hague
52. Blinova I, Chmielewski FM (2008) Subarctic warming and its influence on the growth of orchid populations in the extreme North-East of Europe Murmansk region. J Eur Orch 40:663–680. https://www.pabgi.ru/people/ilona_blinova/paper/t_orchids_08.pdf. Accessed 12 August 2019
53. Light MHS, MacConaill M (1998) Factors affecting germinable seed yield in *Cypripedium calceolus* var. *pubescens* (Willd.) Correll and *Epipactis helleborine* (L.) Crantz (Orchidaceae). Bot J Linn Soc 126:3–26
54. Øien DI, Moen A (2002) Flowering and survival of *Dactylorhiza lapponica* and *Gymnadenia conopsea* in the Solendet nature reserve, Central Norway. In: Kindlmann P, Willems JH,

Whigham DF (eds) Trends and fluctuations and underlying mechanisms in terrestrial orchid populations. Backhuys Publishers, Leiden
55. Inghe O, Tamm CO (1988) Survival and flowering of perennial herbs. V. Patterns of flowering. Oikos 51:203–219
56. Rasmussen HN, Dixon KW, Jersáková J, Těšitelová T (2015) Germination and seedling establishment in orchids: a complex of requirements. Ann Bot 116:391–402
57. Zhang S, Yang Y, Li J, Qin J, Zhang W, Huang W, Hu H (2018) Physiological diversity of orchids. Plant Diversity 40:196–208
58. Carey PD, Farrell L (2002) *Himantoglossum hircinum* (L.) Sprengel (Biological flora of the British Isles, 641.1). J Ecol 90:206–218
59. Wells TCE, Cox R (1989) Predicting the probability of the bee orchid (*Ophrys apifera*) flowering or remaining vegetative from the size and number of leaves. In: Pritchard H (ed) Modern methods in orchid conservation. Cambridge University Press, Cambridge
60. Štípková Z, Romportl D, Černocká V, Kindlmann P (2017) Factors associated with the distributions of orchids in the Jeseníky Mountains, Czech Republic. Eur J Environ Sci 7:135–145. http://www.ejes.cz/index.php/ejes/article/view/318. Accessed 13 August 2019
61. Meekers T, Hutchings MJ, Honnay O, Jacquemyn H (2012) Biological Flora of the British Isles: *Gymnadenia conopsea* s.l. J Ecol 100:1269–1288
62. Rasmussen HN (1995) Terrestrial orchids from seed to mycotrophic plant. Cambridge University Press, Cambridge
63. Wells TCE, Cox R (1991) Demographic and biological studies on *Ophrys apifera*: some results from a 10 year study. In: Wells TCE, Willems JH (eds) Population ecology of terrestrial orchids. SPB Academic Publishing, The Hague
64. Bódis J, Biró É, Nagy T, Takács A, Sramkó G, Bateman RM, Gilián L, Illyés Z, Tökölyi J, Lukács BA, Csábi M, Molnár AV (2019) Biological flora of Central Europe *Himantoglossum adriaticum* H. Baumann. Perspect Plant Ecol Evol Syst 125461. https://doi.org/10.1016/j.ppees.2019.125461
65. Willems JH, Bik L (1991) Population biology of *Orchis simia* in the Netherlands. In: Wells TCE, Willems JH (eds) Population ecology of terrestrial orchids. SPB Academic Publishing, The Hague, pp 1972–1990
66. Brzosko E (2002) The dynamics of *Listera ovata* populations on mineral islands in the Biebrza National Park. Acta Soc Bot Pol 71:243–251
67. Wells TCE (1981) Population ecology of terrestrial orchids. In: Synge H (ed) The biological aspects of rare plant conservation. Wiley, Chichester
68. Jacquemyn H, Brys R, Hutchings MJ (2014) Biological flora of the British Isles: *Epipactis palustris*. J Ecol 102:1341–1355
69. Tsiftsis S (2016) Morphological variability of *Himantoglossum* s.s. (Orchidaceae). Phytotaxa 245:17–30
70. Bailes C, Clements M, Cribb PJ, Muir H (1986) The cultivation of European orchids. Curtis's Bot Mag 3:8–13. JSTOR. http://www.jstor.org/stable/45066409 Accessed 12 August 2019
71. Abernethy A (2002) Light regimes as a control of terrestrial orchid distribution in New Zealand. Dissertation, University of Canterbury
72. Diez JM, Pulliam HR (2007) Hierarchical analysis of species distribution and abundance across environmental gradients. Ecology 88:3144–3152
73. Lõhmus A, Kull T (2011) Orchid abundance in hemiboreal forests: stand-scale effects of clear-cutting, green-tree retention, and artificial drainage. Can J For Res 41:1352–1358
74. Shefferson RP, Kull T, Tali K (2005) Adult whole-plant dormancy induced by stress in long-lived orchids. Ecology 86:3099–3104
75. Zhang SB, Hu H, Xu K, Li ZR, Yang YP (2007) Flexible and reversible responses to different irradiance levels during photosynthetic acclimation of *Cypripedium guttatum*. J Plant Physiol 164:611–620
76. Jacquemyn H, Brys R, Jongejans E (2010) Size-dependent flowering and costs of reproduction affect population dynamics in a tuberous perennial woodland orchid. J Ecol 98:1204–1215

77. Tsiftsis S, Djordjević V (2018) Habitat effects and differences in the reproductive success of *Orchis punctulata* and *Orchis purpurea* (Orchidaceae). Turk J Bot 42:400–411
78. Jacquemyn H, Brys R, Honnay O, Hermy M (2008) Effects of coppicing on demographic structure, fruit and seed set in *Orchis mascula*. Basic Appl Ecol 9:392–400
79. Jacquemyn H, Brys R, Honnay O, Hutchings MJ (2009) Biological flora of the British Isles: *Orchis mascula* (L.) L. J Ecol 97:360–377
80. Jacquemyn H, Brys R (2010) Temporal and spatial variation in flower and fruit production in a food-deceptive orchid: a five year study. Plant Biol 12:145–153
81. Selosse MA, Weiss M, Jany JL, Tillier A (2002) Communities and populations of sebacinoid basidiomycetes associated with the achlorophyllous orchid *Neottia nidus-avis* (L.) L.C.M. Rich. and neighbouring tree ectomycorrhizae. Mol Ecol 11:1831–1844
82. Taylor L, Roberts DL (2011) Biological flora of the British Isles: *Epipogium aphyllum* Sw. J Ecol 99:878–890
83. Preiss K, Adam IKU, Gebauer G (2010) Irradiance governs exploitation of fungi: fine-tuning of carbon gain by two partially myco-heterotrophic orchids. Proc R Soc B 277:1333–1336
84. Hornemann G, Michalski SG, Durka W (2012) Short-term fitness and long-term population trends in the orchid *Anacamptis morio*. Plant Ecol 213:1583–1595
85. Jersáková J, Malinová T, Jeřábková K, Dötteri S (2011) Biological Flora of the British Isles: *Pseudorchis albida* (L.) Á. & D. Löve. J Ecol 99:1282–1298
86. Tsiftsis S, Antonopoulos Z (2011) *Pseudorchis albida*: an enigmatic orchid for the Greek flora. J Eur Orch 43:795–806
87. Jacquemyn H, Brys R, Adriaens D, Honnay O, Roldán-Ruiz I (2009) Effects of population size and forest management ongenetic diversity and structure of the tuberous orchid *Orchis mascula*. Conserv Genet 10:161–168
88. Cruz-Fernández QT, Alquicira-Arteaga ML, Flores-Palacios A (2011) Is orchid species richness and abundance related to the conservation status of oak forest? Plant Ecol 212:1091–1099
89. Hurskainen S, Jäkäläniemi A, Ramula S, Tuomi J (2017) Tree removal as a management strategy for the lady's slipper orchid, a flagship species for herb-rich forest conservation. Forest Ecol Manag 406:12–18
90. Shefferson RP, Kull T, Tali K (2006) Demographic response to shading and defoliation in two woodland orchids. Folia Geobot 41:95–106
91. van der Meer S, Jacquemyn H, Carey PD, Jongejans E (2016) Recent range expansion of a terrestrial orchid corresponds with climate-driven variation in its population dynamics. Oecologia 181:435–448
92. Shefferson RP, Mizuta R, Hutchings MJ (2017) Predicting evolution in response to climate change: the example of sprouting probability in three dormancy prone orchid species. R Soc Open Sci 4:160647
93. Liu H, Feng CL, Luo YB, Chen BS, Wang ZS, Gu HY (2010) Potential challenges of climate change to orchid conservation in a wild orchid hotspot in southwestern China. Bot Rev 76:174–192
94. Reina-Rodríguez GA, Rubiano Mejía JE, Castro Llanos FA, Soriano I (2017) Orchid distribution and bioclimatic niches as a strategy to climate change in areas of tropical dry forest in Colombia. Lankesteriana 17:17–47
95. Ongaro S, Martellos S, Bacaro G, De Agostini A, Cogoni A, Cortis P (2018) Distributional pattern of Sardinian orchids under a climate change scenario. Community Ecol 19:223–232
96. Gaskett AC, Gallagher RV (2018) Orchid diversity: spatial and climatic patterns from herbarium records. Ecol Evol 8:11235–11245. https://doi.org/10.1002/ece3.4598
97. Molnár VA, Tökölyi J, Végvári Z, Sramkó G, Sulyok J, Barta Z (2012) Pollination mode predicts phenological response to climate change in terrestrial orchids: a case study from Central Europe. J Ecol 100:1141–1152
98. Hutchings MJ, Robbirt KM, Roberts DL, Davy AJ (2018) Vulnerability of a specialized pollination mechanism to climate change revealed by a 356-year analysis. Bot J Linn Soc 186:498–509

99. Tsiftsis S, Tsiripidis I, Trigas P (2011) Identifying important areas for orchid conservation in Crete. Eur J Environ Sci 1:28–37
100. Hijmans RJ, Cameron SE, Parra JL, Jones PG, Jarvis A (2005) Very high resolution interpolated climate surfaces for global land areas. Int J Climatol 25:1965–1978
101. Fick SE, Hijmans RJ (2017) WorldClim 2: new 1-km spatial resolution climate surfaces for global land areas. Int J Climatol 37:4302–4315
102. Bowles M, Zettler L, Bell T, Kelsey P (2005) Relationship between soil characteristics, distribution and restoration potential of the federal threatened eastern prairie fringed orchid, *Platanthera leucophaea* (Nutt.) Lindl. Am Midl Nat 154:273–285
103. Dijk E, Willems JH, van Andel J (1997) Nutrient responses as a key factor to the ecology of orchid species. Acta Bot Neerl 46:339–363
104. Tsiftsis S, Tsiripidis I, Papaioannou A (2012) Ecology of orchid *Goodyera repens* in its southern distribution limits. Plant Biosyst 146:857–866
105. Landi M, Frignani F, Lazzeri C, Angiolini C (2009) Abundance of orchids on calcareous grasslands in relation to community species, environmental, and vegetation conditions. Russ J Ecol 40:486–494
106. Lang D (2004) Britain's orchids, a guide to the identification and ecology of the wild orchids of Britain and Ireland. Wild Guides Ltd., Old Basing
107. Djordjević V, Tsiftsis S (2019) Patterns of orchid species richness and composition in relation to geological substrates. Wulfenia 26:1–21
108. Tyler G (2003) Some ecophysiological and historical approaches to species richness and calcicole/calcifuge behaviour – contribution to a debate. Folia Geobot 38:419–428
109. Farrell L (1985) Biological Flora of the British Isles: *Orchis militaris* L. J Ecol 73:1041–1053
110. Tali K, Foley MJY, Kull T (2004) Biological flora of the British Isles, 232. *Orchis ustulata* L. J Ecol 92:174–184
111. Jersáková J, Traxmandlová I, Ipser Z, Matthias K, Pellegrino G, Schatz B, Djordjević V, Kindlmann P, Renner SS (2015) Biological flora of Central Europe: *Dactylorhiza sambucina* (L.) Soó. Perspect Plant Ecol Evol Syst 17:318–329
112. Kotilínek M, Těšitelová T, Jersáková J (2015) Biological Flora of the British Isles: *Neottia ovata*. J Ecol 103:1354–1366
113. Petrova AS, Venkova DY (2006) Epipactis pontica (Orchidaceae): a new species for the Bulgarian flora. Phytol Balcan 12:249–253. http://www.bio.bas.bg/~phytolbalcan/PDF/14_1/14_1_11_Petrova_&_Venkova.pdf. Accessed 12 August 2019
114. van der Ent A, Wood JJ (2013) Orchids of extreme serpentinite (ultramafic) habitats in Kinabalu Park. Malesian Orchid J 12:39–54
115. van der Ent A, van Vugt R, Wellinga S (2015) Ecology of *Paphiopedilum rothschildianum* at the type locality in Kinabalu Park (Sabah, Malaysia). Biodivers Conserv 24:1641–1656
116. Filimonova E, Lukina N, Glazyrina M, Borisova G, Kumar A, Maleva M (2019) A comparative study of *Epipactis atrorubens* in two different forest communities of the Middle Urals, Russia. J For Res 1–10. https://doi.org/10.1007/s11676-019-01010-y
117. Stuckey I (1967) Environmental factors and the growth of native orchids. Am J Bot 54:232–241
118. Knudson MD, Vanlooy JA, Hill MJ (2015) A habitat suitability index (HSI) for the western prairie fringed orchid (*Platanthera praeclara*) on the Sheyenne National Grassland, North Dakota, USA. Ecol Indic 57:536–545
119. Wolken PM, Sieg CH, Williams SE (2001) Quantifying suitable habitat of the threatened western prairie fringed orchid. J Range Manag 54:611–616
120. Sieg CH, King RM (1995) Influence of environmental factors and preliminary demographic analysis of a threatened orchid, *Platanthera praeclara*. Am Midl Nat 134:307–323
121. Jacquemyn H, Waud M, Merckx VS, Lievens B, Brys R (2015) Mycorrhizal diversity, seed germination and long-term changes in population size across nine populations of the terrestrial orchid *Neottia ovata*. Mol Ecol 24:3269–3280

122. Illyés Z, Halász K, Rudnóy S, Ouanphanivanh N, Garay T, Bratek Z (2009) Changes in the diversity of the mycorrhizal fungi of orchids as a function of the water supply of the habitat. J Appl Bot Food Qual 83:28–36
123. Foley M, Clarke S (2005) Orchids of the British Isles. Griffin Press Publishing Limited, Cheltenham
124. Danihelka J, Chrtek J Jr, Kaplan Z (2012) Checklist of vascular plants of the Czech Republic. Preslia 84:647–811
125. Ståhlberg D (2009) Habitat differentiation, hybridization and gene flow patterns in mixed populations of diploid and autotetraploid *Dactylorhiza maculata* s.l. (Orchidaceae). Evol Ecol 23:295–328
126. Molnár A (ed) (2011) Magyarország orchideáinak atlasza. Kossuth kiadó, Budapest
127. Hrivnák R, Hrivnák M, Slezák M, Vlčko J, Baltiarová J, Svitok M (2014) Distribution and eco-coenotic patterns of the forest orchid *Epipactis pontica* in Slovakia. Ann For Res 57:55–69
128. Djordjević V, Jakovljević K, Stevanović V (2016) Three taxa of *Epipactis* (Orchidaceae-Epidendroideae) new for the flora of Serbia. Phyton-Ann Rei Bot 56:77–89
129. Zilioli DM, Bini C, Wahsha M, Ciotoli G (2011) The pedological heritage of the Dolomites (Northern Italy): features, distribution and evolution of the soils, with some implications for land management. Geomorphology 135:232–247
130. Tsiftsis S (2009) The orchids (Orchidaceae) of E. Macedonia: distribution, ecology and high conservation value areas. Dissertation, Aristotle University of Thessaloniki (in Greek with English summary)
131. Sundermann H (1980) Europäische und mediterrane Orchideen. Brücke-Verlag Kurt Schmersow, Hildesheim
132. Breiner R (1979) pH-Messungen an Orchideen-Standorten auf Kreta und Zypern. Mitt Bl Arb Heim Orchid Baden-Wurttemberg 11:54–58. http://biolis.ub.uni-frankfurt.de/search/detail/22028. Accessed 12 August 2019
133. Vakhrameeva MG, Tatarenko IV, Varlygina TI, Torosyan GK, Zagulski MN (2008) Orchids of Russia and adjacent countries (within the borders of the former USSR). ARG Gantner Verlag, Ruggell
134. Syska M (1995) Die Orchideenflora des westlichen Nestos-Deltas und des angrenzenden Berglandes (Nordost-Griechenland): Verbreitung – Ökologie – Gefahrdung. J Eur Orch 27:339–552
135. Wallenwein F, Saad A (2000) Messungen des pH-Wertes an den Wuchsorten mediterraner Orchideen. J Eur Orch 32:375–386
136. Möller O (1985) Die Mineralsalze der Standortböden der europäischer Orchideen. Die Orchidee 36:118–121. https://orchidee.de/gesellschaft/die-orchidee/. Accessed 12 August 2019
137. Tsiripidis I (2001) Plant communities of beech forests in Rodopi mountain range and their environmental assessment for reforestation. Dissertation, Aristotle University of Thessaloniki (in Greek)
138. Kull T (1999) Biological Flora of the British Isles: *Cypripedium calceolus* L. J Ecol 87:913–924
139. Mróz L (1994) Ekologia *Dactylorhiza sambucina* (L.) Soó w Sudetach. Acta Univ Wratisl Prace Botaniczne LXXVI:103–157
140. Molnár AV, Sramkó G (2012) *Epipactis albensis* (Orchidaceae): a new species in the flora of Romania. Biologia 67:883–888
141. Parzych A, Sobisz Z (2013) Preliminary ecology research on *Epipactis atrorubens* (Hoffm.) Besser on the Słowińskie coast (Northern Poland). Ecol Quest 18:21–32
142. Urban D (2013) Characteristics of the locality of *Hammarbya paludosa* (L.) O. Kuntze on the Łęczna-Włodawa plain (West Polesie). Teka Komisji Ochrony i Kształtowania Środowiska Przyrodniczego. OL PAN 10:448–454
143. Procházka F, Velísek V (1983) Orchideje naší přírody. Čekoslovenské Akademie Věd, Praha

144. Brütsch JP (2000) Die Gattung *Nigritella*. Bauhinia 14:21–32. https://botges.ch/bauhinia/bauhinia14(2000)21-32.pdf. Accessed 12 August 2019
145. Molnár AV, Máté A, Sramkó G (2011) An unexpected new record of the Mediterranean orchid, *Ophrys bertolonii* (Orchidaceae) in Central Europe. Biologia 66:778–782
146. Reinhammar L, Olsson EG, Sormeland E (2002) Conservation biology of an endangered grassland plant species, *Pseudorchis albida*, with some references to the closely related alpine *P. straminea* (Orchidaceae). Bot J Linn Soc 139:47–66
147. Jeřábková K (2006) Ecological demands and optimal management of *Pseudorchis albida*. Master thesis in Czech, Czech University of Life Sciences, Prague
148. Beyrle H, Penningsfeld F, Hock B (1991) The role of nitrogen concentration in determining the outcome of the interaction between *Dactylorhiza incarnata* (L.) Soó and *Rhizoctonia* sp. New Phytol 117:665–672
149. Cameron DD, Leake JR, Read DJ (2006) Mutualistic mycorrhiza in orchids: evidence from plant–fungus carbon and nitrogen transfers in the green-leaved terrestrial orchid *Goodyera repens*. New Phytol 171:405–416
150. McKendrick SL, Leake JR, Read DJ (2000) Symbiotic germination and development of myco-heterotrophic plants in nature: transfer of carbon from ectomycorrhizal *Salix repens* and *Betula pendula* to the orchid *Corallorhiza trifida* through shared hyphal connections. New Phytol 145:539–548
151. Hejcman M, Schellberg J, Pavlů V (2010) *Dactylorhiza maculata*, *Platanthera bifolia* and *Listera ovata* survive N application under P limitation. Acta Oecol 36:684–688
152. Dijk E, Olff H (1994) Effects of nitrogen, phosphorus and potassium fertilization on field performance of *Dactylorhiza majalis*. Acta Bot Neerl 43:383–392
153. Silvertown J, Wells DA, Gillman M, Dodd ME, Robertson H, Lakhani KH (1994) Short-term effects and long-term after effects of fertilizer application on the flowering population of green-winged orchid *Orchis morio*. Biol Conserv 69:191–197
154. Dijk E, Grootjans AP (1998) Performance of four *Dactylorhiza* species over a complex trophic gradient. Acta Bot Neerl 47:351–368
155. Stevanović MB, Janković MM (2001) Ecology of plants with the basics of physiological ecology of plants. NNK, Beograd (in Serbian)
156. Moreira ASFP, Isaias RMDS (2008) Comparative anatomy of the absorption roots of terrestrial and epiphytic orchids. Braz Arch Biol Technol 51:83–93
157. Abadie JC, Püttsepp Ü, Gebauer G, Faccio A, Bonfante P, Selosse MA (2006) *Cephalanthera longifolia* (Neottieae, Orchidaceae) is mixotrophic: a comparative study between green and nonphotosynthetic individuals. Can J Bot 84:1462–1477
158. Batty AL, Dixon KW, Brundrett M, Sivasithamparam K (2001) Constraints to symbiotic germination of terrestrial orchid seed in a mediterranean bushland. New Phytol 152:511–520
159. Jurkiewicz A, Turnau K, Mesjasz-Przybylowicz J, Przybylowicz W, Godzik B (2001) Heavy metal localisation in mycorrhizas of *Epipactis atrorubens* (Hoffm.) Besser (Orchidaceae) from zinc mine tailings. Protoplasma 218:117–124
160. Shefferson RP, Kull T, Tali K (2008) Mycorrhizal interactions of orchids colonizing Estonian mine tailings hills. Am J Bot 95:156–164
161. Esfeld K, Hensen I, Wesche K, Jakob SS, Tischew S, Blattner FR (2008) Molecular data indicate multiple colonizations of former lignite mining areas in Eastern Germany by *Epipactis palustris* (Orchidaceae). Biodivers Conserv 17:2441–2453
162. Rewicz A, Bomanowska A, Shevera MV, Kurowski JK, Krasoń K, Zielińska KM (2017) Cities and disturbed areas as man-made shelters for orchid communities. Not Bot Horti Agrobot Cluj-Na 45:126–139
163. Gajić G, Djurdjević L, Kostić O, Jarić S, Mitrović M, Pavlović P (2018) Ecological potential of plants for phytoremediation and ecorestoration of fly ash deposits and mine wastes. Front Environ Sci 6:124. https://doi.org/10.3389/fenvs.2018.00124
164. Adamowski W (2006) Expansion in native orchids in anthropogenous habitats. Pol Bot Stud 22:35–44

165. Rewicz A, Jaskuła R, Rewicz T, Tończyk G (2017) Pollinator diversity and reproductive success of *Epipactis helleborine* (L.) Crantz (Orchidaceae) in anthropogenic and natural habitats. PeerJ 5:e3159
166. Fekete R, Nagy T, Bódis J, Biró É, Löki V, Süveges K, Takács A, Tökölyi J, Molnár VA (2017) Roadside verges as habitats for endangered lizard-orchids (*Himantoglossum* spp.): ecological traps or refuges? Sci Total Environ 607–608:1001–1008
167. Fekete R, Löki V, Urgyán R, Süveges K, Lovas Kiss Á, Vincze O, Molnár VA (2019) Roadside verges and cemeteries: comparative analysis of anthropogenic orchid habitats in the eastern Mediterranean. Ecol Evol. https://doi.org/10.1002/ece3.5245
168. Grant CD, Koch JM (2003) Orchid species succession in rehabilitated bauxite mines in Western Australia. Aust J Bot 51:453–457
169. Norman MA, Koch JM, Grant CD, Morald TK, Ward SC (2006) Vegetation succession after bauxite mining in Western Australia. Restor Ecol 14:278–288
170. Girlanda M, Selosse MA, Cafasso D, Brilli F, Delfine S, Fabian R, Ghignone S, Pinelli P, Segreto R, Loreto F, Cozzolino S, Perotto S (2006) Inefficient photosynthesis in the Mediterranean orchid *Limodorum abortivum* is mirrored by specific association to ectomycorrhizal Russulaceae. Mol Ecol 15:491–504
171. Mucina L, Bültmann H, Dierßen K, Theurillat JP, Raus T, Čarni A, Šumberová K, Willner W, Dengler J, Gavilán García R, Chytrý M, Hájek M, Di Pietro R, Iakushenko D, Pallas J, Daniëls FJA, Bergmeier E, Santos Guerra A, Ermakov N, Valachovič M, Schaminée JHJ, Lysenko T, Didukh YP, Pignatti S, Rodwell JS, Capelo J, Weber HE, Solomeshch A, Dimopoulos P, Aguiar C, Freitag H, Hennekens SM, Tichý L (2016) Vegetation of Europe: hierarchical floristic classification system of plant, lichen, and algal communities. Appl Veg Sci 19:3–264
172. Djordjević V, Tomović G, Lakušić D (2010) *Epipactis purpurata* Sm. (Orchidaceae) – a new species in the flora of Serbia. Arch Biol Sci 62:1175–1180
173. Kovalchuk A (2016) On the occurrence of *Orchis pallens* L. in the Ukrainian Carpathians. J Eur Orch 48:29–36
174. Tsiftsis S, Karagiannakidou V, Tsiripidis I (2007) The orchid flora of East Macedonia (NE Greece). J Eur Orch 39:489–526
175. Djordjević V (2016) *Epipactis muelleri* (Orchidaceae-Neottieae), a species new to the flora of Serbia. Phyton-Ann Rei Bot 56:303–312
176. Djordjević V, Tsiftsis S, Lakušić D, Jovanović S, Stevanović V (2017) Distribution and conservation status of some rare and threatened orchid taxa in the Central Balkans and the southern part of the Pannonian plain. Wulfenia 24:143–162
177. Djordjević V, Jovanović S, Stevanović V (2014) *Dactylorhiza fuchsii* (Orchidaceae), a new species in the flora of Serbia. Arch Biol Sci 66:1227–1232
178. Pierce S, Vagge I, Brusa G, Cerabolini BEL (2014) The intimacy between sexual traits and Grime's CSR strategies for orchids coexisting in semi-natural calcareous grassland at the Olive Lawn. Plant Ecol 215:495–505
179. Slaviero A, Del Vecchio S, Pierce S, Fantinato E, Buffa G (2016) Plant community attributes affect dry grassland orchid establishment. Plant Ecol 217:1533–1543
180. Oberdorfer E (1994) Pflanzensoziologische Exkursionsflora. Ulmer, Stuttgart
181. Krauss J, Klein AM, Steffan-Dewenter I, Tscharntke T (2004) Effects of habitat area, isolation, and landscape diversity on plant species richness of calcareous grasslands. Biodivers Conserv 13:1427–1439
182. Leuschner C, Ellenberg H (2017) Ecology of central European non-Forest vegetation: coastal to alpine, natural to man-made habitats: vegetation ecology of Central Europe. Springer International Publishing, Cham
183. Haraštová-Sobotková M, Jarsáková J, Kindlmann P, Čurn L (2005) Morphometric and genetic divergence among populations of *Neotinea ustulata* (Orchidaceae) with different flowering phenologies. Folia Geobot 40:385–405

184. Budzhak VV, Chorney II, Tokariuk AI, Kuzemko AA (2016) Numeric syntaxonomical analysis of the communities with participation of species from *Molinia caerulea* complex in the southwest of Ukraine. Hacquetia 15:63–78
185. Schrautzer J, Fichtner A, Huckauf A, Rasran L, Jensen K (2011) Long-term population dynamics of *Dactylorhiza incarnata* (L.) Soó after abandonment and re-introduction of mowing. Flora 206:622–630
186. Blinova IV (2016) Spatial population structure of rare orchid species in rich fens in the central part of Murmansk oblast. Russ J Ecol 47:234–240
187. Hrivnák R, Gömöry D, Cvachová A (2006) Inter-annual variability of the abundance and morphology of *Dactylorhiza majalis* (*Orchidaceae-Orchideae*) in two permanent plots of a mire in Slovakia. Phyton-Ann Rei Bot 46:27–44
188. Löki V, Tökölyi J, Süveges K, Lovas-Kiss A, Hürkan K, Gábor S, Molnár AV (2015) The orchid flora of Turkish graveyards: a comprehensive field survey. Willdenowia 45:231–243
189. Molnár VA, Takács A, Mizsei E, Löki V, Barina Z, Sramkó G, Tökölyi J (2017) Religious differences affect orchid diversity of Albanian graveyards. Pak J Bot 49:289–303
190. Rewicz A, Rewers M, Jędrzejczyk I, Rewicz T, Kołodziejek J, Jakubska-Busse A (2018) Morphology and genome size of *Epipactis helleborine* (L.) Crantz (Orchidaceae) growing in anthropogenic and natural habitats. PeerJ 6:e5992
191. Rewicz A, Kolodziejek J, Jakubska-Busse A (2016) The role of anthropogenic habitats as substitutes for natural habitats: a case study on *Epipactis helleborine* (L.) Crantz (Orchidaceae, Neottieae). Variations in size and nutrient composition of seeds. Turk J Bot 40:258–268
192. Scade A, Brundrett MC, Batty AL, Dixon KW, Sivasithamparam K (2006) Survival of transplanted terrestrial orchid seedlings in urban bushland habitats with high or low weed cover. Aust J Bot 54:383–389
193. Lim WH (2015) Aspects of the physiological ecology of the Western-Australian ruderal orchid, *Microtis media* R. Br, with special reference to the functions of its mycorrhizal fungi. Dissertation, the University of Western Australia
194. Grime JP (1979) Plant strategies and vegetation processes. Wiley, Chichester
195. Coates F, Lunt ID, Tremblay RL (2006) Effects of disturbance on population dynamics of threatened orchid *Prasophyllum correctum* D. L. Jones and implications for grassland management in south-eastern Australia. Biol Conserv 129:59–69
196. Willems JH, Melser C (1998) Population dynamics and life-history of *Coeloglossum viride* (L.) Hartm.: an endangered orchid species in The Netherlands. Bot J Linn Soc 126:83–93
197. Smith PH, Cross S (2016) Effect of mowing regime on abundance of green-winged orchid *Anacamptis morio* on coastal grassland in Merseyside, England. Conserv Evid 13:79–81. https://www.conservationevidence.com/individual-study/5910. Accessed 13 August 2019
198. Sletvold N, Øien DI, Moen A (2010) Long term influence of mowing on population dynamics in the rare orchid *Dactylorhiza lapponica*: the importance of recruitment and seed production. Biol Conserv 143:747–755
199. Humphrey JW, Coombs EL (1997) Effects of forest management on understorey vegetation in a *Pinus sylvestris* L. plantation in NE Scotland. Bot J Scotl 49:479–488
200. Duncan M (2012) Response of orchids to bushfire: black Saturday Victoria 2009 – natural values fire recovery program. Department of Sustainability and Environment. Victoria, Heidelberg
201. Fridley JD, Vandermast DB, Kuppinger DM, Manthey M, Peet RK (2007) Co-occurrence based assessment of habitat generalists and specialists: a new approach for the measurement of niche width. J Ecol 95:707–722
202. Boulangeat I, Lavergne S, Van Es J, Garraud L, Thuiller W (2012) Niche breadth, rarity and ecological characteristics within a regional flora spanning large environmental gradients. J Biogeogr 39:204–214
203. Schiebold JMI, Bidartondo MI, Lenhard F, Makiola A, Gebauer G (2018) Exploiting mycorrhizas in broad daylight: partial mycoheterotrophy is a common nutritional strategy in meadow orchids. J Ecol 106:168–178

204. McCormick MK, Jacquemyn H (2014) What constrains the distribution of orchid populations? New Phytol 202:392–400
205. McCormick MK, Whigham DF, Canchani-Viruet A (2018) Mycorrhizal fungi affect orchid distribution and population dynamics. New Phytol 219:1207–1215
206. McCormick MK, Lee Taylor D, Whigham DF, Burnett RK (2016) Germination patterns in three terrestrial orchids relate to abundance to mycorrhizal fungi. J Ecol 104:744–754
207. Dearnaley JDW (2007) Further advances in orchid mycorrhizal research. Mycorrhiza 17:475–486
208. Rasmussen HN, Rasmussen FN (2014) Seedling mycorrhiza: a discussion of origin and evolution in Orchidaceae. Bot J Linn Soc 175:313–327
209. Selosse MA, Faccio A, Scappaticci G, Bonfante P (2004) Chlorophyllous and achlorophyllous specimens of *Epipactis microphylla* (Neottieae, Orchidaceae) are associated with ectomycorrhizal septomycetes, including truffles. Microb Ecol 47:416–426
210. Rasmussen HN, Rasmussen FN (2009) Orchid mycorrhiza: implications of a mycophagous life circle. Oikos 118:334–345
211. Bailarote BC, Lievens B, Jacquemyn H (2012) Does mycorrhizal specificity affect orchid decline and rarity? Am J Bot 99:1655–1665
212. Davis BJ, Phillips RD, Wright M, Linde CC, Dixon KW (2015) Continent-wide distribution in mycorrhizal fungi: implications for the biogeography of specialized orchids. Ann Bot 116:413–421
213. Shefferson RP, Taylor DL, Weiß M, Garnica S, McCormick MK, Adams S, Gray HM, McFarland JW, Kull T, Tali K, Yukawa T, Kawahara T, Miyoshi K, Lee YI (2007) The evolutionary history of mycorrhizal specificity among lady's slipper orchids. Evolution 61:1380–1390
214. Pellegrino G, Luca A, Bellusci F (2016) Relationships between orchid and fungal biodiversity: mycorrhizal preferences in Mediterranean orchids. Plant Biosyst 150:1–10
215. McCormick MK, Lee Taylor D, Juhaszova K, Burnett RK, Whigham DF, O'Neill JP (2012) Limitations on orchid recruitment: not a simple picture. Mol Ecol 21:1511–1523
216. Tremblay RL, Ackerman JD, Zimmerman JK, Calvo RN (2005) Variation in sexual reproduction in orchids and its evolutionary consequences: a spasmodic journey to diversification. Biol J Linn Soc 84:1–54
217. van der Pijl L, Dodson CH (1966) Orchid flowers: their pollination and evolution. University of Miami Press, Coral Gables
218. Jersáková J, Johnson SD, Kindlmann P (2006) Mechanisms and evolution of deceptive pollination in orchids. Biol Rev 81:219–235
219. Vereecken NJ, Dafni A, Cozzolino S (2010) Pollination syndromes in Mediterranean orchids – implications for speciation, taxonomy and conservation. Bot Rev 76:220–240
220. Neiland MRM, Wilcock CC (1998) Fruit set, nectar reward, and rarity in the Orchidaceae. Am J Bot 85:1657–1671
221. Darwin C (1862) On the various contrivances by which British and foreign orchids are fertilised by insects. John Murray, London
222. Jacquemyn H, Brys R, Hermy M, Willems JH (2005) Does nectar reward affect rarity and extinction probabilities of orchid species? An assessment using historical records from Belgium and the Netherlands. Biol Conserv 121:257–263
223. Catling P (1990) Auto-pollination in the Orchidaceae. In: Arditti JE (ed) Orchid biology: reviews and perspectives, vol V. Timber Press, Portland

Which Environmental Factors Drive Distribution of Orchids? A Case Study from South Bohemia, Czech Republic

2

Zuzana Štípková, Dušan Romportl, and Pavel Kindlmann

Contents

Abstract

Species distribution models are a useful tool applied in many branches of biology, especially when dealing with threatened organisms. In combination with GIS techniques, these models are especially important and valuable for predicting

Z. Štípková (✉) · P. Kindlmann
Global Change Research Institute, Academy of Sciences of the Czech Republic, České Budějovice, Czech Republic

Institute for Environmental Studies, Faculty of Science, Charles University, Benátská 2/Prague 2, Czech Republic
e-mail: zaza.zuza@seznam.cz; pavel.kindlmann@centrum.cz

D. Romportl
Department of Physical Geography and Geoecology, Faculty of Science, Charles University, Albertov 6, Czech Republic
e-mail: dusan.romportl@natur.cuni.cz

J.-M. Mérillon, H. Kodja (eds.), *Orchids Phytochemistry, Biology and Horticulture*, Reference Series in Phytochemistry, https://doi.org/10.1007/978-3-030-38392-3_27

occurrence of rare species, for example, orchids. Orchids are an endangered plant group, protected in the whole world. Questions of their conservation are therefore highly discussed, but not all factors affecting their survival and distribution are known so far. Here we show an example of using SDMs for analysis of orchid species occurrence data from the region of South Bohemia in the Czech Republic. Our data were analyzed using the MaxEnt program, which produces species distribution maps and thus allows predicting potential occurrence of orchids in yet unknown localities. This program also determines the environmental factors affecting species distribution. This is important for better protection of orchids, because we can improve management plans that are crucial for maintaining orchid localities to stay alive. We determined the most important factors affecting studied species occurrence and areas, where new sites are most likely to be discovered. This approach can help us to find new localities of orchids and to understand which environmental factors influence the occurrence of these endangered plants.

Keywords

Orchids · Distribution · Environmental variables · Species distribution models · MaxEnt

1 Introduction

Questions concerning species diversity have attracted ecologists for over a century. Recently, this issue became even more important, because the diversity of life on Earth is in rapid decline [1]. Therefore, one of the most pressing tasks facing the global conservation community is trying to understand the main factors determining diversity of species [2] and identifying important areas for their conservation [3]. This effort is often followed by creation of a network of protected areas, wherein negative human influence is considerably limited [4–7]. This especially holds for threatened groups of organisms, such as orchids [8, 9].

The orchid family, with estimates of about 20,000–35,000 species [10–12], is an important group with respect to conservation biology [13], being at the front line of extinction [14]. Many characteristics, such as great species richness, its specific role in ecosystem, or endangered situation, make it crucial to explore the distribution and conservation status of Orchidaceae [15]. Orchids are also known for their sensitivity to environmental changes [16], as well as to their high extinction risk, compared to other plant families, as a result of natural and/or anthropogenic causes [17, 18]. However, decrease of many orchid species occurred in whole Europe, mainly as a result of the loss or even alteration of their natural habitats [8, 19, 20]. The most effective methods for conserving orchids undoubtedly involve protection of their habitats [12, 21].

Species distribution models (SDMs) are a useful tool, which is often applied in many branches of biogeography, conservation biology, and ecology in the last decades [22], especially when threatened species are concerned [23]. These numerical tools combine species occurrence records with environmental data [22]. In combination with GIS techniques, these models are especially important and useful for predicting occurrence of rare species [24]. Despite the fact that the results of species distribution models often suffer from high levels of uncertainty to several factors, concerning biased species distribution data, errors in environmental variables used as predictors, spatial resolution, and the modeling process [25, 26], SDMs have become widely accepted tools to predict species distributions [27].

The maximum entropy algorithm in the MaxEnt application [28–31] is often used for modeling species distributions from presence-only species records [31]. This approach was used by conservation practitioners for predicting the distribution of a species from a set of occurrence records and environmental variables [31, 32]. MaxEnt is one of the most robust approaches of species distribution in terms of successfully estimating the area from only a few records of occurrence [33, 34]. Despite long history of studies on orchids, only a minute part of previous papers concerning distribution, phytogeography, or conservation strategies of this taxonomic group included application of species distribution models (e.g., see [35–38]). Presence-only modeling methods require exclusively a set of known species occurrences together with predictor variables such as topographic, climatic, edaphic, biogeographic, and/or remotely sensed data [29, 30].

Here, we show an example of using the species distribution models and MaxEnt for analyses of orchid species distribution in the region of South Bohemia, Czech Republic. Using MaxEnt analysis, we estimated which environmental factors affect the distribution of selected orchid species and tried to find new suitable localities for orchid occurrence in the area selected.

2 Materials and Methods

This study was conducted in the region of South Bohemia, in the south of the Czech Republic. This area with about 10,057 km^2 stretches from 400 m as the lowest parts to more than 1300 m above sea level as the highest parts of the Šumava National Park. This region is quite rich in orchid flora; it includes also critically endangered species of the Czech Republic such as *Liparis loeselii*, *Neottia cordata*, or *Malaxis monophyllos*.

As a source of information, we used data about orchid occurrence from five databases: (1) the database of the Nature Conservation Agency of the Czech Republic [39]; (2) the Czech National Phytosociological Database and (3) the Floristic Documentation, both deposited at the Department of Botany and Zoology, Faculty of Science, Masaryk University in Brno [40]; (4) the database of the South Bohemian Branch of the Czech Botanical Society [41]; and (5) the database of the inheritance of the late František Procházka (10,000 items, digitized from original cards). The whole database that consists of all data from these five databases is deposited at the Global Change Research Institute, Department of Biodiversity

Research in České Budějovice. However, there is no public access to either of these databases in order to protect the existing localities of orchid species.

Our field studies took place within this region during 2014–2016 (Fig. 1) and were based on information taken from databases described above. During the field studies, we checked localities listed in databases in order to determine whether a selected orchid species is still present there or not. To do this, we visited as many localities as we were able to and recorded any important notes that we were able to. If an orchid species was found, the number of flowering individuals was counted, and other important information was registered, such as accurate GPS position of the locality, notes about how the locality looked like, or if the locality is somehow maintained or not. At the end of our field studies, the total of 428 localities were visited and checked.

As was mentioned above, there are many orchid species occurring in the region of South Bohemia. To describe which environmental factors affect their distribution, we used only species with an adequate number of records [42, 43], which is required to obtain reliable predictions. Two most numerous species in the region, *Dactylorhiza majalis* and *Platanthera bifolia*, have been tested before and described in Štípková et al. [44], and other two abundant species, *Epipactis atrorubens* and *Cephalanthera damasonium*, were analyzed in Kosánová [45]. In this study, other seven orchid species were analyzed:

- *Anacamptis morio*, a representative of thermophilous orchid species
- *Cephalanthera rubra*, preferring forest habitats
- *Dactylorhiza fuchsii*, preferring forest habitats
- *Neottia nidus-avis*, preferring forest habitats
- *Epipactis palustris*, thriving in wet habitats with stable water regime
- *Neottia ovata*, species that can be found in various types of habitats
- *Platanthera chlorantha*, which prefers place in higher altitude

A set of possible important environmental variables was chosen according to our knowledge and available information published in the literature about ecological demands and factors affecting distribution of studied species [46–49]. These factors might be important for individual orchid species distribution, and therefore they were included into the analysis. Environmental factors we used in analyses of all species are summarized in Table 1. One of the factors we used needs more description: the "consolidated layer of ecosystems," KVES [50]. It is a list of 40 habitat types (KVES 1, KVES 2, … KVES 40), such that each KVES number refers to a specific type of habitat. For example, KVES 6 means mesophilic meadows, KVES 10 means oak and oak-hornbeam forests, KVES 30 means mixed forests, and so on. The list and meaning of KVES we used are shown in Table 1. A KVES without a number means that the species occurrence depends on a certain habitat type. KVES is a categorical variable.

It is clear that the intensity of agriculture in the vicinity of an orchid locality affects its distribution in a broad scale. Previous studies about factors affecting orchid distribution [44, 45] indicate that the lower the amount of arable land is in

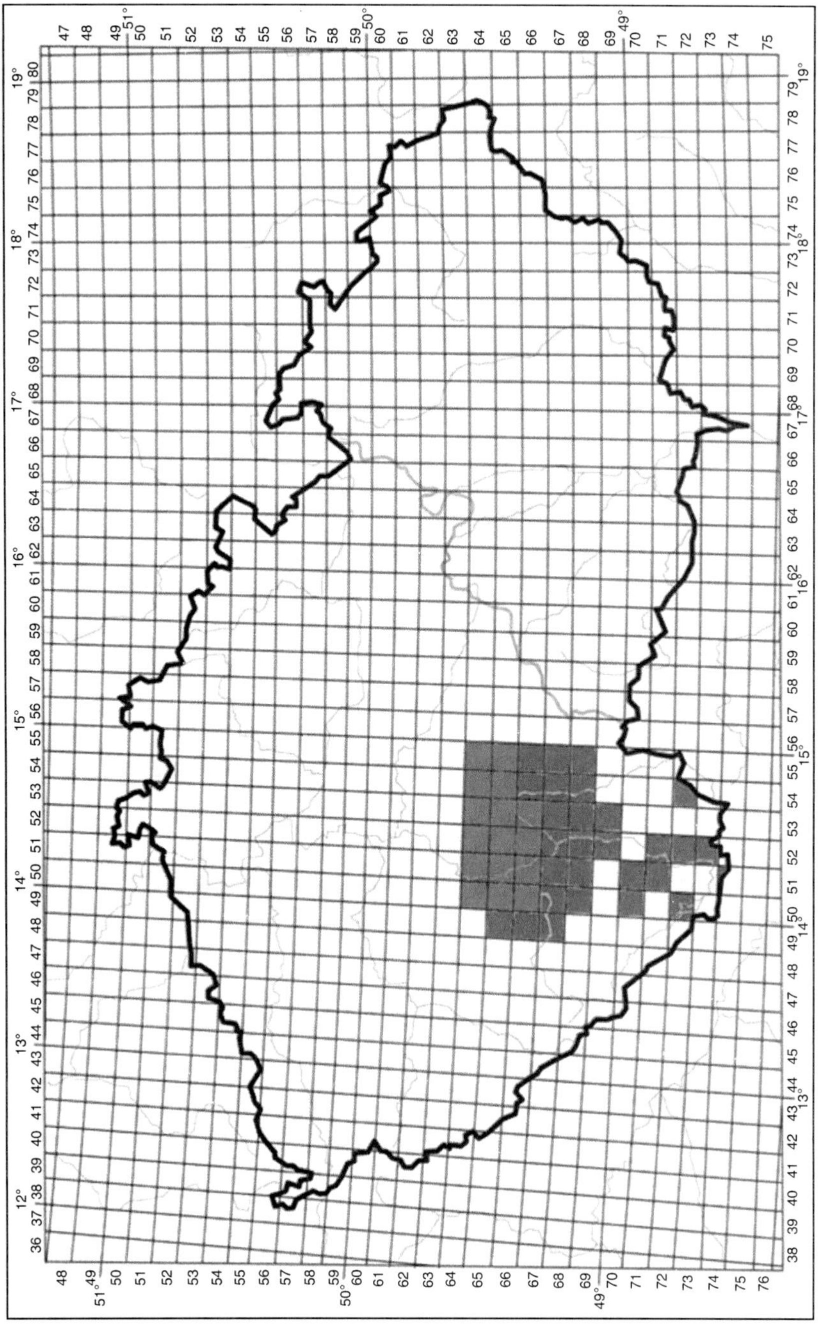

Fig. 1 Map of the study site in region of South Bohemia in the Czech Republic where field study has taken place

Table 1 The list of all environmental variables that were used in our analysis

Code	Description
dem	Altitude
frost_days	Number of freezing days per year
KVES	Consolidated layer of ecosystems
4	Alluvial and wet meadows
5	Dry grasslands
6	Mesophilic meadows
10	Oak and oak-hornbeam forests
12	Beech forests
13	Dry pine groves
17	Natural scrublands
19	Wetlands and coastal vegetation
23	Swamps and marshes
24	Ponds
29	Deciduous forests
30	Mixed forests
31	Coniferous forests
33	Urban green areas, gardens, parks, cemeteries
36	Discontinuous urban development
39	Agricultural meadows
40	Arable land
precipitation	Total precipitation per year (mm)
slope	Slope of terrain (degrees)
solar_rad	Solar radiation – total amount of incoming solar insolation (WH/m^2)
summer_days	Number of summer days (with temperature exceeding 25 °C) per year
temp_1	Mean annual temperature (°C)
temp_2	Temperature variability during year (°C)
trop_days	Number of tropical days (with temperature exceeding 30 °C) per year
veg_season	Duration of vegetation season

the vicinity of an orchid locality, the higher is the probability of presence of an orchid. For this reason, the amount of arable land was not tested separately again in our study, but the amount of arable land is included in the list of KVES under the number 40 (see Table 1).

The climatic data were obtained from the Global Change Research Institute of the CAS, and a climate character from a timeline of 1981–2011 was created. The resolution of these data was 500 × 500 m. Slope of the terrain [51] was also added into the analysis as additional factor that could influence the distribution of studied species.

The ecological niche modeling was conducted using maximum entropy method implemented in MaxEnt program version 3.3.2 [29–31] based on the species presence-only observations. Environmental data (described above) with spatial resolution of 500 m was used. The maximum number of iterations was set to 10,000 and convergence threshold to 0.00001. The "random seed" option, which provided random test partition and background subset for each run, was applied.

As an output from MaxEnt program, jackknife pictures, importance of each environmental factor, and potential distribution maps were obtained. In the resulting figure of jackknife procedure, two different blue bars are always displayed. The length of the dark-blue bar is telling us how large is the impact of selected variable. The length of light-blue bar is showing how much information would be lost if the corresponding factor were excluded from the analysis. The most important thing is to look on the combination of lengths of both blue bars, because these factors play an important role in the distribution of studied species. The most relevant factors are often displayed as at least a bit of dark-blue bar length in combination with shorter light-blue bar in comparison with other factors in the jackknife figure. Thus, deletion of such factor would cause a large loss of explanatory power of the model. Resolution of potential distribution maps was set to 500 × 500 m to make the map precise and detailed for determining possible new localities of studied species with suitable conditions.

In the description of important factors for studied species, percentage contributions of all factors were displayed in a table, but only the first three most important factors were described in more details in the text belonging to each studied species.

3 Results and Discussion

3.1 *Anacamptis morio* (L.) R.M. Bateman, A.M. Pridgeon & M.W. Chase 1997

The results of jackknife procedure in Fig. 2 revealed that the consolidated layer of ecosystems (KVES) is the most important factor that influences the distribution of *Anacamptis morio* in the South Bohemian region. Other important factors for this species are precipitation and slope of terrain (slope).

A closer look at the pictures of the most important environmental variables (Fig. 3) that play an important role in the distribution of *A. morio* reveals some interesting patterns. Consolidated layer of ecosystems (KVES) was determined as the most important factor with the contribution of 61.2%. Figure 3a of the analysis of KVES indicates that the highest probability of presence of this species is in oak and oak-hornbeam forests (KVES 10), mixed forests (KVES 30), discontinuous urban development (KVES 36), and agricultural meadows (KVES 39). According to our personal observation in the field, discontinuous urban development and agricultural meadows may be suitable habitats for *A. morio* when suitable management is applied. However, oak and oak-hornbeam forests and mixed forests are not suitable habitats for this species. These inconsistencies may be caused by border zone of two or more different habitat types. Plants may be present close to a forest border, and this place could have been identified as a forest during a monitoring of habitats and not as mesophilic meadow, agricultural meadow, or dry grassland, which *A. morio* can favor. On one hand, this is not a precise result, but, on the other hand, we can also identify negative aspects of MaxEnt analysis in prediction of a particular species distribution and possible gaps in procedure of habitat or biotope monitoring.

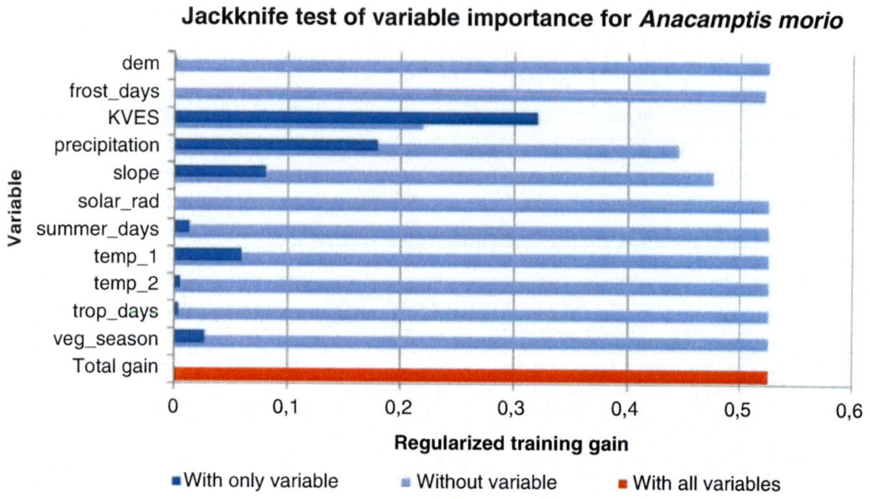

Fig. 2 Graph of the jackknife procedure for *Anacamptis morio*

Another factor that was determined as important is the amount of precipitation per year (mm). It is clearly visible in Fig. 3b that *A. morio* prefers places with low amount of precipitation. It is in agreement with ecological demands of this species mentioned in the literature – we can find it on dry or slightly humid substrates [46–48].

The last factor with an important effect on distribution of *A. morio* is the slope of terrain. Figure 3c shows that this species prefers rather flat or slightly inclined places, which was confirmed by our personal observation on its localities. We would probably not find this species in steep or exposed habitats.

The potential distribution map of *Anacamptis morio* is depicted in Fig. 4. It shows that there are many ecologically suitable places for distribution of this species but not all of them are suitable indeed because of inappropriate management. Other vegetation overgrows many of these suitable places, and it is impossible for *A. morio* to set a new population on such places [52, 53]. In the past, there were many thriving localities of this species in the region of South Bohemia, but the majority of them is now extinct (about 95%) [54] because of inappropriate management (overgrowing or even no management) or conversion of suitable places into agricultural fields that happened in the past [49, 55].

3.2 *Cephalanthera rubra* (Linne) L.C.M. Richard 1818

The jackknife procedure in Fig. 5 revealed that more factors have a similar impact on the distribution of this species. The first one is mean annual temperature (temp_1), then precipitation, and consolidated layer of ecosystems (KVES). Other important factors affecting its distribution are altitude (dem) and slope of the terrain (slope). Using Fig. 5, it may be hypothesized that many factors have a certain impact on

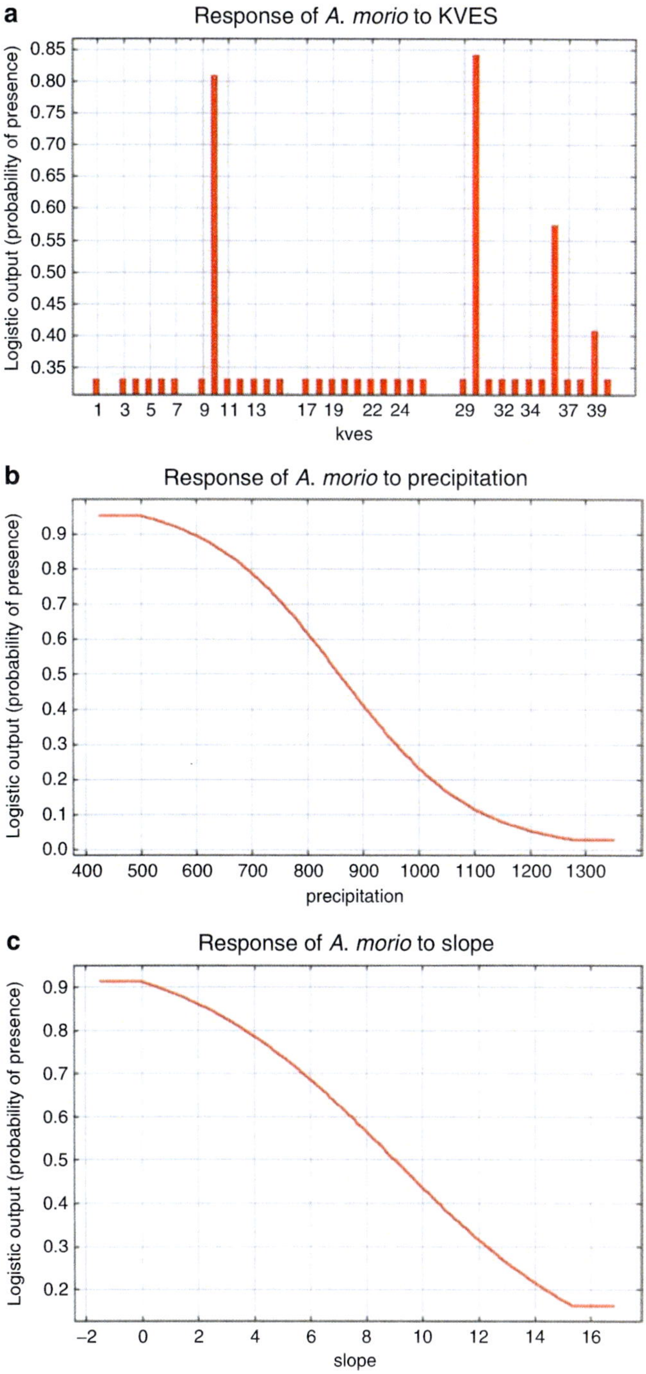

Fig. 3 Responses of *Anacamptis morio* to (**a**) consolidated layer of ecosystems (KVES), (**b**) precipitation, (**c**) slope of a terrain

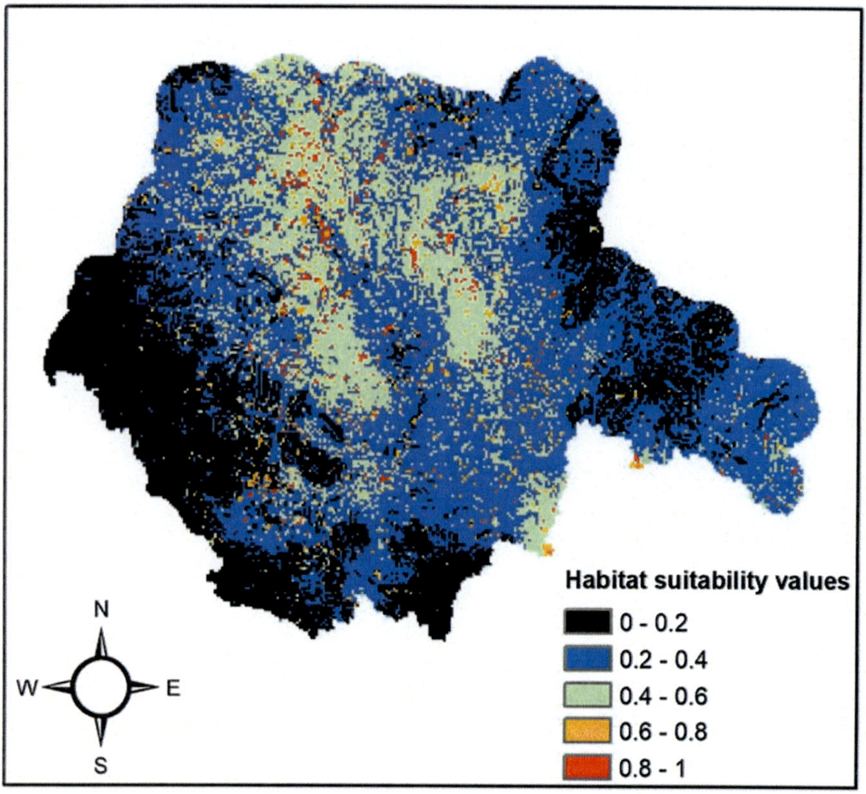

Fig. 4 Potential distribution map of *Anacamptis morio* in the region of South Bohemia

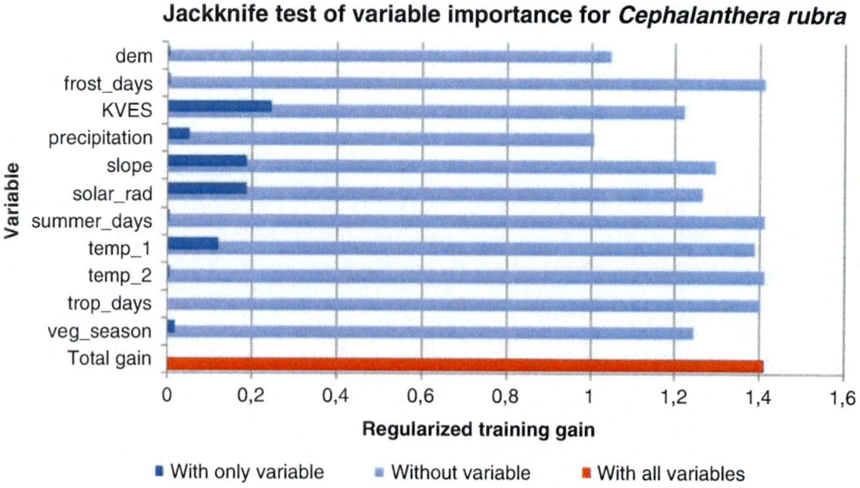

Fig. 5 Graph of the jackknife procedure for *Cephalanthera rubra*

distribution of *C. rubra*; they are co-affecting its presence together to some extent, and no single one factor has such an outstanding impact that KVES had in the case of *Anacamptis morio*.

The most important factor affecting the distribution of *C. rubra* is the mean annual temperature (temp_1). Figure 6a shows that *C. rubra* thrives in places where mean annual temperature is high. Intuitively, this is connected with altitude because high mean annual temperatures are in lowlands. Therefore, this means that this species will flourish in lower altitudes and we should not find it in high mountains. It is confirmed by information in the literature [46, 48] that says that we can find this species from lowlands to middle altitudes in the Czech Republic.

Second factor that was found to be important for distribution of *C. rubra* is the amount of precipitation per year, and its effect is depicted in Fig. 6b. It is clearly visible that this species prefers low amount of precipitation. This is again connected to altitude, as in the case of the previous factor – *C. rubra* should occur in the lower altitudes where there is only a little rain, which is congruent with the literature [46, 48].

Consolidated layer of ecosystems (KVES) was found to be the third most important factor affecting the presence of *C. rubra* in South Bohemian region. Figure 6c indicates that the most suitable habitats for this species are oak and oak-hornbeam forests (KVES 10), mixed forests (KVES 30), and agricultural meadows (KVES 39). It is said in the literature [46–49] that this species grows in bright forests, so both oak/oak-hornbeam forests and open mixed forests could be suitable for this species. To compare with one European country, Hungary, this species also prefers deciduous or mixed forests with *Pinus nigra* (Pacsai, pers. com.). Agricultural meadows may be considered as suitable, if a proper management is applied or if an agricultural meadow neighbors a suitable open forest.

The potential distribution map of *Cephalanthera rubra* (Fig. 7) shows potential suitable places for distribution of this species in the region of South Bohemia. Such places may be found around the city of Český Krumlov and toward the southern borders from this town and in the southeastern part bordering Austria. It may be also present in smaller limestone areas, because *C. rubra* prefers places where limestone is present [46, 48] such as abandoned limestone quarries. Such places were found, for example, around Horažďovice and Sušice, Tábor, Milevsko, and Písek city.

3.3 *Dactylorhiza fuchsii* (Druce) Soó 1962

Figure 8 shows the result of jackknife procedure generated by MaxEnt. It revealed that the most important factor affecting the distribution of *D. fuchsii* is consolidated layer of ecosystems (KVES). The second most important factor was slope followed by annual year temperature (temp_1).

Responses of the most important factors affecting the distribution of *D. fuchsii* are shown in Fig. 9. It represents the most important factor that affects the distribution of this species – consolidated layer of ecosystems (KVES). It is clearly visible that in this region, *D. fuchsii* prefers coniferous forests (KVES 31). It is known from the

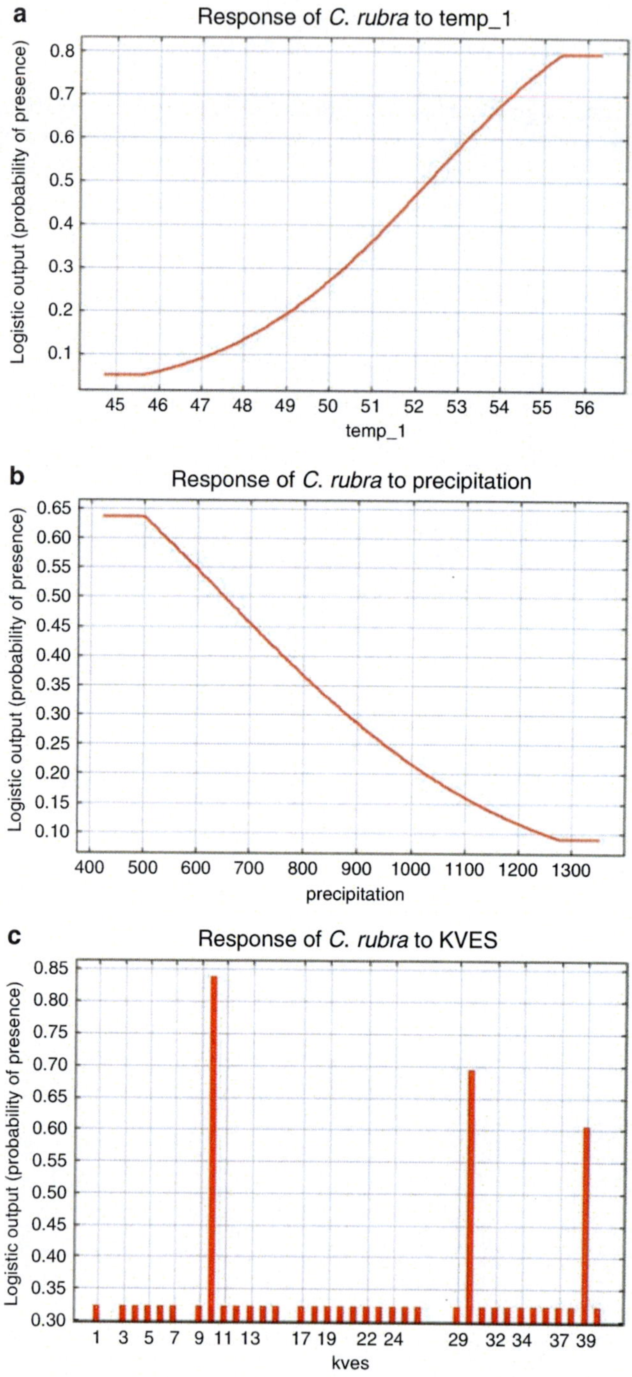

Fig. 6 Responses of *Cephalanthera rubra* to (**a**) mean annual temperature (temp_1), (**b**) precipitation, (**c**) consolidated layer of ecosystems (KVES)

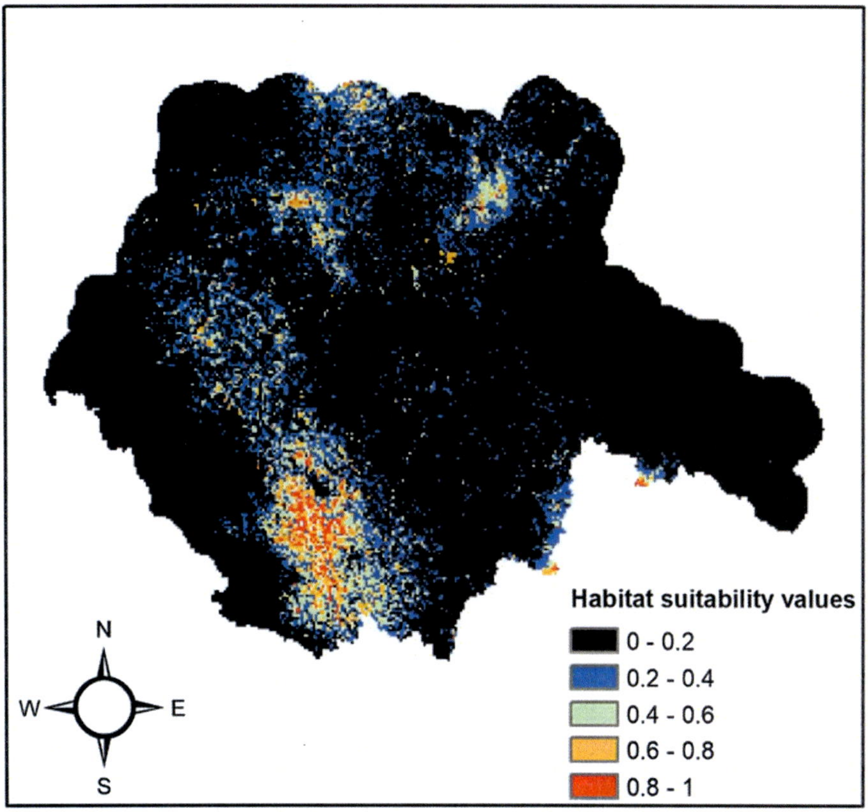

Fig. 7 Potential distribution map of *Cephalanthera rubra* in the region of South Bohemia

literature [46–48] that this species can be found in meadows and pastures as well as in forests. Coniferous forests are suitable [49] especially if they are not dense and the species has enough light to grow. From our personal observations in the field, suitable habitats are mainly on the borders of coniferous forests with other habitats (meadows or pastures) and in sparse coniferous forests, too.

Slope of terrain was determined as the second most important factor for distribution of *D. fuchsii*. From Fig. 9b, it is clearly visible that this species prefers a bit hilly landscape, and we will probably not find it in flat areas. It is in accordance with the literature, because *D. fuchsii* can be found from middle altitudes [46, 48], where the landscape is a bit wavy, not completely flat, and it disappeared from the low altitudes in South Bohemia [49].

The important factor that has an effect on the distribution of this species is mean annual temperature (temp_1). From Fig. 9c, it may be assumed that *D. fuchsii* prefers from middle to higher values of annual mean temperature. It means it will not be probably found in the highest places that are most exposed and cold, but it may be found in middle altitudes, where the temperatures are still high enough for its presence.

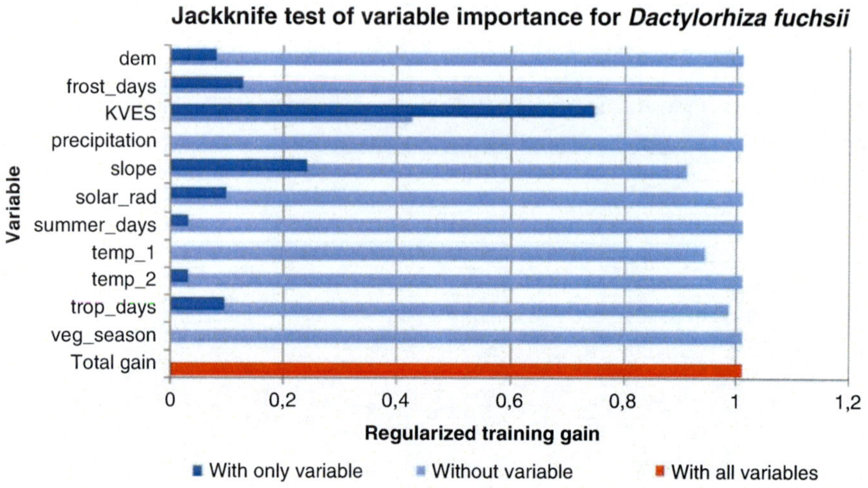

Fig. 8 Graph of the jackknife procedure for *Dactylorhiza fuchsii*

In Fig. 10, the map of potential distribution of *Dactylorhiza fuchsii* is displayed. The most suitable places, where new localities of this species may be found, are distributed mainly in the southernmost part of the South Bohemian region around Vyšší Brod city near the borders of Austria. Other suitable places can be found around Prachatice city toward Boletice Military Training Area and in middle altitudes in Šumava National Park. However, smaller suitable habitats are scattered across the whole region of South Bohemia.

3.4 *Epipactis palustris* (Linne) Crantz 1769

In Fig. 11, the results of jackknife procedure of *Epipactis palustris* are displayed. From this picture, it is clearly visible that consolidated layer of ecosystems (KVES) was revealed as the most important factor affecting its distribution in the region of South Bohemia. Other two factors that have an impact on its distribution were slope of the terrain (slope) and solar radiation (solar_rad).

According to the results of jackknife procedure, the most important factor was the consolidated layer of ecosystem (KVES). In a closer look at picture of this factor (Fig. 12a), we can clearly distinguish that most probably we will find this species in habitats of dry pine forests (KVES 13), mesophilic meadows (KVES 6), and partly in agricultural meadows (KVES 39). The information in the literature says that this species prefers habitats, mainly meadows, with stable water level and regime [49], and it may also be present in secondary habitats that were somehow altered by people in the past [46–48]. According to our personal observations in the field, it was found that this species is often present in the near vicinity of a forest (e.g., at the border zone between meadow and pine forest) – often in a kind of terrain depression

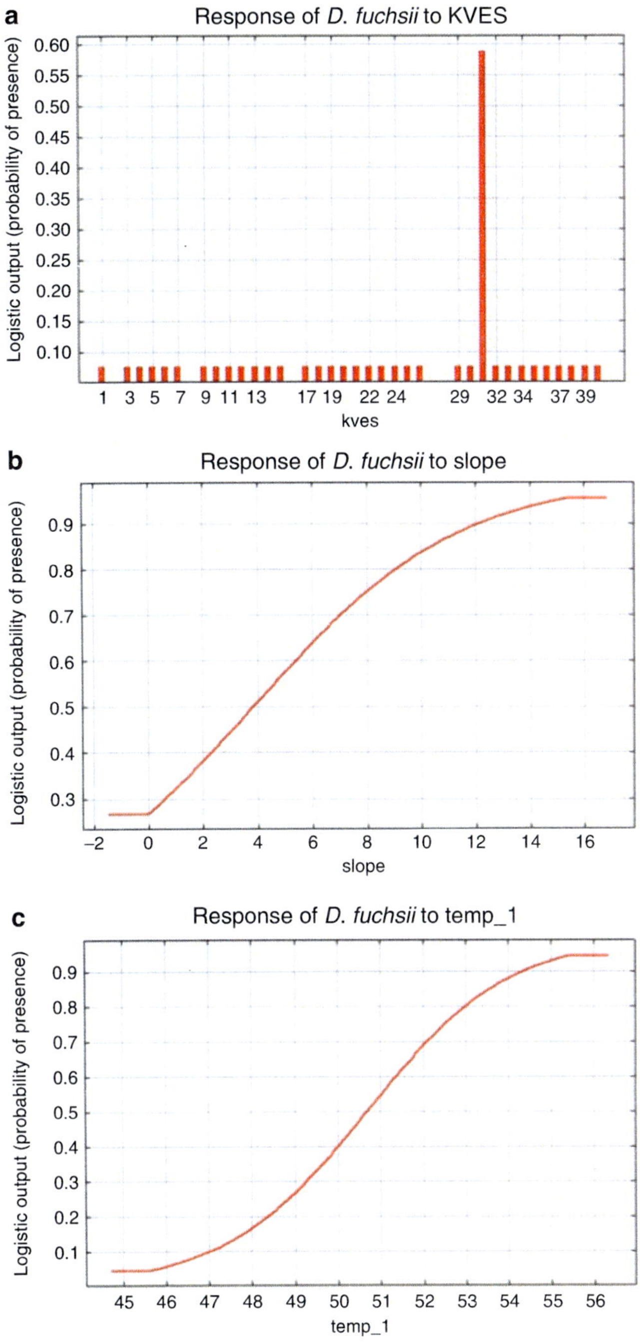

Fig. 9 Responses of *Dactylorhiza fuchsii* to (**a**) consolidated layer of ecosystems (KVES), (**b**) slope of a terrain, (**c**) mean annual temperature (temp_1)

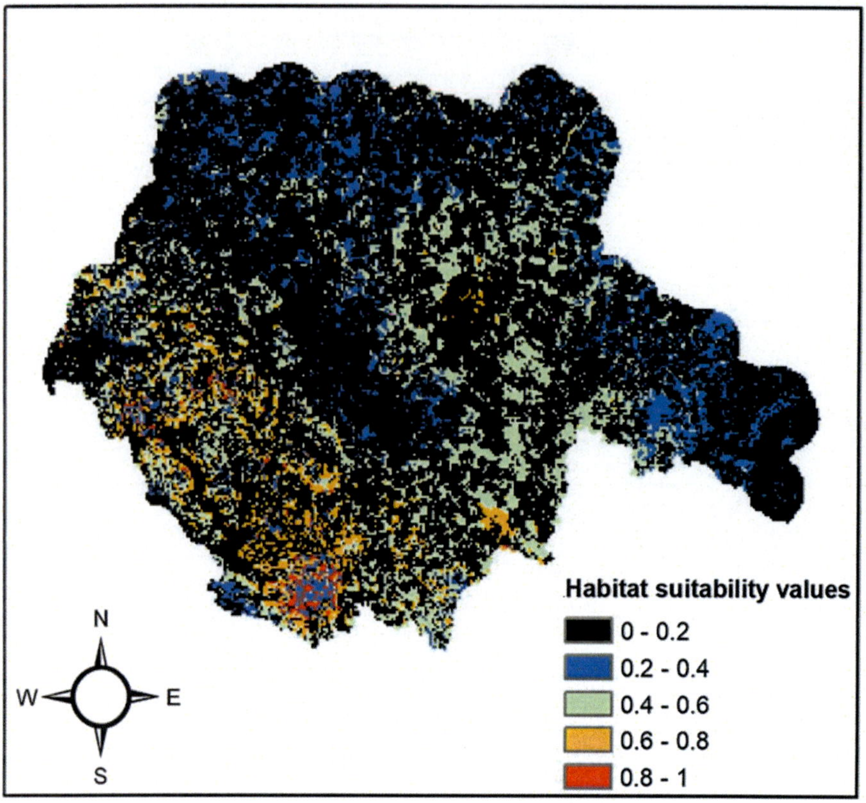

Fig. 10 Potential distribution map of *Dactylorhiza fuchsii* in the region of South Bohemia

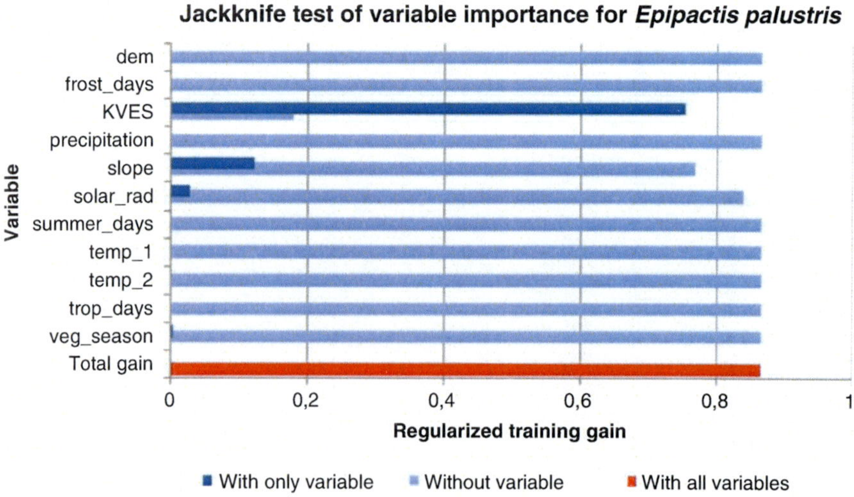

Fig. 11 Graph of the jackknife procedure for *Epipactis palustris*

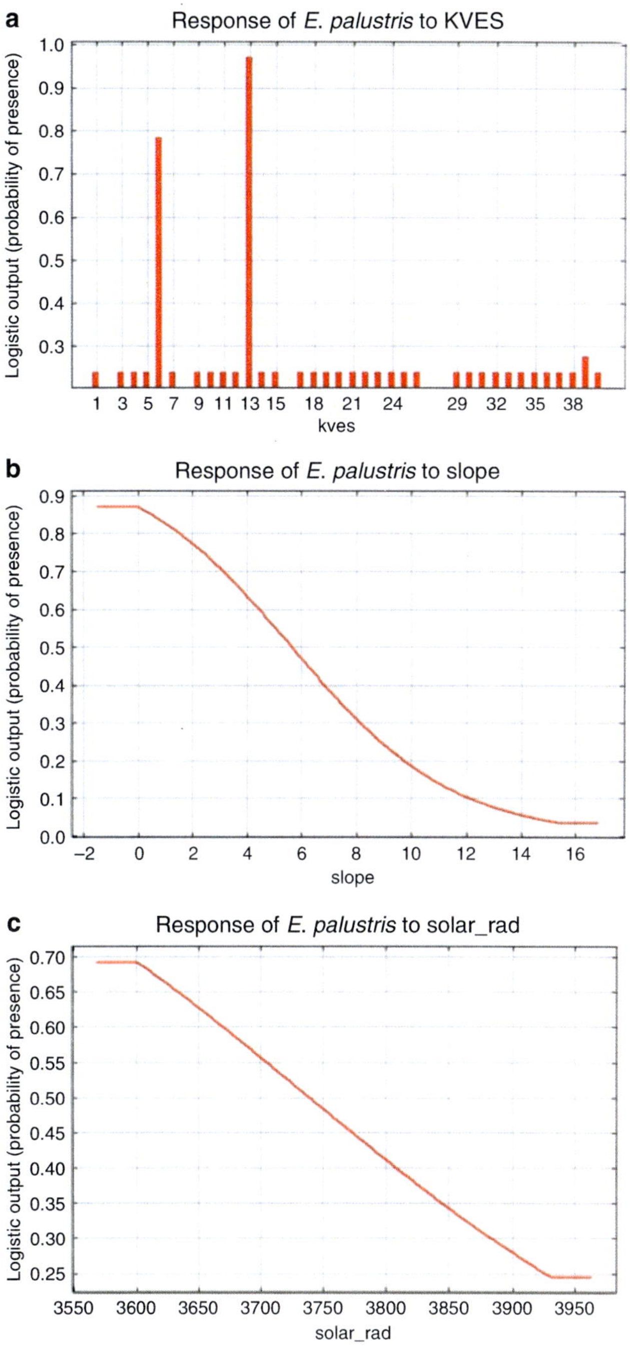

Fig. 12 Responses of *Epipactis palustris* to (**a**) consolidated layer of ecosystems (KVES), (**b**) slope of a terrain (slope), (**c**) solar radiation (solar_rad)

where water regime is stable and higher. Both mesophilic and agricultural meadows are also suitable for distribution of *E. palustris* if a suitable water regime is maintained.

Slope of a terrain (slope) was determined as the second most important factor affecting the distribution of this orchid. Figure 12b shows the probability of presence of *E. palustris* on a different slope. According to the results in Fig. 12b, this species will be most probably found in flat terrain that is in accordance with the previous statements about water regimes and possible terrain depressions. We will not find this species in steep slopes, because water regime varies here a lot during the year and is not stable.

The last factor – the amount of incoming solar radiation (solar_rad) – goes in hand with the first factor. From the graph generated by MaxEnt (Fig. 12c), it is clearly visible that *E. palustris* prefers shady places which is in accordance with its suitable habitats in this region which are pine groves. It was also monitored during our field studies that *E. palustris* was found in humid meadows, where surrounding vegetation was quite high, so orchid plants were not exposed to direct sunbeams in such places. According to Fig. 12c, *E. palustris* can be also found in semi-shaded places.

The potential distribution map of *Epipactis palustris* is depicted in Fig. 13. It shows suitable places, where it is possible to find new localities of this species in the future if their management and climate will not change. Such suitable localities are quite scattered in this region, and the majority of this area is not much suitable for this species. However, few suitable places can be still found mainly in the vicinity of Veselí nad Lužnicí city, where some Special Areas of Conservation (SAC) are present, and toward the southeastern borders with Austria near Suchdol nad Lužnicí and Chlum u Třeboně villages in Třeboňsko Nature Conservation Area. Some smaller scattered suitable places were also found between Dolní Dvořiště and Vyšší Brod city in the southernmost part of this region and in the Šumava National Park.

3.5 *Neottia nidus-avis* (Linne) L.C.M. Richard 1817

Figure 14 shows the effect of various factors tested that influence the distribution of *Neottia nidus-avis* in the South Bohemian region, according to the jackknife procedure. Clearly, consolidated layer of ecosystem (KVES) has the main impact on the distribution of this species. The following two main important environmental variables were slope of a terrain (slope) and mean annual precipitation.

The pictures from the results of the three most important variables affecting the distribution on *N. nidus-avis* in the region of South Bohemia (Fig. 15) revealed some interesting patterns. According to MaxEnt analysis, the most important factor was set to KVES (consolidated layer of ecosystem). From Fig. 15a, it is clearly visible that the most suitable habitats for this species are oak and oak-hornbeam forests (KVES 10); however, it is possible to find it partly also in coniferous forests (KVES 31) and in agricultural meadows (KVES 39). Suitable habitat of oak and oak-hornbeam forests is in accordance with the information from literature [47, 48, 56]; partly also coniferous forests may be suitable but only in case that

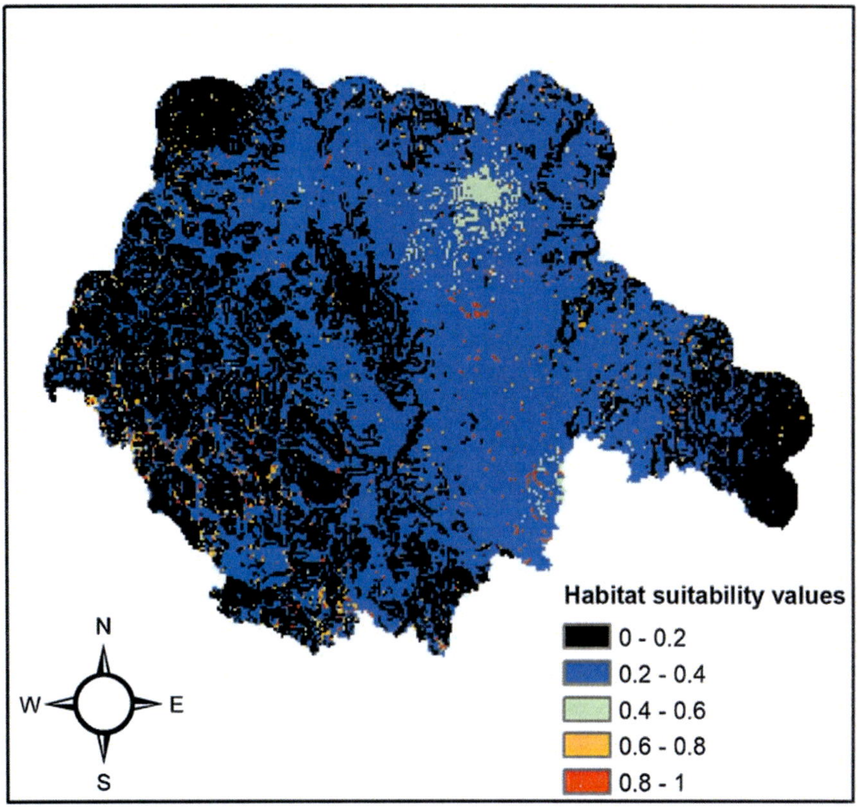

Fig. 13 Potential distribution map of *Epipactis palustris* in the South Bohemian region

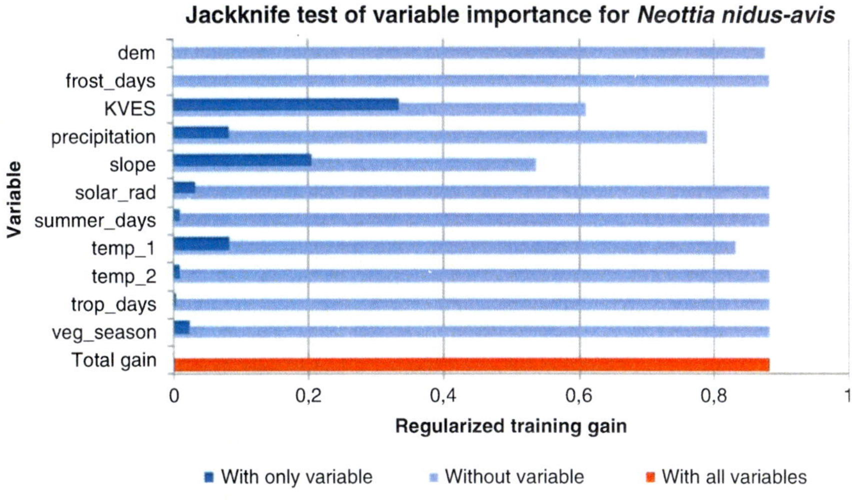

Fig. 14 Graph of the jackknife procedure for *Neottia nidus-avis*

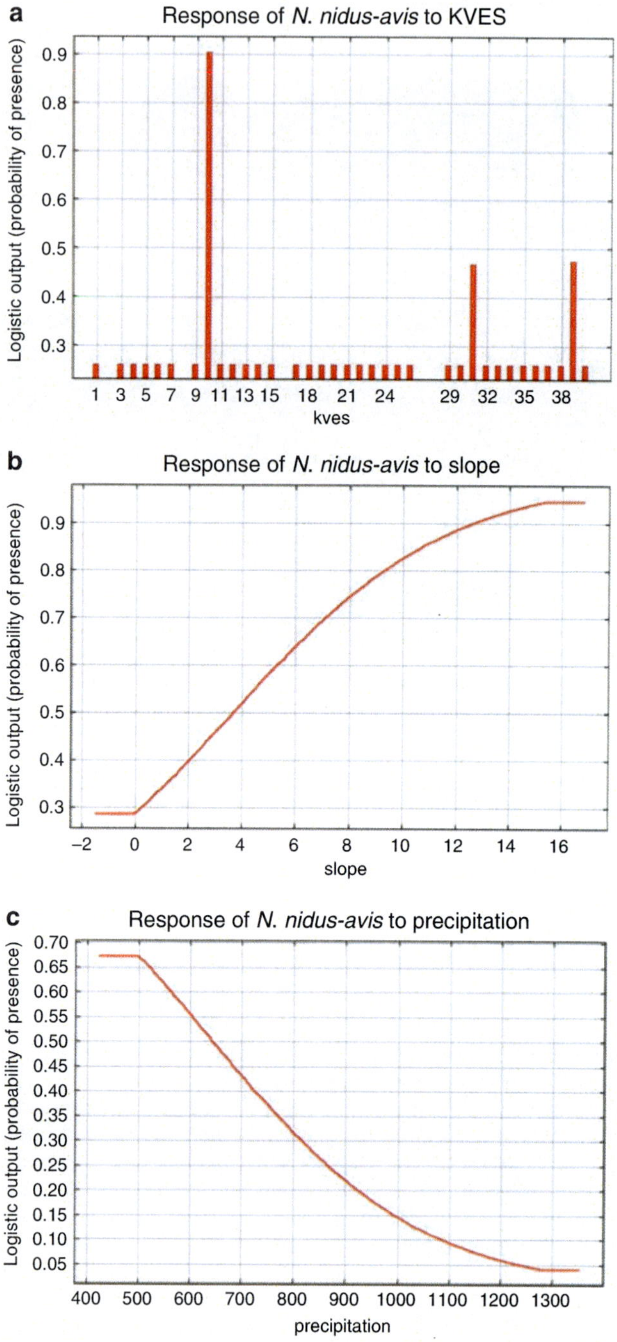

Fig. 15 Responses of *Neottia nidus-avis* to (**a**) consolidated layer of ecosystems (KVES), (**b**) slope of a terrain (slope), (**c**) mean annual precipitation

particular forest is not too dense. But agricultural meadows are not suitable for *N. nidus-avis*. This inconsistency may be caused by a borderline between forest habitat and agricultural meadow. Dykyjová stated that this species is able to reach also atypical habitats from forests edges, such as grass meadows or scrublands [46]. Also Lepší et al. confirmed forest edges as a partly suitable habitat [49].

Slope of a terrain was revealed as the second most important factor affecting the presence of *N. nidus-avis*. According to Fig. 15b, it may be assumed that this species will be found in at least a bit hilly countryside and not in completely flat places. Based on the literature, it prefers the altitudinal range from lower places or foothills to mountains [46, 48].

Another important factor for distribution of *N. nidus-avis* is annual mean precipitation. Figure 15c indicates that this species prefers places with lower amount of precipitation during the whole year. Intuitively, these places can be found in lower altitudes but also in higher altitudes in rain shadow, and these may be suitable for presence of *N. nidus-avis*.

In Fig. 16, the map of potential distribution of *Neottia nidus-avis* is shown. It indicates that there are many places with suitable conditions for this species in the

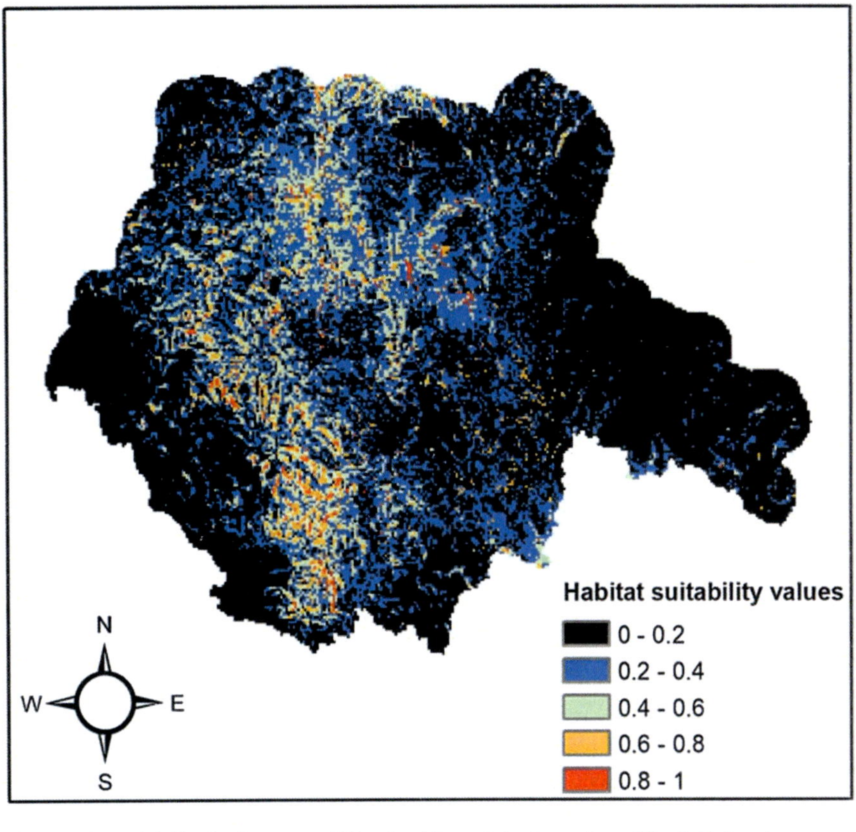

Fig. 16 Potential distribution map of *Neottia nidus-avis* in the region of South Bohemia

South Bohemian region. They can be found in the area between Český Krumlov and Vyšší Brod cities in the southern part of this region, in foothills of Šumava National Park, and in the central part of South Bohemia around Písek city.

3.6 *Neottia ovata* (L.) Bluff & Fingerh. 1838

In Fig. 17, the results of jackknife procedure for *Neottia ovata* are displayed. From this picture, it is clearly visible that the most important factor affecting the distribution of this species is again the consolidated layer of ecosystems (KVES). Other two factors with the second and third highest percent contribution were slope of a terrain (slope) and solar radiation (solar_rad).

A closer look at pictures of environmental variables that had the most important effect on the distribution of *N. ovata* (Fig. 18) reveals certain patterns. Figure 18a shows that most suitable habitats for this species are present in urban green areas, gardens and parks (KVES 33); in oak and oak-hornbeam forests (KVES 10), beech forests (KVES 12); and in agricultural meadows (KVES 39). All of the habitats that were revealed as suitable by the analysis of MaxEnt are places proved by the information from literature. It says that *N. ovata* is one of the species that has broad ecological niche, so it may be found in various types of habitats [46–48, 56].

Slope of a terrain was also determined as important factor that may affect the distribution of *N. ovata*. Figure 18b indicates that we will probably not find it in completely flat places, but it prefers at least a bit hilly countryside. The center of its occurrence in the region of South Bohemia is in a foothill level [49]. However, as it was stated above, this species has no specific ecological demands, so we can find it in various places.

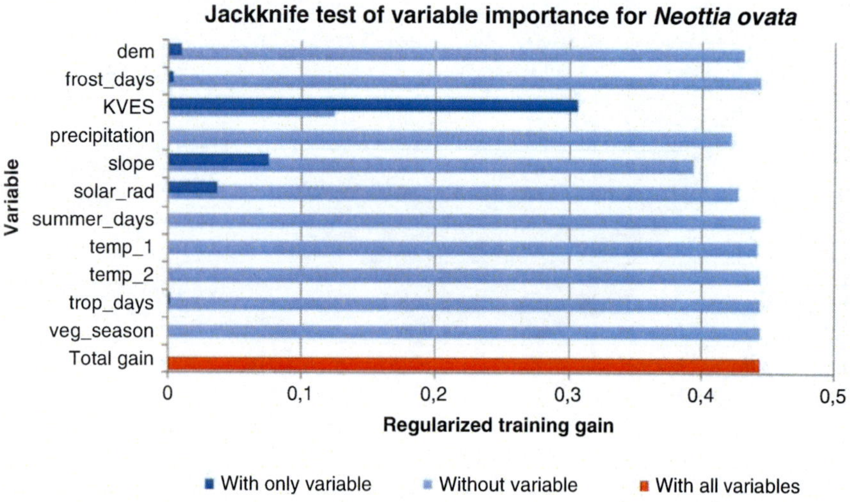

Fig. 17 Graph of the jackknife procedure for *Neottia ovata*

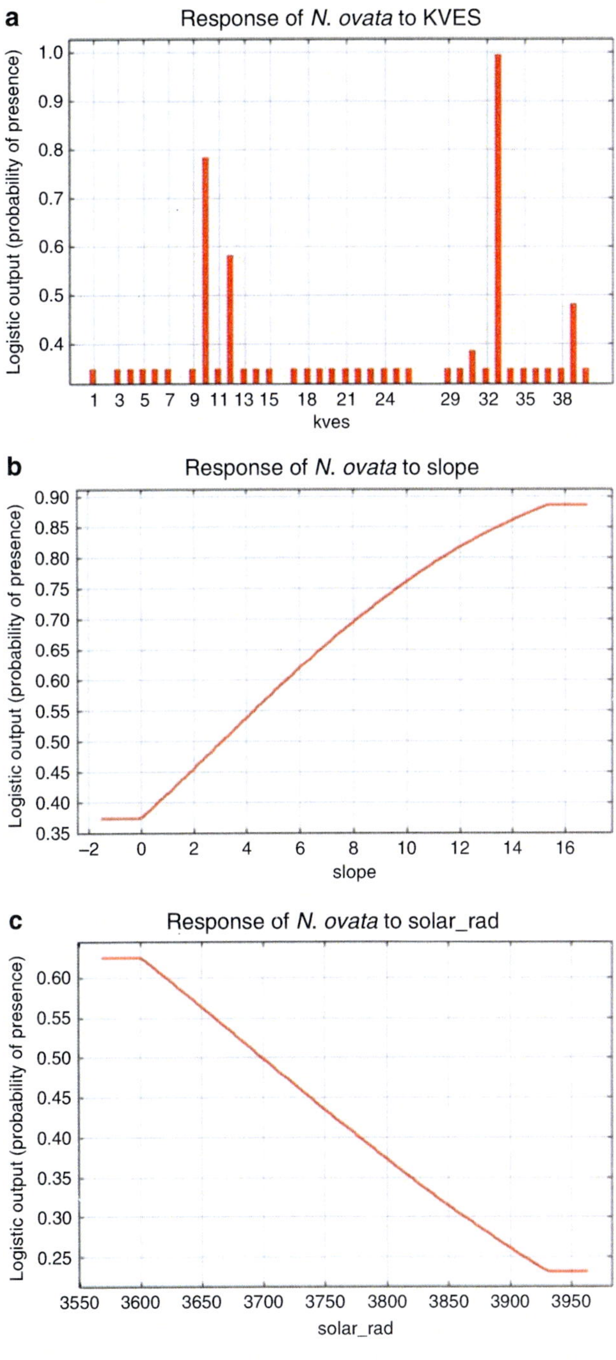

Fig. 18 Responses of *Neottia ovata* to (**a**) consolidated layer of ecosystems (KVES), (**b**) slope of a terrain (slope), (**c**) solar radiation

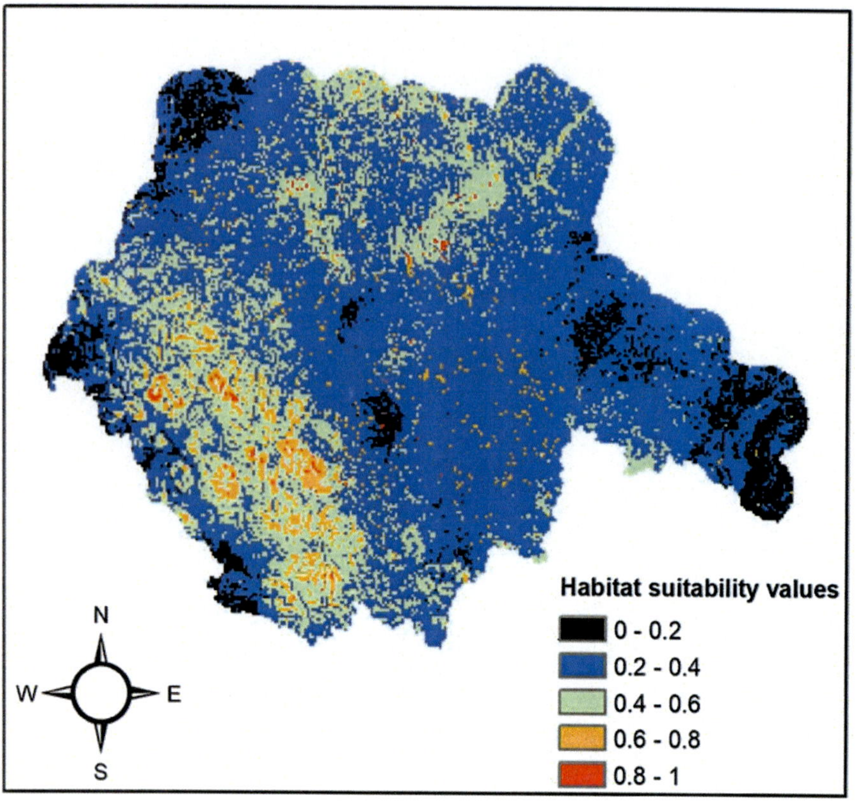

Fig. 19 Potential distribution map of *Neottia ovata* in the region of South Bohemia

Figure 18c shows the impact of solar radiation on the distribution of *N. ovata*. It is clearly visible from the picture that the more solar radiation is present, the higher is the probability of presence of this species on such places. It does not prefer entirely shady places.

The potential distribution map of *Neottia ovata* is displayed in Fig. 19. From this picture, it can be assumed that there are many suitable places in the region of South Bohemia for presence of *N. ovata* and there are almost no places that would be completely unsuitable (dark-blue color). This also proved the statement above that this species may be found in many different types of habitats and climatic conditions; it is not specialized in this term. The most suitable places (red and yellow colors) are present in southwestern part of the South Bohemian region, in the foothills of Šumava National Park in the area from Český Krumlov toward Prachatice city.

3.7 *Platanthera chlorantha* (Custer) Rchb. 1828

The results of jackknife procedure are displayed in Fig. 20. It implies that there are more environmental variables that have the main impact on the distribution of

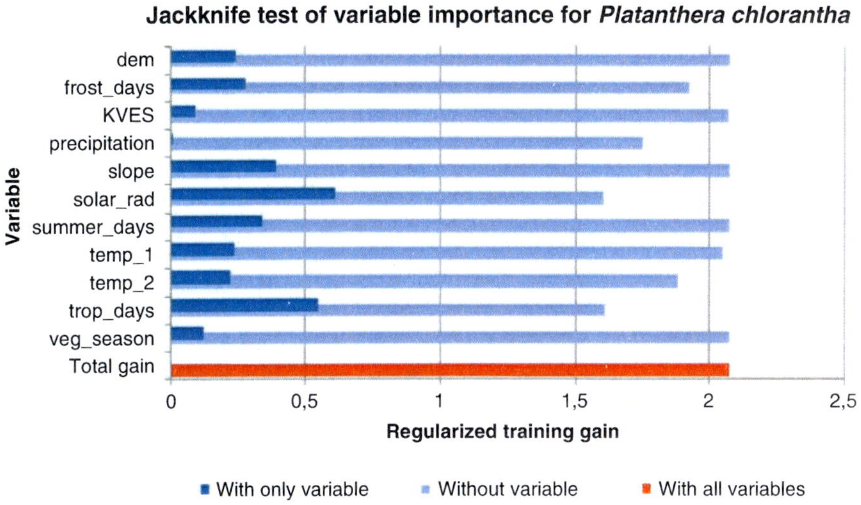

Fig. 20 Graph of the jackknife procedure for *Platanthera chlorantha*

P. chlorantha. There are many factors that cooperate with each other and influence where this species occurs. According to Fig. 20, the most important factors affecting its distribution in the region of South Bohemia are number of tropical days per year (trop_days) followed by total amount of solar radiation that enters a locality (solar_rad) and amount of precipitation per year (precipitation). It is also worth to say that consolidated layer of ecosystems (KVES) was not evaluated as one of the most important factors affecting species distribution, as was the case with species described above, so it is the only species that do not rely strongly on a type of habitat that is present on its localities. Literature says that this species has no clear relation with a particular habitat type [47].

A closer look at pictures of the most important factors that had a significant impact on the distribution of *P. chlorantha* (Fig. 21) shows interesting results. In Fig. 21a, the impact of the number of tropical days per year is displayed. It is visible that there is a high probability of occurrence of this species in places with zero or only a few tropical days per year and almost zero probability in places where many tropical days are present. These findings imply that *P. chlorantha* prefers higher altitudes and we will probably not find this species in lowlands in the studied region. Also according to the literature, it prefers higher and colder places [49].

Figure 21b shows a response of the studied species to solar radiation (solar_rad), a typical mesoclimatic factor. In general, the extent of solar radiation is not different throughout the whole Czech Republic, so this factor tells us whether *P. chlorantha* prefers shady or sunny places. From the graph, it is clearly visible that it is more likely to find this species in shady places and it tries to avoid places in full sunlight. This is in accordance with the literature; it may occur also in mountain meadows that may represent the middle part of the curve presented.

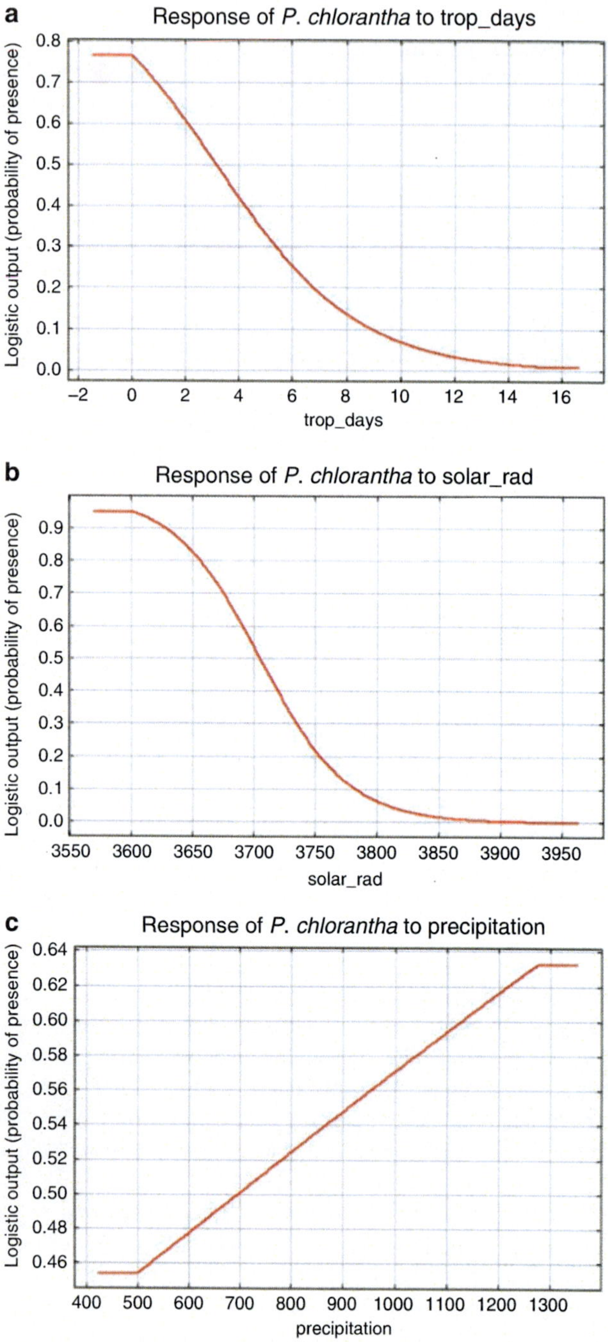

Fig. 21 Responses of *Platanthera chlorantha* to (**a**) amount of tropical days per year (trop_days), (**b**) total amount of solar radiation (solar_rad), (**c**) mean annual precipitation

The impact of mean annual precipitation on the distribution of *P. chlorantha* is displayed in Fig. 21c. This picture shows that the more precipitation per year, the higher is the probability of presence of this species, and it does not prefer places with low amount of rainfalls throughout a year. Generally speaking, high amount of precipitations is a characteristic feature for higher altitudes and mountain areas, so this result indicates that such places are suitable for the presence of *P. chlorantha*. It is similar to the case of the impact of tropical days on this species, and it is also stated in the literature [46, 47, 49].

In Fig. 22, the potential distribution map of *Platanthera chlorantha* is depicted. This map shows that there are a few suitable places for potential occurrence of the studied species, but climatic and environmental conditions are not suitable for *P. chlorantha* in most of the region. Potential suitable places for this species may be found mainly in the southeastern edge at the borders with Austria around Nová Bystřice city, in smaller scattered areas in the south around Vyšší Brod city, in a middle-to-higher-altitude belt stretching from Český

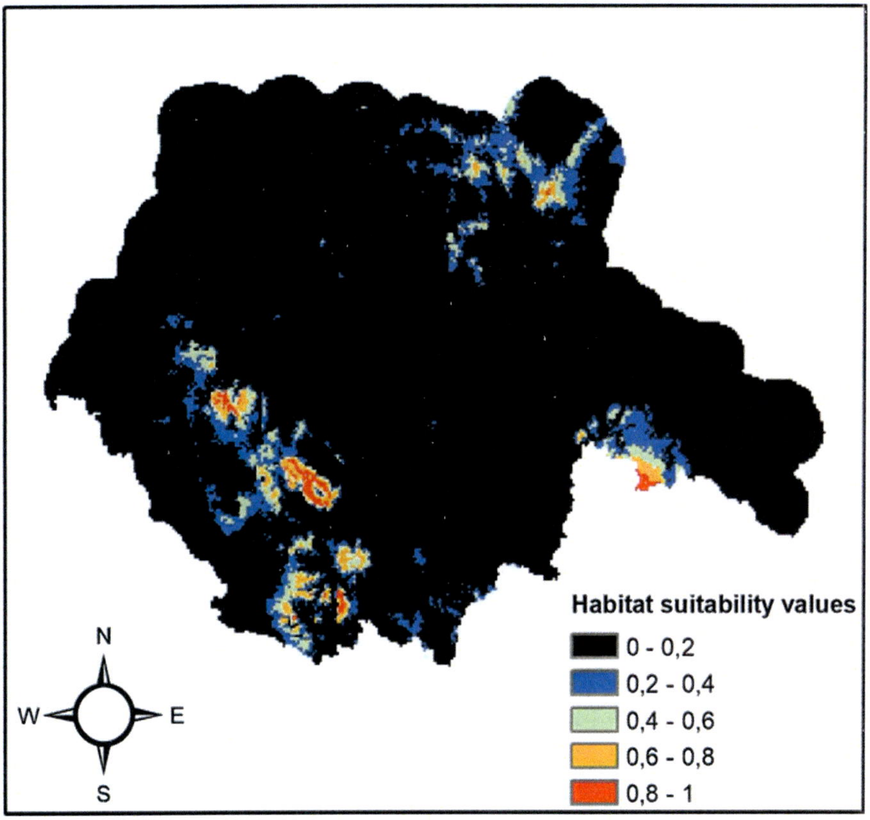

Fig. 22 Potential distribution map of *Platanthera chlorantha* in the South Bohemian region

Krumlov toward Prachatice, and partly also in the northeastern part of the region in the vicinity of Chýnov.

4 Summary

As stated in other papers, the amount of arable land is an important factor that affects the orchid distribution. In our study, no orchid species studied prefer any kind of arable land (not presented here). It proved the previous statements that the high amount of arable land in the vicinity of an orchid locality has a negative effect on its distribution [44, 45].

The most important and most common environmental factor affecting the distribution of numerous orchid species in the region of South Bohemia was the consolidated layer of ecosystems (KVES) as it played the most important role for 10 out of 11 species studied (see Table 2). Only for *Platanthera chlorantha*, KVES was not found as important factor affecting its occurrence. The other two most important variables were mean annual precipitation and slope of a terrain that was important for 7 out of 11 species studied (Table 2). According to the fact that KVES was set as the most important factor that can influence the distribution and presence of many orchids species, evaluation of a particular habitat type (KVES) was also done (see Table 3). Based on our analysis, the most important KVES types (habitat type) are oak and oak-hornbeam forests (KVES 10) followed by agricultural meadows (KVES 39). However, forests as well as meadows in general should be protected as they host many endangered species of orchids. The duration of vegetation season (veg_season) was also added into our analysis, but it has no important effect on the distribution of studied species because the length of the vegetation season does not differ a lot across the whole country. The small differences in the length of vegetation season are more connected with altitude.

5 Conclusions

The MaxEnt program is a useful tool for predicting potential distribution of species in general but is especially effective when working with threatened and endangered species. According to the results of our study, the most important factors for many orchid species in the South Bohemian region are habitat type (represented by consolidated layer of ecosystems, KVES), precipitation, and slope of a terrain.

Our results are important and helpful in determination of possible new localities in the region of South Bohemia but may be also used in larger scale. Without potential distribution maps, searching of new localities would be only a random choice of researchers. Our findings may help in orchid conservation by preserving suitable habitats for chosen orchid species.

Table 2 The list of the most important environmental factors (the first three) affecting distribution of studied species in the region of South Bohemia in the Czech Republic (° impact of a factor is less than 50%, • impact of a factor is 50% and more). Results for species in red are described in Štípková et al. [44] and Kosánová [45]

	dem	KVES	precipitation	slope	solar_rad	temp_1	temp_2	trop_days	veg_season
Anacamptis morio		•	°	°					
Cephalanthera damasonium	°	°		°					
Cephalanthera rubra		°	°			°			
Dactylorhiza fuchsii		•		°		°			
Dactylorhiza majalis		°	°						°
Epipactis atrorubens		•		°		°			
Epipactis palustris		•		°	°				
Neottia nidusavis		°	°	°					
Neottia ovata		•	°	°					
Platanthera chlorantha			°		°			°	
Platanthera bifolia		°	°		°				

Table 3 The most important habitat types (KVES) for studied species (*Platanthera chlorantha* was not included in the table because it does not strongly rely on a specific habitat type)

	KVES 4	KVES 5	KVES 6	KVES 10	KVES 12	KVES 13	KVES 17	KVES 19	KVES 23	KVES 24	KVES 30	KVES 31	KVES 33	KVES 36	KVES 39
Anacamptis morio				•							•			•	•
Cephalanthera damasonium				•							•			•	•
Cephalanthera rubra				•							•				•
Dactylorhiza fuchsii												•			
Dactylorhiza majalis	•	•	•				•	•	•	•			•		
Epipactis atrorubens				•		•				•	•				
Epipactis palustris			•			•									•
Neottia nidus-avis				•								•			•
Neottia ovata				•	•								•		•
Platanthera bifolia				•	•						•			•	•

Acknowledgments This work was supported by the Ministry of Education, Youth and Sports of CR within the National Sustainability Program I (NPU I), grant number LO1415. We also thank the South Bohemian Branch of the Land Office in Ceske Budejovice for their kind cooperation and Kristina Kosánová for her help in the field.

References

1. Dirzo R, Raven PH (2003) Global state of biodiversity and loss. Annu Rev Environ Resour 28:137–167. https://doi.org/10.1146/annurev.energy.28.050302.105532
2. Possingham HP, Wilson KA (2005) Biodiversity – turning up the heat on hotspots. Nature 436:919–920. https://doi.org/10.1038/436919a
3. Tsiftsis S, Tsiripidis I, Trigas R (2011) Identifying important areas for orchid conservation in Crete. Eur J Environ Sci 1(2):28–37
4. Margules CR, Pressey RL (2000) Systematic conservation planning. Nature 405:243–253. https://doi.org/10.1038/35012251
5. Gaston KJ, Pressey RL, Margules CR (2002) Persistence and vulnerability: retaining biodiversity in the landscape and in protected areas. J Biosci 27:361–384. https://doi.org/10.1007/BF02704966
6. Williams PH, Margules CR, Hilbert DW (2002) Data requirements and data sources for biodiversity priority area selection. J Biosci 27:327–338. https://doi.org/10.1007/BF02704963
7. Heywood VH, Iriondo JM (2003) Plant conservation: old problems, new perspectives. Biol Conserv 113:321–335. https://doi.org/10.1016/S0006-3207(03)00121-6
8. Efimov PG (2011) Revealing the decline and expansion of orchids of NW European Russia. Eur J Environ Sci 1(2):7–17
9. Feldman D, Prat D (2011) Conservation recommendations from a large survey of French orchids. Eur J Environ Sci 1(2):8–27
10. Dressler RL (1993) Phylogeny and classification of the orchid family. Cambridge University Press, Cambridge
11. Chase MW, Cameron KM, Barrett RL, Freudebstein JV (2003) DNA data and Orchidaceae systematics: a new phylogenetic classification. In: Dixon KW, Kell SP, Barrett RL, Cribb PJ (eds) Orchid conservation. Natural History Publications, Kota Kinabalu
12. Cribb PJ, Kell SP, Dixon KW, Barrett RL (2003) Orchid conservation: a global perspective. In: Dixon KW, Kell SP, Barrett RL, Cribb PJ (eds) Orchid conservation. Natural History Publications, Kota Kinabalu
13. Pillon Y, Chase M (2006) Taxonomic exaggeration and its effects on orchid conservation. Conserv Biol 21:263–265
14. Swarts ND, Dixon KW (2009) Terrestrial orchid conservation in the age of extinction. Ann Bot 104:543–556
15. Zhang ZJ, Yan YJ, Tian Y, Li JS, He JS, Tang ZY (2015) Distribution and conservation of orchid species richness in China. Biol Conserv 181:64–72. https://doi.org/10.1016/j.biocon.2014.10.026
16. Vakhrameeva MG, Tatarenko IV, Varlygina TI, Torosyan GK, Zagulski MN (2008) Orchids of Russia and adjacent countries (within the borders of the former USSR). ARG Gantner Verlag KG, Ruggell
17. Hutchings MJ (1989) Population biology and conservation of *Ophrys sphegodes*. In: Pritchard HW (ed) Modern methods in orchid conservation: the role of physiology, ecology and management. Cambridge University Press, Cambridge
18. Kull T, Kindlmann P, Hutchings M, Primack B (2006) Conservation biology of orchids: introduction to the special issue. Biol Conserv 129:1–3
19. Wotavová K, Balounová Z, Kindlmann P (2004) Factors affecting persistence of terrestrial orchids in wet meadows and implications for their conservation in a changing agricultural landscape. Biol Conserv 118:271–279

20. Janečková P, Wotavová K, Schödelbauerová I, Jersáková J, Kindlmann P (2006) Relative effects of management and environmental conditions on performance and survival of populations of a terrestrial orchid, *Dactylorhiza majalis*. Biol Conserv 129:40–49
21. Hágsater E, Dumont V (1996) Orchids: status, survey and conservation action plan. IUCN, Gland
22. Elith J, Leathwick JR (2009) Species distribution models: ecological explanation and prediction across space and time. Annu Rev Ecol Evol Syst 40:677–697. https://doi.org/10.1146/annurev.ecolsys.110308.120159
23. Guisan A, Tingley R, Baumgartner JB et al (2013) Predicting species distributions for conservation decisions. Ecol Lett 16:1424–1435. https://doi.org/10.1111/ele.12189
24. Guisan A, Thuiller W (2005) Predicting species distribution: offering more than simple habitat models. Ecol Lett 8:993–1009. https://doi.org/10.1111/j.1461-0248.2005.00795.x
25. Elith J, Graham CH (2009) Do they? How do they? WHY do they differ? On finding reasons for differing performances of species distribution models. Ecography 32:66–77
26. Rocchini D, Hortal J, Lengyel S, Lobo JM, Jimenez-Valverde A, Ricotta C, Bacaro G, Chiarucci A (2011) Accounting for uncertainty when mapping species distributions: the need for maps of ignorance. Prog Phys Geogr 35:211–226
27. Tsiftsis S, Tsiripidis I, Trigas R, Karagiannakidou V (2012) The effect of presence/absence vs. continuous suitability data on reserve selection. Eur J Environ Sci 2(2):125–137
28. Elith J, Graham CH, Anderson RP et al (2006) Novel methods improve prediction of species' distributions from occurrence data. Ecography 29:129–151
29. Phillips SJ, Anderson RP, Schapire RE (2006) Maximum entropy modeling of species geographic distribution. Ecol Model 190:231–259. https://doi.org/10.1016/j.ecolmodel.2005.03.026
30. Phillips SJ, Dudík M (2008) Modeling of species distributions with Maxent: new extensions and a comprehensive evaluation. Ecography 31:161–175. https://doi.org/10.1111/j.0906-7590.2008.5203.x
31. Elith J, Phillips SJ, Hastie T, Dudik M, Chee YE, Yates CJ (2011) A statistical explanation of MaxEnt for ecologist. Divers Distrib 17:43–57. https://doi.org/10.1111/j.1472-4642.2010.00725.x
32. Fourcade Y, Engler JO, Rödder D, Secondi J (2014) Mapping species distributions with MAXENT using a geographically biased sample of presence data: a performance assessment of methods for correcting sampling bias. PLoS One 9:e97122
33. Hernández P, Graham C, Lawrence L, Albert D (2006) The effect of sample size and species characteristics of performance of different species distribution modeling methods. Ecography 29:773–785
34. Yi Y-J, Cheng X, Yang Z-F, Zhang S-H (2016) Maxent modeling for predicting the potential distribution of endangered medicinal plant (*H. riparia* Lour) in Yunnan, China. Ecol Eng 92:260–269
35. Kolanowska M (2013) Glacial refugia and migration routes of the neotropical genus Trizeuxis (Orchidaceae). Acta Soc Bot Pol 82:225–230. https://doi.org/10.5586/asbp.2013.024
36. Wan J, Wang C, Han S, Yu J (2014) Planning the priority protected areas of endangered orchid species in northeastern China. Biodivers Conserv 23:1395–1409. https://doi.org/10.1007/s10531-014-0671-0
37. Reina-Rodríguez GA, Rubiano JE, Llanos FAC, Otero JT (2016) Spatial distribution of dry forest orchids in the Cauca River Valley and Dagua Canyon: towards a conservation strategy to climate change. J Nat Conserv 30:32–43. https://doi.org/10.1016/j.jnc.2016.01.004
38. Vollering J, Schuiteman A, de Vogel E, van Vugt R, Raes N (2016) Phytogeography of New Guinean orchids: patterns of species richness and turnover. J Biogeogr 43:204–214. https://doi.org/10.1111/jbi.12612
39. Nature Conservation Agency of the Czech Republic (2006) http://portal.nature.cz/publik_syst/ctihtmlpage.php?what=3&nabidka=hlavni. Accessed 24 Mar 2014

40. Czech National Phytosociological Database. Vegetation Science Group, Department of Botany and Zoology, Faculty of Science, Masaryk University (2005) http://www.sci.muni.cz/botany/vegsci/dbase.php?lang=cz. Accessed 21 Feb 2014
41. South Bohemian Branch. Czech Botanical Society (2017) https://botanospol.cz/cs/node/42. Accessed 03 Mar 2014
42. Pearson RG, Thuiller W, Araújo MB et al (2006) Model-based uncertainty in species range prediction. J Biogeogr 33:1704–1711
43. Wisz MS, Hijmans RJ, Li J, Peterson AT, Graham CH, Guisan A, NCEAS Predicting Species Distributions Working Group (2008) Effect of sample size on the performance of species distribution models. Divers Distrib 14:763–773
44. Štípková Z, Kosánová K, Romportl D, Kindlmann P (2018) Chapter 8, Determinants of orchid occurrence: a Czech example. In: Şen B, Grillo O (eds) Selected studies in biodiversity. InTechOpen, London
45. Kosánová K (2017) Dynamika výskytu orchidejí ve vybraném modelovém území v jižních Čechách. Mgr. Thesis, Charles University
46. Dykyjová D (2003) Ekologie středoevropských orchidejí. KOPP, České Budějovice
47. Jersáková J, Kindlmann P (2004) Zásady péče o orchidejová stanoviště. KOPP, České Budějovice
48. Průša D (2005) Orchideje České republiky. Computer Press, Brno
49. Lepší P, Lepší M, Boublík K, Stech M, Hans V (2013) Červená kniha květeny jižní části Čech. Jihočeské muzeum v Českých Budějovicích, České Budějovice
50. AOPK ČR (2013) Konsolidovaná vrstva ekosystémů. Agentura ochrany přírody a krajiny ČR. [Electronic geographical data]
51. ČÚZK (2010) Digitální model reliéfu České republiky 4. generace (DMR4G). http://geoportal.cuzk.cz/(S(wbdeojptgceogdtyhisrjzjf))/Default.aspx?head_tab=sekce-00-gp&mode=TextMeta&text=uvod_uvod&menu=01&news=yes&UvodniStrana=yes
52. Kindlmann P, Balounová Z (1999) Flowering regimes of terrestrial orchids: chaos or regularity? J Veg Sci 10:269–273
53. Jersáková J, Kindlmann P, Stříteský M (2002) Population dynamics of *Orchis morio* in the Czech Republic under human influence. In: Kindlmann P, Willems JH, Whigham DF (eds) Trends and fluctuations and underlying mechanisms in terrestrial orchid populations. Backhuys Publishers, Leiden
54. Chán V (1999) Komentovaný červený seznam květeny jižní části Čech. AOPK ČR, Praha
55. Štípková Z, Kindlmann P (2015) Extent and reasons for meadows in South Bohemia becoming unsuitable for orchids. Eur J Environ Sci 5:142–147
56. Baumann H, Künkele S, Lorenz R (2009) Orchideje Evropy a přilehlých oblastí. Academia, Praha

Diversity, Ecology, and Conservation of Mauritius Orchids

3

Cláudia Baider and F. B. Vincent Florens

Contents

Abstract

Mauritius was one of the last places on Earth to be colonized by humans offering one of the most complete history of what native species occurred originally and what was lost, when, and why. This situation can therefore serve as a laboratory to

C. Baider (✉)
The Mauritius Herbarium, Agricultural Services, Ministry of Agro-Industry and Food Security, Réduit, Mauritius
e-mail: cbaider@govmu.org

F. B. V. Florens
Tropical Island Biodiversity, Ecology and Conservation Pole of Research, Department of Biosciences and Ocean Studies, University of Mauritius, Réduit, Mauritius
e-mail: Vin.Florens@uom.ac.mu

J.-M. Mérillon, H. Kodja (eds.), *Orchids Phytochemistry, Biology and Horticulture*, Reference Series in Phytochemistry, https://doi.org/10.1007/978-3-030-38392-3_29

study human impacts in the current age of human-driven species extinction. Mauritius is also one of the most human-impacted places, thereby reflecting what awaits much of the tropics as human impacts intensify. We used available literature, herbarium samples, and personal observations and studies on the Orchidaceae to characterize its diversity, distribution and ecology, and the human-induced threats they face, to better inform their conservation in Mauritius. There are 91 native orchid species from 30 genera recorded on the island. Twenty species (22%) appear extinct, although some may survive undetected. New species and records continue to be added. Only 10% of the species are endemic to Mauritius, and 80% are unique to the south-west Indian Ocean islands. Most species are epiphytic, and the highest diversity occurs in native forests of the wet uplands. Mauritian orchids, particularly the larger ones, face many threats, some inexorably worsening. There exists much room to improve knowledge about Mauritian orchids that would better inform their conservation which is today still very neglected. This includes taxonomic research, detection of ecological patterns and trends, ecology of the species, as well as quantification and hierarchization of threats to prioritize conservation management. Studying Mauritius native orchids helps understand how devastating, sustained, and accelerating the many threats that human activities pose to orchid biodiversity can be and which await other countries currently less human-impacted than Mauritius.

Keywords

Mascarenes · Biodiversity · Invasive alien species · Island · Extinction

1 Introduction

1.1 Mass-Extinction and Island Biodiversity

Biodiversity worldwide is rapidly being lost under a wide array of threats from human activities driven principally by human overpopulation and unsustainable consumption patterns of natural resources [1]. The situation is such that it has been described as the sixth mass extinction [2]. Such an extent and worsening biodiversity loss is particularly damaging to humans and their societies given the dependence on biodiversity in many ways including for the provision of food, water, or a swath of ecosystem services [3, 4]. Biodiversity loss also shares its root causes with and positively feeds back into global climate change, which is the greatest existential threat facing humanity today [5]. These global problems must be tackled to stop the degradation and reverse the trend, and the implementation of solutions often has a local or regional theatre [6, 7]. In turn, the success of locally applied solutions depends on a good understanding of the threats driving degradation at that level because conservation management is as sound as the science on which it is based [8].

In this context, oceanic islands are particularly informative places to study biodiversity, ecology, and conservation [9, 10]. They have played and continue to play an important role in the development and refining of the theory of evolution by natural selection [11] and in understanding species distribution and how communities assemble through time [12], as well as, the biogeographical factors that strongly shape biotas [13–15]. Furthermore, oceanic islands contribute a disproportionately large share of global biodiversity relative to their area by virtue of the often-high endemism that characterizes their biota [16], particularly when they are relatively isolated, high and old, and occur in tropical or subtropical areas. With the advent of the widespread and profound impacts of human activities, humans have now combined the high degree of endemism of oceanic islands to a high degree of threat to biodiversity, resulting in a disproportionately large representation of oceanic island among the world's biodiversity hotspots [17, 18].

Interestingly, oceanic islands are often among the last places on Earth to have been reached by humans and be subjected to their impacts and modifications [19, 20]. This means that oceanic islands can provide us with some of the most accurate accounts and best understanding of how human activities influence nature and drive biodiversity loss, because they enable a better understanding of what biodiversity was initially present when the place was pristine, and what was lost, when and why following human colonization, a situation absent from places which have been inhabited by humans for eons like the continents. A more accurate understanding of human impacts on biodiversity is itself crucial if we are to devise sound responses to the global mass extinction. Oceanic island can therefore serve as useful laboratories to study the impacts and consequences of human activities on biodiversity, and by extension the solutions that are required to reverse biodiversity loss.

Orchids, in particular, make for an interesting model to study biodiversity patterns and threats besetting species and the corresponding conservation solutions, because they form one of the most diverse family of flowering plants present globally (>29,500 species, summing about 8% of all known flowering plants [21]) and on many tropical archipelagos [22], including the Mascarenes (around 166 species [23–28]). The occurrence of many species gives the best chances for any pattern to be more reliably established than would be the case if one were to be dealing with smaller groups of species, as the latter are more highly subjected to possible spurious conclusions caused, for example, by sampling error. Furthermore, orchids have been a group of plants that attracted much interest [29, 30] and since long [31], including being part of culture and iconography of older cultures (e.g., Greek, Roman, Chinese; see [32]). Importantly, today the whole family is listed under Appendix II of CITES, even though some orchids are invasive [33, 34].

1.2 Mauritius

Mauritius (1865 km^2, 828 m maximum elevation) is one of the three main volcanic oceanic islands comprising the Mascarenes along with La Réunion (2512 km^2) some 175 km to the west-south-west and Rodrigues (108 km^2) located about 595 km to the east. Mauritius is centered around 20° 20′S and 57° 35′E some 900 km east of

Madagascar. It emerged some 7.8 million years ago and experienced its last volcanic activity in its north-east region about 20,000 years ago [35]. The bedrock is mainly basaltic with limited calcarenitic areas confined to small patches on the coastline or on lagoonal islets in the south and southeast. Annual rainfall varies from about 800 mm on parts of the western coast to about 4000 in the wettest highlands, with an average of about 2100 for the island [36]. The temperature varies from an average of 16.4 °C at night in July–August to an average of 29.2 °C during the day in January–February. The original native vegetation varied from a palm-rich woodland in the drier coastal areas behind the coastal vegetation fringe, to small mossy forest patches on the highest grounds, and comprised in majority, of a wet forest which covered about half of the island [37]. The vegetation ecosystems did not appear to have shifted or changed majorly from the end of the Pleistocene (38,000 years ago) to the time of human colonization in 1638 but underwent a series of relatively limited reassortment of species dominance instead [38].

Although Mauritius was among the last places on Earth to be settled by humans (in 1638), it underwent one of the most rapid and advanced levels of native habitat destructions that spared only about 4.4% of its original terrestrial habitats [39] within the following 375 years or so of human presence under the Dutch (1638–1710), French (1715–1810), British (1810–1968), and Mauritian (1968–present) sovereignty. Some types of habitat, like the palm-rich drier forests, have been completely destroyed from the mainland and only survive as highly degraded small patches on tiny offshore islets [37] which are fortunately undergoing ecological restoration. The remaining habitats on Mauritius are also highly fragmented [40], and despite their small extent, habitat destruction continues and has been recorded even within Nature Reserves protected by law [41]. Furthermore, the overwhelming majority of the 80 or so km^2 of native habitats that have so far escaped deforestation are currently highly invaded by encroaching alien plants [42]. In effect, Mauritius may arguably be regarded as representing a "window" into the future of many other tropical places as the latter catch up, in line with current trends, with the already advanced levels of habitat destruction and fragmentation, alien species invasion, native species extinction and endangerment rates, human overpopulation, and urban sprawl, among others [39, 41]. Mauritius, therefore, approximates what other places are increasingly approaching and can thus serve as an informative laboratory for them of how biodiversity will be lost, but also of possible solutions to stem this biodiversity loss.

Here we discuss the diversity, ecology, and conservation of biodiversity using the native orchid flora of Mauritius as a model. We used the available literature, herbarium samples, and notes thereon as well as personal observations and studies on the field to characterize the diversity of native orchids in terms of the number of genera and species, the patterns of species discoveries through time, species distribution (whether island endemic, archipelago endemic, or of wider distribution), ecology (pollination, habit, elevational distribution, etc.), the various human-induced threats that they face (e.g., habitat destruction, overcollection, the impact of invasive alien plants and animals, etc.), and the rate of extinction and species rarity that ensued, in order to propose conservation measures to try to reverse the current ongoing loss of the orchid flora of Mauritius.

2 Diversity

2.1 Background

The oldest traceable samples of orchids collected in Mauritius were two native species of *Nervilia*, a genus of ground orchid, placed on the same herbarium sheet, and collected in 1769 by the French botanist Philibert Commerson (1727–1773; https://en.wikipedia.org/wiki/Philibert_Commerson). Although this first collection was made over 130 years after the island was first colonized by humans, it is unlikely that orchid species could have already been lost as a result of human activities by then (in contrast with vertebrates [37]) because 82.5% of the island's native cover still subsisted and none of the broad types of vegetation communities had yet been destroyed [37, 43]. After Commerson, who botanized on Mauritius until 1773, came Louis-Marie Aubert du Petit Thouars (1758–1831; https://en.wikipedia.org/wiki/Louis-Marie_Aubert_du_Petit-Thouars), another French botanist who is most well-known for his work on collecting orchids of Mauritius and also of nearby Réunion island and Madagascar from 1793 to 1802. Du Petit Thouars produced a small book with six paintings (believed to have been published between1804 and 1819 [44]), followed by an article that contained tables with the description of dozens of species present on Mauritius in 1809 [45] and, later, a more comprehensive first book on Mauritius orchids (along with species from Réunion and Madagascar) in 1822 [46] and described 11 orchid genera, four of which still stand today, including *Bulbophyllum*, the largest genus of orchids with over 2000 species [47], which makes it among the largest genera of flowering plants. Du Petit Thouars recognized 51 orchid species for Mauritius. Another notable botanist of the nineteenth century, who contributed significantly to expanding knowledge on Mauritius native orchids, was the Czech Wenceslas Bojer (1795–1856; https://en.wikipedia.org/wiki/Wenceslas_Bojer) who lived on Mauritius for over 30 years [48] and published the first Flora of the island [49].

Additions to the native flora of Mauritius orchids continue to be made despite the very limited extent of native habitats that survive the rapid deforestation experienced [41]. Thus, during the last 16 years alone (which represents 7% of the total period of orchid study on the island), seven species have been added, representing about 8% of the total known Mauritian native orchid flora. This suggests that several species might have been driven extinct before they had a chance of being discovered, while the 95% or so of the island's native habitats were being destroyed. New additions of native species which were already known to exist elsewhere but recorded or confirmed for the first time on Mauritius include the diminutive aphyllous *Taeniophyllum coxii* [50] and the fairly large – by Mauritius standard – epiphytic *Jumellea exilis* and *J. rossi* [51]. In addition, four new species were described including two that are also shared with La Réunion, namely, the epiphytic *Polystachya jubaultii* [52] and *Bulbophyllum mascarenense* [27], and two that are endemic to Mauritius, namely, *Angraecum jeannineanum* [53] and *A. baiderae* [25]. Various taxa (genera or sections) of the orchid flora of Mauritius are also currently undergoing reviews in the context of the updating of the Flora of the Mascarenes [25, 54].

2.2 Diversity and Endemism

To date, 91 native species of orchids belonging to 30 genera have been recorded in Mauritius. The most diverse genera are the predominantly epiphytic *Angraecum* (18 species) and *Bulbophyllum* (17 species) and the terrestrial *Cynorkis* (9 species) (Fig. 1). The flora is dominated by species that are endemic to the biodiversity hotspot region (80%), followed by those endemic to the Mascarene archipelago (41%) (Fig. 2), with a rate of endemism to the island of only 10%, which is relatively low compared to other families of native angiosperms (up to 100% in families like the Ebenaceae, or nearly so for Pandanaceae). About 67% of the island's orchid flora had been discovered within the first 40 years of the 250 or so years of botanical history of Mauritius orchids, with some more (ca. 12% of today's known species) collected later by Bojer and Louis Hyacinthe Boivin (https://plants.jstor.org/stable/10.5555/al.ap.person.bm000000840) over a period of 30 years (1821–1851) (Fig. 3). After a hiatus of more than 70 years, interest in orchids restarted with the arrival in 1924 of the British botanist Reginald E. Vaughan (1895–1987). Vaughan started botanizing as soon as he established himself on the island (sadly part of his early collections were lost in a fire in 1937), and a decade later he re-organized existing herbarium collections, which led to the creation of a centralized national herbarium (The Mauritius Herbarium). He also initiated the ongoing project of updating the account of the Mascarene flora [56]. Vaughan, together with his assistant (and later the first curator of The Mauritius Herbarium, Joseph Guého [57], and the scientific director of the Flore des Mascareignes, Jean Bosser (1922–2013;

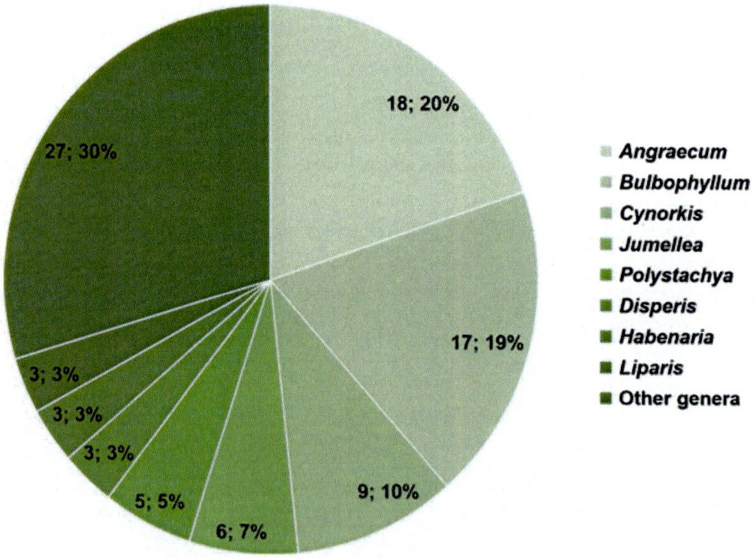

Fig. 1 Number of species and their respective percentage within each native genus of native orchids recorded for Mauritius. Only genera with three or more species are listed by name, others being lumped

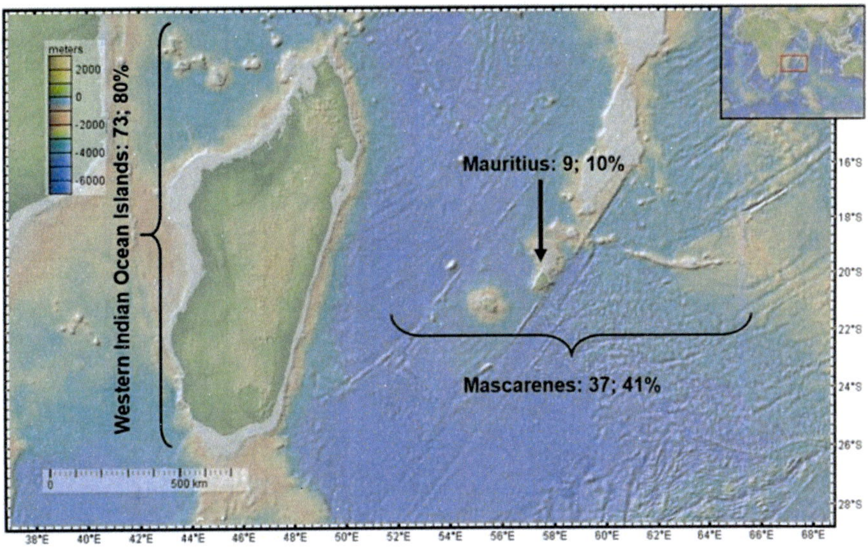

Fig. 2 Number of species and respective percentages of species of orchids present on Mauritius that are endemic to the island, to the Mascarene archipelago and the Western Indian Islands hotspot. Figure made with GeoMapApp (www.geomapapp.org [55])

https://en.wikipedia.org/wiki/Jean_Marie_Bosser), collected an additional 15% of the known species. David Roberts [58], while doing his PhD on the orchids of the Mascarenes, added some new records (Fig. 3).

Although the collection and description of the orchid flora of the Mascarenes started relatively early, several problems have been delaying a greater understanding of its diversity. For example, some closely related species had previously been misidentified (e.g., *Jumellea fragrans* and *J. rossii*, see [51, 59] (but see also [60]), or not recognized as different entities (e.g., *Angraecum cadetii* and *A. jeannineanum* [53]; *Bulbophyllum nutans* and *B. mascarenense* [27]). Also, orchid specimens can easily deteriorate, and many type specimens and other collections were lost, especially those of Thouars, Commerson, Boivin, and Bojer. Finally, a number of species were described from different islands of the region, but only more recently, with better available tools (digitization and imaging of collections, and larger sampling of specimens on phylogenetic studies), many of them are being synonymized and being recognized to have larger distribution than previously thought (e.g., [25, 51, 54], decreasing the so-called "taxonomic inflation" as the group gets better studied and understood [61, 62].

2.3 Types and Distribution

Like is often the case in the wet tropics [63, 64], most orchids of Mauritius are chiefly epiphytic (66%) (e.g., most species of the genera *Angraecum* and

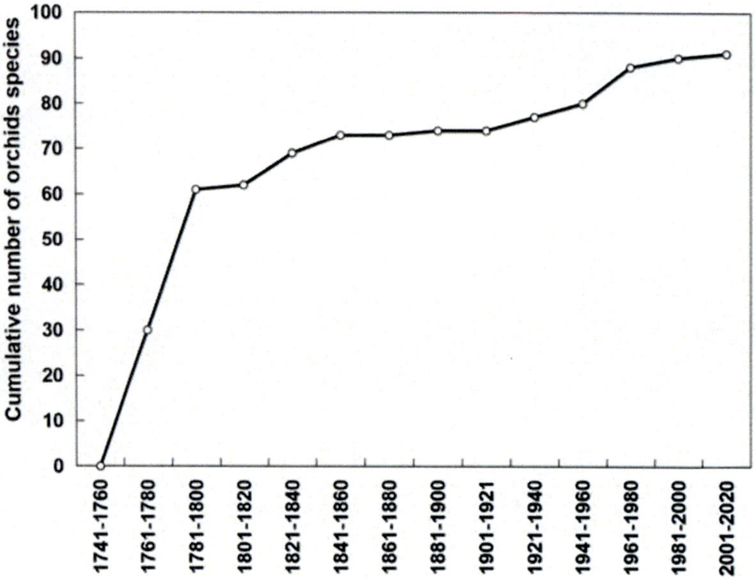

Fig. 3 Cumulative number of first collection of orchid specimens of species native to Mauritius by date (in periods of 20 years). Data is based on available herbarium records, or date of publication of the basionym

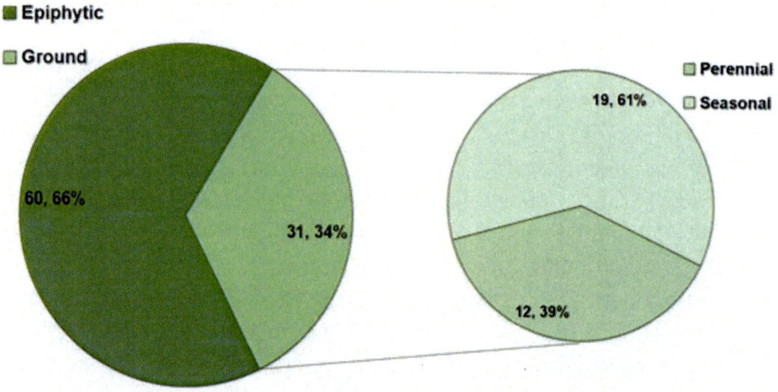

Fig. 4 Habit of Mauritius native orchids by number of species, and their respective percentages

Bulbophyllum), the rest being ground orchids (Fig. 4), which comprise either perennial species (12 species, e.g., *Calanthe* and *Phaius*) or seasonally emerging species (19 species, 61% of the ground orchids, e.g., *Nervilia* and *Disperis*). A number of these typically epiphytic species are, however, often also found growing on rock as lithophytes, good examples of which may be found on the cliff face by the south access path towards the Le Pouce Mountain Nature Reserve (*Angraecum pectinatum*), along Montagne Longue crestline (*Polystachya concreta*) or on the southern steep cliff of Trois Mamelles mountain's highest peak (*Bulbophyllum* sp.). Similarly, some typically ground orchids are often encountered growing in moss patches on trees in very wet regions, like some *Cynorkis* and *Benthamia*. No saprophytic orchid species, those depending on root mycorrhizal fungi for uptake of carbon [65], are known from Mauritius like occurs on nearby island of Réunion (*Gastrodia similis* [66]) or Madagascar. There are two aphyllous species recorded from the island, namely, the predominantly moist to dry forest species *Microcoelia aphylla*, and the more recently discovered *Taeniophyllum coxii*, which has a more restricted distribution in parts of the Black River Gorges National Park [50].

A very marked pattern in Mauritius is the fact that native orchids are almost exclusively confined to native vegetation or their remnants, where, in the case of epiphytic species, they grow predominantly on a variety of native trees. Although no quantitative studies of preferred phorophytes have been done, observations suggest that only one species of native epiphyte is found nearly all the time on a single tree species (the African orchid *Angraecopsis parviflora* on the Mascarene endemic tree *Nuxia verticillata*). In very rare cases, native orchid species may be found growing outside native habitats. An example is the native *Angraecum calceolus* observed growing on planted roadside Jacaranda (*Jacaranda mimosifolia*) or *Aeranthes arachnites* growing on planted camphor (*Cinnamomum camphora*) or litchi (*Litchi chinensis*) trees. These, and other species, may also be found growing naturally in certain private gardens, for example, in Curepipe, but then they would most often be found on native trees or native tree clumps remnants of the original native forest, or on trees planted in their close vicinity.

Of the orchid species with known intra-island distribution, few on Mauritius can grow from sea level (or close to it, at around 100 m), to the island's highest elevation (N = 5, 6%). Similarly, few species are found to be restricted to the upper limits of the island's elevation (> 700 m; N = 2, 2.4%). A majority of species (N = 50, 61%) occur at least at elevations of 300–600 m, and an even greater majority grow between 600 m elevation to the island's culminating point of 828 m (N = 70, 85%) (Fig. 5). Although one could suspect that such patterns could be artifactual because most lowland native vegetation has been destroyed early after human colonization [37], and that specimens with more precisely recorded locality are those that were more recently collected, it has also been shown that orchid diversity (including many of the same species as in Mauritius) drops with elevation on the sister island of Réunion [67], where much more intact elevational gradient of native vegetation subsists and that both elevation and climate do influence distribution of orchids and the epiphytic species [68–70], among other factors (e.g., area in case of islands). In general, on Mauritius, elevation above 300 m would be mostly within transitional and wet forests.

Orchid species richness on Mauritius is strongly related with elevation, with sites at 600 m elevation or more harboring higher species richness despite their smaller area and greater isolation. Because of risks associated with such information, we opted not to disclose the name of species and locations (for more on this see "Threats" below). The most species-rich areas include a variety of habitat types with low to high vegetation canopy. Also, they tend to be areas that are least invaded

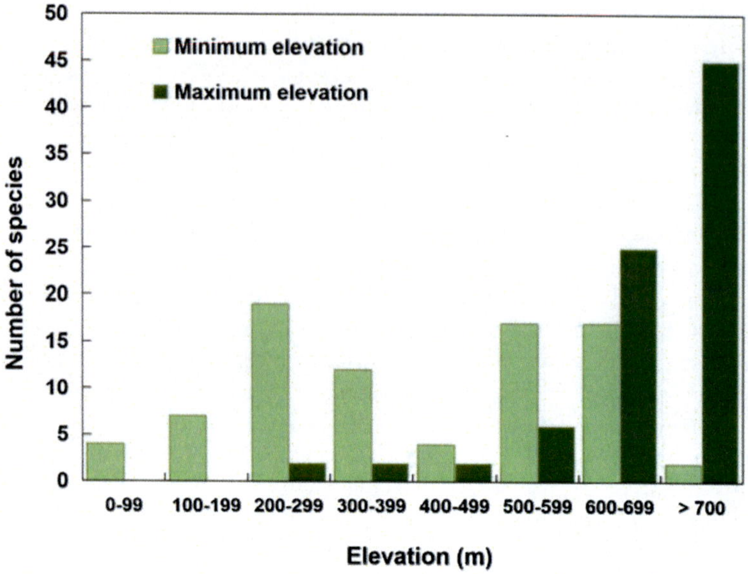

Fig. 5 Number of species of native orchids per bracket of elevation range, in terms of both minimum and maximum known elevation per species

by alien plant species and/or receiving conservation management, sometimes since 50 years. Of the extant species, the majority have at least one locality within legally protected areas – either a Nature Reserve or National Park, which also represent a large part of the surviving habitat remnants. However, the distribution of several species does not coincide with protected areas. Worryingly, a few species or populations appear to have gone extinct in the last 50–60 years, even within protected areas on the island.

3 Ecology

Although orchid seeds tend to be among the smallest and lightest of flowering plants [71], long-distance dispersal might be limited on islands [72], with climatic and environment factors being more important diversification factors [64, 68–70, 72]. On Mauritius, the endemism of orchids is relatively low at island scale (~10%), but higher at the level of the Western Indian Ocean Islands hotspot (~80%, Fig. 2). In fact, at this scale, most species of this region are endemic [26], a pattern resulting from factors operating through time and space, including island hopping facilitated by sea level drops during the Pleistocene [15], as well as regional forcing [13].

Based mostly on literature (e.g., [67], the majority of orchid of Mauritius are pollinator-dependent (N = 52, 57%), with just over a third (N = 32, 35%) being self-pollinated, although a few have unknown pollination systems (N = 7, 8%) as these species are known only from drawings or few old specimens. Nevertheless, the percentage of auto-pollinating species on Mauritius is higher than on Réunion at similar elevation (around 23%) [67]. Among the pollinator-dependent species, one section (*Hadrangis*) of the genus *Angraecum*, which is endemic to the Mascarenes, evolved unique pollination patterns. Indeed, the two Réunion endemic species are bird pollinated [73, 74], while a species shared with Mauritius is pollinated on Réunion by an endemic nocturnal orthopteran [75]. On Mauritius, observations on this same orchid species (and on the sister Mauritius endemic species) were done but no pollinator was seen during the period of the study [53].

4 Threats and Conservation

4.1 Deforestation

The most immediate and damaging impact of human activities on biodiversity is habitat destruction, something that happened in Mauritius at one of the fastest rates recorded worldwide [37, 41]. Indeed Mauritius was one of the last places on Earth to be colonized by humans, when Dutch settlers established on the island in 1638 and, by the 1990s, its native habitat extent had been reduced 20 folds [40, 76]. The island of Mauritius now has one of the lowest extents of surviving original native cover of any country in the tropics. Native habitat destruction has continued beyond the 1990s, although on a much smaller scale given the small fragments of native habitats that survive as a confetti of remnants over parts of the island [77]. Early twenty-first-century deforestation occurred, for example, in what was the best surviving remnant of coastal native forest at Roches Noires on the east coast or in a Nature Reserve harboring a hardwood forest that was supposed to be strictly protected [41], and which contain a fair diversity of orchids. Other small-scale destruction of native habitat also continues on mountain reserves, mostly for ranching of alien deer (for hunting and meat) and for ecotourism facilities, along river reserves (mainly for cultivation), and for road enlargement, as happened at Le Pétrin, one of the hotspots of orchids inside the Black River Gorges National Park and Chamarel in 2021. An often-overlooked threat for epiphytic and terrestrial orchids comes from thinning of trees within native forests for illicit cannabis plantations. The last estimate puts native habitats as covering 4.4% of Mauritius [39].

The rapid loss of habitat cover on Mauritius corroborates relatively well with the last records or disappearance of the 20 species (22%) of Mauritian native orchids that are currently considered to have been driven extinct by human activities, as almost all of them ($N = 17$) were last collected between 1769 and 1888. However, it must also be kept in mind that a swath of other threats, although relatively less severe, as described below, has also been operating alongside deforestation, even within habitats that have not been destroyed. Some of these threats, like the negative impacts of alien long-tailed macaques (*Macaca fascicularis*), likely introduced in 1602 [37], would have started even before humans finally colonized the island (see below).

4.2 Harvesting

Mauritian native orchids are typically unremarkable relative to showy commercial orchids like those of the Neotropical genus *Cattleya* or, the mostly southeast Asian *Dendrobium*. Nevertheless, harvesting of wild plants still appears to pose problems, although it has not been quantified. Some species like *Angraecum eburneum* or *Cryptopus elatus* have some degree of horticultural interest, and indeed the latter is often collected from the wild. Collection is likely to have contributed to decline of species that have been used as beverage enhancer (e.g., *J. rossii* and *J. fragrans* used to give fragrance to rum [54]), or as a medicinal plant (*Angraecum mauritianum* as stated on specimen MAU 0016681 at The Mauritius Herbarium). However, these along with some other species of no real horticultural interest for their few and inconspicuous greenish flowers are still being collected and sold, for example, in the market of the capital city of Port Louis (pers. obs., Z Jhumka, pers. comm. 2019). Some plants seen on sale were clearly recently uprooted from their host and sold as such without potting, possibly benefiting from the buyers' ignorance that the plants would in most likelihood not survive, and that if they do, any flower would be most unremarkably inconspicuous. Rarity increases the willingness for keeping samples of species in personal collections [78, 79], and this might have been the reason why the largest individual of *Taeniophyllum coxii*, which was growing by a forest track inside the Black River Gorges National Park and illustrated in fruit in 2017 [80], ended up being peeled off its host tree as witnessed by the sharp blade marks left behind. This species, recently recorded in the country, is extremely rare on Mauritius, and it would be considered locally as Critically Endangered [50], according to the Red List criteria of the International Union for the Conservation of Nature [81].

Harvesting of other native species also poses a more indirect threat to indigenous orchids. There is abundant evidence in wet forests of Mauritius like at Macchabé and Brise Fer in the Black River Gorges National Park that illegal harvesting of the larger specimens of the native tree fern *Alsophila excelsa* was common in the past [82]. The wide base of the stem of these tree ferns, apart from having some medicinal uses, was commonly harvested to make flower pots, and their fibrous roots often used as a growth medium for orchids since the time Mauritius was a French colony in the eighteenth century [83]. Many cut stumps can be observed especially in the vicinity of access roads, largely because the fern bases are often massive and doubtless hard to carry over longer distances. In the wild, *A. excelsa* is an important host to many orchid species to which its harvest is therefore harmful. Furthermore, *A. excelsa* provides the best microsites of germination to a few native tree species including *Nuxia verticillata*, a large, long-lived tree often host to many epiphytic orchid species including *Angraecopsis parviflora*. The illegal harvesting of tree ferns thus triggers both a direct and an indirect loss of highly favorable hosts to many native epiphytic orchids.

4.3 Alien Plants

The small areas of native vegetation that escaped deforestation so far [39], and which constitute the stronghold or sole habitat of all native orchids, are today for the most part invaded by many alien plant species either in the understory (e.g., the strawberry guava, *Psidium cattleyanum* in tall forests) or in the canopy (e.g., the cinnamon *Cinnamomum verum*, or the emergent travellers palm *Ravenala madagascariensis*), or both [42]. Alien plants exert intense competition for light on native orchids [84], which greatly reduces orchid diversity (expressed as both species richness and abundance) by virtue of their extreme density [42] (but see more below). Forest areas where alien plants have been controlled for about a decade as part of conservation management [85] harbor a much higher diversity of orchids than adjacent alien invaded forests [84]. Monitoring of permanent plots indicated that the epiphytic orchid community progressively recovers within areas cleared of alien plants while simultaneously declining in adjacent areas invaded by alien plants (unpublished data). Control of invasive alien plants on Mauritius is therefore extremely important for the conservation of native epiphytic orchids. However, for each hectare of forest undergoing such ecological restoration, there are about 20 hectares where invasion continues to progress. Consequently, there exists no reasonable doubt today that native epiphytic orchids are declining as a whole over the island particularly that it has been shown that the invasion by alien plants has not stabilized but is in fact worsening with time [42].

Alien woody weeds also harm native orchids in indirect ways. Alien-invaded native forests are losing native host trees at an alarming rate. For instance, native trees with stem of at least 10 cm diameter at breast height (DBH) have diminished by half within 68 years from what were some of the best preserved native forests of Mauritius in the 1930s [86] despite their presence in protected areas [87]. A decline among smaller diameter woody native plants (>2.5 to <10 cm DBH) is also apparent when comparing with studies done in the 1980s [87, 88]. A closer tree by tree monitoring through time between adjacent weeded and nonweeded native forests showed native tree mortality to be increased by the presence of invasive alien plants at both community [89] and population levels [90]. Even native trees that largely overtop the alien plants are dying faster when growing within stands of alien plants [90], indicating that root interactions alone suffice in increasing mortality of native trees. This elevated native tree mortality is generating an inexorable loss of suitable host plants for epiphytic orchids.

Competition for light is not the only direct harm delivered by invasive alien plant on native orchids. As alien plant invasion progressively worsens through time [42], native tree densities progressively drop [87], such that a long-term gradual replacement of native by alien trees is ongoing in invaded native forests. One might think that this would be of little consequence for epiphytic orchids as they need a woody host (phorophyte) to support themselves and that it would not matter much if that host is native or alien. But epiphytes are not randomly distributed because conditions offered by a particular host species matter and, in turn, influence, among other factors, the microclimatic conditions for the orchid development and availability of resources (light and water) which vary through time, for example, when host trees seasonally shed leaves, or if the bark of the host has the right rugosity for the orchid roots to grip, or if the orchid is able to develop as the stem or branch where they rest grows [91, 92]. *Psidium cattleyanum* is by far the main alien plant invading the most orchid-rich forests of Mauritius (79% of sampled alien stems [42]). It is an understory tree of small diameter that frequently sheds its bark, making it an exceptionally poor host for virtually all native epiphytic orchids (Fig. 6) although a few species may still establish and mature on strawberry guava [50]. Several other invasive alien tree species, for example, *C. verum* and *Ligustrum robustum*, also present particularly poor substratum to epiphytic orchids, although not as extreme as *P. cattleyanum*. The latter species also creates a very thick root mass at the soil surface which appears to substantially change soil texture and porosity away from what is conducive for ground orchids. Indeed, only after this weed had been removed from a patch of forest in the Black River Gorges National Park, was a *Nervilia* species that was hitherto thought extinct on Mauritius, relocated as a strongly regenerating small population [93].

The invasion by alien plants has also been shown to be reducing the extent of flowering of native woody plants [90, 94] as well as the production of fruits [90, 94, 95] hence of seeds too, resulting in a much lowered native plant regeneration in forests invaded by alien plants [93]. Invasive plants also reduce growth rate of native trees [90], which in turn is expected to result in smaller trees with fewer opportunities for epiphytic orchids to grow on, as well as feeding back into reduced seed

Fig. 6 (**a**) Stems of native tree species often form suitable substratum for a variety of plants, including lichens, mosses, ferns, and orchids compared to (**b**) the smooth stems of the invasive alien strawberry guava, *Psidium cattleyanum* whose outer surface is frequently shed. (Photos: F. B. Vincent Florens)

production and regeneration because smaller trees typically carry fewer fruits. Invasion also decreases the density of tree ferns [82] and litter basket ferns [96], suggesting that establishment of new orchids could be reduced if mycorrhizal fungi availability is harmed by lower phorophyte diversity or ideal microclimate produced by mosses and ferns that enable this symbiotic relation (fungi-orchid) to occur. Finally, there is evidence also that invasive alien plants greatly harm some insect groups [97] including the native butterfly community [98]. Butterflies are often used as an indicator group of insect diversity, itself important for native plant regeneration as pollinators. In conclusion, through several direct and indirect mechanisms, invasive alien plants are harmful to the establishment, growth, and reproduction of native trees, which themselves are important as hosts for a large number of epiphytic orchids of Mauritius.

4.4 Alien Animals

Invasive alien animals like the long-tailed macaque (*Macaca fascicularis*) whose introduction predates human colonization also threaten orchids. They chew on the meristem of predominantly the larger terrestrial orchids of the genera *Calanthe* and *Phaius*, killing the individuals affected (Fig. 7). Monkeys also chew on meristems of

Fig. 7 (**a**) The invasive alien long-tailed macaque (*Macaca fascicularis*) in the canopy of native vegetation in the highlands of Mauritius, (**b**) disarticulated leaves (about 30 cm long) of a native terrestrial orchid (*Calanthe* sp.) that has been attacked by a long-tailed macaque, (**c**) details of leaves in **b** showing chewed bases of the leaves. The plant's meristem was completely destroyed. (Photos: F. B. Vincent Florens, February 2021)

epiphytic species of *Jumellea* [51], which might have significant parts of the plant removed. Indeed, four on six (67%) *Jumellea* species shared between Mauritius and the nearby Réunion, which is free of alien macaques, are more threatened with extinction on Mauritius than on the latter island [54]. But more interestingly, the larger colonies of Mauritian *Jumellea* are clearly today primarily confined to places that are inaccessible or most difficult of access to macaques, like cliffs or high up on isolated large trees with straight boles that are difficult to climb [51]. Another example of the contrast between macaque-infested Mauritius and macaque-free Réunion is illustrated by *Bulbophyllum longiflorum*, a species known to be chewed by macaques, and down to about seven known individuals at two sites on Mauritius, but very common in drier forests on Reunion [99]. Also noteworthy is that the largest Mascarene orchid, *Angraecum palmiforme* (up to 1 or 1.2 m tall), was last recorded on Mauritius in the early 1800s [49], and although rare [100], it has been seen on Réunion less than three decades ago [101]. Similarly, *A. eburneum* now occurs in the wild in Mauritius in very small numbers and almost exclusively in inaccessible steep cliffs, whereas on Réunion it often grows in highly accessible spots [99, 101].

Macaques also pose substantial indirect threats by hindering the growth and reproduction of host trees of epiphytic orchids. They do so by commonly breaking and chewing on many young shoots of saplings to adults of many native trees and reducing the vitality of the plants [89, 102], thereby incidentally destroying flower buds, flowers, or unripe fruits [90, 95]. Macaques also directly attack large quantities of fruits before they have time to ripen their seeds. For example, 95% of the fruits of the canopy endemic *Sideroxylon grandiflorum* are destroyed by macaques before the seeds within have time to ripen contributing to the tree to virtually stop regenerating in the wild [90]. Similar types of damages are known to happen to various other large native trees and large-fruited species of the genus *Mimusops* (Sapotaceae) and *Diospyros* (Ebenaceae) (pers. obs.; [95, 102], which collectively form a sizeable proportion of trees in Mauritian native moist to wet forests [103]. Some smaller understorey trees are also impacted in a similar fashion, one example being *Eugenia alletiana* (Myrtaceae) [104].

Native orchids also face several threats from other invasive alien species in addition to the ones described above. For example, the invasive feral pigs (*Sus scrofa*) destroy terrestrial orchids by uprooting and smothering them, especially when rooting, which apart from reducing plant cover is known to also change soil physical structure and chemical composition [105]. Damage by feral pigs is also influenced by their density. In Australia, higher density of feral pigs is known to increase damage to ground orchids [106]. On Mauritius, pigs seasonally converge to the upland wet forests during the fruiting of the alien strawberry guava [89] to eat the abundant fallen guavas, indicating that this alien plant species is an important resource for the pig. Some ground orchid species might be more at risk of negative effects from feral pigs, such as species of *Habenaria* (a genus counting an endemic extinct species) which tend to grow in swampy and very wet soils, which are particularly impacted by feral pigs, especially for wallowing.

Several other alien animals also negatively affect orchids. Terrestrial orchids, especially those with soft leaves and succulent stems (e.g., *Platylepis*), have their stem grazed and chopped by the alien invasive Giant African snails (*Lissachatina fulica* and *Achatina immaculata*). The very rare native orchid, *Nervilia*, which seasonally forms a single leaf appearing above ground for 4–5 months during a year, is also attacked by these same mollusks, but also by alien slugs (pers. obs.). The invasive java deer (*Rusa timorensis*), introduced in 1608 [37], impacts orchids by grazing, soil compacting, direct removal of epiphytes during antler rubbing, which itself can kill even large trees by triggering wood rot that causes the tree to eventually snap, reducing the number of available phorophytes hosts for epiphytic orchids. Invasive alien rats (*Rattus rattus* and *R. norvegicus*) can chew on orchids [27], and the alien hare (*Lepus nigricollis*) may graze on ground orchids in more open vegetation.

4.5 Indirect Effects

All large frugivores of Mauritius (>1 kg) were quickly driven extinct following human colonization [37]. Such losses are known to impact the dynamics and composition of forests [107, 108]. Today, the role of maintaining seed dispersal, and consequently, fostering regeneration of adequate phorophyte for orchids, depends almost exclusively, especially for large seeded-species, on the Mauritius flying fox (*Pteropus niger*), a Mascarenes endemic and Endangered species [109]. This single species has an ecological keystone role of seeds dissemination on Mauritius as it is known to feed on fruits of, on average, about 53% of individual woody plants in various Mauritian forests, and particularly those of the larger trees [110], which are known to have important physical ecosystem engineer roles that enable the survival of many other forest species [111], including orchids. Yet, Mauritian authorities have long been planning to cull flying foxes [112, 113] and since 2015, implemented multiple mass-culling campaigns on spurious justifications [114–116]. These recurrent mass-culling campaigns are unnecessarily heightening the risk of extinction of the flying fox [109] and must already be weakening its ecological keystone role given the way such roles are played out by this group of fruit bats [117]. Indeed, the detrimental effects on regeneration of native forest trees that the weakening and losing of the seed dissemination role of *P. niger* bring are emerging and being confirmed [118]. Loss of frugivores, including *P. niger*, is known to have strong detrimental effects on forest resilience [119].

4.6 Other Threats

A rather wide variety of unusual threats to native orchids have also come to light in Mauritius often unexpectedly originating from certain conservation managers themselves. Perhaps the most direct one was the targeted removal of a ground orchid inside conservation management areas as part of the program of maintenance weeding meant to control invasive alien plants. A wide diversity of native plants recover by regenerating much better [93] following the initial weeding of the principally woody invasive alien plants [42] that invade native forests, when conservation management areas are created [85]. However, alien plants tend to reinvade the weeded areas, necessitating their regular maintenance weeding. One of the rare ground orchids of Mauritius, a *Platylepis*, was observed to benefit much from alien plants weeding only to then be mistakenly taken for an alien weed itself by managers during maintenance weeding and weeded (Fig. 8). Many years later, the population of the species appears to have still not yet recovered from this unfortunate event. This situation underscores the importance of conservation managers to develop appropriate levels of identification skills and of implementing effective supervision on the field, so that similar events do not recur on this or other species, the more so that it is not an isolated event because several other native species are often weeded during maintenance weeding, including woody plants like *Cnestis glabra*, *Mussaenda arcuata*, or *Scutia myrtina* themselves serving as phorophytes for orchids.

Another activity that is damaging to native orchids within protected areas, and which is more recurrent this time, concerns indiscriminate maintenance weeding. Indeed, the use of hoes to remove alien grasses also destroys many species of ground orchids growing along the alien plants being removed (Fig. 8). A much better solution would be to remove the alien plants by hand alongside putting a definitive stop to the ongoing practice of cutting native pioneer trees [85] or lopping their

Fig. 8 (**a**) The native ground orchid *Platylepis occulta* was mistakenly weeded by the hundreds during maintenance weeding operations within a protected area receiving conservation management. (**b**) All individuals from small juveniles (on the right) to large adults (on the left) that were spotted ended up being uprooted. (**c**) Maintenance weeding of alien plants being carried out using hoes, an indiscriminate technique that destroys many native plants including ground orchids. (**d**) A pile of freshly weeded alien plants removed with hoes and containing native seedlings, ferns, and ground orchids. (Photos: F. B. Vincent Florens)

branches, because these activities not only, at a cost, set back ecological restoration, but also they themselves increase light at ground level which boosts the growth of the same alien grasses which the managers are trying to control with hoes inside conservation management areas.

Brush cutting is also a common practice along forest tracks within protected areas. Many orchid species grow better at track edges because the high invasion levels of alien plants further inside the vegetation [42] make it too dark for many orchids to grow. While brush cutting has been observed to damage some perennial orchids growing on edges of tracks, such as *Angraecum mauritianum* or *A. ramosum*, along with many other native plants, the activity appears to be much more damaging to seasonal ground orchids like *Disperis* or *Cynorkis* because the brush cutting is often carried out precisely when these orchids are growing leaves, flowers, and fruits above ground. Brush cutting edges would itself become largely redundant if managers allowed overtopping vegetation to link up above the forest tracks thereby shading them, but the opposite management of widening tracks and lopping branches is done instead, which create more influx of light which favors thicker alien plants growth, which is then dealt with by brush cutting. If ever brush cutting is truly necessary, it should be done in seasons where the ground orchids are dormant underground, or else avoiding areas where colonies are growing, flowering, and fruiting.

A new threat has been noted recently within the country's main National Park which is damaging ground orchids particularly the smaller seasonal species that grow on rock on the forest floor, such as *Disperis* spp. and *Cynorkis* spp. During forest track maintenance works, tons of rocks were gathered from the adjacent protected forest floor and used for leveling the tracks. Several species of seasonal orchid grow on these rocks, which are also colonized by various bryophytes and ferns. They are all destroyed when the extracted rocks are used on the tracks. Many orchids, because they are seasonal, are not visible on these rocks for the most part of the year, as they occur as dormant tubers in the rocks' cracks and depressions or within the moss cover. They only do sprout into view in the growing season. These forest floor rocks form an integral part of the forest and are important for native biodiversity both for plants growing on top of them and for animals hiding from predators below them and should therefore not be harvested for building purposes, particularly within protected areas.

Another unusual threat to orchids, which is this time more persistent and long-lasting, concerns the practice of conservation managers cutting native pioneer trees (*Harungana madagascariensis*) that grow predominantly in areas undergoing ecosystem restoration after alien plant weeding [85]. Although this practice endured for many years and was more widespread before subsiding in around 2013, the native pioneer trees are still occasionally cut by conservation managers in areas undergoing restoration as happened in the Conservation Management Area of Pétrin in 2019. Lopping of the branches of this pioneer tree is more commonly practiced in recent years, but still represents inappropriate management because it involves investing resources in an activity that reduces native biomass and increases re-infestation rates of alien plants due to increased influx of light [85]. As far as orchids are concerned, the native pioneer tree that is cut or pruned makes for a good phorophyte for epiphytic species [84], particularly also because it typically grows in forest gaps, where other phorophytes are rare [120].

Many of the unusual threats highlighted above, could be reduced or avoided if decision-makers, managers and technicians working in conservation were better trained in ecology and conservation, in particular of tropical native forest biodiversity, as this would have led to more informed evidence-based management, compared to the current situation, where many staff, for example of the authorities, are primarily trained in agriculture-related fields (https://civilservice.govmu.org/Pages/SOS/SOS/agro.pdf). With the help of foreign institutions and scientists, Mauritius has nonetheless achieved several conservation successes [41], some of which of worldwide acclaim [121]. Sustaining such successes rests on sufficient appropriately trained conservation professionals, an increasing number of whom are now Mauritians compared to three to four decades ago. Worryingly, however, there has recently been a recrudescence of obstacles to capacity building in ecology and conservation in Mauritius. For instance, the country's main university, and only one locally producing graduates in biology, many of whom with a particular interest in conservation [122], has decided in 2015 to replace its biology undergraduate program with an equivalent that effectively reduced the ecology-related component three-folds, removing all of it from the final year of studies and removing conservation biology altogether from the program for at least 4 years (V.F. pers. obs), despite this field counting among the research strengths of the university. The new program contains one of the lowest ecology/conservation components of comparable programs worldwide and incidentally, it coincided with a severe drop in student intake that culminated with just four students graduating in 2019, a ten-fold drop over the long-term average.

To complicate matters, several other barriers to local ecology and conservation training and research have also emerged, including a policy by the authorities to restrict such research in the National Parks to office hours on week days, and selectively applied to the country's main university [122]. Another restriction of the policy, similarly selectively applied, included payments being associated with permission to carry out ecology/conservation research in the National Parks [122]. The policy effectively selectively blocked university students from accessing field stations of government for their research, a situation that has not yet returned to normal, although the policy was shelved after several years. The policy was also followed by an attempt of academic gagging [123], suggesting that conservation authorities sometimes have pursuits other than promoting research into conservation. Another barrier concerns the delays for obtaining permits from the authorities for carrying out research in ecology and conservation, which changed from the officially stated 3 weeks to at least 3 months, often more, effectively crippling many research projects and capacity building. Examples include a foreign student in Mauritius for about 5 months to study native orchids conservation and whose research permission was granted a few days before the end of her stay, and a PhD candidate also seeking to study orchid ecology and conservation, but who could not because his request for research permission was ignored.

5 Conclusion

Even in a small island with a rich and long-lasting heritage of natural history studies and very little native habitat surviving, recent dedicated surveys and research are still revealing new species of orchids and improving the understanding of orchid taxonomy, distribution, ecology, and conservation. It could be expected that further research would reveal more species and useful information for conservation managers including relocation of species currently believed extinct on the island, such as the ground orchid *Nervilia* that was relocated after a lapse of 239 years after it was last collected [93]. The orchids comprise the largest family of native flowering plants in Mauritius, and one with the highest rate of extinction so far and on those bases should be considered the current top priority for such surveys.

Furthermore, given the numerous threats besetting the biodiversity of Mauritius and in particular its native orchids, there is an urgent need to greatly strengthen ecological and conservation research and capacity building to rise to the challenges of improving evidence-based management to appropriate standards, to give Mauritius a chance to not just slow but rather stem and reverse the current trend of loss of native biodiversity, particularly in the context of the worsening and longer-lasting threats imposed by anthropogenic climate change. Local institutions involved in conservation should seek to strengthen their collaboration and work in a more complementary manner than at present as well as build stronger links with international institutions and scientists to encourage and attract more research being done locally. While increasing the training and capacity of many existing conservation professional in fields like taxonomy, ecology, or conservation is urgent, there is also a need for creating a much larger number of stable and sufficiently valorized posts of conservation professionals with clear career paths so as to attract, retain and motivate highly trained and able professionals in the field.

Training and capacity building alone would not suffice to adequately deal with the daunting challenges of the current biodiversity loss in Mauritius. There is also an urgent need to address the numerous and widespread problems posed by invasive alien species at a scale that is ecologically meaningful to ensure long-term conservation. Valorizing biodiversity to the appropriate extent for the diversity of services that it provides, including for the water cycle, carbon sequestration etc. would help mainstreaming conservation action and stimulate ecosystem restoration and reforestation programs to the scale needed to rise to the challenge. More collaboration, dialogue and trust among local stakeholders and actors is essential in the broader interest of the country, as well as, identifying and removing barriers that are currently impeding or distracting from true positive outcomes. The best way by far to conserve native orchids would be to do so in their restored natural habitats where sufficiently large enough populations can subsist that will make extinction risks negligible, and where the orchids ecological function will continue, thereby sustaining the biodiversity associated with the various species like pollinators, herbivores etc.

Apart from serving the interests of Mauritius in better conserving its threatened native orchid biodiversity and the habitats and ecological function that are crucial to it, with the appropriate actions and levels of involvement, Mauritius is poised to play an important role for the world beyond as a laboratory of sorts to better understand threats to biodiversity and experiment corresponding solutions, because the island combines many attributes, such as high human population densities, high degree of habitat destruction and fragmentation or high rates of alien species invasion, that await many other countries with rich biodiversity. If Mauritius is able to take the right actions to turn things around and become an example of successful orchid conservation, like it is known for globally with conservation of endemic birds, the island could well open a path useful for conserving threatened orchid floras elsewhere thereby generate a multiplier effect of its conservation efforts.

Acknowledgment We thank the late Jean Bosser and Thierry Pailler for their insights about Mauritius orchids over the years and the reviewers for their constructive comments on the manuscript. The National Parks and Conservation Service and the Forestry Service of the Ministry of Agro-Industry and Food Security are acknowledged for permission of access and research over the years. Mary-Ann Stanley and Owen L. Griffiths of Bioculture (Mauritius) Ltd. and Ebony Forest Reserve Chamarel Ltd. gave permission to access and carry out research at Mt. Camizard and Chamarel and provided substantial logistical support in terms of transport and on-site camping facilities for some of our surveys.

References

1. Ripple WJ, Wolf C, Newsome TM et al (2017) World scientists' warning to humanity: a second notice. Bioscience 67:1026–1028
2. Ceballos G, Ehrlich PR, Raven PH (2020) Vertebrates on the brink as indicators of biological annihilation and the sixth mass extinction. Proc Natl Acad Sci 117:13596–13602
3. Cardinale BJ, Duffy JE, Gonzalez A et al (2012) Biodiversity loss and its impact on humanity. Nature 486:59–67
4. Mace GM, Norris K, Fitter AH (2012) Biodiversity and ecosystem services: a multilayered relationship. Trends Ecol Evol 27:19–26
5. Ripple WJ, Wolf C, Newsome TM et al (2019) World scientists' warning of a climate emergency. Bioscience 70:8–12
6. Dinerstein E, Joshi AR, Vynne C et al (2020) A "Global Safety Net" to reverse biodiversity loss and stabilize Earth's climate. Sci Adv 6:eabb2824
7. Griscom BW, Adams J, Ellis PW et al (2017) Natural climate solutions. Proc Natl Acad Sci 114:11645–11650
8. Gillson L, Biggs H, Smit IPJ et al (2019) Finding common ground between adaptive management and evidence-based approaches to biodiversity conservation. Trends Ecol Evol 34:31–44
9. Kueffer C, Kinney K (2017) What is the importance of islands to environmental conservation? Environ Conserv 44:311–322
10. Whittaker RJ, Fernandez-Palacios JM (2007) Island biogeography: ecology, evolution, and conservation. OUP, Oxford
11. Darwin C (1859) On the origin of species by means of natural selection, or, the preservation of favoured races in the struggle for life. Murray, London
12. Fridriksson S (2013) Surtsey: evolution of life on a volcanic island. Elsevier, Kent
13. Ibanez T, Keppel G, Baider C et al (2018) Regional forcing explains local species diversity and turnover on tropical islands. Glob Ecol Biogeogr 27:474–486
14. MacArthur RH, Wilson EO (1967) The theory of island biogeography. Princeton University Press, Princeton

15. Norder SJ, Proios K, Whittaker RJ et al (2019) Beyond the last glacial maximum: island endemism is best explained by long-lasting archipelago configurations. Glob Ecol Biogeogr 28:184–197
16. Kier G, Kreft H, Lee TM et al (2009) A global assessment of endemism and species richness across island and mainland regions. Proc Natl Acad Sci 106:9322–9327
17. Jenkins CN, Pimm SL, Joppa LN (2013) Global patterns of terrestrial vertebrate diversity and conservation. Proc Natl Acad Sci 110:E2602–E2610
18. Myers N, Mittermeier RA, Mittermeier CG et al (2000) Biodiversity hotspots for conservation priorities. Nature 403:853–858
19. Keegan WF, Diamond JM (1987) Colonization of islands by humans: a biogeographical perspective. In: Schiffer MB (ed) Advances in archaeological method and theory. Academic Press, San Diego, pp 49–92
20. Rick TC, Kirch PV, Erlandson JM et al (2013) Archeology, deep history, and the human transformation of island ecosystems. Anthropocene 4:33–45
21. COL (2021) Catalogue of life, 2021. Available from: https://www.catalogueoflife.org/data/taxon/DPL
22. Taylor A, Weigelt P, König C et al (2019) Island disharmony revisited using orchids as a model group. New Phytol 223:597–606
23. Hermans J, Rajaovelona L, Cribb P et al (2020) New species and nomenclatural changes in *Cynorkis* (Orchidaceae) from Madagascar, the Comoros and the Mascarenes. Kew Bull 75:50
24. Pailler T (2020) Un nouveau *Cynorkis* (Orchidaceae) pour la flore des Mascareignes. Plant Ecol Evol 153:492–497
25. Pailler T, Verlynde S, Bytebier B et al (2020) Revision of *Angraecum* sect. *Perrierangraecum* (Orchidaceae; Epidendroideae; Vandeae) for the Mascarenes, with a description of a new endemic species for Mauritius. Phytotaxa 442:183–195
26. Pailler T, Andilyat M, Andrianarivo C et al (2018) Guide des orchidées des îles de l'Océan Indien: espèces indigènes et endémiques. Université de La Réunion, St Denis, La Réunion
27. Pailler T, Baider C (2020) *Bulbophyllum mascarenense* Pailler and Baider sp. nov.: a new endemic orchid species from the Mascarenes. Bot Lett 167:417–423
28. Pailler T, Henze F (2020) Orchidées de La Réunion. Editions Orphie, St Denis, La Réunion
29. Chase MW, Cameron K, Barrett RL et al (2003) DNA data and Orchidaceae systematics: a new phylogenetic classification. In: Dixon KW, Kell SP, Barret RL, Cribb PJ (eds) Orchid conservation. Natural History Publications (Borneo), Kota Kinabalu, pp 69–89
30. Fay MF, Chase MW (2009) Orchid biology: from Linnaeus via Darwin to the 21st century. Ann Bot 104:359–364
31. Gesner C (1751) Opera botanica. Mich. Seligmanni, Nuremberg
32. Kumbaric A, Savo V, Caneva G (2013) Orchids in the Roman culture and iconography: evidence for the first representations in antiquity. J Cult Herit 14:311–316
33. Cohen IM, Ackerman JD (2009) *Oeceoclades maculata*, an alien tropical orchid in a Caribbean rainforest. Ann Bot 104:557–563
34. Swarts ND, Dixon KW (2009) Terrestrial orchid conservation in the age of extinction. Ann Bot 104:543–556
35. Montaggioni L, Nativel P (1988) La Réunion, Ile Maurice: Géologie et aperçus biologiques. Masson, Paris
36. Padya BM (1989) Weather and climate of Mauritius. Mahatma Gandhi Institute, Moka
37. Cheke A, Hume J (2008) Lost land of the Dodo: an ecological history of the Mascarene Islands. T and AD Poyser, London
38. de Boer EJ, Hooghiemstra H, Florens FBV et al (2013) Rapid succession of plant associations on the small ocean island of Mauritius at the onset of the Holocene. Quat Sci Rev 68:114–125
39. Hammond DS, Gond V, Baider C et al (2015) Threats to environmentally sensitive areas from peri-urban expansion in Mauritius. Environ Conserv 42:256–267
40. Page W, D'Argent GA (1997) A vegetation survey of Mauritius (Indian Ocean) to identify priority rainforest areas for conservation management. IUCN/MWF report, Port Louis, Mauritius

41. Florens FBV (2013) Conservation in Mauritius and Rodrigues: challenges and achievements from two ecologically devastated oceanic islands. In: Sodhi NS, Gibson L, Raven PH (eds) Conservation biology: voices from the tropics. Wiley-Blackwell, Chichester, pp 40–50
42. Florens FBV, Baider C, Martin GM et al (2016) Invasive alien plants progress to dominate protected and best-preserved wet forests of an oceanic island. J Nat Conserv 34:93–100
43. Vaughan RE, Wiehe PO (1937) Studies on the vegetation of Mauritius: I. A preliminary survey of the plant communities. J Ecol 25:289–343
44. Du Petit-Thouars AA (1804) Cahier de six planches, tirées en couleur, grand in-folio. The author, Paris
45. Du Petit-Thouars AA (1809) Extrait de trois mémoires lus à la première classe de l'Institut, sur l'histoire des plantes orchidées des îles australes d'Afrique. Tableau des espèces des plantes orchidées qui composent le genre *Angorkis*. Nouv Bull Sci Philom Paris 1:314–319
46. Du Petit-Thouars AA (1822) Histoire particulière des plantes orchidées recueillies sur les trois îles Australes d'Afrique, de France, de Bourbon et de Madagascar. The author, Paris, 32 pp + 109 plates
47. Vermeulen JJ, Schuiteman A, de Vogel EF (2018) New taxa in *Bulbophyllum* (Orchidaceae), Epidendroideae: Malaxideae from New Guinea, lifting a mega-genus over the 2000-species mark. Malesian Orchid J 21:31–68
48. Vaughan RE (1958) Wenceslaus Bojer 1795–1856. Proc R Soc Arts Sci Mauritius 2:73–98
49. Bojer W (1837) Hortus Mauritianus: ou énumération des plantes, exotiques et indigènes, qui croissent à l'Ile Maurice, disposées d'après la méthode naturelle. Imprimerie d'Aimé Mamarot et Compagnie, Mauritius
50. Roberts DL, Florens FBV, Baider C et al (2004) *Taeniophyllum coxii* (Summerh.) Summerh. (Orchidaceae): a new record for Mauritius, Indian Ocean. Kew Bull 59:493
51. Baider C, Florens FBV, Rakotoarivelo F et al (2012) Two new records of *Jumellea* (Orchidaceae) for Mauritius (Mascarene Islands) and their conservation status. Phytotaxa 52:21–28
52. Pailler T, Baider C (2012) *Polystachya jubaultii* Pailler (Orchidaceae), une espèce nouvelle endémique de Mascareignes. L'Orchidophile 195:285–289
53. Fournel J, Micheneau C, Baider C (2015) A new critically endangered species of *Angraecum* (Orchidaceae) endemic to the island of Mauritius, Indian Ocean. Phytotaxa 222:211–220
54. Pailler T, Rakotoarivelo FP, Razafimandimbison SG et al (2020) Taxonomic revision of *Jumellea* (Orchidaceae, Angraecinae) in the Mascarenes. Phytotaxa 477:1–34
55. Ryan WBF, Carbotte SM, Coplan JO et al (2009) Global multi-resolution topography synthesis. Geochem Geophys Geosyst 10:Q03014
56. Bosser J, Cadet T, Guého J et al (1976) Flore des Mascareignes: La Réunion, Maurice, Rodrigues. MSIRI/KEW/IRD, Port Louis
57. Baider C (2009) MEL Joseph Guého 1937–2008: in memoriam. Taxon 58:679
58. Roberts DL (2001) Reproductive biology and conservation of the orchids of Mauritius. PhD thesis, University of Aberdeen, 290 p
59. Mallet B, Pailler T, Blambert L et al (2012) Différences morphologiques et identification de *Jumellea rossii* et *Jumellea fragrans* (Orchidaceae) à l'Ile de la Réunion: Implications pour la conservation. Rev Ecol (Terre Vie) 11(Suppl):73–83
60. Bosser J (1965) Contribution à l'étude des Orchidaceae de Madagascar V. A. Révision de quelques sections du genre *Bulbophyllum* à Madagascar. B. Autres espèces nouvelles de *Bulbophyllum*. C. Espèces nouvelles des genres *Cryptopus* et *Angraecum*. Adansonia 5:375–410
61. Fay MF (2018) Orchid conservation: how can we meet the challenges in the twenty-first century? Bot Stud 59:16
62. Isaac NJB, Mallet J, Mace GM (2004) Taxonomic inflation: its influence on macroecology and conservation. Trends Ecol Evol 19:464–469
63. Givnish TJ, Spalink D, Ames M et al (2015) Orchid phylogenomics and multiple drivers of their extraordinary diversification. Proc R Soc B Biol Sci 282:20151553

64. Taylor A, Keppel G, Weigelt P et al (2021) Functional traits are key to understanding orchid diversity on islands. Ecography 4:703–714
65. Martos F, Dulormne M, Pailler T et al (2009) Independent recruitment of saprotrophic fungi as mycorrhizal partners by tropical achlorophyllous orchids. New Phytol 184:668–681
66. Bosser J (2006) Contribution à l'étude des Orchidaceae de Madagascar, des Comores et des Mascareignes. XXXV. Description d'un *Oeceoclades* nouveau de Madagascar, et notes sur trois genres nouveaux pour les Mascareignes. Adansonia 28:45–54
67. Jacquemyn H, Micheneau C, Roberts DL et al (2005) Elevational gradients of species diversity, breeding system and floral traits of orchid species on Réunion Island. J Biogeogr 32:1751–1761
68. Ackerman JD, Trejo-Torres JC, Crespo-Chuy Y (2007) Orchids of the West Indies: predictability of diversity and endemism. J Biogeogr 34:779–786
69. Ding Y, Liu G, Zang R et al (2016) Distribution of vascular epiphytes along a tropical elevational gradient: disentangling abiotic and biotic determinants. Sci Rep 6:19706
70. Keppel G, Gillespie TW, Ormerod P et al (2016) Habitat diversity predicts orchid diversity in the tropical south-west Pacific. J Biogeogr 43:2332–2342
71. Roberts DL, Dixon KW (2008) Orchids. Curr Biol 18:R325–R329
72. Vollering J, Schuiteman A, de Vogel E et al (2016) Phytogeography of New Guinean orchids: patterns of species richness and turnover. J Biogeogr 43:204–214
73. Micheneau C, Fournel J, Humeau L et al (2008) Orchid–bird interactions: a case study from *Angraecum* (Vandeae, Angraecinae) and *Zosterops* (white-eyes, Zosteropidae) on Reunion Island. Botany 86(10):1143–1151
74. Micheneau C, Fournel J, Pailler T (2006) Bird pollination in an angraecoid orchid on Reunion Island (Mascarene Archipelago, Indian Ocean). Ann Bot 97:965–974
75. Micheneau C, Fournel J, Warren BH et al (2010) Orthoptera, a new order of pollinator. Ann Bot 105:355–364
76. Safford RJ (1997) A survey of the occurrence of native vegetation remnants on Mauritius in 1993. Biol Conserv 80:181–188
77. FAO (2020) Global forest resources assessment 2020, report Mauritius. FAO, Rome
78. Angulo E, Deves A-L, Saint Jalmes M et al (2009) Fatal attraction: rare species in the spotlight. Proc R Soc B Biol Sci 276:1331–1337
79. Courchamp F, Angulo E, Rivalan P et al (2006) Rarity value and species extinction: the anthropogenic Allee effect. PLoS Biol 4:e415
80. Baider C (2017) Fireside Science, The smallest orchid of Mauritius, 2017. Available from: https://scifundchallenge.org/firesidescience/2017/03/08/the-smallest-orchid-of-mauritius/
81. IUCN SSC (2001) IUCN Red List categories and criteria: version 3.1. IUCN Species Survival Commission, Gland/Cambridge
82. Thormann B (2010) Threat-analysis of endemic tree fern species of Mauritius – role of invasive plant species. BSc thesis, University of Applied Sciences, Eberswalde
83. Rouillard G, Guého J (1999) Les Plantes et leur histoire à l'île Maurice. MSM Ltd, Mauritius
84. Mawlah U (2007) Ecology of native Mauritian orchids and response to conservation management. BSc thesis, University of Mauritius, Réduit
85. Florens FBV, Baider C (2013) Ecological restoration in a developing island nation: how useful is the science? Restor Ecol 21:1–5
86. Vaughan RE, Wiehe PO (1941) Studies on the vegetation of Mauritius: III. The structure and development of the upland climax forest. J Ecol 29:127–160
87. Florens FBV, Baider C, Seegoolam NB et al (2017) Long-term declines of native trees in an oceanic island's forests invaded by alien plants. Appl Veg Sci 20:94–105
88. Lorence DH, Sussman RW (1986) Exotic species invasion into Mauritius wet forest remnants. J Trop Ecol 2:147–162
89. Florens FBV (2008) Ecologie des forêts tropicales de l'Ile Maurice et impact des espèces introduites envahissantes. PhD thesis, Université de La Réunion

90. Baider C, Florens FBV (2006) Current decline of the 'Dodo-tree': a case of broken-down interactions with extinct species or the result of new interactions with alien invaders? In: Laurance W, Peres C (eds) Emerging threats to tropical forests. Chicago University Press, Chicago, pp 199–214
91. Rasmussen HN, Rasmussen FN (2018) The epiphytic habitat on a living host: reflections on the orchid–tree relationship. Bot J Linn Soc 186:456–472
92. Zarate-García AM, Noguera-Savelli E, Andrade-Canto SB et al (2020) Bark water storage capacity influences epiphytic orchid preference for host trees. Am J Bot 107:726–734
93. Baider C, Florens FBV (2011) Control of invasive alien weeds averts imminent plant extinction. Biol Invasions 13:2641–2646
94. Monty MF, Florens FBV, Baider C (2013) Invasive alien plants elicit reduced production of flowers and fruits in various native forest species on the tropical island of Mauritius (Mascarenes, Indian Ocean). Trop Conser Sci 6:35–49
95. Krivek G, Florens FBV, Baider C et al (2020) Invasive alien plant control improves foraging habitat quality of a threatened island flying fox. J Nat Conserv 54:125805
96. Bindewald A (2011) Threat analysis of the large epiphytic ferns *Asplenium nidus* L. and *Microsorum punctatum* (L.) Copel. in the wet lowland forests of Mauritius: impact of invasive alien plants. BSc thesis, University of Mauritius, Réduit
97. Hugel S (2012) Impact of native forest restoration on endemic crickets and katydids density in Rodrigues island. J Insect Conserv 16:473–477
98. Florens FBV, Mauremootoo JR, Fowler SV et al (2010) Recovery of indigenous butterfly community following control of invasive alien plants in a tropical island's wet forests. Biodivers Conserv 19:3835–3848
99. Bernet P (2010) Les Orchidées de La Réunion, MSM, Mauritius
100. de Cordemoy EJ (1895) Flore de l'île de la Réunion:(phanérogames, cryptogames, vasculaires, muscinées) avec l'indication des propriétés économiques & industrielles des plantes, P. Klinsksieck, Paris
101. Szelengowicz M, Tamon JM (2013) Les orchidées des Mascareignes. Printec, Seychelles
102. Reinegger RD, Oleksy RZ, Bissessur P et al (2021) First come, first served: fruit availability to keystone bat species is potentially reduced by invasive macaques. J Mammal. 102:428–439
103. Florens FBV, Baider C, Martin GM et al (2012) Surviving 370 years of human impact: what remains of tree diversity and structure of the lowland wet forests of oceanic island Mauritius? Biodivers Conserv 21:2139–2167
104. Baider C, Florens FBV (2013) *Eugenia alletiana* (Myrtaceae), a new critically endangered endemic species to the island of Mauritius. Phytotaxa 94:1–12
105. Singer FJ, Swank WT, Clebsch EEC (1984) Effects of wild pig rooting in a deciduous forest. J Wildl Manag 48:464–473
106. Hone J (2002) Feral pigs in Namadgi National Park, Australia: dynamics, impacts and management. Biol Conserv 105:231–242
107. Harrison RD, Tan S, Plotkin JB et al (2013) Consequences of defaunation for a tropical tree community. Ecol Lett 16:687–694
108. Terborgh J, Nuñez-Iturri G, Pitman NCA et al (2008) Tree recruitment in an empty forest. Ecology 89:1757–1768
109. Kingston T, Florens FBV, Oleksy RZ, et al. (2018) Pteropus niger, The IUCN Red List of Threatened Species 2018: https://doi.org/10.1109/e.T18743A86475525
110. Florens FBV, Baider C, Marday V et al (2017) Disproportionately large ecological role of a recently mass-culled flying fox in native forests of an oceanic island. J Nat Conserv 40:85–93
111. Jones CG, Lawton JH, Shachak M (1997) Positive and negative effects of organisms as physical ecosystem engineers. Ecology 78:1946–1957
112. Florens FBV (2015) Flying foxes face cull despite evidence. Science 350:1325
113. Florens FBV (2012) Going to bat for an endangered species. Science 336:1102
114. Florens FBV (2016) Mauritius culls threatened fruit bats. Nature 530:33

115. Florens FBV, Baider C (2019) Mass-culling of a threatened island flying fox species failed to increase fruit growers' profits and revealed gaps to be addressed for effective conservation. J Nat Conserv 47:58–64
116. Vincenot CE, Florens FBV, Kingston T (2017) Can we protect island flying foxes? Science 355:1368–1370
117. McConkey KR, Drake DR (2006) Flying foxes cease to function as seed dispersers long before they become rare. Ecology 87:271–276
118. Albert S, Flores O, Baider C et al (2021) Differing severity of frugivore loss contrasts the fate of native forests on the land of the Dodo (Mascarene archipelago). Biol Conserv 257:109131
119. Albert S, Flores O, Strasberg D (2020) Collapse of dispersal trait diversity across a long-term chronosequence reveals a strong negative impact of frugivore extinctions on forest resilience. J Ecol 108:1386–1397
120. Gouvenain RC d, Kobe RK, Silander JA (2007) Partitioning of understorey light and dry-season soil moisture gradients among seedlings of four rain-forest tree species in Madagascar. J Trop Ecol 23:569–579
121. Sodhi NS, Butler R, Laurance WF et al (2011) Conservation successes at micro-, meso-and macroscales. Trends Ecol Evol 26:585–594
122. Florens FBV (2012) National parks: Mauritius is putting conservation at risk. Nature 481:29–29
123. Florens FBV (2013) Research safeguards protected areas: the important role of governments. Trends Ecol Evol 28:504

Diversity of Orchids from Continental Sub-Saharan Africa

4

Adama Bakayoko, Noufou Doudjo Ouattara, Akoua Clémentine Yao, Djah François Malan, Danho Fursy-Rodelec Neuba, Bi Fézan Honora Tra, and Tanoh Hilaire Kouakou

Contents

Abstract

Africa is a vast continent containing many types of ecological environments. It harbors the second largest forest reserve in the world but its flora is not well known for lack of financial means to carry out botanical prospecting studies. While many families have been well studied, others like the Orchidaceae are little known. This study is intended to contribute to the knowledge of this family through its use and distribution in continental sub-Saharan Africa.

The overall analysis of the orchid flora was done based on the four major regions except North Africa and South Africa (West Africa, Central Africa, Southern Africa, and East Africa). This study is based on the investigation of the literature. We have consulted previous published studies on orchids, floras,

A. Bakayoko (✉) · N. D. Ouattara · A. C. Yao
UFR des Sciences de la Nature (SN), Université NANGUI ABROGOUA, Abidjan, Ivory Coast

Centre Suisse de Recherches Scientifiques en Côte d'Ivoire, Abidjan, Ivory Coast

D. F. Malan · D. F.-R. Neuba · B. F. H. Tra · T. H. Kouakou
UFR des Sciences de la Nature (SN), Université NANGUI ABROGOUA, Abidjan, Ivory Coast

J.-M. Mérillon, H. Kodja (eds.), *Orchids Phytochemistry, Biology and Horticulture*, Reference Series in Phytochemistry, https://doi.org/10.1007/978-3-030-38392-3_38

and distribution maps of the targeted areas. We were able to draw up a list of 1373 species belonging to 88 genera. The results show that of the four phytogeographic zones, Central Africa is the richest with 708 species, followed by Southern Africa and East Africa with 637 and 583 species, respectively. West Africa, with 413 species, is the least rich area. Several uses have been listed. Mostly, orchids are using in pharmacopoeia, in feeding and as ornamental plants. We were also able to establish endemism in several countries (e.g., Tanzania, Zimbabwe, the Democratic Republic of Congo, Ethiopia, Cameroon, Mozambique, Zambia, Malawi, Rwanda, Angola, Kenya, Nigeria, Gabon, Central African Republic, Uganda, and Burundi).

Keywords

Diversity · Distribution · Orchids · Sub-Saharan Africa · Endemic

1 Introduction

Third largest continent in the world in terms of area after Asia and America, African continent has a great diversity of ecological environments ranging from wet evergreen forests to vast desert areas [1]. With this multitude of vegetation, the African continent is rich in more than 62,000 species of flowering plants but still remains largely unknown. Compared to other continents, the diversity is relatively low. Some studies estimate this diversity probably at 25% of the world's species. Several reasons could explain this ignorance. The fragmentation of Africa due to the history of colonization is believed to have encouraged sectarianism. The large surface area of the continent and the lack of financial support to carry out botanical inventories is also a cause of the ignorance of the flora of the African continent. Most of the country or regional floras of Africa are old.

African continents contain the second largest forest reserve in the world. While many botanical families have been well studied, others are very little known. This is the case of Orchidaceae family which has an estimated global richness of 25,000 species [2]. This family has good representation in tropical and equatorial regions of the world but in Africa it is little studied. The richness of the genera of the Orchidaceae is estimated at 700 accord [3]. This family presents the greatest diversity and number of species in flowering plants [4]. In Africa, the richness of Orchidaceae is estimated at 1500 species belonging to 1 87 genera [5, 6].

The Orchidaceae family is important both in botanical terms for its richness, and economically through the use of several species as ornamental plants [7]. Many of these species are sold internationally as an ornamental. This international trade of the orchids species has an impact on the abundance of the species, in their habitats. Some species became rare, and others are on the way of extinction, while others have already disappeared. The main reason for their disappearance is the degradation of their habitat by human activities. Therefore, orchids represent an excellent indicator of the quality of biotopes in which they are found [8]. This work, which aims to contribute to the knowledge of the floristic diversity of the Orchidaceae family, will focus on species from sub-Saharan Africa, except South Africa.

2 Method of Data Collection

This work on the orchids of sub-Saharan Africa was based on the use of literature and on all the flora that we had at our disposal. We have particularly used the books presenting the distribution of Orchidaceae in the countries of sub-Saharan Africa [9, 10]. It should be noted that the work of these authors is based on specimens deposited in different herbaria around the world. We analyzed the flora of orchids according to the four major regions except North Africa and South Africa. Thus the areas considered in this work are: West Africa, Central Africa, Southern Africa (not the country), and East Africa.

3 Brief Presentation of Orchidaceae

3.1 Distribution of Orchids

The Orchidaceae are a cosmopolitan family with a very large ecological environment. Orchids are much more present in warm regions [11]. Herbaceous and perennial, the species of this family have colonized practically all habitats. They can be found in the coastal areas as well as at higher altitudes. However, they are absent from extreme environments such as Antarctica, the sea, the most arid deserts, and the coldest mountain peaks. Orchidaceae are more common in tropical areas where its flora represent nearly 10% of tropical flora. According to Ref. [11], they can be terrestrial plants (tropical, temperate, and boreal environments for geophytes with tubers or rhizomes), epiphytes (generally in tropical environments), and even lianas as species of *Vanilla* genera. In Africa, the Orchidaceae constitute an important part of the epiphytes of tropical forests. Among the 153 species of epiphytes recorded in the evergreen forests of West Africa, 101 belong to the Orchidaceae family [12].

3.2 Botanical Description and Systematic

The plants belonging to the Orchidaceae family are terrestrial or epiphytic herbs, rarely helophytes, never halophytes, aquatic, or parasitic. The terrestrial Orchidaceae are provided with important underground apparatus, in the form of tubers of stem or root origin or of rhizomes, on which are born the adventitious roots and the leafy and flowering aerial stems. The stems of Orchidaceae are simple or branched and erect. The creeping form is rarely observed.

The strongly adherent roots of orchids allow a good attachment to trees and better withstand the stresses of aerial life. In some species, the leafless vegetative apparatus is reduced to a system of green roots, from which the inflorescences arise directly.

In Orchidaceae family, the leaves are simple, alternate, more or less fleshy, with a sheathing base, parallel to the veins. They are often arranged in two rows, rarely opposite or in whorls. They are sessile or petiolate and sometimes reduced to scales. The greatest diversity at the leaf level is observed in terrestrial Orchidaceae, hence their extensive use as decorative plants.

Epiphytic Orchidaceae have both leathery and succulent, green, or spotted leaves with usually long, hanging aerial roots, gray or white, covered with a velamen, with a meristematic apex. This system absorbs the moisture and nutrients necessary for the life of the plant. These thick ligulate leaves reduce water loss and promote photosynthesis. The survival of many epiphytes in environments subject to extreme conditions is due to the development of particularly active photosynthetic tissues in all organs (stems, roots, and even flowers).

Many tropical and subtropical orchids, both terrestrial and epiphytic, have pseudo-bulbs as reserve organs for water and nutrients. Those pseudo-bulbs vary, depending on the species. We can observe from barely swollen stems to very hard, to apple-shaped bright green organs, from which the leaves emerge. A plant may have only one pseudo-bulb, but these may also be gathered in a tight group, or, in some tropical species, be spaced along a creeping or climbing rhizome. Their size varies from that of a pinhead to a thick cylinder up to 3 m high. Pseudo-bulbs are always on the ground or above ground level, but in several groups of terrestrial orchids in temperate regions, there are similar organs growing on roots in the ground.

The Orchidaceae family is constituted by five subfamilies which are the Epidendroideae, the Apostasioideae, the Vanilloideae, the Cypripedioideae, and the Orchidoideae [13]. The classification of African orchids has been a real challenge for a very long time [14]. Taxonomic confusions are not limited only to the specific level but also to the generic level. For example, the generic delineation within the African Angraecoid group resulted in the discovery of 6 new species and 2 new genera [14]. In addition, 30 species were not correctly assigned, from a phylogenetic point of view, to genera.

The taxonomic insufficiencies of this family, in continental Africa, are justified by several reasons including the political instability of the countries (making field missions difficult), logistical problems, the lack of effective protocols for inventories, etc. At the phylogenetic level, the studies showed that the genera *Angraecopsis*, *Diaphananthe*, and *Margelliantha* are polyphyletic while the genera *Aerangis*, *Ancistrorhynchus*, *Bolusiella*, *Campylocentrum*, *Cyrtorchis*, *Dendrophylax*, *Eurychone*, *Microcoelia*, *Nephrangis*, *Podangis*, and *Solenangis* are monophyletic [14].

4 The Orchids of Sub-Saharan Africa

4.1 Richness According to Geographic Areas

The Orchidaceae species inventoried in sub-Saharan Africa are 1373 species grouped into 88 genera. The genera *Habenaria* (391 spp.), *Polystachya* (279 spp.), *Eulophia* (232 spp.), and *Bulbophyllum* (115 spp.) are the important genera. These genera are followed by *Disa*, *Satyrium*, *Tridactyle*, *Angraecum*, *Disperis*, *Rhipidoglossum*, and *Brachycorythis* with more than 50 species. We recorded 108 species common to the 4 major regions (West Africa, Central Africa, East Africa, and Southern Africa). According to the work of Refs. [9, 10], the richness of Central Africa phytogeographic zones is 708 species. This zone is followed by southern Africa and eastern Africa with 637 and 583 species, respectively. These numbers are higher than those given by the work of Ref. [15]. For this author, East Africa is the richest region with 679 species followed by Central Africa with 517 species, South Africa with 425 species, and West Africa with 413 species. The region with the lowest orchid abundance is West Africa. Central Africa appears to be the most studied area. Indeed, several works have been carried out in this area by numerous researchers. After Central Africa, the orchids of East and South Africa which have been the best studied. In West Africa very few studies have been focused on Orchidaceae. However, in this area, the important work on the orchids of Côte d'Ivoire should be highlighted [4].

The most abundant species distributed in most countries are: *Ansellia africana* Lindl., *Bolusiella alinae* Szlach., *Brachycorythis ovata* Lindl., *Brachycorythis pubescens* Harv., *Brachycorythis tenuior* Rchb.f., *Bulbophyllum fuscum* Lindl., *Bulbophyllum maximum* (Lindl.) Rchb.f., *Bulbophyllum scaberulum* (Rolfe) Bolus, *Cyrtorchis arcuata* (Lindl.) Schltr. (Fig. 10), *Eulophia angolensis* (Rchb.f.) Summerh., *Eulophia cucullata* (Afzel. ex Sw.) Steud., *Eulophia guineensis* Lindl., *Eulophia horsfallii* (Bateman) Summerh., *Habenaria ichneumonea* (Sw.) Lindl., *Habenaria schimperiana* Hochst. ex A. Rich., *Habenaria zambesina* Rchb.f., *Liparis nervosa* (Thunb.) Lindl., *Nervilia bicarinata* (Blume) Schltr., *Nervilia kotschyi* (Rchb.f.) Schltr., *Nervilia petraea* (Afzel. ex Sw.) Summerh., *Platylepis glandulosa* (Lindl.) Rchb.f., *Polystachya golungensis* Rchb.f., *Polystachya modesta* Rchb.f., *Polystachya concreta* (Jacq.) Garay & H. R. Sweet., *Rangaeris muscicola* (Rchb.f.) Summerh., *Rhipidoglossum rutilum* (Rchb.f.) Schltr., *Tridactyle anthomaniaca* (Rchb.f.) Summerh., *Tridactyle bicaudata* (Lindl.) Schltr., *Tridactyle filifolia* (Schltr.) Schltr. et *Tridactyle tridactylites* (Rolfe) Schltr.

4.2 Endemism of Orchidaceae Species in the Sub-Saharan Countries

Apart from species widely distributed in all regions, other species seem rather confined to a given country or region. Recent studies showed that 284 species are found in East Africa, 259 species in Central Africa, 201 in South Africa, and only 58 species in South Africa [9, 10]. There is thus an endemism in certain countries or in two border countries.

The richness countries in endemic species based on the distribution map of herbarium samples are Tanzania (78 species), Zimbabwe (41 species), the Democratic Republic of Congo (38 species), Ethiopia (32 species), Cameroon (29 species), Mozambique (27 species), Zambia (24 species), Malawi (23 species), Rwanda (21 species), Angola (20 species), Kenya (18 species), Nigeria (17 species), Gabon

(16 species), Central African Republic (8 species), and Uganda and Burundi with 5 species each [9, 10]. This number of endemic species could vary with taxonomic revisions and sampling effort conducted in several countries. The endemic species found in Tanzania include: *Aerangis confusa* J. Stewart, *Ancistrorhynchus parviflorus* Summerh., *Angraecopsis tenerrima* Kraenzl.; I. & E. la Croix, *Angraecum brevicornu* Summerh., *Bonatea volkensiana* (Kraenzl.) Rolfe, *Brachycorythis tanganyikensis* Summerh, *Cynorkis pleistadenia* (Rchb.f.) Schltr., *Disperis elaphoceras* Verdc., *Epipactis ulugurica* Mansf., *Eulophia amblyosepala* (Schltr.) Butzin, *Habenaria apiculata* Summerh., *Holothrix hydra* P. J. Cribb, *Margelliantha clavata* P. J. Cribb, *Mystacidium nguruense* P. J. Cri*bb, Nervilia similis* Schltr, *Oeceoclades zanzibarica* (Summerh.) Garay & P. Taylor, *Platycoryne ambigua* (Kraenzl.) Summerh., *Polystachya acuminata* Summerh., *Pterygodium ukingense* Schltr, *Rangaeris schliebenii* (Mansf.) P. J. Cribb, *Rhipidoglossum melianthum* (P. J. Cribb) Senghas, *Satyrium comptum* Summerh., *Tridactyle brevifolia* Mansf., *Vanilla zanzibarica* Rolfe, *Zeuxine lunulata* P. J. Cribb & J. Bowden.

Kenya, which borders Tanzania, has 243 Orchidaceae species divided into 47 genera according to Ref. [16]. Among the species we note as endemic to Kenyan flora are *Ancistrorhynchus paysanii* Senghas, *Angraecum decipiens Summerh.*, *A. spectabile* Summerh., *Bilabrella kraenzliniana*, *Polystachya bella* Summerh., *Eulophia endlichiana* (Kraenzl)., *Habenaria altior* Rendle, *H. bonateoides* Ponsie, *H. haareri* Summerh., *H. keniensis* Summerh., *H. thomsonii* Rchb. f., *Holothrix pentadactyla* (Summerh.) Summerh., *Polystachya holstii* Kraenzl., *P. teitensis* P. J. Cribb, *Rhipidoglossum montanum* (Piers) Senghas, *Ypsilopus longifolius* (Kraenzl.) Summerh, *Y. viridiflorus* P. J. Cribb & J. Stewart.

According to the distribution map, these two East African countries have in common 17 endemic species found only on their territory. These are: *Aerangis coriacea* Summerh., *Angraecopsis breviloba* Summerh, *Angraecum viride* Kraenzl., *Bonatea rabaiensis* (Rendle) Rolfe, *Cynorkis uncata* (Rolfe) Kraenzl., *C. usambarae* Rolfe, *Disperis egregia* Summerh., *Habenaria helicoplectrum* Summerh., *H. plectromaniaca* Rchb.f. & S. Moore, *Polystachya confusa* Rolfe, *P. disiformis* J. Cribb, *P. leucosepala* P. J. Cribb, *P. shega* Kraenzl., *Tridactyle cruciformis* Summerh., *T. furcistipes* Summerh., *T. tanneri* P. J. Cribb, *Rhipidoglossum tanneri* (P. J. Cribb) Senghas.

Ethiopia has 160 Orchidaceae species, divided into 37 genera [15]. These authors have recorded 26 endemic species while 35 species have been reported by Refs. [9, 10]. Among the species indicated, we can cite *Angraecopsis trifurca* (Rchb.f.) Schltr, *Habenaria lefebureana* (A. Rich.) T. Durand & Schinz, *Disa pulchella* Hochst. ex A. Rich., *Disperis crassicaulis* Rchb.f., *Eulophia abyssinica* Rchb. f., *Habenaria aethiopica* S. Thomas & P. J. Cribb, *Holothrix praecox* Rchb.f., *Liparis abyssinica* A. Rich., *Polystachya aethiopica* P. J. Cribb, *Rhipidoglossum candidum* (P. J. Cribb) Senghas, *Roeperocharis alcicornis* Kraenzl., *Satyrium aethiopicum* Summerh.

In Central Africa, Cameroon harbors the highest number of endemic species with 29 endemic species recorded [9, 10]. In a previous work in this country, 44 endemic species were identified [17]. These endemic species are, among others,

Ancistrorhynchus constrictus Szlach. & Olsz., *Angraecopsis cryptantha* Cribb, *Angraecum curvipes* Schltr., *Bolusiella zenkeri* (Kraenzl.) Schltr., *Bulbophyllum saltatorium* Lindl., *Chamaeangis letouzeyi* Szlach. & Olsz., *Cheirostylis divina* (Guinea) Summerh. var. *ochyrae* Szlach. & Olsz., *Diaphananthe garayana* Szlach. & Olsz., *Disperis kamerunensis* Schltr., *Disperis nitida* Summerh., *Gastrodia africana* Kraenzl., *Genyorchis macrantha* Summerh., *Habenaria alinae* Szlach., *Liparis kamerunensis* Schltr., *Microcoelia sanfordii* L. Jónss., *Ossiculum aurantiacum* P. J. Cribb & Laan, *Platycoryne alinae* Szlach., *Polystachya albescens* Ridl. subsp. *angustifolia* (Summerh.) Summerh., *Rangaeris trachypus* (Kraenzl.) Guillaumin, *Rhipidoglossum ochyrae* Szlach. & Olsz., *Rhipidoglossum polydactylum* (Kraenzl.) Garay, *Vanilla ochyrae* Szlach. & Olsz.

Orchids species of Gabon can be checked at http://orchideesgabon.e-monsite.com/pages/orchidees-liste.html. This site gives a non-exhaustive list of 284 species divided into 53 genera among which 16 endemic species [9, 10]. These endemics species are: *Angraecopsis hallei* Szlach. & Olszewski, *Angraecum cribbianum* Szlach. & Olszewski, *Bulbophyllum apodum* Hook.f., *B. subligaculiferum* J. J. Vern., *Cyrtorchis henriquesiana* (Ridl.) Schltr., *Dinklageella villiersii* Szlach. & Olszewski, *Rhipidoglossum magnicalcar* Szlach. & Olszewski, *Taeniorrhiza gabonensis* Summerh., *Tridactyle latifolia* Summerh., *T. minutifolia* Stévart & D'Haijère, *T. pentalobata* P. J. Cribb & Stévart, *Vanilla acuminata* Rolfe, *V. chalotii* Finet, *V. hallei* Szlach. & Olszewski, *Veyretella hetaerioides* (Summerh.) Szlach. & Olszewski, *V. flabellata* Szlach.

Gabon and Cameroon have 11 species in common [17]. These species are: *Aerangis gracillima* (Kraenzl.) J. C. Arends & J. Stewart, *Ancistrorhynchus obovata* Stévart inédit., *Angraecum eichlerianum* Kraenzl. var. *curvicalcaratum* Szlach. & Olsz., *Bulbophyllum fayi* J. J. Verm., *Bulbophyllum minutifolium* Stévart in Stévart *& al., Diaphananthe ceriflora* J. B. Petersen, *Eggelingia gabonensis* P. J. Cribb & Laan, *Genyorchis platybulbon* Schltr., *Halleorchis aspidogynoides* Szlach. & Olsz., *Polystachya kubalae* Szlach. & Olsz., *Polystachya letouzeyana* Szlach. & Olsz. However, 19 species have been indicated as common endemic species to these two countries [9, 10].

Cameroon shares with the Republic of Congo, the Democratic Republic of Congo, Equatorial Guinea, and Ethiopia five endemic species which are *Habenaria phantasma* la Croix, *Habenaria stenoceras* Summerh., *Rhipidoglossum montealenense* Descourvières, Stévart & P. J. Cribb and *Stolzia grandiflora* P. J. Cribb. *Habenaria egregia* Summerh. is only observed in Cameroon, Ethiopia, and Kenya.

The species *Orestias micrantha* Summerh., *Polystachya batkoi* Szlach. & Olszewski, *Polystachya letouzeyana* Szlach. & Olszewski, *Polystachya moniquetiana* Stévart & Geerinck, *Polystachya riomuniensis* Stévart & Nguema and *Polystachya stodolnyi* Szlach. & Olszewski are only found in Cameroon, Gabon, and Equatorial Guinea. Cameroon also shares with Gabon and the Republic of Congo the species *Habenaria lisowskii Szlach.* and with Gabon and the Democratic Republic of Congo the species *Tridactyle laurentii* (De Wild.) Schltr. Two species *Polystachya bifida* Lindl. and *Polystachya testuana* Summerh are endemic to Gabon and Equatorial Guinea.

The Democratic Republic of Congo has 38 endemic species according to the distribution maps of Refs. [9, 10]. Some of these species are: *Angraecum mofakoko* De Wild., *Bulbophyllum horridulum* J. J. Vern., *Cynorkis summerhayesiana* Geerinck, *Diaphananthe divitiflora* (Kraenzl.) Schltr., *Disa alinae* Szlach., *Eulophia fernandeziana* Geerinck, *Habenaria garayana* Szlach. & Olszewski, *Polystachya alicjae* Mytnik, *Roeperocharis maleveziana* Geerinck, *Tridactyle fimbriatipetala* (De Wild.) Schltr., *Tridactyle stipulata* (De Wild.) Schltr., *Tridactyle vanderlaaniana* Geerinck.

In République Centrafricaine, we found only eight endemics species which are: *Ancistrorhynchus brevifolius* Finet, *Eulophia falcatiloba* Szlach. & Olszewski, *Habenaria kornasiorum* Szlach. & Olszewski, *Habenaria letestuana* Szlach. & Olszewski, *Microcoelia jonssonii* Szlach. & Olszewski, *Platycoryne lisowskiana* Szlach. & Kras, *Platycoryne ochyrana* Szlach., *Disperis raiilabris* Summerh.

A total of 21 endemic species from Rwanda flora has been noticed [9, 10]. They are: *Bulbophyllum kivuense* J. J. Vern., *Diaphananthe eggelingii* P. J. Cribb, *Diaphananthe liae* Ed. Fischer, Killmann, J.-P. Lebel & Delep., *Disperis natascha-oppeltae* Eb. Fischer, Killmann, *Disperis reklieberae* Eb. Fischer, *Eulophia pocsii* Eb. Fisch., *Gastrodia rwandensis* Eb. Fisch. & Killmann, *Habenaria bequaertii* Summerh., *Margelliantha lebelii* Eb. Fischer & Killmann, *Polystachya anastacialynae* Eb. Fischer, *Polystachya bruechertiae* Eb. Fischer, Killmann & J.-P. Lebel, *Polystachya erica-lanzae* Eb. Fischer, *Polystachya isabelae* Mytnik, *Polystachya lawalreeana* Geerinck, *Polystachya samilae* Eb Fisch., *Rhipidoglossum cuneatum* (Summerh.) Garay, *Rhipidoglossum mildbraedii* (Kraenzl.) Garay, *Stolzia heiligenthalii* Eb. Fisch., Killmann, *Stolzia kalkhof-roseae* Eb. Fischer, Killmann, *Tridactyle nanne-ritzkae* Eb. Fischer, Killmann, Lebel & Delepierre, *Tridactyle stevartiana* Geerinck.

According to the two authors [9, 10], there are five endemic species in Burundi which are: *Habenaria lewallei* Geerinck, *Polystachya editae* Eb. Fisch., *Polystachya lacroixiana* Geerinck, *Polystachya maculata* P. J. Cribb, *Polystachya walravensiana* Geerinck & Arbonn.

Burundi and Rwanda share 16 endemic species which are: *Habenaria coeloglossoides* Summerh., *Habenaria lehae* Eb. Fischer, *Polystachya tridentata* Summerh, *Polystachya troupiniana* Geerinck, *Polystachya undulata* P. J. Cribb & Podz., *Polystachya woosnamii* Rendle, *Rhaesteria eggelingii* Summerh., *Polystachya proterantha* P. J. Cribb, *Habenaria brachylobos* (Summerh.) Summerh., *Rhipidoglossum ovale* (Summerh.) Garay, *Tridactyle eggelingii* Summerh., *Margelliantha burttii* (Summerh.) P. J. Cribb, *Polystachya fabriana* Geerinck, *Polystachya macropoda* Summerh., *Rhipidoglossum arbonnieri* (Geerinck) Eb. Fischer, Killmann, *Rhipidoglossum delepierreanum* (Lebel & Geerinck) Eb. Fisch., Killmann.

In West Africa, the orchids flora of Nigeria and Côte d'Ivoire is well known compared to the other countries of the area. Nigeria is rich with 239 species divided between 34 genera. It is the country with the most endemic species in West Africa [9, 10]. According to these authors, there are 15 endemic species. But other studies based on inventories reveal 7 endemic species which are: *Diaphananthe dorotheae* (Rendle) Summerh., *Genyorchis apertiflora* Summerh., *Habenaria linguiformis*

Summerh., *Habenaria nigerica* Summerh., *Habenaria phylacocheira* Summerh., *Habenaria prionocraspedon* Summerh., *Alectra virgata* Hemsl., which are indicated to be endemic species of Nigeria [18].

By observing the distribution maps of species, Nigeria (West Africa), neighbor of Cameroon (Central Africa), shares with this country 26 endemic species [9, 10]. These are for example: *Aerangis gravenreuthii* (Kraenzl.) Schltr., *Ancistrorhynchus serratus* Summerh., *Angraecopsis cryptantha* P. J. Cribb, *Angraecum angustum* (Rolfe) Summerh., *Diaphananthe lanceolata* (Summerh.) P. J. Cribb & Carlsward, *Disperis mildbraedii* Schltr. ex Summerh., *Genyorchis micropetala* (Lindl.) Schltr, *Habenaria alinae* Szlach., *Liparis letouzeyana* Szlach. & Olszewski, *Platycoryne megalorrhyncha* Summerh., *Polystachya alpina* Lindl., *Tridactyle lagosensis* (Rolfe) Schltr., *Rhipidoglossum obanense* (Rendle) Summerh. *Platylepis xerostele* Ormerod is observed on the border of Nigeria and Cameroon.

The book entitled *Les Orchidées de Côte d'Ivoire* gave for this country a total of 264 species divided into 48 genera [4]. This study increased the number of species recorder in the frame of previous studies [19]. Indeed three new species (*Ancistrorbyncbus akeassiae Perez-Vera, Bulbupbyllum danii Perez-Vera*, and *Cbamaeangis pauciflora Perez-Vera*) and two new varieties (*Bulbopbyllum scaberulum* (Rolfe) Bolus var. *album* Perez-Vera and *Cyrtorcbis broumii* (Rolfe) Schltr. var. *guillaumetii* Perez-Vera) have been described [4].

Côte d'Ivoire shares with Liberia and Sierra Leone three species: *Polystachya pseudodisa Kraenzl., Tridactyle fusifera Mansf.*, and *Bulbophyllum denticulatum Rolfe. Polystachya leonensis Rchb.*f. is found in Guinea and Sierra Leone. This species is not counted among ivorian orchids, although it is collected near the Ivorian border. This species has been also observed in Cameroon flora. The species *Polystachya bancoensis Burg* has only been observed in Côte d'Ivoire and Ghana. The species *Polystachya rivae* Schweinf. encountered in Ethiopia has been reported in Côte d'Ivoire [9, 10]. This species is not included in the list compiled by Ref. [4].

Côte d'Ivoire, Liberia, Sierra Leone, and Ghana share *Bulbophyllum danii* Perez-Vera, *Angraecum modicum* Summerh., *Diaphananthe suborbicularis* Summerh. and *Bulbophyllum parvum* Summerh. The species *Malaxis melanotoessa* Summerh. and *Polystachya bequaertii* Summerh. et *Polystachya elastica* Lindl. are only found in Liberia and Sierra Leone. Species like *Habenaria jacobi* Summerh. and *Habenaria jaegeri* Summerh. are found in Liberia and Sierra Leone and Guinea. The species *Habenaria angustissima* Summerh. is endemic in Senegal and Guinea. The species *Bulbophyllum parvum* Summerh. and *Diaphananthe suborbicularis* Summerh. are found in Ghana and Sierra Leone. The species *Rhipidoglossum laticalcar* (J. B. Hall) Senghas is only found in Ghana and Nigeria. *Polystachya monolenis* Summerh. is an endemic species of Sierra Leone and Ghana. The species *Rhipidoglossum laxiflorum* Summerh. has been reported in Ghana, Togo, and Côte d'Ivoire in the work of Refs. [9, 10]. However, this species has not been reported in Côte d'Ivoire [4]. *Habenaria dinklagei* Kraenzl. has only been observed in Liberia and Nigeria. It has not been found in other regions between the two countries.

The well-known Southern African countries for orchids are Angola, Malawi, Mozambique, Zambia, and Zimbabwe. The flora of Angola is rich of 228 orchids belonging to 40 genera [20]. This Angolan flora has 20 endemic species according to Refs. [9, 10]. For example we can mention *Angraecopsis gassneri* G. Will., *Brachycorythis mixta* Summerh., *Diaphananthe welwitschii* (Rchb.f.) Schltr., *Eulophia aloifolia* Welw. ex Rchb.f, *Eulophia protearum* Rchb.f., *Habenaria decaptera* Rchb.f., *Habenaria robusta* Welw. ex Rchb.f., *Holothrix klimkoana* Szlach. & Marg., *Polystachya angularis* Rchb.f., *Satyrium aciculare* van der Niet & P. J. Cribb et *Satyrium welwitschii* Rchb.f.

The site https://www.malawiflora.com/speciesdata/family.php?family_id=161 gave for Malawi 420 Orchidaceae species grouped into 57 genera. Among these 420 species, 23 are endemic according to the work of Refs. [9, 10]. Three hundred six species distributed among 48 genera have been counted by Ref. [21]. It is understandable that this number may be increasing with the work prior to this study [22]. In this flora of Malawi, 23 species are endemic according to the work of Refs. [9, 10]. Those species are: *Aerangis carnea* J. Stewart, *Angraecopsis lovettii* P. J. Cribb, *Angraecum umbrosum* P. J. Cribb, *Bonatea stereophylla* (Kraenzl.) Summerh., *Disa walleri* Rchb.f., *Disperis decipiens* Verdc.; *Habenaria diselloides* Schltr., *Holothrix buchananii* Schltr., *Polystachya mzuzuensis* P. J. Cribb & la Croix, *Polystachya suaveolens* P. J. Cribb, *Rhipidoglossum oxycentron* (P. J. Cribb) Senghas, *Satyrium afromontanum* la Croix & P. J. Cribb.

Mozambique harbors 232 species and 49 genera of orchids with 7 endemic species [23]. These species are: *Cyrtorchis glaucifolia* Summerh., *Disperis mozambicensis* Schltr., *Eulophia biloba* Schltr., *Eulophia bisaccata* Kraenzl., *Habenaria hirsutissima* Summerh., *Habenaria mosambicensis* Schltr. et *Liparis hemipilioides* Schltr. In addition, 14 other species are shared with Zimbabwe, Tanzania, Malawi, and South Africa. The country shares 10 endemic species with Zimbabwe: *Bulbophyllum ballii* P. J. Cribb, *Cynorkis anisoloba Summerh., Disa chimanimaniensis* (H. P. Linder) H. P. Linder, *Disa zimbabweensis* H. P. Linder, *Neobolusia ciliata* Summerh., *Oligophyton drummondii* H. P. Linder & G. Will., *Polystachya subumbellata* P. J. Cribb & Podz., *Polystachya valentina* la Croix & P. J. Cribb, *Satyrium flavum* la Croix, and *Schizochilus lepidus* Summerh. Mozambique shares, respectively, with Malawi and Tanzania the species *Polystachya songaniensis* G. Will. et *Habenaria stylites* Rchb.f. & S. Moore subsp. johnsonii (Rolfe) Summerh.

In Zambia 412 orchids species divided into 46 genera are recorded on the site https://www.zambiaflora.com/speciesdata/family.php?family_id=161. Among these species, 22 are endemic. These species are: *Bonatea antennifera* Rolfe, *Bonatea cassidea* Sond., Cynorkis clarae Geerinck, *Disa helenae* Buscal. & Schltr., *Disa praecox* (H. P. Linder) H. P. Linder in Linder & Kurzweil, *Disperis bifida* P. J. Cribb, *Eulophia brenanii* P. J. Cribb & la Croix, *Eulophia richardsiae* P. J. Cribb & la Croix; *Habenaria argentea* P. J. Cribb, *Habenaria bertauxiana* Szlach. & Olszewski, *Habenaria binghamii* G. Will., *Habenaria macrotidion* Summerh., *Habenaria orthocentron* P. J. Cribb, *Habenaria pasmithii* G. Will., *Habenaria petraea* Renz & Grosvenor, *Habenaria tubifolia* la Croix & P. J. Cribb, *Habenaria velutina* Summerh., *Holothrix villosa* Lindl. var. villosa, *Polystachya asper* P. J. Cribb & Podz., *Polystachya erythrocephala* Summerh., *Polystachya mafingensis* P. J. Cribb, *Pteroglossaspis corymbosa* G. Will.

Zimbabwe has 358 species of orchids for 54 genera that can be viewed at the site https://www.zimbabweflora.co.zw/speciesdata/family.php?family_id=161.

A list of the plant taxa endemic or nearly endemic to Zimbabwe has not previously been compiled [24]. The list of endemics identified in this flora will be based on Refs. [9, 10]. Zimbabwe comes after Tanzania with 41 endemic species. We have retained *Aeranthes parkesii* G. Will., *Angraecum chimanimaniense* G. Will., *Bulbophyllum chikukwa* Fibeck & Mavi, *Brownleea galpinii* Bolus subsp. galpinii, *Cynorkis anisoloba* Summerh., *Disa chimanimaniensis* (H. P. Linder) H. P. Linder, *Disperis concinna* Schltr., *Eceoclades quadriloba* (Schltr.) Garay & P. Taylor, *Eulophia foliosa* (Lindl.) Bolus, *Goodyera afzelii* Schltr, *Habenaria bicolor* Conrath & Kraenzl., *Holothrix macowaniana* Rchb. f., *Mystacidium gracile* Harv., *Neobolusia ciliata* Summerh., *Oeceoclades pandurata* (Rolfe) Garay & P. Taylor, *Oligophyton drummondii* H. P. Linder & G. Will., *Platycoryne affinis* Summerh., *Polystachya pubescens* (Lindl.) Rchb.f., *Satyrium flavum* la Croix, *Schizochilus calcaratus* P. J. Cribb & la Croix, *Stenoglottis woodii* Schltr., *Tridactyle hurungweensis* W. Fibeck.

5 Uses of Orchidaceae Species in African Countries

Orchids are best known for their beautiful flowers which make them a resource of great economic importance in the horticultural industry.

Very little information related to the indigenous African ornamental orchids are available. However, several orchids from other continents or countries have been introduced and domesticated and constitute important sources of income for African horticulturalists. Among these species *Arachnis hookeriana* (Rchb.f.) Rchb., *Arachnis flos-aeris* (L.) Rchb.f. native to Asia, *Papilionanthe hookeriana* (Rchb.f.) Schltr. native to Malaysia and *Bletilla striata* (Thunb.) Rchb.f. from Japan, China, and Tibet. These species are cultivated and sold in Cameroon [25]. *Ansellia africana* Lindl. (Fig. 1), nicknamed the panther orchid because of its remarkable size of flowers and its yellow color streaked with brown, is an ornamental species in Gabon. This species is also called the queen of African orchids. Its range stretches from West Africa to South Africa. This large epiphytic orchid is arguably the most popular African orchid in the West. In Gabon, we can also see, *Eulophia bouliawongo* (Rchb.f.) J.Raynal (*Eulophia oedoplectron* Summerh.), another ornamental orchid, named the queen of marshes and coastal savannas prone to flooding. It is a large pink orchid over 2 m tall. Its flower stalk can produce up to 35 quite spectacular, deep pink flowers (https://www.lepratiquedugabon.com/les-orchidees-du-gabon/).

Some orchids are also consumed by several communities in Africa. The species of orchids consumed by people are different from one country to another. However, *Disa robusta* N.E.Br. (Fig. 2) and *Satyrium buchananii* Schltr. are eaten in several countries. In Cameroon (Central Africa), the underground organs of *Habenaria keayi* Summerh. and *Habenaria zambesina* Rchb.f. (Fig. 3) are eaten [26]. Orchids are also eaten in East Africa. In Malawi the tubers of several orchids are used for food. These are, among others, *Disa engleriana* Kraenzl. (Fig. 4), *Disa zombica*

Fig. 1 Image of *Ansellia africana* Lindl. (From Aké-Assi)

Fig. 2 Image of *Disa robusta* N.E.Br.

N.E.Br., *Habenaria walleri* Rchb.f., *Satyrium amblyosaccos* Schltr. [27]. In Tanzania these are *Brachycorythis pleistophylla* Rchb.f., *Disa erubescens* Rendle, *Disa ochrostachya* Rchb.f., *Eulophia schweinfurthii* Kraenzl., *Habenaria xanthochlora* Schltr., *Satyrium atherstonei* Rchb.f., *Disa robusta* N.E.Br., *Satyrium robustum* Schltr., *Satyrium sceptrum* Schltr., *Satyrium acutirostrum* Summerh., *Roeperocharis wentzeliana* Kraenzl. which are used for the preparation of several meals [28]. In Benin (West Africa), *Eulophia horsfallii* (Bateman) Summerh. (Fig. 5), *Habenaria cirrhata* (Lindl.) Rchb.f. (Fig. 6), and *Nervilia bicarinata* (Blume) Schltr. are used in the preparation of sauces [29].

An often-overlooked value of plants in the Orchidaceae family is their role in medicine. In many countries of Europe, America, Asia, and Africa, orchids have been used as traditional medicine for a very long time. Particularly in Africa, they are

Fig. 3 Image of *Habenaria zambesina* Rchb.f.

Fig. 4 Image of *Disa engleriana* Kraenzl.

used in human and animal medicine. In Benin, in human health, orchids are used to cure several diseases [29]. *Calyptrochilum christyanum* (Rchb.f.) Summerh. is used in the treatment of dysmenorrhea (painful periods), edema of the lower limbs, malaria, and liver disease (disease of the faith). This species is also used to speed up the walking of a baby. *Eulophia guineensis* Lindl. is known to cure edema of the lower extremities, cough, epigastralgia (stomach pain), and fever. *Habenaria schimperiana* Hochst. ex A. Rich. is used to cure visual disturbances. To treat fever, myalgia (muscle pain), urinary disorders, epigastralgia (stomach pain), people of Benin use *Nervilia bicarinata* (Blume) Schltr. *Nervilia kotschyi* (Rchb.f.) Schltr. is used to treat dysmenorrhea (painful periods), epigastralgia (stomach pain), and

Fig. 5 Image of *Eulophia horsfallii* (Bateman) Summerh.

Fig. 6 Image of *Habenaria cirrhata* (Lindl.) Rchb.f.

cough [29]. In Cameroon, orchids are also used to treat several diseases. *Angraecum angustipetalum* Rendle is used for bone fortification of children. It is also used to perform abortions. This species is also used against snakes. *Ancistrorhynchus serratus* Summerh. is used to treat diabetes. *Bulbophyllum fuscum* var. *melinostachyum* (Schltr.) J. J. Verm., *Bulbophyllum barbigerum* Lindl., *Bulbophyllum intertextum* Lindl., and *Bulbophyllum calyptratum* Kraenzl. are used to cure diseases such as side pain, otodynia (ear pain), burns, wounds, and dermatoses (skin diseases). *Cyrtorchis arcuata* (Lindl.) Schltr. (Fig. 7) is recommended in the treatment of diabetes and skin diseases.

Diaphananthe bidens (Afzel. ex Sw.) Schltr. (Fig. 8) is used for fertility. *Graphorkis lurida* (Sw.) Kuntze is used to treat coughs and tuberculosis. *Habenaria procera* (Afzel. ex Sw.) Lindl. is recommended for the purification of blood, gastritis, and arthritis. *Liparis nervosa* (Thunb.) Lindl. is used to treat ulcers and

Fig. 7 Image of *Cyrtorchis arcuata* (Lindl.) Schltr.

Fig. 8 Image of *Diaphananthe bidens* (Afzel. ex Sw.) Schltr. (From Swiss Center for Scientific Research)

burns. *Listrostachys pertusa* (Lindl.) Rchb.f. (Fig. 9) is recommended for epigastralgia (stomach pain), constipation, and measles. *Polystachya concreta* (Jacq.) Garay & H. R. Sweet, *Polystachya cultriformis* (Thouars) Lindl. ex Spreng., and *Polystachya caloglossa* Rchb.f. are used to treat rheumatism, arthritis, measles, and burns. *Tridactyle tridactylites* (Rolfe) Schltr. is used for the treatment of dementia or mental disorders [25].

In Benin, orchids are used spiritually. *Calyptrochilum christyanum* (Rchb.f.) Summerh. is used to attract good luck, the power of prophecy. It also has a vanishing power. *Eulophia guineensis* Lindl. fights the spirits of twins. *Nervilia bicarinata* (Blume) Schltr. (Fig. 10) is used for the power of prophecy. This species is also used to fight against the spirit of the deceased, witchcraft, and the spirits of twins [29]. In Cameroon, *Ansellia africana* Lindl. is used to ward off evil spirits and promote romantic relationships. *Angraecum birrimense* Rolfe, *Bulbophyllum lupulinum* Lindl., and *Bulbophyllum falcatum* (Lindl.) Rchb.f. are used to fight witchcraft. These species are also used to make predictions. *Calyptrochilum emarginatum* (Afzel. ex Sw.) Schltr. is used for seduction and luck [25].

Fig. 9 Image of *Listrostachys pertusa.* (From the Parc National des Île Ehotilés)

Fig. 10 Image of *Nervilia bicarinata* (Blume) Schltr. (From Sinfra)

In animal health, *Calyptrochilum christyanum* (Rchb.f.) Summerh. and *Habenaria cirrhata* (Lindl.) Rchb.f. are used to treat diseases in chickens [29].

6 Conclusion

Based on this review of the literature, the contribution of the sub-Saharan African orchid flora is estimated to be more than 5% of the world flora of orchids. It is clear that the number of species is underestimated taking into account the fact that several regions of the continent were not highly inventoried. Also, many species are epiphytes, thus difficult to be collected. Orchidaceae is a diverse family and important for communities for the diversity of its uses (ornamental, food, medicine, etc.). The fact that some regions are more studied than others should be highlighted. For instance, Central and East Africa are relatively well known compared to West Africa.

With the high rate of deforestation in Africa, several species could be threatened in the continent. Intensive prospection of understudied areas should be carried out to avoid the loss of many species. Also, it is important to assess the conservation status of the African orchids.

References

1. White F (1986) La végétation de l'Afrique. ORSTOM-UNESCO, Paris, p 391
2. Govaerts R, Cribb PJ, Wood J (2003) Monocot checklist. Royal Botanic Gardens, London
3. Geerinck D (1992) Orchidaceae (seconde partie). In: Bamps P (ed) Flore d'Afrique centrale (Zaïre, Rwanda, Burundi). Spermatophyte. Jardin Botanique National de Belgique, Meise
4. Perez-Vera F (2003) Les Orchidées de Côte d'Ivoire. Collection Parthénope. Biotope, Mèze, p 576
5. Lebrun JP, Stork AL (1995) Enumération des plantes à fleurs d'Afrique tropicale. Monocotylédones: Limnocharitaceae à Poaceae, vol 3. Conservatoire et Jardin Botaniques de la Ville de Genève, Geneva, p 341
6. Lebrun JP, Stork AL (1997) Enumération des plantes à fleurs d'Afrique tropicale. Gamopétales: Clethraceae à Lamiaceae, vol 4. Conservatoire et Jardin Botaniques de la Ville de Genève, Geneva, p 342
7. Hamisy WC (2007) Development of conservation strategies for the wild edible orchid in Tanzania. Progress report for the Rufford Small Grants Foundation. The Rufford Foundation, London
8. Owen J (2011) Kew scientists lead fight to save orchids from extinction. The Independent 15(4): 509–514
9. Lebrun JP, Stork AL (2015) Tropical African flowering plants. Ecology and distribution. Orchidaceae, part one (Genera A–G). Conservatoire et Jardin Botaniques de la Ville de Genève, Geneva
10. Lebrun JP, Stork AL (2017) Tropical African flowering plants. Ecology and distribution. Orchidaceae, part two (Genera H–Z), vol 10. Conservatoire et Jardin Botaniques de la Ville de Genève, Geneva, p 352
11. Schatz B, Geniez P (2001) Les orchidées, un patrimoine naturel à conserver. In: Le génie de la nature. Biotope, Mèze, p 47
12. Johansson D (1974) Ecology of vascular epiphytes in West African rain forest. Acta Phytogeogr Suec 59:1–136
13. Freudenstein JV, Chase MW (2015) Phylogenetic relationships in Epidendroideae (Orchidaceae), one of the great flowering plant radiations: progressive specialization and diversification. Ann Bot 115:665–681
14. Simo-Droissart M, Plunkett GM, Droissart V, Edwards MB, Farminhão JNM, Jecmenica V, D'Haijère T, Lowry PP II, Sonké B, Micheneau C, Carlsward BS, Azandi L, Verlynde S, Hardy OJ, Martos F, Bytebier B, Fischer E, Stévart T (2018) New phylogenetic insights toward developing a natural generic classification of African Angraecoid orchids (Vandeae, Orchidaceae). Mol Phylogenet Evol 126:241–249
15. Cribb PJ, Thomas S, Rasmussen FN (2001) The orchids of Ethiopia and Eritrea. Biol Skr 54: 75–84
16. Zhou Y, Liu B, Mbuni Y, Yan X, Mwachala G, Hu G, Wang Q (2017) Vascular flora of Kenya, based on the Flora of Tropical East Africa. PhytoKeys 90:113–126
17. Droissart V, Sonké B, Stévart T (2006) Les Orchidaceae endémiques d'Afrique centrale atlantique présentes au Cameroun. Syst Geogr Plants 76(1):3–84
18. Borokini TI (2014) A systematic compilation of endemic flora in Nigeria for conservation management. J Threat Taxa 6(11):6406–6426

19. Cribb PJ, Pérez-Vera F (1975) *Bulbophyllum ivorense* P. J.Cribb & Pérez-Vera. Asansonia 15:209
20. Figueiredo E, Smith GF (2008) Plants of Angola. Strelitzia 22:292
21. Morris B (1982) The orchids of Malawi. Soc Malawi J 35(2):7–23
22. White F, Dowsett-Lemaire F, Chapman JD (2001) Evergreen forest flora of Malawi. Royal Botanic Gardens, Kew, p 708
23. Darbyshire I, Timberlake J, Osborne J, Rokni S, Matimele H, Langa C, Datizua C, de Sousa C, Alves T, Massingue A, Hadj-Hammou J, Dhanda S, Shah T, Wursten B (2019) The endemic plants of Mozambique: diversity and conservation status. PhytoKeys 136:45–96
24. Mapaura A (2002) Endemic plant species of Zimbabwe. Nat Herb Bot Gard 18(1):117–149
25. Fonge BA, Essomo SE, Bechem TE, Tabot PT, Arrey BD, Afanga Y, Assoua EM (2019) Market trends and ethnobotany of orchids of Mount Cameroon. J Ethnobiol Ethnomed 15(29):1–11
26. Menzepoh SB (2011) Les orchidées comestibles chez le peuple Bagam au Cameroun. Biotechnologie. Agron Soc Environ 15(4):509–514
27. Kasulo V, Mwabumba L, Cry M (2009) A review of edible orchids in Malawi. J Hortic For 1(7): 133–139
28. Mapunda LND (2007), Edible orchids in Makete district, the Southern Highlands of Tanzania: distribution, population and status. Master thesis. Swedish Biodiversity 553 Centre, Uppsala Universitet, Uppsala, p 68
29. Assédé ESP, Djagoun CAMS, Azihou AF, Kouton MD, Gogan YSC, Geldenhuys CJ, Chirwa PW, Sinsin BA (2017) Folk perceptions and patterns of use of orchid species in Benin, West Africa. Flora et Vegetatio Sudano-Sambesica 20:26–36

Orchid Biodiversity and Genetics

5

Seeja G and Sreekumar S

Contents

Abstract

The angiosperm family Orchidaceae comprises ~900 genera and 30,000–35,000 species, which are distributed all over the world. Besides, more than two lakhs man made hybrids are available. The unparalleled unique characteristic features and adaptations to thrive on almost all habitats on the earth made this group of plants as a distinct one from the rest of the plants in the Plant Kingdom. The diversity in morphology, physiology, and genetic peculiarities induces large level of speciation with remarkable evolutionary significance. The constructions of orchid flowers are always a curiosity to both layman and scientists. The unique flower characters (phenotype) such as wide variations in floral morphology,

S. G
Department of Plant Breeding and Genetics, College of Agriculture, Ambalavayal, India
e-mail: seeja.g@kau.in

S. S (✉)
Biotechnology and Bioinformatics Division, KSCSTE- Jawaharlal Nehru Tropical Botanic Garden and Research Institute, Thiruvananthapuram, India
e-mail: drsreekumar@rediffmail.com

J.-M. Mérillon, H. Kodja (eds.), *Orchids Phytochemistry, Biology and Horticulture*, Reference Series in Phytochemistry, https://doi.org/10.1007/978-3-030-38392-3_2

colors, shapes, enchanting smells to attract pollinators are the expression of genes (genotype). The construction of orchid flowers as well as the efficiency in successfully carrying out the function for which they are intended and underlying genetics is briefly discussed.

Keywords

Orchid · Biodiversity · Genetics · MADS box · Phytomolecules · Pigments · Pollinium · Homeotic · Secondary metabolite

1 Introduction

Orchids are one of the natural wonders with exceptional characteristic features, not only amazing flowers with large diversity but also with remarkable evolutionary significance. It belongs to the angiosperm (flowering plants) family orchidaceae and consists of about 900 genera and 30000–35000 species. Many species are commercially exploited as ornamentals, medicinal, and food additives. For examples, many species under the genus *Dendrobium, Vanda, Cymbidium, Phalaenopsis, Oncidium, Cattleya, Paphiopedilum, Aranda, Renanthera* are highly remunerative floriculture/horticulture crops; *Vanda tessellate, Rhyncostylis retusa*, and *Dendrobium monticola* are used in Indian traditional medicine; *Gastrodia sesamoides* and *Gastridia falconeri* are used as food and vanillin derived from *Vanilla planifolia* as a food additive. In floriculture industry, orchid flowers rank top most position as cut flowers and potted plants. Considering the ornamental value, plant breeders have been explored its significant traits and produced over two lakhs hybrids, which are multiplied in large numbers by the application of tissue/seed culture method for flourishing the floriculture industry in world wide. However, illegal harvesting of the wild species for local, regional, and international trade is ever increasing that leads to the extinction of many orchid species which grow only in specific micro-agro-climatic zones or specific ecological niche and limited population. Besides, the rapid deforestation and urbanization/change in land use pattern results devastation of habitat with interacting factors like pollinating agents and mycorrhizal symbionts which became a great threat to orchid population and extinction of many species. Therefore, conservation of orchid biodiversity and sustainable utilization of wild species attain prime importance.

The incomparable amazing floral diversity of over 30,000 orchid species and its underlying genetic mechanism still remain as a mystery. Even though Orchids' unveiling species-specific variation in the floral whorl initiation and floral whorl development is a multifaceted one, its evolutionary schoolwork is worthy and underlying molecular data is relatively limited. The recently proposed "orchid code" theory explained the genetic mechanism of floral whorl and its diversity development in orchids [1]. The uniqueness and diversity is generated by the combinatorial collaboration of four *DEF*-like MADS-box genes groups with other floral homeotic genes. The changes in the expression of DEF-like or CYC-like genes create floral diversity and promote floral whorl development as the perception of evolutionary developmental biologist ("evo-devo") that the developmental genetic system can influence the direction and pace of evolutionary changes [1]. True changes in organ identity are rare events in the evolution of orchid flowers. Hence, the unprecedented developmental genetic predisposition that originated early in orchid evolution may be the cause of floral diversity in orchids [1].

2 Orchid Biodiversity Versus Adaptation

Orchids belonging to the family Orchidaceae consist of ~900 genera and 30,000–35,000 species with exceptional phenotypic and genotypic diversity. The database, "The Plantlist.org" currently provides accepted names of 27,801 species under 899 genera. Besides, hundreds of synonyms and unassessed species names are in the database for further clarification. Orchids are cosmopolitan in distribution and growing throughout the world from tropical to subtropical and temperate regions. Due to the specific ecological adaptations evolved through evolution, the members of orchidaceae have diverse habitat on earth, such as soil (terrestrial), for example, *Spathoglottis*, rock surfaces (lithophytic), for example, *Paphiopedilum*, and on other plants (epiphytic), for example, *Dendrobium*. Among the terrestrial forms, some species are associated with fungus as endotrophic or as ectotrophic. Majority of the species (70%) are epiphytes and some are saprophytes, for example, *Didymoplexis*. In order to thrive on diverse habitat, orchids have variety of designs and adaptations. Presence of velamen roots, thick, waxy, or leathery leaves, pseudobulbs, Crassulacean acid metabolism (CAM) photosynthesis mechanism, etc., are some of the adaptations found among the orchids to grow under water stress condition. Orchids are popular mainly due to the presence of unique and fascinating flowers with great complexity, incredible diversity, and long floral lifespan as cut flowers, which are unrivalled in the plant world. The unique floral characters such as wide variations in floral morphology, colors, shapes, enchanting smells to attract pollinators are expression of genes (genotype), and these all indicate high degree of speciation among orchids. It is interesting to note that in many orchids, the flowers are constructed in such a way to attract specific insect or bird to pollinate the flowers and in some others produce chemicals to attract specific pollinators. These all features establish floral and reproductive isolation that facilitate genetic diversity and consequent speciation.

The special characteristic features of the orchid flowers include except few species all others have zygomorphic flowers (bilateral symmetry) with trimerous arrangement, outer whorl of three sepals and inner whorl of three petals, both are alike in appearance and known as tepals. The posterior petal is modified to form labellum or lip and twisting of flower on its axis through 180° makes it to the anterior position such that it forms a convenient landing platform for pollinator insects. The labellum is distinctly modified and attractive in each species. It is variously cut, highly colored, may bear warts or out-growths and in various shapes. A spur developed either from the base of the lip or partly from it and axis which produces nectar for attracting the pollinators. The staminal filaments and style fused to form a cylindrical structure, gynostemium, or column. At the top of the column, pollinia are attached. A pollinium is a coherent mass of pollen grains put together by glue-like alkaloid viscidium, a combination of cellulosic strands and mucopolysaccharides. The pollinium has a caudicle with sticky disc, which may attach on the insects' legs when they visit the flower. The column serves the same functions of style and stigma. It has an opening toward the labellum side and stigma is inside the column. These features direct the pollinating insects to the flowers. A innumerable number of minute seeds facilitate the expression of genetic variability and dispersal of seeds in large proportion across geographical/ecological barriers. All these unique characteristic features of orchids facilitate evolution and speciation [2] leading to high diversity among orchids. The genetic mechanism involved in the evolution of different orchid genera and species, and diversity of flowers engenders keenness exclusively to orchid breeders. It led to the production of over 200,000 man-made orchid commercial hybrids through hybridization and that made orchids as a top-ranked cut flower in the global floriculture industry.

3 Floral Development and Genetics in Orchids

In reproductive biology, orchids exhibit distinctive evolutionary mechanism. One is the fused structure of androecium and gynoecium to form single column known as gynostemium which facilitates pollination by pollinators evolved simultaneously naturally through coevolution. Presence of protandry, pollinia, synchronization in microsporogenesis, pollination-triggered megagametogenesis, effective fertilization on entire length of placenta, and production of incalculable small immature embryos known as seeds especially without endosperm in dehiscent capsule are the special features evolved in orchids and its underlying mechanism; particularly the gene regulation processes are investigated by different molecular workers.

Generally at a particular stage of plant development due to the activities of flowering time genes, the apical meristem switches program from a leaf-producing to a flower-producing tissue and becomes an inflorescence meristem and forms numerous primitive primordia that later develops into sepal, petal, stamen, and carpel. This phenomenon is due to homeotic mutations in normal vegetative cells that lead to develop normal organs in inappropriate position. As in other flowering plants, orchids also exhibit two stages, floral transition and flower development. Floral transition is influenced by various factors such as juvenility, ambient temperature, and photoperiod. These factors play pivotal role in determining flower initiation time with respect to ontogeny and season [3]. In *Phalaenopsis* [4] and *Dendrobium* [5], temperature below 26 °C promoted flower induction, but in *Dendrobium* hybrids, namely *Dendrobium* Chao Praya Smile and *Dendrobium* Madame Thong-In, flower induction occurs in high temperature (30 °C) and in some others like *Cypripedium* species vernalization below 5 °C or subzero temperature is a prerequisite for flowering [6]. Photoperiod also regulates flower induction in certain orchids, for example, in *Phalaenopsis pulcherrima*, 30 °C day and 20 °C night condition, flowering is more efficient under 9 h light than 12 h light treatment

[3, 7]. Tropical regions are the treasure house of orchids where the day and night length are almost equal throughout the year; hence, the influence of photoperiod on orchid flowering may be insignificant. The significant effect of phytohormones on flower transition in orchids has been demonstrated, for example, in *Phalaenopsis* and *Doritaenopsis* (hybrid *Doritis* × *Phalaenopsis*) plants sprayed with 6-benzylaminopurine (BAP) produced inflorescence 3–9 days earlier [8]. Similarly, influence of synthetic cytokinins (BAP, gibberellic acid, abscisic acid) on orchid flower induction in in vitro has been demonstrated [3].

Flower transition and inflorescence architecture are modulated by two homologous proteins, FLOWERING LOCUS T (FT) and TERMINAL FLOWER 1 (TFL1). The FT promotes the transition to reproductive development and flowering, while TFL1 represses this transition [9]. FT and TFL1 showed ~60% amino acid sequence similarity but function in opposite manner [9–11]. Photoperiod is a primary trigger of FT expression. FT facilitates the transition to flowering in response to its induction by the transcription factor CONSTANS (CO) [9, 12, 13]. Temperature-responsive FT regulators, which target the FT promoter or noncoding regions, include SHORT VEGETATIVE PHASE (SVP) [13], PHYTOCHROME INTERACTING FACTOR 4 (PIF4) [14], LIKE HETEROCHROMATIN PROTEIN 1 (LHP1) [15], and FLOWERING LOCUS C (FLC) [9, 16]. Once the FT is inducted by CO, it will move from the leaves to shoot apical meristem (SAM) and binds to the bZIP transcription factor FD (basic region/leucine zipper motif) to form a complex that regulates meristem identity genes, resulting in flowering induction [9, 17, 18]. The meristem identity genes induced by the FT-FD complex are APETALA 1(AP1) and FRUITFULL(FUL) that reprogram the primordia to produce reproductive organs instead of vegetative ones [9, 19]. In addition to inducing FT, CO is suggested to induce, directly or indirectly, the expression of TERMINAL FLOWER 1(TFL1), which controls inflorescence meristem identity and delays the transition to the reproductive phase at the SAM [9, 20]. TFL1 represses the expression of several genes downstream of FT such as LEAFY(LFY) and AP1, perhaps by partnering with FD, with which it weakly interacts [9, 21]. The high sequence homology of TFL1 to FT suggests conserved biochemical action, but the action and regulation of these proteins at the molecular level remain unclear [9–11].

In orchids, flower transition study is limited. In *Phalaenopsis aphrodite*, FLOWERING LOCUS T1 *(*PaFT1*)* was isolated and functionally characterized [22]. Its expression depends on the ambient temperature and photoperiod has insignificant role [22]. In *Dendrobium nobile*, two genes homologous to FLOWERING LOCUS (FT) and MOTHER OF FT (MFT) were isolated and designated as DnFT and DnMFT, respectively. Expression and function of these genes in regulating the transition from vegetative to reproductive stage were studied [23]. In accordance with maturation, expression of DnFT was increased in leaves and decreased in buds, while DnMFT level was decreased in leaves and increased in buds. Low-temperature treatment enhanced DnFT expression in leaves and declined in buds when compared to normal. But low temperature treatment has insignificant effect on the expression of DnMFT in both leaves and buds [23]. In *Arabidopsis*, the overexpression of DnFT induced flowering, which indicated that DnFT may act as a promoter of flowering in *D. nobile* [23]. Ding et al. [24] isolated similar type of gene, that is, DOSC1 from *Dendrobium* Chao Praya Smile; DOSC1 has been particularly expressed in emerging floral meristems and created seven 35S:DOSOC1 transgenic *Dendrobium* orchid lines, which consistently exhibited earlier flowering than wild-type orchids, and it can be utilized for genetic manipulation of flowering time in orchids. DOAP1, an ortholog of AP1, was also identified and characterized from *Dendrobium* Chao Praya Smile [25], which also promote early flowering. Study on differential gene expression in shoot apical meristem (SAM) during floral transition in in vitro cultures of *Dendrobium* Madame Thong-In revealed that several transcription factors, including a MADS-box gene of the AP1/AGL2 family, a class I KNOX gene, and a homolog of the *Drosophila* SEVEN-UP gene, were differentially expressed in vegetative and transitional SAM. The KNOX gene plays an important role in the function of SAM and encodes a KNOTTED1-like homeobox (Knox) protein later designated as *Dendrobium* Orchid Homeobox 1 (DOH1) [3, 26]. Overexpression of antisense DOH1 resulted in early flowering in *Dendrobium* [3]. In *Oncidium*, high temperature (30 °C) promotes flowering. Both flowering promoter FT and repressor TERMINAL FLOWER 1 (TFL1) have been identified in *Oncidium* Gower Ramsey [3, 27]. The expression of OnFT is regulated by photoperiod but expression of OnTFL1 is not influenced by photoperiod. The OMADS1, a homolog of *Arabidopsis* AGL6, has also been identified in *Oncidium* Gower Ramsey [3] and its overexpression in *Oncidium* resulted in precocious flowering [3, 28]. According to Chin et al. [29], ascorbate/dehydroascorbate (AsA/DHA) redox ratio may act as one of the endogenous signals that induce the flowering of *Oncidium* in response to high ambient temperature [3]. Reduced GSH redox ratio caused by down-regulation of GSH metabolism-related genes such as glutathione reductase (GR1), γ-glutamylcysteine synthase (GSH1), and glutathione synthase (GSH2) was linked to the decrease in the AsA redo redox ratio for flowering of *Oncidium* orchid [3, 29]. A *FVE* homologue gene was isolated from Doritaenopsis "Tinny Tender" (*Doritaenopsis* Happy smile × Happy valentine) and designated as DhFVE [3, 30], which has pivotal role in flowering time and cold response regulation. Besides, EARLY FLOWERING 4 (EFL4) family genes, DhEFL2, DhEFL3, and DhEFL4, have been identified in *Dortiaenopsis* [3, 31]. In *Cymbidium* and *Erycina*, many putative flowering genes have also been identified through transcriptome analyses [3, 32, 33].

4 Orchid Flower Development and MADS Box Genes

In plants, homeotic genes are called as MADS box, which is a conserved sequence motif found in genes comprises of MADS-box gene family. The acronym MADS box is derived from the initials of four loci, MCMI of *Saccharomyces cerevisiae*, AG of *Arabidopsis thaliana*, DEF of *Antirrhinum majus*, and SRF of *Homo sapiens*, all of which contain the MADS-box domain, a conserved 56 amino acid DNA binding domain [34, 35]. To exemplify the functional activity of floral homeotic, various flower-development models such as ABC, ABCE, ABCED, and floral quartet have been suggested. Coen and Meyerowitz [36] described the genetic interactions controlling flower development in *Arabidopsis* and *Antirrhinum* and proposed ABC model. According to them, the genetic control of meristem behavior has two classes of genes: those control identity of meristems and those determine the identity of organs. The genes of both classes can be considered homeotic. A series of homeotic mutations affect the identity of floral organs, sepals, petals, stamens, and carpels. They define three regions of the floral meristem, each coincides with the domain of action of one of the three classes of floral homeotic genes. Region A comprises whorl 1 (sepal) and 2 (petal), region B comprises whorl 2 (petal) and 3 (stamen), and region C comprises whorl 3 (stamen) and 4 (carpel) [36]. The interactions of these genes express a variety of unique phenotypic expression. For example, genes acting in regions A, B, and C are required for three regulatory functions a, b, and c, respectively, then the combination of functions in the four whorls of wild type would be a, ab, bc, c. In principle, this might provide sufficient information to specify the identity of organs in each whorl [36]. That is, sepals form if a alone is expressed, a and b together direct petal development, b and c together specify stamens, and c expressed alone determines carpel formation [36]. Further genetic analysis in *Arabidopsis thaliana* that revealed the A function is mediated by APETALA1 (AP1) and APETALA2 (AP2), the B function by APETALA3 (AP3) and PISTILLATA (PI), and the C function by AGAMOUS (AG) [37]. All of these genes encode putative transcription factors [38, 39], suggesting that ABC genes may control the transcription of other genes ("target genes") whose products are directly or indirectly involved in the formation or function of floral organs. According to Irish [40] except for AP2, all ABC genes encode MIKC-type MADS-domain proteins [37].

The ABC model has some limitations since it is not sufficient for the speciation of floral organ identity. Subsequently, based on the studies in *Petunia* hybrid, ABC model was extended to an "ABCD" model by addition of a D function specifying ovule identity [37, 41]. In *Arabidopsis thaliana*, three genes closely related to AG, namely, SEEDSTICK (STK; formerly known as AGL11), SHATTERPROOF1 (SHP1; formerly known as AGL1), and SHATTERPROOF2 (SHP2; formerly known as AGL5) [42, 43], were identified as D-function genes; stk shp1 shp2 triple mutants are characterized by conversion of ovules into carpel-like or leaf-like structures [37, 43]. The C-function gene AGAMOUS was also considered as an additional class D gene [37].

The ABCD model was further extended to ABCDE model based on the study in *Arabidopsis thaliana* and *Antirrhinum* [44]. According to this model, the A genes APETALA1 (AP1) and APETALA2 (AP2) control sepal development [28] and together with B genes regulate the formation of petals (LIPLESS 1 & 2). The B genes PISTILLATA (PI) and APETALA3 (AP3), together with C genes, mediate stamen development (e.g., DEFICIENS (DEF) and GLOBOSA (GLO)). The C genes AGAMOUS determine the formation of carpel and the D genes SEEDSTICK and SHATTERPROOF specify the identity of the ovule. The E genes SEPALLATA express in the entire floral meristem and are necessary (SEP1, SEP2, SEP3, and SEP4) [34]. The controlling genes are A sepal and petal, B petal and stamen, C stamen and carpel, D ovule specification, and E all floral organs.

Both ABC and ABCDE models relied on genetic data which cannot properly explain the molecular mechanism of different floral homeotic genes' interaction. To overcome from this shortcoming instead of genes level, encoded proteins level is considered that led to a new model called as floral quartet model (FQM). According to the floral quartet model (FQM), the identity of the different floral organs is specified during development by quaternary (tetrameric) protein complexes composed of MIKC type MADS-domain proteins [37, 44]. These quartets are assumed to function as transcription factors by binding to the DNA of their target genes, which they either activate or repress to control the development of the respective floral organs [37, 44].

Floral development in orchids has been investigated based on the forgoing models by various authors. The ABC model [45] enumerated profiling of gene expression in the transition of flower development in orchids by in vitro flowering of *Dendrobium* species and later examined by the mRNA differential display method [46]. It revealed that genes involved in the regulation of transcription, cell division, and other metabolic processes are exhibited close association with transition process of flower development in orchids [46]. MADS-box genes have been identified in *Aranda* cv Deborah and *Dendrobium* cv Madame Thong-in [46, 47]. Gene expression investigation revealed that these genes have significant role in flowering.

5 Floral Whorl Development in Orchids – Class A MADS Box Genes

It includes the phylogenetically related MADS-box genes of class A and class E [34]. It involved in the floral meristem initiation and development and also determines floral organ development. Based on the ABCDE model, the class A and class E MADS-box genes are necessary for the correct development of sepals, petals, and floral meristem. Class A includes subfamily AP1/SQUA-like (from the APETALA1 and SQUAMOSA locus of *Arabidopsis thaliana* and *Antirrhinum majus*, respectively), and class E includes subfamily SEP-like (from the SEPALLATA locus of *A. thaliana*), subfamily SEP-like again includes SEP3, SEP1/2/4 (previously known as AGL 9 and AGL 2/3/4 clades, respectively), and AGL6 clades [34]. Class A genes have been identified and functionally characterized in many orchids such as *Dendrobium thyrsiflorum*, *Dendrobium* Madame Thong-In, *Oncidium* Gower Ramsey, and *Phalaenopsis amabilis*, and class E includes *Oncidium* Gower Ramsey and *Dendrobium* Madame Thong-In [34].

6 The Orchids Class B MADS-Box Genes

Based on the ABCDE model, the class B MADS-box genes are necessary for the correct development of petals and stamens and include two major lineages, the AP3/DEF-like genes (from the APETALA3 and DEFICIENS loci of *A. thaliana* and *A. majus*, respectively) and the PI/GLO-like genes (from the PISTILLATA and

GLOBOSA loci of *A. thaliana* and *A. majus*, respectively) [34]. Class B MADS-box orchid genes involve in the first whorl of floral organs that may be responsible for the development of petaloid sepals in orchids. In several orchids viz. *Cymbidium* hybrid cultivar, *Dendrobium crumenatum*, *Dendrobium moniliforme*, *Gongora galeata*, *Habenaria radiata*, *Oncidium* Gower Ramsey, *Phalaenopsis equestris*, *Phragmipedium longiflorum*, *Spiranthes odorata*, *Vanilla planifolia*, *Epipactis palustris*, and *Orchis italic*, class B genes have been functionally characterized [34].

7 The Orchid Class C and D MADS-Box Genes

Within the ABCDE model of flower development, the class C genes regulate the development of carpels and, together with the class B genes, of stamens. The class D genes are primarily involved in the development of ovules [34]. Class C includes AG-like and class D includes STK-like (SEEDSTICK) genes. Class C genes have been characterized from the orchid species viz. *Cymbidium ensifolium*, *Dendrobium crumenatum*, *Dendrobium thyrsiflorum*, and *Phalaenopsis* sp. Class D includes *Dendrobium crumenatum*, *Dendrobium thyrsiflorum*, and *Phalaenopsis* sp. [34].

In orchids according to the ABCDE model, class B gene involved in the development of outer whorl of sepals (tepals), inner whorl of petals (tepals), and third whorl known as stamens. But in the development of modified petal (labellum), gene involved has not been satisfactorily explained [34].

The recent theory known as "the orchid code" proposes an elegant model describing the development and evolution of the orchid perianth [1, 34, 48, 49]. This theory explained about the genetic code attributes to the class B AP3/DEF-like genes a pivotal role in tepal and lip identity and leaves unchanged the function of the class B PI/GLO-like genes and the functions of the A, C, D, and E class genes with respect to the modified ABCDE model [34].

In *Arabidopsis*, the identity of petals is realized through the interaction of one AP3/DEF-like and one PI/GLO-like gene product, and the orchid code theory suggests that the identity of orchid tepals and lips is determined by the interactions of the products of four paralogous AP3/DEF-like genes belonging to four different clades with the product of one PI/GLO-like gene. The orchid AP3/DEF-like genes are grouped into four well-defined clades: clade 1 (PeMADS2-like) is sister to clade 2 (OMADS3-like), while clade 3 (PeMADS3-like) is sister to clade 4 (PeMADS4-like). Each clade is characterized by a specific expression pattern [1, 34, 48, 49].

In orchid code theory, the interactions of the clade 1 and clade 2 gene products mediate the development of the outer tepals (whorl 1). The formation of the two lateral inner tepals (whorl 2) is specified by the interaction of high levels of the clade 1 and 2 and low levels of the clade 3 and 4 gene products, whereas the development of the lip, which is a highly modified inner tepal, is determined by the expression of high levels of the clade 3 and 4 gene products, in addition to low levels of clades 1 and 3 gene products. Thus, the expression of clade 3 genes differentiates between the inner and outer tepals, whereas the expression of clade 4 genes distinguishes between the two lateral inner tepals and the lip [1, 34, 48]. This proposed scheme can also explain the evolution of the zygomorphic orchid flower, starting from an actinomorphic flower composed of six nearly identical tepals in which the ancestor of the current AP3/DEF-like genes was equally transcribed. The duplication and evolution of different cis regulatory elements played a fundamental role in the functional diversification of the four AP3/DEF-like orchid clades. An initial duplication event produced the ancestor of the clade 1 and clade 2 genes and the ancestor of the clade 3 and clade 4 genes. At this stage, the evolution of a more specialized expression of the ancestor of the clade 3 and 4 genes, which was excluded from the outer tepals, might have established an intermediate flower structure, with distinctive outer and inner tepals. After a second duplication round, clade 3 and clade 4 genes differentiated, and the modularization of their expression led to the evolution of the lip [34, 48, 49]. These are well explained through the EVO/DEVO molecular approach [34].

Similar to MADS-box genes, class 1 knox genes are transcription factors involved in floral development. During flower development, there is interaction between these two genes [26]. In cv Madame Thong-In, down-regulation of the expression of DOH1 gene, a class 1 knox gene, causes multiple shoot apical meristem formation and early flowering, in association with the expression of DOMADS 1 which is a MADS-box gene involved in the floral transition of orchids [45, 46]. But in a wild-type orchid plant, DOMADS 1 gene expression in shoot apical meristem during floral transition is associated with a marked reduction in DOH1 gene and later both type genes are located at the same region in the inflorescence meristem and the developing floral primordia [45, 46]. Compared to other flowering plants, unique and different developmental programs may be present in orchids due to the highly evolved floral structures, which are being investigated [46]. In orchids MADS-box genes, DOMADS2 and DOMADS3 have shown novel expression patterns in the shoot apical meristem during floral transition [45, 46, 50].

8 Ovule Development

Unlike other flowering plants, in orchids ovule maturation is elicited by pollination. Genes associated with ovule development have been investigated in *Phalaenopsis* spp. Isolation and characterization of several genes involved in ovule differentiation provided that similar function of homologues in *Arabidopsis* also. *Phalaenopsis* O39 gene is a member of a new class of plant home box transcription factors designated HD-GL2 and expressed in ovule from primordium formation at early stages to various late stages of ovule differentiation. So it is an important regulator gene involved in the ovule initiation and development [46, 51]. Role of ethylene in ovary maturation and ovule differentiation in orchids has been extensively investigated by both physiological and molecular methods [46, 51, 52]. Even though there is coordinated regulation of ACC synthase and ACC oxidase gene expression in the ovary [52], the molecular events triggered by pollen-pistil interactions leading to these gene expressions are not well understood [46].

9 Perianth Senescence/Development

Wilting and drying up of calyx and corolla after pollination is a key indication of successful pollination in orchids. The exact reason for this senescence is explained as the resources are directed for ovule development and embryogenesis after fertilization. The physiological and molecular mechanisms of pollination-induced senescence have been studied in orchid species *Phalaenopsis* and *Dendrobium*, regarding ethylene sensitivity and production. Initial event triggering is an increase in ethylene production and the consequent physiological changes of flower [46, 53]. The identification of sensitivity factors such as GTP binding protein [54], short-chain saturated fatty acids [55], and auxin [56] gives way to the elucidation of the mechanisms under the regulation of ethylene sensitivity [46]. In orchid flowers, after pollination, there is an increase in ethylene production. In *Phalaenopsis* spp., the production of abundant ethylene in the perianth up to 72 h after pollination was observed but not the accumulation of ACC synthase. But ACC oxidase expression is up-regulated in the petals and sepals about 48 h after pollination in parallel with the onset of perianth senescence [46]. Both ACC synthase and ACC oxidase are positively regulated by increased ethylene production [46, 52] in orchids.

In some *Phalaenopsis* species (subdivision stauroglottis), the sepals and petals turn green and photosynthetic following successful pollination. These organs become leaf-like and provide photosynthates for the developing ovules/ovary and the embryos subsequent to fertilization over an extended period of many months until the capsules are mature. The molecular genetics for this transformation of the perianth from an energy sink to an energy source during postpollination development is interaction of ABC genes [46, 57].

10 Genetic Variation Versus Secondary Metabolite and Adaptations

Some sudden or gradual changes in traits occurred in individuals, like mutation, as well as the influence of various types of stresses may induce alteration in plant metabolism which leads to the production of various types of plant secondary compounds and those phytomolecules help to overcome the adverse effects or ensure lively functions. Secondary compounds impart flower color, volatile fragrance, food flavor, attract pollinators, interaction with symbiotic microorganisms, and tolerance against pest and diseases (isofalvonoids and phenylpropanoid derivatives). These are also take part in the mechanism of frost tolerance, nutrient storage, structural reinforcement, photo protective, UV–Vis absorption, signaling to mutualist, and in the adaptation of the plant under various environmental conditions and stress [58, 59, 60]. Besides, these phytomolecules have medicinal properties and majority of the currently used lifesaving drugs are derived from the natural source like plants. Production of the metabolites by the plants is regarded an adaptive capacity to cope up with stressful constraints during challenging and changing environment of growth that may involve production of complex chemical types and interactions in the structural and functional stabilization through signaling processes and pathways [61, 62].

The unique adaptations evolved in orchids are a curiosity among the biologist but the underlying mechanism is still unclear. For over a century, it has been debated whether adaptations are likely caused by a large number of mutations of small phenotypic effect or by very few genetic changes of large effect [63]. The vegetative modifications among the orchids to thrive as lithophytes and epiphytes or under drought condition might cause many changes in physiology and biochemistry of plants, including arrest of cell growth and alteration of photosynthesis mechanism (CAM plant) with an enhanced respiration [62, 64], which thus may affect biosynthetic pathways for the production of PSMs through provision of precursors or intermediates from the primary metabolism [62]. Many authors have reported that plants under drought condition produced higher rate of different categories of plant secondary metabolites such as terpenes, complex phenols, and alkaloids during in vitro and in vivo growth through the induction of ionic or osmotic stress [62, 65–69]. Increasing antioxidant enzyme activities and osmolytes play an important role in protecting plants under drought stress [62]. In many orchids, the foregoing phytochemicals are reported, for example, alkaloids, triterpenoids, tannins, phenols, steroids, flavonoids have been reported from the leaves of *Dendrobium panduratum* subsp. villosum [70]. In *Dendrobium ochreatum*, presence of glycosides, alkaloids, flavonoids, tannins, and phytosterol were reported by Banerjee et al. [71]. Leaves and roots of six *Phalaenopsis* hybrids, namely Green Field Sweet Valentine "Montclair," Sakura Hime, Sogo Yukidian "V3," Chian Xen Queen, Fusheng's Bridal Dress "Meidarland," and Younghome Golden Leopard "Peachy," contain high amounts of phenolic compounds with strong antioxidant activity. Ferulic acid, p-coumaric acid, and sinapic acid were accumulated largely in the roots in comparison with the leaves of these hybrids [72]. Hence, the root extracts of these hybrids may be used as a potential source of natural antioxidants [72]. In *Phalaenopsis* Sogo Yukidian "V3" roots, stems and leaves are the rich source of phenolics, flavonoids, and antioxidants, and also isolated five phenolic compounds including caffeic acid, syringic acid, vanillin, ellagic acid, and cinnamic acid and found that the plant parts have potent antioxidant activity and may be utilized as an efficient and safe antioxidant source [73]. In five different species of *Dendrobium*, namely, *D. nobile*, *D. fimbriatum*, *D. moschatum*, *D. chrysanthum*, and *D. chrysotoxum*, content of terpenoids (ursolic acid, β-sitosterol, and lupeol) and a phenolic compound (gallic acid) were analyzed in different plant parts such as stem, leaves, and roots, and it was found that all samples were good source of β-sitosterol with maximum content in *D. fimbriatum* stem and minimum content in *D. chrysanthum* roots. *D. nobile* (roots and stem) and *D. moschatum* (stem) were also found to be a source of ursolic acid and lupeol, respectively [74]. Perusal of the literature revealed that about 100 compounds from 42 *Dendrobium* species including 32 alkaloids, 6 coumarins, 15 bibenzyls, 4 fluorenones, 22 phenanthrenes, and 7 sesquiterpenoids constituents have been reported [75]. Williams [76] conducted a major survey on 142 orchid species belonging to 75 genera and found that the most common constituents were flavone C-glycoside and flavonols [75].

The diversity of floral color, shape, and fragrance is the key factor that makes the orchid flowers a distinctive one. In most species, the sepals are just uniformly green and do not contribute to interesting color patterns. Orchids are exceptional in that the sepals and the lip are usually as colorful as the petals, which results in an unbalanced pigment distribution among different segments of the perianths and lip to show various flower color patterns [77]. The flower color is mainly attributed by pigmentation such as flavonoids, carotenoids, and betalains. Chlorophyll also plays a significant role in floral whorl pigmentation in orchids. Betalains are red colored pigments substituted by anthocyanins exclusively in the Angiosperm order Caryophyllales and certain fungus belonging to the group Basidiomycetes. These pigments are structurally and biosynthetically distinct from flavonoids and anthocyanins [78]. In orchids, presence of betalains is not so far reported. Flavonoids are low molecular weight natural compounds with varying phenolic structures. They are the major floral pigments which provide a wide spectrum of coloration. There are about 6000 flavonoids that contribute to the colorful pigments of fruits, herbs, vegetables, and medicinal plants [79]. Among flavonoids, anthocyanin belongs to the red series and controls pink to blue-violet flower colors. Other flavonoids belong to the pure yellow series, among which chalcone and aurone are deep yellow, and flavones, flavonols, and flavanones are light yellow or nearly colorless [80].

The initial step of flavonoid synthesis is the condensation of three acetate units and a hydroxycinnamic acid unit and formed chalcone, the key intermediate in the synthesis of flavonoids [81]. Chalcone itself acts as a pigment having yellow or orange in color and it may be converted into bright yellow aurone. Usually, however, chalcone is modified to a colorless flavanone, and flavanone may be directly converted into flavones, which vary in color from very pale to bright yellow, depending on their degree of hydroxylation. Alternatively, flavanone can be converted into dihydro flavonol, which can then be modified by flavonol synthase to various flavonols. The flavonols are usually colorless, but act as co-pigments, stabilizing and modifying the color of other pigment molecules. Alternatively, dihydroflavonols may be modified through a number of steps to make anthocyanins [82]. In orchids, anthocyanins are the widely distributed pigments and produce red, pink, purple, black, and blue coloration. Carotenoids are lipophilic isoprenoid compounds responsible for yellow, orange, and red coloration; they also contribute to brown and bronze hues in combination with anthocyanin. Although many study reports are available for describing the role of pigments and floral coloration in many crops, the study in this line in orchids is limited.

The main pigments (anthocyanin, beta-carotene, and chlorophyll) in the petals of six different orchid species such as Mokara Pink, Mokara Aranda, Mokara Gold Nugget, *Ascocenda* Dong Tarn, *Dendrobium* Sonia 17, and *Dendrobium* Shavin White and the phenylalanine ammonia-lyase (PAL) activity were investigated [83]. Petals that have intense color have high amount of anthocyanin content, whereas those pale in color have high amount of chlorophyll content. PAL activity was shown to be positively correlated with the total anthocyanin content in the orchid flower petals [83] since anthocyanin is synthesized throughout phenylpropanoid pathway which is mainly catalyzed by PAL [84]. Matsui and

Nakamura [85] studied the distribution pattern of flower pigments and shape of epidermal cells in perianth tissues in 68 orchid species and reported that the yellow, orange, and purple color of the flowers depend on the differential distribution pattern of anthocyanins and carotenoids in the epidermal and parenchymatous cells. Red flowers were ascribed to the coexistence of carotenoids and anthocyanins. Understanding of pigmentation and its inheritance patterns may contribute to the breeding programs [85].

Besides flower color, nectar and volatile compounds produced from the floral secretary glands/cells and pollen play crucial role in the reproductive biology by attracting pollinators. Nectar and pollen are the food resources of various insects, which consist of sugar, amino acids, sterols, lipids, and vitamins. Other minor secondary metabolites such as organic acids, terpenes, alkaloids, flavonoids, and glycosides are also present at various level [86]. The specificity of floral scent emitted by different species of flowers epitomises key floral signals to the particular insect to direct towards rewarding flower species.

Genes that encode enzymes of the flavonoid pathway affect flower pigmentation in orchids [87, 88]. These genes include chalcone synthase (CHS), chalcone isomerase (CHI), flavanone-3-hydroxylase (F3H), flavonoid 3′-hydroxylase (F3′H) and flavonoid 3′,5′-hydroxylase (F3′5′H), dihydroflavonol 4-reductase (DFR), anthocyanidin synthase (ANS), UDP-glucose: flavonoid 3-o-glucosyltransferase (UFGT) and methyl transferase (MT) [80]. For examples, based on homology analysis of the *Phalaenopsis* genome, a total of 49 genes which include all above category were identified [89]. DFR genes were cloned from *Freesia hybrid* [90], *Bromheadia finlaysoniana* [87], *Cymbidium* "Rosannagirl Mild" [88], *Oncidium* Gower Ramsey [91], *Dendrobium* [92], *Dendrobium* Sonia "Earsakul" and *Dendrobium* Red Bull [93], *Dendrobium helix* x cv. Pomeo Brown hybrid [94]. DFR, CHS, and F3'5'H genes were isolated from *Dendrobium moniliforme* [95], and F3H genes were isolated from *Ascocenda* orchid, which showed homology to F3H from *Bromheadia finlaysoniana* [96]. Identification of additional regulatory genes in the flavonoid pathway is essential. Further investigations of their regulatory mechanisms and pathways will provide strategies toward genetic engineering of color in orchids [57].

Orchid flowers are unique for their specific patterns of colors in sepals, petals, and the modified dorsal petals (lips). These are discrete spots, streaks, or blotches. The patterns on the lips of some species are strikingly contrasting to serve as the landing platform for insect pollinators. The regulation of pigmentation is refined to specific cells of the different floral organs, and the expression of genes involved in flavonoid synthesis is the initial steps in the complex regulation of pigmentation [57].

11 Conclusion

The unparalleled adaptive features and evolutionary developments in orchids particularly to thrive under adverse environmental conditions and its instinctive genetic mechanism evolved through evolutionary process to survive in different habitats are always generate curiosity and give significant insights to the biologist. Knowledge about underlying genetic mechanisms behind the vegetative and reproductive biology particularly in floral development, pollination mechanism, seed production, dissemination, and germination is meager and in-depth studies in this line certainly ensure novel insights in evolutionary biology of this unique ornamental crop.

References

1. Mondragon-Palomino M, Theissen G (2009) Why are orchid flowers so diverse? Reduction of evolutionary constraints by paralogues of class B floral homeotic genes. Ann Bot (Lond) 104: 583–594
2. Schiestl FP, Schluter PM (2009) Floral isolation, specialized pollination, and pollinator behavior in orchids. Annu Rev Entomol 54:425–446
3. Wang SL, Viswanath KK, Tong CG, An HR, Jang S, Chen FC (2019) Floral induction and flower development of orchids. Front Plant Sci 10:1258
4. Blanchard MG, Runkle ES (2006) Temperature during the day, but not during the night, controls flowering of *Phalaenopsis* orchids. J Exp Bot 57(15):4043–4049
5. Campos KO, Kerbauy GB (2004) Thermoperiodic effect on flowering and endogenous hormonal status in *Dendrobium* (Orchidaceae). J Plant Physiol 161(12):1385–1387
6. Zhang Y, Zhao S, Liu D, Zhang Q, Cheng J (2014) Flowering phenology and reproductive characteristics of *Cypripedium macranthos* (Orchidaceae) in China and their implication in conservation. Pak J Bot 46(4):1303–1308
7. Wang WY, Chen WS, Huang KL, Huang L-S, Chen WH, Su WR (2003) The effects of day length on protein synthesis and flowering in *Doritis pulcherrima*. Sci Hortic 97:49–56
8. Blanchard MG, Runkle ES (2008) Benzyladenine promotes flowering in *Doritaenopsis* and *Phalaenopsis* orchids. J Plant Growth Regul 27(2):141–150
9. Wickland DP, Hanzawa Y (2015) The FLOWERING LOCUS T/TERMINAL FLOWER 1 gene family: functional evolution and molecular mechanisms. Mol Plant 8:983–997
10. Hanzawa Y, Money T, Bradley D (2005) A single amino acid converts a repressor to an activator of flowering. Proc Natl Acad Sci U S A 102:7748–7753
11. Ahn JH, Miller D, Winter VJ, Banfield MJ, Lee JH, Yoo SY, Henz SR, Brady RL, Weigel D (2006) A divergent external loop confers antagonistic activity on floral regulators FT and TFL1. EMBOJ 25:605–614
12. Suarez-Lopez P, Wheatley K, Robson F, Onouchi H, Valverde F, Coupland G (2001) CONSTANS mediates between the circadian clock and the control of flowering in *Arabidopsis*. Nature 410:1116–1120
13. Lee JH, Yoo SJ, Park SH, Hwang I, Lee JS, Ahn JH (2007) Role of SVP in the control of flowering time by ambient temperature in *Arabidopsis*. Genes Dev 21:397–402
14. Kumar SV, Lucyshyn D, Jaeger KE, Alo's E, Alvey E, Harberd NP, Wigge PA (2012) Transcription factor PIF4 controls the thermos-sensory activation of flowering. Nature 484: 242–245
15. Adrian J, Farrona S, Reimer JJ, Albani MC, Coupland G, Turck F (2010) Cis-regulatory elements and chromatin state coordinately control temporal and spatial expression of FLOWERING LOCUS T in *Arabidopsis*. Plant Cell 22:1425–1440
16. Searle I, He Y, Turck F, Vincent C, Fornara F, Krober S, Amasino RA, Coupland G (2006) The transcription factor FLC confers a flowering response to vernalization by repressing meristem competence and systemic signaling in *Arabidopsis*. Genes Dev 20:898–912
17. Abe M, Kobayashi Y, Yamamoto S, Daimon Y, Yamaguchi A, Ikeda Y, Ichinoki H, Notaguchi M, Goto K, Araki T (2005) FD, a bZIP protein mediating signals from the floral pathway integrator FT at the shoot apex. Science 309:1052–1056

18. Wigge PA, Kim MC, Jaeger KE, Busch W, Schmid M, Lohmann JU, Weigel D (2005) Integration of spatial and temporal information during floral induction in *Arabidopsis*. Science 309:1056–1059
19. Amasino R (2010) Seasonal and developmental timing of flowering. Plant J 61:1001–1013
20. Simon R, Igeno MI, Coupland G (1996) Activation of floral meristem identity genes in *Arabidopsis*. Nature 384:59–62
21. Hanano S, Goto K (2011) Arabidopsis TERMINAL FLOWER1 is involved in the regulation of flowering time and inflorescence development through transcriptional repression. Plant Cell 23: 3172–3184
22. Jang S, Choi SC, Li HY, An G, Schmelzer E (2015) Functional characterization of *Phalaenopsis aphrodite* flowering genes PaFT1 and PaFD. PLoS One 10(8):e0134987
23. Li R, Wang A, Sun S, Liang S, Wang X, Ye Q et al (2012) Functional characterization of FT and MFT ortholog genes in orchid (*Dendrobium nobile* Lindl.) that regulate the vegetative to reproductive transition in *Arabidopsis*. Plant Cell Tissue Organ Cult 111(2):143–151
24. Ding L, Wang Y, Yu H (2013) Overexpression of DOSOC1, an ortholog of *Arabidopsis* SOC1, promotes flowering in the orchid *Dendrobium* Chao Parya Smile. Plant Cell Physiol 54(4):595–608
25. Sawettalake N, Bunnag S, Wang Y, Shen L, Yu H (2017) DOAP1 promotes flowering in the orchid *Dendrobium* Chao Praya Smile. Front Plant Sci 8:400
26. Yu H, Yang SH, Goh CJ (2000) DOH1, a class 1 knox gene, is required for maintenance of the basic plant architecture and floral transition in orchid. Plant Cell 12(11):2143–2160
27. Hou CJ, Yang CH (2009) Functional analysis of FT and TFL1 orthologs from orchid (*Oncidium* Gower Ramsey) that regulate the vegetative to reproductive transition. Plant Cell Physiol 50(8): 1544–1557
28. Hsu HF, Huang CH, Chou LT, Yang CH (2003) Ectopic expression of an orchid (*Oncidium* Gower Ramsey) AGL6-like gene promotes flowering by activating flowering time genes in *Arabidopsis thaliana*. Plant Cell Physiol 44(8):783–794
29. Chin DC, Hsieh CC, Lin HY, Yeh KW (2016) A low glutathione redox state couples with a decreased ascorbate redox ratio to accelerate flowering in *Oncidium* orchid. Plant Cell Physiol 57(2):423–436
30. Sun X, Qin Q, Zhang J, Zhang C, Zhou M, Paek KY et al (2012) Isolation and characterization of the FVE gene of a *Doritaenopsis* hybrid involved in the regulation of flowering. Plant Growth Regul 68(1):77–86
31. Chen W, Qin Q, Zhang C, Zheng Y, Wang C, Zhou M et al (2015) DhEFL2, 3 and 4, the three EARLY FLOWERING4-like genes in a *Doritaenopsis* hybrid regulate floral transition. Plant Cell Rep 34(12):2027–2041
32. Li X, Luo J, Yan T, Xiang L, Jin F, Qin D et al (2013) Deep sequencing-based analysis of the *Cymbidium ensifolium* floral transcriptome. PLoS ONE 8(12):e85480. https://doi.org/10.1371/journal.pone.0085480
33. Lin CS, Hsu CT, Liao DC, Chang WJ, Chou ML, Huang YT et al (2016) Transcriptome-wide analysis of the MADS-box gene family in the orchid Erycina pusilla. Plant Biotechnol J 14(1): 284–298. https://doi.org/10.1111/pbi.12383
34. Aceto S, Gaudio L (2011) The MADS and the beauty: genes involved in the development of orchid flowers. Curr Genomics 12(5):342–356
35. Schwarz-Sommer Z, Huijser P, Nacken W, Saedler H, Sommer H (1990) Genetic control of flower development by homeotic genes in *Antirrhinum majus*. Science 250:931–936
36. Coen ES, Meyerowitz EM (1991) The war of the whorls: genetic interactions controlling flower development. Nature 353:31–37
37. Theißen G, Melzer R, Rümpler F (2016) MADS-domain transcription factors and the floral quartet model of flower development: linking plant development and evolution. Development 143(18):3259–3271
38. Yanofsky MF, Ma H, Bowman JL, Drews GN, Feldmann KA et al (1990) The protein encoded by the *Arabidopsis* homeotic gene agamous resembles transcription factors. Nature 346:35–39

39. Ó'Maoiléidigh DS, Graciet E, Wellmer F (2014) Gene networks controlling *Arabidopsis thaliana* flower development. New Phytol 201:16–30
40. Irish VF (2010) The flowering of *Arabidopsis* flower development. Plant J 61:1014–1028
41. Angenent GC, Colombo L (1996) Molecular control of ovule development. Trends Plant Sci 1: 228–232
42. Favaro R, Pinyopich A, Battaglia R, Kooiker M, Borghi L, Ditta G, Yanofsky MF, Kater MM, Colombo L (2003) MADS-box protein complexes control carpel and ovule development in *Arabidopsis*. Plant Cell 15:2603–2611
43. Pinyopich A, Ditta GS, Savidge B, Liljegren SJ, Baumann E, Wisman E, Yanofsky MF (2003) Assessing the redundancy of MADS-box genes during carpel and ovule development. Nature 424:85–88
44. Theißen G (2001) Development of floral organ identity: stories from the MADS house. Curr Opin Plant Biol 4:75–85
45. Yu H, Goh C (2000) Differential gene expression during floral transition in an orchid hybrid *Dendrobium* Madame Thong-In. Plant Cell Rep 19:926–931
46. Yu H, Goh CJ (2002) Molecular genetics of reproductive biology in orchids. Plant Physiol 127: 1390–1393
47. Lu ZX, Wu M, Loh CS, Yeong CY, Goh CJ (1993) Nucleotide sequence of a flower-specific MADS box cDNA clone from orchid. Plant Mol Biol 23:901–904
48. Mondragon-Palomino M, Theissen G (2008) MADS about the evolution of orchid flowers. Trends Plant Sci 13:51–59
49. Mondragon-Palomino M, Theissen G (2011) Conserved differential expression of paralogous DEFICIENS- and GLOBOSA-like MADS-box genes in the flowers of Orchidaceae: refining the 'orchid code'. Plant J 66:1008–1019
50. Yu H, Goh CJ (2000) Identification and characterization of three orchid MADS-box genes of the AP1/AGL9 subfamily during floral transition. Plant Physiol 123(4):1325–1336
51. Nadeau JA, Zhang XS, Li J, O'Neill SD (1996) Ovule development: identification of stage-specific and tissue-specific cDNAs. Plant Cell 8:213–239
52. O'Neill SD, Nadeau JA, Zhang XS, Bui AQ, Halevy AH (1993) Interorgan regulation of ethylene biosynthetic genes by pollination. Plant Cell 5:419–432
53. Porat R, Halevy AH, Serek M, Borochov A (1995) An increase in ethylene sensitivity following pollination is the initial event triggering an increase in ethylene production and enhanced senescence of *Phalaenopsis* orchid flower. Physiol Plant 93:778–784
54. Porat R, Borochov A, Halevy AH (1994) Pollination induced senescence in *Phalaenopsis* petals. Relationship of ethylene sensitivity to activity of GTP-binding proteins and protein phosphorylation. Physiol Plant 90:679–684
55. Halevy AH, Porat R, Spiegelstein H, Borochov A, Botha L, Whitehead CS (1996) Short chain saturated fatty acid in the regulation of pollination-induced ethylene sensitivity of *Phalaenopsis* flowers. Physiol Plant 97:469–474
56. Zhang XS, O'Neill SD (1993) Ovary and gametophyte development are coordinately regulated by auxin and ethylene following pollination. Plant Cell 5:403–418
57. Yu H, Goh CJ (2001) Molecular genetics of reproductive biology in orchids. Plant Physiol 127: 1390–1393
58. Neelam SS, Jyoti SJ, Sandhya T, Bhawana M, Lokesh KN, Rajender SS (2018) Plant metabolic engineering. In: Barh D, Azevedo V (eds) Omics technologies and bio-engineering towards improving quality of life, volume 2: microbial, plant, environmental and industrial technologies. Academic, Elsevier, London, pp 143–175. https://doi.org/10.1016/B978-0-12-815870-8.00009-7
59. Singer AC, Crowley DE, Thompson IP (2003) Secondary plant metabolites in phytoremediation and biotransformation. TRENDS Biotechnol 21(3):123–130
60. Verpoorte R, Contin A, Memelink J (2002) Biotechnology for the production of plant secondary metabolites. Phytochem Rev 1:13–25
61. Edreva A, Velikova V, Tsonev T et al (2008) Stress-protective role of secondary metabolites: diversity of functions and mechanisms. Gen Appl Plant Physiol 34(1–2):67–78

62. Isah T (2019) Stress and defense responses in plant secondary metabolites production. Biol Res 52:39
63. Xu S, Schlüter PM, Grossniklaus U, Schiestl FP (2012) The genetic basis of pollinator adaptation in a sexually deceptive orchid. PLoS Genet 8(8):e1002889. https://doi.org/10.1371/journal.pgen.1002889
64. Mashilo J, Odindo AO, Shimelis HA, Musenge P, Tesfay SZ, Magwaza LS (2017) Drought tolerance of selected bottle gourd [*Lagenaria siceraria* (Molina) Standl.] landraces assessed by leaf gas exchange and photosynthetic efficiency. Plant Physiol Biochem 120:75–87
65. de Matos NJ, Bertodo LOO, Da Rosa LMG, Von Poser GL, Rech SB (2014) Stress induction of valuable secondary metabolites in *Hypericum polyanthemum* acclimatized plants. South Afr J Bot 94:182–189
66. Niinemets Ü (2015) Uncovering the hidden facets of drought stress: secondary metabolites make the difference. Tree Physiol 36(2):129–132
67. Quan NT, Anh LH, Khang DT et al (2016) Involvement of secondary metabolites in response to drought stress of rice (*Oryza sativa* L.). Agriculture 6(2):23
68. Afzal SF, Yar AK, Ullah RH et al (2017) Impact of drought stress on active secondary metabolite production in *Cichorium intybus* roots. J Appl Environ Biol Sci 7(7):39–43
69. Piasecka A, Sawikowska A, Kuczyńska A et al (2017) Drought-related secondary metabolites of barley (*Hordeum vulgare* L.) leaves and their metabolomic quantitative trait loci. Plant J 89 (5):898–913
70. Johnson M, Janakiraman N (2013) Phytochemical and TLC studies on stem and leaves of the orchid *dendrobium panduratum* subsp. villosum Gopalan & AN Henry. Indian J Nat Prod Resour 4:250–254
71. Banerjee J, Chauhan N, Dey BK (2018) Pharmacognostical, physiochemical and phytochemical evaluation of leaf, stem and root of orchid *Dendrobium ochreatum*. J Appl Pharma Res 6(1):16–25
72. Minh TN, Khang DT, Tuyen PT, Minh LT, Anh LH, Quan NV, Ha PT, Quan NT, Toan NP, Elzaawely AA, Xuan TD (2016) Phenolic compounds and antioxidant activity of *Phalaenopsis* orchid hybrids. Antioxidants (Basel) 5(3):31. https://doi.org/10.3390/antiox5030031
73. Minh TN, Tuyen PT, Khang DT, Quan NV, Ha PTT, Quan NT, Andriana Y, Xinyan F, Van TM, Khanh TD, Xuan TD (2017) Potential use of plant waste from the moth orchid (*Phalaenopsis* Sogo Yukidian "V3") as an antioxidant source. Foods 27(6–10):85
74. Sunita S, Suman K, Deepti G, Mayuresh J, Prasenjit P, Nasheman K (2015) Variation in the marker content of five different *Dendrobium* species: comparative evaluation using validated HPTLC technique. J App Pharm Sci 5(10):32–38
75. Lam Y, Tzi Bun NG, Yao RM, Shi J, Xu K, Sze SCW, Zhang KY (2015) Evaluation of chemical constituents and important mechanism of pharmacological biology in *Dendrobium* plants. Evid Based Complement Alternat Med. https://doi.org/10.1155/2015/841752. Accessed 10 July 2021
76. Williams (1979) The leaf flavonoids of the Orchidaceae. Phytochemistry 18(5):803–813
77. Li BJ, Zheng BQ, Wang JY, Tsai WC, Lu HC, Zou LH et al (2020) New insight into the molecular mechanism of colour differentiation among floral segments in orchids. Commun Biol 3(89):1–13. https://doi.org/10.1038/s42003-020-0821-8
78. Brockington SF, Walker RH, Glover BJ, Soltis PS, Soltis DE (2011) Complex pigment evolution in the Caryophyllales. New Phytol 190(4):854–864
79. Panche AN, Diwan AD, Chandra SR (2016) Flavonoids: an overview. J Nutr Sci 5(e47):1–15. https://doi.org/10.1017/jns.2016.41
80. Zhao D, Tao J (2015) Recent advances on the development and regulation of flower color in ornamental plants. Front Plant Sci 6:261. https://doi.org/10.3389/fpls.2015.00261
81. Martin C, Gerats T (1993) Control of flower coloration. In: Jordan BR (ed) The molecular biology of flowering. CAB International, Wallingford
82. Glover BJ (2011) The diversity of flower color: how and why?. WIT Transactions on State of the Art in Science and Engineering, WI T Press
83. Tan SH, Manap SA, Karim MR, Rashid SS, Mahmood M, Ling Ma N (2014) Comparative flower pigment study of orchid plants. Adv Environ Biol 8(20):20–24

84. Vagas FD, Jimenez AR, Paredes-Lopez O (2000) Natural pigments: carotenoids, anthocyanins and betalains-characteristics, biosynthesis, processing and stability. Crit Rev Food Sci 40: 173–289
85. Matsui S, Nakamura M (1988) Distribution of flower pigments in perianth of *Cattleya* and genera I. species. J Japan Soc Hort Sci 57(2):222–232
86. Honda K, Ômura H, Hori M, Kainoh Y (2010) Allelochemicals in plant-insect interactions. In: Comprehensive natural products II: chemistry and biology, vol 4. Elsevier Ltd, Boston, pp 563–594
87. Liew CF, LOH CS, Goh CJ, Lim SH (1998) The isolation, molecular characterization and expression of dihydroflavonol 4-reductase cDNA in the orchid, *Bromheadia finlaysoniana*. Plant Sci 135(2):161–169
88. Johnson ET, Yi H, Shin B, Oh BJ, Cheong H, Choi G (1999) *Cymbidium* hybrida dihydroflavonol 4-reductase does not efficiently reduce dihydrokaempferol to produce orange pelargonidin-type anthocyanins. Plant J 19(10):81–85
89. Wang LM, Zhang J, Dong XY, Fu ZZ, Jiang H, Zhang HC (2018) Identification and functional analysis of anthocyanin biosynthesis genes in *Phalaenopsis* hybrids. Biol Plant 62(1):45–54
90. Li Y, Liu X, Cai X, Shan X, Gao R, Yang S, Han T, Wang S, Wang L, Gao X (2017) Dihydroflavonol 4-reductase genes from *Freesia hybrida* play important and partially overlapping roles in the biosynthesis of flavonoids. Front Plant Sci 8:428. https://doi.org/10.3389/fpls.2017.00428
91. Hieber AD, Mudalige-Jayawickrama RG, Kuehnle AR (2006) Color genes in the orchid *Oncidium* Gower Ramsey: identification, expression and potential genetic instability in an interspecific cross. Planta 223:521–531
92. Mudalige-Jayawickrama RG, Champagne MM, Hieber AD, Kuehnle AR (2005) Cloning and characterization of two anthocyanin biosynthetic genes from *Dendrobium* orchid. J Am Soc Hortic Sci 130:611–618
93. Pitakdantham W, Sutabutra T, Chiemsombat P, Pitaksutheepong C (2011) Isolation and characterization of dihydroflavonol 4-reductase gene in *Dendrobium* flowers. J Plant Sci 6:88–94
94. Anggraini R, Febriani AL, Mazieda MN, Al-Yamini TH, Listyorini DD (2017) Isolation of dihydroflavonol-4-reductase (DFR) gene in *Dendrobium* helix cv. Pomeo Brown. KNE Life Sci 3(4):213–218. https://doi.org/10.18502/kls.v3i4.707
95. Whang SS, Um WS, Song IJ, Lim PO, Choi K, Park KW, Kang KW, Choi MS, Koo JC (2011) Molecular analysis of anthocyanin biosynthetic genes and control of flower coloration by flavonoid 3′,5′-hydroxylase (F3′5′H) in *Dendrobium moniliforme*. J Plant Biol 54:209–218
96. Khumkarjorn N, Thanonkeo S, Yamada M, Thanonkeo P (2017) Cloning and expression analysis of a flavanone 3-hydroxylase gene in *Ascocenda* orchid. J Plant Biochem Biotechnol 26(2):179–190

Part II

Biology

Orchid-Associated Bacteria and Their Plant Growth Promotion Capabilities

6

Héctor Herrera, Alejandra Fuentes, Javiera Soto, Rafael Valadares, and Cesar Arriagada

Contents

Abstract

Orchids are one of the most widespread plants inhabiting lands in almost all terrestrial ecosystems. Most plants from the Orchidaceae family depend on specific mycorrhizal fungi to germinate their tiny and dust-like seeds. There, mycorrhizal fungi provide the resources that the embryo needs to advance to further developmental stages. However, at seed germination and the plantlet stage orchids need other nutrients to sustain the plantlet development and completion of its life cycle. There, other soil-borne beneficial microorganisms, such as free-living and symbiotic bacteria, can associate with orchid roots establishing a special interaction that has scarcely been explored. Most of the beneficial

H. Herrera · A. Fuentes · J. Soto · C. Arriagada (✉)
Laboratorio de Biorremediación, Facultad de Ciencias Agropecuarias y Forestales, Departamento de Ciencias Forestales, Universidad de La Frontera, Temuco, Chile
e-mail: hector.herrera@ufrontera.cl; alejandra.fuentes@ufrontera.cl; cesar.arriagada@ufrontera.cl

R. Valadares
Instituto Tecnologico Vale, Belém, PA, Brazil

J.-M. Mérillon, H. Kodja (eds.), *Orchids Phytochemistry, Biology and Horticulture*, Reference Series in Phytochemistry, https://doi.org/10.1007/978-3-030-38392-3_35

root-associated bacteria can help plants solubilize essential nutrients such as phosphorous and nitrogen, as well as to produce diverse beneficial metabolites that can contribute to improving biomass production, stress tolerance, and biocontrol of potential phytopathogenic fungal species. This chapter will summarize the principal studies performed in orchid-associated bacteria and analyze their potential plant growth–promoting capabilities.

Keywords

Bacteria · Endosphere · Rhizosphere · Plant growth promotion · Symbiosis

1 Introduction

Orchids belong to the widespread Orchidaceae family, which includes about 899 genera and 27,801 species distributed in almost all terrestrial ecosystems [1–3]. Orchids can grow in soil (terrestrial), rocks (lithophytes), or tree trunks (epiphytes) [4]. According to their physiological mode of carbon nutrition, orchids can be classified as: (i) photoautotrophic, which can obtain their own carbon throughout photosynthesis; (ii) mycoheterotrophic, which lack a photosynthetic apparatus and fully depend on the carbon obtained via a compatible mycorrhizal fungus; and (iii) partially mycoheterotrophic, which can obtain carbon from both photosynthesis and mycorrhizal fungi [5–7]. A common characteristic of orchids is the production of a capsule with thousands of dust-like seeds, with a nutrient-poor endosperm which lacks the essential nutrients to sustain the initial plant growth [8, 9]. These seeds are dispersed into the environment and once they reach the growth substrate they must find a compatible mycorrhizal fungus able to provide the carbon necessary to start the seed germination process [10, 11]. After that, a mycoheterotrophic organ known as protocorm is formed. At the protocorm stage, the mycoheterotrophic processes are completely active and the fungi provide almost all the nutrition for the embryo [12, 13]. Once the plantlet stage is reached, the mycoheterotrophic processes can be progressively reduced, especially in autotrophic or partially mycoheterotrophic species. At seed germination, plantlets develop and in the adult mature plant, orchids must associate with other fungal and bacterial taxa that can synergistically contribute to plant growth and development, providing essential nutrients such as nitrogen (N), protection against disease and phytopathogens, stress tolerance, and others [14–16].

The orchid mycorrhizal fungi live inside root tissues, inhabiting specific symbiotic structures known as pelotons [17, 18]. Such structures are essential for orchid nutrition, playing key roles in resource mobilization and orchid development [19]. Commonly, the mycorrhizal fungi associated with orchids are included in the *Rhizoctonia*-like fungi complex, which can involve Basidiomycetes from the genera *Ceratobasidium*, *Tulasnella*, *Thanatephorus*, *Sebacina*, and *Serendipita* [20, 21]. However, nowadays our understanding of mycorrhizal fungi associated with orchids has grown considerably and they can include different ectomycorrhiza-forming fungi, wood- or litter-decomposing fungi, and a broad spectrum of other different endophytic fungi without a specific role for seed germination or development [4, 22, 23]. Despite our understanding about fungi-associated microorganisms and their role in the orchid metabolism having been largely addressed [24–26], the study of orchid-associated bacteria and their benefits for orchids are mainly unknown [14]. Bacteria are commonly isolated from soils and can also live in association with other organisms, playing essential roles in the ecosystems [27, 28]. Specifically, several studies have demonstrated that plants are commonly associated with free-living or symbiotic bacteria and these microorganisms can have positive effects on the metabolisms of the associated plant [28–30]. In this chapter, we will summarize the principal studies performed on orchid-associated bacteria and discuss their putative function on orchids, with emphasis on plant growth–promoting mechanisms.

2 Plant Growth–Promoting Bacteria

Bacteria that positively influence plant growth are categorized as plant growth–promoting bacteria (PGPB) and include those that are free-living and those that form specific symbiotic relationships with plants (e.g., *Rhizobia* spp. and *Frankia* spp.) [31]. They can have beneficial effects on plant growth via direct and indirect mechanisms [32, 33]. Despite the differences between symbiotic and free-living bacteria, they all utilize the same mechanisms to promote plant growth: (i) directly by facilitating the supply of nutrients to the plant (N fixation or phosphate solubilization), increasing iron (Fe) bioavailability (through siderophore production), producing phytohormones like auxins, or modulating hormone levels such as ethylene through enzyme 1-aminocyclopropane-1-carboxylate (ACC) deaminase; or (ii) indirectly acting as biocontrol agents by controlling phytopathogenic soil-borne microbes, for instance, by producing antibiotics, hydrogen cyanide, or enzymes with the ability to hydrolyze the fungal cell wall, and stimulating mycorrhizal processes [34, 35].

PGPB should be capable of supplying host plants with additional nutrients or facilitating the acquisition of existing ones, such as potassium (K), N, phosphorus (P), or Fe from the growth substrate. Some bacteria are able to fix N_2 (conversion of atmospheric N to ammonium) and supply N to plants. Diverse genera such as *Azotobacter*, *Azospirillum*, and *Rhizobium* have known roles in plant growth promotion by N fixation [36–38]. Among N-fixing bacteria, some are free-living and do not require a host to perform the process, whereas others are symbiotic and only fix N in association with certain plants in specific root structures, such as leguminous [39]. P is usually found strongly adhered to the soil colloids in insoluble forms, thus not available for plants. Some bacteria have the ability of dissolving inorganic phosphates helping plants to cope their P needs. Such bacteria comprise genera such as *Burkholderia* and *Serratia*, among others [40–42]. Several genera of rhizobacteria (*Bacillus*, *Pseudomonas*, *Burkholderia*, *Agrobacterium*, and *Rhizobium*) are involved in the solubilization of K in the soil by converting it to soluble forms through release of organic acid [43–45]. These microbes play a vital role in the release of nutrients from soil adhered to mineral surfaces, living in association with plants or beneficial soil-borne fungi and contributing to growth and development [46, 47].

Siderophores are low-weight molecules (between 500 and 1500 Da) that possess great affinity and selectivity to binding and complex Fe [48]. Bacterial siderophores make Fe available to plants via two methods: one is by a ligand exchange reaction and the other is by direct uptake of siderophore-Fe complexes [49]. The primary function of these compounds is to chelate the ferric Fe [Fe(III)] from the environment and thereby make it available for microbial and plant cells [50]. Bacterial siderophores excreted by various genera such as *Pseudomonas*, *Rhizobium*, *Streptomyces*, *Azotobacter*, *Bacillus*, *Burkholderia*, and *Serratia* are known to sequester Fe and benefit plant growth [51–53]. Studies in orchid nutrition reporting on the interaction among nutrients and their ideal balance within this plant are scarce [54]. However, it is known that Fe is the third most limiting nutrient for plant growth and metabolism, primarily due to the low solubility of the oxidized ferric form in aerobic environments [55]. As a critical component of proteins and enzymes, Fe plays a significant role in basic biological processes such as photosynthesis, chlorophyll synthesis, respiration, N fixation, uptake mechanisms, and DNA synthesis [56, 57]. It is a cofactor of many enzymes that are necessary for plant hormone synthesis, such as ethylene, lipoxygenase, ACC oxidase, or abscisic acid [58].

The phytohormones auxins control and model several aspects on plant, such as cell expansion and division, cell elongation and differentiation, and a variety of physiological responses that result in improved plant growth [59]. Among auxins, the most common is indole-3-acetic acid (IAA), which is known to be produced not only by plants but also by bacteria [60]. Other phytohormones are salicylic acid and abscisic acid, which are particularly known for regulating the defense response in plants against pathogens [61, 62]. Another important plant growth regulator is ethylene, which controls the growth of roots, leaves, flowers, and fruits [63]. Ethylene is derived from ACC and is produced in plants as a response to multiple stresses [63]. Thus, different environmental stresses such as metals, chemicals, water, and extreme temperatures induce ethylene biosynthesis through the induction of ACC synthase [63]. When in excess, ethylene may be toxic to the plant, causing defoliation and other harmful effects [64]. ACC deaminase (ACCD) is an enzyme produced by some bacteria such as *Bacillus*, *Pseudomonas*, and *Azotobacter*, among others [65–67] and is considered as a key characteristic for PGPB strains [68]. ACCD is beneficial for plants as it decreases ACC levels by cleaving it into ammonia and α-ketobutyrate. This helps to reduce the adverse effects of ethylene by reducing its levels. Therefore, PGPB capable of producing ACCD are beneficial for plants grown under stress conditions such as drought, salt, and heavy metals by regulating the plant's ACC levels [65, 69, 70] and thus reducing the ethylene to nontoxic levels.

PGPB can grow in most plant structures (flowers, fruits, stems, leaves, and roots) [71], and can be localized inside or outside the plant cells (vascular tissues or seeds) [72, 73]. In the soil, microorganisms play important roles in the ecosystem, interacting closely with plant roots and soil nutrients [74]. The rhizosphere is an area of intensive molecular interaction between plant roots and soil colloids. It is defined as a narrow zone of soil that surrounds the plant roots and which is rich in nutrients compared to bulk soils, showing intense biological and chemical activities [75]. Over the years, the rhizosphere definition has been refined to include three zones which are defined based on their relative proximity to the root [76]: (i) the endorhizosphere, which includes the endodermis and cortical layers; (ii) the rhizoplane, known as the root surface where soil particles and microbes are adhered, including the epidermis, cortex, and a mucilaginous polysaccharide layer; and (iii) the ectorhizosphere, the outermost zone which extends from the rhizoplane out into the bulk soil. The volume of the soil which is not a part of the rhizosphere and is not influenced by any plant root is known as bulk soil [77]. During the course of growth and development of plants, a variety of organic compounds are released from the roots by exudation, secretion, and deposition making the rhizosphere rich in nutrients as compared to the bulk soil [75]. This acts as a driving force for the setup of active and enhanced microbial populations in the root zone, much higher than the bulk soil [78, 79]. Plants secrete low-molecular-weight compounds, such as amino acids, sugars, phenolics, terpenoids, and lipids, and high-molecular-weight compounds, including proteins, polysaccharides, and nucleic acids, depending on the growth stage and environmental conditions [80]. Upon secretion into the rhizosphere, most metabolites are rapidly degraded by soil microbes, but some, particularly specialized metabolites, remain in the soil and mediate biological communication [81, 82]. Microbial diversity is reduced near the roots, with further reduction in the endosphere [83, 84]. In this sense, the communities of microorganisms which reside in plants for at least a specific part of their life cycle are referred to as endophytes [85, 86]. Endophytes are often classified as obligate or facultative. Obligate endophytes are inherently dependent on their host, and their transmission to other plants occurs vertically or via vectors [87], whereas facultative endophytes have a biphasic lifestyle and can also survive in other habitats, such as the plant surface or soil [88]. Inside the host plant, endophytes can colonize both intercellular or intracellular spaces [89]. Endophytic communities are affected by several biotic and abiotic factors, such as plant species, developmental stage, or interactions with other plant-associated microorganisms, temperature, radiation, and physical or chemical attributes of the soil [90, 91]. The roles of endophytes inside the plant are varied and include nutrient acquisition, such as N fixation and phosphate solubilization, phytohormone and siderophore production, and protection against abiotic stresses (i.e., salinity, drought, or pollution) or phytopathogen control [87].

3 Mycorrhiza Helper Bacteria

Other indirect beneficial bacteria associated with plants are mycorrhiza helper bacteria (MHB). Soil microorganisms have significant effects on the physiological state of plants and productivity. Root-associated fungi and bacteria can establish mutualistic associations with plants, the most studied being mycorrhizal symbiosis, an association between filamentous fungi and most plant roots [92]. More than 90% of plants can establish these root-fungus associations, being established as the most widespread and ancient in the biosphere [93, 94]. As a general rule, the host plant provides the fungal associate carbohydrates from photosynthesis, while the fungus improves the acquisition of essential nutrients and water for the plants [95]. According to the root colonization strategy, two main types of mycorrhizae are commonly recognized: (i) ectomycorrhizae, in which the fungus colonizes the intercellular spaces of the root tissues, basically growing around it [96]; and (ii) endomycorrhizae, in which the fungus develops within the cortical cells [97]. Endomycorrhizae are further subdivided into specific types, the most recognized being arbuscular, ericaceous, and orchid mycorrhizae [98].

A group of specific soil bacteria colonizes the mycorrhizosphere, defined as the soil zone influenced by the roots and hyphae of external mycelium of the symbiotic fungus. These bacterial communities are known as MHB, establishing a complex physical and metabolic interaction producing beneficial effects for the mycorrhizal symbiosis [99]. The main effects through which MHB can favor the establishment of mycorrhizal symbiosis can be summarized as: (i) increasing survival and germination of fungal spores or other reproductive structures; (ii) stimulating presymbiotic growth mycelium; (iii) increasing root receptivity since bacteria proliferate in the rhizosphere prior to the development of the symbiotic fungus; (iv) stimulating the root-mycelium recognition; (v) changing physicochemical properties in the mycorrhizosphere; and (vi) reducing soil-mediated stress by biotic and abiotic factors through detoxification and phytopathogen control [100–102]. Although the role of MHB associated with orchid remains largely unexplored, the high diversity of root-associated bacteria growing in the orchid-growing substrates may represent an opportunity to study and design strategies to explore the beneficial potential of such microorganisms for mycorrhizal orchid plants.

MHB can produce diverse bioactive compounds including phytohormones and enzymes. The enzyme ACCD prevents the ACC produced by the plant from being transformed into ethylene, a key regulator in the stress response in plants [103]. In addition to their role in tolerance to environmental stress, ACC deaminase-producing bacteria improve nodulation and mycorrhizal colonization in the plant [104]. Other enzymes produced by MHB seem to have a positive role in the colonization of the root cortex due to their ability to secrete enzymes such as endoglucanases, cellobiose hydrolases, pectatoliases, and xylanases, which softens the cell wall, improving the receptivity of the plant tissues [105].The phytohormone IAA is the most important signal molecule and is involved in the processes of cell division, elongation, and differentiation, promoting root elongation and proliferation of secondary roots [106]. Some MHB have the ability to produce large amounts of IAA, which promotes the development and growth of the hypha [107].

MHB can be found in many genera of gram-negative and gram-positive bacteria [100]. Many of them belong to Proteobacteria (*Azotobacter*, *Klebsiella*, and *Pseudomonas*), α-Proteobacteria (mainly *Azospirillum*, *Bradyrhizobium*, and *Rhizobium*), *β*-Proteobacteria (*Burkholderia*), Firmicutes (mainly *Bacillus*, *Brevibacillus*, and *Paenibacillus*), and Actinobacteria (mainly *Streptomyces*, *Rhodococcus*, and *Arthrobacter*) [105]. These genera are frequently recognized as having a direct effect on plant growth-promoting solubilization and mineralization of phosphate, N fixation, secondary metabolite production, phytohormone synthesis, and secretion of siderophores, among others [31].

The tripartite interaction between plant-mycorrhizal fungi-bacteria is still scarcely explored. The study of the functions of MHB, their cultivation, and determination of their potential can be of great utility to improving the growth and development of the mycorrhizal fungus and the plant. This issue can be relevant for sustainable productive practices as well as in ecological restoration programs based on an efficient management of the microbial components of the soil, biodiversity, and the symbiotic interactions that occur there.

4 Methods for the Study of Orchid-Associated Bacteria

For the isolation and identification of culturable bacteria, it is necessary to have pure cultures of the strains, for which the technique of serial dilution and spreading onto agar nutrient media is the most standard [108]. For the isolation of rhizospheric microorganisms only the soil firmly attached to the roots is considered [109]. The procedures for isolating endophytic bacteria involve a crucial step of surface disinfection to eliminate saprophytes or other microorganisms associated with the surface of the tissue, including leaves, stems, bark, roots, fruits, and seeds [110]. Common culture media used for the isolation of plant-associated bacteria are Luria-Bertani agar, tryptic soy agar, yeast extract sucrose agar, nutrient agar, and others. To limit fungal growth, the media should be amended with antifungal agents such as cycloheximide, benomyl, benzalkonium chloride, captan, and others [111, 112]. The study of the diversity of root-associated bacteria has been conducted using mainly these culture-dependent techniques. The identification of the isolated strains can be carried out based on the morphological and biochemical characterization. However, the culture-dependent methods present discrepancies between the observable morphological and/or phenotypic characteristics of the isolates, which means that the same strain can generate different results in repeated tests [113]. Currently for the identification of bacteria, complementary genotypic methods are used and can be divided mainly into two categories: (i) pattern or fingerprint-based techniques (such as phospholipid fatty acid, denaturing gradient gel electrophoresis, single-strand conformation polymorphism, terminal restriction fragment length polymorphism, and amplified fragment length polymorphism) and (ii) sequence-based techniques (such as metagenomics, pyrosequencing, metatranscriptomics, and transcriptomics). Both techniques generate a database of fingerprints or specific DNA sequences against which the test organisms or sequences can be compared. The degree of similarity or coincidence is a measure of how related the two organisms are to each other [114]. For the molecular identification of bacterial species based on the nucleotide sequence, a wide variety of genes have been used, such as 16S–23S rRNA intergenic space (*ITS*), rRNA *23S*, *RpoB* (β subunit of RNA polymerase), and *Gyr*B (ß subunit of DNA gyrase) [115–117]. However, the 16S rRNA analysis has proven to be the most efficient gene for identification; due to its highly conserved analysis, it is possible to obtain information on phylogenetic relationships to classify and identify different prokaryotic species [118].

The new proteomic tools, based mainly on mass spectrometry, have made it possible to complement the classical techniques of microbiology and genomics for the classification, identification, and phenotypic characterization of bacteria [119]. Bacterial identification using ribosomal protein profiling using matrix-assisted laser desorption/ionization time-of-flight mass spectrometry (MALDI-TOF MS) has been commonly applied to accurately identify microorganisms given that the mass spectrum is specific for each species [120]. The results are expressed as a homology percentage for sequencing and in an absolute value for the MALDI TOF and allows the proximity of the proposed profile to the data banks to be determined. Among the main advantages of this technique is speed and accuracy, in addition to the low operating cost [121].

In recent decades, the greatest advances in molecular techniques have shown that only a small fraction of microorganisms can grow in culture media [122] and 95–99% of microorganisms in any environment are uncultured. Regarding soil bacteria, only 1% can be cultured by traditional techniques, which is not representative of the total phylogenetic diversity [123]. Culture-independent techniques are used to investigate the microbial diversity of a variety of environmental niches including the root-associated bacteria (rhizosphere and endosphere). Denaturing gradient gel electrophoresis (DGGE) allows as much as 95–99% of the bacterial community to be detected. This technique separates genes of the same size that differ in their profile of denaturation due to differences in the nucleotide sequence [124]. However, it presents disadvantages at the time of analysis such as the presence of artifacts which have been a serious problem for genetic diversity analyses [125]. Currently, next-generation sequencing techniques (NGS) allow massive parallel sequencing with a high throughput of amplicons at low cost compared to those based on the Sanger method. Metagenomic approaches analyze the composition and diversity of a microbial community with the aim of detecting genetic diversity, metabolic routes, and access to complete genes or sequences of the sample [109, 126]. For a bacterial community study, the 16 rRNA ribosomal subunit gene is used as a target. It consists of nine hypervariable regions flanked by conserved regions, of which the hypervariable regions are specific to each genus and species [127]. In addition, other advantage of 16S rRNA gene is the availability of several databases of reference sequences and taxonomy [128]. On the other hand, shotgun metagenomics and pyrosequencing technologies are alternatives that provide more complete data about diversity and metabolic functional features present in microbial communities [129]. Additionally, they describe the corresponding genes in an ecosystem, facilitating the characterization of the potential functionality of the microbiome in specific environments [130]. Thus, these methodologies can be important for the study of orchid-associated bacteria and their metabolic effects on the plant.

5 Diversity of Orchid-Associated Bacteria

Like most land plants, orchids also have associated bacteria inside their structures. Although non-culture-dependent methods have increased our understanding of root-associated bacteria in several plants, these NGS methodologies have been scarcely explored in plants from the Orchidaceae family. Recently, Alibrandi, Schnell [131] analyzed the endophytic bacterial diversity of the Mediterranean orchids *Neottia ovata* (L.) Bluff and Fingerh., *Serapias vomeracea* (Burm.f.) Briq., and *Spiranthes spiralis* (L.) Chevall. using metabarcoding of the 16S rRNA gene. That study identified Proteobacteria and Actinobacteria as principal phyla associated with the analyzed orchids, with less richness and diversity in the aerial biomass than in the roots. Similarly, Li, Xiao [132] used a metagenome pyrosequencing-based approach to identify bacterial communities associated with *Dendrobium catenatum* Lindley, identifying the genera *Pseudomonas*, *Escherichia/Shigella*, *Delftia*, and *Burkholderia* as the dominant taxa in roots, leaves, and stems. Additionally, Yu, Zhou [133] used a nested PCR-DGGE approach to analyze bacterial endophytes in *Dendrobium officinale* Kimura et Migo plants, identifying the genus *Burkholderia* as the principal orchid-associated bacteria that can contribute mainly to N fixation. Pei, Mi [134] also studied the diversity of endophytic bacteria associated with *D. officinale* through metagenomics, showing that the phylum Proteobacteria, Actinobacteria, Firmicutes, and Bacteroidetes were commonly identified in the analyzed structures (roots, stems, and leaves). Lin, Xiong [135] showed that the terrestrial orchid *Gymnadenia conopsea* (L.) R.Br. has a particular bacterial root-associated microbiome including mainly the taxa Proteobacteria, Bacteroidetes, Acidobacteria, Actinobacteria, Verrucomicrobia, Chloroflexi, and Planctomycete, showing that geographical location was the main factor affecting the diversity of bacterial taxa associated with the roots. Overall, such studies involving culture-independent methods have provided essential information to understand the huge diversity of orchid-associated bacteria with the ability to grow inside plant structures and for a better understanding of the ecology of orchids in relation to their bacterial associates. However, many such genera cannot be cultured using standard laboratory procedures, limiting the exploration of plant growth–promoting mechanisms, their use in productive systems. and conservation strategies involving commercial and threatened species. Therefore, it is essential to study the diversity of culturable microorganisms in order to assign a putative beneficial role to orchid-associated bacteria in seed germination and/or plant growth promotion at advanced developmental stages. Consequently, we will summarize some of the principal studies performed on orchid-associated bacteria involving culture-dependent methods and their possible plant growth–promoting traits detected (Table 1).

5.1 Seed-Associated Bacteria

An orchid interacts with bacteria throughout its life cycle. Several plants harbor a selective and specific group of bacteria that can colonize pollen and seeds [2]. Such bacteria are associated with the initial plant developmental stages, contributing to processes such as disease prevention, phytohormone production, stress tolerance, and biological control of phytopathogenic microorganisms. In this sense, the study of pollen-/seed-associated bacteria with orchids may provide valuable data for the sustainable management of orchid cultivars.

Table 1 Diversity of culturable bacteria isolated from orchid plants and with potential plant growth–promoting capabilities

Isolated bacteria	Host	Isolation source	Plant growth–promoting capabilities	Location	Reference
Pseudomonas sp., *Pantoea* sp., *Rahnella* sp., *Staphylococcus* sp., *Microbacterium* sp., *Fictibacillus* sp., *Streptomyces* sp., *Sphinomonas* sp., *and Bacillus* sp.	*Spiranthes spirali, Serapias vomeracea, and Neottia ovata*	Root, stem, leaf, and capsule endosphere	Phosphate solubilization ACC deaminase activity IAA production Siderophore production potassium solubilization antimicrobial activity	Italy	Alibrandi, Lo Monaco [140]
Bacillus sp., *Flavobacterium* sp., *Nocardia* sp., *Pseudomonas* sp., *Curtobacterium* sp., *Rhodococcus* sp., *Xanthomonas* sp., *Acinetobacter* sp., *Aquaspirillum* sp., *Micrococcus* sp., *Streptomyces* sp., *Cellulomonas* sp., *Gluconobacter* sp., *and Mycobacterium* sp.	*Dendrobium moschatum and Acampe papillosa.*	Aerial and substrate root endosphere	Phosphate solubilization[a] ACC deaminase activity[a] IAA production[a] Siderophore production[a] potassium solubilization[a]	Southeast Asia	Tsavkelova, Cherdyntseva [152]
Caulobacter vibrioides, *Roseomonas cervicalis*, *Streptomyces* sp., *Azospirillum irakense*, *Enterobacter cloacae*, *Agrococcus iejuensis*, *Sphingomonas* sp., *and Bacillus pumilus*	*Dendrobium moschatum*	Root endosphere, rhizoplane	IAA production	Southeast Asia	Tsavkelova, Egorova [138]
Pseudomonas fluorescens-putida	*Thelymitra crinita, Lyperanthus nigricans, and Pterostylis recurva*	Root endosphere	Seed germination	Western Australia	Wilkinson, Dixon [162]

Bacillus toyonensis, *Bacillus mobilis*, *Pseudomonas fluorescens*, *Acinetobacter calcoaceticus*, *Bacillus simplex*, *Pseudomonas baetica*, *agrobacterium tumefaciens*, *Bacillus cereus*, *Stenotrophomonas maltophilia*, *Bacillus mobilis*, *Pseudomonas frederiksbergensis*, and *Bacillus thuringiensis*	*Anacamptis pyramidalis*, *Himantoglossum caprinum*, *Limodorum abortivum*, *Platanthera bifolia*, *Serapias vomeracea*, *Spiranthes spiralis*, *Ophrys apifera*, *Ophrys sphegodes*, *Orchis coriophora*, *Orchis laxiflora*, *Orchis provincialis*, and *Orchis tridentata*	Ectorhizosphere, endorhizosphere, and rhizoplane	IAA production Phosphate solubilization ACC deaminase activity	Turkey	Altinkaynak and Ozkoc [143]
Streptomyces sp., *Bacillus* sp., *Pseudomonas* sp., *Burkholderia* sp., *Erwinia* sp., *Nocardia* sp., *Flavobacterius* sp., *Stenotrophomonas* sp., *Pantoea* sp., *Chryseobacterium* sp., *Agrobacterium* sp., and *Paracoccus* sp.	*Paphiopedilum appletonianum*, and *Pholidota articulata*	Endosphere and rhizoplane	IAA production	Vietnam	Tsavkelova, Cherdyntseva [145]
Sphingomonas paucimobilis	*Dendrobium officinale*	Endosphere	Salicylic acid, IAA, zeatin, abscisic acid production. Nitrogen fixation	China	Yang, Zhang [153]
Bacillus sp., *Sporosarcina* sp., *Paenibacillus* sp., *Burkholderia* sp., *Methylobacterium* sp., *Brevibacterius* sp., and *Curtobacterium* sp.	*Dendrobium officinale*	Root, stem, and leaf endosphere	Phosphate solubilization[a] ACC deaminase activity[a] IAA production[a] Siderophore production[a] potassium solubilization[a]	China	Pei, Mi [134]

(continued)

Table 1 (continued)

Isolated bacteria	Host	Isolation source	Plant growth–promoting capabilities	Location	Reference
Bacillus amyloliquefaciens	*Vanilla phaeantha, V. planifolia x V. pompona*	Shoot tip	Fungal inhibitor production Seedling growth promotion	United States of America	White, Torres [141]
Bacillus sp., *Gemella* sp., *Staphylococcus* sp., *Streptococcus* sp., *Paracoccus* sp., *Achromobacter* sp., *Acidobacter* sp.	*Stanhopea connata*	Root fungal hyphae	IAA production[a] Siderophore production[a] antifungal activity [a] nitrogen fixation[a]	Ecuador	Novotna and Suárez [154]
Collimonas pratensis, Pseudomonas sp., *Pandoraea oxalativorans, Pseudomonas koreensis, Exiguobacterium aurantiacum, Dyella marensis, Luteibacter rhizovicinus, Bacillus* sp., *Pseudomonas azotoformans, Chryseobacterium* sp., *Pseudomonas costantinii*	*Chloraea barbata, Chloraea collicensis, Chloraea gavilu, Chloraea magellanica, Gavilea araucana, and Gavilea lutea*	Root endosphere	Phosphate solubilization Siderophore production IAA production antifungal activity	Chile	Herrera, Sanhueza [14]
Paenibacillus lentimorbus, Paenibacillus macerans	*Cymbidium eburneum*	Meristem	Indole compound production Phosphate solubilization	Not specified	Faria, Dias [142]

Sphingomonas sp., *Sinorhizobium* sp., *Bacillus* sp., *Nocardia cerradoensis. Bacillus megaterium, and Burkholderia phytofirmans*	*Guarianthe skinneri*	Root	IAA production phosphate solubilization nitrogen fixation	Mexico	Aguilar Díaz, Bertolini [146]
Bacillus thuringiensis, Burkholderia cepacia, Burkholderia gladioli, Herbaspirillum frisingense, Pseudomonas stutzeri, rhizobium cellulosilyticum, rhizobium radiobacter, and Stenotrophomonas maltophilia	*Cymbidium* sp.	Root and leaf endosphere	Nitrogen fixation, phosphate solubilization, and zinc oxide. Indolic compound production	Brazil	Gontijo, Andrade [15]

[a]Potential plant growth promotion assigned based on the plant growth–promoting capabilities detected in other plant species

In orchids, the diversity of endophytic bacteria associated with pollen/seeds have been scarcely explored. Pavlova, Leontieva [136] reported that the rhizobacterial strains *Pseudomonas fluorescens* and *Klebsiella oxytoca* can colonize root tissues and the surface of protocorms and seeds of the orchid *Dendrobium nobile* Lindl. Additionally, the same study demonstrated plant growth–promoting capabilities of the tested rhizobacterial strains mainly through auxin production. Similarly, Tsavkelova, Cherdyntseva [137] demonstrated that bacteria isolated from the rhizoplane of the epiphytic tropical orchid *Dendrobium moschatum* Buch.-Ham., such as the genera *Sphingomonas* and *Mycobacterium*, can enhance seed germination despite the absence of compatible mycorrhizal fungi, demonstrating the crucial role of certain bacterial taxa at the same level as orchid mycorrhizal fungi. Additionally, Tsavkelova, Egorova [138] showed that endophytic or rhizoplane bacterial isolates belonging to the genera *Mycobacterium*, *Bacillus*, *Agrococcus*, and *Sphingomonas* can have positive roles in orchid growth, colonizing the seed surface and inner tissues of protocorms and rhizoids, influencing the seed germination process and embryo growth. Colonization of endophytic bacteria was also detected in mature capsules of the terrestrial orchid *S. spiralis* [88]. In this study, the genera *Bacillus*, *Sphingomonas*, and *Staphylococcus* were effectively isolated and characterized showing intense plant growth–promoting attributes such as phosphate solubilization, K solubilization, and IAA production. More recently, Alibrandi, Schnell [131] described the bacterial microbiome associated with orchids seeds after superficial disinfection of the capsule, describing seed microbiota associated with the *S. spiralis* as sharing a core of taxa with roots, stems, and leaves, suggesting a vertical transfer of the core microbiota, which can be essential to stimulating the early colonization of endophytic-associated bacteria. Strictly speaking, the diversity of culturable orchid seed–endophytic bacteria still remain unexplored, but our understanding of seed-associated bacteria has provided essential information to understand the beneficial role of these microorganisms at the first developmental stages of the orchid.

5.2 Phyllosphere-Associated Bacteria

Microorganisms inhabiting the phyllosphere can be beneficial for the associated plants, contributing to resistance to biotic/abiotic stresses and production of antimicrobial compounds, among others [139]. Alibrandi, Lo Monaco [140] analyzed the diversity of endophytic bacteria associated with leaves and stems of *S. spiralis*, *N. ovata*, and *S. vomeracea* orchids, exploring their plant growth–promoting traits. That study mainly described to the genera *Pseudomonas*, *Staphylococcus*, *Pantoea*, and *Rahnella* as principal bacterial associates to orchid aboveground organs. The analysis of plant growth–promoting capabilities revealed that most of the isolated strains have potential beneficial effects on the associated orchid, including nutrient solubilization, ACC deaminase activities, and IAA biosynthesis. In the same vein, Pei, Mi [134] identified *Burkholderia* sp., *Bacillus* sp., *Methylobacterium* sp., and *Curtobacterium* sp. as principal bacterial associates of the leaf and stem of *D.*

officinale. Most of these bacterial genera involve species with clear roles in plant growth promotion including phosphate solubilization, ACC deaminase activity, IAA production, siderophore production, and K solubilization. White, Torres [141] analyzed the colonization pattern of the endophytic bacterial strain *Bacillus amyloliquefaciens* in shoot meristems and stomatal areas of stems and leaves of *Vanilla phaeantha* Rchb. f., in which a protective and defensive role was expected for vanilla orchids. Gontijo, Andrade [15] analyzed leaf-associated bacteria of *Cymbidium* sp. plants, identifying several genera with beneficial effects on plant growth including the species *Burkholderia cepacia*, *Burkholderia gladioli*, and *Rhizobium radiobacter*. Similarly, Faria, Dias [142] identified *Paenibacillus* spp. associated with meristems of *Cymbidium eburneum* Lindl., showing that most of these isolates can promote plant growth and can be effectively used as a strategy for improving the acclimatization of orchid plantlets.

The studies conducted on phyllosphere-associated bacteria have increased our understanding about the potential benefits of these bacteria for orchids plants. Several of the isolated strains have been identified as PGPB (i.e., *Bacillus* spp., *Burkholderia* spp., *Pseudmonas* spp., and *Rhizobium* spp.), which can have a potential beneficial role at the first developmental stage, where improved growth rates, acclimatization, and biocontrol against phytopathogens are the main benefits expected.

5.3 Rhizosphere-Associated Bacteria

Currently, few studies have characterized the diversity of culturable bacteria associated with the rhizosphere of orchid roots. In the rhizosphere, several bacteria live under the influence of the plant root metabolism and can have essential roles in nutrient solubilization, influencing the growth of the associated plants, some of which can also colonize the inner root tissues. Altinkaynak and Ozkoc [143] characterized bacteria inhabiting the rhizosphere of native orchids from Turkey, identifying the genera *Bacillus*, *Paenibacillus*, *Pseudomonas*, *Acinetobacter*, *Agrobacterium*, and *Stenotrophomonas*. That study isolated almost 16 PGPB with the ability of phosphate solubilization, ACC deaminase activity, and IAA production. Similarly, Tsavkelova, Cherdyntseva [144] isolated and characterized auxin-producing bacteria from the rhizoplane of *D. moschatum* (*Rhizobium*, *Microbacterium*, *Sphingomonas*, and *Mycobacterium*), which were further analyzed for their potential role in seed germination, showing that some of these isolates were able to improve the seed germination rates of the inoculated seeds [137]. Similarly, Tsavkelova, Cherdyntseva [145] analyzed the diversity of bacteria colonizing the rhizoplane of the terrestrial orchid *Paphiopedilum appletonianum* (Gower) Rolfe and the epiphytic orchid *Pholidota articulate* Lindl. The main results showed that the culturable bacterial diversity was different in the two analyzed orchids, identifying *Streptomyces*, *Bacillus*, *Pseudomonas*, *Burkholderia*, *Erwinia*, and *Nocardia* as principal associates of *P. appletonianum*, whereas *Pseudomonas*, *Flavobacterium*, *Stenotrophomonas*, *Pantoea*, *Chryseobacterium*, *Bacillus*, *Agrobacterium*, *Erwinia*, *Burkholderia*, and *Paracoccus* were associated with *P.*

articulate roots. Similarly, Aguilar Díaz, Bertolini [146] identified rhizospheric bacteria are associated with *Guarianthe skinneri* (Bateman) Dressler and W.E. Higgins in Mexico, identifying *Sphingomonas* sp., *Sinorhizobium* sp., *Bacillus* spp., *Nocardia cerradoensis*, and *Burkholderia phytofirmans* as the main bacterial isolates.

Among the isolated bacteria, several genera with demonstrated roles in plant growth promotion and nutrient solubilization have been identified. Therefore, rhizosphere-associated bacteria can contribute to vital processes such as auxin production (i.e., *Sphingomonas* spp. and *Rhizobium* spp.), nutrient solubilization (i.e., *Sphingomonas* spp., *Sinorhizobium* spp., and *Nocardia* spp.), siderophore production (i.e., *Pseudomonas* spp.), volatile organic compounds (i.e., *Bacillus* spp.), protection against disease (i.e., *Bacillus* spp., and *Pseudomonas* spp.), and control of phytopathogens (i.e., *Rhizobium* spp., *Pseudomonas* spp., and *Bacillus* spp.).

5.4 Root Endosphere–Associated Bacteria

Mycoheterotrophic species like orchids commonly associate with a broad range of soil-borne fungi, including those accepted as orchid mycorrhizae as well as a broad range of free-living or endophytic species [112]. Some of these microorganisms can have a saprophytic or phytopathogenic lifestyle, including species from the genera *Fusarium*, *Thanatheporus*, and dark septate endophytes [147–149]. Therefore, endophytic bacteria associated with orchids may provide essential information about the mechanisms of control of intracellular fungal hyphae to avoid overspread in vital plant tissues [150, 151].

Endophytic bacteria associated with orchids have a strong influence on orchid metabolism, interacting directly with the cells of the host plant. In this sense, endophytic bacteria may represent an opportunity to improve the yield of orchid plants. Tsavkelova, Cherdyntseva [145] identified endophytic bacteria associated with roots of *P. appletonianum*, showing *Streptomyces*, *Bacillus*, *Erwinia*, and *Pseudomonas* genera as the principal endophytic taxa, whereas *Pseudomonas*, *Bacillus*, and *Flavobacterium* were the endophytic bacteria isolated from *P. articulata* roots. In the same way, Tsavkelova, Egorova [138] explored the role of the PGPB bacterial genera isolated from *D. moschatum* on the growth of *D. nobile*, showing that some endophytic isolates, such as *Agrococcus*, *Sphingomonas*, *Mycobacterium*, and *Bacillus pumilus*, can have a beneficial effect on seed germination. Alibrandi, Lo Monaco [140] also isolated endophytic bacteria colonizing the roots of *S. spiralis*, *N. ovata*, and *S. vomeracea*. The main endophytic taxa associated with the roots of the analyzed species were *Pseudomonas* sp., *Staphylococcus* sp., *Microbacterium* sp., *Streptomyces* sp., *Fictibacillus* sp., and *Bacillus* sp., some of which showed strong plant growth–promoting traits including ACC deaminase activity, phosphate solubilization, IAA production, siderophore production, and K solubilization. Tsavkelova, Cherdyntseva [152] studied the diversity of endophytic bacteria isolated from the root of the epiphytic orchids *Acampe papillosa* (Lindl.) Lindl. and *D. moschatum*, identifying the bacterial taxa *Bacillus* sp., *Flavobacterium* sp., *Pseudomonas* sp., *Rhodococcus* sp., *Xanthomonas* sp., *Alcaligenes* sp., and

Gluconobacter sp. as the principal associated taxa. Similarly, Pei, Mi [134] also analyzed the diversity of endophytic bacteria colonizing roots of *D. officinale*, showing to *Bacillus* sp., *Sporosarcina* sp., *Paenibacillus* sp., and *Brevibacterium* sp. as principal associates. In the same vein, Yang, Zhang [153] isolated and characterized a PGPB isolates from the root endosphere of *D. officinale*. The authors identified *Sphingomonas paucimobilis* as the main bacterial associate, which showed strong plant growth–promoting traits such as N fixation and secretion of plant growth regulators such as salicylic acid, IAA, and abscisic acid, which were involved in an improved plant growth of the inoculated seedlings. A recent study isolated and characterized endophytic bacterial strains by directly colonizing the mycorrhizal structures (pelotons) of some terrestrial Andean orchids [14]. That study identified the main endophytic bacterial taxa associated with fungal hyphae of mycorrhizal fungi-colonizing roots of native orchids from the genera *Chloraea* spp. and *Gavilea* spp. in southern Chile. The bacterial genera *Collimonas* sp., *Pseudomonas* spp., *Pandoraea* sp., *Exiguobacterium* sp., *Dyella* sp., *Luteibacter* sp., *Bacillus* sp., and *Chryseobacterium* sp. were commonly detected. Plant growth–promoting capabilities were detected in almost all of the isolates as well as a restriction of fungal growth, especially the isolates *Collimonas pratensis*, *Dyella marensis*, *Exiguobacterium aurantiacum*, and *Pseudomonas azotoformans*, an attribute that can contribute to the intraradical control of fungal hyphae of mycorrhizal fungi as well as potential phytopathogenic species.

Until now, several genera have been described as common associates of the endosphere of plants from the Orchidaceae family. These endophytic isolates include several taxa also identified as rhizospheric-associated microorganisms, such as *Pseudomonas* spp., *Bacillus* spp., *Sphingomonas* spp., and *Burkholderia* spp., showing also plant growth-promoting capabilities such as IAA production and phosphate/K solubilization. Additionally, the potential capabilities of some bacterial isolates to inhibit the growth of potential phytopathogenic species (i.e., *Collimonas* sp., *Pseudomonas* spp., *Exiguobacterium* sp., *Dyella* sp., and *Chryseobacterium* sp.) points to essential roles in the control of fungal spread inside the root tissues of symbiotic orchids.

5.5 Fungi-Associated Bacteria

Bacteria associated with fungal hyphae have been commonly detected in several terrestrial plants. Similarly, they can be indirectly involved in orchid mycorrhizal processes, especially helping the extraradical fungal hyphae inhabiting the growth substrate. In this sense, only one study has reported the direct association of orchid mycorrhizal fungi mycelia with bacteria. Specifically, Novotna and Suárez [154] characterized bacteria associated with living hyphae of the mycorrhizal fungi *Serendipita* sp., isolated from roots of the epiphytic orchid *Stanhopea connata* Klotzsch in southern Ecuador. Several genera such as *Bacillus* sp., *Gemella* sp., *Staphylococcus* sp., *Streptococcus* sp., *Paracoccus* sp., *Achromobacter* sp., and *Acidobacter* sp. were detected. Some of the detected species match with other bacterial isolates with the ability to live inside fungal hyphae, such as *Bacillus* spp., *Paracoccus* spp. and *Achromobacter* spp. [155–157]. Therefore, this study has provided essential information on the diversity of bacteria living in association with fungal hyphae of orchid mycorrhizal fungi.

6 Management of Root-Associated Bacteria in Cultural Practices

As reviewed, several studies have reported that orchid species establish specific symbiotic interactions with soil bacteria. In this sense, identifying microhabitats in which threatened orchid populations grow naturally may be essential to knowing the reproduction mechanisms in nature and applying the information obtained to seed germination strategies and plant growth promotion, with a focus on orchid-associated bacteria. Although most soil root–associated bacterial communities (endophytic and rhizospheric) cannot be cultured, identifying potential PGPB with a positive influence on orchid metabolism, especially at the first developmental stages, can be useful to improving the germination and survival rates of wild orchids, improving growth rates, plant establishment, and development. Therefore, developing a potential inoculant involving a single bacterium or consortium may be essential for cultural management in orchids, especially for plantlets obtained by asymbiotic seed germination procedures. In this sense, key developmental stages for orchid bacterization can be seed germination, initial plantlet development, and acclimatization. A compatibility test is necessary because recent studies have shown that bacterization can have different effects depending on the host [138]. The methods for inoculation of bacteria can include direct inoculation of the bacterial cells or through different microencapsulation methodologies [158, 159]. Most studies on orchid-associated bacteria have tested the plant growth–promoting capabilities under controlled conditions, but information about the in vivo effects of inoculation on plants are limited. Therefore, studies regarding in vivo effects are needed for a better understanding of the role of orchid-associated bacteria.

Soil microorganisms can establish synergistic interactions with soil-borne microbes that usually result in improved plant growth. Therefore, it is necessary to explore the potential of the noncultured microbiota for a better understanding of the symbiotic interactions of orchids. The soil microbiome, involving all microbiological populations inhabiting a soil, certainly influences the orchid seed germination processes, because in the microbiome the compatible mycorrhizal fungi and effective orchid-associated bacteria can colonize the growth substrate. Identifying such hotspots of orchid reproduction and developing efficient management methodologies involving the respective associated microbiome can contribute to improving the seed germination rates in both, nature and controlled conditions. In this sense, recent studies have shown that the use of soil microbiomes can be an effective strategy to improve plant growth [160, 161]. Therefore, the use of the original soil microbiome able to sustain the germination and growth of orchid plants may be essential to improving reproduction and yield of orchids plants.

7 Conclusion

The presence of diverse bacterial genera in orchid structures denotes that bacteria are common associates at different life cycle stages (seed, protocorm, plantlet, and mature plant). Additionally, the isolation and characterization of orchid-associated bacteria have provided evidence of the benefits for the associated plants. Commonly, orchid mycorrhizal fungi have been described as essential microorganism associated with orchids, but nowadays knowledge of orchid-associated bacteria has provided clues about the beneficial role of bacteria for stress tolerance, plant growth promotion, and protection against disease or phytopathogens. Therefore, orchid-associated bacteria must be considered essential to orchid development, and some can effectively colonize inner plant structures, having direct contact with the plant cells. However, further evidence is required to define the optimal developmental stage and specific strains for an adequate inoculation of plant growth–promoting bacteria, as well as to test their synergistic or antagonistic effect with orchid mycorrhizal fungi.

Acknowledgments The authors would like to thank to Fondo Nacional de Desarrollo Científico y Tecnológico (FONDECYT), grant numbers 1211857 to Cesar Arriagada and 3200134 to Hector Herrera.

References

1. Roberts DL, Dixon KW (2008) Orchids. Curr Biol 18(8):R325–R329
2. Herrera H et al (2019) Orchid mycorrhizal interactions on the Pacific side of the Andes from Chile. A review. J Soil Sci Plant Nutr 19(1):187–202
3. Clements M et al (1999) Genera Orchidacearum Vol. 1: General introduction, Apostasioideae, Cypripedioideae. Oxford University Press, Oxford
4. Herrera H et al (2017) Mycorrhizal compatibility and symbiotic seed germination of orchids from the coastal range and Andes in south Central Chile. Mycorrhiza 27(3):175–188
5. Merckx V et al (2013) Mycoheterotrophy, vol 10. Springer, New York, pp 978–971
6. Herrera H et al (2018) Adaptation and tolerance mechanisms developed by mycorrhizal Bipinnula fimbriata plantlets (Orchidaceae) in a heavy metal-polluted ecosystem. Mycorrhiza 28(7):651–663
7. Selosse M-A, Roy M (2009) Green plants that feed on fungi: facts and questions about mixotrophy. Trends Plant Sci 14(2):64–70
8. Valadares RB et al (2012) Narrow fungal mycorrhizal diversity in a population of the orchid Coppensia doniana. Biotropica 44(1):114–122
9. Fan X-L et al (2020) Transitions between the terrestrial and epiphytic habit drove the evolution of seed-aerodynamic traits in orchids. Am Nat 195(2):275–283
10. Liu H, Luo Y, Liu H (2010) Studies of mycorrhizal fungi of Chinese orchids and their role in orchid conservation in China—a review. Bot Rev 76(2):241–262
11. Fuji M et al (2020) Relative effectiveness of Tulasnella fungal strains in orchid mycorrhizal symbioses between germination and subsequent seedling growth. Symbiosis 81:53–63
12. Yeung EC (2017) A perspective on orchid seed and protocorm development. Bot Stud 58(1): 1–14
13. Valadares R et al (2014) Proteome changes in Oncidium sphacelatum (Orchidaceae) at different trophic stages of symbiotic germination. Mycorrhiza 24(5):349–360

14. Herrera H et al (2020) Isolation and identification of endophytic bacteria from mycorrhizal tissues of terrestrial orchids from southern Chile. Diversity 12(2):55
15. Gontijo JB et al (2018) Bioprospecting and selection of growth-promoting bacteria for Cymbidium sp. orchids. Sci Agric 75(5):368–374
16. Sisti LS et al (2019) The role of non-mycorrhizal fungi in germination of the mycoheterotrophic orchid Pogoniopsis schenckii Cogn. Front Plant Sci 10:1589
17. Dearnaley J, Perotto S, Selosse MA (2016) Structure and development of orchid mycorrhizas. Mol Mycorrhizal Symbiosis 2016:63–86
18. Peterson RL et al (1996) The interface between fungal hyphae and orchid protocorm cells. Can J Bot 74(12):1861–1870
19. Kuga Y, Sakamoto N, Yurimoto H (2014) Stable isotope cellular imaging reveals that both live and degenerating fungal pelotons transfer carbon and nitrogen to orchid protocorms. New Phytol 202(2):594–605
20. Herrera H et al (2020) Mycorrhizal Fungi isolated from native terrestrial orchids from region of La Araucanía, Southern Chile. Microorganisms 8(8):1120
21. Weiß M et al (2016) Sebacinales–one thousand and one interactions with land plants. New Phytol 211(1):20–40
22. Jiang J et al (2019) Fusarium oxysporum KB-3 from Bletilla striata: an orchid mycorrhizal fungus. Mycorrhiza 29(5):531–540
23. Meng Y-Y et al (2019) Are fungi from adult orchid roots the best symbionts at germination? A case study. Mycorrhiza 29(5):541–547
24. Valadares R et al (2020) Proteomic and transcriptomic analyses indicate metabolic changes and reduced defense responses in mycorrhizal roots of Oeceoclades maculata (Orchidaceae) collected in nature. Journal of Fungi 6(3):148
25. Dearnaley JD, Cameron DD (2016) Nitrogen transport in the orchid mycorrhizal symbiosis-further evidence for a mutualistic association. New Phytol 213(1):10–12
26. Sathiyadash K et al (2020) Orchid mycorrhizal fungi: structure, function, and diversity. In: Orchid biology: recent trends & challenges. Springer, Singapore, pp 239–280
27. Gnanamanickam SS (2006) Plant-associated bacteria, vol 1. Springer, Dordrecht
28. Yu AO, Leveau JH, Marco ML (2020) Abundance, diversity and plant-specific adaptations of plant-associated lactic acid bacteria. Environ Microbiol Rep 12(1):16–29
29. Eida AA et al (2018) Desert plant bacteria reveal host influence and beneficial plant growth properties. PLoS One 13(12):e0208223
30. Dellagi A, Quillere I, Hirel B (2020) Beneficial soil-borne bacteria and fungi: a promising way to improve plant nitrogen acquisition. J Exp Bot 71(15):4469–4479
31. Basu S, Rabara R, Negi S (2017) Towards a better greener future-an alternative strategy using biofertilizers. I: plant growth promoting bacteria. Plant Gene 12:43–49
32. Soto J et al (2019) Enhanced arsenic tolerance in Triticum aestivum inoculated with arsenic-resistant and plant growth promoter microorganisms from a heavy metal-polluted soil. Microorganisms 7(9):348
33. Olanrewaju OS, Glick BR, Babalola OO (2017) Mechanisms of action of plant growth promoting bacteria. World J Microbiol Biotechnol 33(11):1–16
34. Glick BR (1995) The enhancement of plant growth by free-living bacteria. Can J Microbiol 41 (2):109–117
35. Bhattacharyya PN, Jha DK (2012) Plant growth-promoting rhizobacteria (PGPR): emergence in agriculture. World J Microbiol Biotechnol 28(4):1327–1350
36. Remigi P et al (2016) Symbiosis within symbiosis: evolving nitrogen-fixing legume symbionts. Trends Microbiol 24(1):63–75
37. Ejaz S et al (2020) Effects of inoculation of root-associative Azospirillum and Agrobacterium strains on growth, yield and quality of pea (Pisum sativum L.) grown under different nitrogen and phosphorus regimes. Sci Hortic 270:109401
38. Sumbul A et al (2020) Azotobacter: a potential bio-fertilizer for soil and plant health management. Saudi J Biol Sci 27(12):3634

39. Madigan MT, Martinko JM, Parker J (1997) Brock biology of microorganisms, vol 11. Prentice hall, Upper Saddle River
40. Din M et al (2019) Production of nitrogen fixing Azotobacter (SR-4) and phosphorus solubilizing Aspergillus niger and their evaluation on Lagenaria siceraria and Abelmoschus esculentus. Biotechnol Rep 22:e00323
41. You M et al (2020) Isolation and characterization of Burkholderia cenocepacia CR318, a phosphate solubilizing bacterium promoting corn growth. Microbiol Res 233:126395
42. Ludueña LM et al (2017) Effects of P limitation and molecules from peanut root exudates on pqqE gene expression and pqq promoter activity in the phosphate-solubilizing strain Serratia sp. S119. Res Microbiol 168(8):710–721
43. Zeng X et al (2012) Characterization and potassium-solubilizing ability of Bacillus Circulans Z 1–3. Adv Sci Lett 10(1):173–176
44. Rajawat MVS et al (2016) A modified plate assay for rapid screening of potassium-solubilizing bacteria. Pedosphere 26(5):768–773
45. Etesami H, Emami S, Alikhani HA (2017) Potassium solubilizing bacteria (KSB):: mechanisms, promotion of plant growth, and future prospects a review. J Soil Sci Plant Nutr 17(4): 897–911
46. Lian B et al (2002) A comprehensive review of the mechanism of potassium releasing by silicate bacteria. Acta Mineral Sin 22(2):179–183
47. Sugumaran P, Janarthanam B (2007) Solubilization of potassium containing minerals by bacteria and their effect on plant growth. World J Agr Sci 3(3):350–355
48. Ahmed E, Holmström SJ (2014) Siderophores in environmental research: roles and applications. Microb Biotechnol 7(3):196–208
49. Thomine S, Lanquar V (2011) Iron transport and signaling in plants. In: Transporters and pumps in plant signaling. Springer, Berlin, pp 99–131
50. Braud A et al (2009) Enhanced phytoextraction of an agricultural Cr-and Pb-contaminated soil by bioaugmentation with siderophore-producing bacteria. Chemosphere 74(2): 280–286
51. Sabet H, Mortazaeinezhad F (2018) Yield, growth and Fe uptake of cumin (Cuminum cyminum L.) affected by Fe-nano, Fe-chelated and Fe-siderophore fertilization in the calcareous soils. J Trace Elem Med Biol 50:154–160
52. Kumar P et al (2018) Inoculation of siderophore producing rhizobacteria and their consortium for growth enhancement of wheat plant. Biocatal Agric Biotechnol 15:264–269
53. El Attar I et al (2019) Screening of stress tolerant bacterial strains possessing interesting multi-plant growth promoting traits isolated from root nodules of Phaseolus vulgaris L. Biocatal Agric Biotechnol 20:101225
54. Novais SV et al (2016) Phosphorus-zinc interaction and iron and manganese uptake in the growth and nutrition of phalaenopsis (Orchidaceae). Rev Bras Ciênc Solo 40:1–10
55. Zuo Y, Zhang F (2011) Soil and crop management strategies to prevent iron deficiency in crops. Plant Soil 339(1):83–95
56. Kim J, Rees D (1992) Structural models for the metal centers in the nitrogenase molybdenum-iron protein. Science 257(5077):1677–1682
57. Reichard P (1993) From RNA to DNA, why so many ribonucleotide reductases? Science 260 (5115):1773–1777
58. Camprubi E et al (2017) Iron catalysis at the origin of life. IUBMB Life 69(6):373–381
59. Li J, Li C, Smith SM (2017) Hormone metabolism and signaling in plants. Academic
60. Duca D et al (2014) Indole-3-acetic acid in plant–microbe interactions. Antonie Van Leeuwenhoek 106(1):85–125
61. Großkinsky DK et al (2016) Cytokinin production by Pseudomonas fluorescens G20-18 determines biocontrol activity against Pseudomonas syringae in Arabidopsis. Sci Rep 6(1): 1–11
62. Akhtar SS et al (2020) Role of cytokinins for interactions of plants with microbial pathogens and pest insects. Front Plant Sci 10:1777

63. Dubois M, Van den Broeck L, Inzé D (2018) The pivotal role of ethylene in plant growth. Trends Plant Sci 23(4):311–323
64. Desbrosses G et al (2009) PGPR-Arabidopsis interactions is a useful system to study signaling pathways involved in plant developmental control. Plant Signal Behav 4(4):319–321
65. Gowtham H et al (2020) Induction of drought tolerance in tomato upon the application of ACC deaminase producing plant growth promoting rhizobacterium Bacillus subtilis Rhizo SF 48. Microbiol Res 234:126422
66. Win KT et al (2018) The ACC deaminase expressing endophyte Pseudomonas spp. enhances NaCl stress tolerance by reducing stress-related ethylene production, resulting in improved growth, photosynthetic performance, and ionic balance in tomato plants. Plant Physiol Biochem 127:599–607
67. Pandey S, Gupta S (2020) Diversity analysis of ACC deaminase producing bacteria associated with rhizosphere of coconut tree (Cocos nucifera L.) grown in Lakshadweep islands of India and their ability to promote plant growth under saline conditions. J Biotechnol 324:183–197
68. Glick BR (2014) Bacteria with ACC deaminase can promote plant growth and help to feed the world. Microbiol Res 169(1):30–39
69. del Carmen Orozco-Mosqueda M, Glick BR, Santoyo G (2020) ACC deaminase in plant growth-promoting bacteria (PGPB): an efficient mechanism to counter salt stress in crops. Microbiol Res 235:126439
70. Belimov AA et al (2019) Rhizobial ACC deaminase contributes to efficient symbiosis with pea (Pisum sativum L.) under single and combined cadmium and water deficit stress. Environ Exp Bot 167:103859
71. Berg G et al (2016) The plant microbiome explored: implications for experimental botany. J Exp Bot 67(4):995–1002
72. Backer R et al (2018) Plant growth-promoting rhizobacteria: context, mechanisms of action, and roadmap to commercialization of biostimulants for sustainable agriculture. Front Plant Sci 9:1473
73. Zhang R, Vivanco JM, Shen Q (2017) The unseen rhizosphere root–soil–microbe interactions for crop production. Curr Opin Microbiol 37:8–14
74. Zhang Y et al (2017) Soil bacterial and fungal diversity differently correlated with soil biochemistry in alpine grassland ecosystems in response to environmental changes. Sci Rep 7(1):1–10
75. Prashar P, Kapoor N, Sachdeva S (2014) Rhizosphere: its structure, bacterial diversity and significance. Rev Environ Sci Biotechnol 13(1):63–77
76. Pinton R (2001) The rhizosphere as a site of biochemical interactions among soil components, plants, and microorganisms. In: Willig S, Varanini Z, Nannipieri P (Eds) The rhizosphere: Biochemistry and Organic Substance at the Soil-Plant Interface, 1st ed. CRC Press, Boca Raton
77. Gobat J-M, Aragno M, Matthey W (2004) The living soil: fundamentals of soil science and soil biology. Science Publishers, Enfield
78. Grayston S, Vaughan D, Jones D (1997) Rhizosphere carbon flow in trees, in comparison with annual plants: the importance of root exudation and its impact on microbial activity and nutrient availability. Appl Soil Ecol 5(1):29–56
79. Beckers B et al (2017) Structural variability and niche differentiation in the rhizosphere and endosphere bacterial microbiome of field-grown poplar trees. Microbiome 5(1):1–17
80. Massalha H et al (2017) Small molecules below-ground: the role of specialized metabolites in the rhizosphere. Wiley Online Library
81. Cesco S et al (2012) Plant-borne flavonoids released into the rhizosphere: impact on soil bio-activities related to plant nutrition. A review. Biol Fertil Soils 48(2):123–149
82. Sugiyama A, Yazaki K (2014) Flavonoids in plant rhizospheres: secretion, fate and their effects on biological communication. Plant Biotechnol 31(5):431–443
83. Edwards J et al (2015) Structure, variation, and assembly of the root-associated microbiomes of rice. Proc Natl Acad Sci 112(8):E911–E920

84. Bulgarelli D et al (2012) Revealing structure and assembly cues for Arabidopsis root-inhabiting bacterial microbiota. Nature 488(7409):91–95
85. Petrini O (1991) Fungal endophytes of tree leaves. In: Microbial ecology of leaves. Springer, pp 179–197
86. Hardoim PR et al (2015) The hidden world within plants: ecological and evolutionary considerations for defining functioning of microbial endophytes. Microbiol Mol Biol Rev 79 (3):293–320
87. Papik J et al (2020) The invisible life inside plants: deciphering the riddles of endophytic bacterial diversity. Biotechnol Adv 44:107614
88. Hardoim PR, van Overbeek LS, van Elsas JD (2008) Properties of bacterial endophytes and their proposed role in plant growth. Trends Microbiol 16(10):463–471
89. Reinhold-Hurek B, Hurek T (2011) Living inside plants: bacterial endophytes. Curr Opin Plant Biol 14(4):435–443
90. Yu X et al (2015) Effects of growth stage and fulvic acid on the diversity and dynamics of endophytic bacterial community in Stevia rebaudiana Bertoni leaves. Front Microbiol 6:867
91. Yandigeri MS et al (2012) Drought-tolerant endophytic actinobacteria promote growth of wheat (Triticum aestivum) under water stress conditions. Plant Growth Regul 68(3):411–420
92. van der Heijden MGA et al (2015) Mycorrhizal ecology and evolution: the past, the present, and the future. New Phytol 205(4):1406–1423
93. Prasad R et al (2017) Introduction to mycorrhiza: historical development. In: Varma A, Prasad R, Tuteja N (eds) Mycorrhiza – function, diversity, state of the art. Springer International Publishing, Cham, pp 1–7
94. Malhi GS et al (2020) Arbuscular mycorrhiza in combating abiotic stresses in vegetables: an eco-friendly approach. Saudi J Biol Sci 28(2):1465
95. Bonfante P, Genre A (2010) Mechanisms underlying beneficial plant–fungus interactions in mycorrhizal symbiosis. Nat Commun 1(1):1–11
96. Martin F et al (2016) Unearthing the roots of ectomycorrhizal symbioses. Nat Rev Microbiol 14(12):760–773
97. Teotia P et al (2017) Mobilization of micronutrients by mycorrhizal fungi. In: Varma A, Prasad R, Tuteja N (eds) Mycorrhiza – function, diversity, state of the art. Springer International Publishing, Cham, pp 9–26
98. Peterson RL, Massicotte HB, Melville LH (2004) Mycorrhizas: anatomy and cell biology. NRC Research Press, Ottawa
99. Garbaye J (1994) Tansley review no. 76 helper bacteria: a new dimension to the mycorrhizal symbiosis. New Phytol 128(2):197–210
100. Frey-Klett P, Garbaye J, Tarkka M (2007) The mycorrhiza helper bacteria revisited. New Phytol 176(1):22–36
101. Deveau A, Labbé J (2016) Mycorrhiza helper bacteria. In: Martin F (ed) Molecular Mycorrhizal Symbiosis. Wiley, Hoboken
102. Rigamonte TA, Pylro VS, Duarte GF (2010) The role of mycorrhization helper bacteria in the establishment and action of ectomycorrhizae associations. Brazilian J Microbiol 41(4):832–840
103. Singh R et al (2015) Biochemistry and genetics of ACC deaminase: a weapon to "stress ethylene" produced in plants. Front Microbiol 6:937
104. Barnawal D et al (2014) ACC deaminase-containing Arthrobacter protophormiae induces NaCl stress tolerance through reduced ACC oxidase activity and ethylene production resulting in improved nodulation and mycorrhization in Pisum sativum. J Plant Physiol 171(11):884–894
105. Lies A et al (2018) Using mycorrhiza helper microorganisms (MHM) to improve the mycorrhizal efficiency on plant growth. In: Meena VS (ed) Role of rhizospheric microbes in soil: volume 1: stress management and agricultural sustainability. Springer Singapore, Singapore, pp 277–298
106. Myo EM et al (2019) Indole-3-acetic acid production by Streptomyces fradiae NKZ-259 and its formulation to enhance plant growth. BMC Microbiol 19(1):155

107. Krause K et al (2015) Biosynthesis and secretion of Indole-3-acetic acid and its morphological effects on Tricholoma vaccinum-spruce ectomycorrhiza. Appl Environ Microbiol 81(20): 7003–7011
108. Pettipher GL, Jay JM, Wang HH (2005) Microbiological techniques. In: Worsfold P, Townshend A, Poole C (eds) Encyclopedia of analytical science, 2nd edn. Elsevier, Oxford, pp 16–25
109. Fuentes A et al (2020) Fungal and bacterial microbiome associated with the rhizosphere of native plants from the Atacama Desert. Microorganisms 8(2):209
110. Fouda AH et al (2015) Biotechnological applications of fungal endophytes associated with medicinal plant Asclepias sinaica (Bioss.). Ann Agric Sci 60(1):95–104
111. Verma VC et al (2009) Endophytic actinomycetes from Azadirachta indica a. Juss.: isolation, diversity, and anti-microbial activity. Microb Ecol 57(4):749–756
112. Herrera H et al (2019) Root-associated fungal communities in two populations of the fully mycoheterotrophic plant arachnitis uniflora phil.(corsiaceae) in Southern Chile. Microorganisms 7(12):586
113. Bodor A et al (2020) Challenges of unculturable bacteria: environmental perspectives. Rev Environ Sci Biotechnol 19(1):1–22
114. Emerson D et al (2008) Identifying and characterizing bacteria in an era of genomics and proteomics. Bioscience 58(10):925–936
115. Gürtler V, Stanisich VA (1996) New approaches to typing and identification of bacteria using the 16S-23S rDNA spacer region. Microbiology (Reading) 142(Pt 1):3–16
116. Hunt DE et al (2006) Evaluation of 23S rRNA PCR primers for use in phylogenetic studies of bacterial diversity. Appl Environ Microbiol 72(3):2221–2225
117. Karah N et al (2011) Species identification and molecular characterization of Acinetobacter spp. blood culture isolates from Norway. J Antimicrob Chemother 66(4):738–744
118. Lau SKP et al (2015) Chapter 12 – Gene amplification and sequencing for bacterial identification. In: Sails A, Tang Y-W (eds) Methods in microbiology. Academic Press, pp 433–464
119. Welker M (2011) Proteomics for routine identification of microorganisms. Proteomics 11(15): 3143–3153
120. Risch M et al (2010) Comparison of MALDI TOF with conventional identification of clinically relevant bacteria. Swiss Med Wkly 140:w13095
121. Rahi P, Prakash O, Shouche YS (2016) Matrix-assisted laser desorption/ionization time-of-flight mass-spectrometry (MALDI-TOF MS) based microbial identifications: challenges and scopes for microbial ecologists. Front Microbiol 7:1359–1359
122. Stewart EJ (2012) Growing unculturable bacteria. J Bacteriol 194(16):4151–4160
123. Pham VHT, Kim J (2012) Cultivation of unculturable soil bacteria. Trends Biotechnol 30(9): 475–484
124. Muyzer G, Smalla K (1998) Application of denaturing gradient gel electrophoresis (DGGE) and temperature gradient gel electrophoresis (TGGE) in microbial ecology. Antonie Van Leeuwenhoek 73(1):127–141
125. Al-Mailem DM, Kansour MK, Radwan SS (2017) Capabilities and limitations of DGGE for the analysis of hydrocarbonoclastic prokaryotic communities directly in environmental samples. Microbiology 6(5):e00495
126. Ravin NV, Mardanov AV, Skryabin KG (2015) Metagenomics as a tool for the investigation of uncultured microorganisms. Russ J Genet 51(5):431–439
127. Baker GC, Smith JJ, Cowan DA (2003) Review and re-analysis of domain-specific 16S primers. J Microbiol Methods 55(3):541–555
128. Muhamad Rizal NS et al (2020) Advantages and limitations of 16S rRNA next-generation sequencing for pathogen identification in the diagnostic microbiology laboratory: perspectives from a middle-income country. Diagnostics 10(10):816
129. Kaul S, Sharma T, Dhar MK (2016) "Omics" Tools for better understanding the plant-endophyte interactions. Front Plant Sci 7:955–955

130. Marchesi JR, Ravel J (2015) The vocabulary of microbiome research: a proposal. Microbiome 3(1):31
131. Alibrandi P et al (2020) Diversity and structure of the endophytic bacterial communities associated with three terrestrial orchid species as revealed by 16S rRNA gene metabarcoding. Front Microbiol 11:604964
132. Li O et al (2017) Bacterial and diazotrophic diversities of endophytes in Dendrobium catenatum determined through barcoded pyrosequencing. PLoS One 12(9):e0184717
133. Yu J et al (2013) Design and application of specific 16S rDNA-targeted primers for assessing endophytic diversity in Dendrobium officinale using nested PCR-DGGE. Appl Microbiol Biotechnol 97(22):9825–9836
134. Pei C et al (2017) Diversity of endophytic bacteria of Dendrobium officinale based on culture-dependent and culture-independent methods. Biotechnol Biotechnol Equip 31(1):112–119
135. Lin M et al (2020) The effect of plant geographical location and developmental stage on root-associated microbiomes of Gymnadenia conopsea. Front Microbiol 11:1257
136. Pavlova A et al (2017) Colonization strategy of the endophytic plant growth-promoting strains of Pseudomonas fluorescens and Klebsiella oxytoca on the seeds, seedlings and roots of the epiphytic orchid, Dendrobium nobile Lindl. J Appl Microbiol 123(1):217–232
137. Tsavkelova EA et al (2007) Orchid-associated bacteria produce indole-3-acetic acid, promote seed germination, and increase their microbial yield in response to exogenous auxin. Arch Microbiol 188(6):655–664
138. Tsavkelova EA et al (2016) Dendrobium nobile Lindl. Seed germination in co-cultures with diverse associated bacteria. Plant Growth Regul 80(1):79–91
139. Vorholt JA (2012) Microbial life in the phyllosphere. Nat Rev Microbiol 10(12):828–840
140. Alibrandi P et al (2020) Plant growth promoting potential of bacterial endophytes from three terrestrial mediterranean orchid species. Plant Biosyst https://doi.org/10.1080/11263504.2020.1829731
141. White JF et al (2014) Occurrence of B acillus amyloliquefaciens as a systemic endophyte of vanilla orchids. Microsc Res Tech 77(11):874–885
142. Faria DC et al (2013) Endophytic bacteria isolated from orchid and their potential to promote plant growth. World J Microbiol Biotechnol 29(2):217–221
143. Altinkaynak H, Ozkoc I (2020) Isolation and molecular characterization of plant growth promoting bacteria from the rhizosphere of orchids in Turkey. Rhizosphere 16:100280
144. Tsavkelova E, Cherdyntseva T, Netrusov A (2005) Auxin production by bacteria associated with orchid roots. Microbiology 74(1):46–53
145. Tsavkelova EA et al (2007) Bacteria associated with orchid roots and microbial production of auxin. Microbiol Res 162(1):69–76
146. Aguilar Díaz T et al (2018) Rizobacterias promotoras de crecimiento en Guarianthe skinneri (Orchidaceae). Rev Biol Trop 66(3):953–968
147. Ortiz J et al (2019) The endophytic fungus Chaetomium cupreum regulates expression of genes involved in the tolerance to metals and plant growth promotion in Eucalyptus globulus roots. Microorganisms 7(11):490
148. Srivastava S, Kadooka C, Uchida JY (2018) Fusarium species as pathogen on orchids. Microbiol Res 207:188–195
149. Hou XQ, Guo SX (2009) Interaction between a dark septate endophytic isolate from Dendrobium sp. and roots of D. nobile seedlings. J Integr Plant Biol 51(4):374–381
150. Herrera H et al (2020) Isolation and identification of plant growth-promoting bacteria from rhizomes of Arachnitis uniflora, a fully mycoheterotrophic plant in southern Chile. Appl Soil Ecol 149:103512
151. Adamo M et al (2020) The dark side of orchid Symbiosis: can Tulasnella calospora decompose host tissues? Int J Mol Sci 21(9):3139
152. Tsavkelova E, Cherdyntseva T, Netrusov A (2004) Bacteria associated with the roots of epiphytic orchids. Microbiology 73(6):710–715

153. Yang S et al (2014) Growth-promoting S phingomonas paucimobilis ZJSH 1 associated with D endrobium officinale through phytohormone production and nitrogen fixation. Microb Biotechnol 7(6):611–620
154. Novotna A, Suárez J (2018) Molecular detection of bacteria associated with Serendipita sp., a mycorrhizal fungus from the orchid Stanhopea connata Klotzsch in southern Ecuador. Botany Lett 165(2):307–313
155. Pakvaz S, Soltani J (2016) Endohyphal bacteria from fungal endophytes of the Mediterranean cypress (Cupressus sempervirens) exhibit in vitro bioactivity. For Pathol 46(6):569–581
156. Bravo D et al (2013) Isolation of oxalotrophic bacteria able to disperse on fungal mycelium. FEMS Microbiol Lett 348(2):157–166
157. Deng Z-S et al (2011) Paracoccus sphaerophysae sp. nov., a siderophore-producing, endophytic bacterium isolated from root nodules of Sphaerophysa salsula. Int J Syst Evol Microbiol 61(3):665–669
158. Young CC et al (2006) Encapsulation of plant growth-promoting bacteria in alginate beads enriched with humic acid. Biotechnol Bioeng 95(1):76–83
159. Vemmer M, Patel AV (2013) Review of encapsulation methods suitable for microbial biological control agents. Biol Control 67(3):380–389
160. Basu S et al (2021) Role of soil microbes in biogeochemical cycle for enhancing soil fertility. In: New and future developments in microbial biotechnology and bioengineering. Elsevier, pp 149–157
161. Chaudhary DR, Rathore AP, Sharma S (2020) Effect of halotolerant plant growth promoting rhizobacteria inoculation on soil microbial community structure and nutrients. Appl Soil Ecol 150:103461
162. Wilkinson K, Dixon K, Sivasithamparam K (1989) Interaction of soil bacteria, mycorrhizal fungi and orchid seed in relation to germination of Australian orchids. New Phytol 112(3): 429–435

Mycorrhiza in Orchids

7

Saranjeet Kaur

Contents

Abstract

The orchids are a highly medicinal and floriferous assemblage of flowering plant species. Each orchid fruit encloses thousands of dust like minute and highly reduced seeds. Due to lack of endosperm and presence of complex carbohydrates, these seeds require specific fungal partner to accomplish germination in nature. The fungus plays a crucial role in the germination of the orchid seeds and their growth and development in mature orchid plants. In this manuscript, some intricacies pertaining to mycorrhizal interactions are being discussed.

S. Kaur (✉)
Department of Chemistry, University Institute of Sciences, Chandigarh University, Mohali, Punjab, India
e-mail: sarana_123@rediffmail.com

J.-M. Mérillon, H. Kodja (eds.), *Orchids Phytochemistry, Biology and Horticulture*, Reference Series in Phytochemistry, https://doi.org/10.1007/978-3-030-38392-3_7

Keywords

Endomycorrhiza · Monocot · Orchid seeds · Root cortex · Symbiosis

1 Introduction

The orchids are highly evolved group of angiosperm plants belonging to monocot family Orchidaceae. In the world, a total of 28,000 species are confined to 763 orchid genera [1]. Despite of being cosmopolitan in distribution, they do not appear as dominant vegetation in any part of the world.

The orchids produce a large number of structurally and functionally highly reduced, microscopic seeds (Fig. 1). Although, the seeds are produced in enormous quantities in a single capsule, merely 0.2–0.3% germinate in nature, whereas countless perish away. Moreover, the seeds lack endosperm tissue and possess little amount of complex carbohydrates as reserve food material which the seeds are unable to utilize. The undifferentiated embryos of orchid seeds lack distinct root and shoot meristem. Therefore, orchid seeds necessarily require specific mycobionts which could convert complex carbohydrates into simple molecules and provide them to the germinating entities. Various orchid mycobionts establish a symbiotic relationship with the roots of orchid plants. All orchid species are myco-heterotrophic at certain stage in their life cycle. In symbiotic association, the specific fungus invades into the seed as well as roots. The fungus forms loosely coiled structures called pelotons in the parenchymatous cells of the cortical region of the roots. The orchids maintain symbiotic association with fungal endophytes throughout their lifecycle to obtain nutrients, sugars, and minerals [2]. Conversely, there are some orchid species that establish symbiotic association with their specific fungal endophytes only in severe conditions [3]. The orchid seeds germinate into a pyriform structure called protocorm; these get associated with specific mycorrhizal partners. The achlorophyllous orchid species such as *Corallorhiza maculata* and *Rhizanthella* species known as achlorophyllous myco-heterotrophs maintain their fungal symbionts throughout their life cycle to get nutrition.

Fig. 1 Minute orchid seeds

Table 1 Shows endophytic fungi promote enzyme activity in orchid species

Host Plant	Name of the endophytic fungus	Activity	Reference
Anoectochilus formosanus	*Epulorhiza* sp.	Enhances activities of following the enzymes: chitinase, β-1,3-glucase, phenylalanine ammonia-lyase, polyphenol oxidase	[7]
Anoectochilus roxburghii	*Epulorhiza* sp., *Mycena anoectochila*	Enhance enzyme activities	[8, 9]
Cymbidium sinense	*Mycena orchdicola*	Secretes phytohormones	[10]
Dendrobium candidum	*Mycena dendrobii*	Secretes phytohormones	[10]
Dendrobium nobile, *D. chrysanthum*	*Epulorhiza* sp., *Mycena* sp., Tulasnellales, Sebacinales, Cantharellales	Enhance the absorption of nutrients in plants, promotes seed germination of the host plant	[11]
Pecteilis susannae	*Epulorhiza* sp., *Fusarium* sp.	Enhance the absorption of N, P, and K elements in plants promoting the seed germination of host	[12]

The importance of a variety of phytohormones in the regulation of entire life process including plant development and their defense response has long been established in orchids. Besides providing nutrients such as nitrogen and phosphorus for various life processes in orchid plants, the fungal partners are also reported to be secreting certain growth hormones, for instance, cytokinins and auxins mainly (indole-3-acetic acid, indole-3-acetonitrile) [4, 5]. A study in *Gastrodia elata* demonstrated that the fungal *Mycena dendrobii* secreted IAA which promoted seed germination and growth of its host plant [6]. Literature studies also reported that endophytic fungi promote the growth of the host plants by enhancing the enzyme activity (Table 1).

2 Geological Location and Environment

Some orchids are extremely specific in selecting their symbionts, as they prefer a single genus of fungi. *Corallorhiza maculata*, a myco-heterotroph, associate only with *Russulaceae* irrespective of their geological location and presence of other orchids in its vicinity [13].

Certain orchid species change their symbiotic fungal partner in response to environmental stress with regard to variations in altitude from tropical to temperate regions [3]. *Goodyera pubescens* associates only with one fungal mycobiont, if not subjected to changes in the environment, like drought, etc. The orchids with high degree of specialization have fewer fungal associations [14]. Some other orchid species, for instance, *Chloraea collicensis* and *Chloraea gavilu*, make symbiotic association with only *Rhizoctonia* [15].

3 Mycobiont Invasion in Orchid Tissues at Different Stages of Development

The fungus invades into the tissues of orchid at its different stages of development. Fungal hyphae penetrate inside the testa of seed through the opening and into the parenchymatous cells of germinating orchid seeds, protocorms, and cortex of the roots. Certain physiological and cytological changes also take place at the time of the invasion of fungus into parenchyma cells. Increased number of mitochondria and few vacuoles enhances metabolic activity of the parenchyma cells of embryo. Fungal mycobiont enter into the protocorms through the chalazal end of the embryo [16, 17].

4 Root Cortex and Fungal Pelotons

In the present study, it was observed that the fungus penetrated into roots mainly through root hair tips. Once the fungus enters into the parenchymatous cells of cortex of the orchid root, its hyphae start coiling loosely inside the cells; these coiled structures are known as pelotons [18, 19]. Pelotons inhabit the parenchymatous region of the cortex of roots (Fig. 2), which is an important anatomical feature of orchid mycorrhiza that clearly distinguishes it from the other forms of fungi [20].The pelotons vary in their size, packaging, and arrangement of their hyphal mass [21]. The pelotons remain functionally active for certain period. Later, these disintegrate inside the cortical root cell forming a sort of round clumped or disc-like structure. In present study, it was also observed that the pelotons of adjacent cells make seen interconnection with each other (Fig. 3).

As soon as the fungus invades into the root cell, it undergoes biochemical changes. The cells with disintegrated pelotons lack starch grains, whereas the newly invaded cortical root cells possess large starch grains, which indicate the hydrolysis of starch grains after the fungus colonization [14]. When pelotons

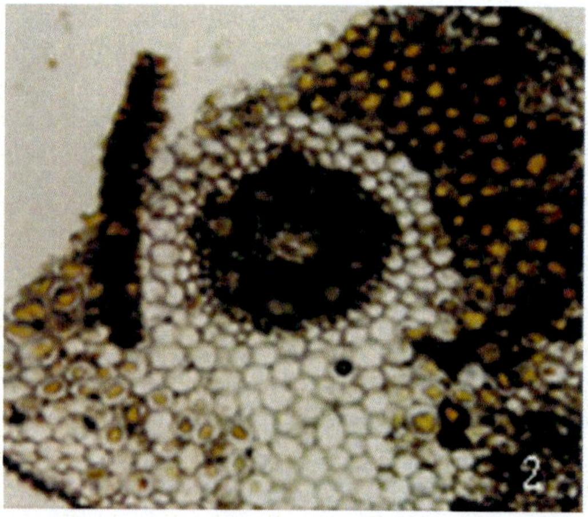

Fig. 2 Transverse hand-section of orchid root showing fungal invasion in the parenchyma cells of cortex

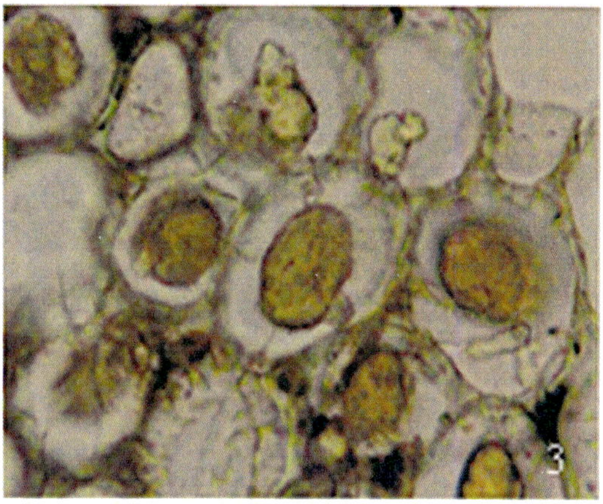

Fig. 3 Pelotons interconnected with the pelotons of neighboring cells

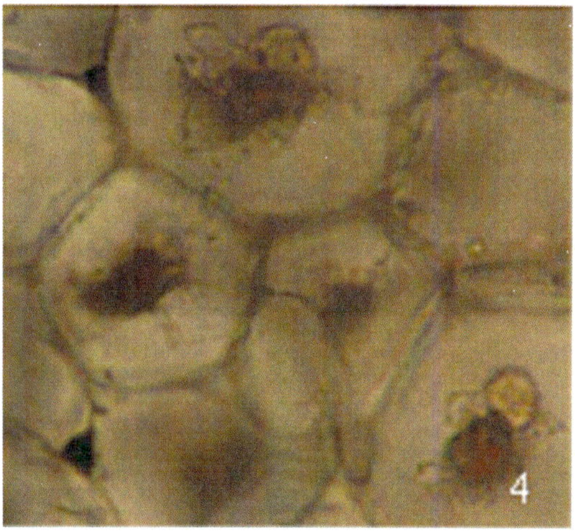

Fig. 4 Completely lysed pelotons

disintegrate or are lysed, they appear as brown or yellow clumps in the orchid cells [22] as also observed in the present study (Fig. 4). At the time of peloton disintegration, certain structural and physiological changes also occur [23]. Certain cytological changes also occur in the invaded cell. The nucleus becomes conspicuous, it enlarges in size considerably, and shows increased amount of DNA content [21]. The increased DNA content is correlated with the differentiation of parenchyma cells suggesting its role in orchid growth [24].

In orchids, two types of host cells such as digestive cells and host cells are involved in nutrient transfer [25]. The digestive cells are engaged in dense peloton development followed by digestion and subsequent reinvasion, whereas the host cells contain live hyphae in pelotons, which are not digested, or at least not as readily [25]. There is an additional mode of nutrient transfer known as phytophagy. It involves lysis of fungal cells. The complete fungal hyphae are not digested, but only the growing end (tip region) is degraded, and the cell contents are released into interfacial space between the plant and hyphal membrane and are provided to the plant [20].

5 Fungal Members as Orchid Mycorrhiza

5.1 *Rhizoctonia* Fungi

The fungi that act as orchid mycorrhizae belong to class basidiomycetes. These basidiomycetous fungi include certain genera of fungus, namely, *Rhizoctonia*, *Sebacina*, *Tulasnella*, and *Russula* species. Most of the orchid species build up their association with saprotrophic or pathogenic fungi, whereas a few show preference for ectomycorrhizal fungal species. The orchid species show association with different fungal partners at different developmental stages in their life cycle. These associations could be at the time of either seed germination or protocorm development or could be throughout the life cycle of an orchid plant species [26]. Moreover, different orchid species show preference for their symbiotic fungi depending on the type of environmental niches in which they thrive, whether terrestrial or growing on other plants as an epiphyte [27].

5.2 Basidiomycetous and Ascomycetous Ectomycorrhiza

Few species of basidiomycetous fungi are also reported to be making symbiotic association with orchid species, but they do not belong to *Rhizoctonia*. Specific myco-heterotrophic orchids are also reported to be associated with ectomycorrhizal basidiomycetes that belong to genera such as *Thelephora*, *Tomentella*, and *Russula*. Ascomycetous fungi are rarely seen establishing symbiotic connections with orchid species. A terrestrial orchid species *Epipactis helleborine* has a specific association with ectomycorrhizal ascomycetes in the *Tuberaceae*.

6 Orchid Specificity for a Symbiont

At successive developmental stages, orchid species show preference for their fungal mycobionts. Terrestrial orchid species build symbiotic association with members of family *Tulasnellaceae*, yet a few autotrophic and saprophytic orchids make association with several ectomycorrhizal fungi also [28, 29]. Few clades of fungus *Rhizoctonia* also show association with some epiphytic species [30]. *Rhizoctonia* fungi can form symbiotic relationship with either an epiphytic or terrestrial orchid, but very rarely, they associate with both [30]. It has been shown through seed baiting techniques that the seeds of *Dendrobium aphyllum* germinate when they make symbiotic association with *Tulasnella*, but do not germinate when treated with *Trichoderma* isolated from orchid plant, which strongly advocates about the specificity of symbionts to different developmental stages. The preference for symbiosis with a fungal partner differs at various developmental stages of an orchid species [31, 32]. With the advancing age of an orchid plant, the fungal associations also become more complex. *Cephalanthera longibracteata*, a mixotrophic orchid species, symbiotically associates with numerous fungal species belonging to family *Russulaceae*, *Tricholomataceae*, *Sebacinales*, and *Thelephoraceae* [33].

7 Peloton Formation and Mycophagy or Necrotrophy

Metabolically active and live pelotons facilitate the transfer of nitrogen and carbon; however, when fungal pelotons get digested, most of the nitrogen and carbon is absorbed by the plant itself by the process of mycophagy [25, 34]. Shortly, after the invasion of fungus into the cortical tissues and peloton formation, their lysis starts [35]. Pelotons are formed in a unique way. At the time of their development, a thin membrane surrounds them, which eventually acts as endoplasmic reticulum surrounded by Golgi apparatus. Afterward, inside this cover, digestive enzymes are secreted into the space between the plant membrane and peloton to digest them [36]. Further, the digestion of pelotons starts; at the same time, a secondary membrane is also formed around the fungal peloton which is a large vacuole and allows the degradation of the peloton in isolated manner [36]. Additionally peroxisomes accumulate within the digestive plant cells and undergo exocytosis into the newly formed vacuole; a number of enzymes concentrate such as chitinases, uricases, peptidases, oxidases, and catalases are secreted which ultimately, breaks down the peloton [35, 36].The fungal remnants are consumed by the plant itself, thus, transferring the nutrients to the host plant [25].

Few experimental studies using stable isotope imaging technique reveal that C13 and N15 when applied to mycorrhizal hyphae got readily transported to the host plant through fungal pelotons, leading to an inconsistent quantity of these isotopes inside the peloton containing plant cell and the peloton itself. It had also been observed that the senescing pelotons contain higher concentrations of C13 and N15 isotopes than in live pelotons [34].The fungal hyphae also undergo morphological change; they swell before collapsing, most probably due to the increased load of nutritive compounds [34]. Once the pelotons are completely digested, reinvasion into the digestive cell occurs shortly after, and a new peloton starts forming again [25]. Reinvasion and digestion occur cyclically throughout the entire life span of the symbiotic association [35].

8 Orchid Mycorrhiza and Nutrient Transport

Orchid mycorrhizal associations involve a variety of nutrient transport systems, structures, and phenomenon which are specifically found in the family Orchidaceae. These interactions are observed between basidiomycetes and almost all species of Orchidaceae [37]. In basidiomycete, the fungi *Rhizoctonia* is generally found associated with orchid species. *Rhizoctonia* is known for its saprophytic abilities also and establishes anomalous associations [37]. The orchid plant species which inhabit dense and highly shaded forest areas depend completely on their mycorrhiza for nutrition such as carbon [35, 38]. Orchid seeds being extremely reduced and non-endospermic show an obligatory parasitic stage during germination and draw nutrients with the help of mycorrhizal fungus [39]. After the orchid seed germination, the orchid fungal interactions become specific to utilize the carbon and available nutrients. These associations are often governed by the orchid plant itself [40]. Orchid mycorrhizal interactions could either be completely parasitic on the fungal associate or show mutualistic interaction, thus, establishing bidirectional nutrient flow between the plant and mycorrhizal fungus [40]. In the natural habitats, orchid mycorrhiza shows a specific mycorrhizal nutrient transfer interaction upon which the diversity of the orchid genera depends [35].

9 Nutrient Transfer Mechanism in Orchids

Orchid mycorrhizal interactions show unique flow of nutrients. In the arbuscular mycorrhizal associations, it is observed that plant species channelize unidirectional supply to fungus with carbon swapping with either or both, phosphorus or nitrogen, depending on the environment [41, 42]. There occurs bidirectional flow of carbon between the fungus and plant, besides flow of nitrogen and phosphorus from the fungus to plant. Nearly 400 plant species are not able to provide carbon to their system. In fact, all of the nutrients of the plant are supplied by the fungus [40]. However, in these interactions, carbon gain by the plant is positive in the majority of the observed interactions [25]. Peloton starts forming shortly 20–36 hours, after the invasion of fungus [35]. The plasma membrane of invaded parenchyma cells also assists with the fungal infection and its further growth [43]. For exchange of nutrient materials between pelotons and surrounding plasma membrane, extensive surface area is created. The invaginated plasma membrane surrounds the growing pelotons and creates vast surface area from which nutrients can be exchanged. The pelotons are highly coiled fungal hyphal mass as compared to endomycorrhiza of arbuscular mycorrhiza [35, 44, 45]. During fungal invasion, increase in ribosomes also takes place. Plasma membrane participates in the exchange of nutrients between plant and fungus besides the enzyme excretion inside the space termed interfacial apoplast [40, 46].

The pelotons are not permanent structures; these are swiftly digested within a few hours of their formation in orchid parenchyma cells. The digestion of pelotons is a universal feature observed in almost all endomycorrhizal associations; in case of orchid species, these coiled structures get digested sooner after their formation [34].

10 Transport of Phosphorus

Phosphorus is a macroelement which is required by all plants. Phosphorus is taken up by the mycorrhizal fungus from the soil particles in three distinct forms: inorganic phosphorus, organic phosphorus, and phosphate. Deficiency of phosphate in soil allows the formation of a symbiotic relationship between plant and fungus. Mycorrhizal fungi are capable of increasing soil surface area besides initiating the secretion of a variety of enzymes [41, 42, 47]. Inorganic phosphate is transferred either through active transport as phosphate through Pi transporters (inorganic phosphorus) and is moved out of the fungal hyphae into the interfacial apoplast, where it forms dihydrogen phosphate and then subsequently gets transferred by active Pi transporters into the plant cell or it depends upon passive efflux of Pi out of the fungus and active absorption by the plant [41, 47]. For efficient transport of phosphorus, these pathways depend on comparative high concentrations of Pi inside fungal cell and low concentration of Pi inside the plant cell. The second method is more dependent on this condition. Certain genes which are Pi transporter genes such as MtPT4 and StPT3 are known to regulate the exchange of phosphorus in orchid plants along with H+ ATPase transporters [42]. In orchid species, once mycorrhizal symbiotic connections are established afterward phosphorus is obtained by the plant only through the metabolically active pelotons of fungal tissue; once the digestion of pelotons starts degrading simultaneously, the flow of phosphorus also ceases [47].

11 Transfer of Other Micronutrients

Mostly, passive transport helps in the transfer of micronutrients across the cell membranes, both during absorption from soil by fungi and further from fungi to the host plant [41]. But under specific conditions, active transport of micronutrients takes place as well [48]. The upregulation of cation transporters is seen in orchid *D. officinale* symbioses, suggesting that fungi make possible the transfer of nutrients from fungi to plant [49]. Cation, such as iron, mostly found adhered tightly to the organic substrates and remains out of reach of plants and fungi. There are certain compounds, for instance, siderophores (small molecule which have high affinity for Fe^{3+} utilized by fungal species), which are secreted into the soil by fungi to acquire these cations [50]. These cations are liberated into the soil around the hypha and absorb iron from the soil. These siderophore molecules are reabsorbed into the fungal mycelium where the iron has to be dissociated from the siderohore and quickly utilized [50]. The orchid species possess siderophores in association with mycorrhizal fungi within the genus *Rhizoctonia* which can utilize the siderophore "basidiochrome" as the major iron-chelating compound [48]. Apart from these known chelating compounds, other vital nutrients may also be transferred between mycorrhizal fungi and orchid plants through specialized methods also.

12 Transport of Nitrogen

Nitrogen transport is an equally important and essential process that often occurs through mycorrhizal associations [51]. Nitrogen is abundant and much easy to obtain as compared to phosphorus. Mycorrhizal interactions give a significant benefit in the allocation of nitrogen. Bioavailable nitrogen (nitrate and ammonium) is absorbed from the soil media by the mycorrhizal fungi and further assimilated into the amino acids [42]. There are a few proposed mechanisms by which nitrogen is transferred to the host plant. These pathways are biotrophic; a significant amount of nitrogen may also be transferred necrotropically but through a distinct process [51].

In pathway-1, the amino acids get transferred into the extra-radical mycelium where these amino acids are broken down. The amino acid such as arginine is synthesized and transferred intra-fungally and later on is catabolized into ammonium ions and is moved into the interfacial space between peloton and surrounding plant membrane and later on transported into the plant cell through ammonium transporters and incorporated into the plant [46].

In the transportation of nitrogen primarily as ammonium, certain ammonium transporter genes get regulated in the plant and mycorrhizal fungal associate which further regulate a class of genes called "protease genes" along with other three transporters such as an external amino acid permeases, nitrate transporters, and ammonium transporters [41, 46]. Nitrogen may also be transferred in the form of other amino acids, namely, arginine, glycine, and glutamine into the cell through specialized amino acid transporters. The mycorrhizal fungus *T. Calospora* in symbiotic association with orchid plant is able to regulate the expression of SvAAP1 and SvAAP2. These genes encode amino acid permeases that strongly support amino acids to be important molecules involved in nitrogen transport [34, 46, 47]. The transport of inorganic nitrogen in the form of ammonium and the transport of organic nitrogen as amino acids occur simultaneously [35]. In fact, it has been proved through isotope (C13 and N15) studies that amino acids may be the primary nitrogen compound transferred in the orchid [46].

13 Transport of Carbon

As soon as the fungus gets carbon, it converts it into sugar mainly trehalose and assimilates into the fungal mycelium. The transport of carbon from fungal partner to plant cells occurs in one of two forms primarily trehalose, but carbon can also get converted into glucose and sucrose or as an amino acid arginine but can also be converted into glycine and glutamine [34, 40, 52].The transport of these molecules occurs through specialized amino acid permeases and carbohydrate transporter protein molecules. These molecules are fixed into the fungal peloton membrane, into the interfacial space where they get absorbed into the plant cell by similar transporter protein molecules in the orchid cell endoplasmic reticulum membrane surrounding the hyphal coils [34, 52].

The active transport of carbon from symbiotic fungal partner to plant cell is a biotrophic phenomenon. A significant amount of carbon gets transferred inside plant cell when the fungal pelotons are degraded and digested. Genes encoding the transporter proteins get regulated, simultaneously, both in plant and fungi, in the similar fashion as nitrogen and phosphorus compound transporter genes during symbiosis [46].

Orchid mycorrhizal interactions consist of various symbiotic fungi, ranging from myco-heterotrophic plants (*Monotropa uniflora*) to chlorophyllous orchid species such as *Goodyera repens* [40, 52, 53]. As mentioned earlier in the text that during symbiotic interactions in orchid mycorrhiza carbon is translocated readily from fungi to the plant tissues. Conversely, this may or may not occur with the transfer of carbon from plant to fungi [25, 52, 54]. Orchid mycorrhiza is more or less considered to be showing partial myco-heterotrophic interactions [35]. In myco-heterotrophic orchids, carbon is taken up, by the fungi (basidiomycetes), in the form of molecules of peptide and carbohydrate [35, 55, 56]. Specific genes that codes for proteases and cellulose, lignin digestive enzymes, as well as oligopeptide and carbohydrate transporters get regulated mycelium present in soil to support enhanced carbon uptake [46].

Interestingly, it is also observed that apart from orchids, myco-heterotrophic fungi interact with roots of beech trees as well [57]. Few studies report that myco-heterotrophic fungi are also involved in the translocation of photosynthates from tree to the fungus and then to orchid, but a thorough investigation is required to unravel such intricacies in orchid fungus interactions [58].

14 Conclusion

The orchids have meticulously evolved to survive in nature in spite of having highly reduced seed structure. It becomes imperative to understand orchid fungus intricacies for conservation purpose. A few studies have been reported about orchid fungus interactions. The knowledge of species-specific as well as developmental stage-specific fungal symbionts would be a step toward saving them from getting extinct in nature. More morphological and molecular studies are required for the identification of fungal symbionts that inhabit at every stage of their development. It would prove useful in replenishing their natural resources, thus saving them for future use.

Acknowledgments English language assistance from Dr. H.S. Sekhon is acknowledged with deep gratitude.

References

1. Christenhauz MJM, Byng JW (2016) The number of known plants species in the world and its annual increase. Phytotaxa 261:201–217
2. Pecoraro L, Girlanda M, Kull T, Perini C, Perotto S (2012) Molecular identification of root fungal associates in Orchis pauciflora Tenore. Plant Biosys 146:985–991

3. McCormick MK, Whigham Dennis F, Sloan DM, Hodkinson K, Brendan H (2006) Orchid–fungus Fidelity: a marriage meant to last? Ecol 87:903–911
4. Zhang J, Jia W, Yang J, Ismail AM (2006) Role of ABA in integrating plant responses to drought and salt stresses. Field Crop Res 97:111–119
5. Hartley SE, Gange AC (2009) Impacts of plant symbiotic Fungi on insect herbivores: mutualism in a multitrophic context. Ann Rev Ento 54:323–342
6. Guo SX, Wang QY (2001) Character and action of good strain on stimulating seed germination of *Gastrodia elata*. Mycosystema 3:408–412
7. Tang MJ, Meng ZX, Guo SX (2008) Effect of endophytic fungi on the content of sugars and inorganic elements of cultivated seedling of *Anoectochilus*. Chin Tradit Herb Drugs 39:1565–1568
8. Yu XM, Guo SM (2000) Establishment of symbiotic system for Anoectochilus roxburghii (wall.) Lindl. And endophytic fungi. J Chin Materia Med 25:81–82
9. Chen X, Wu CH, Tang JJ, Hu SJ (2005) Arbuscular mycorrhizae enhance metal lead uptake and growth of host plants under a sand culture experiment. Chemosphere 60:665–671
10. Zhang X, Lund AA, Sarath G, Cerny RL, Roberts DM, Chollet R (1999) Soybean nodule sucrose synthase (Nodulin-100): further analysis of its phosphorylation using recombinant and authentic root-nodule enzymes. Arch Biochem Biophys 371:70–82
11. X-m C, Guo S-X (2005) Effects of four species of endophytic fungi on the growth and polysaccharide and alkaloid contents of *Dendrobium nobile*. China Materia Med 30:253–257
12. Chutima R, Dell B, Suyanee V, Boonsom B (2011) Endophytic fungi from *Pecteilis susannae* (L.) Rafin (Orchidaceae), a threatened terrestrial orchid in Thailand. Mycorrhiza 21:221–229
13. Taylor D, Lee B, Thomas D, Hodges SA (2004) Evidence for mycorrhizal races in a cheating orchid. Proceedings of biological sciences. R Soc Pub 271(1534):35–43
14. Jacquemyn H, Honnay O, Cammue Bruno PA, Brys R, Lievens B (2010) Low specificity and nested subset structure characterize mycorrhizal associations in five closely related species of the genus Orchis. Mol Ecol 19:4086–4095
15. Pereira G, Romero C, Suz LM, Atala C (2014) Essential mycorrhizal partners of the endemic Chilean orchids Chloraea collicensis and C. gavilu. Flora Morph Dist Funct Ecol Plants 209:95–99
16. Peterson RL, Currah RS (1990) Synthesis of mycorrhizae between protocorms of Goodyera repens (Orchidaceae) and Ceratobasidium cereale. Can J Bot 68:1117–1125
17. Leake JR (1994) The biology of myco-heterotrophic ('saprophytic') plants. New Phytol 127:171–216
18. Pecoraro L, Girlanda M, Kull T, Perini C, Perotto S (2012) Molecular identification of root fungal associates in *Orchis pauciflora*. Plant Biosys 146:985–991
19. Smith SE, Read DJ, Harley JL (1997) Mycorrhizal symbiosis, 2nd edn. Academic Press, San Diego
20. Rasmussen HN (2002) Recent developments in the study of orchid mycorrhiza. Plant Soil 244:149–163
21. Hadley G, Williamson B (1971) Analysis of the post-infection growth stimulus in orchid mycorrhiza. New Phytol 70:445–455
22. Selosse M-A, Minasiewicz J, Boullard B (2017) On the germination of *Neottia nidus-avis*. Mycorrhiza 27:611–618
23. Peterson RL, Uetake Y, Bonfante P, Faccio A (1996) The interface between fungal hyphae and orchid protocorm cells. C J Botany 74:1861–1870
24. Alvarez MR (1968) Quantitative changes in nuclear DNA accompanying post germination embryonic development in Vanda (Orchidaceae). Am J Bot 55:1036–1041
25. Rasmussen HN (2009) Orchid mycorrhiza: implications of a mycophagous life style. Oikos 118:334–345
26. Shefferson RP, WeiB M, Kull T, Taylor DL (2005) High specificity generally characterizes mycorrhizal association in rare lady's slipper orchids, genus *Cypripedium*. Mol Ecol 14:613–626
27. Alghamdi SA (2017) Influence of mycorrhizal fungi on seed germination and growth in terrestrial and epiphytic orchids. Saudi J Biol Sci 26:495–502

28. Bidartondo MI, Burghardt B, Gebauer G, Bruns TD, Read DJ (2004) Changing partners in the dark: isotopic and molecular evidence of ectomycorrhizal liaisons between forest orchids and trees. Proc Biol Sci 271(1550):1799–1806. The Royal Society Publishers
29. McKendrick SL, Leake JR, Read DJ (2000) Symbiotic germination and development of myco-heterotrophic plants in nature: transfer of carbon from ectomycorrhizal *Salix repens* and *Betula pendula* to the orchid *Corallorhiza trifida* through shared hyphal connections. New Phytol 145:539–548
30. Martos F, Munoz F, Pailler T, Kottke I, Gonneau C, Selosse M-A (2012) The role of epiphytism in architecture and evolutionary constraint within mycorrhizal networks of tropical orchids. Mol Ecol 21:5098–5109
31. Zi X-M, Sheng C-L, Goodale UM, Shao S-C, Gao J-Y (2014) *In situ* seed baiting to isolate germination-enhancing fungi for an epiphytic orchid, *Dendrobium aphyllum* (Orchidaceae). Mycorrhiza 24:487–499
32. McKendrick SL, Leake JR, Taylor DL, Read DJ (2002) Symbiotic germination and development of the myco-heterotrophic orchid *Neottia nidus-avis* in nature and its requirement for locally distributed *Sebacina* spp. New Phytol 154:233–247
33. Sakamoto Y, Yokoyama J, Maki M (2015) Mycorrhizal diversity of the orchid Cephalanthera longibracteata in Japan. Myco Sci 56:183–189
34. Kuga Y, Sakamoto N, Yurimoto H (2014) Stable isotope cellular imaging reveals that both live and degenerating fungal pelotons transfer carbon and nitrogen to orchid protocorms. New Phytol 202:594–605
35. Dearnaley JDW, Martos F, Selosse MA (2012) Orchid mycorrhizas: molecular ecology, physiology, evolution and conservation aspects. In: Esser K (ed) The mycota, vol IX – fungal associations, 2nd edn. Springer, Berlin, pp 207–230
36. Barroso J, Casimiro A, Carrapico F, Pais M, Salome S (1988) Localization of uricase in mycorrhizas of *Ophrys lutea*. New Phytol 108:335–340
37. Selosse M-A, Weiss M, Jany J-L, Tillier A (2002) Communities and populations of sebacinoid basidiomycetes associated with the achlorophyllous orchid Neottia nidus-avis (L.) L.C.M. Richb. And neighbouring tree ectomycorrhizae. Mol Ecol 11:1831–1844
38. Girlanda M, Segreto R, Cafasso D, Liebel HT, Rodda M, Ercole E, Cozzolino S, Gebauer G, Perotto S (2011) Photosynthetic mediterranean meadow orchids feature partial mycoheterotrophy and specific mycorrhizal associations. Am J Bot 98:1148–1163
39. Alexander C, Hadley G (1985) Carbon movement between host and mycorrhizal endophyte during the development of the orchid *Goodyera repens*. New Phytol 101:657–665
40. Zimmer K, Hynson NA, Gebauer G, Allen EB, Allen MF, Read DJ (2007) Wide geographical and ecological distribution of nitrogen and carbon gains from fungi in pyroloids and monotropoids (Ericaceae) and in orchids. New Phytol 175:166–175
41. Smith SE, Gianinazzi PV, Koide R, Cairney JWG (1994) Nutrient transport in mycorrhizas: structure, physiology and consequences for efficiency of the symbiosis. Plant Soil 159:103–113
42. Balestrini R, Nerva L, Sillo F, Girlanda M, Perotto S (2014) Plant and fungal gene expression in mycorrhizal protocorms of the orchid *Serapias vomeracea* colonized by *Tulasnella calospora*. Plant Signal Behav 9:e977707
43. Fochi V, Falla N, Girlanda M, Perotto S, Balestrini R (2017) Cell-specific expression of plant nutrient transporter genes in orchid mycorrhizae. Plant Sci 263:39–45
44. Imhof S (1998) Subterranean structures and mycotrophy of the achlorophyllous *Triuris hyalina* (Triuridaceae). Can J Bot 76:2011–2019
45. Imhof S (1999) Anatomy and mycotrophy of the achlorophyllous Afrothismia winkleri. New Phytol 144:533–540
46. Fochi V, Chitarra W, Kohler A, Voyron S, Singan VR, Lindquist EA, Barry KW, Girlanda M, Grigoriev IV, Martin F, Balestrini R, Perotto S (2017) Fungal and plant gene expression in the - symbiosis provides clues about nitrogen pathways in orchid mycorrhizas. New Phytol 213:365–379
47. Cameron DD, Johnson I, Leake JR, Read DJ (2007) Mycorrhizal acquisition of inorganic phosphorus by the green-leaved terrestrial orchid *Goodyera repens*. Ann Bot 99:831–834

48. Haselwandter K, Passler V, Reiter S, Schmid DG, Nicholson G, Hentschel P, Albert K, Winkelmann G (2006) Basidiochrome – a novel siderophore of the orchidaceous mycorrhizal fungi *Ceratobasidium* and *Rhizoctonia* spp. Bio Met 19:335–343
49. Zhao M-M, Zhang G, Zhang D-W, Hsiao Y-Y, Guo S-X (2013) ESTs *analysis reveals putative genes involved in symbiotic seed germination in Dendrobium officinale*. PLoS One 8:e72705
50. Neilands JB (1995) Siderophores: structure and function of microbial iron transport compounds. J Biol Chem 270:26723–26726
51. Dearnaley JDW, Cameron DD (2016) Nitrogen transport in the orchid mycorrhizal symbiosis – further evidence for a mutualistic association. New Phytol 213:10–12
52. Cameron DD, Leake JR, Read DJ (2006) Mutualistic mycorrhiza in orchids: evidence from plant-fungus carbon and nitrogen transfers in the green-leaved terrestrial orchid *Goodyera repens*. New Phytol 171:405–416
53. Hadley G, Purves S (1974) Movement of 14 carbon from host to fungus in orchid mycorrhiza. New Phytol 73:475–482
54. Cameron DD, Bolin JF (2010) Isotopic evidence of partial mycoheterotrophy in the Gentianaceae: and as case studies. Am J Bot 97:1272–1277
55. Smith SE (1967) Carbohydrate translocation in orchid mycorrhizas. New Phytol 66:371–378
56. Whigham D, McCormick MK, Smith CR, Rasmussen HN, Marie S, Caldwel P, Bruce DT, O'Neill J (2012) Interactions between decomposing wood, mycorrhizas and terrestrial orchid seeds and protocorms. In: Kindlmann P, Willems JH, Whigham DF (eds) Trends and fluctuations and underlying mechanisms in terrestrial orchid populations. Backhuys, Leiden, pp 117–131
57. www.psu.edu
58. Yang S, Pfister DH (2006) Monotropa uniflora plants of eastern Massachusetts form mycorrhizae with a diversity of russulacean fungi. Mycologia 98:535–540

Phytoalexins in Orchids

8

Saranjeet Kaur

Contents

Abstract

The plant kingdom harbors a number of chemicals called phytoalexins, as natural products, which are secreted temporarily whenever plants are attacked by any kind of microbe or pathogen. The family Orchidaceae synthesizes phytoalexins in their system as defensive compounds. These bioactive compounds act as antimicrobials upon attack by any kind of microbe or fungi in orchids. There are few reports in literature regarding the phytoalexins. This chapter reviews phytoalexins in the monocot family Orchidaceae.

Keywords

Defense mechanism · Monocot · Bioactive · Orchid root · Symbiosis

S. Kaur (✉)
Department of Chemistry, University Institute of Sciences, Chandigarh University, Mohali, Punjab, India
e-mail: sarana_123@rediffmail.com

J.-M. Mérillon, H. Kodja (eds.), *Orchids Phytochemistry, Biology and Horticulture*, Reference Series in Phytochemistry, https://doi.org/10.1007/978-3-030-38392-3_28

1 Introduction

Since the dawn of civilization, the plants have been resonating with human life on earth. Various diseases of human are treated by the plants. The plants harbor a variety of bioactive chemical compounds including alkaloids, flavonoids, sesquiterpenes, etc. At one hand where the plants have the ability to cure human system, they are efficient enough to combat any type of pathogen attack or disease if that has invaded into their system.

The plant kingdom harbors a number of chemicals called phytoalexins, as natural products, which are secreted temporarily whenever plants are attacked by any kind of microbe or pathogen. These inhibitory compounds are biologically active against a variety of pathogens such as insects, fungi, bacteria, or nematodes and sometimes toxicity produced by plants themselves and animals as well [1]. A perusal of literature reveals that phytoalexins are secondary metabolites having low molecular weight [2, 3]. Phytoalexins are lipophilic in nature and are capable of traversing through the plasma membrane. They exhibit their toxicity because of their acidic character. As soon as the fungus attacks the plant tissues, these phytoalexins starts disorganizing the cellular contents, ruptures plasma membrane, granulation of cytoplasm occurs, and later these phytoalexins inhibit the fungal enzymes thus reducing or inhibiting the growth of the mycelium [4]. In plant system, their synthesis occurs de novo and gets accumulated rapidly at the site when and wherever there is any kind of pathogenic attack or some kind of abiotic cause. Accumulation of phytoalexins at the site of pathogenic attack itself triggers resistance response to that particular pathogen. Although these "defense compounds" are synthesized in minute quantities; however, they are quite efficient as toxins to the infecting agent/s. Phytoalexins are often found in dicots but hardly ever in monocots and in few gymnosperms.

The Orchids: The orchids belong to one of the highly evolved monocot family Orchidaceae which is very well known for its floriferous blooms and therapeutically important herbs. The orchid species for instance *Dendrobium chrysanthum* is known to harbor chemical compounds that are highly cytotoxic to cancerous cells. The orchids are highly priced ornamental plants as species like *Cymbidium*, *Dendrobium Oncidium*, *Phalaenopsis*, etc. find special place in cut-flower trade world over and are able to generate handsome revenue.

The orchid species of diverse habits and habitats, similar to other plant species, are also known to synthesize bioactive chemical compounds identified as "phytoalexins." These are self-defense allelo-chemicals synthesized in plants. Phytoalexins are low molecular weight compounds that provide resistance to plant system against microbes. The phytoalexin Pisatin was first discovered in *Pisum sativum* [1]. The main phytoalexin discovered in orchid is "orcinol" which is a dihydrophenanthrene from *Orchis militaris* when infected with an endophytic fungus *Rhizoctonia* repens [5, 6]. The tubers of the orchid species *Loroglossum hirsinum*, upon infection with another species of genus *Rhizoctonia*, that is, *Rhizoctonia versicolor* synthesized two potential phytoalexins namely loroglossal which is an isomer of orcinol and the other phytoalexin from *Loroglossum* is hircinol [6]. These two compounds were observed to be structurally related to orcinol. The compound loroglossal was found to be active against *Phytophthora infestans* and *Monilinia fruticola* whereas hircinol was reported for its antifungal activity at ultra-low concentration of 6×10^{-6} M [7].

Phytoalexins are the secondary metabolites that are lipophilic in nature. Many angiosperm families such as Malvaceae, Solanaceae, and Orchidaceae report the secretion of secondary metabolites. The class of phytoalexins secreted by members of Malvaceae and Solanaceae are sesquiterpenes, and in family Leguminosae the phytoalexins found are isoflavonoids or polyacetylenes, and that of family Orchidaceae are dihydrophenanthrenes (http://www1.biologie.uni-hamburg.de/b-online/e33/33d.htm).

Some plant species like the potato produce several similar substances at the same time. Little is known about their mode of action. Some results point an effect that changes the membrane properties of the fungus cell. Other substances seem to block the oxidative phosphorylation, and still others can link up DNA molecules. Phytoalexins provide no absolute protection against fungus infections. They are mainly directed at "non-pathogenic" species. Many parasites are able to protect themselves against these substances or to develop defense mechanisms of their own.

2 Biosynthetic Pathways

A survey of literature indicates that biosynthesis of phytoalexins in plants is regulated by certain compounds synthesized endogenously, for instance, phytohormones (cytokinins, auxins, abscisic acid, ethylene, salicylic acid, jasmonic acid, and gibberellins), and by other transcriptional regulators, defense-related genes, phosphorylation relays and cascades [8, 9].

In general, there are pathways that help in synthesizing different phytoalexins. These pathways are: (a) the phenylpropanoic-polymalonic acid pathway, (b) the methylerythritol phosphate and geranyl-geranyl diphosphate pathway, and (c) the indole phytoalexin pathway. Phytoalexins deriving from phenylpropanoic-polymalonic acid pathway: All kinds of flavonoid phytoalexins such as isoflavonoids, isoflavones, pterocarpans, isoflavans, coumestans and arylbenzofurans, stilbenes and its derivatives (dihydrophenanthrenes) are synthesized by the common phenylpropanoic-polymalonic acid pathway.

The pathway starts with phenylalanine/phenylalanine ammonia lyase (PAL)/ tyrosine ammonia lyase (TAL)/tyrosine precursors. Subsequently, para-coumaric acid gets activated in para-coumaroyl-CoA through its ligation to coenzyme A by 4-coumaroyl: CoA ligase (C4L). Later, chalcone synthase (CHS), stilbene synthase (STS) using same substrate condensed it with three successive units of malonyl-CoA, further producing naringenin chalcone, the first intermediate which is C15 in flavonoid pathway and other compound resveratrol, the precursor molecule of stilbenes [10, 11, 12]. Another pathway is mevalonoid-derived phytoalexin pathway: In this pathway, sesquiterpene, monoterpene, carboxylic sesquiterpene, and diterpene phytoalexins are synthesized. In this pathway the gene transcripts (OsDXS3, OsDXR, OsCMS, OsCMK, OsMCS, OsHDS, and OsHDR) are involved which were observed in *Oryza sativa* cells [13].

Indole phytoalexins pathway: The third type of pathway of phytoalexin synthesis is well studied in *Arabidopsis*. The pathway starts from chlorismate. This pathway involves production of camalexin from tryptophan which shows the involvement of two cytochrome P450 homologs CYP79B2 and CYP79B3, which further form indole-3-acetaldoxime; it is converted into indole-3-acetonitrile (IAN) through cytochrome P450, CYP71A13. The last step involved in the synthesis of camalexin is under the control of a gene named Phytoalexin Deficient 3, Pad 3 (CYP71B15), which codes for camalexin through dihydrocamalexic acid [13].

3 Role of Phytoalexins in Plants

The term phytoalexin was coined by K.O. Müller in 1940. Phytoalexins are considered to be plant antibiotics that are synthesized de novo when the plant part gets some microbial infection [14]. Phytoalexins, an antimicrobial compound, could not be preformed in the tissue or released from preexisting plant constituents [14]. Thus, in contemporary terms, these antibiotics are produced in response to a microbial elicitor, and their production requires the expenditure of plant energy, generally in the form of new transcriptional and or translational activity. The species of family Brassicaceae show piling up of metabolites such as sulfur-containing tryptophan-derived alkaloids whenever there is a pathogen attack [15, 16].

Phytoalexin compounds are synthesized in minute quantities over the surface of plants; if consumed by human system in excess, they can prove harmful to human beings and animals. For instance, in other plant species, the phytoalexin named "pisatin" which are produced by garden pea and phaseolin by green bean are capable of lysing bovine red blood cells at 200 ppm and 17.5 ppm concentration, respectively. Likewise, carrots contain phytoalexin "myristicin," an insecticidal, which can induce cerebral excitation in human beings. Damaged *Ipomaea batata* show the elevated levels of phytoalexin called Ipomeamarone which can cause damage to lungs and liver in cattle and human beings.

Blighted white potato causes poisoning and ultimately death due to the formation of two glyco-alkaloids: alpha-chaconine and alpha-solanine. From consumer point of view, it becomes essential to trim down the production of phytoalexins through improved agricultural practices. In family Leguminosae, the phytoalexins produced by the members belong to six iso-flavonoid classes such as iso-flavones, iso-flavanones, pterocarpans, pterocarpenes, iso-flavans, and coumestans.

Different plant species harbor a variety of phytoalexins. Some of the types of phytochemicals synthesized by different genus of angiosperm families are summarized in Table 1.

The compound phytoalexins are low molecular weight antimicrobial compounds, which gather at site of infection, releases enzymes such as proteases, chitinases, beta-1,3-glucanases which destroy the pathogens, and also release certain defensive biopolymers (peroxidases and phenol oxidases) which restrict the spread of pathogens for instance lignin, callose, hydroxyproline-rich glycoproteins, and some other compounds which regulate the induction and/or activity of the protective compounds. A kind of phytoalexin named *trans*-resveratrol has been reported, when a fungus

Table. 1 Shows various types of phytoalexins found in angiosperm families [3]

S. No.	Angiosperm Family	Phytoalexin Type
1	Amaryllidaceae	Flavans
2	Brassicaceae	Indole phytoalexins/camalexin Sulfur-containing phytoalexins/ brassinin
3	Chenopodiaceae	Flavanones/betagarin Isoflavones/betavulgarin
4	Convolvulaceae	Furanosesquiterpenes/Ipomeamarone
5	Compositae	Polyacetylenes/safynol
6	Poaceae	Diterpenoids: Kauralexins; Momilactones; Phytocassanes Oryzalexins; Zealexins Deoxyanthocyanidins/luteolinidin and apigeninidin Flavanones/ sakuranetin Phenylamides
7	Leguminosae	Isoflavones, Isoflavanones, Isoflavans, Coumestans Pterocarpans/ pisatin, phaseollin, glyceollin and maiackiain Furanoacetylenes/ wyerone Stilbenes/resveratrol Pterocarpens
8	Malvaceae	Terpenoids naphtaldehydes/gossypol
9	Linaceae	Phenylpropanoids/coniferyl alcohol
10	Moraceae	Gossypol/Terpenoids naphtaldehydes
11	Orchidaceae	Loroglossol/Dihydrophenanthrenes
12	Rutaceae	Methylated phenolic compounds/xanthoxylin
13	Umbelliferae	Falcarinol Phenolics: xanthotoxin 6-methoxymellein/ Polyacetylenes
14	Solanaceae	Coumarins/Sesquiterpenoids
15	Rosaceae	Auarperin Dibenzofurans/Cotonefurans
16	Vitaceae	Resveratrol/Stilbenes

Botrytis cinerea infects *Vitis vinifera* [17], and another phytoalexin called delta-viniferin is synthesized when *Plasmopara viticola* infects the grapevine [18]. A phytoalexin 6-methoxymellein is known to be induced in carrot slices by UV-C, [19], and it induces resistance to *Botrytis cinerea* [20] and other microorganisms [21].

A survey of available reports reveals that the plant *Sorghum* produces two discrete phytoalexin molecules of 3-deoxyanthocyanidin chemical group, apigeninidin and luteolinidin. The precursor molecule for biosynthesis is flavanone naringenin which differ from marginally from that of the anthocyanin pathway. In *Zea mays*, the phytoalexins such as zealexins and kauralexins belong to terpenoid class [22]. Whenever any pathogenic fungi attacks the plant, the phytoalexins are produced. These phytoalexins interact with invaded pathogen and start to neutralize its noxious effects. *Alternaria brassicicola* which is a necrotrophic fungus can detoxify brassinin, which is a phytoalexin synthesized in Brassicaceae family [22].

4 Role of Plant Hormones

Plant hormones are also known to activate defense-response related genes to synthesize defensive metabolites such as sesquiterpenoids. Plant hormones play an important role in controlling plant stress and various diseases. Literature study reveals that external application of abscicic acid (ABA) in certain way elevates the resistivity of plant to pathogen attack [23]. In dicot family Solanaceae, the synthesis of phytoalexin known as capsidiol, a sesquiterpenoid, is regulated by ABA [24].

4.1 Phytoalexins in Gymnosperms

Apart from Angiosperms, Gymnosperms are also known to produce certain phytoalexins. A toxin named Pinosylvin which is a stilbenoid toxin is synthesized before fungal invasion. The heartwood tissues of members of Pinaceae show phytoalexin synthesis, and it is a fungitoxin protecting the wood from fungal infection [25].

5 Phytoalexins in Orchids

In family Orchidaceae, there exists an inverse relationship between mycorrhiza and the amount of phytoalexin synthesized. It is noticed that when mycorrhiza invades into parenchyma cells of roots, the level of endogenous amount of phytoalexin declines. Phytoalexins are synthesized in almost every part of orchid plant such as roots, rhizomes, and bulbs as soon as they are invaded by any fungus as observed in *Bletila striata*, which showed 100% in enzyme bibenzyl synthase [26]. Literature survey indicates the isolation of more than 40 dihydrophenanthrene phytoalexins in orchid species of diverse habits and habitats [27]. Radioactive experimental study indicated that L-phenylalanine is a precursor molecule for the biosynthesis of 9,10-dihydrophenanthrenes with intermediate compounds *m*-coumaric acid, dihydro-*m*-coumaric acid, and 3,3′,5-trihydroxybibenzyl [28]. Since these are derivatives of dihydrostilbenes or bibenzyls so these dihydrophenanthrenes are categorized as stilbenoids [29]. When mycorrhiza is digested completely in orchid root cell, its production increases again inside the cell.

Molecular investigations reveal that whenever an orchid is attacked or invaded by any kind of pathogen (fungi or bacteria) intense activation of specific genes that encodes for the synthesis of phytoalexin enzymes takes place [27]. The key enzyme, bibenzyl synthase, that is primarily involved in the synthesis of phytoalexins gets activated as soon as fungus infects the orchid roots [27]. The phytoalexin molecule precursors were synthesized in *Epipactis palustris* rhizomes, when these got infected by *Rhizoctonia* strain [28]. Literature studies reveal an interesting observation about stem tissues showing more fungicidal activity than tubers of *Dactylorhiza incarnata* and less in roots [29].

6 Phytoalexins and Human Health

The phytoalexins are synthesized in minute quantities over the surface of plants; if consumed by human system in excess, can prove harmful. For instance, in other plant species, the phytoalexins named "pisatin," which is produced by garden pea, and phaseolin, by green bean, are capable of lysing bovine red blood cells at

200 ppm and 17.5 ppm concentration, respectively. Likewise, carrots contain phytoalexin "myristicin," an insecticidal, which can induce cerebral excitation in human beings. Damaged *Ipomaea batata* show the elevated levels of phytoalexin called Ipomeamarone which can cause damage to lungs and liver in cattle and human beings. Likewise, blighted white potato causes poisoning and ultimately death due to the formation of two glyco-alkaloids: alpha-chaconine and alpha-solanine. From consumer point of view, it becomes essential to trim down the production of phytoalexins through improved agricultural practices. On the contrary, there are reports that indicate the positive influence of phytoalexins on human health. A few of phytoalexins have shown health-promoting effects in human beings [22]. They are reported to be acting as antioxidants, showing anticarcinogenic and acting as cardiovascular protective. The topical applications of phytoalexin called resveratrol shows antitumor activities against skin cancer in vivo. Another phytoalexin named as maslinic acid, synthesized in olives, exerts a broad spectrum of biological activities such as antitumoral, antidiabetic, neuroprotective, cardioprotective, antiparasitic, and growth stimulating. Synthesis of maslinic acid in olive makes it a food having nutraceutical value [22].

7 Conclusion

The orchids have smartly evolved themselves to survive in nature in spite of having highly reduced seed structure. It is crucial to understand orchid fungus intricacies for conserving them. Reports have been cited in orchid literature about orchid fungus interactions in relation to phytoalexins. Studies are required to unravel the mode of interaction in which these biomolecules show their activity against microorganisms. Furthermore, research in molecular studies would be a step toward understanding the relationship of symbiont and phytoalexin that is synthesized by various orchid species upon getting infected by any pathogen attack.

Acknowledgments English language assistance from Dr. H.S. Sekhon is acknowledged with deep gratitude.

References

1. Braga MR et al (1991) Phytoalexins induction in Rubiacea. J Chem Ecol 17:1079–1090
2. Grayer RJ, Kokubun T (2015) Plant-fungal interactions: the search for phytoalexins and other antifungal compounds from higher plants. Phytochemistry 56:253–263
3. Arruda RL, Santana Paz AT, Freitas Bara MT, Côrtes MVCB, Filippi MCC, Conceição EC (2016) An approach on phytoalexins: function, characterization and biosynthesis in plants of the family *Poaceae*. Ciência Rural 46(7):1206–1216. https://doi.org/10.1590/0103-8478cr20151164
4. Cavalcanti LS et al (2005) Aspectos bioquímicos e moleculares da resistência induzida. In: Cavalcanti LS et al (eds) Indução de resistência em plantas a patógenos e insetos, vol 81. FEALQ, Piracicaba, p 124
5. Boller AH, Corrodi F, Gaumann E et al (1957) Uber induzierte Abwehrstoffe bei Orchideen Pt. 1. Helv Chim Acta 40:1062–1066

6. Hardegger E, Schellenbaum M, Corrodi H (1963) Uber induzierte Abwehrstoffe bei Orchideen II. Helv Chim Acta 46:1171–1180
7. Ward EWB, Urwin CH, Stoessl A (1975) Loroglossal: an orchid phytoalexin. Phytopathology 65(5):632–633. Jeandet P, Delaunois B, Conreux A, Donnez D, Nuzzo V, Cordelier S, Clément C, Courot E (2010) Biosynthesis, metabolism, molecular engineering and biological functions of stilbene phytoalexins in plants. Biofactors 36:331–341
8. Jeandet P, Clément C, Courot E, Cordelier S (2013) Modulation of phytoalexin biosynthesis in engineered plants for disease resistance. Int J Mol Sci 14:14136–14170
9. Ahuja I, Kissen R, Bones AM (2012) Phytoalexins in defence against pathogens. Trends Plant Sci 17(2):73–90
10. Jeandet P, Delaunois B, Conreux A, Donnez D (2010) Biosynthesis, metabolism, molecular engineering, and biological functions of stilbene phytoalexins in plants. Biofactors 36(5): 331–341
11. Deavours BE, Dixon RA (2005) Metabolic engineering of isoflavonoid biosynthesis in alfalfa. Plant Physiol 138:2245–2259
12. Kaimoyo E, VanEtten HD (2008) Inactivation of pea genes by RNAi supports the involvement of two similar O-methyltransferases in the biosynthesis of (+)-pisatin and of chiral intermediates with a configuration opposite that found in (+)-pisatin. Phytochemistry 69:76–87
13. Schmelz EA, Huffaker A, Sims JW, Christensen SA, Lu X, Okada K, Peters RJ (2014) Biosynthesis, elicitation and roles of monocot terpenoid phytoalexins. Plant J 79:659–678
14. Müller K, Borger H (1940) Experimentelle Untersuchungen uber die Phytophthora-Resistenz der Kartoffel. Arb. Biol. Reichsanstalt, Berlin. Land-u Forstwirtsch. 23:189–231
15. Pedras MSC, Yaya EE, Glawischnig E (2011) The phytoalexins from cultivated and wild crucifers: chemistry and biology. Nat Prod Rep 28:1381–1405
16. Bednarek P (2012) Sulfur-containing secondary metabolites from Arabidopsis thaliana and other Brassicaceae with function in plant immunity. Chembiochem 13:1846–1859
17. Favaron F, Lucchetta M, Odorizzi S, Cunha AT, Sella L (2009) The role of grape polyphenols on trans-resveratrol activity against *Botrytis cinerea* and of fungal laccase on the solubility of putative grape PR proteins. J Plant Pathol 91(3):579–588
18. Timperio AM, Alesndro AD, Fagioni M, Magro P (2012) Production of the phytoalexins trans-resveratrol and delta-viniferin in two economy-relevant grape cultivars upon infection with *Botrytis cinerea* in field conditions. Plant Physiol Biochem 50(1):65–71
19. Mercier J, Arul J, Ponnampalam R, Boulet M (1993) Induction of 6-methoxymellein and resistance to storage pathogens in carrot slices by UV-C. J Phytopathol 137:44–54
20. Hoffman R, Heale JB (1987) Cell death, 6-methoxymellein accumulation, and induced resistance to *Botrytis cinerea* in carrot root slices. Physiol Mol Plant Pathol 30:67–75
21. Kurosaki F, Nishi A (1983) Isolation and antimicrobial activity of the phytoalexin 6-methoxymellein from cultured carrot cells. Phytochemistry 22(3):669
22. Jeandet P (2015) Phytoalexins: current progress and future prospects. Molecules 20(2): 2770–2774
23. Lopez MA, Bannenberg G, Castresana C (2008) Controlling hormone signaling is a plant and pathogen challenge for growth and survival. Curr Opin Plant Biol 11:420–427
24. Mialoundama AS, Heintz D, Debayle D, Rahier A, Camara B, Bouvier F (2009) Abscisic acid negatively regulates elicitor-induced synthesis of capsidiol in wild tobacco. Plant Physiol 150(3):1556–1566
25. Lee SK, Lee HJ, Min HY, Park EJ, Lee KM, Ahn YN, Cho YJ, Pyee JH (2005) Antibacterial and antifungal activity of pinosylvin, a constituent of pine. Fitoterapia 76(2):258–260
26. Reinecke T, Kindl H (1993) Characterization of bibenzyl synthase catalysing the biosynthesis of phytoalexins of orchids. Phytochemistry 35(1):63–66
27. Stoessl A, Arditti J (1984) Orchid phytoalexins. In: Arditti J (ed) Orchid biology, reviews and perspectives, III. Cornell University Press, Ithaca/London
28. Reinecke T, Kindl H (1994) Characterization of bibenzyl synthase catalysing the biosynthesis of phytoalexins of orchids. Phytochemistry 35:63–66
29. Burges A (1939) The defensive mechanism in orchid mycorrhiza. New Phytol 38(3):273–283

Part III

Horticulture

Micropropagation of Some Orchids and the Use of Cryopreservation

9

Kanchit Thammasiri, Nipawan Jitsopakul, and Sasikarn Prasongsom

Contents

Abstract

Orchid micropropagation is an indispensable tool for seed and clonal propagation, especially for commercial purposes. It has been interestingly researched and used for enormous species and hybrids. In contrary, many wild orchid species are in endangered and extinction because of deforestation and natural disaster. Only outstanding horticultural characteristic orchids are cultivated. Therefore, orchid conservation is

K. Thammasiri (✉)
Department of Plant Science, Faculty of Science, Mahidol University, Bangkok, Thailand
e-mail: kanchitthammasiri@gmail.com

N. Jitsopakul
Department of Plant Science, Textile and Design, Faculty of Agriculture and Technology, Rajamangala University of Technology Isan, Surin, Thailand

S. Prasongsom
Pathum Wan District, Bangkok, Thailand

J.-M. Mérillon, H. Kodja (eds.), *Orchids Phytochemistry, Biology and Horticulture*, Reference Series in Phytochemistry, https://doi.org/10.1007/978-3-030-38392-3_10

urgently needed by various means, such as living collection, seed and pollen storage, in vitro conservation, and cryopreservation for small aseptic materials, such as protocorms, protocorm-like bodies (PLBs), buds, root tips, meristem, callus, and cell suspension. In this review, orchid cryopreservation which is a long-term storage with genetic stability is focused for the development and successful use.

Keywords

Orchid · Micropropagation · Protocorms · Cryopreservation · Vitrification · Cryo-plate

1 Introduction

Orchids are in the family *Orchidaceae* which is one of the most diverse, advance, and largest with about 25,000 species. They are unique and fascinating in characteristics that make them popular worldwide for plant studies, hobbies, and commercial uses. They are used as ornamentals, medicine, and cosmetics. Orchid micropropagation is an indispensable tool for seed and clonal propagation, especially for commercial purposes; otherwise, orchid industry will not be rapidly progressed. Orchid micropropagation has been interestingly researched and used for enormous species and hybrids. In contrary, many wild orchid species are in endangered and extinction because of deforestation and natural disaster. Only outstanding horticultural characteristic orchids are cultivated. Therefore, orchid conservation is urgently needed by various means, such as living collection, seed and pollen storage, in vitro conservation, and cryopreservation for small aseptic materials, such as protocorms, protocorm-like bodies (PLBs), buds, root tips, meristem, callus, and cell suspension. Cryopreservation is a long-term storage with genetic stability. It was developed about 60 years ago from using sophisticated equipment with time-consuming protocols to simple equipment with fast, efficient, and high survival. For orchids, cryopreservation was developed for many orchid species for over 20 years.

2 Orchid Micropropagation

Asexual propagation by division or cutting, which is a conventional method for general plant propagation, as well as for orchids, is also practiced but mainly as a hobby and not for large-scale production because multiplication is slower in such cases. This method is; however, unavoidably used when tissue culture (micropropagation) fails to work. The beginning of orchid tissue culture experiment is in the nineteenth century, scientists reported the successful use for both seeds [1] and explants [2]. Many studies have focused on asymbiotic propagation of orchids from seeds [3, 4]. Success for propagating orchids through tissue culture was first achieved in 1949 in *Phalaenopsis* species. Later in 1967, the successful development of techniques for tissue culture of *Dendrobium* provided a breakthrough, especially for the Thai orchid cut-flower business, since this genus has contributed the most to cut-flower production. Many types of culture media were used for orchid tissue culture. Medium composition is an important factor for growth and morphogenesis. It consists of water, a solid or semisolid support, macronutrients, micronutrients, vitamins, carbon sources, plant growth regulators or hormones, amino acids, sorbents, and other undefined organic supplements [5]. Plant tissue culture is one of the most important techniques that used for growing plant cells under sterile conditions. The suitable medium compositions and environment are required for producing clones of a plant. At present, successful propagation through tissue culture or micropropagation has been achieved in over 80 genera of orchids [6].

2.1 Protocorm and Protocorm-Like Body Formation in Orchids

The orchid seeds are very small and dust-like because they contain only the small embryo and without any associated endosperm storage tissue. Most range in length from 0.05 mm to 6.0 mm. Seed weight extends from 0.31 μg to 24 μg [7]. Single fruit or pod contains millions of seeds that are suitable for dispersing by wind [8]. Enormous numbers of seeds produced but only a few seeds germinate in nature [9]. Other reasons for seed germination, seeds lack enzyme to metabolize polysaccharides but utilize liquid as a major nutrient source and the embryos also lack enzyme to convert liquid to soluble sugar [10]. Orchid seeds require a symbiotic with a suitable fungus [11]. Under natural conditions, germination of the mature orchid seeds, partially those from the terrestrial orchids, is dependent on the association of a mycorrhizal fungus. Since Knudson's discovery in 1946 [12], orchid seeds can also germinate in culture medium without mycorrhiza fungus [13]. For germination, the embryo enlarges to form a small, corm-like structure, called a protocorm, which possesses a quiescent shoot and root meristem at opposite poles. In nature, a protocorm becomes green and accumulates carbohydrate reserves through photosynthesis. Normal seedling growth then continues utilizing the stored protocorm food reserves. These somatic protocorms can appear to be similar to seedling protocorms, and many workers on orchid propagation, have used terms such as "protocorm-like bodies" (PLBs) to describe them. When a shoot tip of an orchid is transferred to culture on a suitable medium, PLBs can grow and develop as a mature shoot apex. PLBs also arise directly on some other orchid explants and proliferate from other PLBs in a fashion which is exactly comparable to the direct formation of somatic embryos. Champagnat and Morel [14] and Norstog [15] considered the appearance of protocorms to be a manifestation of embryogenesis is because they represent a specialized stage in embryo development and are normally derived directly from zygotic embryos.

2.2 The Components of Orchid Tissue Culture Media

Artificial media used in plant tissue culture are copied from natural nutrients in soil for supporting plant growth. Essential inorganic elements can be divided into two types, first is macro-elements, such as nitrogen (N), potassium (K), calcium (Ca), phosphorus (P), magnesium (Mg), and sulfur (S). The second is micro-elements, such as iron (Fe), nickel (Ni), chlorine (Cl), manganese (Mn), zinc (Zn), boron (B), copper (Cu), and molybdenum (Mo) [16]. Other elements, such as cobalt (Co), aluminum (Al), sodium (Na), and iodine (I) are essential for some plant species. Other components are carbon sources, plant growth regulators, nondefined organic additives (banana homogenate, potato extract, coconut water, etc.), chitosan, and activated charcoal.

2.2.1 Culture Media

The most commonly used media are Murashige and Skoog (MS) [17], Vacin and Went (VW) [18], and Knudson (KC) [12]. For orchid growth media, MS is often used in many researches because MS was the best liquid medium than VW and KC for inducing PLBs of *Dendrobium* [19]. Half-strength MS (½MS) was popularly used in many orchid species, such as somatic embryogenesis of *Coelogyne cristata* formed embryo on ½MS (100%) medium better than cultured on MS (0–30%) [20]. PLBs of *Grammatophyllum speciosum* were cultured on ½MS supplemented with 2 mg/l naphthalene-1-acetic acid (NAA) and 1 mg/l BAP induced shoot and root formation [21]. Kishor and Sharma [22] reported that ½MS was suitable for germinating the hybrid of *Renanthera imschootiana* × *Vanda coelurea* seeds when compared with VW medium supplemented with 15% coconut water. The results had significant difference on growth responses of leaf and root between two media. Kishor and coworkers [23] reported that the best basal medium for culturing immature embryo of *Aerides vandarum* and *Vanda stangeana* was ½MS medium supplemented with 20% (v/v) coconut water. Seedlings of *Paphiopedilum callosum* were cultured on ½MS supplemented with 5 μM thidiazuron (TDZ) gave the highest number of shoots per explant [24]. VW basal medium was used for inducing the PLBs in hybrid *Cymbidium* Twilight Moon "Day Light" [25]. Even though MS contains more elements and concentration than VW but also some orchids grow the best on VW, such as seeds of *Paphiopedilum villosum* var. *densissimum* germinated better on VW medium than ½MS and ¼MS media [26]. In different media, SH [27] medium was developed for culturing calli of monocots and dicots. This medium was studied in some orchids. Luo et al. [28] reported that SH supplemented with 30 g/l sucrose was the best medium for proliferating PLBs of *Dendrobium huoshanense* when compared with MS, B5 [29], and N6 [30] media. PLBs of *Cymbidium* that cultured on SH medium can induce the small of both calli and PLBs [31].

2.2.2 Sugar

Carbohydrates are very important component in in vitro cultures because they have energy and carbon source, as well as an osmotic agent. They were added in any nutrient media that is essential for in vitro growth and development because photosynthesis is insufficient in in vitro culture [32]. Sucrose is mostly used but glucose, fructose, sorbital, maltose, and other sugars are also used for comparing the efficacy with sucrose. Germination of *Paphiopedilum ciliolare* seeds was the best on a medium containing 5 g/l fructose plus 5 g/l glucose; a mixture of 7.5 g/l of each sugar was optimal for seedling growth [33]. Glucose showed the best frequency of germination in *Paphiopedilum villosum* var. *dentissimum* seeds on ¼MS; while, sucrose, maltose, and mannitol provided lower on seed germination [26]. Hong et al. [34] reported that 30 g/l sucrose for cultivar Grower Ramsey and 20 g/l glucose for cultivar Sweet Sugar in free PGRs ½MS media for 60 days was suitable for inducing direct embryo for *Oncidium*. Maltose (2%) in ½liquid MS medium was suitable carbon source for enhancing growth rate of *Grammatophyllum speciosum* PLBs when compared with the same concentration of sucrose, glucose, trehalose, sorbitol, and mannitol [21]. Including the results of Luo et al. [35], PLBs of *Dendrobium huoshanense* cultured on ½MS medium supplemented with 10 g/l maltose was the best for inducing number of new shoots and had significant difference from the results of sucrose and glucose. Different concentrations of sugar gave different results on orchid growth, such as *Dendrobium* Second Love explants were increased in the number of shoot and root length when cultured on modified VW medium supplemented with 2–4% sucrose under light and dark condition [36].

2.2.3 Plant Growth Regulators

Auxins and cytokinins are two main groups of plant growth regulators that often used in several plants, including orchids [37, 38]. Auxins are used for inducing plant root or enhance the cytokinin effects. Cytokinins are popularly used for shoot proliferation and induce shoot formation. Plant growth regulators (PGRs) with different types and concentrations play an important role during in vitro propagation including multiple of shoots, root and shoot formation, elongation of root tips and callus formation of many orchid species. PGRs have been considered to accelerate the growth and development of PLBs and plantlets from root-derived calli of *Oncidium* [38]. The presence of auxins in the medium is generally essential for embryo initiation [39]. Auxins play two main roles in plant tissue culture. Firstly, they can be used to induce root formation, such as indole acetic acid (IAA) and indole butyric acid (IBA). Secondly, they can initiate callus formation, such as 2,4-dichlorophenoxy acetic acid (2,4-D) and NAA [40]. Jheng et al. [41] showed that a lower level of 2,4-D seemingly stimulated the increase of cell masses, and a higher level of TDZ induced granular structure under dark condition. But the combination of exogenous 2,4-D with TDZ was found very crucial for long-term maintenance of callus cultures without loss of vigorous growth and regenerative capability. Cytokinins are widely used for their ability to induce either shoot proliferation by breaking dormancy in lateral buds or adventitious meristem (shoot formation), such as kinetin, 6-benzylaminopurine (BAP) [42–44] and thidiazuron (TDZ) [37, 45, 46]. TDZ is a substituted phenyl urea with cytokinin-like activity. It is generally believed that TDZ is more active in stimulating shoot formation than BAP or kinetin, even at extremely low concentration [47]. For some years, TDZ has been generally used to culture orchid tissue, which could induce organogenesis and somatic embryo formation [40, 48]. Disadvantage of TDZ in regeneration is the difficulty in elongation and rooting of the regenerated shoots. This problem was overcome by transferring regenerated shoots to MS medium supplemented with BAP and NAA. Wasiksiri [49] studied the effects of six different liquid media including, modified Vacin and Went [18], Murashige and Skoog [17], Schenk and

Hildebrandt [27], half strength MS (½MS), half strength SH (½SH), and half strength KNO_3 in SH (½KNO_3SH) which contained 100 ml/l coconut water and without sucrose on multiple shoot formation from lateral buds of *Vanda coerulea*. After 8 weeks of culture, the survival of lateral buds (100%) was observed on ½SH and ½KNO_3SH but the protocorm-like structure was found only in ½KNO_3SH. For increasing multiple shoots or PLBs, lateral buds were cultured on ½SH and ½KNO_3SH media supplemented with different concentrations of plant growth regulators (TDZ, BAP, and NAA). The highest fresh weight of PLBs (total fresh weight of 1.4 g) were observed on ½KNO_3SH medium supplemented with 1 mg/l BAP and 0.1 mg/l NAA (Fig. 1). Development of PLBs was established by organic additives (activated charcoal, potatoes, bananas, and sucrose). The results revealed that PLBs which cultured on ½KNO_3SH supplemented with 1 g/l activated charcoal, 50 g/l potatoes, 50 g/l bananas, and 20 g/l sucrose produced the highest shoots (434 shoots/lateral bud). These new shoots were able to develop into plantlets. New shoots were cultured on ½KNO_3SH supplemented with 1 g/l activated charcoal, 50 g/l bananas, and 20 g/l sucrose. After 16 weeks of culture, the best growth of plantlets was found (Fig. 2).

2.2.4 Vitamins

All media used for orchid micropropagation have vitamins. Some are in organic materials, such as banana, coconut water, potato, etc. Most added vitamins in orchid culture media are niacin (nicotinic acid), pyridoxin (vitamin B6), and thiamine (vitamin B1). Biotin, folic acid, and pantothenic acid are also used in some media.

2.2.5 Amino Acids

Glycine which is a component of the MS medium is the most used in orchid culture media. Other amino acids are also used in some media. All organic.materials already have various amino acids.

2.2.6 Banana Homogenate

Homogenized banana fruit is sometimes added to media of orchid culture and is often reported to promote plant growth. The reason for its stimulatory effect has not been explained. One suggestion mentioned earlier is that it might help to stabilize the pH of the medium. Banana pulp is a rich source of natural cytokinins which inhibit culture initiation but promote differentiation and growth of shoots at later stages [33, 50]. Vyas et al. [51] reported that KC medium supplemented with 10% (v/v) banana homogenate provided high percentage germination of *Dendrobium lituiflorum* seeds and gave maximum on regeneration of leaves and roots within 60 days of culture. Lo et al. [52] studied the effect of various additives in MS medium on seedling growth of *Dendrobium tosaense*. The results showed that 8% banana and 8% potato juice gave the same on fresh weight, plant height, stem diameter, and root length after 20 weeks of culture. When compared to the different concentrations of banana homogenate (5–30%), 10% gave the best proliferation of PLBs on *Dendrobium* hybrid on ½MS medium for 4 weeks and all 4 banana cultivars and banana powder that were added into the agar media gave significant difference on fresh weight of PLBs [53]. The addition of 2.5 g/l banana powder

Fig. 1 Micropropagation of a lateral bud of *Vanda coerulea* in ½KNO_3SH liquid medium after 8 weeks of culture in medium; (**a**) without PGR, (**b**) 0.1 mg/l TDZ, (**c**) 0.5 mg/l TDZ, (**d**) 1 mg/l TDZ, (**e**) 1.5 mg/l TDZ, (**f**) 2 mg/l TDZ, (**g**) 1 mg/l BAP, (**h**) 2 mg/l BAP, (**i**) 0.1 mg/l NAA and 1 mg/l BAP, (**j**) 0.5 mg/l NAA and 1 mg/l BAP, (**k**) 1 mg/l NAA and 1 mg/l BAP, (**l**) 0.1 mg/l NAA and 2 mg/l BAP, and (**m**) 0.5 mg/l NAA and 2 mg/l BAP. Bar = 0.5 cm

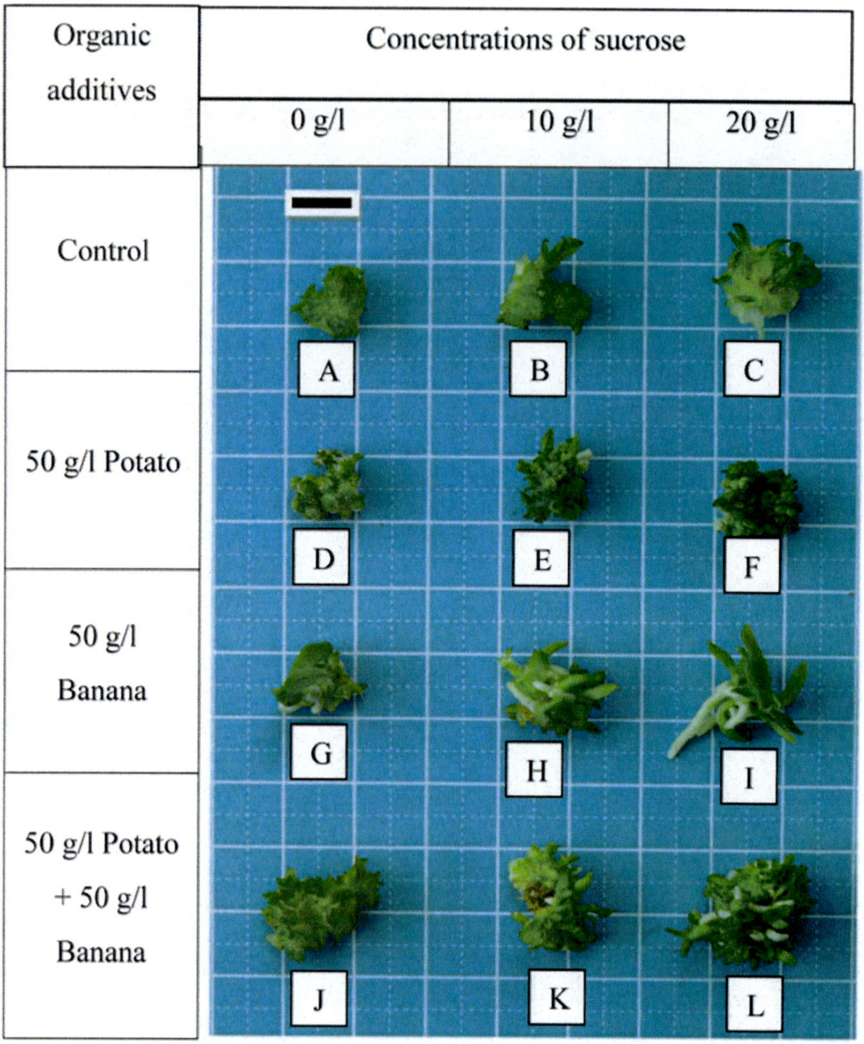

Fig. 2 Effects of organic additives and combinations with concentrations of sucrose and organic additives on embryogenesis on PLBs of *Vanda coerulea* under culture in ½KNO_3SH contained 100 ml coconut water and 1 g/l activated charcoal after 8 weeks of culture. (**a**) ½KNO_3SH agar medium without organic additives and sucrose. (**b**) ½KNO_3SH agar medium supplemented with 10 g/l sucrose. (**c**) ½KNO_3SH agar medium supplemented with 20 g/l sucrose. (**d**) ½KNO_3SH agar medium in combination with 50 g/l banana without sucrose. (**e**) ½KNO_3SH agar medium in combination with 50 g/l banana and supplemented with 10 g/l sucrose. (**f**) ½KNO_3SH agar medium in combination with 50 g/l banana and supplemented with 20 g/l sucrose. (**g**) ½KNO_3SH agar medium in combination with 50 g/l potato without sucrose. (**h**) ½KNO_3SH agar medium in combination with 50 g/l potato and supplemented with 10 g/l sucrose. (**i**) ½KNO_3SH agar medium in combination with 50 g/l potato and supplemented with 20 g/l sucrose. (**j**) ½KNO_3SH agar medium in combination with 50 g/l banana and 50 g/l potato without sucrose. (**k**) ½KNO_3SH agar medium in combination with 50 g/l banana and 50 g/l potato and supplemented with 10 g/l sucrose, and (**l**) ½KNO_3SH agar medium in combination with 50 g/l banana and 50 g/l potato and supplemented with 20 g/l sucrose. Bar = 1 cm

was the best to enhance the number and length of roots of *Phalaenopsis amabilis* PLBs after 4, 6, and 8 weeks on ½MS supplemented with 10% coconut water, 2 g/l peptone, and 1 g/l activated charcoal [54]. The best proliferated PLBs of *Phalaenopsis violacea* (80%) was induced when cultured on ½MS medium supplemented with 10% banana pulp extract; while, other organic additives (tomato, coconut water, and papaya) provided lower effect for inducing PLBs (≤35%) in the same condition [55]. Half strength MS medium containing100 g/1 banana homogenate facilitated root development and shoot growth of *Rhynchostylis gigantea* PLBs [56]. The in vitro plantlets of *Renanthera* Tom Thumb "Qilin" provided the highest number of roots and root length when cultured on VW medium supplemented with 1 g/l peptone, 1 g/l NAA, 1 g/l activated charcoal, and 100 g/l banana homogenate. While other concentrations of banana homogenate (50 and 200 g/l), coconut water, and potato homogenate had low effect to regenerate plantlet roots [50]. The seedlings of *Dendrobium lituiflorum* were cultured on KC medium with 20% (v/v) banana homogenate gave elongated leaves and well-developed roots compared to the same medium without banana homogenate after 30 days of the fourth subculture [57]. Hyponex N016 medium supplemented with 1.0 mg/1 NAA, 1.0 g/1 peptone, 100 g/1 banana homogenate, and 1.0 g/1 AC was suitable for 2 cm *Paphiopedilum wardii* plantlet growth when compared with 50, 150, and 200 g/l banana homogenate in the same medium [58]. The PLBs of *Vanda coerulea* which cultured on ½KNO_3SH supplemented with 1 g/l activated charcoal, 50 g/l potatoes, 50 g/l bananas, and 20 g/l sucrose produced the highest shoots (434 shoots/lateral bud). These new shoots were able to develop into plantlets. New shoots were cultured on ½KNO_3SH supplemented with 1 g/l activated charcoal, 50 g/l bananas, and 20 g/l sucrose [49].

2.2.7 Potato Extract

Potato is a carbohydrate-rich food which contains about 80% water and 20% dry matter, with 60–80% of the latter composed of starch [59]. In orchid seed germination, potato extract showed 100% seed germination in *Dendrobium hamaticalcar* after 50 days of culture on ½MS supplemented with 1 g/l potato extract that is the same result as on ½MS supplemented with 1 g/l yeast extract [60]. VW medium supplemented with coconut water and 5–20% potato extract gave high proliferation percentage of *Vanda* Kasem's Delight PLBs that was the same effect as supplemented with 5–20% tomato extract [59]. Regenerated shoots of *Cypripedium formosanum* produced roots when cultured on the basal medium with 1 g/1 activated charcoal and 20 g/1 potato extract for 60 days [61]. The PLBs of *Vanda coerulea* which cultured on ½KNO_3SH supplemented with 1 g/l activated charcoal, 50 g/l potatoes, 50 g/l bananas, and 20 g/l sucrose gave the highest shoots (434 shoots/ lateral bud) [49].

2.2.8 Coconut Water

Coconut water (CW) is a natural growth promoter which contains higher levels of zeatin, zeatin riboside, 1,3-diphenylurea (contains cytokinin-like activity), auxins, nitrogenous compounds, inorganic elements, organic acids, sugars and their alcohols, peptides, vitamins, amino acids, and many other unknown components in its composition. CW was demonstrated that physiologically active substances presented in CW promote the cell division which further enhance shoot multiplication. Amino acids increase the number of shoots by inducing cell division [62, 63]. In orchid seed germination, CW showed 100% seed germination in *Dendrobium tetrachromum* after 50 days of culture on ½MS supplemented with 15% CW that is the same result as on ½MS supplemented with 2 g/l peptone [60]. The plantlets of *Calanthe* hybrids (Bukduseong × Hyesung) that cultured on modified Hyponex medium supplemented with 50 ml/l CW showed significant difference and gave good results in fresh weight and dry weight of plantlets after 8 weeks of culture [64]. CW was reported to enhance the regeneration of *Paphiopedilum rothschildianum* PLBs to the new plantlets on ½MS medium supplemented with 20% CW. Ng and Saleh [65] compared the effect of CW with banana homogenate, potato homogenate, and tomato homogenate. The results showed that CW was the best organic additive on the regeneration of plantlet from secondary PLBs of *Paphiopedilum rothschildianum*. Asghar et al. [66] studied the effects of CW on *Dendrobium nobile* var. Emma White. Plantlet regeneration on phytotechnology medium (O753) supplemented with 100 ml/l CW gave the highest number of shoots per explant, shoot length, fresh weight shoot, and dry weight shoot when compared with other concentrations of CW (50–300 ml/l). The immature seeds of *Rhynchostylis retusa* and *Aerides maculosa* on VW, MS, and Mitra media that added 15% of CW increased the percentage of callus induction when compared with control (VW) [67]. CW was the best organic additive to increase PLBs of *Dendrobium* Alya Pink on ½MS after 4 weeks of culture when compared with the same percentage of banana homogenate, tomato extract, and control group [53].

2.2.9 Chitosan

Chitin, consisting of a copolymer of *N*-acetyl-D-glucosamine and D-glucosamine residues linked by β-1,4 glycosidic bonds, is a natural polysaccharide. It is presented in broad range of species: in shells of crustaceans, in cuticles of insects, and in the cell wall of fungi and some algae. The deacetylated form of chitin is chitosan. Chitosan has been extensively used in agriculture, such as in seed, leaf, fruit, and vegetable coating, as fertilizer and in controlled agrochemical release, to increase plant product, to stimulate the immunity of plants, to protect plants against microorganisms, and to stimulate plant growth [68]. Very diluted chitosan solution was sprayed on orchid roots showing stimulation of growth, renewed flower production, and enhanced resistance against fungi and virus [69]. Nge et al. [68] suggested that fungal chitosan is an attractive candidate to be used as growth stimulator in VW liquid medium for orchid tissue culture. Sopalun et al. [21] studied effects of chitosan on growth rate, number of new PLBs per explant, number of shoots per explant, number of roots per explant, and number of leaves per explant of *Grammatophyllum speciosum* PLBs in in vitro culture. The 0, 5, 10, 15, 20, 25, 50, or 100 mg/l of chitosan were supplemented in ½MS liquid or on solid medium containing 2% (w/v) sucrose. The results showed that liquid medium supplemented with 15 mg/l chitosan showed the highest growth at 756% after 1 month of culture. The supplementation of chitosan more than 25 mg/l reduced the growth of

Grammatophyllum speciosum PLBs. Relative growth of *Grammatophyllum speciosum* PLBs were significantly reduced when chitosan concentrations were more than 25 mg/l. Moreover, the supplement of 100 mg/l chitosan caused the necrosis of PLBs and released some browning compounds into the medium. Various concentrations of chitosan supplemented in 1/2 MS solid medium showed the differences in the development of *Grammatophyllum speciosum* PLBs. The 5, 10, and 15 mg/l of chitosan supplemented in medium promoted leaf development. The highest number of leaves per explant was found in the medium supplemented with 15 mg/l chitosan (1.8 leaves per explant). Moreover, supplementation with 5, 10, and 15 mg/l chitosan promoted a higher number of new PLBs and shoots. However, chitosan supplementation did not promote rooting.

2.2.10 Activated Charcoal

Activated charcoal (AC) is composed of carbon that has numerous pores on the surface. AC was used as a culture component for adsorption of toxic plant metabolites, such as phenolic compound and plant growth regulators [70]. AC provided a dark environment during in vitro culture that affected shoot and root production. Dark condition when added the AC in the media can enhance to prolong the activity of some PLBs that can be degraded by light, such as IAA and IBA in culture media [71]. For seed germination, three different agar media (Fast, Knudson C, and 0.1 MS) that added 2 g/l AC increased the percentage of seed germination of *Encyclia* aff. *oncidioides*. Znaniecka et al. [72] concluded that activity of AC to accumulate phenolic compounds and oxidative products promoted the growth of *Encyclia* seedling. *Calanthe* hybrids are popular orchids in Japan and Korea but wild orchids of this genus had low percentage of seed germination (2%). Shin et al. [73] reported that Hyponex medium supplemented with 0.1 ppm NAA or 0.5 ppm BAP combined with 0.1 g/l AC enhanced germination percentage of *Calanthe* hybrid seeds when compared with other treatments. For plantlet growth, in vitro cultures of endangered orchid, *Anoectochilus formosanus* on H3 medium with different concentrations of AC showed that 0.5 mg/dm^3 gave the highest plant height, root length, fresh mass, and dry mass. Addition of AC (0.5, 1, 3, and 5 mg/dm^3) in culture media gave stronger plantlets that produced more roots and leaves than control [74]. Thomus and Michael [75] reported that MS medium supplemented with BAP, NAA, and 1 g/l of AC showed more multiple shoots per explant of *Rhynchostylis retusa* seedlings than in medium without AC. The regeneration of *Rhynchostylis rubrum* plantlets from callus after 8 weeks of culture on New Dogashima (ND) and VW medium supplemented with 15% CW and 0.2% AC showed greener somatic embryo than the same non-AC additive media [76]. Mass propagation and seedling of *Vanda coerulea* on both Phytamax medium and MS medium supplemented with 3 g/l AC showed healthier plantlets from PLBs than the same medium that added lower concentration of AC [77]. In vitro plantlets of endangered herb orchid, *Malaxis acuminata* on MS medium supplemented with 30 g/l sucrose and 3 g/l AC showed the highest number of shoots per subculture and number of roots per plant when compared with different concentrations of AC (0, 1, 2, 4, and 5 g/l) [78]. PLBs of *Phalaenopsis cornu-cervi* that cultured on ND medium supplemented with 0.2% AC and 4% sucrose gave the best result in 100% survival and promoted the plantlet growth after 5 months of culture [79]. Khatun et al. [80] reported that MS medium supplemented with different concentrations of PGRs had significant difference to promote the root production (root numbers and root length) of *Dendrobium* hybrid explants when compared with the same medium without AC after 120 days of culture. In callus culture of *Coelogyne cristata*, AC could effectively stimulate formation of somatic embryos on ½MS medium [20]. Graphite and AC have carbon structures, odorless, and non-toxic substances; therefore, they were used to darken the culture medium of plant tissue culture. In *Cattleya bicolor* seedlings, free PGRs KC agar medium supplemented with 6 or 7.5 g/l of graphite gave the highest number of buds; while, 6 g/l of AC gave the largest roots; however, in double hybrid "BLC Pastoral Innocence" seedlings, 4.5 g/l of AC gave the highest number of buds and roots [81].

2.2.11 Solid or Semisolid Supports

Micropropagation protocols have been established for many orchid species and hybrids by different media, carbon sources, plant growth regulators, and have been employed for several orchid species [82]. Knudson [1, 12] developed the culture media for asymbiotic seed germination by using Knudson B and C media. After that, other tissue culture media with other supplementations (carbon source, organic supplementation, etc.) were suggested for orchid germination, such as Vacin and Went medium [18], Murashige and Skoog medium [17], Malmgren Modified Terrestrial Orchid Medium [83], tomato culture medium [84], Mitra medium [85], banana culture medium [84], and PDA medium (potato dextrose agar). Various species of orchids have different requirements for nutrient and other conditions for growth. Rasmussen et al. [86] suggested that specific orchid species may have different limiting factors for growth and development.

However, the relationship between plant growth regulators and plant are investigated by using tissue culture technique. The supplementation of different ratios of plant growth regulators was observed by measuring the growth rate [16]. Normally, five major classes are accepted as plant hormones, including auxins, cytokinins, gibberellins, ethylene, and abscisic acid. The chemicals of hormone are different based on structure of each group that showed their effects on plant physiology [87]. Tissue culture technique was used as a model to study the effects of plant growth regulators, usually cytokinins, auxins, and the combination of both. However, it has been considered that plant growth regulators may cause somaclonal variation in plant. Tissue culture in orchids remains an indispensable tool for the commercial production of elite selections in many orchid producing countries (Table 1) because of low cost, uniformity, fast propagation, and high yield in a short period of time [88]. Most cut-flower orchids, *Dendrobium*, *Oncidium*, *Mokara*, *Aranda*, *Ascocenda*, and *Cattleya* alliances are propagated successfully through tissue culture using mainly modified VW, MS, and ½MS media. Within 1–2 years, lateral buds and apical bud from one young pseudobulb multiply to over 10,000 plantlets from the laboratory and are ready to grow in the greenhouse. Figures 3 and 4

Table 1 Micropropagation of some orchid species and hybrids

Orchid species and hybrids	Explant	Regenerants	Medium composition	References
Phalaenopsis amboinensis	Seeds	Protocorms	VW agar medium	[89]
Dimorphorchis lowii	Seeds	Protocorms	Knudson C agar medium + 150 ml/l coconut water	[90]
Dendrobium lasianthera	Seeds	Protocorms	VW agar medium + 2 g/l peptone	[91]
Cymbidium iridioides D.Don	Seeds	Protocorms	MS agar medium + 1.0 mg/l NAA	[92]
Aerides odorata Lour.	Seeds	Protocorms	MS agar medium	[93]
Coelogyne cristata Lindl.	Seeds	Protocorms	MS agar medium + 0.5 mg/l NAA + 1.0 mg/l BAP	[94]
Dendrobium densiflorum Lindl.	Seeds	Protocorms	MS agar medium + 1.0 mg/l BAP	[95]
Cymbidium elegans Lindl.	Seeds	Protocorms	MS agar medium + 1.0 mg/l BAP	[95, 96]
Phaius tancarvilleae (L'Her) Blume.	Seeds	Protocorms	MS agar medium + 0.5 mg/l BS	[97]
Cymbidium dloifolium Lindl.	Seeds	Protocorms	MS agar medium + 0.1% casein hydrolysate + 10% CW	[98]
Coelogyne fuscescens Lindl.	Seeds	Protocorms	MS agar medium + 1 mg/l BAP + 0.5 mg/l NAA	[99]
Cymbidium aloifolium Lindl. SW.	Seeds	Protocorms	MS agar medium + 0.5 mg/l BAP + 0.5 mg/l NAA	[100]
Cymbidium devonianum Paxton	Seeds	Protocorms	MS agar medium + 2 mg/l BAP + 0.5 mg/l NAA	[101]
Phaius tancarvilleae (L'Her.)	Shoot tips	Shoots	MS medium + 1.0 mg/l BAP	[102]
Grammatophyllum speciosum	Shoot tips	PLBs	½Murashige and Skoog (MS) liquid medium + 2% (w/v) sucrose	[21]
Rhynchostylis gigantea (Lindl.) Ridl.	Shoot tips	Adventitious shoots	VW liquid medium + 10 g/l sucrose	[103]
Dendrobium cruentum Rchb. f.	Shoot tips	PLBs	½Murashige and Skoog (MS) liquid + 1% (w/v) sucrose + 1.0 mg/l NAA	[104, 105]

(continued)

Table 1 (continued)

Orchid species and hybrids	Explant	Regenerants	Medium composition	References
Vanda coerulea	Shoot tips	Adventitious shoots	Vacin and Went (VW) medium + 10 g/l sucrose	[106]
Spathoglottis eburnea Gagnep	Shoot tips	Shoots	½MS medium + 1 mg/l NAA + 2 mg/l BAP	[107]
Cattleya sp.	Shoot tips	Rooting	Half strength MS medium + 2.4 mg/l IBA + 3% sucrose + 60 PPFD	[108]
Anacamptis pyramidalis Lindl. Rich.	Shoot tips	PLBs	MS medium + NAA/IBA/IAA; 0.5–1 mg/l + CW	[109]
Anoectochilus formosanus Hay.	Shoot tips	Shoot buds	Hyponex medium + 1 mg/l BAP	[74]
Arundina bambusifolia Lindl.	Shoot tips	Shoots	Raghavan and Torrey's medium, N and N medium	[110]
Cymbidium aloifolium Lindl. Sw.	Shoot tips	PLBs	VW medium + 5.0 mg/l NAA	[111]
Cymbidium atropurpureum (Lindley) Rolfe.	Shoot tips	PLBs	MS medium + 2.5 mg/l BAP	[112]
Dendrobium wardianum R. Warner	Shoot tips	PLBs	VW medium + 1 mg/l BAP + 1.5 mg/l NAA	[113]
Dendrobium cv. Sonia	Shoot tips	Shoot buds	½MS medium + 1 mg/l BAP + 7.5%CW	[114]
Dendrobium Joannie Ostenhault	Shoot tips	PLBs	VW medium + 15% CW	[115]
Phaius tankervilleae (Banks ex Aiton) Blume.	Shoot tips	PLBs	Raghavan and Torrey's (1964) basal medium	[110]
Vanilla planifolia Andr.	Shoot tips	Shoots	MS medium + 1 mg/l BAP + 150 ml/l CW	[116]
Aerides crispum Lindl.	Leaf explant	PLBs	MS medium + 2.0 mM BAP	[117]
Aerides multiflora Roxb.	Leaf explant	PLBs	MPR medium + 2 mg/l BAP + 0.5 mg/l NAA	[118, 119]
Dendrobium Cheingmai Pink	Leaf explant	Somatic embryos	½MS medium + 18.16 mM TDZ	[120]
Dendrobium hybrids (Sonia 17 and 28)	Leaf explant	PLBs	MS medium + 44.4 mM BAP	[121]

(continued)

Table 1 (continued)

Orchid species and hybrids	Explant	Regenerants	Medium composition	References
Micropera pallida Lindl.	Leaf explant	PLBs	½MS medium + 2 mg/l NAA+ 2 mg/l BAP	[122]
Phalaenopsis Little Steve	Leaf explant	Somatic embryos	½MS medium + 4.54 mM TDZ	[123]
Vanilla planifolia Andr.	Leaf explant	Callus Shoots from the callus	MS medium + 4.52 mM 2,4-D + 2.22 mM BAP	[124]
Phalaenopsis amabilis var. 'Manila'	Leaf explant	PLBs	MS medium + 15 mg/l BAP + 3 mg/l NAA	[125]
Aranda Deborah (*Arachis hookeriana* Rchb. f. × *Vanda lamellate* Lindl.)	Inflorescence explants	PLBs	KC medium + 1 mg/l BAP + CW	[126]
Epidendrum radicans Pav. Lindl.	Inflorescence explants	PLBs/shoot buds	½MS medium + 0.1 mg/l TDZ	[127]
Oncidium Sweet Sugar	Inflorescence explants	PLBs	½MS medium + 5 mg/l BAP + 5 mg/l NAA	[128]
Ponerorchis graminifolia Rchb. f.	Inflorescence explants	Shoot buds	½MS medium + 4.44 mM BAP + 0.54 mM NAA	[129]
Vanda hybrid "Dr. Anek"	Inflorescence explants	Direct shooting	½MS medium + 10 mg/l BAP+ 2 mg/l TDZ+ 30 g/l sucrose +7.5 g/l agar + 250 mg/l cefotaxime	[130]

showed the establishment of *Grammatophyllum speciosum* micropropagation and transfer plantlets to the saranhouse, respectively [21].

3 Orchid Cryopreservation

Cryopreservation is the process of freezing living materials using liquid nitrogen (LN) at −196 °C or −320 °F that physical and metabolic cellular processes are effectively stopped, and then they can be recovered and grown to regenerate a whole plant after cryopreservation. Cryopreservation is an ideal method for a long-term conservation of plant genetic resources because it demands the least space, labor, and maintenance; reduces susceptibility to disease and mutation. Orchid materials can be used for cryopreservation including pollen, seeds, protocorms, PLBs, apical bud, lateral bud, and meristem.

Fig. 3 Establishment of *Grammatophyllum speciosum* micropropagation; (**a**) A shoot tip was excised under a stereo microscope, (**b**) PLBs were induced in ½MS liquid medium (2 months after culture), (**c**) PLBs were subcultured for multiplication (1 month after culture), (**d**) PLBs were cultured on ½MS solid medium supplemented with 0.05% (w/v) of activated charcoal, 2.0 mg/l NAA and 1.0 mg/l BA, (**e**, **f**) Shoots and roots were induced, respectively (3 months after culture). Bar = 1 mm

Cryopreservation methods are based on the phenomenon of slow freezing (cell dehydration due to cold temperature) and vitrification (glass formation). Slow freezing was applied for cryopreservation of dormant buds and slow freezing methods are concerned. In vitrification-based methods, cells are dehydrated by exposure to concentrated chemicals or air drying before freezing, followed by rapid warming to avoid intracellular ice formation. Six different vitrification-based methods are: (1) vitrification, (2) encapsulation-dehydration, (3) encapsulation-

Fig. 4 Transfer of *Grammatophyllum speciosum* plantlets to the saranhouse. (**a**) Plantlets (6.0–8.0 cm long) were obtained after 3 months of culture, (**b**) Plantlets were removed from the bottle and washed with tap water, (**c**) Plantlets were transplanted to pots filled with small pieces of coconut husk, (**d**) Plants grown 6 months in the saranhouse, (**e**) Plants grown 1 year in the saranhouse, and (**f**) Plants grown 2 years in the saranhouse. Bar = 1 cm (**a**–**d**)

vitrification, (4) droplet-vitrification, (5) vitrification cryo-plate method (V cryo-plate), and (6) dehydration cryo-plate (D cryo-plate).

3.1 Dormant Bud Method

This is the first cryopreservation method that dormant buds were collected during winter when buds were dehydrated at −10 °C to −30 °C before storage in liquid nitrogen (−196 °C). This method was successful in willows (*Salix koriyanagi*) and poplar (*Populus sieboldii*) [131]. Later, dormant bud method was studied in apple [132–134], cherry [135], blueberry [136], etc. There is no research on orchid dormant bud method.

3.2 Slow Freezing Method

Slow freezing was the standard method in the initial stages [137], but need programmable freezing controller to reduce the temperature at the rate of 0.5–2 °C/min depending on the plant type and the growth stage until about −40 °C, then store in liquid nitrogen. The protocol takes many hours to complete and require expensive tools. This method was popular during 1980–1990 to store undeveloped plant tissues, such as suspended cells and calli of various plants including ornamental plants [138] except orchids.

3.3 Vitrification Method

Vitrification is a method developed to reduce the temperature quickly and to warm quickly, all parts of cells and tissues are in a glass state. Vitrification method involves treatment of explants with plant vitrification solution (PVS), such as PVS1 (consisted of 22% glycerol + 15% EG + 15% PEG + 7% DMSO + 0.5 M sorbitol) [139], PVS2 (consisted of 30% glycerol, 15% ethylene glycol, and 15% dimethyl sulfoxide) [140], PVS3 (consisted of 50% glycerol + 50% sucrose in medium) [141], and PVS4 (consisted of 35% glycerol + 20% EG + 20.5% sucrose) [142] to induce dehydration of explants during cooling and warming to avoid intracellular ice-crystal formation [139, 140]. The key to successful cryopreservation by vitrification method is to prevent injury by optimizing exposure time to PVS for dehydration [143] because over exposure time to PVS may result in cell injury and intracellular ice formation during cooling. The optimum exposure time to PVS depends on explant size and species specific. The suitable dehydration duration was related to the sample size, the composition, and loading solution [144]. Sakai et al. [140] succeeded to cryopreserve nucellar cells of naval orange using PVS2 solution, after that many plants were experimented with success including orchids (Table 2). Thammasiri [145] succeeded to cryopreserve seeds, using orchid seeds of *Doritis pulcherrima*. The results showed that 50 min exposure time to PVS2 solution gave 62% survival.

3.4 Encapsulation-Dehydration Method

Encapsulation-dehydration method is developed from artificial seed production that explants are encapsulated in alginate beads, precultured with high sucrose, desiccated by air-drying in a laminar air-flow cabinet or with silica gel, and then plunged into liquid nitrogen [142, 146]. The advantages of this method are easy for manipulation of encapsulated explants [147], preventing direct contact of toxic chemicals

Table 2 Successful cryopreservation of some orchid species and hybrids

Material	Species	Method	Dehydration	Survival/ regrowth	References
Zygotic embryo	*Bletilla striata*	Vitrification	3 h PVS2 at 0 °C	60% germination	[181]
Seeds	*Dendrobium candidum*	Drying	Drying to <12%	95% survival	[182]
Seeds	*Doritis pulcherrima*	Vitrification	50 min PVS2 at 25 ± 2 °C	92% germination	[145]
Immature seeds	*Bletilla striata*	Vitrification	60 min PVS2 on ice	81% survival	[183]
Seeds	*Ponerochis graminifolia* var. *suzukiana*	Vitrification	60 min PVS2	86% survival	[184]
Seeds	*Vanda coerulea*	Vitrification	70 min PVS2 at 25 ± 2 °C	67% germination	[185]
Seeds	*Vanda tricolor*	Vitrification	180 min PVS2 on ice	13.6% germination	[186]
Seeds	*Oncidium bifolium*	Encapsulation-dehydration	5 h silica gel	4.8% of the cryopreserved seeds produced complete plants	[187]
Mature seeds	*Bletilla striata*	Droplet-vitrification	60 min PVS2 at 25 °C	93% germination	[167]
Seeds	*Dendrobium cruentum*	D cryo-plate	60 min in laminar air-flow cabinet	57.8% regrowth	[105]
Protocorms	*Dendrobium candidum*	Vitrification	15 min PVS2	88% survival	[182]
Cell suspension	*Doritaenopsis*	Vitrification	PVS2	64% survival	[188]
Protocorms	*Dactylorhiza fuchsii*	Vitrification	PVS2	9% survival	[189]
Protocorms	*Vanda coerulea*	Encapsulation-dehydration	8 h in laminar air-flow cabinet	40%	[150]
Protocorms	Grammato-phyllum speciosum	Encapsulation-dehydration	8 h in laminar air-flow cabinet	24% regrowth	[168]

(continued)

Table 2 (continued)

Material	Species	Method	Dehydration	Survival/ regrowth	References
Protocorms	*Dendrobium cariniferum*	Encapsulation-vitrification	60 min PVS2 at 25 ± 2 °C	15% survival	[190]
Protocorm like-bodies	*Dendrobium candidum*	Encapsulation-vitrification	50 min PVS2 at 0 °C	85% survival	[191]
Protocorms	*Grammato-phyllum speciosum*	Encapsulation-vitrification	60 min PVS2	14% regrowth	[168]
Protocorms	*Bletilla striata*	Droplet-vitrification	40 min PVS2 at 25 °C	66% survival	[167]
Protocorms	*Grammato-phyllum speciosum*	Droplet-vitrification	30 min PVS2	38% regrowth	[168]
Protocorms	*Arundina graminifolia*	D cryo-plate	30 g drying beads in laminar air-flow cabinet	77% regrowth	[180]
Protocorms	*Vanda lilacina* Teijsm. & Binn	V cryo-plate	20 min PVS2	33.33% survival	[177]
Protocorms	*Vanda lilacina* Teijsm. & Binn	D cryo-plate	Silica gel for 1 h	83.78% survival	[177]
Shoot primordia	*Vanda pumila*	Desiccation	Precultured in 1 mg/l ABA for 3 days	65% survival	[192]
Shoot tips	*Dendrobium* Walter Oumae	Encapsulation-dehydration	6–8 h in laminar air-flow cabinet	13.33% survival	[193]
Pollinia	*Dendrobium* hybrids	Vitrification	3 h PVS2 on ice	60%	[194]
Pollinia	*Luisia macrantha*	Vitrification	10 min PVS2	56% germination	[195]

with plant materials, and nontoxic cryoprotectants are applied to protect during dehydration [148]. However, this method is the longer dehydration procedure than vitrification method [145]. Fabre and Dereuddre [149] succeeded to cryopreserve tomato apical meristems by using this method. There are many plant species including orchids to be cryopreserved by encapsulation-dehydration.

Fig. 5 Regrowth of *Vanda coerulea* protocorms after cryopreservation by encapsulation-dehydration in combination with loading solution. (**a**) Precultured beads after sterile air-flow dehydration for 10 h. (**b**) Cryopreserved protocorms after 20 days of regrowth. (**c**) 3 months of culture on modified VW agar medium showing shoot growth. (**d**) Plantlets after 8 months of culture on modified VW agar medium (left: noncryopreserved, right: cryopreserved plantlet). (**e**) Plantlets derived from cryopreserved protocorms after 5 months, and (**f**) 15 months of culture in the greenhouse. Bar for **a**–**c** = 1 mm, for **d** = 0.5 cm, and for **e**, **f** = 1 cm

Jitsopakul et al. [150] successfully cryopreserved protocorms of *Vanda coerulea* by encapsulation-dehydration in combination with a loading solution (Fig. 5). Protocorms were selected 70 days after sowing seeds, harvested from 7-month-old fruits. After encapsulation in an alginate matrix composed of 2% Na-alginate, 2 M glycerol plus 0.4 M sucrose (loading solution), the protocorms were precultured in modified VW [18] liquid medium supplemented with 0.7 M sucrose on a shaker (100 rpm) at 25 ± 3 °C for 20 h. Encapsulated protocorms were then dehydrated in a sterile air-flow in a laminar air-flow cabinet at 25 ± 3 °C for 1–10 h, and then directly plunged into liquid nitrogen for 1 day. After thawing at 49 °C for 2 min, cryopreserved beads were cultured on modified VW agar medium for regrowth. The highest regrowth of 40% was observed with cryopreserved bead with 35% water content for 8 h dehydration. No morphological variation was detected between noncryopreserved and cryopreserved plantlets, and ploidy level was unchanged as a result of cryopreservation.

3.5 Encapsulation-Vitrification Method

Encapsulation-vitrification method is a combination of encapsulation-dehydration method and vitrification method. The explants are encapsulated in alginate bead, and then subjected to dehydration by concentrated plant vitrification solutions, such as PVS1, PVS2, PVS3, and PVS4. This method gives higher survival than encapsulation-dehydration [151–153]. This method is used for many orchid species and hybrids [154].

3.6 Droplet-Vitrification Method

This recent method is the fast freezing from small drops of plant vitrification solution (PVS) on aluminum foil. Droplet-vitrification method is a combination of droplet-freezing and solution-based vitrification, then placed on aluminum foil strip in droplet of vitrification solution and then frozen by rapidly immersion in liquid nitrogen [155]. Rapid warming was done by dipping the aluminum foil strips in unloading solution without using a water bath [156]. During the cooling and warming procedures, rapid heat transfer is needed to avoid freezing injury [156–158]. Aluminum foil has an efficient thermal conductivity, resulting in quick and uniform heat distribution among tissue [156, 158, 159]. A high warming rate was employed to avoid recrystallization of intracellular ice or addition cell dehydration by extracellular ice [156]. The first success of droplet-vitrification method was studied in potatoes [160]. Later, successes in papaya [161], prunes [162], jam [163], chrysanthemum [164], banana [165], rose [166], *Bletilla striata* [167], *Grammatophyllum speciosum* [168], *Vanda coerulea* [150], Sugar cane [169], *Vanilla* orchid [170, 171], Oil Palm [172], Grapes [173], Orange [174], and etc. Jitsopakul et al. [150] studied droplet-vitrification method for cryopreservation of *Bletilla striata* mature seeds (0 day after sowing), zygotic embryos (3 days after sowing), and protocorms (6, 9, and 12 days after sowing). Mature seeds were surface-sterilized and sown on solidified ND medium supplemented with 3% sucrose and cultured under illumination provided at an intensity of 62.0 μmol m^{-2} s^{-1} for 16 h/day at 25 °C for preparation of zygotic embryos and protocorms. Mature seeds, zygotic embryos, and 6-day-old protocorms were precultured in liquid ND medium supplemented with 0.3 M sucrose for 3 h on a shaker (110 rpm) and then dehydrated with 2 M glycerol and 0.4 M sucrose in ND liquid medium (loading solution) for 15 min, followed by exposure to PVS2 solution for 60 min at 25 °C. Then, plant materials were soaked in liquid nitrogen by droplet-vitrification method, and then were cultured on solidified ND medium supplemented with 3% sucrose, germination rates of cryopreserved mature seeds, cryopreserved zygotic embryo, and survival rate of cryopreserved 6-day-old protocorms were 93%, 91%, and 84%, respectively. Cryopreserved 9-day-old protocorms gave the highest survival rate of 66% when precultured with 0.5 M sucrose for 3 h on a shaker, dehydrated with loading solution for 15 min, followed by exposure to PVS2 solution for 40 min at 25 °C and cultured on solidified ND medium supplemented with 480 mg/l ammonium nitrate and 3% sucrose. No survival was observed in cryopreserved 12-old-day protocorms (Fig. 5). Figure 6a–e showed development of mature seeds. Mature seeds developed into zygotic embryos at 3 days after sowing (Fig. 6b). Protocorms formed green spots on zygotic embryos at 6 days of sowing (Fig. 6c). Protocorms

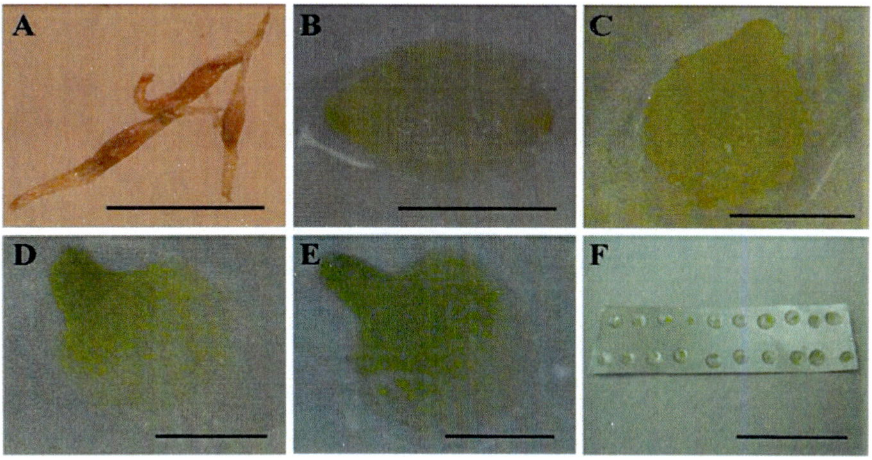

Fig. 6 Mature seeds of *Bletilla striata*. (**a**) Development of mature seeds sown on solidified ND medium supplemented with 3% sucrose under illumination provided at an intensity of 62 μM m^{-2} s^{-1} for 16 h/day at 25 °C for 3 days, (**b**) 6 days, (**c**) 9 days, (**d**) 12 days, (**e**) and Twenty droplets of PVS2 solution (2 μl) with protocorms, placed on sterilized aluminum foil strip (7 × 20 mm^2) (**f**). Bar: **a**–**e** = 0.5 mm, **f** = 1 cm

formed apical meristems at 9 days of sowing (Fig. 6d) and formed primary leaves at 12 days of sowing (Fig. 6e).

3.7 V Cryo-Plate Method

A new method for preserving plant varieties in cold conditions was developed 9 years ago by Yamamoto et al. [175] is the V cryo-plate method. It is the combination of encapsulation-vitrification and droplet-vitrification methods to make artificial seeds to attach well (high quality) on aluminum sheets (cryo-plate) size 7 mm × 37 mm × 0.5 mm, which has 10–12 holes (hole size 1.5 mm × 0.75 mm) in which the aluminum sheet is thicker than the aluminum foil (aluminum foil) used with the droplet-vitrification method. The thermal conductivity of the aluminum sheet is better than the aluminum foil which has a cooling rate of approximately 4,000 °C/min and an increase of temperature of around 3,000 °C/min; while, aluminum sheets have a temperature reduction of around 5,000 °C/min and an increase of temperature around 4,500 °C/min. Therefore, V cryo-plate method gives higher survival than droplet-vitrification method because the cooling and heating rates are faster. The advantages of the V cryo-plate method are: (1) The cooling and heating rates are very fast. Therefore, providing high survival. (2) The protocol is easy and convenient because the plant materials are attached to the aluminum sheet throughout the operation, not in a suspended state in a super cool solution, such as PVS2 and PVS3. Using this method does not need special skills. Jitsopakul et al. [176]

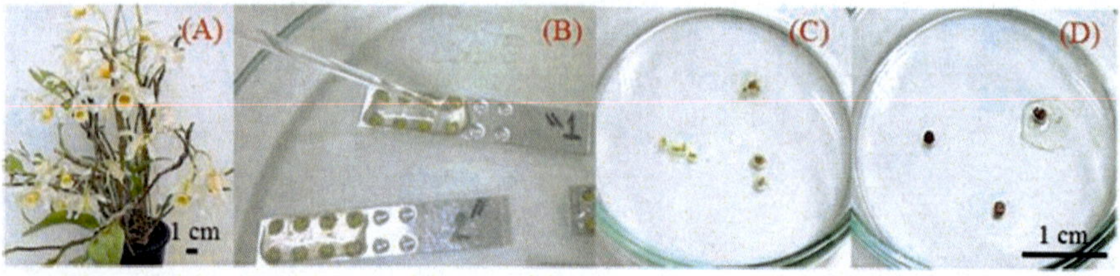

Fig. 7 *Dendrobium signatum* Rchb. f. (**a**) blooming flowers, (**b**) pollinia embedded with 3% sodium alginate gel on aluminum cryo-plates, (**c**) survival of pollinia, and (**d**) no survival of pollinia after cryopreservation using V cryo-plate method

studied the effect of V cryo-plate and D cryo-plate methods for cryopreservation of pollinia of Dendrobium signatum Rchb. f., a Thai orchid for pollination. Pollinia were collected from flowers and then placed on the aluminum cryo-plate containing 12 wells embedded with 3% sodium alginate gel (Fig. 7). In V cryo-plate method, cryo-plates with pollinia were immersed in 0.6 M sucrose + 2 M glycerol loading solution (LS) for 15 min, then dehydrated with PVS2 solution for 40 min at room temperature (29 ± 2 °C). In D cryo-plate method, cryo-plates with pollinia were treated with LS for 15 min, then dehydrated in a laminar air-flow cabinet for 3 h at room temperature (29 ± 2 °C). In both methods, cryo-plates were directly plunged into liquid nitrogen for 40 min, and rapidly warmed in 1.2 M sucrose for 15 min. The cryopreserved and non-cryopreserved pollinia were used to pollinate flowers of the same species. The results showed that cryopreserved pollinia retained fertilizing ability, and had similar fruit formation as those pollinated with noncryopreserved pollinia. The fruit formation from cryopreserved pollinia in V cryo-plate and D cryo-plate methods were 55.6% and 50%, respectively. Seeds from noncryopreserved and cryopreserved pollinia were successfully produced and germinated into plantlets with well-formed leaves and roots. Multiple stems (4.8 stems/plant) produced from one plantlet when cultured on modified VW agar medium [18] supplemented with 100 g/l banana, 150 ml coconut water, 50 g/l potato, and 20 g/l sucrose for 120 days at 25 ± 2 °C.

Imsomboon et al. [177] studied two new cryopreservation methods, the V cryo-plate and D cryo-plate methods of PLBs of *Vanda lilacina* Teijsm. and Binn. PLBs were dehydrated with PVS2 solution (0, 20, 40, 60, and 120 min) and silica gel (0, 0.5, 1, 1.5, 2, 2.5, and 3 h) for the V cryo-plate and D cryo-plate methods, respectively. The results showed that the V cryo-plate method, dehydration with PVS2 solution for 20 min, was a suitable method. It gave a survival score of 0.31, higher than control (+LN; 0 min) and gave a percentage of survival of 33.33%. For the D cryo-plate method, the suitable method for cryopreserved PLBs of *Vanda lilacina* was dehydration with silica gel for 1 h. It gave the highest survival score of 0.78 and percentage of survival was up to 83.78%. Then, an unpaired t-test was used to compare data of survival scores of cryopreserved *Vanda lilacina* PLBs between the V cryo-plate and the D cryo-plate method using silica gel for 1 h in dehydration step (Figs. 8 and 9).

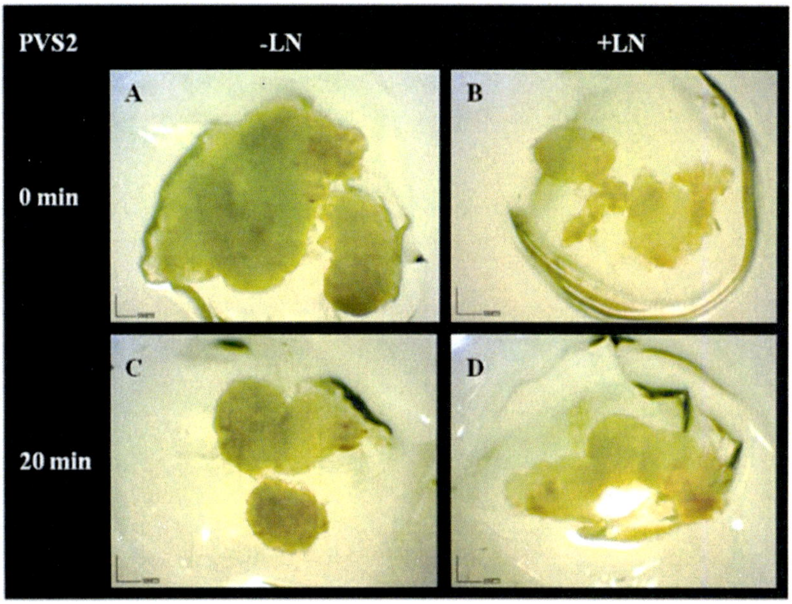

Fig. 8 PLBs of *Vanda lilacina* recovered from V cryo-plate method on ½MS agar medium at the 4th week. (**a**, **c**) Noncryopreserved (−LN) and (**b**, **d**) cryopreserved (+LN), (**a**, **b**) without PVS2 solution (0 min), and (**c**, **d**) with PVS2 solution (20 min) (bar = 0.2 mm)

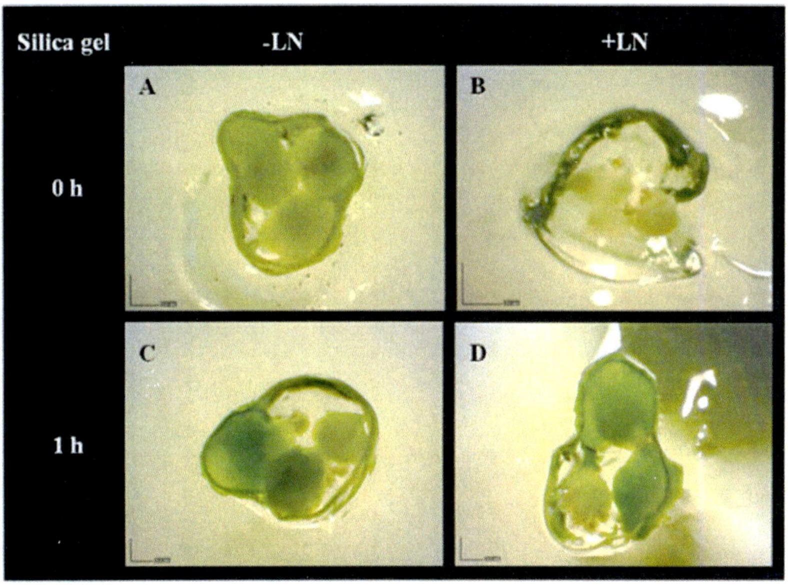

Fig. 9 PLBs of *Vanda lilacina* recovered from D cryo-plate method on ½MS agar medium at the 4th week. (**a**, **c**) Noncryopreserved (−LN) and (**b**, **d**) cryopreserved (+LN), (**a**, **b**) without silica gel (0 min), and (**c**, **d**) with silica gel (1 h) (bar = 0.5 mm)

3.8 D Cryo-Plate Method

Niino et al. [178] developed the D cryo-plate method, which is a combination of encapsulation-dehydration with the V cryo-plate method [179]. As some plants may be affected by the use of PVS2 solutions, there is low survival. Therefore, in order to avoid this damage, cells were dried by blowing air from a laminar air-flow cabinet. In addition, the cryopreserved plant materials can be larger than those using the V cryo-plate method because there is no problem of dehydration and the toxicity of the PVS2 solution after prolonged immersion. Both V cryo-plate and D cryo-plate methods have the same protocol except for the cell dehydration by air-drying, silica gel, or drying beads, instead of using concentrated chemical solutions, such as PVS2 solution. Later, Cordova II and Thammasiri [180] developed D cryo-plate method using silica gel and drying beads for dehydrating cells.

In this experiment, protocorms were placed in the preculture solution consisting of 0.7 M sucrose on a shaker (110 rpm) at 25 ± 3 °C for 1 day. After that, protocorms were placed one by one in the wells which filled before with the alginate solution containing 2% (w/v) sodium alginate in calcium-free ½MS basal medium with 0.4 M sucrose. The cryo-plates were hardened for 20 min by slowly dispensing the calcium chloride solution containing 0.1 M calcium chloride in ½MS basal medium with 0.4 M sucrose. Then the cryo-plates were surface dried using sterile filter paper, placed in Petri dishes containing silica gel or drying beads in a laminar air-flow cabinet (Fig. 10). Cryo-plates were dehydrated for 5 h until 25% moisture content was achieved. Dehydrated cryo-plates were placed in 2 ml cryotubes and plunged directly into liquid nitrogen for 1 day. Cryo-plates were removed from cryotubes and warmed in unloading solution (1.2 M sucrose solution) for 20 min. Protocorms were then removed from the cryo-plate and placed on ½MS agar medium for regrowth. Growth conditions were conducted using 16 h light at 25 ± 3 °C. For the cryo-plate method, regrowth of control treatments dehydrated using silica gel was observed to be 90%. Regrowth of control treatments dehydrated using drying beads was observed to be 92.1%. In all other treatments, regrowth was observed to be 73.8% using silica gel for dehydration. Regrowth for all other treatments dehydrated using drying beads was observed to be 76.5%. Regrowth was observed at the 2nd week of transfer to ½MS media. Dehydration using silica gel or drying beads did not significantly affect regrowth rate. Protocorms dehydrated using silica gel or drying beads developed into normal plantlets (Table 2).

4 Conclusions

In conclusion, plant tissue culture is a useful tool to increase and conserve orchid under sterile conditions. But their problems are time-consuming, quite expensive, and labor intensity and costly on growth of orchids in sterile condition. Furthermore, the higher somaclonal variation, phenotypically change and losing materials by microbial contamination were observed by using plant tissue culture technique.

Fig. 10 Cryo-plate method dehydrated with silica gel or drying beads. (**a**) Protocorm development, (**b**) Preculture of protocorms in ½MS liquid medium with 0.7 M sucrose for 1 day, (**c**) Pour the alginate solution containing 2% (w/v) sodium alginate in calcium-free ½MS basal medium with 0.4 M sucrose in the wells, (**d**) Place the precultured protocorms in the wells one by one, (**e**) Pour the calcium chloride solution containing 0.1 M calcium chloride in ½MS basal medium with 0.4 M sucrose, (**f**) Dehydration with 50 g silica gel, (**g**) Dehydration with 30 g drying beads, (**h**) Put each cryo-plate in a 2 ml cryotube, (**i**) Plunge 2 ml cryotubes into liquid nitrogen for 1 day, (**j**) Warming in 1.2 M sucrose solution for 20 min, (**k**) Plate on ½MS agar medium, (**l**) Regrowth, and (**m**) Regrowth after 60 days

Plant cryopreservation methods have been developed from the beginning in 1960. The methods changed from time-consuming protocols to more simple and more efficient. For orchids, the methods are developed for various plant materials, such as seeds, embryo, protocorm, protocorm-like bodies, apical bud, lateral bud, root tip, and meristem. The developed methods are vitrification, encapsulation-dehydration, encapsulation-vitrification, droplet-vitrification, V cryo-plate, and D cryo-plate, respectively.

References

1. Knudson L (1922) Symbiosis and asymbiosis relative to orchids. New Phytol 26:328–336
2. Arditti J (1984) An history of orchid hybridization, seed germination and tissue culture. Bot J Linn Soc 89:359–381
3. Arditti J (1967) Factors affecting the germination of orchid seeds. Bot Rev 33:1–96
4. Sheehan TJ (1983) Recent advances in botany, propagation and physiology of orchids. In: Janick J (ed) Horticultural reviews, vol 5. AVI Publishing Company, Westport, pp 279–315
5. Stephen R, Jayakoalen N (2000) Artificail seed production in coriander (*Coriandrum sativum* L.). Plant Tissue 10(1):45–49
6. Arditti A, Ernst R (1993) History. In: Micropropagation of orchids. Wiley, New York, pp 3–23

7. Arditti J, Ghani AKA (2000) Numerical and physical properties of orchid seeds and their biological implications. New Phytol 145:367–421
8. Seaton P, Ramsay M (2005) Growing orchids from seed. Mondadori Printing, Surrey
9. Kalimuthu K, Senthilkumar R, Vijayakumar S (2007) In vitro micropropagation of orchid, *Oncidium* sp. (Dancing Dolls). Afr J Biotechnol 6(10):1171–1174
10. Manning JC, Staden J (1987) The development and mobilization of seed reserves in some African orchids. Aust J Bot 35:343–353
11. Rasmussen HN, Anderson TF, Johansen B (1990) Temperature sensitivity of in vitro germination and seedling development of *Dactylorhiza majalis* (Orchidaceae) with and without a mycorrhizal fungus. Plant Cell Environ 13:171–177
12. Knudson L (1946) A nutrient for germination of orchid seeds. Am Orchid Soc Bull 15:214–217
13. Lee YI, Yeung EC, Chung MC (2007) Embryo development of orchids. In: Orchid biotechnology. Stallion Press, Singapore, pp 23–44
14. Champagnat M, Morel G (1972) La culture in vitro des tissues tubercules d'Ophrys. Comptes Rendus de I'Academie de Sciences, Paris 274:3379–3380
15. Norstog K (1979) Embryo culture as a tool in the study of comparative and developmental morphology. In: Plant cell and tissue culture: principles and applications. Ohio State University, Columbus, pp 179–202
16. Georage EF, Hall MA, Klerk GJD (2008) Plant propagation by tissue culture, 3rd edn. Springer, Dordrecht
17. Murashige T, Skoog F (1962) A revised medium for rapid growth and bioassays with tobacco tissue cultures. Physiol Plant 15:473–497
18. Vacin EF, Went FW (1949) Some pH changes in nutrient solutions. Bot Gaz 110:605–613
19. Puchooa D (2004) Comparison of different culture media for the in vitro culture of *Dendrobium* (Orchidaceae). Int J Agric Biol 6:884–888
20. Naing AH, Chung JD, Lim KB (2011) Plant regeneration through indirect somatic embryogenesis in *Coelogyne cristata* orchid. Am J Plant Sci 2:262–267
21. Sopalun K, Thammasiri K, Ishikawa K (2010) Micropropagation of the Thai orchid *Grammatophyllum speciosum* Blume. Plant Cell Tissue Organ Cult 101:143–150
22. Kishor R, Sharma GJ (2009) Intergeneric hybrid of two rare and endangered orchids, *Renanthera imschootiana* Rolfe and *Vanda coerulea* Griff. ex L. (Orchidaceae): synthesis and characterization. Euphytica 165(2):247–256
23. Kishor R, Devi HS, Jeyaram K et al (2008) Molecular characterization of reciprocal crosses of *Aerides vandarum* and *Vanda stangeana* (Orchidaceae). Plant Biotechnol Rep 2:145–152
24. Wattanawikkit P, Bunn E, Chayanarit K, Tantiwiwat S (2011) Effect of cytokinins (BAP and TDZ) and auxin (2,4-D) on growth and development of *Paphiopedilum callosum*. Kasetsart J (Nat Sci) 45:12–19
25. Jaime ATS (2012) New basal media for protocorm like body and callus induction of hybrid *Cymbidium*. J Fruit Ornam Plant Res 20(2):127–133
26. Long B, Niemiera AX, Cheng ZY et al (2010) In vitro propagation of four threatened *Paphiopedilum* species (Orchidaceae). Plant Cell Tissue Organ Cult 101:151–162
27. Schenk RU, Hildebrant AC (1972) Medium and techniques for induction and growth of monocotyledonous and dicotyledonous plant cell cultures. Can J Bot 50:199–204
28. Luo JP, Zha XQ, Fan XJ (2006) Effects of carbon and nitrogen sources on protocorm-like bodies proliferation of *Dendrobium huoshanense* by liquid culture. Acta Hortic 725(1): 197–202
29. Gamborg OL, Miller RA, Ojima K (1968) Nutrient requirements of suspension cultures of soybean root cells. Exp Cell Res 50(1):151–158
30. Chu CC (1978) The N6 medium and its applications to another culture of cereal crops. In: Proceedings of symposium on plant tissue culture. Science Press, Peking, pp 43–50

31. Silva JAT, Yam TW, Fukai S et al (2005) Establishment of optimum nutrient media for in vitro propagation of *Cymbidium* sw. (Orchidaceae) using protocorm-like body segments. Propag Ornam Plants 5(3):129–136
32. Pierik RLM (1988) In vitro culture of higher plants as a tool in the propagation of horticultural crops. Acta Hortic 226:25–40
33. Pierik RLM, Sprenkels PA, Harst BV et al (1988) Seed germination and further development of plantlets of *Paphiopedilum ciliolare* Pfitz. in vitro. Sci Hortic 34:139–153
34. Hong PI, Chen JT, Chang WC (2008) Promotion of direct somatic embryogenesis of *Oncidium* by adjusting carbon sources. Biol Plant 52(3):597–600
35. Lou JP, Wawrosch C, Kopp B (2009) Enhanced micropropagation of *Dendrobium huoshanense* C.Z. Tang et S.J. Cheng through protocorm-like bodies: the effects of cytokinins, carbohydrate sources and cold pretreatment. Sci Hortic 123:258–262
36. Ferreira WM, Suzuki RM, Pescador R et al (2011) Propagation, growth, and carbohydrates of *Dendrobium* Second Love (Orchidaceae) in vitro as affected by sucrose, light, and dark. In Vitro Cell Dev Biol 47:420–427
37. Chen TY, Chen JT, Chang WC (2004) Plant regeneration through direct shoot bud formation from leaf cultures of *Paphiopedilum* orchids. Plant Cell Tissue Org Cult 76:11–15
38. Wu IF, Chen JT, Chang WC (2004) Effects of auxins and cytokinins on embryo formation from callus of *Oncidium* 'Gower Ramsey'. Plant Cell Tissue Org Cult 77:107–109
39. Steward FC, Mapes MO, Mears K (1958) Growth and organised development of cultured cells: II. Organisation in cultured grown from freely suspended cells. Am J Bot 45:705–707
40. Chen JT, Chang WC (2000) Efficient plant regeneration through somatic embryogenesis from callus culture of *Oncidium* (Orchidaceae). Plant Sci 160:87–93
41. Jheng FY, Do YY, Liauh YW et al (2006) Enhancement of growth and regeneration efficiency from embryogenic callus cultures of *Oncidium* 'Gower Ramsey' by adjusting carbohydrate sources. Plant Sci 170:1133–1140
42. Nayak NR, Sahoo S, Patnaik S, Rath SP (2002) Establishment of thin cross section (TCS) culture method for rapid micropropagation of *Cymbidium aloifolium* (L.) Sw. and *Dendrobium nobile* Lindl. (Orchidaceae). Sci Hortic 94:107–116
43. Tokuhara K, Mii M (1993) Micropropagation of *Phalaenopsis* and *Doritaenopsis* by culturing shoot tips of flower stalk buds. Plant Cell Rep 13:7–11
44. Tokuhara K, Mii M (2001) Induction of embryogenic callus and cell suspension culture from shoot tips excised from lower flower stalk buds of *Phalaenopsis* (Orchidaceae). In Vitro Cell Dev Biol Plant 37:457–461
45. Chen Y, Piluek C (1995) Effect of thidiazuron and N6-benzylaminopurine on shoot regeneration of *Phalaenopsis*. Plant Growth Regul 16:99–101
46. Roy J, Banerjee N (2003) Induction of callus and plant regeneration from shoot-tip explants of *Dendrobium fimbriatum* Lind. var *oculatum* Hk.f. Sci Hortic 97:333–340
47. Huetteman CA, Preece JE (1993) Thidiazuron: a potent cytokinin for woody plant tissue culture. Plant Cell Tissue Org Cult 33:105–119
48. Ernst R (1994) Effect of thidiazuron on in vitro propagation of *Phalaenopsis* and *Doritaenopsis* (Orchidaceae). Plant Cell Tissue Org Cult 39:273–275
49. Wasiksiri W, Thammasiri K, Chuenboonngarm N et al (2010) Medium development for micropropagation of *Vanda coerulea* Griff. ex Lindl. from lateral buds. In: Abstract of 36th Congress on science and technology of Thailand
50. Wu K, Zeng S, Silva JAT et al (2012) Efficient regeneration of *Renanthera* Tom Thumb 'Qilin' from leaf explants. Sci Hortic 135:194–201
51. Vyas S, Guha S, Bhattacharya M et al (2009) Rapid regeneration of plants of *Dendrobium lituiflorum* Lindl. (Orchidaceae) by using banana extract. Sci Hortic 121:32–37
52. Lo SF, Narawade SM, Kou CL et al (2004) Asymbiotic germination of immature seeds, plantlet, development and ex vitro establishment of plant of *Dendrobium tosaense* Makino – a medicinally important orchid. In Vitro Cell Dev Biol Plant 40:528–535

53. Nambiar N, Tee CS, Maziah M (2012) Effects of organic additives and different carbohydrate sources on proliferation of protocorm-like bodies in *Dendrobium* Alya Pink. Plant Omics J 5(1):10–18
54. Sinha P, Jahan MAA, Munshi JL et al (2010) High frequency regeneration of *Phalaenopsis amabilis* (L.) Bl. Cv. Lovely through in vitro culture. Plant Tissue Cult Biotechnol 20(2): 185–193
55. Gnasekaran P, Rathinam X, Sinniah UR et al (2010) A study on the use of organic additives on the protocorm-like bodies (PLBS) growth of *Phalaenopsis violacea* orchid. J Phytology 2:29–33
56. Li Z, Xu L (2009) In vitro propagation of white-flower mutant of *Rhynchostylis gigantea* (Lindl.) Ridl. through immature seed-derived protocorm-like bodies. J Hortic For 1(6):93–97
57. Vyas S, Kapoor-Pandey P, Guha S et al (2011) Synchronous plantlet formation by using banana extract and in vitro hardening in orchid, *Dendrobium lituiflorum* Lindl. J Ornam Hortic Plants 1(3):175–184
58. Zeng S, Wu K, Silva JAT et al (2012) Asymbiotic seed germination, seedling development and reintroduction of *Paphiopedilum wardii* Sumerh., an endangered terrestrial orchid. Sci Hortic 138:198–209
59. Gnasekaran P, Poobathy R, Mahmood M et al (2012) Effects of complex organic additives on improving the growth of PLBs of *Vanda* Kasem's delight. Aust J Crop Sci 6:1245–1248
60. Ali MJ, Murdad R, Latip MA et al (2011) In vitro seed germination of Bornean endemic orchids *Dendrobium tetrachromum* and *Dendrobium hamaticalcar*. In: Empowering science, pp 770–777
61. Lee YI (2010) Micropropagation of *Cypripedium formosanum* Hayata through axillary buds from mature plants. HortScience 45(9):1369–1372
62. Dix L, Van SJ (1982) Auxin and gibberellins-like substances in coconut milk and malt extract. Plant Cell Tissue Org Cult 1:239–245
63. Staden JV, Zazimalova E, George EF (2008) Plant growth regulators II: cytokinins, their analogues and antagonists. In: Plant propagation by tissue culture, 3rd edn. Springer, Dordrecht, pp 206–217
64. Baque MA, Shin YK, Elshmari T et al (2011) Effect of light quality, sucrose and coconut water concentration on the microporpagation of *Calanthe* hybrids ('Bukduseong' × 'Hyesung' and 'Chunkwang' × 'Hyesung'). Aust J Crop Sci 5(10):1247–1254
65. Ng CY, Saleh NM (2011) In vitro propagation of *Paphiopedilum* orchid through formation of protocorm-like bodies. Plant Cell Tissue Organ Cult 105:193–202
66. Asghar S, Ahmad T, Hafiz IA et al (2011) In vitro propagation of orchid (*Dendrobium nobile*) var. Emma white. Afr J Biotechnol 10(16):3097–3103
67. Parab G, Krishnan S (2012) Rapid in vitro mass multiplication of *Aerides maculosa* and *Rhynchostylis retusa* from immature seeds. Indian J Biotechnol 11:288–294
68. Nge KL, Nwe N, Chandrkrachang S et al (2006) Chitosan as a growth stimulator in orchid tissue culture. Plant Sci 170:1185–1190
69. Chanrkrachange S (2002) The applications of chitin and chitosan in agriculture in Thailand, In: Suchiva K, Chanrkrachange S, Methacanon P and Peter MG. (Eds). Advances in Chitin Science Volume 5. Bangkok, Thailand:458–462
70. Thomus TD (2008) The role of activated charcoal in plant tissue culture. Biotechnol Adv 26:631–618
71. Pan MJ, Staden JV (1998) The use of charcoal in vitro culture – a review. Plant Growth Regul 26:155–163
72. Znaniecka J, Królicka A, Sidwa-Gorycka M et al (2005) Asymbiotic germination, seedling development and plantlet propagation of *Encyclia* aff. oncidioides – an endangered orchid. Acta Soc Bot Pol 74(3):193–198
73. Shin YK, Baque AM, Elghamedi S et al (2011) Effects of activated charcoal, plant growth regulators and ultrasonic pre-treatments on in vitro germination and protocorm formation of *Calanthe* hybrids. Aust J Crop Sci 5(5):582–588

74. Ket NV, Hahn EJ, Park SY et al (2004) Micropropagation of an endangered orchid *Anoectochilus formosanus*. Biol Plant 48(3):339–344
75. Thomus TD, Michael A (2007) High-frequency plantlet regeneration and multiple shoot induction from cultured immature seeds of *Rhynchostylis retusa* Blume., an exquisite orchid. Plant Biotechnol Rep 1:243–249
76. Te-chato S, Kongruk S, Khaimuk W (2010) Micropropagation of Chang Daeng (*Rhynchostylis rubrum*) by embryogenic callus. J Agric Technol 6(3):589–597
77. Roy AR, Patel RS, Patel VV et al (2011) Asymbiotic seed germination, mass propagation and seedling development of *Vanda coerulea* Griff ex. Lindl. (Blue *Vanda*): an in vitro protocol for an endangered orchid. Sci Hortic 128:325–331
78. Arenmongla T, Deb CR (2012) Germination of immature embryos and multiplication of *Malaxis acuminata* D. Don, an endangered therapeutically important orchid, by asymbiotic culture in vitro. Indian J Biotechnol 11:464–469
79. Rittirat S, Thammasiri K, Te-chato S (2012) Effect of media and sucrose concentrations with or without activated charcoal on the plantlet growth of *P. cornu-cervi* (Breda) Blume & Rchb. f. J Agric Technol 8(6):2077–3087
80. Khatun MM, Khatun H, Khanam D et al (2010) In vitro root formation and plantlet development in *Dendrobium* orchid. Bangladesh J Agric Res 35(2):257–265
81. Prizão EC, Gonçalves LM, Gutierre MAM et al (2012) Activated charcoal and graphite for the micropropagation of *Cattleya bicolor* Lindl. and an orchid double-hybrid 'BLC Pastoral Innocence'. Acta Sci 34(2):157–161
82. Arditti J (1977) Clonal propagation of orchids by means of tissue culture – a manual. In: Arditti J (ed) Orchid biology: reviews and perspectives, vol I. Cornell University Press, Ithaca, pp 203–293
83. Malmgren S (1996) Orchid propagation: theory and practice. In: Allen C (eds) North American native orchids: propagation and production. North American Native Terrestrial Orchid Conference, Germantown, Maryland, pp 63–71
84. Barbery R, Morales I (2011) Manual para el cultivo in vitro de la orquídea *Cattleya nobilior*. Flor símbolo de Concepción. CEPAD, Editorial El País, Santa Cruz, 46 pp
85. Mitra GC, Prasad RN, Roychowdhary A (1976) Inorganic salts and differentiation of protocorms in seed-callus of an orchid and correlated changes in its free amino acid content. Indian J Exp Biol 14:350–351
86. Rasmussen HN, Dixon KW, Jersáková J et al (2015) Germination and seedling establishment in orchids: a complex of requirements. Ann Bot 116:391–402
87. Bajqūz A, Piotrowska A (2009) Conjugates of auxin and cytokinin. Phytochemistry 70(8):957–969
88. Thammasiri K (1997) Orchids in Thailand: a success story. APAARI, FAO, Angkor Publishers (P.) Ltd., New Delhi
89. Utami ESW, Hariyanto S (2019) In vitro seed germination and seedling development of a rare Indonesian native orchid *Phalaenopsis amboinensis* J.J Sm. Hindawi Scientifica 2019:1–6
90. Jainol JE, Jualang AG (2015) In vitro shoot multiplication and rooting of shoot tip explants of *Dimorphorchis lowii*: an endemic orchid of Borneo. Plant Physiol 7:14–25
91. Utami ESW, Hariyanto S, Sri Y et al (2017) In vitro propagation of the endangered medicinal orchid, *Dendrobium lasianthera* J.J. Sm through mature seed culture. Asian Pac J Trop Biomed 7(5):406–410
92. Pant B, Swar S (2011) Micropropagation of *Cymbidium iridioides*. Nepal J Sci Technol 12:91–96
93. Pant B, Gurung R (2005) In vitro seed germination and seedling development in *Aerides odorata* Lour. J Orchid Soc India 19:51–55
94. Pant B, Swar S, Karanjeet A (2008) Micropropagation of *Coelogyne cristata* Lindl. J Orchid Soc India 22(1–2):45–48
95. Pradhan S, Pant B (2009) In vitro seed germination in *Cymbidium elegans* Lindl. and *Dendrobium densiflorum* Lindl. ex Wall (Orchidaceae). Bot Orient J Plant Sci 6:100–102

96. Pant B, Pradhan S (2010) Micropropagation of *Cymbidium elegans* Lindl. through protocorm and shoot tip culture. Role of biotechnology in food security and climate change. In: Proceedings of 6th international plant tissue culture and biotechnology conference, Dhaka, Bangladesh, December, pp 3–5
97. Pant B, Shrestha S, Pradhan S (2011) In vitro seed germination and seedling development of *Phaius tancarvilleae* (L'Her.) Blume. Sci World 9(9):50–52
98. Rajkarnikar KM (2011) Propagation of *Cymbidium aloifolium* (L.) Sw. in vitro by seeds. Bull Dept Plant Res 33:27–30
99. Koirala D, Pradhan S, Pant B (2013) Asymbiotic seed germination and plantlet development of *Coelogyne fuscescens* Lindl., a medicinal orchid of Nepal. Sci World 11(11):97–100
100. Pradhan S, Regmi T, Parmar G et al (2013b) Effect of different media on in vitro seed germination and seedling development of *Cymbidium aloifolium* (L.) Sw. Nepal J Sci Technol 14(1):51–56
101. Parmar G (2014) In vitro seed germination and seedling development of *Cymbidium devonianum* Paxton (Orchidaceae). Bull Dept Plant Res 36:61–64
102. Pant B, Shrestha S (2011) In vitro mass propagation of a ground orchid – *Phaius tancarvilleae* (L'Her.) Blume through shoot tip culture. Plant Tissue Cult Biotechnol 21(2):181–188
103. Prasongsom S, Thammasiri K, Chuenboonngarm N (2016) Efficient adventitious shoot regeneration from shoot tip culture of *Rhynchostylis gigantea* (Lindl.) Ridl. (amethyst-purple), a rare Thai orchid species. Walailak J Sci Technol 13(9):757–767
104. Sangdum S, Thammasiri K, Chuenboonngarm N et al (2017) Micropropagation of *Dendrobium cruentum* Rchb. f., a rare Thai orchid species. Acta Hortic 1167:69–74
105. Prasongsom S, Thammasiri K, Narangajavana J et al (2020) Cryopreservation of Dendrobium cruentum Rchb. f. seeds by D cryo-plate and V cryo-plate techniques. Walailak Journal of Science and Technology 17(3):181–191
106. Jitsopakul N, Thammasiri K, Ishikawa K (2013) Efficient adventitious shoot regeneration from shoot tip culture of *Vanda coerulea*, a Thai orchid. ScienceAsia 39:449–455
107. Pornchuti W, Thammasiri K, Chuenboonngarm N et al (2017) Micropropagation of *Spathoglottis eburnea* Gagnep, a Thai orchid species, through shoot tips. Acta Hortic 1167: 87–94
108. Dewir YH, El-Mahrouk EM, Murthy HN et al (2015) Micropropagation of *Cattleya*: improved in vitro rooting and acclimatization. Hortic Environ Biotechnol 56(1):89–93
109. Morel G (1970) Neres ouf dem Gebiet der Meristem-Forchung. Die Orchidee 20:433–443
110. Nagaraju V, Parthasarathy VS (1995) In vitro propagation of *Phaius* and bamboo orchid by shoot tip culture. Ann Plant Physiol 9:102–104
111. Devi JB, Borthakur B, Deka PC (1997) Clonal propagation of *Dendrobium moschatum* and *Cymbidium aloifolium* through shoot tip culture. J Orchid Soc India 11:19–21
112. Subramanium G, Taha RM (2003) Morphogenesis of *Cymbidium atropurpureum* in vitro. Malays J Sci 22:1–5
113. Sharma A, Tandon P (1992) In vitro culture of *Dendrobium wardianum* Warner: morphogenetic effects of some heterogenous adjuvants. Indian J Plant Physiol 35:80–85
114. Sheela VL, Rajmohan K, Anita S et al (2004) Effect of growth regulators on development and multiplication of protocorm like bodies in *Dendrobium* cv Sonia. J Orchid Soc India 18:21–23
115. Sharon M, Vasundhara G (1990) Micropropagation of *Dendrobium* Joannie Ostenhault. J Orchid Soc India 4:145–148
116. Kalimuthu K, Senthikumar R, Murugalatha N (2006) Regeneration and mass multiplication of *Vanilla planifolia* Andr. – a tropical orchid. Curr Sci 91:1401–1403
117. Sheelavanthmath SS, Murthy HN, Hema BP et al (2005) High frequency of protocorm like bodies (PLBs) induction and plant regeneration from protocorm and leaf sections of *Aerides crispum*. Sci Hortic 106:395–401
118. Vij SP, Aggarwal S, Pathak P (2004) Regeneration competence of *Cymbidium* Great Waltz Valley flower roots: a study in vitro. J Orchid Soc India 18:109–115

119. Vij SP, Sembi JK, Verma J et al (2004) In vitro rapid mass multiplication of *Aerides multiflora*, a floriculturally significant species. J Orchid Soc India 17:63–68
120. Chung HH, Chen JT, Chang WC (2005) Cytokinins induce direct somatic embryogenesis of *Dendrobium* Chiengmai Pink and subsequent plant regeneration. In Vitro Cell Dev Biol Plant 41:765–769
121. Martin KP, Madassery JP (2006) Rapid in vitro propagation of *Dendrobium* hybrids through direct shoot formation from foliar explants and protocorm like bodies. Sci Hortic 108:95–99
122. Bhadra SK, Hossain MM (2004) Induction of embryogenesis and direct organogenesis in *Micropera pallida* Lindl., an epiphytic orchid of Bangladesh. J Orchid Soc India 18:5–9
123. Kuo HL, Chen JT, Chang WC (2005) Efficient plant regeneration through direct somatic embryogenesis from leaf explants of *Phalaenopsis* 'Little Steve'. In Vitro Cell Dev Biol Plant 41:453–456
124. Janarthanam B, Seshadri S (2008) Plantlet regeneration from leaf derived callus of *Vanilla planifolia* Andr. In Vitro Cell Dev Biol Plant 44:84–89
125. Balilashaki K, Ghehsareh MG (2016) Micropropagation of *Phalaenopsis amabilis* var. 'Manila' by leaves obtained from in vitro culturing the nodes of flower stalks. Not Sci Biol 8(2):164–169
126. Goh CJ, Wong PF (1990) Micropropagation of the monopodial orchid hybrid *Aranda* Deborah using inflorescence explants. Sci Hortic 44:315–321
127. Chen LR, Chen JT, Chang WC (2002) Efficient production of protocorm like bodies and plant regeneration from flower stalk explants of the sympodial orchid *Epidendrum radicans*. In Vitro Cell Dev Biol Plant 38:441–445
128. Chen JT, Chang WC (2002) Effects of tissue culture characteristics on direct somatic embryogenesis in *Oncidium* 'Gower Ramsey'. Plant Cell Tissue Org Cult 69:41–44
129. Mitsukuri K, Mori G, Johkan M et al (2009) Effects of explant position and dark treatment on bud formation in floret culture of *Ponerorchis graminiflolia* Rchb. f. Sci Hortic 121:243–247
130. Baby R, Valsala PA, Doddamani MB (2019) In vitro micropropagation protocol for *Vanda* hybrid 'Dr. Anek'. Int J Curr Microbiol App Sci 8(4):2073–2084
131. Sakai A (1960) Survival of the twig of woody plants at −196 °C. Nature 185:393–394
132. Sakai A, Nishiyama Y (1978) Cryopreservations of winter vegetative buds of hardy fruit trees in liquid nitrogen. HortScience 13:225–227
133. Forsline PL, Stushnoff C, Towill LE et al (1998) Recovery and longevity of cryopreserved apple buds. J Am Soc Hortic Sci 123:365–370
134. Towill LE, Bonnart R (2003) Cracking in a vitrification solution during cooling or warming does not affect growth of cryopreserved mint shoot tips. CryoLetters 24:341–346
135. Towill LE, Forsline PL (1999) Cryopreservation of sour cherry (*Prunus cerasus* L.) using a dormant vegetative bud method. Cryo-Letters 20:215–222
136. Jenderek MM, Reed BM (2017) Cryopreserved storage of clonal germplasm in the USDA national plant germplasm system. In Vitro Cell Dev Biol Plant 53:299–308
137. Panis B, Lambardi M (2005) Status of cryopreservation technologies in plants (crops and forest trees). In: The role of biotechnology. Villa Gualino, Turin, Italy, 5–7 March
138. Uemura M, Sakai A (1980) Survival of carnation (*Dianthus caryophyllus* L.) shoot to the temperature of liquid nitrogen. Plant Cell Physiol 21:85
139. Uragami A, Sakai A, Nagai M et al (1989) Survival of cultured cells and somatic embryos of *Asparagus officinalis* cryopreserved by vitrification. Plant Cell Rep 8:418–421
140. Sakai A, Kobayashi S, Oiyama I (1990) Cryopreservation of nucellar cells of navel orange (*Citrus sinensis* Osb. var *brasiliensis* Tanaka) by vitrification. Plant Cell Rep 9:30–33
141. Nishizawa S, Sakai A, Amano Y et al (1993) Cryopreservation of asparagus (*Asparagus officinalis* L.) embryogenic suspension cells and subsequent plant regeneration by vitrification. Plant Sci 88:67–73
142. Sakai A, Matsumoto T, Hirai D et al (2000) Newly developed encapsulation-dehydration protocol for plant cryopreservation. CryoLetters 21:53–62

143. Niino T, Tanaka D, Tantely RR et al (2007) Cryopreservation of basal stem buds of in vitro growth mat rush (*Juncus* spp.) by vitrification. CryoLetters 28:197–206
144. Chen Y, Wang JH (2002) Cryopreservation of carrot (*Daucus carota* L.) cell suspensions and protoplasts by vitrification. CryoLetters 24:57–64
145. Thammasiri K (2000) Cryopreservation of seeds of Thai orchid (*Doritis pulcherrima* Lindl.) by vitrification. CryoLetters 21:237–244
146. Matsumoto T, Sakai A (1995) An approach to enhance dehydration tolerance of alginate-coated dried meristems cooled to −196 °C. CryoLetters 16:299–306
147. Hirai D, Shirai K, Shirai S et al (1998) Cryopreservation of in-vitro grown meristems of strawberry (*Fragaria* X ananassa Duch.) by encapsulation- vitrification. Euphytica 101: 109–115
148. Niino T, Sakai A (1992) Cryopreservation of alginate – coated in-vitro grown shoot tips of apple, pear and mulberry. Plant Sci 87:199–206
149. Fabre J, Dereuddre J (1990) Encapsulation-dehydration: a new approach to cryopreservation of *Solanum* shoot-tips. CryoLetters 11:413–426
150. Jitsopakul N, Thammasiri K, Ishikawa K (2008) Cryopreservation of *Vanda coerulea* protocorms by encapsulation-dehydration. CryoLetters 29:253–260
151. Hirai D, Sakai A (1999) Cryopreservation of in vitro-grown axillary shoot-tip meristems of mint (*Mentha spicata* L.) by encapsulation vitrification. Plant Cell Rep 19(2):150–155
152. Sakai A, Hirai D, Niino T (2008) Development of PVS-based vitrification and encapsulation-vitrification protocols. In: Reed BM (ed) Plant cryopreservation: a practical guide. Springer, New York, pp 33–58
153. Matsumoto T (2017) Cryopreservation of plant genetic resources: conventional and new methods. Rev Agric Sci 5:13–20
154. Thammasiri K, Prasongsom S, Kongsawadworakul P et al (2019) Cryopreservation of *Arundina graminifolia* (D. Don) hochr. seeds using D cryo-plate method. Acta Hortic 1234:301–308
155. Sakai A, Englemann F (2007) Vitrification, encapsulation-vitrification and droplet-vitrification: a review. CryoLetters 28:151–172
156. Kim HH, Yoon JW, Park YE et al (2006) Cryopreservation of potato cultivated varieties and wild species: critical factors in droplet vitrification. CryoLetters 27(4):223–234
157. Agrawal A, Swennen R, Panis B (2004) A comparision of four methods for cryopreservation of meristems in banana (*Musa* spp.). CryoLetters 25:101–110
158. Yoon JW, Kim HH, Ko HC et al (2006) Cryopreservation of cultivated and wild potato varieties by droplet- vitrification: effect of subculture of mother-plants and of preculture of shoot tips. CryoLetters 27:211–222
159. Halmagyi A, Deliu C, Coste A (2005) Plant regrowth from potato shoots tips cryopreserved by a combinated vitrification-droplet method. CryoLetters 26:313–322
160. Pennycooke JC, Towill LE (2000) Cryopreservation of shoot tips in vitro plants of sweet potato [*Ipomoea batatas* (L.) Lam.] by vitrification. Plant Cell Rep 19:733–737
161. Ashmore S, Azimi M, Drew RA (2001) Cryopreservation trials in *Carica* papaya. Acta Hortic (ISHS) 560:117–120
162. De Boucaud MT, Brison M, Helliot B et al (2002) Cryopreservation of *Prunus*. In: Towill LE, YPS B (eds) Biotechnology in agriculture and forestry 50. Cryo-preservation of plant germplasm II. Springer, Berlin/Heidelberg, pp 287–311
163. Leunufna S, Keller ERJ (2003) Investigating a new cryo-preservation protocol for yams (*Dioscorea* spp.). Plant Cell Rep 21:1159–1166
164. Halmagyi A, Fischer-Kluver G, Mix-Wagner G et al (2004) Cryopreservation of *Chrysanthemum morifolium* (*Dendranthema grandiflora* Ramat.) using different approaches. Plant Cell Rep 22:371–375
165. Panis B, Piette B, Swennen R (2005) Droplet vitrification of apical meristem: a cryopreservation protocol applicable to all Musaceae. Plant Sci 168:45–55

166. Halmagyi A, Pinker I (2006) Plant regeneration from Rosa shoot tips cryopreserved by a combined droplet vitrification method. Plant Cell Tissue Org Cult 84:145–153
167. Jitsopakul N, Thammasiri K, Ishikawa K (2008) Cryopreservation of *Bletilla striata* mature seeds, 3-day germination seeds and protocorms by droplet-vitrification. CryoLetters 29:517–526
168. Sopalun K, Thammasiri K, Ishikawa K (2010) Vitrifcation-based cryopreservation of *Grammatophyllum speciosum* protocorms. CryoLetters 31:347–357
169. Barraco G, Sylvestre I, Engelmann F (2011) Comparing encapsulation-dehydration and droplet-vitrification for cryopreservation of sugarcane (*Saccharum* spp.) shoot tips. Sci Hortic 130:320–324
170. Gonzalez-Arnao MT, Lazaro-Vallejo CE, Engelmann F et al (2009) Multiplication and cryopreservation of vanilla (*Vanilla planifolia* 'Andrews'). In Vitro Cell Dev Biol Plant 45:574–582
171. Hernandez-Ramírez F, Gonzalez-Arnao MT, Cruz-Cruz CA et al (2014) Comparison of different preconditioning and loading treatments with vanilla (*Vanilla planifolia* Jack.) apices cryopreserved using the droplet-vitrification procedure. Acta Hortic 1039:173–180
172. Gantait S, Sinniah UR, Suranthran P et al (2015) Improved cryopreservation of oil palm (*Elaeis guineensis* Jacq.) polyembryoids using droplet-vitrification approach and assessment of genetic fidelity. Protoplasma 252:89–101
173. Pathirana R, McLachlan A, Hedderley D et al (2015) Removal of leafroll viruses from infected grapevine plants by droplet vitrification. Acta Hortic 1083:491–498
174. Volk G, Bonnart R, Krueger R et al (2012) Cryopreservation of citrus shoot tips using micrografting for recovery. CryoLetters 33:418–426
175. Yamamoto S, Rafique T, Priyantha WS et al (2011) Development of a cryopreservation procedure using aluminum cryo-plates. CryoLetters 32:256–265
176. Jitsopakul N, Sangyojarn P, Homchan P et al (2019) Efficiency of aluminum cryo-plates for cryopreservation of *Dendrobium signatum* Rchb. F. pollinia. Acta Hortic 1234:279–286
177. Imsomboon T, Thammasiri K, Kosiyachinda P et al (2020) Cryopreservation of protocorm-like bodies of *Vanda lilacina* Teijsm. & Binn., a Thai Orchid species, by V-cryo-plate and D-cryo-plate methods. Walailak J Sci Technol 17(4):369–379
178. Niino T, Yamamoto S, Fukui K et al (2013) Dehydration improves cryopreservation of mat rush (*Juncus decipiens* Nakai) basal stem buds on cryo-plates. CryoLetters 34:549–560
179. Niino T, Watanabe K, Nohara N et al (2014) Cryopreservation of mat rush lateral buds by air dehydration using aluminum cryo-plate. Plant Biotechnol 31(3):281–287
180. Cordova LB II, Thammasiri K (2016) Cryopreservation on a cryo-plate of *Arundina graminifolia* protocorms, dehydrated with silica gel and drying beads. CryoLetters 37:68–76
181. Ishikawa K, Harata K, Mii M et al (1997) Cryopreservation of zygotic embryos of a Japanese terrestrial orchid (*Bletilla striata*) by vitrification. Plant Cell Rep 16:754–757
182. Wang JH, Ge JG, Liu F et al (1998) Cryopreservation of seeds and protocorms of *Dendrobium candidum*. CryoLetters 19:123–128
183. Hirano T, Godo T, Mii M et al (2005) Cryopreservation of immature seeds of *Bletilla striata* by vitrification. Plant Cell Rep 23:534–539
184. Hirano T, Ishikawa K, Mii M (2005) Cryopreservation of immature seeds *Ponerochis graminifolia* var. *suzukiana* by vitrification. CryoLetters 26:139–146
185. Thammasiri K, Soamkul L (2007) Cryopreservation of *Vanda coerulea* Griff. ex Lindl. seeds by vitrification. ScienceAsia 33:223–227
186. Jitsopakul N, Thammasiri K, Yukawa T et al (2012) Effect of cryopreservation on seed germination and protocorm development of *Vanda tricolor*. ScienceAsia 38:244–249
187. Flachsland E, Terada G, Scocchi A et al (2006) Cryopreservation of seeds and in vitro-cultured protocorms of *Oncidium bifolium* Sims. (RCHIDACEAE) by encapsulation-dehydration. CryoLetters 27:235–242
188. Tsukazaki H, Mii M, Tokuhara K et al (2000) Cryopreservation of *Doritaenopsis* suspension culture by vitrification. Plant Cell Rep 19:1160–1164

189. Nikishina TV, Popov AS, Vakhrameeva MG et al (2007) Cryopreservation of seed and protocorms of rare temperate orchids. Russ J Plant Physiol 54:121–127
190. Pornchuti W, Thammasiri K (2008) Cryopreservation of protocorms of *Dendrobium cariniferum* Rchb. f. Acta Hortic 788:63–68
191. Yin M, Hong S (2009) Cryopreservation of *Dendrobium candidum* Wall. ex Lindl. protocorm like-bodies by encapsulation-vitrification. Plant Cell Tissue Org Cult 98(2):179–185
192. Na HY, Kando K (1996) Cryopreservation of tissue-cultured shoot primordial from shoot apices of cultured protocorms in *Vanda pumila* following ABA preculture and desiccation. Plant Sci 118:195–201
193. Lurswijidjarus W, Thammasiri K (2004) Cryopreservation of shoot tips of *Dendrobium* Walter Oumae by encapsulation/dehydration. ScienceAsia 30:293–299
194. Vendrame WA, Carvalho V, Dias MM et al (2008) Pollination of *Dendrobium* hybrids using cryopreserved pollen. HortScience 43(1):264–267
195. Ajeeshkumar S, Decruse SW (2013) Fertilizing ability of cryopreserved pollinia of *Luisia macrantha*, an endemic orchid of Western Ghats. CryoLetters 34:20–29

Cymbidium: Botany, Production, and Uses 10

Ram Pal, N. K. Meena, R. P. Pant, and M. Dayamma

Contents

R. Pal (✉) · M. Dayamma
ICAR-National Research Centre for Orchids, Darjeeling Campus, Darjeeling, West Bengal, India
e-mail: rampal_nrco@yahoo.com

N. K. Meena
ICAR-National Research Center on Seed and Spices, Tabiji, Ajmer, India

R. P. Pant
ICAR- Indian Agricultural Research Institute, Pusa Campus, New Delhi, India

J.-M. Mérillon, H. Kodja (eds.), *Orchids Phytochemistry, Biology and Horticulture*, Reference Series in Phytochemistry, https://doi.org/10.1007/978-3-030-38392-3_8

Abstract

Cymbidiums enjoy a prime position in the floriculture industry for heavy substance of the flowers and high flower longevity. They are cultivated for cut flowers as well as pot plants for house decorations. Besides, they are used in medicine, cosmetics, and food. The successful cymbidium cultivation requires effective management of several factors controlling production and productivity. Various aspects of cymbidium orchid cultivation including genetics and breeding and plant health management have been discussed in the chapter.

Keywords

Genetic diversity · Breeding · Cultivation · Disease and pest management · Propagation · Agroclimatic requirements

1 Introduction

Cymbidiums are in cultivation since the last several centuries in China. The early Chinese scholars, Wang Kuei Hseh in 1247, Zhang Shi Geng in 1233, Lu Ting Weng in 1368, and others, wrote about orchids and their cultivation practices [1]. This indicates the popularity of cymbidium orchids during the period. Orchids in China were not hybridized but collected from the forests, alba forms, inter- and intraspecific variations, and plants with variegated leaves and fragrance which had the highest regard. The first artificial cymbidium hybrid, *Cymbidium eburnolowianum*, was evolved by crossing *Cymbidium eburneum* with *Cymbidium lowianum* in 1889 in England [2]. Over the past decades, many efforts have been made to improve its flower size, color, shape, length and nature of inflorescence, and flower longevity and to extend the period of cultivation, free-flowering habit, etc. by transferring genes from different species as well as genera. Nowadays, cymbidium cultivars are available in a range of colors except blue and black for cut flowers as well as a house plant. The technological advancement made in the field of biotechnology, production technologies, disease and pest management, and postharvest management has greatly benefitted orchids, including cymbidiums.

Orchid growers in the Netherlands, Australia, New Zealand, and America cultivate cymbidium varieties that are cool growing and bear large attractive flowers. However, growers in China, Japan, Taiwan, and Korea cultivate cymbidiums only for pot plants and appreciate it for flower, foliage, and fragrance. Cymbidiums for flowers are preferred for clear colors, suberect or arching foliage, and round petals with no markings or suffusion. The plants for foliage with various types of variegation are also preferred. The combined character, variegation and flower, have an added advantage. The two species of this region, *C. floribundum* and *C. ensifolium*, are very useful in the breeding of cymbidiums. *C. floribundum* transfers its trait for dwarfing, free-flowering, clear coloration, and earliness and C. *ensifolium* for compact growing, fragrance, and warm tolerance to their progenies [3]. The genus, *Cymbidium*, has vast potential in the field of horticulture and medicine. However, the natural population of some species like *C. whiteae* in India and *C. tortisepalum* in China is dwindling very fast. The species need to be protected and conserved using modern conservation techniques under in situ and ex situ.

2 Genetic Diversity

The genus *Cymbidium* shows a large variation in habitat ecology, plant architecture, and floral morphology. The species of this genus are widely distributed from Sri Lanka, Nepal, and India to China and Japan south through Malay Archipelago to the Philippines, New Guinea, and north and east Australia [2]. The number of flower in this genus has been reported to vary from 1–3 in *C. eburneum* to 20–60 in *C. madidum* or 22–60 in *C. canaliculatum* and flower size from 2.5 cm in *C. suave* to 15 cm in *C. tracyanum*. Likewise, the leaf length ranges from 22 cm in *C. tigrinum* to 125 cm in *C. atropurpureum* and pseudobulb length from 50 cm in *C. suave* to barely 1 cm in *C. tortisepalum*. These few morphological characters indicate that a significant variation exists within the genus *Cymbidium*. Traditionally, the diversity in *Cymbidium* is evaluated for morphological and physiological characters. The members of the genus were grouped into 3 subgenera, 15 sections, and 53 species [4]. The classification of species is based on morphological observations primarily floral morphology that cause taxonomic confusion. These are further supplemented by anatomical, micromorphological, biochemical, and molecular markers. Obara-Okeyo et al. [5] demonstrated that isozyme markers are helpful in resolving the ambiguities arising through the morphological characterization of *Cymbidium* species. In their study, they found the placement of different species under different sections agreed with the morphological and karyological classification of species proposed by [3] except *C. floribundum* which is currently under section *Cymbidium*. Choi et al. [6] differentiated subtropical species from temperate species, single terminal flower bud species with many-flowered species, broad-leaved species with narrow-leaved species, and cultivars from 21 species and cultivars of *Cymbidium* native to Korea, China, Taiwan, and Japan. The classification using RAPD supported ecological-, physiological-, and morphological-based classifications. Wang et al. [7] reported that ISSR markers provided valuable data on the origin and biology of *C. ensifolium* cultivars. The markers successfully identified the cultivars and helped in explaining their origin. It was also revealed that about 69% of total genetic variation in this species is due to genetic divergence inside geographical groups suggesting both germplasm collection and in situ conservation are essential for conserving the species. Sharma et al. [8] reported low intraspecific variation in *C. aloifolium* and *C. tigrinum* compared to *C. mastersii*, *C. elegans*, and *C. eburneum*, and these were closely related. Working on *C. goeringii* population structure, [9] recorded genetic similarity among the individuals within population up to 14 m distance is partly due to limited pollen and seed dispersal in *C. goeringii*. He suggested sampling should be conducted at 14–16-m interval to optimize the genetic diversity.

3 Cytology

3.1 Chromosome Number

Majority of species under genus *Cymbidium* have 40 diploid chromosomes except for *Cymbidium macrorhizon* Lindl., *C. lancifolium* Hook., and *C. javanicum* Blume. These species have 38 diploid chromosomes. Thus, there are two groups in the genus *Cymbidium* [10]. The occurrence of aneuploid has also been noticed in different *Cymbidium* species like *C. kanran* Makino (2n = 40 and 41), *C. faberi* Rolfe (2n = 40, 42, 43, and 44), *C. javanicum* Blume (2n = 38, 43), *C. lancifolium* Hook. (2n = 38 and 39), and *C. cyperifolium* Wall. ex. Lindl. (2n = 36, 40). Leoonhardt [11] reported chromosomal counts of 119 clones of 88 species and hybrids of *Cymbidium* in which 89 were diploid, 15 triploid, 12 tetraploid, 1 hexaploid, and 2 aneuploid. *C. insigne* "Bierii" and *C. pumilum* "Yashima" were found to be triploid with 60 chromosomes each. The other famous triploid grex were *Cymbidium* Alexette, *Cymbidium* Alexfrida, *Cymbidium* Arabian Nights, *Cymbidium* Bengal Bay, and *Cymbidium* Cassandra. The triploid arose from the breeding of tetraploids with diploid. The tetraploid clones produce large flower and have heavier flower substance bred out of crossing tetraploid with a tetraploid. Some of important tetraploid grexes or clones are *Cymbidium* Early Bird "Pacific," *Cymbidium* Alexanderi "Westonbirt," *Cymbidium* Babylon "Carpentier," *Cymbidium* Babylon "Castle Hill," *Cymbidium* Balkis, *Cymbidium* Cleo's Sherman, and *Cymbidium* Dorchester "Alpha." Nowadays, *Cymbidium* breeders use tetraploid lines in their breeding program.

3.2 Pre- and Postzygotic Barriers

In addition to the delivery of the male nucleus to the ovule for the formation of zygote, the pollen tube in orchids also serves as a stimulus for megasporogenesis. Megasporogenesis occurs when the pollen tube reaches the ovarian cavity, after assurance of fertilization. After delivery of pollens on the stigma, if it fails to germinate or the pollen tube fails to reach the ovarian cavity, the flowers fall off within a week [12]. The pre- and postzygotic barriers in flowering plant affect successful seed formation. The prezygotic barriers include asynchronous flowering, floral rewards for pollinators, and nonadherence of pollen to the stigmatic surface and abnormal pollen tube growth. The postzygotic barriers are effective after successful fertilization, and these include seed abortion and debility or sterility of F1 hybrids and future generations. The prezygotic barriers like asynchronous flowering could be overcome by growing the crops in a controlled environment or storing the pollens for future use. Stort [13] studied microsporogenesis, pollen germination, and fertility of male gametes in 24 artificial intergeneric and interspecific F1 hybrids of orchids and found that irregular microsporogenesis in parental species has similar chromosomes due to lack of homology of chromosomes in parental species. The lack of homology in chromosome resulted in the formation of tetrads without micronuclei, triads with 1–8 micronuclei, triads, dyads with or without micronuclei, and monads. He encountered three situations: first pollens do not germinate on the stigma, second the pollen tube grows down to the style yet do not penetrate the ovary, and third the pollen tube grows normally though the style and reaching the ovule. No fruit set occurred when the pollen tube failed to penetrate the ovary. Thus, in vitro pollen germination and pollen tube growth test may be insufficient to predict the fertilization. Wimber [14] reported normal cycle of meiotic and mitotic division and pollen formation in the clones of *Cymbidium schroederi*, *C. erythraeum*, *C. grandiflorum*, *C. tracyanum*, and *C. lowianum* except the clones of *C. insigne* and *C. lowianum* var. *concolor* that showed varying levels of abnormalities during reduction division. These included formation of univalents and fragments leading to formation of polyads rather than tetrads. Du Puy [15] observed occasional disruption in the meiotic cycle of *C. lowianum* and *C. tracyanum* by forming univalents or fragments during telophase I of the reduction division. Stort and Morales [12] reported break down of megasporogenesis at various stages of megasporogenesis. He observed degeneration of megaspore mother cell at the beginning of the first meiotic division and attributed it to lack of chromosome homology that prevented full cell division and, consequently, degeneration of megasporocyte during megasporogenesis. However, integument developed normally and ovule without embryo sack formed. Leoonhardt [11] reported production of apparently viable yet not germinating seeds from 11 intergeneric crosses of cymbidium. He suggested that this could be due to accumulation of inhibitors; however, an attempt to leach out the inhibitors in liquid germination media did not encourage germination. Another kind of endogenous incompatibility was suggested by [11] where the progenies of complex hybrid grow normally but are incapable of further hybridization due to chromosomal nonhomology during meiosis. The fertility of such clones can be restored by doubling of chromosomes. The diploid *Cym* Peter Pan (*C. ensifolium* × *Cym* Miretta) is a good example of such incompatibility. For restoring the fertility, *Cym* Peter Pan was converted to tetraploid using colchicines. The *Cym* Peter Pan has now produced 914 descendants.

4 Natural Hybridization

Plants maintain reproductive isolation through pre- and postzygotic barriers. The prezygotic barriers include asynchronous flowering, rewards for different pollinators, the inability of pollen to adhere on the surface of stigma, and failure of the pollen tube to fertilize the ovules. The postzygotic barriers cause seed abortion or sterile seed production. Despite the prevalence of these barriers, several natural hybrids are known to occur in the Orchidaceae family. Sympatric species in the family of Orchidaceae maintain reproductive isolation through pollinator specificity.

Table. 1 List of natural hybrids of *Cymbidium*

Sl no.	Natural hybrid	Female parent	Male parent	Area of occurrence	Year of report
1.	*C.* × *ballianum* Rolfe	*C. eburneum*	*C. mastersii*	Myanmar	1904
2.	*C.* × *baoshanense* F.Y. Liu & Perner	*C. lowianum*	*C. tigrinum*	China to N. Thailand	2001
3.	*C.* × *florinda* Rolfe	*C. erythrostylum*	*C. iridioides*	Vietnam	1913
4.	*C.* × *gammieanum* King & Pantl.	*C. elegans*	*C. erythraeum*	Northeast Himalayas and Nepal	1895
5.	*C.* × *glebelandense* Rolfe	*C. insigne*	*C. schroederi*	Vietnam	1911
6.	*C.* × *hillii* F.Muell. ex Regel	*C. canaliculatum*	*C. suave*	N. Queensland	1879
7.	*C.* × *monanthum* J.M. H.Shaw	*C. qiubeiense*	*C. serratum*	China	2012
8.	*C.* × *nishiuchianum* Makino ex J.M.H. Shaw	*C. goeringii*	*C. kanran*	S. Japan	2002
9.	*C.* × *nomachianum* T. Yukawa & Nob.Tanaka	*C. kanran*	*C. lancifolium*	Japan (Shikoku)	2008
10.	*C.* × *nujiangense* X.P. Zhou, S.P.Lei & Z.J. Liu	*C. cyperifolium* var. szechuanicum	*C. tortisepalum*	China	2007
11.	*C.* × *oblancifolium* Z.J. Liu & S.C.Chen	*C. ensifolium*	*C. lancifolium*	China	2000
12.	*C.* × *purpuratum* L.J. Chen, Li.Q.Li & Z.J. Liu	*C. kanran*	*C. qiubeiense*	China	2007
13.	*C.* × *rosefieldense* Rolfe	*C. insigne*	*C. tracyanum*	Vietnam	1912
14.	*C.* × *woodlandense* Rolfe	*C. mastersii*	*C. tracyanum*	Myanmar	1913

According to [16], there are 14 natural hybrids of *Cymbidium* all over the world. The first natural hybrid in the genus *Cymbidium*, *C.* × *gammieanum*, was described by [17]. It is a cross of *C. erythraeum* and *C. elegans* flowering at the same time in Himalayan region. The natural hybrids reported in the genus *Cymbidium* are mentioned in Table 1.

5 Cymbidium Breeding

Artificial hybridization of cymbidium orchids began in the late nineteenth century in England. The first artificial *Cymbidium* hybrid, *C. eburno-lowianum* (*C. eburneum* × *C. lowianum*), flowered in Messrs James Veitch & Sons Nursery,

England, in 1889. In the early years, the gene pool consisted of a few species like *C. tracyanum*, *C. eburneum*, *C. lowianum C. hookerianum*, *C. mastersii*, and *C. giganteum* (=*C. iridioides*). These species limited the improvement over color and size, but the introduction of *C. insigne* brought a great change in the breeding of cymbidiums. This species passed on characters like long inflorescence well above the leaves, floriferousness, and large and long-lasting flowers to its progenies [2]. C. Alexanderi (*C. insigne* × *C. eburno-lowianum*) is one of the most significant contributions of this species. Further, various color forms of *C. insigne* and *C. erythrostylum* contributed significantly in evolving new hybrids of cymbidium. Occurrence of natural tetraploids, C. Alexanderi "Westonbirt" and C. Rosanna "Pinkie," in two diploid grexes C. Alexanderi and C. Rosanna, respectively, made the beginning of polyploidy breeding in cymbidiums. Soon after, the importance of polyploidy in the breeding of cymbidium was realized, and a number of superior hybrids were converted into tetraploids for using in the breeding program. The gene pool was expanded further by introducing miniature species, viz., *C. pumilum* (=*C. floribundum*), *C. devonianum*, *C. madidum*, and *C. ensifolium*, in the breeding program, and many desirable traits such as perfume, cascading inflorescence, color, and size were improved. The breeders are now utilizing less known species and related genera for introducing variability in the cultivated hybrids. Over the years, about 16,000 hybrids of cymbidium have been registered with International Orchid Registration Authority, Royal Horticultural Society, London.

The important traits for cymbidium breeding include compact in growth habit, free flowering, fast growing, multiple flower spikes, upright and self-supporting flower spikes, round flower shape, orientation of flowers on the spike, flower longevity, and attractive colors including pure color without sustaining. The flowers should have attractive color, and the traits like fragrance, off-season flowering, extension of the flowering season, and earliness should also be taken into consideration.

5.1 Breeding Strategies in Cymbidium

5.1.1 Hybridization and Selection

The cultivars of cymbidium are highly heterozygous, and crossing between them usually leads to the production of fertile seeds that provides a range of genotypes. The segregating F1 population in these crosses gives a source of useful gene recombinants for selection and cloning of new, desired individuals. In hybridization of cymbidiums, selection of parents, direction of cross, time of crossing, and F1 population size are the points taken into consideration. Seed germination and raising of seedlings in tissue culture, acclimation of seedling and growing them to flowering size in greenhouses, evaluation of seedlings for desired characters, selection, multiplication of selected clones, testing the elite clones in genotype x environment under trials for testing their stability, and uniformity are the major steps required for the development of the superior variety through hybridization and selection. It takes about 12–14 years from crossing to release a commercial variety.

The primary and secondary crosses are developed through crossing between the species and cultivars or variety. For large for flowers, *C. eburneum*,

C. lowianum, *C. mastersii*, *C. tracyanum*, *C. iridioides*, *C. hookerianum*, *C. insigne*, and *C. erythrostylum* and, for breeding of miniatures, *C. pumilum* (= *C. floribundum*), *C. ensifolium*, *C. devonianum*, *C. madidum*, *C. tigrinum*, and *C. canaliculatum* were used for developing primary and secondary hybrids. Since the last few years, Australian species *C. canaliculatum*, *C. suave*, and *C. madidum*, and Oriental cymbidium species *C. faberi* and *C. goeringii* have also been included in breeding program. The superior diploid and tetraploid clones are used further to improve desirable traits of the cultivar. Intergeneric hybrids involving two, three, and four genera in crossing with cymbidium are mentioned below.

Bigeneric
Ansellia × *Cymbidium* = Ansidium (Asdm.)
Bifrenaria × *Cymbidium* = Bifrenidium (Bifdm.)
Bolbidium × *Cymbidium* = Bolbicymbidium (Bby.)
Catasetum × *Cymbidium* = Cymasetum (Cymst.)
Eulophia × *Cymbidiella* = Cymbidilophia (Cdl.)
Cymbidiella × *Grammangis* = Cymbidimangis (Cdn.)
Cymbidium × *Promenaea* = Cymbidinaea (Cbn.)
Cymbidium x *Zygopetalum* = Cymbipetalum (Cbp.)
Ansellia × *Cymbidiella*= Cymbisellia (Cml.)
Cymbidium × *Eulophiella* = Cymphiella (Cymph.)
Cymbidium × *Maxillaria* = Maxidium (Mxd.)
Cymbidium × *Phaius* = Phaiocymbidium (Phcym.)
Trigeneric
Catasetum × *Clowesia* x *Cymbidium* = Cymaclosetum (Cma.)
Cyrtopodium × *Cymbidium* x *Grammatophyllum* = Cyrtogramcymbidium (Cgc.)
Cymbidium × *Grammangis* x *Grammatophyllum* = Gramcymbimangis (Gbi.)
Cymbidium × *Eulophia* x *Grammatophyllum* = Gramcymbiphia (Gyp.)
Cymbidium × *Grammatophyllum* = Grammatocymbidium (Grcym.)
Catasetum × *Cymbidium* x *Grammatophyllum* = Thompsonara (Thmpa.)
Quadrageneric
Catasetum × *Clowesia* x *Cymbidium* x *Grammatophyllum* = Kalakauara (Kal.)

5.1.2 Polyploidy Breeding

The importance of polyploidy in cymbidium orchids was recognized soon after the flowering of *Cymbidium* Alexanderi "Westonbirt," a clone of *Cymbidium insigne* × *Cymbidium eburno-lowianum* registered in 1911. It was a chance tetraploid with a large flower, heavy substance, superior color, and full shape. This clone was repeatedly used to develop superior varieties of cymbidium. *Cymbidium* Rosanna, a cross of *Cymbidium* Alexander × *Cymbidium* Kittiwake, was registered in 1927. *Cymbidium* Alexander was a tetraploid, while *Cymbidium* Kittiwake was a diploid, but a clone "Pinkie" was a tetraploid. It is likely that *Cymbidium* Alexanderi "Westonbirt" and *Cymbidium* Rosanna "Pinkie" have had developed through the fertilization of unreduced gametes of their parents. Menninger [18] produced tetraploid cymbidiums by applying colchicine on shoots produced through the pseudobulbs. Wimber and Cott [19] cultured cymbidium protocorms in liquid nutrient culture added with colchicine and converted in tetraploid seedlings (40%). Kim et al. [20] produced triploid and tetraploid of *Cymbidium* Silky by treating PLBs with 0.05% concentration colchicines for a week. The popular cultivars are often tetraploid, while the wild species are diploid. Therefore, it is challenging to transfer favorable genes from wild species to advanced varieties or hybrids. This problem can be solved through the conversion of diploid species or primary hybrids to tetraploid for using with superior cultivars in cymbidium breeding program. Conversion of diploid to different ploidy levels in cymbidiums is generally realized through using colchicine. The fertility in the *Cymbidium* Peter Pan "Greensleeves" (*Cymbidium ensifolium* × *Cymbidium* Miretta) was restored by doubling the chromosome number, and this clone produced many beautiful hybrids [2]. The change in ploidy level is also associated with change in morphological and physiological characteristics of plant like increase in number and size of stomata, increase in cell size and flower size, etc. According to [21], tetraploid flowers have a stronger scent than diploid flowers in *Cymbidium* Golden Elf "Sundust." The micropropagation characteristics such as rate of proliferation, shoot bud and root differentiation changes after chromosome conversion. The diploid has a higher rate of proliferation and shoot and root differentiation than tetraploids.

6 Propagation

6.1 Conventional Propagation

Division of mother plant and sprouting of back bulbs are the two conventional methods practiced for the propagation of cymbidiums. The conventional propagation methods are slow and take a reasonably long time to produce planting material in a sufficient quantity. Viral diseases in cymbidium can be transmitted through the use of the unsterilized cutting tool for division of plant and removal of pseudobulb hence, sterilise the cutting tools before the division of a cymbidium plant.

6.1.1 Division of Plant

The division of the mother plant is the most familiar method for propagation of sympodial orchids like *Cattleya*, *Dendrobium*, *Cymbidium*, and *Epidendrums*. It involves the division of large clumps into smaller segments. The divided segment should retain at least 1 back bulb, 2–3 leafy pseudobulbs, and 2–4 vegetative shoots. The plants should be divided when new leads sprout and new roots have begun to emerge from the base of the lead. The best time to divide orchids is from spring into early fall because during this time, the plants are in active growth and are likely to establish quickly with a little shock [22]. For dividing the plant, observe for natural divisions of the plants and divide from the natural division by pulling them apart; remove leafless pseudobulbs, dead roots, and leaf from the plant; and repot the division in growing media. The plant establishes when the new roots anchor the growing medium. Always use sterilized cutting tools as CymMV and ORSV spread through mechanical means [23].

6.1.2 Propagation Through Back Bulbs

Pseudobulbs are the enlarged and swollen part of the stem from which all leaves and inflorescences arise. After flowering, leaves on the pseudobulb gradually dry up and fall from the pseudobulbs. These leafless pseudobulbs are back bulbs and employed for the regeneration of new plants. Soon [24] described a method for producing plantlets from the back bulbs of cymbidium where back bulbs hung under the shade and watered periodically until new shoots sprout and develop into plantlet. If the bottommost bud is healthy, it will develop into new plantlet; otherwise, any bud along the length of pseudobulb may develop into a new plantlet. A slightly different method was proposed by [25]. In this method, washed back bulbs were placed in plastic bags, and then the bags were hooked under the growing benches in a cool and shady place. The bags were checked periodically for the emergence of shoots, upon the emergence of shoot bud planted in small pots. All the above-described methods do not reuse the pseudobulb in the process of propagation. Pal and Medhi [26] suggested that back bulbs should be cleaned, washed, and soaked in fungicide solution (Bavistin 2 g/l) for 30 min. Thereafter, back bulbs are immersed in a solution containing 100 mg BAP (6-benzylamino purine) for 12 h, removed, air-dried, and placed layer by layer in polyethylene bag filling empty spaces with sawdust of *Cryptomeria japonica* (Dhuppi) and stored for 45 days in a cool and dark place and monitored periodically for sprouting. After sprouting of pseudobulbs, they are planted in the sawdust covering 3/4 the portion of the pseudobulb in beds or planting trays. The sprouted buds develop shoots within 2–3 months. When the plantlets attain 2–3 leaves and 2–3 roots, these are severed from the back bulb using a sterilized knife and planted separately. The back bulbs with no sprouts are again soaked in BAP solution and stored in a cool and shady place as described above. In this process, back bulbs can be used 4–6 times to produce 6–8 plantlets in their life span.

6.2 Nonconventional Propagation

Clonal multiplication through traditional methods of propagation is not rapid enough to meet the demand for disease-free quality planting materials. By micropropagation, it has become possible to propagate quickly the cultivars that are slow to increase through conventional means. Because of its importance, micropropagation of cymbidiums has been widely studied [27–31]. The technique is based on in vitro culture of explants capable of plant regeneration through culturing of various organs and tissues on the media which are suitable for plant regeneration.

6.2.1 Seeds

Orchid seeds are very tiny and produced in thousands per capsule. The seeds lack metabolic machinery and functional endosperm, with the result of only 0.2–0.3% of the seeds germinate in nature [32]. Bernard [33] discovered that in nature, orchid seeds would germinate if infected by a fungus. He described the role of mycorrhiza in the germination of orchid seeds [34]. Knudson [35], for the first time, demonstrated that seeds of orchid could germinate bypassing the requirement of fungus. He germinated orchid seeds on sugar-enriched mineral medium in aseptic conditions. Later, success in asymbiotic germination of orchid seeds and seedling formation on Knudson's medium was reported by [36–39]. Arditti et al. [40] opined that nutritional requirement, pH, source of nitrogen, and light requirement vary with genus, with species, and even with ecotypes. Besides, the age of the capsule also affects the seed germination in orchids [41].

A number of *Cymbidium* species often have been successfully propagated in vitro through the seeds. Oriental cymbidiums are difficult to propagate than epiphytic cymbidiums. The seeds of *C. sinense* (And.) Willd. germinate well on MS medium, and the one supplemented with 0 mg BA, 1.0 mg l^{-1} NAA, 8.0 g agar, and 30 g sucrose is best for shoot development. However, a half-strength medium added with 3.0 mg l^{-1} NAA is best for root development [42]. Cymbidium seeds take about 8–10 months to mature, and thus culturing at an early stage can save time. [43] cultured 50-day-old *C. faberi* Rolfe on hormone-free medium, and the rhizome developed within 4 months that can be multiplied about five times by subculturing on the same medium supplemented with 1.0 mg l^{-1} NAA at 40-day interval. Rhizomes induced shoot buds on medium containing 1.0 mg l^{-1} NAA + 2.0 or 5.0 mg l^{-1} BA and complete plantlet on medium added with 1.0 g l^{-1} activated charcoal within 50 days. Seeds of *C. giganteum* Wall ex. Lindl. germinate well (100%) on [44] medium (M) or Phytamax medium (PM) supplemented with 2.0 g l^{-1} peptone or 1.0 mg l^{-1} BAP. The highest protocorm multiplication occurs in liquid medium supplemented with 2.0 mg l^{-1} BAP and 1.0 mg l^{-1} NAA. BAP and NAA 2.0 mg l^{-1} each induces multiple shoots, and roots develop in half-strength PM or M medium [45]. The seeds obtained through the crossing of two cultivars or species are highly heterozygous and unsuitable for utilization in commercial production of cut flowers or house plants. The commercial orchid production industry requires identical and uniform plants for the production of cut flowers or house plant. The seed culture is only resorted to multiply a particular species or for raising the plants of crosses by the breeders.

6.2.2 Induction of Protocorm-Like Bodies (PLBs)

Arditti and Ernst [46] reported that like many other higher plants, regeneration of orchids by tissue culture method takes two main pathways, namely, (i) direct differentiation of protocorm-like bodies (PLBs) from cultured tissues and organs and their subsequent development into plants and (ii) indirect differentiation of PLBs from explant tissues through an intermediary callus phase.

Shoot Tip Culture

Morel [27] produced virus-free cymbidium by culturing shoot tips on modified Knudson C medium that gave rise to 1–2 mm size colorless PLBs within 2 months of culture that turned green and produced rhizoids and leaf primordium. These bodies were sectioned to produce more PLBs. Wimber [28] published a detailed protocol for in vitro production of cymbidium. He could produce the PLBs within a month on liquid Tsuchiya medium. Sagawa et al. [47] cultured the explants first in liquid-modified KC medium for 4 weeks and then transferred them on solid medium to induce PLBs from the explants. Gu et al. [29] produced PLBs from lateral buds on VW medium added with 1.0 mg BA l^{-1}. Likewise, [48] also induced PLBs from the shoot apices of *Cymbidium* cv. Horunoumi on MS medium supplemented with 0.1 mg l^{-1} BAP and 0.5 mg l^{-1} NAA. Subramaniam and Taha [49] found that VW medium and 5.0 mg l^{-1} NAA are the best for PLB induction from the shoot tips of *C. atropurpureum* (Lindl.) Rolfe.

Leaf Segment Culture

The foliar explants are readily available and do not require the sacrifice of the mother plant, and they are available throughout the year. Wimber [50] reported the production of PLBs from the leaf explants of cymbidium. It is the only report on the propagation of cymbidium using a leaf as explants. Probably, leaf explants do not respond well to callus or PLB induction.

Root Segment Culture

Beechy [51] suggested the possibility of using aerial roots for micropropagation of orchids. The root meristem consists of highly determinate cells which have limited morphogenic competence for bud formation [52]. Yasugi et al. [53] induced PLBs from basal root segments of in vitro-derived root segments of Kenny "Wine Color" on MS medium supplemented with NAA and BA at 0 or 1 mg l^{-1} under light conditions. Nayak et al. [54] induced multiple shoots from root segments of *C. aloifolium* (L.) Sw., on MS medium, supplemented with 2.2–4.5 μM thidiazuron. In all the published reports, the explant has been derived from the in vitro-grown plants, not from the ex vitro-grown plants.

6.3 Thin Section Culture

Thin cell layer (TCL) system refers to excision of small explants excised from plant organs like stem, leaves, floral inflorescence, or flower primordial either longitudinally (lTCL) or transversally (tTCL). The lTCL consists of only one type of tissue such as monolayer of epidermal cells, whereas tTCL comprises a small number of cells from different tissue types, epidermal, cortical, cambium, perivascular, and medullary tissue, and parenchyma cells [55]. Thin cell layer culture is highly efficient as compared to the standard tissue culture technique [56]. Recently, thin cross section of actively growing plant parts such as shoots, leaves, inflorescence stalk, and developing PLBs has effectively been used by some workers for plantlet generation in some orchids [57].

Begum et al. [30] reported that outer tissues (OT) from the PLBs of *Cymbidium* × Thanksgiving cv. Nativity when cultured on medium supplemented with 0.1 mg NAA and 0.5 mg l^{-1} BA showed highest PLB formation. Culturing

TCS of C. *aloifolium* PLBs on MS medium supplemented with ZR at 14.0 μM produced 28.2 PLBs per explant 80% shoot formation frequency [31]. Huan et al. [58] reported protocorm segments of *Cymbidium* Great Flower "Rainbow" when cultured on a medium containing 0.5 mg l^{-1} NAA and 0.1 mg l^{-1} TZD produced callus that converted to PLBs by culturing on hormone-free medium. Chang and Chang [59] induced calli in the sections of pseudobulb, rhizome, and the roots of seed-derived plantlets of *C. ensifolium* var. *misericors* on medium supplemented with 10 mg l^{-1} 2,4-D and 0.1 mg l^{-1} TDZ. Le et al. [60] also induced callus on VW medium containing 0.1 mg l^{-1} NAA and 0.01 mg l^{-1} TDZ within 1 month of culture that can proliferate at a 4-week interval and can convert into PLBs on culturing at hormone-free medium. The callus-derived PLBs form complete plant within 4 months.

6.4 Shoot and Root Development

Skoog and Miller [61] reported that a balance between cytokinin and auxin controls organogenesis. They found a relatively high auxin/cytokinin ratio induced root formation in tobacco callus, whereas a low ratio of the same hormone favored shoot production. Paek and Kozoi [62] reported that the growth pattern and regeneration ability of rhizomes derived from asymbiotic seeds or shoot tip culture varied according to media composition, kind and concentration of growth regulators, culture conditions, and species and varieties. The BA is the best for inducing shoot formation, switching rhizome in PLBs, and directly forming shoot from branched rhizomes.

The rhizome tips of *C. faberi* Rolfe. produced shoots in modified Kyoto solution incorporated with 0.1 mg l^{-1} BA; however, the higher concentration of BA (10.0 mg l^{-1}) prevented the shoot formation [63]. On the contrary, [64] reported that shoot formation in *C. kanran* occurred best when the basal medium (MS) was supplemented with 10 mg l^{-1} BA. Both the species might have varying requirements for cytokinins for shoot bud development. Kinetin concentration 10 mg l^{-1} in culture medium inhibited adventitious root formation in *C. goeringii* and dependent on the intensity of light [65]. Ogura and Okubo [66] achieved proper shoot development in rhizome apices of *C. ensifolium* (L.) Sw. by supplementing the MS medium with 1 or 10 mg l^{-1} NAA and by reducing ammonium and potassium nitrate to 25% and 50%, respectively, from their original medium. However, in *C. kanran*, shoot development occurred by adding 1.0 mg l^{-1} BA and 0.1 mg l^{-1} NAA to the basal medium. TDZ 0.1–1.0 mg l^{-1} in culture medium initiated shoot bud development in the rhizomes of *C. sinense* Willd. but retarded rhizome proliferation. Culturing regenerated shoot buds on basal medium with 0.5 mg l^{-1} NAA and 1 g l^{-1} activated charcoal developed complete plant within 4 months. Kusumoto [67] achieved reliable shoot bud formation and their consistent development to shoots in cymbidium protocorms with kinetin 0.1–1.0 mg l^{-1} and 0.01–0.1 mg l^{-1} NAA or with 0.1–1.0 mg l^{-1} GA3 and 0.01–0.1 mg l^{-1} NAA, but the combination of GA3 (1.0 mg l^{-1}) and NAA (0.1 mg l^{-1}) was most effective for root formation.

Compared to terrestrial cymbidiums, the epiphytic cymbidiums are easy to propagate. Siddique and Paswan [68] observed the best growth and differentiation in PLBs of *C. longifolium (=Cymbidium erythraeum)* on NC medium supplemented with 0.5 mg l^{-1} BA. Nayak et al. [69] reported the effect of growth regulators on proliferation and organogenesis of rhizomes derived from seeds of *C. aloifolium* Sw. They found that auxins increase the proliferation of rhizomes, but NAA 5 mg l^{-1} is most effective. The highest shoot formation frequency (91.5%) and maximum number shoot bud/rhizome (3.5) were obtained when the medium was supplemented with 1 mg l^{-1} BA.

Apart from growth regulators, altering the compositions of defined culture media or adding organic supplements is useful in inducing shoot buds from the PLBs. Kukulczanka and Jastrezebska [70] reported that addition of Heller's minor elements to Tsuchiya medium stimulates shoot and root formation in the PLBs. However, Mg at 250–400 mg l^{-1} retarded the protocorm differentiation but increased protocorm number. Paek and Chun [71] obtained shoot formation in cymbidium protocorms by supplementing KC medium with 4 mM NH_4 N or 8 mM Ca^{2+} or 1 mM Mg^{2+} and the root formation on medium supplemented with of 4 mM NH_4^+ or 4 mM Ca^{2+}. Shimasaki et al. [72] studied the effect of sucrose and trehalose on organogenesis of PLBs of *C. finlaysonianum* Wall. ex Lindl. and *C. kanran* and found that trehalose at 10 g l^{-1} enhanced shoot bud differentiation, while 5 g l^{-1} increased root formation in both the species of *Cymbidium*. Ueda and Torikata [73] studied the effects of growth substance on organogenesis of *C. insigne* Rolfe., *C. pumilum* Rolfe, and *C. goeringii* Rchb. f. and found that *C. insigne* and *C. pumilum* protocorms develop into plants in basal medium (KC with NC microelements) within 2 months of culture. The organic supplements NAA, bacto-tryptone, L-arginine, and L-aspartic acid promoted growth and development, whereas yeast extract inhibited the growth. In *C. goeringii* protocorm developed in rhizome rather than shoots. The basal medium supplemented with 0.1 mg l^{-1} NAA produced shoots, while yeast extract and 1 g l^{-1} and bacto-tryptone (1–5 g l^{-1}) promoted the growth of shoots remarkably.

6.5 Acclimatization and Hardening

The acclimatization of plantlets from in vitro to ex vitro environment is concerned with the transfer of plantlets in compots and then placing them in high humidity structures and high shade levels and then slowly reducing the humidity and increasing the intensity of light so that the plantlets could withstand the outside environment. Successful ex vitro acclimatization of micropropagated plants determines the quality of end products and its commercial introduction and the economic viability of enterprise [74]. When the shoots or plantlets are transplanted from culture to greenhouse, they may desiccate or wilt rapidly and can die as a result of the change in the environment unless substantial precautions are taken to accommodate them [75]. The in vitro environment is characterized by a saturating atmosphere with water vapor, with very low vapor pressure deficit [76, 77], low CO_2

concentration during photoperiod [78], high CO_2 concentration during dark period [79], and high ethylene concentration [80]. These characteristics except for light intensities are primarily due to the low number of air exchanges of the vessel and relatively small air volume of the vessel headspace. High relative humidity and low CO_2 concentration in the vessel limit the gas exchange [81]. Mani and Nagaraju [82] suggested that cocopeat alone or in combination with sand is suitable for hardening micropropagated plantlets of cymbidium.

7 Cultivating Cymbidiums

The commercial cultivation of cymbidium begins with the acquisition of mericlones or seedlings from the plant propagators. The mericlones are used for the production of cut flowers or pot plants as they are genetically identical and uniform in size. The seedling is the product of seeds obtained through artificial hybridization of two dissimilar plants for evolving a new variety. These plants are grown under optimum conditions for a certain period to attain maturity. Thus, plants defer sexual reproduction until they acquire the capacity to produce flower and seed. In cymbidium orchids, this period varies from 3 to 4 years from the deflasking and differs with species or variety. Miniature cymbidium cultivars require a shorter period than cultivars of standard cymbidiums.

7.1 Agroclimatic Requirements

7.1.1 Temperature

Based on climatic requirements, cymbidiums are divided into two groups: standard and miniature. The standard cymbidium cultivars have been evolved from the species occurring in high hills of Himalayas. They bear large flower and require a long cool period to induce flowering. The two views are often proposed for flower induction in large-flowered cymbidiums: (i) cooler nights in the Himalayan region (10–14 °C) during summer months and (ii) a temperature difference of 10–12 °C between day and night helps in induction of flowering in large-flowered cymbidiums [83]. Went [84] reported that no flowering occurred in *C. giganteum* (=*Cymbidium iridioides*) when grown for 1–3 months at 26/14 °C and 23/14 °C day/night temperatures, and little flowering was induced in plants held at 20/10 °C and 20/14 °C. The highest number of plants induced flowering when the plants were grown under 20/10 °C and 20/14 °C temperature regime. He concluded that both day and night temperatures have an effect on the induction of flowering in large size or cool growing cymbidium. Powell et al. [83] reported that *Cymbidium* cultivar Astronaut "Rahah" when exposed to day/night temperatures of 26/12 °C and 14-h photoperiod produced an average of 5.9 inflorescences per plant. The numbers of inflorescence declined to 0.8 and 1.7 at day/night temperature of 20/12 °C or 16/18 °C, respectively. The large-flowered cymbidiums grown today have many species in their ancestry and are likely to influence the flowering behavior of hybrid cultivars. It is likely that contribution of the species to the genome of the hybrid influences the temperature requirement of the hybrids.

The temperature influences flowering and quality of flowers in cymbidium. Li et al. [85] transferred the plants of *Cymbidium* Sazanami in a cold room (8–28 °C), warm room (15–30 °C), and hot room (20–30 °C) after flower bud initiation. The flowering in cymbidium was delayed by 1 month in a cold room and advanced by 1 month in the hot room as compared to the warm room. The best quality of flowers was in warm rooms. The high temperature (>25 °C) at an early stage of flower development causes bud drop in miniature and intermediate cymbidium cultivars. To get rid of this disorder, cymbidium growers in Japan shift their plants from original sites to high hills where the temperature varies between 20 °C and 30 °C [86]. Higher temperature (18 °C) during winters brings to the rapid growth of shoots in *Cymbidium* and results in early flowering (March–April), but plants suffer from bud blast due to higher summer temperature. Whereas the low temperature during winters (6 °C) suppresses the growth during winters that get recovered during spring, shoots mature in June and initiate flower buds in August [86].

The vegetative buds grow normally at 30/25 °C day/night temperature. During winters low night temperature (2 °C) encourages the production of leads, and higher light intensity helps in accumulation of assimilates. The higher temperature during winter reduces the production of leads [87]. According to [88], warm day/night temperature (30/25 and 25/20 °C) accelerates pseudobulb development and flower formation in *Cymbidium ensifolium* var. *misericors*. Plants with 1-year-old pseudobulb produced 2.3 and 1.6 inflorescences, respectively.

7.1.2 Light

In cymbidium orchids, both vegetative and reproductive stages are affected by temperature and light. Komori [87] reported that both light and temperature influence the lead production in *Cymbidium* cv. Lovely Angel "The Two Virgins," a cultivar with difficulty to induce leads. Growing the plants at 2 °C in without shade increased production of leads (1.7 per plant), but these were only 0.9 per plant under 50% shade. The plants cultivated at 15 °C without shade had more fresh weight of shoots and roots compared to plants cultivated under the 50% shade at the same temperature. Low light intensity in flowering years results in a marked reduction in flower stalk emergence because of reduced accumulation of assimilatory products [88, 89]. NI has been effective for accelerating the growth and development of LD herbaceous plants. Kim et al. [90] reported that though night interruption with low light intensity at 3–7 mol m^{-2} s^{-1} for 4 h (22:00–02:00) promotes flower induction with increased growth rate during the juvenile stage in *Cymbidium* Red Fire and *Cymbidium* Yokini, but night interruption with high light intensity (120 mol m^{-2} s^{-1}) produced high-quality plants. The increase of starch in leaves during vegetative growth and soluble sugars in pseudobulbs and roots during reproductive growth of *Cymbidium* "Red Fire" under night interruption is crucial for increasing plant size and thereby promoting flowering [90].

7.1.3 Humidity

Humidity ranging from 50% to 80% in the growing environment is necessary for proper growth and flowering of *Cymbidium*. The excess humidity would likely to increase the incidence of diseases and, therefore, dehumidify the environment by providing proper air circulation by controlling the vents of the greenhouse. Humidity below 40% can be managed by installing overhead sprayers or foggers.

7.2 Growing Media

The growing media for cymbidium should have good water holding capacity; a porous, pH value between 5.5 and 6.5; and electrical conductivity of about 1.05 mhos/cm which is relatively stable. Cymbidium producers use various kinds of growing media like composted bark, vermicompost, carbonized rice husk, pumice, groundnut shells, leafmold, coconut husk, coir dust, leafmold, and alone or in combination with inert growing media constituents like perlite, rock wool granules, expanded clay, and vermiculite. Incorporating 30–40% vermicompost in a growing medium containing 50% pumice, 30% charcoal, 10% vermiculite, and 10% peat moss improves growth and flowering in cymbidium orchid [91]. Yamane and Sakuamoto [92] reported that dusting of the roots of young cymbidium plants with hydrophilic polymer and then planting in composted bark enhances the growth of plantlets, whereas plants older than 1 year grow and flower well in a medium containing composted bark and granular soil containing hydrophilic polymer in 1:1 ratio. The growing substrate that dries quickly would reduce water and nutrient availability for plants, and high water retentivity in substrate reduced oxygen availability and encouraged rotting of the roots.

7.3 Nutrient Management

Usage of fertilizers depends on the plant growth stage, cultivar, growing media, and growing environment. Plants require high nitrogen during the vegetative stage and less during the reproductive stage. Well-balanced fertilizer is needed for optimum growth and flowering and can be supplied through organic and inorganic sources. Naik et al. [93] analyzed growing media (leafmold/coco chips/vermiculite/bricks, 4:2:1:1) at different intervals of growing period and found that NPK content decreases with the growing period. NP was present in the range of sufficiency. K was present, but its deficiency symptoms were not observed may be due to its mobilization to leaves from the storage organs. The required amount of calcium, magnesium, and sulfur was available from the media content. The microelements manganese, copper, iron, and zinc also decreased with the advancing of crop growth. The content of iron was found 30–40% below in young plants. The young plants show the deficiency by manifesting the symptom chlorosis between the veins in extreme case necrosis of the leaf margins and tip of the leaves. The deficiency was corrected by applying 50 ppm iron sulfate at 15-day interval in young plants and

100 ppm iron sulfate in 2-year-old plants. Naik et al. [93] reported the NPK requirement for young plants (1 year old) of *Cymbidium* Pine Clash "Moon Venus" and found that NPK in the ratio of 30:10:10 @ 0.1% is best in terms of growth attributes like leaf length, leaf girth, pseudobulb length, pseudobulb girth, and number of pseudobulb per clump, and for intermediate size of plants (2 years old), the NPK ratio of 20:20:20 @ 0.1% is suitable for enhancing the growth attributes.

Poole and Seeley [94] found that *Cymbidium* hybrids in nutrient culture required 100 ppm N, 50–100 ppm K, and 25 ppm Mg for optimal growth. The nitrogen below 50 ppm shows nitrogen deficiency symptoms. Magnesium at 100 ppm decreased growth in comparison to 50 ppm, and potassium has little response on the growth of cymbidium plants. Increasing nitrogen concentration (4, 6, and 8 mM/l) increased shoot production but reduced spike/shoot ratio in miniature *Cymbidium* cv. "Pendragon Sikkim." Further, withholding of nutrient application during May to June resulted in a reduced shoot formation, but a higher spike/shoot ratio along with earlier flowering, as compared with a continuous fertilizer supply in the nutrient solution [95]. Pascale et al. [96] studied nitrogen in three cultivars of cymbidium, Floripink, Pendragon Irene, and Traceredway, in soilless culture system and reported that N uptake was 8.0, 7.2, and 3.1 g per plant per year. The leachate volume was 40% of the given nutrient solution in Floripink and Traceredway, while it was inconsistent and less in Pendragon Irene (18%), with a mean concentration of 1.68 mmol of NO_3^-, 0.055 of PO_4^{3-}, and 0.24 of K+ per liter of leachate. High EC (1.4 dS m^{-1}) produced more flower spikes per m^2, as compared with low EC (0.6 dS m^{-1}), but fewer per shoot, and the spikes were shorter in *Cymbidium* Mary Pinchess "Del Rey." There was a positive correlation between shoot formation and spike production per shoot, whereas it was negative with the nitrogen content of young leaves. The shoots produced during autumn and winter were more productive than shoots produced during spring and summer [97]. Song et al. [98] investigated nitrogen and potash absorption by 2-year-old plants of *Cymbidium* Jungfrau. Nitrogen absorption was higher in full sunlight than 60% light, whereas the phosphorus absorption was higher under 60% of sunlight. The 67% of N absorbed in plants was redistributed to the bulbs (39%) and leaves (28%), while 46% of P was absorbed and distributed in bulbs (36.2%) and leaves (10.2%). Accumulation of P in leaves is threefold lower than that of N. N and P absorption in 1.5- or 1-year-old daughter plants were greater than in immature daughter plants of the mother plant. The solution culture had more NPK and Mg content in leaves than pot culture.

Naik et al. [99] studied the different concentrations of Panchgavya on the growth and flowering of *Cym* "Sleeping Nymph." Panchgavya is an organic product made out mixing five components, namely, cow dung, cow urine, cow milk, curd, and ghee, in a definite proportion and believed to promote growth and immunity in plant system. Drenching of growing media content or spraying with 1:30 (Panchgavya/water) improved vegetative and reproductive characters of *Cymbidium*.

7.4 Watering

The growing media of cymbidiums should always be moist, but not too wet. Soggy growing medium would reduce oxygen in the container and encourage the pathogens to decay the roots. Cymbidium requires sufficient water during the season of active growth, particularly during spring and summer. Water thoroughly, once or twice a week, more often when the weather is warm. A light misting each day will keep cymbidiums moist enough. Pascale et al. [96] studied water consumption of 6-year-old plants of three cultivars of three cymbidium, Floripink, Pendragon Irene, and Traceredway, in a growing medium containing (v/v) urea-formaldehyde (48%), polyurethane (48%), polyurethane (4%) and pH and EC of the nutrient solution adjusted at 6.0 and below 0.6 dS m^{-1}. The mean water consumption was least in cv. Pendragon Irene (0.6 L per plant per day) compared to Traceredway (1.6 L per plant per day) and Floripink (1.4 L). The water consumption was related to leaf area of the cultivars.

8 Plant Health Management

Cymbidium orchids grow in subtropical to temperate regions of Indian subcontinent. It produces a unique and high-quality flower in diverse color and pattern. The quality and quantum of *Cymbidium* flowers get affected by numerous insect-pests and diseases at different growth stages, both in artificial natural habitation and inside protected structures under controlled conditions. Insect-pests associated with *Cymbidium* are described below.

8.1 Insects and Pests

8.1.1 Aphids

Aphids are the major problem in *Cymbidium* orchid. There are two species of aphid, i.e., black aphid, *Toxoptera aurantii* (Boyer de Fonscolombe), and yellow aphid, *Macrosiphum luteus* (Buckton), which mainly cause damage to orchids.

Black aphid, *Toxoptera aurantii* (Hemiptera: Aphididae): Black aphid (*T. aurantii*) recorded to infests on 120 host plants that include orchids, *Hibiscus*, coffee, cocoa, ficus, citrus, mango, pomelo, etc. [100]. It is also known as black citrus aphid found in South America, Africa, eastern Asia, India, Australia, and the Mediterranean region. Adults are oval, shiny black or brownish-black in color, winged as well as wingless, small in size, and measuring about 2–3 mm in length. Nymphs are wingless and smaller than adults but similar in shape. A pair of black and white banded antennae is reflected well toward the abdomen. This species formed colonies on the young shoots, flowers buds, and flowers. *T. aurantii* completes a life cycle in 6–20 days depending on existing environmental conditions. Lower temperature increases the developmental period of aphid.

Yellow aphid, ***Macrosiphum luteum*** (Hemiptera: Aphididae): Yellow aphid, *M. luteum*, is reported to feed on *Dendrobium*, *Epidendrum*, *Vanda*, *Oncidium*, *Phalaenopsis*, *Cattleya*, and *Arundina* other than *Cymbidium* under protected conditions. Adult of this species is yellow to greenish-yellow in color, while nymphs are pale green. Pest is small-sized, about 2–3 mm length span; oval-shaped with two blackish cornicles are present on the tip of the abdomen. Adults are winged or wingless and the wingless form has a brownish patch on the top of the abdomen.

Damage

Aphids feed on host plants by sucking the cell sap. In *Cymbidium* orchids, nymphs as well as adults colonize on tender shoots or flower buds or new flower spike and even opened flower and suck the cell sap. This kind of damage often causes the plants to become deformed, and tender shoots curled and shrivelled, and sometimes, gall is formed on an affected portion. They also secrete honeydew like other soft-bodied insects, resulting in the development of sooty mold that affects the photosynthesis process. High humidity and cloudy weather fasten the buildup of aphid population. The affected plants retard growth and ultimately deteriorate the quality of flowers. *T. aurantii* is a vector of some viral diseases like citrus tristeza virus on citrus and little leaf and lemon-ribbing virus of lemon and also believed to transmit some viral diseases in orchids, i.e., *Cymbidium mosaic virus* from infested plant to healthy plants.

Management

For producing quality flowers, aphid infestation should be kept under control. Several natural enemies of aphids are present in the orchid ecosystem, which play a vital role in maintaining aphid population below a harmful level. These are *Coccinella septempunctata*, *Menochillus sexmaculata*, *Cryptolaemus montrouzieri*, *Adonia variegata*, and *Chrysoperla carnea* that feed on aphid's nymph and adult stages. On account of the artificial need of control, apply Vertilec (*Verticillium lecanii*, an entomopathogenic fungus formulation) @ 4 g/l of water or azadirachtin 0.03% EC @ 3 ml/l of water alternatively as a foliar spray to reduce aphid population.

If chemical control becomes necessary, plants should be treated with any one of the following insecticides like fipronil 5% SC @ 0.03% or malathion 50 EC @ 0.05% or acephate 75 SP @ 0.05% or imidacloprid 17.8 SL @ 0.003% on the appearance of aphids on the new spikes or flower buds before opening the flowers and repeat the spray at 10–15-day interval to control the pest and quality production of flowers.

8.1.2 Scale Insects

Scales are the major pest problem in orchids in the forest and artificial natural habitation. There are five species of scale insect recorded on various orchid species and hybrids in Indian subcontinent, i.e., boisduval scale, *Diaspis boisduvalii*; ti-scale, *Pinnaspis buxi*; soft brown scale, *Coccus hesperidum*; Florida red scale, *Chrysomphalus aonidum*; and lecanium scale, *Lecanium* sp. These scales cause damage to orchids particularly *Cymbidium aloifolium* in the tropical region to *Cymbidium gammieanum*, *Cymbidium tracyanum*, and *Cymbidium tigrinum* and many hybrids in the temperate region around the year. Scales feed on roots inside potting media, pseudobulbs, leaves, flower spikes, and flowers, causing economic loss.

Boisduval Scale, *Diaspis boisduvalii* (Signoret)

Boisduval scale, *Diaspis boisduvalii* (Signoret), is an economically important pest of orchids including various species and the hybrid of *Cymbidium* in India. It is also reported as an important pest of orchids in Florida [101]. This scale is circular to oval in shape, thin and flat, whitish to light yellow in color, and semitransparent on the removal of waxy material. When insect developed in complete growth, the body is entirely covered with white waxy growth which protects the insect from its natural enemies. It caused a 44% extent of damage to the plants and flowers of *Dendrobium nobile* [102].

Ti Scale, *Pinnaspis buxi* (Bouche)

Ti scale, *Pinnaspis buxi* (Bouche), is a serious pest of *Cymbidium* orchid. It is sticky, pear-shaped, small in size which is about 1–2 mm long, flat-bodied, and elongated without any permanent body organs like wings, legs, or eyes. Ti-scale completes its life cycle 20–30 days depending on existing temperature. Pests attack on all aerial parts and prefer old leaves that look like dead ones, brown to dark brown, and dried rather than plump. Infested plants produce small-sized, deformed flower spikes with inferior quality flowers like miniatures. Other than *Cymbidium*, it also feeds on *Dendrobium*, anthurium, banana, *Hibiscus*, coconut, and palm [101].

Soft Brown Scale, *Coccus hesperidum* Linn.

Coccus hesperidum is a soft scale insect that feeds on the midrib of leaves. It is oval to round in shape and more flattened than either the black or hemispherical scales. Adults are pale brown to dirty white or sometimes grayish mottled with dark brown in color on the back. This species reported on many orchids, citrus, papaya, rubber, etc. [103].

Florida Red Scale, *Chrysomphalus aonidum* Linn.

Chrysomphalus aonidum commonly feeds on nursery plants of *Cymbidium* and few other orchids and ornamentals in greenhouses. It is round or moderately convex in shape, is dark reddish brown to almost black or somewhat ash gray in color, and has a size of almost 2–2.5 mm in diameter. The exuviae are approximately central, reddish brown or brick red sometimes covered with grayish secretion, surrounded by a reddish-brown ring.

Damage

Scale insects are generally found on leaves particularly in the midrib, stems, and pseudobulb and suck the plant sap, resulting in deterioration of plant growth and causing loss of vigor and deformation of infected plants. Substantial scale infestations, however, can reduce overall plant health and cause yellow leaves and leaf drop. If infestation reached the flower buds, flowers do not open properly. Scale insects also excrete sticky honeydew which attracts sooty mold, affecting the average rate of photosynthesis.

Management

Scales are the major challenge in the harvesting of quality flowers of orchids. It is challenging to control the pest or keep the plants free from scales by application of only insecticides; hence, it is indispensable to apply various management approaches such as cultural, biological, and chemical control for effective management of scales.

Cultural Practices

Exclusion is the first measure to be taken to evade scale infestation; thus, select scale-free planting material by carefully examining all plant parts while establishing new orchid farm to prevent early buildup of pest. Keep cleanliness inside the pot and potting media. Need base pruning and burning of infested plant parts should be done to reduce the further spread of scales. Keep a proper distance between plants and isolate infested plants from healthy ones to prevent the scales from moving from one plant to another. If scale infestation is found on the roots, repotting should be done to eradicate harboring eggs and crawlers.

Biological Control

Promote natural enemies of scales. Coccinellid beetles feed on larvae of coccids that help in the suppression of its infestation. *Coccidencyrtus* sp. (Hymenoptera: Encyrtidae) has been reported as a parasite of boisduval scale [104].

Chemical Control

Scales can be removed by rubbing the scurf encrustation with toothbrush or cotton swab and killing them by dipping in 70% isopropyl alcohol or methylated spirit, or after gentle cleaning, roots should be sprayed with any of these insecticides, i.e., malathion 50 EC @ 0.05%, acephate 75 SP @ 0.05%, or carbaryl 0.2%, which would help to reduce scale infestation. A spray of these insecticides should also be done at the time of crawler emergence for effective control of scales on orchids.

8.1.3 Thrips, *Dichromothrips nakahari* (Mound)

Thrips, *Dichromothrips nakahari* Mound (Thysanoptera: Thripidae), is an important pest of *Cymbidium* orchid in all orchid-growing region. It is also reported on *Dendrobium* sp., *Oncidium*, *Arundina*, *Aerides*, *Calanthe*, *Coelogyne*, and *Thunia*. It feeds on leaves, flower buds, and full-bloomed flowers of orchids. Adults are slender, are dark brown to black in color, have apically pointed wings, and measure about 1–2 mm in length. Nymphs are resembling adults but are small in size and have pale yellow color. Wings are absent in the nymphal stage.

Damage

Thrip infestation on *Cymbidium* is reported in hardened plants in the nursery. In the later stage, *D. nakahari* feeds on flower buds and flowers by sucking the cell sap. Both stages by adult and by the nymphs cause damage. They also suck the cell sap from the tender portion of plants and on leaves, resulting in plants becoming discolored and shrivelled. In case of severe infestation, there are malformation in leaves, flower buds, and flowers. If thrip infestation occurs on young plants, they may finally dry up.

Management

For effective management of thrips, regular monitoring is recommended. Keep orchid plants free from weeds; hence, remove weeds and old plant debris and growing medium. If thrips are observed, immediately remove the infected plant parts and discard away to reduce the incidence. Spray orchid plants with neem oil 5 ml/l. of water. In case of severe infestation, apply any of these insecticides, i.e., malathion 50 EC and fipronil 5% SC @ 0.05% or imidacloprid 17.8 SL 0.003%.

8.1.4 Red Spider Mite, *Tetranychus urticae* Koch (Acari: Tetranychidae)

Mites are not insects; they are members of the arachnid family, but in India, spider mite is a most severe problem in *Cymbidium* orchids under controlled conditions. It is also reported on many other orchids, viz., *Dendrobium*, *Epidendrum*, *Arundina*, *Luisia*, *Pholidota*, and *Thunia*. Meena et al. and Nagrare [105, 106] reported that *T. urticae* is a serious pest of *Cymbidium*. It is noticed on the ventral surface of leaves and flowers. Mite incidence on *Cymbidium* is active throughout the year under greenhouses and open polyhouses. Adult mites are oval in shape, about 0.4–0.6 mm in length with well-marked four pairs of legs, and pale greenish to yellow with pairs of distinct dark lateral patches from orange to brick red; nymphs are pale green with darker margins [105].

Damage

The damage is caused by both nymphs and adult stages. Just after hatching, nymphs feed on the undersurface of leaves near the midrib by sucking the cell sap. The loss of cell sap causes yellowing of leaves. The injuries due to feeding can be seen as silvery marks left on the surface of leaves which usually turn brown to black. In case of substantial damage, plant growth is stunted, vigor is lost, and whole plants are covered with webs. Infested plants produce inferior quality flower buds which are not open properly, and flowers become abortive, turn brown, and fall before maturation.

Management

The mite is a serious problem on orchids under controlled conditions. It is more severe during warm and dry weather; hence, increasing humidity and leaf wetness and lowering the temperature help in the suppression of its infestation. Remove the infested plant parts (leaves/flowers) and destroy them to reduce further multiplication of mite. Clean cultivation and proper ventilation are also helpful in mite control.

On heavy incidence, immediately spraying the crop with plain water reduces mite population. If required, spray the plants with miticides, i.e., dicofol or propargite or ethion in recommended dose or concentration, to control mite infestation in orchids.

8.2 Diseases of Cymbidium

There are about 130 plant diseases affecting one or more orchid genera caused by fungi, bacteria, nematodes and viruses. Cymbidium orchids succumb to a number of diseases and pests under protected conditions. Several diseases have attained significant status and can lead to severe economic losses. Plants infected by diseases lose their vigor and production capacity and affect their market value considerably. Among fungal diseases, black rot (*Phytophthora palmivora*, *P. parasitica*, *Pythium ultimum*, and *P. splendens*), anthracnose (*Colletotrichum cymbidicola* and *C. clivae*), orchid wilt (*Sclerotium rolfsii*), petal blight (*Botrytis cinerea*), rust (*Uredo* sp.) and leaf blight (*Fusarium oxysporum*), *Sclerotinia* white rot (*Sclerotinia sclerotium*), and leaf spots caused by species of *Fusarium*, *Cercospora*, and *Haplosporella* are the most common. The bacterial soft rot, caused by *Erwinia* sp., has been reported on many *Cymbidium* hybrids and orchid species. An ectoparasite nematode, *Helicotylenchus microcephalus*, causing root necrosis has also been reported on many *Cymbidium* hybrids. The most serious problem of orchids is viral diseases. More than 50 viruses are known to infect orchids all over the world, but in India, at least 9 viruses have been reported. CymMV and ORSV are the most important and prevalent viruses. These viruses are widely distributed on all commercial hybrids and species.

8.2.1 Anthracnose

Causal Pathogens: *Colletotrichum cymbidicola* and *C. clivae*

Symptoms

The pathogen attacks all the aboveground parts of the plants, but leaves are more frequently attacked. Initially, small oblong to circular, oval, sunken, and reddish-brown to dark brown and gray-colored spots appear at the tip or middle of the leaf lamina which gradually enlarges and covers a large area of the leaf surface. It produces dieback symptoms which start from the tip and proceed downward. It produces conidia within black acervuli. It also affects leaf sheaths and floral spikes. It is found in nature mostly in conidial stage and can overwinter as mycelium or conidia.

Epidemiology

The pathogen perpetuates when phytosanitary measures are not adequate. It spread through infested compost, media, or leaf mold and through old contaminated pots. The disease usually occurs throughout the year. However, during June–September, when the temperature reaches over 30 °C and relative humidity is above 80%, the incidence is very high.

Management

Immediate removal of the diseased plants from the polyhouse and manually cutting of infected leaf portions with sterilized scissors are necessary. Repotting of the plant with properly sterilized potting mixture and fungicidal treatment is recommended. Contaminated pots, potting mix, and the wooden benches should be sterilized with 2% formalin. Keep the surrounding free from other host plants. Plant should not be exposed to direct sunlight as direct exposure to sunlight acts as a precursor of the disease. Spraying of Blitox @ 2.5–3.0 g/l at 10-day interval or spraying of carbendazim + mancozeb @ 1g each in one liter of water at 7-day interval has shown good control of the disease.

8.2.2 Black Rot

Causal Pathogens: *Phytophthora palmivora* and *Phytophthora parasitica*

Symptoms

Black rot is the most destructive disease of orchids. The disease appears as water-soaked small brown patches on the aerial parts of plants. Black necrotic lesions develop on pseudobulbs and roots which later spread upward, resulting in complete defoliation of the plant. The disease later migrates to other potted plants kept within the vicinity of the disease beds. New shoots also show black rot symptoms, which starts from the portion attached with the mother plants/pseudobulbs. Several *Cymbidium* hybrids in pots also get infected with black rot causing initial water-soaked small leaf spot, which later get transformed into blight symptoms covering larger leaf area. Black rot symptoms on leaves might be initiated by secondary airborne inoculum. These pathogens also cause damping-off of seedlings. The disease appears from last week of May or first week of June and continues to occur up to September under Darjeeling and Sikkim conditions when temperature is around 30 °C and humidity is above 90%.

Epidemiology

The disease spreads through contaminated potting media or water splash from adjacent infected plants or even through irrigation water. Plants grown in soil are more infected than raised bed. Soil bed over saturated with more than 90% water holding capacity favors the disease. Rotting occurs mostly on the *Cymbidium* plants grown in clayey soil beds, which are practically observed lower in elevation and easily get moistened with excessive water by rain or overhead irrigation. Besides, where rainwater was continuously dropped on the plants make the plant surface wet for longer period. Temperature in the range of 24–30 °C and relative humidity of 80–95% and continuous rainfall coupled with misty foggy weather favor the disease.

Management

Use of unsterilized potting media and pots should be avoided. Remove the infected plants and also destroy infected parts to check further spread of the disease. A good aeration in the nursery is essential. Reduce watering when the disease is expected to occur (June–September). It is advisable to keep the orchid plants on benches 90–120 cm above the ground level to avoid contamination through water splash. In terrestrial orchids proper drainage should be provided. For effective control, Matco MZ or metalaxyl @ 1 g/l or mancozeb @ 2 g/l can be used as spray or soil drenching. Application of contact fungicide, e.g., Captain, Thiram, or mancozeb, alternately with a systemic base fungicide metalaxyl is recommended.

8.2.3 Bacterial Soft Rot

Causal Pathogen: *Erwinia sp.*

The disease is caused by *Erwinia* sp. Severe infestation of bacterial soft rot has been reported on some cymbidium hybrids and *Eria pubescens* during rainy season. The infected plants initially showed water-soaking lesions and grayish-green lesions that rapidly enlarged. The affected areas are soft, decayed, and brown in color. With increasing severity, the internal tissues disintegrate and produce foul smell, and subsequently the bulb gets completely rotted and the entire plant collapses. When the disease portion is surface sterilized and plated on nutrient agar, bacterial colonies are formed after 48 h. Bacterial colonies on nutrient agar are creamish white, about 1.0 mm in diameter, circular, and raised with entire edge.
Management: Proper sanitation and sterilization of cutting tools with alcohol is recommended. Only disease-free plant material should be planted. Watering frequency needs to be less to minimize leaf wetness. Application of copper-based fungicides is effective against the bacterium.

8.2.4 Bacterial Brown Rot

Causal Pathogen: *Pseudomonas cattleya*
Symptoms: Small soft, water-soaked sunken spots are found on leaves that later become black/brown. The disease advances rapidly, resulting in immediate death of plants.
Epidemiology: High temperature and high humidity favor the disease. The bacterium spreads very fast by rain splash and overhead irrigation water.

Management

Acquire disease-free plants and isolate new stocks for at least 4 weeks before integration with existing stock. It is advised to reduce prolonged wetness by increasing air circulation and the water retention capability of the growing medium. The infected leaves may be cut off to check further spread of the disease. Overhead irrigation should be avoided. Copper-based fungicides are effective against the bacteria. The infected plants should be drenched or sprayed with 8-hydroxyquinoline at a dilution of 1:2000 in water.

8.2.5 Nematode Disease

Causal Organism: *Helicotylenchus microcephalus*

An ectoparasitic nematode has been reported from *Cymbidium* hybrids in different localities in Sikkim [107]. The disease was observed mostly on imported cymbidium hybrids.

Symptoms

Roots of infected *Cymbidium* hybrids develop severe necrosis, swelling, and fluffy root system. The leaves of infected plants exhibited typical bending, twisting, and abnormal growth. The associated nematode, *Helicotylenchus microcephalus*, is spiral and the most frequent plant parasitic nematode found worldwide in temperate and tropical countries. The species is considered to be migratory ectoparasitic feeders and feed from outside the roots by inserting their stylet into the epidermis of young succulent roots. Eggs are laid in the soil close to the roots or on the root surface and hatch in 2 or 3 days under favorable temperature conditions.

There is no evidence of plant parasitic nematode on orchids from India except few interceptions at the plant quarantine check posts. In India, four nematode species were intercepted from Madras Airport during 1989–1991 from the orchid consignment imported from East Asian countries [108]. Again, five nematode species, namely, *Aphelenchoides bicaudatus*, *A. besseyi*, *Helicotylenchus dihystera*, *Rotylenchulus reniformis*, and *Xiphinema elongatum*, were intercepted from *Oncidium*, *Cattleya*, *Dendrobium*, and *Vanda* species imported from Thailand and Singapore [109]. Species of *Helicotylenchus pseudorobustus*, *Aphelenchoides besseyi*, *A. composticola*, *and A. aligarhensis* have been reported on *Phalaenopsis* and *Cattleya* sp. imported from Taiwan to Shanghai [110]. *Aphelenchoides besseyi* has also been reported from *Dendrobium* "Lady Fay."

Management

Control of plant parasitic nematode is achieved with nematode-free planting material. Tissue-cultured plants are the best option. Proper cultural practices limit the spread of nematode. Hot water treatment of propagative material has limited the survival of plant parasitic nematode. Mustard oil cake and neem oil cake can be incorporated with the planting media. For heavy infestation, carbofuran 3G can be used.

8.2.6 Viral Diseases of Orchids

Virus diseases are serious threat to orchid industry. They not only reduce the general vigor of the plant but also lower the flower quality, thereby reducing the marketability of orchids and incurring serious economic losses. The first evidence of virus disease on orchid was described by [111] from Australia as flexuous or rigid rod-shaped particles associated with mosaic disease of orchids.

Cymbidium Mosaic Virus (CymMV)

The virus was first reported by [112] from California on *Cymbidium* sp. The virus produces symptoms like mosaic, necrosis, chlorotic flecks, water-soaked lesions, and flower necrosis on different orchid hosts. Cymbidium mosaic virus is filamentous with a model length of 480 nm and width of 13 nm. The axial canal is obscure.

The basic helix is obvious and the pitch of the basic helix is 2.8 nm. The sedimentation coefficient of CymMV is 121 S. The virus contains 5.6% nucleic acid and 94% protein. Genome of CymMV is unipartite and consists of linear ss RNA with the total genome size of 8.1 Kb.

Odontoglossum Ringspot Virus (ORSV)

The virus was first reported by [113] on *Odontoglossum grande* from the USA. The virus produces ringspot on *O. grande* and diamond mottle on *Cymbidium* spp. It produces color breaking, chlorotic streaking, mosaic, and necrosis. Some orchid cultivars are also symptomless. The virus particles are rod-shaped, not enveloped, and straight having a model length of about 300 nm and width of 18 nm. Axial canal is obvious. The purified preparations have two sedimentation components: 212 S and 119 S. The virions contain 5% nucleic acid, and the molecular weight of the coat protein is 17.5 kDa. In ultrathin sections, virions are found in the mesophyll, epidermis, and vascular parenchyma. The infected cell has crystal inclusion in the cytoplasm. The major ORSV hosts are *Cattleya*, *Cymbidium*, *Epidendrum*, *Odontoglossum*, and *Oncidium*, but it infects many orchid species and hybrids. CymMV and ORSV are generally found in mixed infections causing severe damage to the orchid species. Such plants showed severe necrosis, large sunken patches, and cracking of leaves. In India, the incidence of *Odontoglossum* ringspot virus (ORSV) has been reported in 42 different species of orchids from Sikkim by slot-blot hybridization method [114] and by ELISA on 71 species [115].

Calanthe Mild Mosaic Virus (CalMMV)

Calanthe mild mosaic virus has been reported on *Cymbidium pendulum* and *C. tigrinum* from Sikkim [116]. The infected samples have been tested by ELISA, RT-PCR, and Northern blot analysis and confirmed the presence of potyvirus. The sequencing of an RT-PCR-amplified amplicon using potyvirus-specific primers revealed that the virus is closely related to *Calanthe mild mosaic virus*.

Orchid Fleck Virus (OFV)

Orchid fleck virus was first reported in Japan on *Cymbidium* sp. showing chlorotic and necrotic fleck symptoms [117]. Later on, it was reported from Australia, Brazil, Denmark, Germany, Korea, and the USA producing chlorotic and necrotic spots and rings in many genera of Orchidaceae [118–120]. It is an important virus of orchids and reported on many orchid species from different parts of the world. The virus is sap transmitted to *Dendrobium* sp. and few species of family of Chenopodiaceae, Solanaceae, Leguminosae, and Aizoaceae [117, 121]. This virus has been reported from Sikkim and Kalimpong area of Darjeeling districts on *Coelogyne elata*, *C. flaccida*, and many *Cymbidium* species. Electron microscopy of negatively stained preparations showed bacilliform particles measuring 32–40 nm in diameter and 100–150 nm in length.

8.2.7 Detection and Management

During the last two decades, there has been great advancement in the field of detection technology. There are a number of diagnostic techniques available for the detection of orchid viruses which are accurate, reliable, and less time-consuming and can detect up to attogram (ag) level. Some of the diagnostic techniques like enzyme-linked immunosorbent assay (ELISA), immunosorbent electron microscopy (ISEM), dot immunobinding assay (DIBA), tissue blot immunoassay (TBIA), reverse transcription-polymerase chain reaction (RT-PCR), immune capture PCR, multiplex PCR, DIG-labeled cRNA probes, matrix-assisted laser desorption-ionization (MALDI), TaqMan real-time PCR, and molecular beacons are some of the highly sensitive and specific tests used for the detection of orchid viruses. A number of companies like Mukoyama Orchids in Japan, Forsite Diagnostics Ltd in Britain, and Agdia Incorporation in the USA have developed easy and rapid techniques for diagnosis of orchid viruses which can be performed in the field and provide quick results within 5–10 min. Extensive surveys of orchid nurseries in Sikkim and Darjeeling hills and detection of samples by ELISA revealed that most of the commercial hybrids and species grown in this region are contaminated with CymMV and ORSV. These two viruses are often found as mixed infections [115]. Recently, polyclonal antibodies against CymMV and ORSV have been developed using bacterially expressed recombinant coat protein as immunogen [122]. The antisera are being used for the detection of both viruses from the planting materials.

Orchids are vegetatively propagated crops, and cultural practices play a very significant role particularly as orchid viruses generally spread by inadvertent propagation. Hence, strict sanitation practices are essential for the control of CymMV and ORSV. Clean cultivation by avoiding the sources of inoculums is the best option requiring minimum inputs, but unfortunately this is not practiced by orchid growers. Morel [27] first developed revolutionary techniques of meristem culture to produce virus-free cymbidiums through meristem culture which established the techniques as a clonal propagation procedure. He standardized meristem tip culture to obtain virus-free clonal material of *Cymbidium* based on the concept that majority of plant viruses do not infect the meristematic dome despite severe systematic infection of the plants, and therefore, this region can be excised and regenerated to plantlet [123]. Meristem culture is an important method to produce virus-free plants, and it should be attempted only with valuable breeding plant or valuable unique clone [124].

Chemotherapy using antiviral compounds like ribavirin (Virazole) and dithiouracil is being used in tissue culture media to get virus-free plants. Loi et al. [125] reported that Virazole was found effective in obtaining virus-free *Dendrobium* shoot tips and callus from parent plants infected by CymMV and ORSV from in vitro cultures of *Cymbidium* [126]. Genetic engineering coupled with tissue culture techniques offers a useful way to introduce specific genes into plants. The development of a suitable gene transfer system for *Dendrobium* orchids would provide the breeder with greater opportunity to produce commercially desirable hybrids.

9 Other Uses

Cymbidiums are primarily cultivated for cut flower and pot plant production, but they also possess value for medicine, food, and other uses. Many cymbidium species are used in herbal medicine for curing various kinds of body ailments. Isolation of phytochemicals in a few species has shown encouraging results and can lead to development of useful drugs.

9.1 Medicinal Use

In herbal medicine roots, leaves or whole plant cymbidiums are used for curing various body ailments particularly in India, China, and Australia. Herbalist uses the plant parts fresh, dried, or in powdered form. The choice of herbal plant for curing the disease depends upon its availability, effectiveness, and other substitutes available in the surrounding locality. *Cymbidium aloifolium* is most abundantly used India, Nepal, and Bangladesh, whereas *C. ensifolium*, *C. goeringii*, and *C. faberi* are amply used cymbidiums in Chinese medicine system. The survey of published reports indicates that cymbidiums can cure from minor ailments, viz., cut, boils, and earache, to severe illnesses like paralysis, traumatic injuries, and lung- and kidney-related problems. *C. aloifolium*, *C. hookerianum*, *C. devonianum*, *C. lancifolium*, and *C. bicolor* are used for treating cuts and boils, curing cracks in feet, and healing fractures. The roots of species such as *C. aloifolium*, *C. devonianum*, *C. ensifolium*, *C. faberi*, *C. goeringii*, *C. kanran*, *C. sinense*, and *C. wilsonii* cure reparatory-related diseases like cough, asthma, and bronchitis. *C. ensifolium* and *C. lancifolium* improve the blood circulation in the body. The roots of *C. faberi*, *C. goeringii*, and *C. kanran* are used to clear out stomach worms. *C. aloifolium* and *C. ensifolium* are used in treating menstrual-related disorders in women. *C. aloifolium*, *C. wilsonii*, and *C. goeringii* are used for treating body weaknesses. Cymbidium species are also reported useful in treating for dysentery (*C. madidum*), gastroenteritis (*C. kanran*), tumor (*C. aloifolium*), rheumatism (*C. lancifolium*), inflammation (*C. ensifolium*), headache (*C. faberi*), fever (*C. macrorhizon*), and diarrhea (*C. iridioides*). The seeds of *C. madidum* are used as oral contraceptive.

The activity of the species against a particular disease is due to the presence of phytochemicals in the plant. A few species have been surveyed for phytochemical present in them. *C. aloifolium* contains several phenanthrene aloifol I, alifol II, coelolin, 6-methoxycoelonin, pendulin, and denthyrsinin. Watanabe et al. [127] isolated cymbidine, a monomeric peptidoglycan-related compound with hypotensive and diuretic activities, from a *C. goeringii*. Gigantol was isolated from the whole plant *C. goeringii*. Kim et al. [128] reported that fragrant compounds isolated from *C. goeringii* have either cytotoxic or antibacterial properties. α-Bergamotene has a cytotoxic effect on cancerous cells, while nerolidol has antibacterial property. Nerolidol and β-bisabolene were isolated from *C. forestii*. β-Bisabolene is cytotoxic.

α-Pinene inhibits the growth of glioblastoma cells, and 1,8-cineole that functions as pheromone is found in *C. faberi*.

9.2 Cosmetics

Cymbidium flowers are used in perfume, skin cream, and antiaging cosmetics [129].

9.3 Food

Olatshe is a popular Bhutanese dish made from the flower buds of *Cymbidium hookerianum*. People harvest the flower buds from the inflorescence just before opening and then wash and boil them in water until they become soft. After draining the water, they add a mixture of spices, melted cheese, and salt to it and cook further for 5 min, and the dish is ready. *Olatshe* is served with rice and noodles or used as a dip. The orchid flowers add bitterness, and the addition of spices offset the bitter taste. In China, the flowers of cymbidium orchids are used as herbal tea and drinks [129]. Aboriginals in Australia eat pseudobulbs of *C. canaliculatum* and *C. madidum* as mucilaginous food [2]. In Darjeeling, India monkeys often take out and eat the tender part of stem base of leaves of *C. lowianum*, *C. tracyanum*, and *Cymbidium* hybrids.

9.4 Other Uses

The juice of the pseudobulbs of *C. canaliculatum* is used for making glue. In western Jawa, roasted pseudobulbs of *C. lancifolium* are grounded to stick substance, and the sticky substance is made into handle to fasten Sudanese knives [2]. In India the senesced leaves of cymbidium are collected from the field and woven into baskets.

10 Conclusion

Cymbidiums rank on the top among all the orchids cultivated for cut flower production because of heavy substance and high keeping quality of the flowers. Only a few species of cymbidium are utilized for evolving about 16,000 hybrids from a gene pool of 52 species. Thus, a broad scope still exists for using the other suitable species in the breeding program. Cymbidium breeders mostly use hybridization and selection in the improvement of the cultivars. However, the different breeding methods, such as mutation breeding, are rarely applied in the varietal improvement of cymbidiums, despite promising results in other ornamental crops. Two mechanically viral diseases of cymbidium, CyMV and ORSV, cause severe limitation in the cultivation of cymbidium; no source of resistance is reported so far, and the results have shown that transgenic breeding can help in developing the resistant cultivars. Since cultivars of cymbidium are highly heterozygous, induction and use of haploid in the breeding program would be useful in developing new varieties.

References

1. Hew CS (2001) Ancient Chinese orchid cultivation: a fresh look at an age-old practice. Sci Hortic 87:1–10
2. Puy D, Cribb DJP (2007) The genus Cymbidium. Kew Publishing, Kew
3. Du Puy DJ, P Cribb (1988) The classification of Cymbidium. In: Du Puy D, Cribb P (eds) The genus *Cymbidium*. Kew Publishing, Royal Botanic Garden, Kew, United kingdom, pp 50–194
4. Seth CJ, Cribb PJ (1984) An assessment of sectional limits in the genus Cymbidium Swartz. In: Arditti J (ed) Orchid biology reviews and perspectives, vol 3. Cornell University Press, Ithaca, pp 411–483
5. Obara-Okeyo P, Fujii K, Kako S (1998) Isozyme variation in *Cymbidium* species (Orchidaceae). Hortic Sci 33:133–135
6. Choi SH, Kim MJ, Lee JS, Ryu KH (2006) Genetic diversity and phylogenetic relationships among and within species of oriental cymbidiums based on RAPD analysis. Sci Hortic 108:79–85
7. Wang H-Z, Lu J-J, Hu X, Liu J-J (2011) Genetic variation and cultivar identification in *Cymbidium ensifolium*. Plant Syst Evol 293:101–110
8. Sharma SK, Kumaria S, Tandon P, Rama Rao S (2013) Assessment of genetic variation and identification of species-specific ISSR markers in five species of Cymbidium (Orchidaceae). J Plant Biochem Biotechnol 22:250–255
9. Chung MY, Chung GM, Chung MG, Epperson B (1998) Spatial genetic structure in populations of *Cymbidium goeringii*. Genes Genet Syst 73:281–285
10. Aoyama M (1989) Karyomorphological studies in Cymbidium and its allied genera, Orchidaceae. Bull Hiroshima Botanical Garden 11:1–121
11. Leoonhardt KW (1977) Chromosome numbers and cross compatibility in the genus cymbidium and some related tropical genera (Orchidaceae). Dissertation, University of Hawaii
12. Stort MNS, Morales MAM (1987) Sterility barriers in some artificial F1 orchid hybrids: female sterility. Rev Brasil Genet 10:109–118
13. Stort MNS (1984) Sterility barriers of some artificial hybrids Male sterility I. Microsporogenesis and pollen germination. Am J Bot 71:309–319
14. Wimber DE (1956) Cytogenetic studies in the genus Cymbidium dissertation, Claremont Graduate University
15. Du Puy DJ (1986) A taxonomic revision on of the genus *Cymbidium Sw.* (Orchidaceae). PhD thesis, University of Birmingham and Royal Botanic Gardens, Kew
16. Govaerts R (2019) World checklist of Orchidaceae. Facilitated by the Royal Botanic Gardens, Kew. Published on the Internet. http://wcsp.science.kew.org/. Retrieved
17. King G, Pantling R (1898) The Orchids of Sikkim Himalaya. Ann Roy Bot Gard, Calcutta
18. Menninger ED (1963) Diary of a colchicine induced tetraploid Cymbidium. Am Orchid Soc Bull 32:885–887
19. Wimber DE, Van Cott A (1967) Artificially induced polyploidy in Cymbidium. In: Proceedings of 5th World Orchid Conference, pp 27-32
20. Kim MS, Kim JY, Eun JS (2003) Chromosome doubling of a *Cymbidium* hybrid with colchicine treatment in meristem culture. In: Proceedings of NIOC2003, Nagoya, Japan
21. Hwang SH, Kim MS, Park PH, Park SY (2016) Scent analysis using an electronic nose and flowering period of potted diploid and Tetraploid Cymbidium. Korean J Hortic Sci Technol 34:163–171

22. Fitch CM (2004) Propagating your Orchids: vegetative propagation. In: The gardener's guide to growing of orchids. Botanic Garden, Brooklyn
23. Bag TK (2006) Orchid disease and their management. Tech Bull, ICAR-NRC Orchid Sikkim 3:15–17
24. Soon TE (2005) Orchid Asia. Marshall Cavendish International (Asia) Private Ltd., Singapore
25. Tibs, M. 2007. Orchids: A practical guide to care and cultivation, New Holland, pp. 52-63.
26. Pal R, Medhi RP (2011) Technoguide for propagation of Cymbidiums through backbulbs. National Research Centre for Orchids, Sikkim
27. Morel GM (1960) Producing viruses free Cymbidium. Am Orchid Soc Bull 29:495–497
28. Wimber DE (1963) Clonal multiplication of Cymbidium through tissue culture of the shoot meristem. Am Orchid Soc Bull 32:105–107
29. Gu Z, Higaki PC, Nishidha MM, Arditti J, Nyman LP (1987) The effects of benzyladenine, 2, 4-dichlorophenoxyacetic acid and indole acetic acid on shoot-tip culture of Cymbidium. Lindleyana 2:88–90
30. Begum AA, Tamaki M, Kako S (1994) Formation of protocorm-like-bodies (PLB) and shoot development in vitro culture of outer tissue of Cymbidium PLB. J Jpn Soc Hortic Sci 63:663–673
31. Nayak NR, Sahoo S, Pathak S, Rath SP, Pathak S (2002) Establishment of thin section (TCS) method for rapid micropropagation of *Cymbidium aloifolium* (L) Sw. and *Dendrobium nobile* Lindl. (Orchidaceae). Sci Hortic 94:107–116
32. Singh F (1992) In vitro propagation of orchids 'state of the art'. J Orchid Soc India 6:11–14
33. Bernard N (1899) Sur la germination du Neottia nidusavis. Compt Rend Acad Sci Paris 128(1253):1255
34. Bernard N (1909) Levolution dans la symbiose, les orchidees et leurs champignons commensaux. Ann Sci Nat Bot 9:1–196
35. Knudson L (1922) Non symbiotic germination of orchid seeds. Bot Gaz 73:1–25
36. Ballion G, Ballion M (1924) The non-symbiotic germination of orchid seed in Belgium. Orch Rev 32:305–308
37. Ballion G, Ballion M (1928) A symbiotic germination of orchid seed. Orch Rev 36:103–112
38. Clement E (1925) A resume of the non-symbiotic germination of orchid seed. Orch Rev 33:199–200
39. Smith F (1932) Raising orchid seedlings asymbiotically. Gard Chron 91:9–11
40. Arditti J, Clements MA, Fast G, Hadley G, Nashimura G, Ernst R (1982) Orchid seed germination and seedling culture -A manual. In: Arditti J (ed) Orchid biology: reviews and perspectives, vol II. Cornell University, London
41. Sharma J (1996) Orchids of India commercialization and conservation. Daya Publishing House, New Delhi
42. Xiang Y, Yu FA, Peng ZH (2003) Tissue culture of Cymbidium sinense. Forest Res Beijing 16:434–438
43. Chang C, Chen YC, Hsu YH, Wu JT, Hu CC, Chang WC, Lin NS (2005) Transgenic resistance to *Cymbidium mosaic* virus in *Dendrobium* expressing the viral capsid protein gene. Transgenic Res 14:41–46
44. Mitra GC, Prasad RN, Roychowdhury A (1976) Inorganic salt and differentiation of protocorms in seed callus of orchid and correlative changes in its free amino acid content. Indian J Exp Biol 14:350–351
45. Hossain M, Sharma M, DaSilva JAT, Pathak P (2010) Seed germination and tissue culture of Cymbidium giganteum Wall ex. Lindl. Sci Hortic 123:479–487
46. Arditti J, Ernst R (1993) Micro propagation of Orchids. Willey, New York
47. Sagawa Y, Shoji T, Shoji T (1966) Clonal propagation of Cymbidiums through shoot meristem culture. Am Orchid Soc Bull 35:118–122
48. Kim KW, Kako S (1984) Morphological and histological studies on the formation of protocorm-like bodies and explants development in Cymbidium shoot apex culture in vitro. J Korean Soc Hortic Sci 25:156–163
49. Subramaniam G, Taha RM (2003) Micro propagation of *Cymbidium atropurpureum* in vitro Malaysian. J Sci 22:310–316

50. Wimber DE (1965) Additional observations on clonal multiplication of Cymbidium through culture of shoot meristem. Cymbidium Soc News 20:7–10
51. Beechy N (1970) Propagation of orchids from aerial roots. Am Orchid Soc Bull 39:1085–1088
52. Peterson RL (1975) The initiation and development of root buds. In: Torrey JG, Clarkson DJ (eds) The Development and Function of Roots. Academic, New York
53. Yasugi S, Sukamoto K, Onodera K, Tamashiro M (1994) Plantlet regeneration in root segment culture of Cymbidium Kenny 'Wine Color'. Plant Tiss Cult Lett 11:150–152
54. Nayak NR, Rath SP, Patnaik S (1997) In vitro propagation of three epiphytic orchids *Cymbidium aloifolium* (L.) Sw. *Dendrobium aphyllum* (Roxb.) Fisch and *Dendrobium moschatum* (Buch – Han) Sw through thidiazuron induced high frequency shoot formation. Sci Hortic 71:243–250
55. Tran Thanh Van K (1980) Control of morphogenesis by inherent and exogenously applied factors in thin layers. Int Rev Cytol 32:291–311
56. Rout GR, Mahapatra A, Jain MS (2006) Tissue culture of ornamental pot plant: a critical review on present scenario and future prospects. Biotechnol Adv 24:531–560
57. Chugh S, Guha S, Usha Rao I (2009) Micro propagation of Orchids: a review on the potential of different explants. Sci Hortic 122:507–520
58. Huan LVT, Takamura T, Tanaka M (2004) Callus formation and plant regeneration from callus through somatic embryo structures in *Cymbidium* orchid. Plant Sci 166:1443–1449
59. Chang C, Chang WC (1998) Plant regeneration from callus culture of *Cymbidium ensifolium* var. misericors. Plant Cell Rep 17:251–255
60. Le Van TH, Tanaka T, Tanaka M (2004) Callus formation and plant regeneration from callus through somatic embryo structures in *Cymbidium* orchid. Plant Sci 166:1443–1449
61. Skoog F, Miller CO (1957) Chemical regulation of growth and organ formation in plant tissue culture in vitro. Symp Soc Exp Biol 11:118–130
62. Paek KY, Kozoi T (1998) Micro propagation of temperate Cymbidium via rhizome culture. Hortic Tech 8:283–288
63. Hasegawa A, Oshashi H, Goi M (1985) Effect of BA, rhizome length, mechanical treatment and liquid shaking culture on shoot formation from rhizome in *Cymbidium faberi* Rolfe. Acta Hortic 166:25–40
64. Lee JS, Kwack BA, Lee BK, Chung JD (1986) Studies on *Cymbidium kanran*, a native of Korea. I. Rhizome culture in vitro. J Korean Soc Hortic Sci 25:129–135
65. Ueda H, Torikata H (1970) Organogenesis in the meristem culture Cymbidium: VI. Effects of light and culture medium and root formation. J Jpn Soc Hortic Sci 39:363–368
66. Ogura Y, Okubo H (2002) In vitro shoot formation from rhizome apices of *Cymbidium ensifolium* and *Cymbidium kanran*. J Fac Agric Kyushu Univ 47:301–306
67. Kusumoto M (1978) Effects of combination of growth regulating substances and organic matter on the proliferation and organogenesis of cymbidium protocorms cultured in vitro. J Jpn Soc Hortic Sci 47:391–400
68. Siddique AB, Paswan L (1998) Effect of growth regulators and organic supplements on differentiation Cymbidium longifolium protocorms in vitro. J Hill Res 11:234–236
69. Nayak NR, Chand PK, Rath SP, Patnaik SN (1998) Influence of plant growth regulators on the growth and organogenesis of *Cymbidium aloifolium* (L) Sw. seed derived in vitro. In Vitro Cell Dev Biol 34:185–188
70. Kukulczanka K, Jastrezebska KA (1976) The effect of minor elements and magnesium on the early developmental stage of orchids Cymbidium in vitro culture. Prace-Instytut Sadownictwa-w Skierniewich-B 77:77–86
71. Paek KY, Chun CK (1985) Effect of Ca^{2+} and Mg^{2+} concentration on organogenesis and nutrient content in Cymbidium protocorm culture. J Korean Soc Hortic Sci 26:193–199
72. Shimasaki K, Tanibuchi Y, Fukumoto Y (2003) The effects of sucrose and trehalose on organogenesis in protocorm-like body and rhizome of *Cymbidium* species. J Soc High Tech Agri 15:120–134

73. Ueda H, Torikata H (1968) Organogenesis in meristem culture Cymbidium: I studies on the effect of growth substances added to culture media under continuous illumination. J Jpn Soc Hortic Sci 37:240–248
74. Conner LN, Conner AJ (1984) Comparative water loss from leaves of *Solanum lancilatum* plants cultured in vitro. Plant Sci Lett 36:241–246
75. Ticha IB, Rodociiova B, Kadlecek P (1999) Stomatal morphology during acclimatization of tobacco plantlets to ex vitro condition. Biol Plant 42:469–474
76. Barinerd KE, Fuchigami LH (1981) Acclimatization of aseptically cultured apple plants to low relative humidity. J Am Soc Hortic Sci 106:515–518
77. Fujiwara K, Kozai T (1995) Physical microenvironment and its effects. In: Aitken-Christie J., Kozai T., Smith M.A.L. (eds) Automation and environmental control in plant tissue culture. Springer, Dordrecht, pp 319–369
78. Pospisilova J, Solanova M, Ondej T, Opatryny Z (1988) The photosynthetic characteristics during the micropropagation of tobacco and potato plants. Photosynthetica 22:205–213
79. De Proft MPMML, Debergh P (1985) Carbon dioxide and ethylene evolution in the culture atmosphere of Magnolia cultured in vitro. Plant Physiol 65:375–379
80. Kozai T, Kubota C (2005) In vitro aerial environments and their effects. In: Kozai T, Afreen F, Zobayed SMA (eds) Photoautotrophic (sugar-free medium) micropropagation as a new micropropagation and transplant production system. Springer, Dordrecht, pp 31–52
81. Pospisilova J, Solavova J, Catasky J (1992) Photosynthetic response to stress during in vitro cultivation. Photosynthetica 26:3–18
82. Mani SK, Nagaraju V (2004) Influence of potting media on acclimatization and growth of in vitro plantlets of *Cymbidium* hybrids. J Orn Hort 7:23–70
83. Powell CL, Caldwell KI, Littler RA, Warrington I (1988) Effect of temperature regime and nitrogen fertilizer level on vegetative and reproductive bud development in *Cymbidium* orchids. J Am Soc Hortic Sci 113:552–556
84. Went FW (1957) The experimental control of plant growth. Cron Bot 17:148–152
85. Li J, Zhao X, Matsui S (2001) Effect of temperature on rate of nutrient concentrations in pseudobulbs and leaves and flowering in a *Cattleya* and a *Cymbidium* hybrid. J Soc High Tech Agric 13:85–90
86. Ichihashi S (1997) Orchid production and research in Japan. In: Arditti J, Pridgeon AM (eds) Orchid biology: reviews and perspectives, vol 7. Kluwer Academic Publishers, Dordrecht
87. Komori T, Yoneda K (2002) Effects of autumn and winter light and temperature management on lead emergence in mericloned Cymbidium lovely angel 'The Two Virgins'. Environ Control Biol 40:383–387
88. Lee N, Lee CZ (1993) Growth and flowering of *Cymbidium ensifolium* var. misericors as influenced by temperature. Acta Hortic 30:223–231
89. Komori T, Niitsu A, Murakami T (1990) Effects of light intensity on the growth, flowering and carbohydrate contents of *Cymbidium* orchids. (Japanese text with English abstract) Bull. Yamanashi Agric Res Cent 4:17–26
90. Kim YJ, Lee HJ, Kim KS (2011) Night interruption promotes vegetative growth and flowering of Cymbidium. Sci Hortic 130:887–893
91. Hatamzadeh A, Shafyii Masouleh SS (2011) The influence of vermicompost on the growth and productivity of cymbidiums. Caspian J Env Sci 9:125–132
92. Yamane M, Sakuamoto (1993) Effect of a hydrophilic polymer and granular soil containing it on the growth and flowering at first potting of mericlones young cymbidium growing bark compost. Bull Fac Agric Tottori Univ 46:7–11
93. Naik SK, Barman D, Medhi RP (2010) Response of Cymbidium Pine Clash Moon Venus to major nutrient at vegetative growth stage. J Ornam Hortic 13:182–188
94. Poole HA, Seeley JG (1978) Nitrogen, potassium and magnesium nutrition of three orchid genera. J Am Soc Hortic Sci 103:485–488
95. Bik AR, van den Berg TJM (1984) Effect of substrate and nitrogen supply on yield and quality of mini-cymbidium. Acta Hortic 150:289–296

96. De Pascale S, Maturi T, Paradiso R, Barbieri G (2001) Comparison of water and nutrients use of three cultivars of cymbidium in a soilless culture system. Acta Hortic 559:535–542
97. de Kreij C, van den Berg TJM (1990) Effect of electrical conductivity of the nutrient solution and fertilization regime on spike production and quality of Cymbidium. Sci Hortic 44:293–300
98. Song S-J, Boo C-H, Zhang-Kual U (1999) Mineral absorption by Cymbidium Jungfrau in the solution culture. Korean J Exp Agric 18:35–40
99. Naik SK, Barman D, Devadas R, Ushabharathi T (2013) Evaluation of Panchgavya as source of nutrient for *Cymbidium* orchids. Afr J Agric Res 8:5728–5732
100. Carver M (1978) The black citrus aphid, *Toxoptera citricidus* (Kirkaldy) and *T. aurantii* (Boyer de Fonscolombe) (Homoptera: Aphididae). J Aust Entomol Soc 17:263–270
101. Dekle GW (1965) Arthropods of Florida and neighboring land areas: Florida armored scale insects, vol 3. Florida Department of Agriculture & Consumer Services, Division of Plant Industry, Gainesville
102. Nagrare VS, Pal R, Barman D (2009) Pests associated with orchid *Dendrobium nobile* under mid-Altitude of Sikkim. Environ Ecol 27:560–562
103. Zimmerman EC (1948) Insects of Hawaii. vol 5. Homoptera: Sternorhyncha. Honolulu, Univ. Hawaii Press p 464
104. Miller DR, Devidson JA (2005) Armored scale insect pests of trees and shrubs (Hemiptera: Diaspididae). Cornell University Press, Ithaca
105. Meena NK, Pal R, Barman D, Medhi RP (2013) Biology and seasonal abundance of two-spotted spider mite, *Tetranychus urticae* on orchids and rose. Phytoparasitica 41:597–609
106. Nagrare VS (2001) Pests of orchids and their management in Sikkim- a survey. J Orchid Soc India 15:65–68
107. Pant RP, Das M, Khan MR, Pun KB, Medhi RP (2012) Association of an ecto-parasite nematode- *Helicotylenchus microcephalus* Sher, with poor growth of *Cymbidium* hybrids in Sikkim. Indian Phytopath 65:196–197
108. Johnson M, Babu DVN, Latha S, Muthaiyan MC (1993) Interception from imported orchid plants. Plant Prot Bull 45:26–27
109. Latha S, Babu DVN, Sathyanarayana N, Reddy OR, Sharma R (1999) Plant parasitic nematodes intercepted from imported ornamental plants. Indian Phytopath 52:283–284
110. Lin YY, Wang KM, Tsay TT (1992) The occurrence of *Aphelenchoides besseyi* Christie on *Dendrobium* Lady Fay. Plant Prot Bull- Taipei 34:202–215
111. Magee CJ (1943) Orchid mosaic. Aust Orchid Rev 8:51–52
112. Jenson DD (1950) Mosaic of *Cymbidium* orchids. Phytopathology 40:966–967
113. Jenson DD, Gold HA (1951) A virus ringspot of *Odontoglossum* orchid: symptoms, transmission and electron microscopy. Phytopathology 41:648
114. Sherpa AR, Hallan V, Pathak P, Zaidi AA (2006a) Coat protein gene of *Cymbidium mosaic* virus: characterization of geographical isolates from India. J Phytopathol 154:275–280
115. Pant RP, Das M, Pun KB, Ramachandran P, Medhi RP (2010) Occurrence of *Cymbidium mosaic* and *Odontoglossum ringspot* viruses in orchid germplasm of Sikkim and Darjeeling hills, their identification and diagnosis. Indian Phytopath 63:326–332
116. Singh MK, Sherpa AR, Hallan V, Zaidi AA (2007) A potyvirus in *Cymbidium* spp. in Northern India. Aust Plant Dis Notes 2:11–13
117. Doi Y, Chang MU, Yora K (1977) Orchid fleck virus. In: CMIIAAB descriptions of plant viruses, vol. 183. Commonwealth Mycological Institute
118. Blanchfield AL, Mackenzie AM, Gibbs A, Kondo H, Tamada T, Wilson CR (2001) Identification of orchid fleck virus by reverse transcriptase-polymerase chain reaction and analysis of isolate relationships. J Phytopathol 149:713–718
119. Gibbs A, Mackenzie A, Cross P, Wilson C, Kitajima EW, Nightingale M, Clements M (2000) Viruses of orchids in Australia; their identification, biology and control. Aust Orchid Rev 65:10–21
120. Kitazima EW, Kondo H, Mackenzie A, Rezende JAM, Gioria R, Gibbs A, Tamada T (2001) Comparative cytopathology and immunochemistry of Japanese, Australian and Brazillian isolates of orchid fleck virus. J Gen Plant Pathol 67:231–237

121. Chang MU, Arai K, Doi Y, Yora K (1976) Morphology and intracellular appearance of orchid fleck virus. Ann Phytopathol Soc Jpn 42:156–167
122. Rani P, Pant RP, Jain RK (2010) Serological detection of *Cymbidium mosaic* and *Odontoglossum ringspot* viruses in orchids with polyclonal antibodies produced against their recombinant coat proteins. J Phytopathol 158:542–545
123. Morel GM (1964) A new means of clonal propagation of orchids. Am Orchid Soc Bull 31:437–477
124. Retheesh ST, Bhat AI (2010) Simultaneous elimination of *Cucumber mosaic* virus and *Cymbidium mosaic* virus infecting *Vanilla planifolia* through meristem culture. Crop Prot 29:1214–1217
125. Loi JS, Lee SM, Can Lam LT, Fan S, Yiang WH (1991) Eradication of orchid viruses in *Dendrobium* Sonia using virazole. Singap J Prim Ind 19:16–22
126. Toussaint A, Kummert J, Maroquin C, Lebrun A, Roggemans J (1993) Use of Virazole to eradicate ORSV from in vitro cultures of *Cymbidium Sw.* Plant Cell Tissue Organ Cult 32:303–309
127. Watanabe K, Tanaka R, Sakurai H, Iguchi K, Yamada Y, Hsu CS, Sakuma C, Kikuchi H, Shibayama H, Kawai T (2007) Structure of Cymbidine A, a monomeric peptidoglycan-related compound with hypotensive and diuretic activities, isolated from a higher plant, *Cymbidium goeringii* (Orchidaceae). Chem Pharm Bull 55:780–783
128. Kim SM, Jang EJ, Hong JW, Song SH, Pak CH (2016) A comparison of functional fragrant components of Cymbidium (Oriental orchid) species. Korean J Hortic Sci Technol 34:331–341
129. Shengji P, Zhiwei Y (2018) Orchids and its uses in Chinese medicine and health care products. Med Res Innov 2:1–3

Biotechnology Approaches on Characterization, Mass Propagation, and Breeding of Indonesian Orchids *Dendrobium lineale* (Rolfe.) and *Vanda tricolor* (Lindl.) with Its Phytochemistry

11

Endang Semiarti, Aziz Purwantoro, and Ika Puspita Sari

Contents

Abstract

Mass propagation of orchids is very important to do extensively due to their slower reproduction and growth naturally. Besides this, many growers do hunting expansionally on orchids especially orchid species which is having uniqueness and exclusivity characters. Moreover, orchids such as *Dendrobium* and *Vanda* also have phytochemistry such as sesquiterpenoid, alkaloid, bibenzyl, and phenolic that could be used as medicinal properties. Mass propagation could be conducted through biotechnology approaches using either conventional *in vitro* technique with adding plant growth regulator or orchid breeding using genetic engineering revealed by *Agrobacterium*-mediated transformation techniques. All of these methods are proposed in high embryogenesis of orchids. The

E. Semiarti (✉)
Faculty of Biology, Universitas Gadjah Mada, Yogyakarta, Indonesia
e-mail: endsemi@ugm.ac.id

A. Purwantoro
Faculty of Agriculture, Universitas Gadjah Mada, Yogyakarta, Indonesia
e-mail: azizp@ugm.ac.id

I. Puspita Sari
Faculty of Pharmacy, Universitas Gadjah Mada, Yogyakarta, Indonesia

J.-M. Mérillon, H. Kodja (eds.), *Orchids Phytochemistry, Biology and Horticulture*, Reference Series in Phytochemistry, https://doi.org/10.1007/978-3-030-38392-3_12

combination of 2,4-D and ascorbic acid was absolutely required in the *Agrobacterium*-mediated transformation. Insertion of *KNAT1* gene produced more buds in micropropagation compared to the wild-type plants. The efficiency of transformation was increased caused by acetosyringone application.

Keywords

Mass propagation · Orchid breeding · Orchid biotechnology · *Dendrobium* · *Vanda*

1 Introduction

Orchid is a kind of an ornamental plant. There are more than 5000 species of orchids that could be found in Indonesia [1]. Indonesia is rich with natural orchid species because of its complete agroclimate from the lowlands to the highlands for tropical climate. As [2] stated that approximately 5000 orchid species in the major divisions of Indonesia (Table 1). Of the main islands, only the orchid flora of Java is probably almost completely catalogued. It is well known that orchids are valuable in many ways as follows.

Orchid is preferred because of the beauty of the flowers that were used for decoration in offices and receptions in certain places and as a collection plant in the garden. Some of them have a specific character which gives a special value due to their uniqueness and exclusivity. A common feature of orchids is due to their beauty of the flowers including the last long a time without wilting. In addition to the beauty of the flowers, it turns out that the body parts of orchids also contain phytochemicals, in the form of flavonoids, polysaccharides, bibenzyls, phenanthrenes, coumarins, sesquiterpenoids, alkaloids, and steroids. In general plants produce various compounds for survival by producing organic compounds called primary and secondary metabolites. Primary metabolites are compounds produced by plants that function directly in the process of photosynthesis, respiration, and growth. Secondary metabolites are intermediate or side compounds of primary metabolites. Secondary metabolites are specific, only produced by plants in certain families because they have different functions. Therefore, the secondary metabolites which are produced by

Table 1 The richness of orchid flora of the major divisions of Indonesia[a]

Island (group)	Number of orchid species	Percentage of endemic orchids
Sumatra	1185	42
Java	799	29
Lesser Sunda Islands	208	26
Borneo	1571	56
Sulawesi	585	66
Moluccas	390	46
New Guinea	2824	86

[a]Taken from [2]

orchids could be used in medicinal purposes. Some orchids that are known to have medicinal properties include *Spathoglottis plicata* seeds for itching, *Dendrobium crumenatum* for anticancer medicine, *Dendrobium lineale* for anticancer drugs, and *Vanda tricolor* fragrance for aroma therapy, while the roots contain metabolites for anticancer drugs (Fig. 1). Therefore, these orchids are now being hunted, taken from the forest to be traded. This will certainly threaten the existence of these orchids in nature, besides the threat of natural disasters and conversion of forest land to settlements or roads.

Most of the orchids are endemic in certain areas in Indonesia, and because of their uniqueness and exclusivity, many growers do hunting expansionally. On the other hand, the reproduction and growth of an orchid are very slow due to its natural biology. Seeds of the orchid were difficult to germinate since it has no endosperm. Naturally the seeds will germinate after getting source of nutrients which is supplied by mycorrhiza in the form of symbiosis mutualism. This causes the population of orchids in nature not too much. Air humidity, sufficient light, wind, and environmental conditions are natural mechanisms for regulating the abundance of orchids in their natural habitat. The destruction of habitat and difficulties in cultivation, however, are threatening these orchid species [3]. Rare and threatened orchid species are propagated by seeds rather than by vegetative methods [4]. Therefore, it is necessary to develop a technique for sowing the seed of orchid in order to reach the highest germination rate to be a protocorm. Since the orchid seeds are having no endosperm, the germination of those seeds was established on the artificial medium. On the other hand, micropropagation efforts are an important way to maintain the population of

Fig. 1 Potential Indonesian orchids for phytochemistry. (**a** and **b**) *Spathoglottis plicata* Blume.; (**c**–**d**) *Vanda tricolor* Lindley; (**e**–**f**) *Dendrobium crumenatum* Sw.; (**g**–**h**) *Dendrobium lineale* Rolfe.

these orchids in nature and the existence of these orchids for biopharmaceutical stock. Thus, a biotechnological approach can be applied to conserve them both *in situ* and *ex situ* by using micropropagation methods.

2 Breeding Programs on Orchids Revealed by Biotechnology Approaches

Plant breeding is often considered as an interaction of art and science to alter the natural variation of crop plants in the direction of human needs [5]. However, at present plant breeding is not only included on art and science but also considered on the technology. Therefore, plant breeding is the art, science, and technology of improving important agricultural plants for the benefit of humankind.

In Fig. 2, it can be explained that basically the plant breeding programs include three kinds of activities which could be started from any point of view as given below.

In general, the main purpose of plant breeding with agricultural crops is to improve both qualities and quantities of traits with the profitable values such as yields, nutritional qualities, and other characters. The qualitative characters, or traits, are the easiest to deal with since most of them are discontinuous traits that are governed by one or a few major genes. On the contrary, quantitative characters are much more difficult for the breeder to control since these traits are governed by many genes, each having a small effect. Unfortunately, many traits of economic

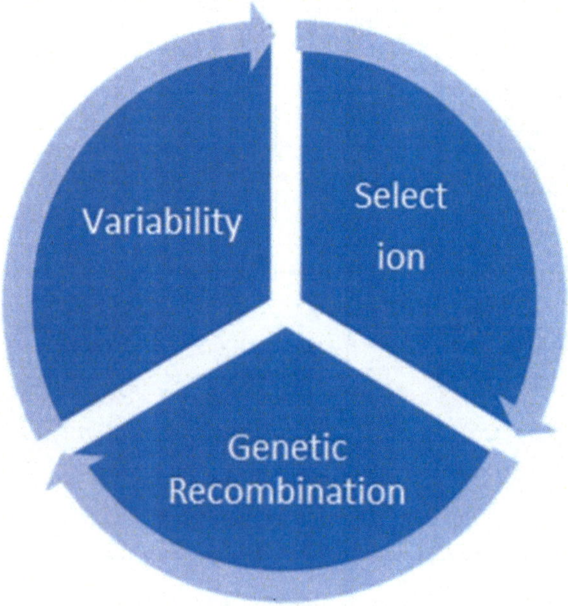

Fig. 2 Basic activities on the plant breeding program

importance are of this type, e.g., height, cold and drought tolerance, time to maturity, and, in particular, yield. However, plant breeding is a rapidly advancing science. It is able to make use of genetic and biotechnological innovations to efficiently develop better crop varieties.

Orchids belong to the ornamental plants which are highly priced in the international market due to their designed spectacular flowers; brilliant colors; delightful appearances; myriad sizes, shapes, and forms; and long-lasting qualities. In breeding ornamental plants, attention is paid to such factors as longer blooming periods, improved keeping qualities of flowers, general thriftiness, and other features that contribute to usefulness and aesthetic appeal. Novelty itself is often a virtue in ornamentals, and the spectacular, even the bizarre, is often sought.

Breeding orchids can be a highly challenging endeavor. Most of the purposes on orchid breeding are to produce commercially important hybrids that have a market demand and are liked by the consumers. The concept behind the development of hybrids in orchids may vary according to the genus and species. The generalized objectives as stated by Bhattacharjee and Das [6] are given below:

- To breed for better color, size, and substance of the flower
- To introduce a perfect blending of colors in sepals, petals, and lips
- To create round and full form of sepals and petals with minimum fenestrations and twists
- To increase the length of inflorescence
- To increase the number of flowers/inflorescence
- To achieve compactness in flower facing on the spike
- To develop hybrids showing a correct mode of display
- To extend the blooming period
- To produce miniature forms
- To produce fragrant varieties
- To produce flowers with longer vase life
- To develop types suitable as pot plants
- To develop hybrids insensitive to strict climatic regime
- To develop hybrids resistant to biotic stress like diseases particularly to viruses

Orchid breeding programs can be divided into two groups including classical and modern plant breeding. In the first one, plants are selected with desirable characters, and elimination of undesirable characters occurs. Otherwise, the modern plant breeding programs called biotechnology have used molecular biology techniques and omics technologies [7]. Biotechnology approaches are more efficient compared to the classical one. It could be due to its most rapid, easiest, cheapest approach, and the important things are not influenced by the environment. Furthermore, biotechnological developments are helping breeders make the desired genetic changes with much greater precision. Therefore, applications of biotechnology on orchid breeding have significantly shortened the time. Another purpose using biotechnology method on orchid breeding is the production of a huge number of plantlets.

Therefore, improvement in orchid breeding is aimed to get stable production of high-quality clones by micropropagation.

Purposing of mass propagation of orchid plants in addition to *in vitro* seed germination, it can also be done with various *in vitro* techniques, namely, organ culture, cell culture, protoplast culture, callus culture, and somatic embryogenesis. Even the current genetic engineering approach is not only used to improve the specific trait of the orchid; it is also used to increase its mass propagation by embryo gene insertion, bud formation genes, etc. Embryogenesis is the one method which could be applied on mass propagation of orchid either zygotic embryogenesis or somatic embryogenesis (Table 2).

The addition of 2 g/L peptone in all culture media accelerated the growth and increased the percentage of seed germination of up to 100%, enlarged the size, and greenish the protocorms and shoot initiation [14].

The most suitable method of biotechnology conducted on orchid breeding is genetic engineering. In this case the modern genetic techniques can insert desirable traits into plants, and the new plants are called transgenic or genetically modified organisms (GMOs). Whatever the purpose on orchid breeding, the method of *Agrobacterium*-mediated transformation system is the popular method.

There are several transformation methods which have been employed for the transfer of exogenous genes into orchid tissues. The most widely used method is particle bombardment, followed by *Agrobacterium*-mediated transformation [22]. Semiarti et al. [10] used *Agrobacterium tumefaciens* for doing the transformation of the wild orchid species *Phalaenopsis amabilis*. Moreover, [22] mentioned that both cells of *Vanda tricolor* and *P. amabilis* transformant carrying *KNAT1* gene were more meristematic compared to the wild-type plants as the organ produced more buds in micropropagation. During working on orchid transformation, much effort has been spent on testing different chemical compounds and selectable markers and their corresponding selection agents for the efficacy and efficiency of

Table 2 Some techniques that have been used for mass propagation of Indonesian orchid species through *in vitro* embryogenesis and *Agrobacterium*-mediated transformation

Orchid species	Techniques conducted	References
Phalaenopsis amabilis	*In vitro* embryogenesis (zygotic and somatic) Genetic transformation by using *KNAT1* gene Genetic transformation by using *AtRKD4* gene	[8–11]
Vanda tricolor	*In vitro* embryogenesis (zygotic and somatic) Genetic transformation by using *KNAT1* gene	[12, 13]
Dendrobium lineale	*In vitro* embryogenesis (zygotic and somatic) Genetic transformation by using *AtRKD4* gene	[14, 15]
Dendrobium phalaenopsis	*In vitro* embryogenesis (zygotic and somatic) Genetic transformation by using *AtRKD4* gene	[16, 17]
Coelogyne pandurata	*In vitro* embryogenesis (zygotic and somatic) Genetic transformation by using *KNAT1* gene	[18, 19]
Dendrobium capra	*In vitro* embryogenesis (zygotic and somatic) Genetic transformation by using *AtRKD4* gene	[20, 21]

transgenic orchids. Pre-culture treatment and application of ascorbic acid were absolutely required in the *Agrobacterium*-mediated transformation of *V. tricolor*. Moreover, acetosyringone had been added in order to improve the efficiency of *Vanda* transformation [12]. Finally, the choice of an appropriate method, a chemical compound, and also a selectable marker is vital to the success of orchid transformation protocol.

3 Phytochemical Compounds of *Dendrobium lineale* and *Vanda tricolor*

In general, secondary metabolites are produced by plants for various purposes, including:

1. As a plant protector against the threat of herbivore animals or microbial infections, fungi, or pathogenic viruses
2. As an attraction for pollinators and seed-dispersing animals
3. As a reproductive hormone

Generally, secondary metabolites are classified based on their biosynthetic pathways, namely, phenolics, terpenes and steroids, alkaloids, and flavonoids. It has been clearly demonstrated that secondary metabolites are the plant adaptation agents to their environment [24]. Secondary metabolites are unique sources for pharmaceuticals. Since the ancient times in Indonesia, the oldest references regarding the use of medicinal herbs are found in temple relief. While the term Jamu (Jampi Oesada) may also be traced to the relics of ancient writing, some say it might be in the manuscripts of Madhawapura, Ghatotkacasraya (MPU Panuluh), Serat Centhini, and Serat Kawruh Bab Jampi-Jampi Jawi. In 1775, Rumphius, a Dutch botanist, conducted a research on herbal medicine in Indonesia. He published a book called *Herbarium Amboinense* possibly in which there are parts of the orchid plant as herbal medicine [25]. In India and China, *Dendrobium* and *Vanda* are already used in Ayurveda and Chinese pharmacopoeia [26, 27]. Pharmacopoeia is a drug standard book issued by a government official body that outlines the ingredients of a drug, the chemicals in a drug and its properties, the medicinal properties, and the usual dosage. In Indonesia there is an Indonesian Herbal Pharmacopoeia since 2008, but unfortunately until now there has been no description of the herbs that come from orchids while Indonesia's wealth in the form of orchids is very abundant.

4 *Dendrobium* Chemical Compounds

The use of plant parts from *Dendrobium* as herbal medicine has been going on for a long time such as tonic, astringent, analgesic, and anti-inflammatory agents. With the development of content analysis techniques in herbal medicine, chemical compound research has been carried out from various materials derived from *Dendrobium*.

Table 3 *Dendrobium* compounds

No	Group	Compounds	*Dendrobium* species
1	Polysaccharide	Polysaccharide	*D. nobile*, *D. fimbriatum*
2	Bibenzyl	Gigantol, dendrocandin	*D. amoenum*, *D. candidum*, *D. aphyllum*, *D. densiflorum*, *D. nobile*
3	Phenanthrene	Phenanthrenedione, phenanthrene, moscatin, moscatilin, denbinodin	*D. chrysotoxum*, *D. densiflorum*, *D. nobile*
4	Coumarin	Ayapin	*D. densiflorum*
5	Sesquiterpenoid	Dendronobilin	*D. nobile*
6	Alkaloid	Scopoletin	*D. densiflorum*
7	Steroid	Sitosterol	*D. thyrsiflorum*

The results of research from [28] found that constituents that are commonly found in *Dendrobium* are flavone C-glycosides and flavonols. This research was conducted on 142 species. A few years later, many researchers found about 100 compounds from 42 *Dendrobium* species including 32 alkaloids, 6 coumarins, 15 bibenzyls, 4 fluorenones, 22 phenanthrenes, and 7 sesquiterpenoids (Table 3).

Nearly all of the *Dendrobium* studied were found to be compounds that emerged from almost all of them, polysaccharide, gigantol, dendrocandin, and moscatin/moscatilin. Some appear dendronophenol [29]. Given the abundance of content in *Dendrobium* species, research has also been carried out on many pharmacological activities of all of these compounds.

5 *Vanda* Compounds

Compounds commonly found in *Vanda* include bibenzyl derivatives (gigantols), phenanthrene derivatives, phenolic compounds, anthocyanins, alkaloids, steroids, and triterpenoids [30] (Table 4).

Research on compound and pharmacological activities of *D. lineale* and *V. tricolor* has never been done. However, there are several ways to estimate the compound and pharmacological activities of these two orchids, one of which is by understanding genetic similarity and chemotaxonomy studies. In chemotaxonomy it is known that plants that are taxonomically close also have genetic similarities. This genetic similarity will give rise to the suspicion that similar compounds will be found (although in varying degrees). By knowing the possibility that the compounds owned by *D. lineale* and *Dendrobium* species have similarities, it will be able to be used to estimate the pharmacological activities that are also not far away (although there are variations in intensity). This also applies to *V. tricolor* with close *Vanda* species [23].

Research conducted by [31] on Papua's endemic orchid including *D. lineale* found several similarities among *Dendrobium*. The study was conducted by isolation and extraction of genomic DNA from Papua's dried leaf orchids (26 types). In this

Table 4 *Vanda* compound

No	Group	Compounds	*Vanda* species
1	Bibenzyl	Gigantol	*V. coerulea*, *V. roxburghii*
2	Phenanthrene	Tessalatin, oxo-tessalatin, parviflorin, imbricatin, methoxycoe lonin, coelonin	*V. tessellate*, *V. parviflora*, *V. coerulea*
3	Phenolic	Parishin, gallic acid, tetracosylferulate	*V. parishii*, *V. tessellata*, *V. roxburghii*
4	Anthocyanins	Delphinidin and cyanidin derivatives	*Vanda* hybrid
5	Steroid and triterpenoids	β-Sitosterol-D-glucoside, β-sitosterol, ϒ-sitosterol, stigmasterol	*V. roxburghii*
6	Alkaloid	Laburnine acetate	*V. hindsii*

study, markers that are commonly used in researching genetic variation are random amplified polymorphism DNA (RAPD). From the analysis using the unweighted pair-group method arithmetic average (UPGMA) with the sequential agglomerative hierarchical nested cluster analysis (SAHN-clustering) and TREEE programs, it is known that *D. lineale* has similarities with *D. mirbelianum* and *Dendrobium* species (variation score 0.23–0.30). Based on chemotaxonomy, it can be assumed that *D. lineale* compound includes polysaccharide, gigantol, dendrocandin, moscatin/moscatilin, and dendronophenol [29].

V. tricolor has similarities with several *Vanda* from research results based on morphological characters (*V. tricolor* is similar to *V. limbata*, *V. celebica*, *V. retusa*, *V. scanen*, and *V. foetida* [32]. Based on the RAPD analysis, *V. tricolor* has genetic similarities with *V. lamellata*, *V. limbata*, *V. luzonica*, *V. sumatrana*, and *V. merrillii* with percentage similarity of all >50% [33].

6 Pharmacological Activities

Polysaccharides isolated from *Dendrobium* have antioxidant and free radical scavenging activities. Environmental stress on *Dendrobium* will result in increased nitric oxide (NO) radicals, namely, catalase (CAT), peroxidase (POD), ascorbate peroxidase (APX), 2,2-azinobis (3-ethylbenzothiazoline-6-sulphonic acid) or ABTS, hydroxyl, and 1,1-diphenyl-2-picrylhydrazyl (DPPH) radicals, and enzymatic antioxidant superoxide dismutase (SOD). Stress-fighting activity is shown in *D. nobile* polysaccharide (scavenging 20–40% of DPPH and 40-0.60% of ABTS). *D. fimbriatum* polysaccharide has the highest scavenging activity of ABTS among *Dendrobium* species at 90% [34, 35]. Furthermore, other studies confirm the antitumor action of *Dendrobium* stem fraction in inhibiting sarcoma 180 *in vivo* and HL-60 cells *in vitro* [36].

Phenanthrene and dihydrophenanthrene from *Dendrobium* species can inhibit tumor necrosis factor alpha (TNFa), interleukin 8 (IL-8), and IL-10 during lipopolysaccharide (LPS) stimulation [37, 38]. Upon lipopolysaccharide (LPS) stimulation, macrophages produce a large amount of inflammatory factors, such as tumor necrosis factor α (TNF-α), interleukin-1 beta (IL-1β), interleukin 6 (IL-6), interferon (IFN), and other cytokines [39]. In addition, denbinobin is a major phenanthrene isolated from the stems of *D. nobile*. The antitumor action of denbinobin in SNU-484 cells decreased the expressions of matrix metalloproteinase-2 and matrix metalloproteinase-9 [40].

Sesquiterpene glycosides (dendrosides) significantly promoted cell proliferation and stimulation of mouse B lymphocytes [41]. B lymphocytes are important to immune cells and play a key role for responses in the immune system. Gigantol inhibits lung cancer cell line due to downregulating caveolin-1 (Cav-1), activating ATP-dependent tyrosine kinase, and regulating cell division cycle 42 (Cdc42) [42].

Moscatilin, moscatin, and moscatin diacetate inhibit arachidonic acid (AA) platelet aggregation-induced in rabbit and collagen induced (50 μg/mL and 100 μg/mL) [43]. Studies on the aggregation of platelets are conducted through thrombin, arachidonic acid (AA), thrombin, or collagen induction method [44]. Polysaccharide from *D. officinale*, *D. nobile*, and *D. chrysotoxum* showed a hypoglycemia activity in diabetic rats [45]. Besides *D. chrysotoxum* is also able to inhibit the occurrence of diabetic retinopathy [46] (Table 5).

The ethanolic extract of stem *V. coerulea* exhibited significant antioxidant and anti-inflammatory activities due to reduction in the production of COX-2 enzymes and PGE2 in irradiated HaCaT cells [47]. The petroleum ether extract of leaves of *V. tessellata* showed potential DPPH and nitric oxide (NO) radical scavenging activities [48], while the hydro-alcoholic extracts of leaves of *V. testacea* reduced axotomy-induced peripheral neuropathy in rats due to antioxidant activity [49]. The extracts of *V. coerulea* showed a skin-hydrating property [50]. *V. teres* stem extract exhibited antiaging effects in immortalized keratinocyte cell lines of human origin (HaCaT) [51]. The methanolic extract of the flowers of *V. spathulata* exhibited an antidepressant activity in mice using the forced swim test and the tail suspension test [52]. Anwar et al. [53] had reported that the petroleum ether extract of leaves of *V. tessellata* showed a dose-dependent hepato-protective activity in rats.

Table 5 Pharmacological activity of *Dendrobium* species

No	Pharmacological activity	Compounds	Orchid species
1	Antioxidant	Polysaccharides	*D. nobile, D. fimbriatum, D. huoshanense, and D. chrysotoxum*
2	Antitumor	Polysaccharides	*D. nobile*
3	Anti-inflammatory	Phenanthrene	*D. nobile and D. chrysanthum*
4	Immunomodulatory	Dendrosides	*D. nobile, D. moniliforme*
5	Anticancer	Gigantol	*D. draconis*
6	Antiplatelet aggregation	Moscatilin, moscatin, and moscatin diacetate	*D. chrysotoxum*
7	Antidiabetes	Polysaccharides	*D. officinale, D. nobile, and D. chrysotoxum*

Table 6 Pharmacological activity of *Vanda* species

No	Pharmacological activity	Compounds	Orchid species
1	Anti-inflammatory	Imbricatin, gigantol, and methoxycoe lonin	*V. coerulea*
2	Antiaging	Vandateroside and eucomic acid; imbricatin, methoxycoe lonin, and gigantol	*V. coerulea and V. teres*
3	Antidepressant	Phenol compounds	*V. spathulata*
4	Antioxidant	Phenanthrene	*V. coerulea*
5	Hepatoprotective	Phenanthrene	*V. tessellata*
6	Aphrodisiac	Phenanthrene	*V. tessellata*

Alcoholic extracts of flowers of *V. tessellata* showed an aphrodisiac activity in male mice [54] (Table 6).

Khan et al. [55] reported that using *in vitro* and *in vivo* models, some members of *Vanda* species have bioactive chemical constituents and pharmacological activities from the aerial parts of the plants and often used as a traditional medicine for ethnopharmacology in Asian countries. The stem and aerial parts of *V. tessellata* and *V. coerulea* have potential anti-inflammatory and anti-microbial activities. Based on the review, it is necessary to prove the pharmacology activity of *D. lineale* and *V. tricolor* by using a guided bioassay of the pharmacological effects that have been done on *Dendrobium* and *Vanda*.

7 Conclusions

Orchids have distinguishable, unique, and exclusive characters. Phytochemical on orchids could be useful for medicinal purposes, such as antitumor, anticancer, anti-inflammatory, etc. At present, the achievement of orchid breeding is mass propagation or micropropagation through biotechnology approaches, namely, either *in vitro* techniques or genetic engineering by using *Agrobacterium*-mediated transformation.

Acknowledgments We thank Mrs. Hj. Sri Suprih Lestari, the owner of "TITI" Orchid Nursery; Pakem-Yogyakarta and Mrs. Titik Handayani, the owner of "Keboen Kita" Orchid Nursery; and Godean, Sleman-Yogyakarta for the permission taking the photograph of *Dendrobium lineale*, and special thanks are given to Mr. Aries Bagus Sasongko for his assistance to take the photo of *Dendrobium crumenatum*.

References

1. Irawati (2002) Pelestarian jenis anggrek Indonesia [En: Preservation of Indonesian orchids]. Buku panduan Seminar Anggrek Indonesia 2002:34–45
2. Schuiteman A, Vermeulen JJ, de Vogel EF (2010) Flora Malesiana: orchids of New Guinea, vol VI; genus *Bulbophyllum*. (CD-ROM). ETI/National Herbarium Nederland, Amsterdam/Leiden

3. Arditi J (1992) Fundamentals of orchid biology. Wiley, New York
4. Kumar A, Nandi SK, Bag N, Palni LMS (2002) Tissue culture studies in two important orchid taxa: *Rhynchostylis retusa* (L.) Bl. and *Cymbidium elegans* Lindl. In: Nandi SK, Palni LMS, Kumar A (eds) Role of plant tissue culture in biodiversity conservation and economic development. Gyanodaya Prakashan, Nainital, pp 113–124
5. Sleper DA, Poehlman JM (2006) Breeding field crops. Blackwell Publishing, Oxford. 424 pp
6. Bhattacharjee SK, Das SP (2008) Orchids- botany, breeding, cultivation, uses and post-harvest management. Aavishkar Publishers, Distributors, Jaipur, p 396
7. Baloglu MC (2018) In: Gyana Ranjan Rout, Peter KV (eds) Genomics of cucurbit in genetic engineering of horticultural crops. Academic Press, London, pp 413–432
8. Utami ESW, Soemardi I, Taryono T, Semiarti E (2007) Embriogenesis Somatik Anggrek Bulan *Phalaenopsis amabilis* (L.) Bl: Struktur dan Pola Perkembangan [En: Somatic embryogenesis of the moth orchid *Phalaenopsis amabilis* (L.) Bl: structure and pattern of development]. Berk Penel Hayati 13: 33–38
9. Mercuriani IS, Purwantoro A, Moeljopawiro S, Jang S, Semiarti E (2012) Selection of *Phalaenopsis amabilis* L. blume orchid resistance to hygromycin. Indian J Biotechnol 17(2):107–113
10. Semiarti E, Indrianto A, Purwantoro A, Isminingsih S, Suseno N, Ishikawa NT, Yoshioka Y, Machida Y, Machida C (2007) Agrobacterium mediated transformation of wild orchid species *Phalaenopsis amabilis*. Plant Biotechnol 24:265–272
11. Mose W, Indrianto A, Purwantoro A, Semiarti E (2017) The influence of thidiazuron on direct somatic embryo formation from various types of explant in *Phalaenopsis amabilis* (L.) blume orchid. Hayati J Biosci 24:201–205
12. Dwiyani R, Purwanto A, Indriarto A, Semiarti E (2012) Konservasi anggrek alam Indonesia *Vanda tricolor* Lindl. varietas Suavis melalui kultur embrio secara *in vitro* [En: Conservation of Indonesia's natural orchid *Vanda tricolor* Lindl. Suavis varieties through *in vitro* embryo culture]. J Bumi Lestari 12(1):93–98
13. Semiarti E, Purwantoro A, Indrianto A, Sasongko AB, Milasari AF, Herawati O (2018) Innovation of natural orchid cultivation technology for tourism development in Banyunganti Village, Jatimulyo Village, Girimulyo Sub-District, Kulon Progo District, Yogyakarta. Proceeding ICCEESD, Yogyakarta
14. Semiarti E, Ningrum WDA (2016) The role of *Dendrobium* Orchid Homeobox 1 (*DOH1*) gene during *in vitro* shoot development of wild orchid *Dendrobium lineale* Rolfe. ABS NAT SEM INTL CONF SOC INDON BIODIV, Yogyakarta, 19–20 Maret, pp 55–58
15. Febryanti NPK, Nurlianab S, Gutierrez-Marcosc J, Semiarti E (2020) The expression analysis of *AtRKD4* transgene in *Dendrobium lineale Rolfe* transgenic orchid carrying *35S::GR::AtRKD4* for micropropagation. AIP Conference Proceeding. Melville, NY, USA. In press
16. Setiari N, Purwantoro A, Moeljopawiro S, Semiarti E (2018) Micropropagation of *Dendrobium phalaenopsis* orchid through overexpression of embryo gene *AtRKD4*. Agrivita 40(2):284–294
17. Zulwanis N, Setiari, Gutierrez-Marcosc J, Semiarti E (2020) The expression of *AtRKD4* transgene during induction of somatic embryogenesis in transgenic *Dendrobium phalaenopsis* orchid carrying 35S::GR::AtRKD4. AIP Conference Proceeding. Melville, NY, USA. In press
18. Semiarti E, Indrianto A, Purwantoro YH, Martiwi INA, Feroniasanti YML, Nadifah F, Mercuriana IS, Dwiyani R, Iwakawa H, Yoshioka Y, Machida Y, Machida C (2010) High-frequency genetic transformation of *Phalaenopsis amabilis* orchid using tomato extract-enriched medium for the pre-culture of protocorms. J Hortic Sci Biotechnol 85(3):205–210
19. Semiarti E (2011) Agrobacterium-mediated transformation of Indonesian orchids for micropropagation (A.285 Indrianto, Ed.). https://doi.org/10.5772/24997
20. Lawrie MD, Indrianto A, Purwantoro A, Semiarti E (2018) Isolation and characterization of *LEAFY* homologous gene partial sequences from *Dendrobium capra* J.J. Smith Endemic Orchid of Java Island. J Biol Res 23(2):53–56
21. Lawrie MD, Zulfa, Layina D R, Ningtias FN, Alifianto A, Indrianto A, Purwantoro A, Semiarti E (2020) *In vitro* germination and flowering of Dendrobium capra J.J. Smith an Endemic Orchid of Java. HAYATI J Bioscience, Bogor, Indonesia, In press

22. Mudalige RG, Kuehnle AR (2004) Orchid biotechnology in production and improvement. HortScience 39:11–17
23. Dwiyani R, Purwantoro A, Indrianto A, Semiarti E (2010) Improvement of genetic transformation technique in *Vanda tricolor* orchid using acetosyringone. Ann Bogoriense 14(2):27–32
24. Kossel A (1891) Ueber die chemische Zusammensetzung der Zelle. Arch Phys 4:181–186
25. Aditama TY (2014) Jamu dan Kesehatan. Lembaga Penerbit Balitbangkes, Jakarta
26. Kirtiker KR, Basu BD (1975) Indian medicinal plants, IV. Bishen Singh Mohendra Pal Singh, Dehradun
27. Das SP, Bhattacherjee SK (2006) Orchids. In: Bhattacherjee SK (ed) Herbaceous perennials and shade loving foliage plants. Pointer Publishers, Jaipur
28. Williams CA (1979) The leaf flavonoids of the Orchidaceae. Phytochemistry 18(5):803–813. https://doi.org/10.1016/0031-9422(79)80019-9
29. Lam Y, Ng TB, Yao RM, Shi J, Xu K, Sze SCW, Zhang KY (2015) Evaluation of chemical constituents and important mechanism of pharmacological biology in *Dendrobium* Plants. Evid Based Complement Alternat Med 2015:841752. https://doi.org/10.1155/2015/841752. Published online 2015 Apr 6
30. Khan H, Tarun Belwal M, Tariq M, Atanasov AG, Devkota HP (2019) Genus Vanda: A review on traditional uses, bioactive chemical constituents and pharmacological activities. J Ethnopharmacol 229:46–53
31. Abbas B, Dailami M, Listyorini FH, Munarti (2017) Genetic variations and relationships of Papua's endemic orchids based on RAPD markers. Nat Sci 9(11):377–385. https://doi.org/10.4236/ns.2017.911035377
32. Kasutjianingati K, Firgiyanto R (2018) Characterization of Morphology from Orchid Vanda sp. as a Genetic Information Source for Preservation and Agribusiness of Orchids in Indonesia. In IOP Conference Series: Earth and Environmental Science (207(1):012006). IOP Publishing
33. Lim SH, Teng PC, Lee YH, Goh CJ (1999) RAPD analysis of some species in the genus *Vanda* (Orchidaceae). Ann Bot 83:193–196
34. Lo SF, Nalawade SM, Mulabagal V et al (2004) *In vitro* propagation by asymbiotic seed germination and 1,1-diphenyl-2-picrylhydrazyl (DPPH) radical scavenging activity studies of tissue culture raised plants of three medicinally important species of *Dendrobium*. Biol Pharm Bull 27(5):731–735. https://doi.org/10.1248/bpb.27.731
35. Luo AX, He XJ, Zhou SD, Fan YJ, He T, Chun Z (2009) *In vitro* antioxidant activities of a water-soluble polysaccharide derived from *Dendrobium nobile* Lindl. extracts. Int J Biol Macromol 45(4):359–363. https://doi.org/10.1016/j.ijbiomac.2009.07.008
36. Wang J-H, Luo J-P, Zha X-Q, Feng B-J (2010) Comparison of antitumor activities of different polysaccharide fractions from the stems of *Dendrobium nobile* Lindl. Carbohydr Polym 79(1):114–118. https://doi.org/10.1016/j.carbpol.2009.07.032
37. Yang L, Qin LH, Bligh SWA et al (2006) A new phenanthrene with a spirolactone from *Dendrobium chrysanthum* and its anti-inflammatory activities. Bioorg Med Chem 14(10):3496–3501. https://doi.org/10.1016/j.bmc.2006.01.004
38. Zhang Y, Liu S, Che Y, Liu X (2007) Epicoccins A-D, epipolythiodioxopiperazines from a Cordyceps-colonizing isolate of *Epicoccum nigrum*. J Nat Prod 70(9):1522–1525. https://doi.org/10.1021/np070239u
39. Chen RM, Chen TG, Chen TL et al (2005) Anti-inflammatory and antioxidative effects of propofol on lipopolysaccharide activated macrophages. Ann N Y Acad Sci 1042:262–271
40. Song JI, Kang YJ, Yong HY, Kim YC, Moon A (2012) Denbinobin, a phenanthrene from *Dendrobium nobile*, inhibits invasion and induces apoptosis in SNU-484 human gastric cancer cells. Oncol Rep 27(3):813–818
41. Ye Q, Qin G, Zhao W (2002) Immunomodulatory sesquiterpene glycosides from *Dendrobium nobile*. Phytochemistry 61(8):885–890. https://doi.org/10.1016/s0031-9422(02)00484-3
42. Charoenrungruang S, Chanvorachote P, Sritularak B, Pongrakhananon V (2014) Gigantol, a bibenzyl from *Dendrobium draconis*, inhibits the migratory behavior of non-small cell lung cancer cells. J Nat Prod 77(6):1359–1366. https://doi.org/10.1021/np500015v

43. Chen C-C, Huang Y-L, Teng C-M (2000) Antiplatelet aggregation principles from *Ephemerantha lonchophylla*. Planta Med 66(4):372–374. https://doi.org/10.1055/s-2000-8553
44. Okada Y, Miyauchi N, Suzuki K et al (1995) Search for naturally occurring substances to prevent the complications of diabetes. II. Inhibitory effect of coumarin and flavonoid derivatives on bovine lens aldose reductase and rabbit platelet aggregation. Chem Pharm Bull 43(8):1385–1387. https://doi.org/10.1248/cpb.43.1385
45. Pan L-H, Li X-F, Wang M-N et al (2014) Comparison of hypoglycemic and antioxidative effects of polysaccharides from four different *Dendrobium* species. Int J Biol Macromol 64:420–427
46. Gong CY, Yu ZY, Lu B et al (2014) Ethanol extract of *Dendrobium chrysotoxum* Lindl ameliorates diabetic retinopathy and its mechanism. Vasc Pharmacol 62(3):134–142. https://doi.org/10.1016/j.vph.2014.04.007
47. Simmler C, Antheaume C, André P, Bonté F, Lobstein A (2011) Glucosyloxybenzyl eucomate derivatives from *Vanda teres* stimulate HaCaT cytochrome c oxidase. J Nat Prod 74:949–955
48. Vijaykumar K (2013) *In vitro* anti-oxidant activity of pet-ether extract of *Vanda tessellata* Roxb. Int Ayur Med J 1:1–4
49. Thaakur SR, Pokkula S (2013) Ameliorative effects of *Vanda testacea* in sciatic nerve transection-induced neuropathy in rats. Int J Pharm Bio Sci 4:271–284
50. Andre P, Archambault JC, Renimel I (2011) Use of an extract of the orchid Vanda coerulea as a skin hydrating agent. US Patent 8,039,028
51. Bonté F, Simmler C, Lobstein A, Pellicier F, Cauchard JH (2011) Action d'un extrait de *Vanda coerulea* sur la sénescence de fibroblastescutanés. In Annales pharmaceutiques francaises (69 (3):177–181). Elsevier Masson
52. Dasari R, Sathyavathi D, Belide SK, Soumy BR (2013) Pharmacological evaluation for antidepressant activity of *Vanda spathulata* in mice. Int J Pharm Bio Sci 4:866–872
53. Anwar M, Kumar SN, Mahendran B (2013) Hepatoprotective activity of pet-ether extract of *Vanda tessellata* Roxb. Int Ayur Med J 1:1–4
54. Kumar PS, Subramoniam A, Pushpangadan P (2000) Aphrodisiac activity of *Vanda tessellata* (Roxb.) Hook exdon extract in male mice. Indian J Pharm 32:300–304
55. Khan H, Marya, Belwal T, Tariq M, Atanasov AG, Devkota HP (2019) Genus *Vanda*: a review on traditional uses, bioactive chemical constituents and pharmacological activities. J Ethnopharmacol 229:46–53

Preferences of Orchid Consumers and the Substitute Products' Influences

12

Adilson Anacleto and Luciane Scheuer

Contents

Abstract

The global floriculture market is very varied, but it is worth mentioning the orchid trade, being among the most important species of this market. Despite the economic relevance of the orchids, the consumer preferences about these flowers are unknown. Thus, an exploratory-descriptive survey was conducted between August and December 2019 with 150 consumers, in order to investigate the profile and behavior of this kind of consumer. The study found that women had a higher average purchase per year than men. As consumers advanced their education, there was a tendency to increase the consumption. Roses were the main choice of the consumers when they did not buy orchids, and the main substitute products for orchids were perfumes and chocolates. Further studies on the use of the orchid as a means of seduction are recommended, given the consumer's availability in these cases to pay for the desired orchid up to 50% more than the market price.

Keywords

Flowers · Floriculture · Orchidaceae · Marketing · Relationship marketing · Gardens · Ornamental plants · Home decor · Orchidists · Flower retail market

A. Anacleto (✉) · L. Scheuer
State University of Paraná, Paranaguá, Brazil
e-mail: adilson.anacleto@unespar.edu.br; luciane.scheuer@unespar.edu.br

J.-M. Mérillon, H. Kodja (eds.), *Orchids Phytochemistry, Biology and Horticulture*, Reference Series in Phytochemistry, https://doi.org/10.1007/978-3-030-38392-3_9

1 Introduction

The floriculture market in the world has revenues of approximately US $ 67.3 billion per year and is expected to grow by an average around 5% in countries where this market is already consolidated, and up to 12% per year in developing countries, especially in Latin America and Asia [1, 2].

The flowers trade other than traditional ones has revealed a new market configuration, which is expected to boost the market over the next decade, with Europe being the main destination, representing the largest share ($n = 30\%$) of the global floriculture market [2, 3].

The global floriculture market is very varied, but it is important to mention the orchid trade, which generates a marketed amount close to US$ 20 billion per year, being among the most important species of this market [4].

Despite the economic relevance of orchids to the global floriculture market, Anacleto et al. [1] report that consumer preferences are unknown about the most flower species, and for orchids this lack of information is even greater given the economic importance of this group of flower species.

According to Anacleto et al. [1] and Bornancin et al. [3] in a globalized scale of trade experienced by the market today, it is essential to know the factors that influence consumer preference, and then delimit strategic actions aimed at the market positioning, resulting in greater trading capacity, expanding the presence of the product, and consequently increasing the consumption.

According to Hirschman [5], the consumption is a continuous process and goes far beyond exchanging a financial amount for a good. This process involves issues that influence the consumer before, during, and after the purchase. According to Kotler et al. [6], the consumption is linked to the marketing, which is a social and managerial process whereby individuals and groups get what they need and want by creating, offering, and freely exchanging valuable products and services with other individuals. The domain of the consumption is when people and objects are brought into contact, acquiring sense, producing social meanings and distinctions.

Solomon [7] explains that when talking about consumption, not only tangible products are cited but also intangible experiences, ideas, and characteristics, many consumer experiences, such as fantasies, feelings, and fun, are behind the buying decisions and are important considerations for the consumption phenomenon.

For Churchill et al. [8], consumers are people who buy goods and services for themselves or others, and every buying process must first have passed for the recognition of a need. The needs may have different stimuli as well as the inner sensations, which are characterized by desires. When the need arises inside the consumer, the impulse to meet it is called motivation. Anacleto et al. [1] describes that the motivation in the flower consumption is a relevant factor to the success of this type of product.

The flower trade has been growing all over the world, but the demand for flowers is still quite irregular and concentrated on the holidays, so it is strongly related to seasonality. So, the growth of flower consumption is highly dependent on the country economic development and the cultural increment, and according to

Churchill et al. [8], it is essential to know the consumers, identify their profile, their preferences and the factors that influence the decision at the time of purchase, as well as understand the perceptions of consumers in front of many possibilities available [8].

Thus, given this context, the research results are presented, which aimed to answer the following questions:

(i) What were the factors that determined the profile and behavior of orchid consumers?
(ii) What products were the main competitors of the orchids in the market?
(iii) What were the main reasons for buying orchids?
(iv) What was the client's willingness to pay more when they found the orchid they wanted?

2 Material and Methods

An exploratory-descriptive survey was conducted between August and December 2019 with 150 consumers who traditionally promoted the purchase of orchids for consumption or for gift giving. The research met the marketing research guidelines related to the consumer preferences, considering also when the population is unknown as proposed by Malhotra [9], and the sample was directed to 150 consumers.

The sampling as proposed by Anacleto et al. [1] required that the consumer had bought orchids at least once in the last 6 months and had agreed in participating in an unidentified research.

The data analysis sought to identify the existence of correlation between the annual consumption averages of orchids among the investigated class considering the following variables: gender, education, income, age, and marital status that were considered as the explanatory factors [1, 9]. The qualitative variables were characterized through absolute and relative frequencies (%) and quantitative variables through the average and standard deviation.

The evaluation of the influence of gender, age, education, and economic condition on the orchid consumption levels was evaluated according to Anacleto et al. [1]. For this, the nonparametric Mann–Whitney and Kruskal–Wallis tests were applied, followed by the multiple comparison test of Dunn averages, at a significance level of 5% ($p < 0.05$), in order to identify the pairs of groups with statistically significant differences [10].

The measurement of the most important species, the factors that limited the increase of orchid consumption, and the influence of substitute products were obtained by calculating the relevance index, obtained from the sum resulting from the hierarchical categorization that allowed each interviewee to issue three indications categorizing the most important with 3 (three) and the least important with 1 (one) among the range of alternatives which were highlighted as the most important ones.

3 Results and Discussion

The research related to the orchid consumption revealed that most consumers were women ($n = 58\%$), and when compared to males, they had higher buying frequency for both their own consumption and gift giving (Mann–Whitney Test = $p < 0.001$) (Table 1).

The predominant age group of purchases (70.1%) was between 20 and 40 years old, and no statistically significant correlation was observed between age and orchid consumption, as well as no significant variations related to marital status of the consumers were observed. In both cases, there were no significant differences in the frequency of orchids purchase for their own use or for gift giving (Table 2).

Regarding education, as schooling progressed, there was a slight tendency to increase consumption, both in purchases for own use and for gift (Table 3).

Related to the monthly family income, significant statistical differences were observed in the number of orchids purchased per year, both for use ($p = 0.080$) and for gift giving ($p = 0.005$), having a correlated relation to the increase in the number

Table 1 Gender comparison on the number of times consumers have bought orchids in the last 12 months ($N = 150$)

		Annual frequency of buying orchids (average ± standard deviation)	
		Own use	For gift
Evaluated criterion	Total of respondents	Average ± standard deviation	Average ± standard deviation
Female	87	$2{,}72 \pm 3{,}46^{a}$	$2{,}37 \pm 2{,}14^{a}$
Male	63	$1{,}29 \pm 2{,}36^{b}$	$2{,}43 \pm 1{,}77^{b}$
Mann-Whitney Test p – Significance value		$p < 0{,}001$	$p = 0{,}513$

Averages followed by the same letter do not differ statistically

^{a,b,c}equal letters do not differ statistically from each other (p > 0.05 in the Mann-Whitney Test)

Table 2 Comparison by marital status on the number of times consumers have bought orchids in the last 12 months ($N = 150$)

		Annual frequency of buying orchids (average ± standard deviation)	
		Own use	For gift
Evaluated criterion	Total of respondents	Average ± standard deviation	Average ± standard deviation
Married	73	$2{,}11 \pm 3{,}31^{a}$	$2{,}09 \pm 1{,}79^{a}$
Single	56	$2{,}20 \pm 2{,}38^{a}$	$2{,}57 \pm 2{,}14^{a}$
Divorced	9	$2{,}22 \pm 3{,}42^{a}$	$2{,}14 \pm 1{,}91^{a}$
Widow(er)	8	$2{,}29 \pm 2{,}37^{a}$	$1{,}82 \pm 1{,}99^{a}$
Kruskal–Wallis Test		$p = 0{,}028$	$p = 0{,}070$

A average, *SD* standard deviation, *p* significance value, *a* groups with no significant differences ($p > 0.05$ in the multiple comparison test by Dunn)

Table 3 Comparison by level of education on the number of times consumers have bought orchids in the last 12 months ($N = 150$)

Evaluated criterion	Total of respondents	Annual frequency of buying orchids (average ± standard deviation)	
		Own use	For gift
		Average ± standard deviation	Average ± standard deviation
Elementary school	17	1,61 ± 1,99[a]	1,72 ± 1,08[a]
High school	68	1,73 ± 1,91[a]	1,79 ± 1,08[a]
University graduation	50	2,36 ± 2,69[b]	2,02 ± 1,69[b]
Postgraduate	15	3,55 ± 2,99[c]	3,23 ± 1,70[c]
Kruskal–Wallis Test		$p = 0{,}032$	$p = 0{,}025$

p significance value of Kruskal–Wallis Test
[a,b,c]Equal letters do not differ statistically from each other ($p > 0.05$ in the multiple comparison test by Dunn)

Table 4 Comparison by the degree of monthly family income on the number of times consumers have bought in the last 12 months ($N = 150$)

Evaluated criterion	Total of respondents	Annual frequency of buying orchids (average ± standard deviation)	
		Own use	For gift
		Average ± standard deviation	Average ± standard deviation
Up to US$ 329	22	1,74 ± 1,87[a]	1,69 ± 1,34[a]
From US$ 330 to US$ 549	47	2,01 ± 1,90[a]	1,99 ± 1,54[a]
From US$ 550 to US$ 975	50	2,55 ± 2,34[b]	2,34 ± 1,72[b]
From US$ 976 to US$ 1902	20	4,31 ± 2,88[c]	3,54 ± 1,70[c]
Over US$ 1903	11	5,30 ± 2,60[c]	4,59 ± 1,88[c]
Kruskal–Wallis Test		$p = 0{,}080$	$p = 0{,}005$

p significance value of Kruskal–Wallis Test
[a,b,c]Equal letters do not differ statistically from each other ($p > 0.05$ in the multiple comparison test by Dunn)

of orchids purchased per year according to the increase in monthly family income (Table 4).

Mother's Day and Valentine's Day were the main dates of the year (Table 5) when there was greater demand for orchids, similar to data reported in other studies [1], but the study revealed that the use of flowers in the seduction processes and birthdays are also periods that motivate orchid consumers to make new purchases.

Regarding the main flowers found in the market and which competed in the flower shops with the orchids (Table 6), it was observed that roses were the main

Table 5 Top dates when consumers used to buy orchids

Classification	Dates	Relevance index
1	Mother's Day	236
2	Valentine's Day	122
3	Seduce a person	82
4	Celebrate birthdays	49
5	Woman's Day	47
6	Own use (decoration)	47
7	All Soul's Day	26
8	Graduation Tribute	16
9	Wedding	15
10	Gifts for various special dates	10

Multiple choice questions

flowers in the consumer's choice. However, studies developed by Anacleto et al. [1] analyzing the flower sales on the shelves of the flower shops, they found that orchids are the flower group favored by consumers for their own use and they are the second option for gift.

The consumer's decision about the flower to be purchased, and whether or not to buy it, depends on factors such as value, species, color, size, payment terms, durability, discounts obtained and especially the easy access to the product, where and when buying the orchids. The orchid consumption tends to be made easier if this set of factors gives the consumer the sense of a routine decision.

The main substitute products for orchids were perfumes and chocolates (Table 7) but other flower species were also considered by consumers as substitutes. The absolute majority of the respondents ($n = 83\%$) admitted that when they did not find the desired orchid for his own use, he/she searched nearby and considered buying a replacement product. However, when the acquisition was for gift, the interviewees ($n = 54\%$) admitted that they searched in up to three different flower shops or did previous research on the Internet in order to facilitate the purchase.

Although the study on the orchid consumer profile reveals data similar to that already reported by Anacleto et al. [1] related to the consumption levels by age, marital status, and education, this study also revealed a greater influence of the substitute products in relation to the orchids, as well as a high willingness of the customers in replacing the orchids if they could not find the desired flower when it was for their own use.

The substitute products for flowers have presented better performance in terms of global market in relation to the flowers. Anacleto et al. [1] describe that the perfume and chocolate industries are better structured and promote massive investment in marketing, especially on seasonal dates when gift giving is a habit. The clothes have a high level of ability to act as a substitute product, in part due to the fact that in the daily life, clothes are more easily perceived by the consumer, and it is easier to get indications from the people close to person that will receive the gift, getting easier to please the receiver [6], as well as the clothes come in a variety of types, shapes, and with wide price range.

Table 6 Flower species preferred by consumers ($N = 150$) when they do not buy orchids

Classification (for gift)	Flower species	Relevance index	Classification (own use)	Flower species	Relevance index
1	Rose	232	1	Rose	172
2	Violet	53	2	Tulip	44
3	Lisianthus	26	3	Cyclamen	28
4	Cyclamen	20	4	Lisianthus	24
5	Astromeliad	15	5	Violet	18
6	Tulip	15	6	Bromeliad	16
7	Bromeliad	10	7	Astromeliad	15
8	Begonia	9	8	Begonia	12
9	Daisy	9	9	Gerbera daisy	11
10	Carnation flower	8	10	Dahlia	8

Multiple choice questions

Table 7 Main products to replace orchids when the consumer did not find the desired orchid ($N = 150$)

Classification	Substitute products	Relevance index
1	Perfumes	131
2	Chocolates	114
3	Other flower species	103
4	Clothes	26
5	Wines	25
6	Books	23
7	Cosmetics/makeup	22
8	Stuffed animals	14
9	Semi jewelry or imitation jewelry	14
10	Decoration items/ornaments	12

Multiple choice questions

The trade of orchids must consider two factors, if the level of desire and will of the customer in having the product is high and if the accessibility to the product is facilitated, because if the consumers can easily purchase the product of their need, naturally occurs the predisposition to pay the values asked when he/she finds the flower object of his/her desire, otherwise the space of the substitute products increases.

The substitute products, according to Porter [11], can be classified as those that can replace the desired product targeted by the consumer under free market conditions. The biggest part of the consumers, facing a great desire of having a product, have a price limit or purchase effort that they are willing to pay in order to access that product. The substitute products reduce the potential returns of a good to a producer or manufacturer, because the more attractive the price in relation to the performance of the substitute product, the greater the pressure on the product profits that can replace them in the case of orchids, as well as in the consumer decision-making process.

Most customers, when faced with conditions that make it difficult to buy any product, tend to consider replacing it, and according to Kotler et al. [6], this pre-purchase tension depends intrinsically on the main reasons that drive the consumer to buy. The process of buying orchids depends basically on the internal or external motivational factors that lead the consumer to search for the product. After this phase of the buying process, there is a decision whether or not to purchase the product.

A product can be classified as a substitute for another when both can perform the same function, similar function or even cause a similar effect as desired by the consumer, such as orchids as internal decoration in homes, if they have high cost, other lower priced species may have consumer preference as it would perform the same function in the residence decorating. The similar function would occur if the consumer in this case made the purchase of plastic flowers, and the similar effect could be obtained by the consumer with the purchase of another product, as the consumer wishes to provide the residence decoration. All commercial products, in general, have a substitute product, which are manufactured or produced by other companies. The ability of a substitute product for orchids and their efficiency are linked to satisfaction and the sense of pleasure or disappointment that results of the expected performance of the replacement product in relation to the expectations of the orchid consumer, which will impact in the decision process of the consumer in the future.

Most of the time, the decision-making process is imperceptible to the consumer, who first feels the desire to have the flower due to the visual, aesthetic, and/or smell interest, and then recognizes the need to own it.

The need to buy flowers in a general context according to Muraro et al. [12] and Anacleto et al. [1] originate from several motivational factors but they point out that the consumer is inserted in which people tend to seek the behavior of repetition of the collective habit in the case of decoration with the use of flowers in their homes or workplaces; however, Kotler et al. [6] describes that the need and desire for consumption may be linked to situation came from childhood memories, loved ones, psychological needs in relation to the environment, or factors such as social and cultural influence.

People make routine decisions every day, and they become consumers as they acquire customary things without having to think and analyze the purchase process more deeply. But the orchid consumption is classified as an extensive decision which degree of difficulty requires the consumer take a longer time to make the decision, which can go from minutes to hours; in a general context, the more relevant the extensive decision, the longer the analysis time. The orchids are classified in this group of purchases as they are not essential products, so the consumer should consider whether they can afford those resources to buy or not.

Orchid consumers, in a general context, showed high price sensitivity when used for their own use; the predisposition in this case to pay more expensive averaged an increase of 5.94% of the market value of the flower, and if the cost was higher for the desired flower, this would be replaced by other products. However, the consumers were less sensitive to the price when buying for the purpose of seducing a person or when orchids were the chosen form of gift for Valentine's Day; in both the cases, they were more likely to spend up to eight times higher than those compared to flowers purchased for their own use (Table 8).

Specifically, in the case of orchids that if the importance of nonmonetary values is increased, the wide range of substitute products disposed to trade tend to have less power of influence for the consumer, and this perception is more pronounced when buying as a gift when cost is less relevant to the consumer.

The preference for consumption of an orchid type can be determined by two sets of factors, first of all by the intrinsic factors of each individual that are able to make the purchase decision; in this respect, it is highlighted the set of feelings, thoughts, and factors that the individual who will give the gift wants to provoke in the receiver, either by remembering colors or smells, which tied to the various types of arrangements, exposure, price, and others which can influence the internal decision-making process and result in an easier behavior of buying. In this case, orchids may have a strong appeal over competing products or even other competing flowers since the orchids have a strong relationship to two relevant organoleptic properties in the

Table 8 Value that customers are predisposed to overpay for the desired orchid according to motivation, need and desire ($N = 150$)

Classification	Dates	How much would pay more for the desired orchid (%)
1	Loving seduction	52.5
2	Valentine's Day	34.8
3	Celebrate Birthday	32.0
4	Mother's Day	20.1
5	Woman' Day	20.0
6	All Soul's Day	10.1
7	Graduation Tribute	7.02
8	Wedding	6.60
9	End of the year	6.23
10	Own use (decoration)	5.94

Multiple choice questions

buying process, especially smell and color, which may arouse the consumer's unconscious to the consumption preference.

Effectively a person's buying action depends on the motivation he/she receives, as well as the perception of a possible internal situation, and whether this process will make him/her take action to seek this satisfaction of desire. The perception combines sensation with the meaning that a previous experience attributes to each individual, yet it depends on one's memory and thought, and the symbolic values that permeate it. According to Schewe and Smith [13], the symbolic perception is a process by which we value what individuals feel, and that value influences and adds to our sensory impressions bringing our past experiences to influence them by giving meaning of completeness and continuity, which were constructed from fragmentary stimuli collected by the sense organs. The sensation depends on the stimulus and the ability of the each individual to record it in previous events that involved the same stimulus and that will affect the interpretation of the sensation by the brain, and thus influencing the consumer's behavior in a situation of a revived sensation, and it is in this context that occurs the motivation for buying which is the inner impulse to fulfill the need.

The substitute products may have reduced strength, and one way to achieve this is by building sales relationships between sellers and buyers. The flower merchants, especially those who have direct contact with the end consumers, develop relationships that bring many benefits to the companies, and then the buyers believe in the sellers, considered partners by establishing a lasting relationship that generates a competitive advantage. Such partnerships between establishments and consumers generate win-win relationship, resulting in what both parties seek; for Kotler et al. [6], the key to reduce the power of the substitute products, especially in the case of nonessentials such as orchids, is the relationship marketing that is a developmental process that must happen in these cases. First of all, the company has to identify the potential customers who are all people able to consume its products. Once identified, the next step is to identify potential customers by analyzing their purchase profile, the interest by the product and whether they can afford it and if they can promote repeated purchases.

Identifying regular and prospective customers is important in the process, because precisely this class of customers can make purchases of substitute products, so the differentiated service can reduce this impact then these regular customers can become preferred customers, who are those who have identification with the product and recommend them to other ones [6].

In this context, the more intensive the relationship marketing, the greater the number of loyal orchids customers and, therefore, the higher the consumption. Companies typically have in their records the market share and the number of potential customers. Within the market, there is a subgroup of clients that have different needs, and for the company to know these clients better, it is necessary to collect more data, through the these information collected, it is possible to identify the profile of the clients, which can be separated by: gender, education, marital state, and with this information, it is possible to identify which type of orchid will be targeted to meet the wants and needs of each individual or each client group.

Fulfilling the consumers' desires in relation to the product is an important factor in the loyalty of a product consumption, Kotler et al. [6] describe that the satisfaction of the consumer desires can be understood as the feeling of pleasure or disappointment resulting from the comparison of the expected performance of the product (or result) in relation to the expectations of the person.

The consumers when looking for a product to buy mentally form an expectation about the quality of this product. The expectations are based on the needs and desires of each individual, his/her past personal experiences, and the influence of others. Then after consumption, the consumer makes a comparison between what he/she wanted and what he/she actually got after purchase, as also described by Bornancin et al. [3], if the performance exceeds expectations, the consumer will be highly satisfied, leading to the possibility of new consumption and the customer loyalty in the orchid consumption.

The relationship marketing actions point to a very viable form of customer loyalty, it also points to organizations the advantages of retaining their current customers over gaining new customers, due to it is usually cheaper to keep customers on track then to conquer new ones. In addition, customer loss can be disastrous, so customer loyalty based on genuine and continuing satisfaction of desires and needs is one of the greatest assets that a flower-trading company can acquire.

4 Conclusion

The study revealed that women, in relation to orchid consumption, had a higher average purchase per year than men.

The predominant age group of purchases (70.1%) was between 20 and 40 years old, but with no difference in consumption between age groups.

It was also observed that as consumers advanced their educations level, there was a slight tendency towards increasing the consumption.

Roses were the main flowers in the choice of consumers when they did not buy orchids, and the main substitute products for the orchids were perfumes and chocolates.

The study showed that there is an increase in the number of orchids purchased per year according to the increase in monthly family income. In this context, orchids, due to the high cost, when compared to the decision-making factors for purchase apparently fits into nonmonetary valuation decision-making processes, so this concept can be realized as, with the exception of roses, all other flowers classified as competitor of the orchids on the shelves of the flower shops have much lower purchase costs to the consumer. Thus, considering that in some cases, this cost reaches less than 50% of the value of an orchid, apparently the nonmonetary appreciation of orchids may be constituting one of the most important factors in the consumer's perception for the motivation to buy. This tendency should be used as a consumer loyalty alternative by relationship marketing to be developed by merchants with their consumers.

Finally, it is recommended to carry out further studies regarding the use of the orchid as a means of seduction between people. It is important to mention that in the seduction processes, the extensive decision has less influence, and the price ceases to be the main issue in the consumption preference and consumer willingness to pay for the orchid he/she wants can reach 50% of the market price.

References

1. Anacleto A, Negrelle RRB, Cuquel FL, Muraro D (2017) Profile and behavior of flower consumer: subsidies for marketing actions. Ceres 64(6):557–566. https://doi.org/10.1590/0034-737x201764060001
2. Tathe S (2019) Global floriculture market research gain impetus due to the growing demand over 2019–2028 with a CAGR of 5.00%. Relatório Técnico
3. Bornancin AAP, Anacleto A (2018) Perfil e comportamento do consumidor de bromélias: orientação a produção rural. Revista Brasileira de Planejamento e Desenvolvimento 7(1):51–66. https://doi.org/10.3895/rbpd.v7n1.7055
4. ITC, International Trade Centre (2019) Market dynamics, annual report. Disponível em http://www.intracen.org. Acesso em 05/11/2019
5. Hirschman E, Holbrook M (1982) Hedonic consumption: emerging concepts, methods and propositions. J Mark 26 (4 ed):138–149. https://doi.org/10.1177/002224298204600314
6. Kotler P, Keller KL, Ancarani F, Costabile M (2014) Marketing management, 14th edn. Pearson, Upper Saddle River
7. Solomon MO (2002) Comportamento do consumidor: comprando, possuindo, sendo, 5th edn. Porto Alegre, Bookman
8. Churchill GA, Peter JP (1995) Marketing: creating value for customers. Irwin, Boston
9. Malhotra NK (2010) Marketing research: an applied orientation. Pearson Education Australia, Boston
10. Hair JF, Black WC, Babin BJ, Anderson RE, Tatham RL (2009) Análise multivariada de dados. Bookman, Porto Alegre

11. Porter ME (2008) Competitive strategy: techniques for analyzing industries and competitors. Simon and Schuster, New York
12. Muraro D, Negrelle RR, Cuquel FL, AnacletoA (2016) Market management: the impact on the development of an ornamental plants supply chain in Curitiba, Brazil. Ciencia e Investigación Agraria 42(3):453–460. https://doi.org/10.4067/S0718-16202015000300013
13. Schewe CD, Smith RM (1979) Marketing: concepts and applications. McGraw-Hill, New York

Part IV

Agri-food Applications

Vanilla: Culture, Reproduction, Phytochemistry, Curing, Pest, and Diseases

13

Keshika Mahadeo, Tony L. Palama, Bertrand Côme, and Hippolyte Kodja

Contents

K. Mahadeo (✉)
Qualisud, Université de La Réunion, Université Montpellier, CIRAD, Institut Agro, Avignon Université, Montpellier, France

Laboratoire de Chimie des Produits Naturels, Université de la Réunion, Faculté des Sciences et Technologies, 15 Avenue René Cassin, CS 92 003, 97 744 St Denis Cedex 9, La Réunion, France
e-mail: keshika.mahadeo@univ-reunion.fr

T. L. Palama
Université Sorbonne Paris Nord, Laboratoire de Chimie, Structures, Propriétés de Biomatériaux et d'Agents Thérapeutiques, CSPBAT, CNRS, UMR 7244, Villetaneuse, France
e-mail: tony.palama@univ-paris13.fr

B. Côme
La Vanilleraie, Sainte-Suzanne, La Réunion, France
e-mail: la.vanilleraie@orange.fr

H. Kodja
Qualisud, Université de La Réunion, Université Montpellier, CIRAD, Institut Agro, Avignon Université, Montpellier, France
e-mail: hippolyte.kodja@univ-reunion.fr

J.-M. Mérillon, H. Kodja (eds.), *Orchids Phytochemistry, Biology and Horticulture*, Reference Series in Phytochemistry, https://doi.org/10.1007/978-3-030-38392-3_13

Abstract

This chapter describes relevant information on vanilla (*Vanilla planifolia*) including the plant description, flowering, and phytochemistry of green pods and leaves. The harvesting and post-harvesting processes are also described along with the cultivation methods. The chapter closes by looking at fungal and viral diseases of vanilla plants.

Keywords

Vanilla · *Vanilla planifolia* · Vanilla cultivation · Curing process · Vanilla diseases

1 Introduction

Vanilla is considered to be the best contribution of the Americas to the world of flavors. The orchid *Vanilla planifolia* is indigenous to Central America. In Mexico, vanilla fruit was called "tlilxochitl," which comes from the words "tlilli" and "xochitl" and means "black" and "pod," respectively [1, 2]. When the Spaniards discovered America, they called it "vanilla" which comes from the Spanish word "vaina" meaning "pod." The Latin word "planifolia" described the flatness ("plani") of the leaves ("folia") [3].

The genus *Vanilla* belongs to the largest family of flowering plants, the Orchidaceae. Among the 800 genera and the 25,000 species within the family, the genus *Vanilla* contains by itself 110 species [4]. Among the 110 species, only 15 species develop an aromatic fruit. *V. planifolia*, *V. tahitensis*, and *V. pompona* (Fig. 1) are the only three species cultivated and commercialized for their relevant flavors

Fig. 1 Cultivation of *V. tahitensis* (left), *V. planifolia* (middle), and *V. pompona* (right). *V. tahitensis*: thin and elongated leaves; *V. planifolia*: flat and large leaves; *V. pompona*: the leaves exhibit dimorphism (large and small leaves)

and aroma compounds. However, the vanilla market is dominated by Bourbon like vanilla (*V. planifolia*). It represents more than 95% of the world production of vanilla.

2 Cultivation Methods

Vanilla is a tropical plant that grows best in warm and moist climate. Therefore, natural growth is obtained between latitude 15° north and 27° south on all continents except Australia [1, 3, 5]. The optimal conditions for Vanilla plants growth are temperature comprised between 21–32 °C, while precipitation required falls between 2000 and 2500 mm per year [1].

Vanilla is a hemi-epiphytic orchid that needs a tree to provide physical support, shade, and organic material. The cultivation is performed through three cultural practices: in forest-type land, intensively in deforested land, and in shade houses (Fig. 2).

In forest-type land, trees are used as tutors that support the vanilla plants. The aerial roots help the vines to climb on the trees to reach the canopy in order to accumulate the maximum of sunlight before flowering. The aerial roots also contribute to absorb water from the air humidity. Ideal soil for vanilla is light, rich in humus and porous allowing the roots to spread without molds development [1, 3]. The plant nutrition comes from the roots found in the soil that feed on decaying organic matter (decomposition of dead leaves from the tutor's trees). A relatively low density of vanilla plants is cultivated in forest-type land leading to low yields of green pod/ha.

In contrast, the production of vanilla in deforested land is an intensive system. This cultural practice consists of planting support trees with sufficient distance for ventilation between the trees in order to avoid molds development. After a year,

Fig. 2 Cultivation of vanilla in forest-type land (left), deforested land (middle), and in shade houses (right)

when there is sufficient shade (60%), the vanilla is planted. The plant feeding is ensured by input of sugar cane mulch, coconut fiber, and organic compost.

The cultivation in shade houses is another intensive system of vanilla production. It is performed in semi-controlled conditions with shade accumulation of 60%. In this intensive system, dead wood trunks are used as tutors and the plant feeding is provided by natural organic compost.

Those two last systems of vanilla cultivation have the advantage of relatively high yields.

Vanilla vines grow around 10 m/year. Crops are established from cuttings of vigorous vines. Higher temperatures or insufficient shading can cause the death of the plant. On the contrary, excessive humidity and shading can induce susceptibility of the plant to diseases (mildew and root rot) [1]. A water stress of about 50 days and sunlight exposure are necessary to induce flowering.

3 Flowering

In general, the first flowers appear 3 years following planting. The physiological cue to flower is promoted by climatic (dry season) or mechanical stress. In Reunion Island, the flowering season is September to December. The flowers (8–10 flowers) are grouped into an inflorescence located underneath the leaf. During the blooming season, the flowers open a few at a time and lasts a single day (Fig. 3). Vanilla flowers are hermaphroditic. The rostellum, a cap like structure, hangs exactly in between the anther sac (male organ) and the stigma (female organ) and acts as a physical barrier to prevent from self-fertilization. Thus, vanilla flowers are either naturally pollinated by the Melipona bee (found in Mexico), or by hand pollination. *V. planifolia* plants were introduced to Reunion Island in 1822. However, without the specific pollinators on the island, no vanilla pods were produced. The biggest advancement in vanilla growth and pollination happened in 1841, when Edmond Albius, a slave on Reunion Island, discovered a practical method of vanilla hand pollination [6]. Albius discovered that the rostellum could be lifted out of the way (using a small stick), so that the anther sac can hang down unimpeded over the stigma lobes. Now practically all vanilla is produced by hand pollination.

Fig. 3 Flower of *V. planifolia*

Fig. 4 Mature fruit of *V. planifolia*

Immediately after hand pollination, pollen tubes begin their germination and fertilization of the ovules. The ovary begins to enlarge and develops a dark green pod. The maximum length (approximately 20 cm) and diameter of the fruit is achieved within 2 months after pollination. Afterward, the fruit enters into a long period of maturation of 7–8 months. When the fruit is mature, the distal tip of the bean changes color from green to yellow (Fig. 4). This senescence zone is the starting point of the fruit dehiscence that can continue after harvesting. At this stage, the vanilla pod starts to lose their aromatic compounds and flavors. Therefore, all mature fruits are harvested 9 months after pollination before the dehiscence appears.

4 Phytochemistry

The leaves and green pods of *V. planifolia* contain various glucosides including the glucoside A (bis[4-(β-D-glucopyranosyloxy)-benzyl]-2-isopropyltartrate) and the glucoside B (bis[4-(β-D-glucopyranosyloxy)-benzyl]-2-(2-butyl) tartrate) [3, 7–9]. These two glucosides were reported in vanilla leaves and may contribute to discriminate the development stage of the leaves. Young leaves were found to have higher level of glucose, glucoside A, and glucoside B whereas older leaves have more sucrose, acetic acid, homocitric acid, and malic acid [10]. As glucosides A and B are thought to be precursors of glucovanillin, there is maybe a connection between their accumulation in green pods and young leaves.

The ^{1}H NMR metabolomic analysis of developing green pods (between 3 and 8 months after pollination) had showed that younger pods contain more glucose, malic acid, homocitric acid, glucosides A, and glucoside B. In the green mature beans, vanillin, the main flavoring constituent of vanilla is present only in very low trace amounts. However, it is mainly present in the uncured pods in the form of a glucoside called glucovanillin. In a recent study, Odoux and Brillouet [11] unambiguously proved that glucovanillin is stored in the placental laminae (92%) of mature green pods whereas only a small amount of glucovanillin (7%) is present in the trichomes [3]. Compared to younger pods, older pods contain more sucrose, glucovanillin, vanillin, *p*-hydroxybenzaldehyde glucoside, and *p*-hydroxybenzaldehyde [12]. In green mature pods, free vanillin can reach 10–20% of the total vanillin content. Besides, the lipid content of vanilla mature beans have also been investigated and the authors reported various fatty acids such as lauric acid (12:0), myristic acid (14:0), pentadecanoic acid (15:0), palmitic acid (16:0), stearic acid (18:0), oleic acid (18:1 n-9), and linoleic acid (18:2 n-6) [13]. The major fatty acids were found to be oleic acid, palmitic acid, and linoleic acid.

Various other minor glycosides were identified in mature green pods after enzymatic hydrolysis including glucosides of vanillic acid, vanilly alcohol, 4-vinylguaiacol, acetovanillon, caffeic acid, ferulic acid, methyl-3,4-dihydroxycinnamate, 2-methoxy-4-cresol, homovanillyl alcohol, 3,4-dihydroxybenzoic acid, ethyl-4-hydroxy-3-methoxyphenylacetate, *p*-cresol, 4-vinylphenol, methylsalicylate, *p*-hydroxybenzylethyl ether, *p*-hydroxybenzaldehyde, *p*-hydroxycinnamic acid, cinnamic alcohol, cinnamic acid, phenethyl alcohol, 3-phenylpropanol [3, 8].

The phenolic compounds detected in mature pods were mostly present as glucosides, underlining the need of the curing process to ensure the release of the aglycones to increase the vanilla flavor.

5 Curing

After harvesting, mature green pods of *V. planifolia* are almost odorless and they develop their characteristic flavor and aroma after a “curing” process. Several procedures have been developed for curing vanilla. Every vanilla growing country has developed its own curing process, but they all generally involve four common steps: “killing,” “sweating,” “drying,” and “conditioning” [3, 14–18]. The whole process takes about 9–12 months. In course of the “killing” process the natural physiological processes in the harvested beans are stopped and thus avoid any dehiscence of the fruit. This can be carried out by hot water scalding, sun or oven heating, and scarification. Killing methods allow cell structure disruption and thus various enzymes can come into contact with their substrates like glucovanillin. During the killing process, the glucovanillin is hydrolyzed to form vanillin and glucose through the action of a β-glucosidase [17]. The “sweating” step allows the moisture to escape rapidly to reach a level which will avoid microbial spoilage during the subsequent operations. This step is also crucial because it lets indigenous enzymes taking effect to develop the characteristic vanilla aroma and flavor. During the “sweating” step the beans get their characteristic brown color and develop a proper texture and flexibility. They are wrapped up in sheets and put in a container overnight. During the

Fig. 5 Curing steps of vanilla

"drying" step, decrease of the moisture content reduces undesirable enzyme activities and biochemical changes. This step usually lasts for 2–4 weeks and is performed indoor and outdoor. The final step of vanilla curing process is the "conditioning" step. Vanilla pods are stored in closed boxes for several months (6 months) to refine its flavor and develop new compounds from Maillard reactions. The full curing process last at least 6 months but can go for 1 year or more (Fig. 5) [3].

6 Pest and Diseases

Vanilla cultivation tends to develop into a more intensive practice. Thus, occurrence of pest and diseases affecting *Vanilla* is more and more increasing. In the last decades, vanilla growers had to face several fungal and/or virus diseases [3].

6.1 Fungal Diseases

Favorable conditions for the development of fungal diseases are drought, deficient soil drainage, poor ventilation, excessive watering of the roots and over-pollination. Fungal diseases are frequently opportunistic infections caused by poor management practices.

Among them Fusarium, a cryptogamic infection, is the most damaging disease of vanilla. The root and stem rot (RSR) of vanilla is a disease caused by *Fusarium oxysporum f. sp. radicis-vanillae* (Forv) [19]. It is a soil-borne fungus found in all vanilla producing countries causing severe damage and yield losses. In the early stage of the disease, there is browning and death of underground roots, followed by the death of aerial roots. Subsequently, the plant stops its shoot growth suggesting a lack of nutrients and the leaves and stems begin to shrivel [3]. When the fungus infects the plant, it is difficult to eradicate the disease. Chemical fungicides, essential oils (clove and cinnamon oil), or biological control such as *Pseudomonas* and *Trichoderma* are used as control methods for RSR [20]. However, none of these methods were efficient enough to restore productivity of vanilla plots. The best management of the disease can be an approach in which the biological control can play a major role in complement of the varietal selection for resistance. However, cultural practices should be the first methods to be used to prevent the development of the pathogen (equipment in good sanitary conditions, managing water in an appropriate manner and use of land with good drainage, keeping plant well-nourished, avoid overcrowding, and excess shade). Besides, identifying genotypes resistant to Fusarium can also be considered as an alternative [21]. Indeed, several species including *V. pompona*, *V. phaeantha*, and *V. bahiana* showed resistance to the pathogen [2]. Most of the *V. planifolia* accessions, *V. tahitensis* and *V. odorata* were susceptible to RSR. However, Koyyappurath et al. [19] identified new accessions of *V. planifolia* resistant to Fusarium.

Anthracnose, caused by *Colletotrichum* sp., is another fungal pathogen that attacks leaves, fruits, stems, and flowers of vanilla plant. Characteristic of the disease are irregular brown spots on the leaves, shoots, and fruits [22]. The spots develop afterward into elliptic necrotic brown spots (Fig. 6). In general, the symptoms appear on the first young leaves of the apical part of the plant followed by fruit damage during the humid and warm months. Infected fruits fall prematurely before reaching their maturity. Anthracnose occurs on poorly maintained plantations with proper

Fig. 6 Brown spots on vanilla pods caused by *Colletotrichum sp* [23]. Reprinted with permission from John Wiley & Sons

shading control. Indeed, an excess of shade and high density of plants are favorable conditions for anthracnose development. In 2011, *Colletotrichum orchidophilum* has been identified as the causal agent of anthracnose in Reunion Island. The annual pod production was reduced by 10–30% due to this pathogen [23].

Vanilla stems and pods can also be infected by mildew. It generally occurs after a significant rainfall event. The propagation of mildew can be rapid in the field resulting in important loss of mature pods.

In these conditions, infected plants have to be eliminated before spread of the disease to the entire plot. The causal agent of mildew is either *Plasmopara palmivora*, or *Phytophtora*. *Phytophthora jatrophae* is one of the sources of vanilla mildew with direct economic impact since it attacks the developing fruits [1, 2]. The disease starts to one extremity and then spreads to the whole pod. Affected parts of the pods turn into a brown chocolate color. Diseased pods lose their swelling and rapidly fall to the ground [1, 3].

6.2 Viral Diseases

Vanilla is also affected by viral diseases due to intensification of cultivation. The damage caused by viruses can be difficult to distinguish: some plants do not show clear symptoms or are asymptomatic. Cymbidium Mosaic Virus (CymMV, Potexvirus) and Odontoglossum RingSpot Virus (ORSV, Tobamovirus) are reported as the most prevalent viruses having economical incidences [3, 24]. Plants infected by CymMV and ORSV are generally asymptomatic, but occasionally they exhibit mild chlorosis or small spots on the leaves and stems of *V. planifolia* and *V. tahitensis* [25]. These two viruses are transmitted through the sap and are dispersed through contaminated cutting tools. They may also be transmitted by contaminated pollen [26]. Potential destructive impact and rapid spread of this virus made it of interest for vanilla growers. The virus was first reported in South Pacific in French Polynesia. Farreyrol et al. [27] reported for the first time infection of Vanilla by Cucumber Mosaic Virus (CMV, Cucumovirus) in the Indian Ocean in La Réunion. Although the authors did not observe severe symptoms, leaves of CMV-infected plants were slightly elongated. It has since been found in vanilla vines growing in other countries in the Indian Ocean, such as Madagascar and India [3, 28, 29]. Another virus, the Vanilla Mosaic Virus (VanMV, Potyvirus) occurs mainly in the islands of French Polynesia. The virus causes leaf distortion and mosaic lesions in *V. planifolia*, *V. pompana* and *V. tahitensis* vines [30, 31]. This virus may be transmitted by aphids [22]. Other viruses have also been reported to infect vanilla vines: the Watermelon Mosaic Virus (WMV, Potyvirus), Bean Common Mosaic Virus (BCMV, Potyvirus), Bean Yellow Mosaic Virus (BYMV, Potyvirus), Cowpea Aphid-Borne Mosaic Virus (CABMV, Potyvirus), Ornithogalum Mosaic Virus (OrMV, Potyvirus) and Wisteria Vein Mosaic Virus (WVMV, Potyvirus) [3, 32–36].

Nowadays, efficient techniques have been developed to detect virus [33, 35, 37], and virus spread can be control with good practice cultivation [3]. In prevention of virus diseases, it is important to use healthy certified cuttings, to install insect-proof netting around the shade houses to avoid the insect vectors (aphids), to eliminate crops around the plantation that can be reservoirs of the virus and to eliminate diseased vanilla vines as soon as they are detected [22, 38].

7 Conclusion

Vanilla is the principal source of natural vanilla of commerce and it is mainly used in food, perfumery, and pharmaceutical preparations. The quality of the bean depends on several aspects including the species used, the volatile constituents, the curing process adopted, and the vanillin content. Hence, this chapter is elaborated on the cultivation methods of vanilla vines, the curing process, and how to avoid the main pest and diseases of vanilla cultivation.

References

1. Bouriquet G (1954) Le vanillier et la vanille dans le monde. In: Lechevalier P (ed) Encyclopédie biologique, vol XLVI. Editions Paul Lechevalier, Paris VI, p 748
2. Purseglove JW, Brown EG, Green CI, Robbins SRJ (1981) Vanilla. In: Spices, vol 2. Longman Inc., New York
3. Palama TL (2010) NMR-based metabolomic characterization of *Vanilla planifolia*. Doctoral Thesis. Leiden University, Netherlands
4. Bory S, Grisoni M, Duval M-F, Besse P (2008) Biodiversity and preservation of vanilla: present state of knowledge. Genet Resour Crop Evol 55:551–571
5. Soto Arenas MA (2003) Vanilla. In: Pridgeon AM, Cribb PJ, Chase MW, Rasmussen FN (eds) Genera Orchidacearum, vol 3. Oxford University Press, USA, pp 321–334
6. Arditti J, Rao AN, Nair H (2009) Hand-pollination of vanilla: how many discoverers? Orchid Biol: Rev Perspect X:233–249
7. Kanisawa T (1993) Flavor development in Vanilla beans. Kouryou 180:113–123
8. Kanisawa T, Tokoro K, Kawahara S (1994) Flavor development in the beans of *Vanilla planifolia*. In: Kurihara K, Suzuki N, Ogawa H (eds) Proceedings of the International Symposium Tokyo, Japan, pp 268–270
9. Tokoro K, Kawahara S, Amano A, Kanisawa T, Indo M (1990) Glucosides in vanilla beans and changes of their contents during maturation. In: Bessiere Y, Thomas AF (eds) Flavour science and technology. Wiley, Chichester, New York, Brisbane, Toronto, Singapore, pp 73–76
10. Palama TL, Fock I, Choi YH, Verpoorte R, Kodja H (2010) Biological Variation of *Vanilla planifolia* leaves metabolome. Phytochemistry 71(5):567–573
11. Odoux E, Brillouet JM (2009) Anatomy, histochemistry and biochemistry of glucovanillin, oleoresin and mucilage accumulation sites in green mature vanilla pod (*Vanilla planifolia*; Orchidaceae): a comprehensive and critical reexamination. Fruits 64:221–241
12. Palama TL, Khatib A, Choi YH, Côme B, Fock I, Verpoorte R, Kodja H (2011) Metabolic characterization of green pods from Vanilla planifolia accessions grown in La Réunion. Metabolic characterization of green pods from *Vanilla planifolia* accessions grown in La Réunion. Environ Exp Bot 72(2):258–265

13. Garros-Patin J, Hahn J (1954) La chimie de la vanille. In: Le Vanillier et la Vanille dans le Monde
14. Balls AK, Arana FE (1941) The curing of vanilla. Ind Eng Chem 33:1073–1075
15. Dignum MJW, Kerler J, Verpoorte R (2002) Vanilla curing under laboratory conditions. Food Chem 79:165–171
16. Havkin-Frenkel D, French JC, Graft N, Joel DM, Pak FE, Frenkel C (2004) Interrelation of Curing and Botany in Vanilla (*Vanilla planifolia*) Bean. Acta Hortic 629:93–102
17. Odoux E (2000) Changes in vanillin and glucovanillin concentrations during the various stages of the process traditionally used for curing *Vanilla fragrans* beans in Réunion. Fruits 55:119–125
18. Pérez-Silva A, Odoux E, Brat P, Ribeyre F, Rodriguez-Jimenes G, Robles-Olvera V, García-Alvarado MA, Günata Z (2006) GC-MS and GC-olfactometry analysis of aroma compounds in a representative organic aroma extract from cured vanilla (*Vanilla planifolia* G. Jackson) beans. Food Chem 99:728–735
19. Koyyappurath S, Conéjéro G, Dijoux JB, Lapeyre-Montès F, Jade K, Chiroleu F, Gatineau F, Verdeil JL, Besse P, Grisoni M (2015) Differential responses of *Vanilla* accessions to root rot and colonization by *Fusarium oxysporum* f. *sp. radicis-vanillae*. Front Plant Sci 6:1125
20. Tombe M, Liew ECY (2010) Fungal diseases of *Vanilla*. In: Odoux E, Grisoni M (eds) Vanilla. CRC Press, Boca Raton, FL
21. Fravel D, Olivain C, Alabouvette C (2003) Research review: *Fusarium oxysporum* and its biocontrol. New Phytol 157:493–502
22. Havkin-Frenkel D, Belanger FC (2018) Vanilla diseases. In: Handbook of vanilla science and technology. Wiley, Oxford, pp 27–40
23. Charron C, Hubert J, Minatchy J, Wilson V, Chrysot F, Gerville S, Loos R, Jeandel C, Grisoni M (2018) Characterization of *Colletotrichum orchidophilum*, the agent of black spot disease of vanilla. J Phytopathol 166(7–8):525–531
24. Zettler FW, Ko NJ, Wisler GC, Elliott MS, Wong SM (1990) Viruses of orchids and their control. Plant Dis 74:621–626
25. Leclercq-Le Quillec F, Rivière C, Lagorce A (2001) Spread of cymbidium mosaic potexvirus and potyviruses in vanilla plants grown in shade houses in Reunion Island. Fruits 56:249–260
26. Hu JS, Ferreira S, Xu MQ, Lu M, Iha M, Pflum E, Wang M (1994) Transmission, movement and inactivation of cymbidium mosaic and odontoglossum ringspot viruses. Plant Dis 78: 633–636
27. Farreyrol K, Pearson MN, Grisoni M, Leclercq-Le Quillec F (2001) Severe stunting of *Vanilla tahitensis* in French polynesia caused by *Cucumber mosaic virus* (CMV), and the detection of the virus in *V. fragrans* in Reunion Island. Plant Pathol 50:414–414
28. Grisoni M, Pearson M, Farreyrol K (2010) Virus diseases of vanilla. In: Odoux E, Grisoni M (eds) Vanilla. CRC Press Taylor & Francis Group, USA, pp 97–123
29. Madhubala R, Bhadramurthy V, Bhat AI, Hareesh PS, Retheesh ST, Bhai RS (2005) Occurencre of *Cucumber mosaic virus* on vanilla (*Vanilla planifolia* Andrews) in India. J Biosci 30:339–350
30. Farreyrol K, Pearson MN, Grisoni M, Cohen D, Beck D (2006) Vanilla mosaic virus isolates from French polynesia and the Cook Islands are Dasheen mosaic virus strains that exclusively infect vanilla. Arch Virol 151:905–919
31. Wisler GC, Zettler FW, Mu L (1987) Virus infections of vanilla and other orchids in French Polynesia. Plant Dis 71:1125–1129
32. Grisoni M, Davidson F, Hyrondelle C, Farreyrol K, Caruana ML, Pearson M (2004) Nature, incidence, and symptomatology of viruses infecting *Vanilla tahitensis* in French polynesia. Plant Dis 88:119–124
33. Grisoni M, Moles M, Farreyrol K, Rassaby L, Davis R, Pearson M (2006) Identification of potyviruses infecting vanilla by direct sequencing of a short RT-PCR amplicon. Plant Pathol 55: 523–529
34. Pearson MN, Brunt AA, Pone SP (1990) Some hosts and properties of a potyvirus infecting *Vanilla fragrans* (Orchidaceae) in the kingdom of Tonga. J Phytopathol 128:46–54

35. Wang YY, Beck DL, Gardner RC, Pearson MN (1993) Nucleotide sequence, serology and symptomatology suggest that vanilla necrosis potyvirus is a strain of watermelon mosaic virus II. Arch Virol 129:93–103
36. Wang YY, Pearson MN (1992) Some characteristics of potyvirus isolates from *Vanilla tahitensis* in French Polynesia and the Cook Islands. J Phytopathol 135:71–76
37. Bhat AI, Bhadramurthy V, Siju S, Hareesh PS (2006) Detection and identification of Cymbidium mosaic virus infecting vanilla (*Vanilla planifolia* Andrews) in India based on coat protein gene sequence relationships. J Plant Biochem Biotechnol 15:33–37
38. Richard A, Farreyrol K, Rodier B, Leoce-Mouk-San K, Wong M, Pearson MN, Grisoni M (2009) Control of virus diseases in intensively cultivated vanilla plots of French Polynesia. Crop Prot 28:870–877

Vanillin: Biosynthesis, Biotechnology, and Bioproduction

14

Shahnoo Khoyratty, Rob Verpoorte, and Hippolyte Kodja

Contents

Abstract

Further understanding of *Vanilla* flavor compounds synthesis, in *Vanilla* plants, is lacking in literature. This can aid in better production of flavor compounds. Additionally, commercial importance of *Vanilla* pods can then be improved. Although vanillin amounts in pod, predominates among flavor compounds, the natural flavor is made of over 200 chemicals. Despite being low cost to produce, synthetic pure vanillin alone does not match the much sought complex notes of the natural product, used in products with high-quality attributes. This, despite the

S. Khoyratty (✉)
Natural Products Laboratory, Institute of Biology, Leiden University, Leiden, The Netherlands

Qualisud, Université de La Réunion, Université Montpellier, CIRAD, Institut Agro, Avignon Université, Montpellier, France
e-mail: shahnookhoyratty@gmail.com

R. Verpoorte
Natural Products Laboratory, Institute of Biology, Leiden University, Leiden, The Netherlands
e-mail: verpoort@chem.leidenuniv.nl

H. Kodja
Qualisud, Université de La Réunion, Université Montpellier, CIRAD, Institut Agro, Avignon Université, Montpellier, France
e-mail: hippolyte.kodja@univ-reunion.fr

J.-M. Mérillon, H. Kodja (eds.), *Orchids Phytochemistry, Biology and Horticulture*, Reference Series in Phytochemistry, https://doi.org/10.1007/978-3-030-38392-3_14

natural product being more than 200 times the price of the synthetic version, fungal biotransformation reactions of vanillin precursors yield "natural" vanillin. Endophytic fungi form symbiotic relationships with asymptomatic plants, where they produce secondary metabolites. In some cases, such metabolites were previously thought to be produced by the plant, but later revised to originate from the fungi. In this view, *Vanilla* flavor metabolites may be due to fungal participation, within the plant. Those fungi may either participate partially or completely, on the vanillin biosynthetic pathway, within the plant. This potential participation requires to be elucidated, to gain a better understanding of the still debatable vanillin biosynthesis, within the plant. The occurrence of different fungal endophyte species, across plant culture regions, may contribute to the observed terroir effect on pod flavor. The study of fungal endophytes within *Vanilla* plants, and their biosynthetic potential, is thus warranted.

Keywords

Vanilla · Vanillin · Flavor · Endophyte · Biosynthesis

1 Introduction

Vanilla is a member of the *Orchidaceae* family, itself comprising 110 species. Within this family, *Vanilla planifolia* Jacks. ex Andrews, *Vanilla pompona* Schiede., and *Vanilla tahitensis* J.W. Moore, all originating from Central America, are important in terms of their flavors. For this reason, the three species have commercial values and thus cultivated. *Vanilla* spp. were introduced in Europe by the Spanish, about thc year 1520, whereas Reunion Island initiated its cultivation around 1819 (Table 1).

Given substrate specificity is lacking during vanillin synthesis, the process is inefficient and expensive as more substrates are then required [5]. Additionally, this synthesis is not environmental friendly [77]. There is also the issue of slow growth of *Vanilla* plants produced through tissue culture, and a low expression of vanillin biosynthesis. As such, vanillin production from those tissues is also inefficient and expensive. Better alternatives, which are more efficient and more cost-effective, are then preferred. Such an alternative is in the synthesis of vanillin by microorganisms.

Among different microorganisms, some fungi have the highest amounts of vanillin synthesis (Table 1). A possible characteristic, for a microorganism to achieve higher amounts of vanillin synthesis, is in the ability to not be affected, from toxicity due to vanillin. Generally, this characteristic can be attributed to the ability to convert vanillin into chemicals of lower toxicity. These chemicals include vanillyl alcohol and vanillic acid. The latter conversion process would reduce the total amount of vanillin obtained, which is not desirable [5]. There is, however, one method to decrease vanillin toxicity, which does not reduce vanillin amount, by the glycosylation of vanillin. With this view, yeast has been genetically modified to obtain this glycosylation ability [5]. It is worth noting that vanillin occurs as glucovanillin, a glycosylated form, in green pods of *Vanilla*. Of interest, fungal endophytes may reside within pods, a *Vanilla* plant organ, and may be involved with vanillin synthesis.

Table 1 Selected landmark events in relation to *Vanilla* flavor

Year	Events
Ca. 1520	***Vanilla* plants imported to Europe. Given absence of required pollinators from Mexico, no pods formed in Europe**
1819	***Vanilla* plants imported to Reunion Island, where artificial pollination method developed. This method opened possibilities of commercialization**
1874	Nicholas-Theodore Gobley isolates and purifies vanillin from *Vanilla* pods
After 1920s	From lignin, vanillin synthesized. However, procedure is not environmental friendly
1970	Green vanillin biotransformation: Vanillin synthesized by *Pseudomonas acidovorans* (bacteria), from ferulic acid precursors
1998 2002 2007	Production of green vanillin: Rice bran oil and wheat bran are inexpensive agricultural by-products. These are sources of ferulic acid, as precursors towards vanillin synthesis, by *Aspergillusniger* (fungus) combined with *Pycnoporuscinnabarinus* (fungus). The later yields a maximum vanillin of 300–2800 mg L^{-1} of liquid media. This amount depends on the fungal strains, addition of glucose, and adsorbent resins[a], concentrating ferulic acid amount. Synthetic vanillin was shown to differ from this green vanillin, through isotope ratio $\delta^{13}C_{PDB}$ analysis
2009	Production of de novo green vanillin: Genetically modified yeasts *Schizosaccharomycespombe* and *Saccharomyces cerevisiae* convert glucose to vanillin. This is only possible after a modification that introduced a novel biosynthetic pathway. The yield of vanillin is 45–65 mg L^{-1} of liquid media
2011	Supply cannot match demand, for *Vanilla* flavoring from *Vanilla* pods. This prompted flavor companies, like International Flavors and Fragrances, Inc. and Givaudan, like to find alternative methods to produce ***substitutes***
2014	Green ripe*Vanilla* beans incubated with ***microorganism*** yields vanillin. This process is part of a patent application by Givaudan (flavor company). Vanillin amounts equal to, or higher than in cured pods, and has additional consistent ***complex sensory profiles***, devoid of off-notes
2015	Production of de novo green vanillin: Genetically modified bacteria *Escherichia coli* convert glucose to vanillin.This is only possible after a modification that introduced a novel biosynthetic pathway. The yield of vanillin is 97.2 mg L^{-1} from l-tyrosine precursor, 19.3 mg L^{-1} from glucose precursor, 13.3 mg L^{-1} from xylose precursor, and 24.7 mg L^{-1} from glycerol precursor
2017	Novel natural *Vanilla* products, by Firmenich. These are less expensive and come from sustainable production practices

Adapted from [1–11]
[a]Vanillin synthesized by microorganisms is removed from media by adsorbent resins. This allows for more vanillin production [5]. In high amounts, vanillin is toxic, prompting the microorganism to degrade vanillin. The latter reduces the yield of vanillin

A literature search shows that de novo vanillin synthesis has not been reported for natural microorganisms. The ability towards de novo synthesis has only been possible after genetically engineering microorganisms e.g. biosynthesis of vanillin from glucose in recombinant yeast [76]. The process has several associated problems, e.g., genetically modified *E. coli* production of vanillin suffers from isovanillin synthesis, a contaminating chemical that reduces vanillin yield [5]. To prevent a decrease in vanillin yield, gene knockout technology was applied to genetically modified yeasts [5]. The gene knockout targeted a gene, for which enzyme (alcohol dehydrogenase ADH6) is responsible for the conversion of vanillin to vanillyl alcohol. Despite this, another issue occurred in accumulation of vanillin, and hence toxicity, as well as decrease in solubility, within the culture medium.

Flavors within cured *Vanilla* pods are of commercial importance. A flavor component of *Vanilla* cured pods is vanillin, generally occurring at 1–2% w/w [12]. In terms of amount, vanillin is thus the major flavored component. Still, other minor components present in lesser amounts than vanillin in the pods also contribute to *Vanilla* flavor. Those minor components are responsible for specific flavors, which give value to natural *Vanilla* flavor, over synthetic vanillin. Flavor descriptors for natural *Vanilla* may range across floral, woody, and spicy [13].

When vanillin and *Vanilla* flavor are compared in terms of organoleptic properties, the former is one-dimensional, whereas the latter is multidimensional [14]. This makes *Vanilla* flavor complex, although subtle as well [14, 15]. *Vanilla* pods thus qualify as spice [16]. One parameter that influences *Vanilla* flavor, is the conditions under which plants were grown [15]. The types of combination of flavor-related chemicals within the pods confer specific flavors to those pods [17]. This is a quality feature which affects *Vanilla* trade and prices.

The type and amount of metabolite, as well as the ratio at which these occur, affect the final flavor of the pods. This, then, constitutes the organoleptic properties of the pod. Given the contribution of minor components (in terms of quantity), to flavor, it is essential to find the minimum amount of such flavor components that can be detected in terms of sensory attributes, by humans. Vanillyl alcohol is an example of this, given olfactory-GC analysis shows the latter registers as intense as vanillin, despite being at an amount 1000 times less than vanillin, in cured pods [18]. Given over 250 components contribute to flavor in cured pods, vanillin accounts to only less than 12% of variability in organoleptic qualities, when different pods, with varying flavors, are compared [18]. Some flavor components are volatiles detected as aromas. These evaporate the moment pods are heated. Thus, baking can lead to flavor alteration [19].

Other minor flavor components include vanillin putative precursors, given these also contribute to flavor qualities of pods. A nonexhaustive list includes volatiles such as monoterpenes, sesquiterpenes, and phenolics [20, 21]. Nonvolatiles include vanillic acid with creamy flavors [22], vanillyl alcohol with balsamic flavors [23], *p*-hydroxybenzoic acid having phenolic tones [22], and *p*-hydroxybenzaldehyde with a sweet flavor [22]. Volatiles are detected with the nose, whereas nonvolatiles with the mouth, i.e., both have differing contributions to flavor.

Vanilla production requires significant labor intervention, compared to other cultivated crops [24, 25]. The cost of labor involved in its production thus makes the pod expensive. Labor availability is also dependent on political stability, which then affects the price of the pods. Climatic conditions also affect the pod supply, leading to price fluctuations. The volatility of pod price, on the international market, has encouraged the production of cheaper synthetic vanillin and nonnatural *Vanilla* flavors, which are then incorporated in food and beverages. The latter production is also more economical than attempting to extract *Vanilla* flavor from pods.

As a comparison, natural extracts are at least 200 times more expensive than synthetic vanillin. This price difference has sometimes led to imitation natural flavors being sold, thus requiring quality control, to prevent this practice [26].

Global demand to supply ratio for *Vanilla* cured pods is at least at 10:1 [2]. Due to this, synthetic vanillin competes with natural *Vanilla* flavor, despite the flavor attributes of natural flavor being better. With time, 95% of *Vanilla* flavor world consumption is that of synthetic vanillin. Through habit, consumers have got used to vanillin over natural *Vanilla* [26].

Still, market trends are shifting towards natural vanillin, synthesized by green chemical reactions, instead of the synthetic version. Even so, natural vanillin is not set to substitute natural *Vanilla* flavor, rather it will take the place of synthetic vanillin at a more accessible price [27]. High-quality products, e.g., fine chocolates, will always contain real *Vanilla* flavor though. Other products using *Vanilla* flavor are Cola beverages, ice cream, and baked foods [26].

As consumer trend shift towards a greater demand for natural products, demand for biotechnology-derived natural flavors are also increasing [28]. Two main methods of flavor compound synthesis are either through biotransformation or through de novo synthesis. The difference between both production methods is with regards to the number of steps in reactions, from precursor, leading to the same vanillin [28]. In biotransformation reactions, single reactions would biotransform precursor into vanillin, whereas in de novo synthesis a precursor is biotransformed into vanillin after going through complex metabolic pathways.

The search for microorganisms capable to produce "natural" vanillin de novo is sought after. Finding such microorganisms could partially address the problem of not having enough natural *Vanilla* flavor extract. Some fungi are good candidates towards this end, given their ability to synthesize vanillin from ferulic acid precursors.

Endophytes are microorganisms that produce secondary metabolites and are sometimes responsible for the source of those metabolites, and in several cases, previously thought to be produced by the plants [29]. Additionally, metabolites from the same endophytes can participate in biosynthetic pathways of the plants, to yield novel secondary metabolites. These include pharmacological metabolites ergoline alkaloids [30–33] and taxol [34, 35].

Endophytes (bacteria and fungi) form symbiotic relationships with plant organs while, unlike pathogens, the plant shows no visible disease symptoms [36]. Still, the definition of an endophyte is not clear-cut. For instance, if a plant pathogen loses its virulence, this microorganism can then be termed as an endophyte [37]. The presence of the endophytes in the plant is due to the metabolome of the plant [38]. Endophyte function within the plant is varied and so far is reported as participating in conferring to the host, survival tools, within the environment in which the plant grows. Endophyte assistance to the plant encompasses stress conditions, e.g., pathogenic attack [39–41], under low water regimes [42] and consumption by herbivores [43–45].

In a similar way, it is reasonable to assume fragrances and flavors traditionally thought to be produced by plants, may either be produced only, or partially, by endophytes instead. For this reason, an in-depth study into endophytes residing in plants, those important for their flavors, is warranted. From there, to study the biosynthetic abilities or biotransformation reactions of those endophytes, on precursors and towards flavor metabolites. Given endophyte composition generally varies across environments, it may also be possible that such differences are responsible for the terroir effects of *Vanilla*.

1.1 *Vanilla* Species as Source of Vanillin

Flavor metabolites are organ specific within *Vanilla* plants. Specifically, mature pods, but not the leaves, contain flavors. The age of the pod is important, given immature pod are devoid of vanillin. During the maturation period, vanillin content increases within the pod, as glucovanillin. The highest yield of vanillin is reached at the end of maturation. *Vanilla* pods take about 8 to 9 months after pollination to mature. Most of the glucovanillin within the mature pods would then be hydrolyzed with the enzyme glucosidase, during the curing process, into vanillin [24, 25]. Although *Vanilla* pods harbor the highest vanillin content, the presence of vanillin is not limited to *Vanilla* species. Other plants also contain vanillin (Table 2).

The synthesis of vanillin, in terms of all steps on the biosynthetic pathway, and the identity of participating vanillin precursors are still uncertain. One issue is that studies conducted to elucidate this pathway, or part of it, used varying biological materials, e.g., *Vanilla* pods, cell culture, and plants [47]. Each of these materials may have different vanillin synthesis pathways which increase uncertainty, in any

Table 2 Vanillin presence and amounts in different plants

Plant species	Vanillin percentage with respect to plant dry weight
Syzygium aromaticum (L.) Merr. and L.M. Perry) Common name: Clove	Trace
Solanum tuberosum L. Common name: Potato	0.01
Proboscidea louisianica (Mill.) Thell. Common name: Ram's horn	0.01
Narcissus tazetta L. Common name: Narcissus	0.01–0.60
Hyacinthus orientalis L. Common name: Hyacinth	0.20–0.50
*Vanilla pompona*Schiede Common name: Vanilla	0.01–2.00
Vanilla tahitensis J.W. Moore Common name: Vanilla	0.50–2.00
Vanilla planifolia Jacks. ex Andrews* Common name: Vanilla	2.00–3.00 (Highest vanillin content)

Adapted from [46]

direct comparison between results for different biological materials. Given the simplicity of vanillin structure, enzyme promiscuity on phenolic pathways, it is hypothesized that synthesis of vanillin can thus occur through several potential pathways, which is a further complication [48]. Due to these issues, it is not quite possible to reconstruct a single reliable pathway for vanillin synthesis, based on data from different sources.

Figure 1 illustrates vanillin putative synthesis network. Precursors include tyrosine and phenylalanine.

Vanillin can be converted into vanillic acid and vanillyl alcohol, both of which also contribute to *Vanilla* flavor as well. As a result, vanillin can either be the product of the vanillin biosynthesis pathway or a precursor to other intermediates.

Through the shikimic acid pathway, tyrosine and phenylalanine precursors (phenylpropanoids) lead to the synthesis of vanillin [52], which is generally accepted. Despite this, the steps through which this conversion occurs vary based on two hypotheses, i.e., either the ferulate or benzoate pathways. C6C3 methylation and hydroxylation yield ferulic acid [52], followed by chain shortening, to give vanillin. The latter reaction describes the ferulate pathway hypothesis. In the alternative benzoate pathway, phenylalanine or tyrosine go through a chain shortening reaction [52], then the aromatic ring undergoes methylation or hydroxylation to yield vanillin. It is also possible for another vanillin precursor, such as*p*-hydroxybenzoic acid, to originate directly from the shikimate pathway as well. In this case, vanillin synthesis is not related to phenylalanine or to tyrosine, making this possibility, different from the ferulate and benzoate pathway hypotheses, for vanillin synthesis.

Enzymes tyrosine or phenylalanine ammonia lyase are two enzymes that intervene at the entry point of the phenylpropanoid pathway. Cinnamic acid derivatives are synthesized through the reactions catalyzed by both aforementioned enzymes. The precursor *p*-hydroxybenzaldehyde can lead to vanillin synthesis. Metabolites from the latter pathway are present in *Vanilla* pods, which suggests that this pathway may be possible towards vanillin synthesis. Such metabolites include *p*-coumaric acid, vanillyl alcohol [18], *p*-hydroxybenzaldehyde, protocatechuic aldehyde, and *p*-hydroxybenzoic acid [13]. And yet, there is the contradictory result that radiolabeled *p*-hydroxybenzaldehyde, after feeding within pod material, is not integrated into glucovanillin [53]. Still, this is not a conclusive result, since highly reactive aldehydes, like the radiolabeled *p*-hydroxybenzaldehyde, may already be used in other reactions before being able to participate within vanillin synthesis. As such, this still opens the possibility that other nonradiolabeled *p*-hydroxybenzaldehyde, already present in the pods, may have been used in vanillin synthesis.

Another controversy is with regards to the function of enzymes that intervene on pathways, leading to vanillin synthesis. An enzyme reported by a research work to participate in vanillin synthesis from ferulic acid as precursor (ferulate pathway) [53] is instead reported by another research work to participate in *p*-hydroxybenzaldehyde synthesis from *p*-coumaric acid (US patent application) [55]. It is possible that an enzyme has broad substrate specificities, as hypothesized by Dixon. For instance, cinnamic aldehydes may be synthesized from the related acid derivatives by an enzyme. Also, aldehyde reduction may occur through enzymes that react

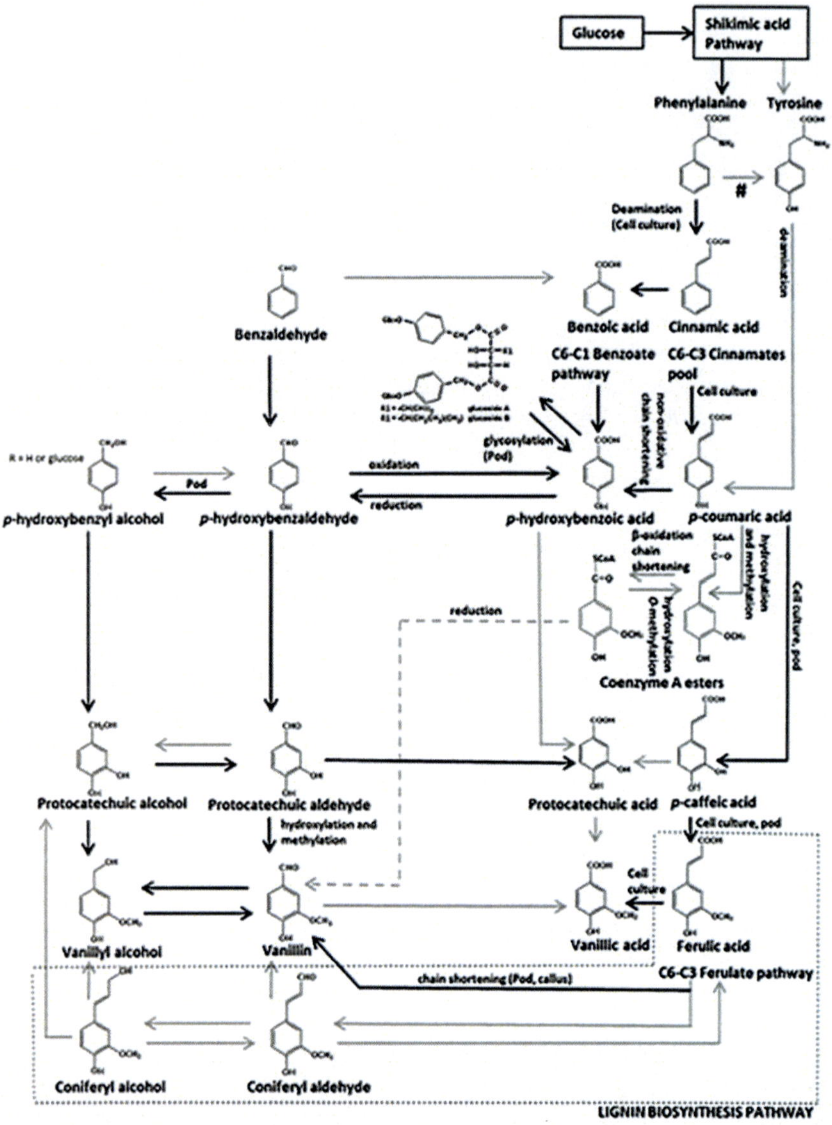

Fig. 1 Scheme of potential pathways to vanillin in relation to confirmed pathways in the plant *Vanilla planifolia* (Bold, Dark lines). *Vanilla planifolia* reactions adapted from [47–53]. The hydroxylation reaction marked with # has not been found in plants and fungi, as phenylalanine and tyrosine are synthesized separately from different precursors [54]. The system (cell culture, pod, and callus) in which the step was shown experimentally to occur is indicated but only for those confirmed steps and not those proposed to occur by the authors without experimental results

similarly to aromatic ring substituents on several substrates. Both vanillin and one of its precursor (ferulic acid, a cinnamic acid derivative) have comparable aromatic ring substitution patterns. Additionally, ferulic acid occurs in large amounts either freely or linked in plant cell walls [56, 57], making it a good candidate towards vanillin synthesis.

1.2 Vanillin Sources: Plants and Microorganisms

Variation in the microbial species within *Vanilla* pods depends on the treatment conditions in the curing process of pods post-harvest. In this way, microbial species within Indonesian pods differed significantly after being subjected to scalding (part of the curing process) [58]. The exact temperature and length of time, through which pods are treated during scalding, vary depending on the region of the world. In Indonesia, the treatment is to put pods in water at 65–70 °C for a period of 2 min. With a decrease in the number of microbial species was associated the proliferation of fungi. Furthermore, *Fusarium* spp. were also identified within Indonesian pods [59]. The latter were identified through sequencing of elongation factor genes, combined with morphological similarities against *Fusarium* spp. references.

Other fungi are also reported to occur within *V. planifolia*. In this way, *Tulasnella* spp., *Ceratobasidium* spp., and *Thanatephorus* spp. are all mycorrhizal fungi, found in roots, while not causing any plant symptoms [60].

Other research work focused, instead, on actinomycetes and bacterial microorganisms, and the species of these present during the curing process [15, 61]. Some bacterial endophytes, living inside *Vanilla* pods, were found to raise vanillin amounts after curing. Those bacteria are *Bacillus subtilis* and *Bacillus vanillea* [62]. It may also be possible that other *Bacillus* species would have similar properties.

The world demand towards natural products at the expense of synthetic ones has pushed for a higher consumption of natural vanillin [57]. To satisfy this increase in demand, a production of natural vanillin by making use of microorganisms has gained favor. As early as the 1940s through the 1990s, some research work was already conducted, focused on producing vanillin, by fungal microorganisms [63]. However, this type of research was not seen as important over the years and is thus lacking. Instead, work concerned with biotransformation of cinnamic acid and ferulic acid was promoted. For this reason, more work is now warranted in the usage of fungi towards vanillin synthesis. In fact, fungal species are known to synthesize vanillin through ferulic acid as precursor [47, 57, 63] (Fig. 2).

What is also lacking is that no work is reported on *Vanilla* pod fungal endophytes in connection to synthesis of *Vanilla* flavor metabolites, whether through biotransformation reaction of precursors or through de novo synthesis. By knowing which fungal species occur within pods that can intervene in the elaboration of *Vanilla* flavors, it may be possible to gain greater control on fine-tuning towards the best *Vanilla* flavor in pods, while these are still growing on the plant. This can be done by encouraging the growth, on the plant, of those specific fungal species.

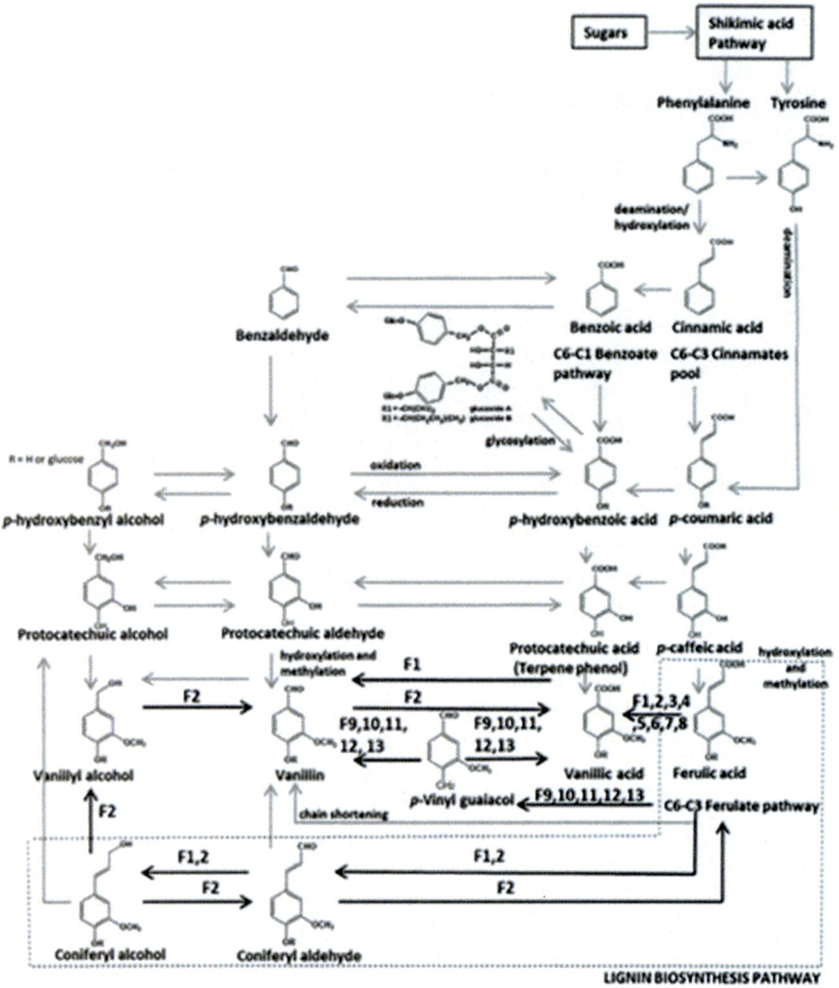

Fig. 2 Scheme of potential pathways to vanillin in relation to confirmed pathways in fungi (Bold, Dark lines). Fungus **F1**: *Pycnoporuscinnabarinus*, **F2**: *Trametes* sp., **F3**: *Aspergillusniger*, **F4**: *Botrytis* sp., **F5**: *Cephalosporium* sp., **F6**: *Penicillium* sp., **F7**: *Trichoderma* sp., **F8**: *Verticillium* sp., **F9**: *Schizophyllum commune*, **F10**: *Paecilomycesvariotii*, **F11**: *Fusariumsolani*, **F12**: *Sporotrichum thermophile*, **F13**: *Debaromyceshansenii*. Fungal reactions adapted from [63]. R = H

1.3 Volatiles Related to *Vanilla* Flavor

Despite the importance of vanillin to *Vanilla* flavor, the desirability of pods does not depend only on vanillin amount within the pod [64]. The problem is that vanillin has been given too much importance at the detriment of other nonvanillin *Vanilla* flavor metabolites. A contributing factor to this is that vanillin is the first metabolite related to *Vanilla* flavor which was isolated (Table 1). Aroma detected by the nose is also an essential component that adds value to pods. For this reason, and since research is lacking in this direction, more attention should be dedicated to volatiles related to flavor in the pods. A technique to detect odor-active volatiles is through the use of GC-olfactory analysis. There are, however, complications given it is not easy to relate single metabolites from complex combinations to specific flavor components.

Generally, nonvanillin *Vanilla* flavor metabolites occur in amounts much lower than vanillin, within the pods. And yet, the former can be appreciated by the senses as a flavor component, at the same title as vanillin. As such, detection of amounts through analytical methods has never been an issue with vanillin but can be difficult with other *Vanilla* flavor metabolites. As a comparison, the *Vanilla* flavor metabolite guaiacol typically occurs 20 times lower in amounts than vanillin [65]. Still, sensory organs can detect guaiacol at 50 times a lower threshold than that of vanillin (detection threshold guaiacolis 13 ppb, vanillin is 680 ppb [66]).

Despite research generally lacking on this theme, some research work has been previously conducted on finding volatile metabolites present in *Vanilla* pods. In this way, a number of pod-derived volatiles are known [20, 21, 62]. Of interest, volatiles detected in pods can be grouped into phenols, sesquiterpenes, monoterpenes, lactones, arenes, and ethers.

Within the group of volatiles from pods, a further step was done in some work, identifying those that can be detected by olfactory senses [20]. Within the context of *Vanilla* flavor, this is still not enough, given not all volatiles from pods detected by olfactory senses are relevant to *Vanilla* flavor. For this reason, it is also essential to do research work, to find which volatiles from pods are related to *Vanilla* flavor. Sesquiterpenes volatiles, for instance, found in pods may contribute to woody notes, which also define *Vanilla* flavor [67]. References such as [22] are available that describe organoleptic tones produced by specific metabolites. These references are then an important first step for this type of work of associating volatile metabolites in pods to *Vanilla* flavors.

Some fungi are reported in literature to produce terpenoids with flavor-related properties [63, 68]. The same could be the case with pod-derived fungal endophytes, producing volatiles with *Vanilla* flavor–related properties. As yet, literature is absent on potential participation of endophytes, in volatile *Vanilla* flavor synthesis. This, then, represents a research avenue of interest.

1.4 Endophyte Species Occurrences Depending on Geographical Location

Depending on the cultivation site of a plant, fungal endophyte species tend to differ for the same plant species [69]. In conjecture, a similar occurrence is possible for fungal endophytes in *Vanilla* pods, depending on the cultivation sites across Reunion Island. Due to this diversity, it is essential to obtain samples which reflect the general fungal endophyte community, in *Vanilla* and across Reunion Island. Thus, the step forward is to sample plants from different cultivation regions in Reunion Island, in view of isolating and identifying endophytes from these (Fig. 3).

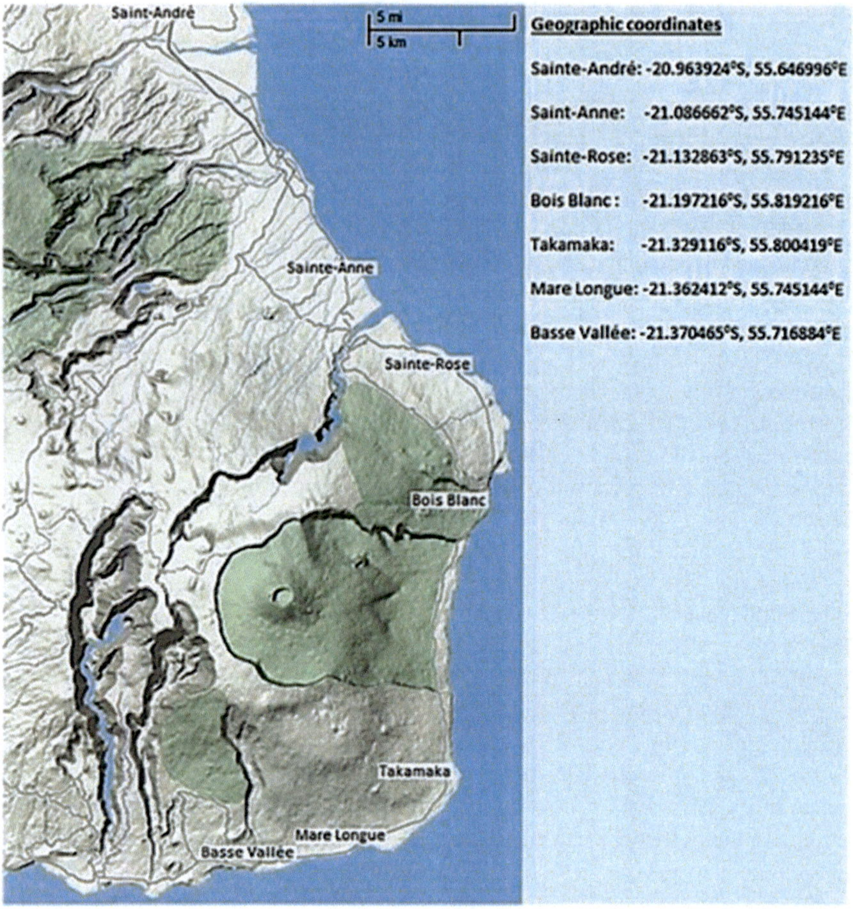

Fig. 3 East and southeastern regions in Reunion Island where *Vanilla* cultivation is performed

This difference in *Vanilla* flavor across cultivation sites is not limited to Reunion Island. As such, for the same plant species (*Vanilla planifolia*), flavor also varies across different cultivation zones in the world [13] (Table 3). This type of difference in flavor is known as the terroir effect, which is also popular in grapes used in wine making. Diversity in microorganisms in grapes has been observed, based on the culture region of the plant [70]. Additionally, it was also shown for those grapes that the metabolome differs due to microorganism species present within the plant. This then contributes to the terroir effect of the final wine product. Additionally, from their results, the authors speculate wine terroir of a location may be obtained across any other locations, by inoculating the same fungal species from that location onto any grape plants. That fungal endophytes may contribute to *Vanilla* flavor, while the fungal species may vary across regions may at least partially explain the observed terroir effect. This then forms a plausible hypothesis to investigate.

Table 3 Region-specific flavors depending on cultivation region

Cultivation region	Common name	Species	Region-specific flavor properties
Indian ocean islands including Reunion Island	Bourbon	*Vanilla planifolia*	Deeply balsamic notes
India	Indian	*Vanilla planifolia*	Lacks balsamic tones, creamy, sweetness but more body compared to Bourbon Vanilla. Mildly pungent sour flavor
Mexico	Mexican	*Vanilla planifolia*	Lesser body than Bourbon Vanilla
Indonesian Islands	Indonesian	*Vanilla planifolia*	Lacks creamy and sweet notes compared to Bourbon Vanilla. Pencil tones, intense woody, and mildly smoky

Adapted from: [13]

Even if all cultivated *Vanilla planifolia* plants are genetically similar clones in Reunion Island, pod flavor differences have still been observed depending on cultivation zones, in the same Island. Additionally, the same curing process is used for all pods (Bertrand Côme – La Vanilleraie, *pers. comm.*) (Fig. 3). As such, observed differences in flavor notes cannot be attributed to plant genetic factors, nor to pod post-harvest processing methods. This opens the possibility to other factors that need to be elucidated and that contribute to differences in pod flavor properties. One such possibility may be fungal endophyte diversity within *Vanilla planifolia*, depending on culture regions for the plant.

Vanilla flavor does not occur in all organs of *Vanilla* plants. For instance, this flavor occurs in the pods but not the leaves. For this reason, it may be informative to find fungal endophytes species occurring within the pods, and to compare against those occurring in the leaves. It is only through such a comparison that fungal endophyte species specific to the pods only, but absent from the leaves, can be found. These endophytes then are likely to be associated more to flavor metabolite synthesis. By performing such an experiment, but including a further parameter, i.e., for plants coming from different cultivation regions in Reunion Island, it may be possible to make a correlation between fungal species and the terroir effect. Also of interest is to find the capacity of fungal endophyte species to synthesize *Vanilla* flavor metabolites and the associated precursors.

1.5 Plant and Endophyte Cohabitation

Interest in endophytes has been growing over time [71]. What was found is that every plant species harbors endophytes [72]. The number of endophytic species varies depending on the plant host. On some hosts, only one species was found, and in other cases, hundreds of species were isolated. While the endophytes reside within the plant host, a physiological, metabolic interaction is formed between both types of organisms. This interaction differs to that, when a pathogen infects a plant, where the plant mounts a defense reaction. The defense reaction translates in the synthesis of phytoalexins, whereas for endophytes, the plant does not mount a defense reaction. Rather, a constitutive synthesis of metabolites is obtained. Another possible reaction within this interaction is the endophyte that biotransforms plant-derived precursors. In fact, the resulting secondary metabolites from endophyte interaction with the plant can protect the plant against pathogens, through constitutive defense. Overall, the metabolic implication of a plant–endophyte interaction is quite complex, given there are many endophyte species, each with its own interaction, within the plant [73]. This complexity might add to terroir effects, within the context of *Vanilla* flavor.

2 Conclusions

The review here elaborates on vanillin and *Vanilla* flavor synthesis, as well as points to the research direction taken, so far, on the same themes. A discussion of the research direction to follow in the future themes on those flavors where literature as research work is lacking is also mentioned. The vanillin biosynthetic network occurs in plants, especially *Vanilla* spp., and in different microorganisms. That some microorganisms participate in vanillin synthesis has led to their usage for this synthesis. By introducing selected genes in some microorganisms, it has even been possible for those to synthesize vanillin de novo, when this was not possible before. It is now well documented that endophytes contribute to the synthesis of economically valuable secondary metabolites [73–75]. It may be possible that microorganisms present inside *Vanilla* plants assist in *Vanilla* flavor synthesis, during pod maturing or during the curing process. Fungal endophyte variability, across plant culture regions, may also contribute to the observed terroir effect in *Vanilla* pods. Given variability in endophyte species within a plant, this variability may, at least partially, help to understand the conflicting studies about vanillin synthesis, within *Vanilla* pods. More studies on bacterial species within *Vanilla* plants are warranted.

References

1. Firmenich (2017) Firmenich enriches *Vanilla* palette with sustainable and cost-effective natural tonalities. http://www.firmenich.com/en_INT/company/news/FIRMENICH-ENRICHES-VANILLA-PALETTE.html. Accessed 09 Sept 2017
2. Givaudan (2016) Anything but Plain *Vanilla*. Givaudan Taste Essentials *Vanilla*. https://www.givaudan.com/flavours/world-flavours/tasteessentials/vanilla. Accessed 29 July 2016
3. Ni J, Tao F, Du H, Xu P (2015) Mimicking a natural pathway for *de novo* biosynthesis: natural vanillin production from accessible carbon sources. Sci Rep 5:13670. https://doi.org/10.1038/srep13670

4. Perfumer and Flavorist (2014) Givaudan submits patent application for *Vanilla* fermentation process. http://www.perfumerflavorist.com/networking/news/company/Givaudan-Submits-Patent-Application-For-Vanilla-Fermentation-Process-238896611.html. Accessed 29 July 2016
5. Kaur B, Chakraborty D (2013) Biotechnological and molecular approaches for vanillin production: a review. Appl Biochem Biotechnol 169:1353–1372
6. Hansen EH, Møller BL, Kock GR, Bünner CM, Kristensen C, Jensen OR, Okkels FT, Olsen CE, Motawia MS, Hansen J (2009) De novo biosynthesis of vanillin in fission yeast (*Schizosaccharomycespombe*) and Baker's yeast (*Saccharomyces cerevisiae*). Appl Environ Microbiol 75:2765–2774
7. Zheng L, Zheng P, Sun Z, Bai Y, Wang J, Guo X (2007) Production of vanillin from waste residue of rice bran oil by *Aspergillusniger* and *Pycnoporuscinnabarinus*. Bioresour Technol 98:1115–1119
8. Lesage-Meessen L, Lomascolo A, Bonnin E, Thibault JF, Buleon A, Roller M, Asther M, Record E, Ceccaldi BC, Asther M (2002) A biotechnological process involving filamentous fungi to produce natural crystalline vanillin from maize bran. Appl Biochem Biotechnol 103:141–153
9. Thibault FJ, Asther M, Colonna-Ceccaldi B, Couteau D, Delattre M, Duarte JC, Craig F, Heldt-Hansen PH, Kroon PA, Lesage-Meessen L, Micard VC, Renard MT, Van Hulle S, Williamson G (1998) Fungal bioconversion of agricultural by-products to vanillin. Lebenson Wiss Technol 31:530–536
10. Rasoanaivo P (1998) Essential oils of economic value in Madagascar: present state of knowledge. HerbalGram 43:31–59
11. Toms A, Wood JM (1970) Degradation of trans-ferulic acid by *Pseudomonas acidovorans*. Biochemist 9:337–343
12. Sinha AK, Sharma UK, Sharma N (2008) A comprehensive review on *Vanilla* flavor: extraction, isolation and quantification of vanillin and others constituents. Int J Food Sci Nutr 59:299–326
13. Ranadive AS (2011) Quality control of *Vanilla* beans and extracts. In: Havkin-Frenkel D, Belanger FC (eds) Handbook of *Vanilla* science and technology. Wiley-Blackwell, Oxford, pp 141–160
14. Gleason-Allured J (2011) *Vanilla*: anything but plain. Perfumer and Flavorist. http://www.perfumerflavorist.com/flavor/application/vanilla/132347233.html. Accessed 30 July 2016
15. Dunphy P, Bala K (2012a) The role of microorganisms in *Vanilla* curing. Part 1: evidence for microbial involvement. Perfum Flavor 37:24–29
16. Cameron KM (2011) *Vanilla* phylogeny and classification. In: Havkin-Frenkel D, Belanger FC (eds) Handbook of *Vanilla* science and technology. Wiley-Blackwell, Oxford, p 243
17. Alwahti AY (2003) A taste of *Vanilla*. The TED case studies. Number 686. http://www1.american.edu/TED/vanilla.htm. Accessed 30 July 2016
18. Hoffman PG, Zapf CM (2011) Flavor, quality, and authentication. In: Havkin-Frenkel D, Belanger FC (eds) Handbook of *Vanilla* science and technology. Wiley-Blackwell, Oxford, pp 162–179
19. Kennedy CR (2015) The flavor rundown: natural vs. artificial flavors: What's in a flavor? Harvard University. http://sitn.hms.harvard.edu/flash/2015/the-flavor-rundown-natural-vs-artificial-flavors/. Accessed 29 July 2016
20. Zhang S, Mueller C (2012) Comparative analysis of volatiles in traditionally cured Bourbon and Ugandan *Vanilla* bean (*Vanilla planifolia*) extracts. J Agric Food Chem 60:10433–10444
21. Toth S, Lee KJ, Havkin-Frenkel D, Belanger FC, Hartman TG (2011) Volatile compounds in *Vanilla*. In: Havkin-Frenkel D, Belanger FC (eds) Handbook of *Vanilla* science and technology. Wiley-Blackwell, Oxford, pp 183–218
22. Sigma-Aldrich (2014) Ingredients catalog. Flavour Frag J
23. Burdock GA (2010) Fenaroli's handbook of flavor ingredients, 6th edn. CRC Press, New York
24. Dignum M, Kerler J, Verpoorte R (2001) *Vanilla* production: technological, chemical and biosynthetic aspects. Int Food Res J 17:199–219

25. Dignum MJW, Kerler J, Verpoorte R (2002) *Vanilla* curing under laboratory conditions. Food Chem 79:165–171
26. Royal Society of Chemistry (2016) Chemistry in its element – vanillin. http://www.rsc.org/chemistryworld/podcast/CIIEcompounds/transcripts/vanillin.asp. Accessed 29 July 2016
27. Labuda I (2011) Biotechnology of vanillin: vanillin from microbial sources. In: Havkin-Frenkel D, Belanger FC (eds) Handbook of *Vanilla* science and technology. Wiley-Blackwell, Oxford, pp 301–327
28. Dionisio AP, Molina G, Carvalho DS, Pastore GM (2012) Natural flavourings from biotechnology for foods and beverages. In: Baines D, Seal R (eds) Natural food additives, ingredients and flavourings. Woodhead Publishing, Cambridge, pp 231–259
29. Gimenez C, Cabrera R, Reina M, Gonzalez-Coloma A (2007) Fungal endophytes and their role in plant protection. Curr Org Chem 11:707–720
30. Kucht S, Groß J, Hussein Y, Grothe T, Keller U, Basar S, König W, Steiner U, Leistner E (2004) Elimination of ergoline alkaloids following treatment of *Ipomoea asarifolia* (Convolvulaceae) with fungicides. Planta 219:619–625
31. Schardl CL, Young CA, Hesse U, Amyotte SG, Andreeva K, Calie PJ, Fleetwood DJ, Haws DC, Moore N, Oeser B, Panaccione DG, Schweri KK, Voisey CR, Farman ML, Jaromczyk JW, Roe BA, O'Sullivan DM, Scott B, Tudzynski P, An Z, Arnaoudova EG, Bullock CT, Charlton ND, Chen L, Cox M, Dinkins RD, Florea S, Glenn AE, Gordon A, Güldener U, Harris DR, Hollin W, Jaromczyk J, Johnson RD, Khan AK, Leistner E, Leuchtmann A, Li C, Liu J, Liu J, Liu M, Mace W, Machado C, Nagabhyru P, Pan J, Schmid J, Sugawara K, Steiner U, Takach JE, Tanaka E, Webb JS, Wilson EV, Wiseman JL, Yoshida R, Zeng Z (2013) Plant-symbiotic Fungi as chemical engineers: multi-genome analysis of the Clavicipitaceae reveals dynamics of alkaloid loci. PLoS Genet 9:e1003323
32. Steiner U, Hellwig S, Leistner E (2008) Specificity in the interaction between an epibioticclavicipitalean fungus and its convolvulaceous host in a fungus/plant symbiotum. Plant Signal Behav 3:704–706
33. Steiner U, Leibner S, Schardl CL, Leuchtmann A, Leistner E (2011) Periglandula, a new fungal genus within the Clavicipitaceae and its association with Convolvulaceae. Mycologia 103:1133–1145
34. Stierle A, Strobel G, Stierle D (1993) Taxol and taxane production by *Taxomyces andreanae*, an endophytic fungus of Pacific yew. Science 260:214–216
35. Strobel GA, Hess WM, Li JY, Ford E, Sears J, Sidhu RS, Summerell B (1997) Pestalotiopsis guepinii, a Taxol-producing endophyte of the Wollemi pine, *Wollemia nobilis*. Aust J Bot 45:1073–1082
36. Souza JJD, Vieira IJC, Rodrigues-Filho E, Braz-Filho R (2011) Terpenoids from endophytic Fungi. Molecules 16:10604–10618
37. Hardoim PR, Overbeek LSV, Berg G, Pirttila AM, Compant S, Campisano A, Doring M, Sessitsch A (2015) The hidden world within plants: ecological and evolutionary considerations for defining functioning of microbial endophytes. Microbiol Mol Biol R 79:293–320
38. Bálint M, Tiffin P, Hallström B, O'Hara RB, Olson MS, Fankhauser JD, Piepenbring M, Schmitt I (2013) Host genotype shapes the foliar fungal microbiome of balsam poplar (*Populusbalsamifera*). PLoS One 8:e53987
39. Álvarez-Loayza P, White JF, Torres MS, Balslev H, Kristiansen T, Svenning JC, Gil N (2011) Light converts endosymbiotic fungus to pathogen, influencing seedling survival and niche-space filling of a common tropical tree, Iriarteadeltoidea. PLoS One 6:e16386
40. Arnold AE, Mejía LC, Kyllo D, Rojas EI, Maynard Z, Robbins N, Herre EA (2003) Fungal endophytes limit pathogen damage in a tropical tree. Proc Natl Acad Sci USA 100:15649–15654
41. Herre EA, Mejía LC, Kyllo DA, Rojas E, Maynard Z, Butler A, Van Bael SA (2007) Ecological implications of anti-pathogen effects of tropical fungal endophytes andmycorrhizae. Ecology 88:550–558
42. Bayat F, Mirlohi A, Khodambashi M (2009) Effects of endophytic fungi on some drought tolerance mechanisms of tall fescue in a hydroponics culture. Russ J Plant Physiol 56:510–516

43. De Sassi C, Müller CB, Krauss J (2006) Fungal plant endosymbionts alter life history and reproductive success of aphid predators. Proc R Soc Biol Sci 273:1301–1306
44. Rudgers JA, Afkhami ME, Rúa MA, Davitt AJ, Hammer S, Huguet VM (2009) A fungus among us: broad patterns of endophyte distribution in the grasses. Ecology 90:1531–1539
45. Saari S, Sundell J, Huitu O, Helander M, Ketoja E, Ylönen H, Saikkonen K (2010) Fungal-mediated multitrophic interactions – do grass endophytes in diet protect voles from predators? PLoS One 5:e9845
46. Havkin-Frenkel D, Belanger FC (2016) Biotechnological production of vanillin. In: Havkin-Frenkel D, Dudai N (eds) Biotechnology in flavor production, 2nd edn. Wiley, West Sussex, p 168
47. Korthout H, Verpoorte R (2007) Vanilla. In: Berger RG (ed) Flavours and fragrances: chemistry, bioprocessing and sustainability. Springer Science + Business Media, Leipzig, pp 203–217
48. Dixon RA (2011) Vanillin biosynthesis – not as simple as it seems? In: Havkin-Frenkel D, Belanger FC (eds) Handbook of *Vanilla* science and technology. Wiley-Blackwell, Oxford, pp 292–298
49. Kundu A (2017) Vanillin biosynthetic pathways in plants. Planta 245:1069–1078
50. Yang H, Barros-Rios J, Kourteva G, Rao X, Chen F, Shen H, Liu C, Podstolski A, Belanger F, Havkin-Frenkel D, Dixon RA (2017) A re-evaluation of the final step of vanillin biosynthesis in the orchid *Vanilla* planifolia. Phytochemistry 139:33–46
51. Funk C, Brodelius PE (1990) Phenylpropanoid metabolism in suspension cultures of *Vanilla planifolia*. Andr Plant Physiol 94:95–101
52. Havkin-Frenkel D, Belanger FC (2007) Application of metabolic engineering to vanillin biosynthetic pathways in *Vanilla* planifolia. In: Verpoorte R, Alfermann AW, Johnson TS (eds) Applications of plant metabolic engineering. Springer, Dordrecht, pp 175–196
53. Gallage NJ, Hansen EH, Kannangara R, Olsen CE, Motawia MS, Jørgensen K, Holme I, Hebelstrup K, Grisoni M, Møller BL (2014) Vanillin formation from ferulic acid in *Vanilla planifolia* is catalysed by a single enzyme. Nat Commun 5:4037
54. Dewick PM (2002) The shikimate pathway: aromatic amino acids and phenylpropanoids. In: Dewick PM (ed) Medicinal natural products. Wiley, West Sussex, pp 121–166
55. Havkin-frenkel D, Podstolski A, Dixon RA (2003) Vanillin biosynthetic pathway enzyme from *Vanilla planifolia*. US Patent US20030070188 A1 (US Patent Office: United States, 2003)
56. Mathew S, Abraha TE (2004) Ferulic acid: an antioxidant found naturally in plant cell walls and feruloylesterases involved in its release and their applications. Crit Rev Biotechnol 24:59–83
57. Lesage-Meessen L, Stentelaire C, Lomascolo A, Couteau D, Asther M, Moukha S, Record E, Sigoillot J, Asther M (1999) Fungal transformation of ferulic acid from sugar beet pulp to natural vanillin. J Sci Food Agric 79:487–490
58. Roling WFM, Kerler J, Braster M, Apriyantono A, Stam H, Verseveld HWV (2001) Microorganisms with a taste for *Vanilla*: microbial ecology of traditional Indonesian *Vanilla* curing. Appl Environ Microbiol 67:1995–2003
59. Pinaria AG, Liew ECY, Burgess LW (2010) *Fusarium*species associated with *Vanilla* stem rot in Indonesia. Australas Plant Pathol 39:176–183
60. Porras-Alfaro A, Bayman P (2007) Mycorrhizal fungi of *Vanilla*: diversity, specificity and effects on seed germination and plant growth. Mycologia 99:510–525
61. Dunphy P, Bala K (2012b) The role of microorganisms in *Vanilla* curing. Part 2: Microbial transformation of phenols and other compounds. Perfum Flavor 37:22–27
62. Gu F, Chen Y, Fang Y, Wu G, Tan L (2015) Contribution of Bacillus isolates to the flavor profiles of *Vanilla* beans assessed through aroma analysis and chemometrics. Molecules 20:18422–18436
63. Mäkelä MR, Marinović M, Nousiainen P, Liwanag AJ, Benoit I, Sipilä J, Hatakka A, de Vries RP, Hildén KS (2015) Aromatic metabolism of filamentous fungi in relation to the presence of aromatic compounds in plant biomass. Adv Appl Microbiol 91:63–137
64. Reineccius G (1998) Natural flavoring materials. In: Reineccius G (ed) Source book of flavors, 2nd edn. Chapman & Hall, New York, p 357

65. Takahashi M, Inai Y, Miyazawa N, Kurobayashi Y, Fujita A (2013) Key odorants in cured Madagascar *Vanilla* beans (*Vanilla planifolia*) of differing bean quality. Biosci Biotech Bioch 77:606–611
66. Leffingwell & Associates (2017) Odor & flavor detection thresholds in water (in parts per billion). http://www.leffingwell.com/odorthre.htm. Accessed 5 Mar 2017
67. Buccellato F (2011) *Vanilla* in perfumery and beverage flavors. In: Havkin-Frenkel D, Belanger FC (eds) Handbook of *Vanilla* science and technology. Wiley-Blackwell, Oxford, pp 235–240
68. Grice K, Lu H, Atahan P, Asif M, Hallmann C, Greenwood P, Maslen E, Tulipani S, Williford K, Dodson J (2009) New insights into the origin of perylene in geological samples. Geochim Cosmochim Acta 73:6531–6543
69. David AS, Seabloom EW (2016) Plant host species and geographic distance affect the structure of aboveground fungal symbiont communities, and environmental filtering affects belowground communities in a coastal dune ecosystem. Microb Ecol 71:912–926
70. Zarraonaindia I, Owens SM, Weisenhorn P, West K, Hampton-Marcell J, Lax S, Bokulich NA, Mills DA, Martin G, Taghavi S, van der Lelie D, Gilbert JA (2015) The soil microbiome influences grapevine-associated microbiota. Microb Cell 2:171–173
71. Varma VC, Gange AC (2014) Advances in endophytic research. Springer, New Delhi
72. Faeth SH, Fagan WF (2002) Fungal endophytes: common host plant symbionts but uncommon mutualists. Integr Comp Biol 42:360–368
73. Pusztahelyi T, Holb IJ, Pócsi I (2015) Secondary metabolites in fungus-plant interactions. Front Plant Sci 6:573. https://doi.org/10.3389/fpls.2015.00573
74. Ludwig-Müller J (2015) Plants and endophytes: equal partners in secondary metabolite production? Biotechnol Lett 37. https://doi.org/10.1007/s10529-015-1814-4
75. Hongsheng Y, Zhang L, Li L, Zheng C, Guo L, Li W, Sun P, Qin L (2010) Recent developments and future prospects of antimicrobial metabolites produced by endophytes. Microbiol Res 165:437–449
76. Gallage NJ, Møller BL (2015) Vanillin–Bioconversion and Bioengineering of the Most Popular Plant Flavor and Its De Novo Biosynthesis in the Vanilla Orchid. Mol Plant 8:40–57
77. Khoyratty S, Kodja H, Verpoorte R (2018) Vanilla flavor production methods: a review. Ind Crops Prod 125:433–442

Part V

Phytochemistry and Medicinal Properties

Ethnobotany and Recent Advances in Indian Medicinal Orchids

15

Ram Pal, N. K. Meena, M. Dayamma, and D. R. Singh

Contents

R. Pal (✉) · M. Dayamma
ICAR-National Research Centre for Orchids, Darjeeling Campus, Darjeeling, West Bengal, India
e-mail: rampal_nrco@yahoo.com

N. K. Meena
ICAR-National Research Center on Seed and Spices, Tabiji, Ajmer, India

D. R. Singh
ICAR-National Research Centre for Orchids, Pakyong, East Sikkim, Sikkim, India

J.-M. Mérillon, H. Kodja (eds.), *Orchids Phytochemistry, Biology and Horticulture*,
Reference Series in Phytochemistry, https://doi.org/10.1007/978-3-030-38392-3_26

Abstract

India is very rich in orchid genetic resource. It is estimated that about 1300 species occur within the political boundaries of India, of which nearly 150 species are used in Ayurvedic, Siddha, Unani, and tribal system of medicine. The country has been paying attention to ornamental orchids for the use in floriculture industry, but the orchid therapeutics did not catch such attention which also has similar potential to contribute in economy and health of the people. The present chapter takes a look on uses of orchids in traditional medicine system as well as progress made for their utilization in healthcare system.

Keywords

Orchids · Ayurveda · Medicinal · Phytochemicals

1 Introduction

Traditional medicines have a long history of catering the healthcare needs of people all over the world. In India, traditional medicine system includes Ayurveda, Siddha, Unani and Homeopathy, Yoga, and Naturopathy. Of these, the first four are heavily dependent on plant-base formulations for curing various kinds of illnesses. The two famous treatises on Ayurveda, *Charaka Samhita* and *Sushruta Samhita*, were composed during 600 B.C., the former listed 341 types of plants and plant products and the latter described 1120 types of illnesses, 700 medicinal plants, and 121 preparations. India possesses 47,000 species of plants distributed in 15 agroclimatic zones: 15,000 of them have medicinal value. Of these, 7000 used in Ayurveda, 700 in Unani, 600 in Siddha, and 30 in modern medicine [1]. Ashtavarga is a group of eight plant species and have prodigious role in Ayurvedic system of medicine. In Sanskrit they are called as Kakoli, Kshirakakoli, Meda, Mahameda, Jeevaka, Rishbhaka, Riddhi, and Vriddhi [2]. The plants in Ashtavarga are called Jeevaniya (strengthens vitality and immunity), Brhnayiya [activates cell regeneration system], and Vayasthapan (activates metabolic and anabolic process leading to youthfulness) [3]. The scientific identity of Ashtavarga plants has been established Kakoli as *Roscoea procera* Wall. (*Roscoea purpurea* Sm), Kshirakakoli as *Lilium polyphyllum* D.Don, Meda as *Polygonatum verticillatum* (L.) All., Mahameda as *Polygonatum cirrhifolium* (Wall.) Royle Jeevak as *Malaxis acuminate* D. Don. (*Crepidium acuminatum*), Rishbhaka as *Microstylis muscifera* Ridl. (*Malaxis muscifera*), Riddhi as *Habenaria edgeworthii* Hook f. ex Colt (*Platanthera edgeworthii*), and Vriddhi as *Habenaria intermedia* (*H. arietina*). The last four are terrestrial orchids found in the Himalayas. The other orchids mentioned in Ayurvedic literature are Jivanti, *Flickingeria macraei* (*Dendrobium alpestre*), shwethuli, and rasna (*Acampe papillosa* and *Vanda tessellata*). Additionally, the tribal medicine system is practiced by 645 tribes who live in isolation and practice their system of medicine. The tribal have passed down their knowledge and practices of curing illnesses generation after generation by word of mouth. With other plants, tribal also use several species of orchids in their healthcare system.

Ethnobotanical studies indicate about 150 species of orchids are used in traditional medicinal system in India. Despite their known potential value in medicine and perennial use, a very few species have been subjected to phytochemical analysis and clinical evaluation. Orchids produce a large number of phytochemical like alkaloids, flavonoids, carotenoids, anthocyanins, and sterols. Aeridin was isolated from *Aerides crispum* Lindl: Agrostophyllin, Agrostophyllinone, and Agrostophyllinol from *Agrostophyllum brevipes* King & Pantl.; Arundinan from *Arundina graminifolia* (D. Don) Hochr; Bulbophythrin A and Bulbophythrin B from *Bulbophyllum odoratissimum* (Sm.); and Coeloginanthrin and Combretastatin C-1 from *Coelogyne cristata* Lindl. The study of phytochemical constituents is essential to validate their usage against a specific disease. The clinically tested biomolecules would have a usage in the medical science. The integration of traditional medicines with western medicine would not only bring down the cost of medication but also increase the safety from side effects of western medicines. The People's Republic of China has successfully incorporated traditional herbal medicine into a modern healthcare system. The unique blend herbal medicine, acupuncture, and western medicine cater the healthcare needs of its people [4].

2 Orchids in Indian System of Medicine

India has six, Ayurveda, Siddha, Unani, Yoga, Naturopathy, and Homeopathy, well-acknowledged systems of medicine. Ayurveda, Siddha, Yoga, and Naturopathy originated in India. Unani system of medicine has its roots in Greece and introduced by Arabs in India. It spread, enriched, and assimilated with Indian culture during the Mughal period. Homeopathy was introduced in India during the eighteenth century and became part of the Indian medicine system. Thus, the systems of medicine that originated in India and the systems of medicine that came from outside into the country got assimilated with Indian culture are called Indian System of Medicine [5]. Though not officially recognized, there are tribal or folk medicines in indigenous healthcare which plays a vital role in meeting the healthcare needs of tribal in India.

2.1 Ayurveda

Ayurveda is one of the most acknowledged holistic healthcare systems based on varying medicinal uses of plants to prevent and cure a variety of illnesses. *Charaka Samhita* and *Sushruta Samhita*, the two compilations are dating back to 600 BC, the primary source of knowledge in healthcare in ancient India. The *Charaka Samhita* consists of 8 sections and 120 chapters dealing with various aspects of medicine and associated subjects. Around 600 drugs of plant, animal, and mineral origin are mentioned in this treatise [6]. *Sushruta Samhita*, a treatise on surgery, describes surgical instruments, operative procedures, and presents a description of 650 drugs. The medical books were written during medieval India (seventh to the eleventh century) mention Ashtavarga which consists of eight plant species with their Sanskrit names as Kakoli, Kshirakakoli, Jeevaka, Rishbhaka, Meda, Mahameda, Riddhi, and Vriddhi [2]. The plants in Ashtavarga are called as Jeevaniya (strengthens vitality and immunity), Brhnayiya (activates cell regeneration system), and Vayasthapan (activates metabolic and anabolic process leading to youthfulness) [3]. There had been a problem of correct identification of plants mentioned in Ashtavarga. However, the researches carried out after the independence of the country, the majority of Indian scientists agree with the plants mentioned in the Ashtavarga. The Jeevak is *Malaxis acuminate* D. Don. (*Crepidium acuminatum*), Rishbhaka is *Microstylis muscifera* Ridl. (*Malaxis muscifera*), Riddhi is *Habenaria edgeworthii* Hook f. ex Colt (*Platanthera edgeworthii*), and Vriddhi is *Habenaria intermedia* (*H. arietina*), respectively. The plants mentioned in Ashtavarga occur in the Himalayas between 1200 and 4000 m. In Himachal Pradesh, an Indian state in Western Himalayas, all the four orchids occur between 1800 and 2800 m altitude and flower during monsoon [7].

Chyavanprash is formulation made up of about 50 herbs, sugar, and honey and very popular in the northern part of India. It maintains physique, vigor, and vitality and also delays the process of aging. Approximately 15,000 tons of Chyavanprash produced annually in India. Adults and children consume it as a remedy for cold, cough, and respiratory infections. Ashtavarga plants were indispensable constituents of Chyavanprash and collectively increased the antioxidant activity of Amla (*Phyllanthus emblica*), one of the principal constituent of the Chyavanprash. The unregulated use of these plants might have led to decline, and replacement of these herbs was proposed in Ayurvedic text Bhavprakash (sixteenth century), Yogaratnakara (seventeenth century), and Vaidyachintamani (eighteenth century). The Ayurvedic Pharmacopoeia of India published by Ministry of AYUSH, Government of India, permits replacement of *Malaxis acuminata* (*Crepidium acuminatat*) and *Malaxis muscifera* with *Pueraria tuberosa* and *Habenaria edgeworthii* (*Platanthera edgeworthii*) and *Habenaria intermedia* (*Habenaria arietina*) with *Dioscorea bulbifera*. *Malaxis acuminata* used for the preparation of Chyavanprash and contains Beta-sitosterol, piperitone, citronellal, eugenol, Limonene, 1,8-cineole, *p*-cymene, *O*-Methylbatatasin, and cetyl alcohol [8]. The ethyl acetate extract of Chyavanprash exhibited a higher level of scavenging activity, i.e., close to ascorbic acid (IC50 20.693 μg/ml) and thus inhibits formation free radicals in the body [9]. The other orchids used in Ayurvedic system of medicine include Munjataka (*Dactylorhiza hatagirea*), Jeevanti (*Flickingeria macraei*) (*Dendrobium macraei*), and Rasna (*Vanda tessellata*). In India, medicinal orchids are under threat due to various human-made and natural calamities. All four orchids of Ashtavarga group have been placed under the Red Data Book of Indian Plants. The causes of the rarity of medicinal orchids vary with the region. In Himachal Pradesh, Ashtavarga orchids are threatened due construction of houses and hydroelectric projects, grazing, and collection fodder from natural habitats [7]. In Niyamgiri hills of Odisha, they are facing endanger because of mining operations [10]. Shifting cultivation, upcoming industries, use of forests for human settlement, and felling of trees are a major threat for the survival of the orchids in other parts of Odisha [11]. *Acampe praemorsa*, *Geodorum densiflorum*, and *Habenaria marginata* are endangered in Harda district of Madhya Pradesh [12]. Unregulated collection of orchids for various purposes including medicines is endangering orchids in their natural habitats. Therefore, it is necessary to ease the pressure on natural habitats by cultivating orchids propagated through tissue culture rather than collecting from natural habitats.

2.2 Siddha

Siddha system of medicine is one of the ancient traditional healthcare system practiced mostly in Tamil-speaking areas of India. Ayurveda system of medicine is relatively more popular than the Siddha system of medicine. Unlike Ayurveda, in Siddha system of medicine, orchids do not find much favor for medicines. However, tribal living in Tamil Nadu uses orchids to cure the illnesses.

2.3 Unani

Unani system of medicine originated in Greece and introduced in India by Arabs and Persians in the eleventh century. It flourished under the patronage Mughal rulers and spread all over the country. The Unani system medicine suffered a setback during British rule in India, but the efforts of Nizam of Hyderabad, Azizi family of Lucknow, and Sharifi family of Delhi brought it to life in a short time [13]. There are several institutions to undertake research and education in the country: the Central Council for Research in Unani Medicine (CCRUM) to promote research; the National Institute of Unani Medicine (NIUM) for postgraduate, teaching, training, and research; and the Central Research Institute on Unani Medicine for research and education in Unani medicine. In Turkey, the salep is prepared by grinding of tuberous orchids to a fine powder to use as food and drug. It is used as a thickening agent in Turkish ice cream and also mixed with milk to make the hot drink known as salep served sprinkled with cinnamon [14]. In Unani system of medicine Salam Panja (palm-like), Salam lahsunia (garlic type), Salam Mishri (translucent and globular), and Salam Badshah (like a king) are common. These names were given to the appearance of tubers used in the preparation of salep. In India, several species of orchids such as *Dactylorhiza hatagirea*, *Eulophia nuda*, *Habenaria commelinifolia*, and *Satyrium nepalense* are used in preparation of salep.

2.4 Tribal Medicine

The tribal system of medicine in India developed based on the experiences and experimentation of tribal people living in isolation in the want of healthcare facilities. This tradition of knowledge passed down from one generation to other. Tribal in India use many plant species, including orchids to get rid of many diseases. Dongaria Kondha tribes of Niyamgiri hills in southwest Odisha, India, use 16 species of orchids to treat 33 kinds of diseases [10]; in another study from Odisha, it was found that Bonda, Dongaria Kondha, Juang, and Saora tribes use 26 species orchids to cure various kinds of body ailments [11]. Tribals in the neighboring state, Andhra Pradesh, in the Eastern Ghats use 23 species of orchids to cure various kinds of diseases. Pragada et al. [15] surveyed 53 families of the tribals in Andhra Pradesh and reported that *Geodorum densiflorum* is used to treat ephemeral fever. Hill-Korwa tribe of Chhattisgarh uses 30 medicinal herbs belonging to 18 families. Among them, there are two orchids, *Saccolabium papillosum* (*Acampe praemorsa*) used for bone fracture and body ache and *Bulbophyllum leopardinum* used against sunstroke and diabetes [16]. Northeastern Himalayan state, Nagaland, is rich in orchid genetic resources; 396 species belonging to 92 genera inhabit in this state. There are 15 species of orchids used by local practitioners to treat various diseases like rheumatism, cholera, nervous disorder, and tuberculosis that are also used as antimicrobial agent and antidotes to snake and insect bites [17]. According a survey of 198 respondents consisting of 57 females and 141 males in Arunachal Pradesh, 101 plant species belonging to 50 families are used in ethnomedicine [18]. It is quite surprising that a state with the highest species of orchids in India did not document orchids species in part of their ethnobotanic medicines, whereas in neighboring state, Nagaland, 15 species of orchids have been recorded for medicinal uses. In the Kashmir Himalayas, 7 species of orchid are used to cure 24 types of different diseases [19]. Not all the orchids with medicinal value have been documented in the country, and those need to be documented earliest. Most of the studies do not focus on the complete preparation and use of herbal medicines. For realizing the benefits of herbal-based medicines used in tribal or folk medicine, a complete formulation and its use should be documented.

3 Orchids in Indian Medicine

3.1 *Acampe* Lindley

Epiphytic or lithophytic, monopodial perennial herb, stem erect or branched, leaves fleshy thick, leathery, bilobulate at apex. Inflorescence axillary, erect sometimes branched, small-medium sized, non-resupinating flowers. The genus is consisting of about 12 species distributed in India, Southwest China, Malaysia, and Africa. In India, this genus is represented by six species, and two of them, viz., *A. praemorsa* and *A. carinata*, are of medicinal interest. *A. praemorsa* is the most widely and intensively used species of this genus.

3.1.1 *Acampe carinata* (Griff.) Panigrahi

Found in India, Nepal, Bhutan, Thailand, Cambodia, and Vietnam. Monopodial epiphyte, plants robust; thick coriaceous leaves; inflorescence, umbel with 5–12 small flowers.

Roots [10] and leaf paste [20] are used to alleviate abdominal pain. Root paste of *A. carnata* is applied locally on scorpion sting and snakebite [10].

3.1.2 *Acampe praemorsa* (Roxb.) Blatt & McCann

Found in northeastern states of India, Nepal, Sri Lanka, and Myanmar at an elevation ranging from 600 to 1600 m. Plants are robust monopodial epiphyte; leaves thick, fleshy, leathery oblong, notched at tip; inflorescence umbel; flowers small, fragrant.

Ethnobotanical surveys conducted in the country showed that roots, leaves, whole plant, seed capsule rind, and seeds are used for curing various body ailments. The paste made from the roots of *A. praemorsa* and *Asparagus racemosus* is taken orally on an empty stomach for 15 days to alleviate arthritis [20]. The decoction made from the fresh roots consumed orally (2 tsp) along with honey (5 tsp) taken twice a day for 5 days for curing cough [21]. Hossain [20] recorded that roots are used for treating *asthma* and have expectorant properties, and the root paste is also applied externally on scorpion and snakebite. Shanavaskhan et al. [22] recorded that tribes of Kerala use whole plant to treat rheumatism. The root paste is applied to fractured organ of the cattle [23]. Leaf paste along with a piece of garlic is taken daily for 7 days to get relief from chest pain and stomach disorder caused by hyper acidity [10]. The juice from leaves is applied over nipple for relief from abdominal pain, and dispensing two to three drops in the ear relieves earache [22]. The fractured bones can also be treated by applying leaf paste [24, 25]. The fibers from the capsule are used to treat the wounds. The capsules fibers are placed on the wound and tied with a piece of cloth. The old fibers and cloth are replaced with new one every day till wound is healed. The seeds are directly applied on the old wound as an antibiotic. The plant is used as tonic [26] and in rheumatism [26]. The leaves contain flavonoids and cyanogenic glycosides [27]. Anuradha and Rao [28] isolated a new phenanthropyran derivative [1,7-dihydroxy-3-methoxy-9,10-dihydrophenanthropyran] from the whole plant.

3.2 *Acanthephippium* Lindl.

Terrestrial, medium- to small-sized herbs; pseudobulb ovoid, ovoid cylindric, cylindric pseudobulb with a few nodes and internodes and usually clasped with membranous sheath at base. The genus consists of about 12 species and is distributed in tropical Asia, Malaysia, to Fiji Islands. Three species inhabited in India, only one used in medicine.

3.2.1 *Acanthephippium bicolor* Lindl.

Hot- to warm-growing, small-sized, terrestrial herb found in southern India and Sri Lanka at an elevation of 650 m. Pseudobulbs clustered, large, smooth, ovate, deeply sulcate tapering above into thin stem; flowers during winter on a short, erect 3.0–4 cm long, up to six-flowered inflorescence. Flowers are pale purple with 9–11 veins.

Tribes in Kolli Hills in Namakkal district of Tamil Nadu (Eastern Ghats), use leaf extract *A. bicolor* for treating urinary tract infections mainly caused by *Escherichia coli*, *Staphylococcus saprophyticus*, *Klebsiella*, *Enterococci bacteria*, and *Proteus mirabilis*. Kala and Senthilkumar [29] tested antimicrobial activity of leaf extract of *A. bicolor* against 35 microbial agents and found Gram-negative bacterium, like *Staphylococcus aureus*, *Streptococcus faecalis*, and *Bacillus cereus*, among Gram-positive bacteria. *Proteus vulgaris*, *Proteus mirabilis*, *Enterobacter aerogenes*, *Shigella dysenteriae*, *Klebsiella pneumoniae*, and *Escherichia coli* and pathogenic fungi, viz., *Microsporum audouinii*, *Microsporum fulvum*, *Candida albicans*, and *Trichophyton rubrum*, are sensitive to leaf extract. This might be the reason for using this plant in treating the urinary tract infection. Further study may lead to development of drug suitable for combating urinary tract infections.

3.3 *Aerides* Lour.

Epiphytic medium-sized monopodial epiphytic herbs with many thick roots; stem with many nodes, short and enclosed by leaf sheaths. Leaves are distichous leathery, fleshy, and apex bilobed. Inflorescence arises from the axils of leaves densely flowered. The genus consists of about 20 species distributed from Sri Lanka, India, Nepal, Bhutan, Myanmar, China, Thailand, Indochina, and Malaysia to the Philippines and Indonesia. Eight species under this genus have been reported from India. Three species, namely, *A. crispum*, *A. emericii*, and *A. maculosa* are endemic. *A. rosea*, *A. odorata*, *A. multiflora*, and *A. crispum*, have been grouped as medicinal orchids. The paste made from the leaves of *A. multiflorum* is applied as poultice on cuts and wound [30]. The root infusion of *A. maculosa* is given once a day for 1–2 months to a patient suffering from tuberculosis. Anuradha and Prakash [31] isolated Aeridin (2,7-dihydroxy-1,3-dimethoxy-9,10-dihydrophenanthropyran), a derivative of phenanthropyran from whole plant part of *A. crispum*. Powder of *A. crispum* is boiled in neem oil and filtered, and 2–3 drops is poured into ears once during night for 3 days to cure pain and ear deafness [32]. The premedical studies conducted by Ghanakash and Kaushik [33] have shown that leaf extract of *A. multiflora* has antibacterial property.

3.4 *Agrostophyllum* Blume

3.4.1 *Agrostophyllum callosum* Rchb. f.

Epiphytic herbs, pseudostems are clustered bilaterally flattened with many internodes; leaves are twisted and almost lie in the same plane; inflorescence terminal, globular, densely flowered; flowers usually white, resupinating, often self-pollinating. Not much known about ethnobotanical use of this species. Two stilbenoids (agrostophyllol and isoagrostophyllol) and two diastereomeric 9,10-dihydrophenanthropyran derivatives from *A. callosum* Rchb. f. [34–37] and two terpenoids (agrostophyllinol and agrostophylline) from *A. brevipes* Ridley and *A. callosum* Rchb. f. [38].

3.5 Arundina Blume

3.5.1 *Arundina graminifolia* (D. Don) Hochr.

A. graminifolia is commonly known as a bamboo orchid. It is herbaceous terrestrial plant found in India, Nepal, Thailand, Malaysia, Singapore, South China, through the Pacific Islands. Terrestrial, thin and tall stem, leaves alternating acuminate, distributed from sea level to 1200 m, flowers similar to *Cattleya* and short-lived usually 5–6 days.

Flower stalk paste and rhizome are used for treating ear pain and rheumatism, whereas the roots are used in snakebite and intestinal biliary colic in Bangladesh [39].

It is used for healing the cracks on the skin [40]. *A. graminifolia* contains numerous stilbenoids, namely, arundin and its analogues: triterpenoid, Arundinol, and hydroxybenzaldehyde [41–43]. The plant contains tannins, saponin, and heptacosane [44]. Benzyldihydrophenanthrene, arundinaol (7-hydroxy-1-(*p*-hydroxybenzyl)-2,4-dimethoxy-9,10-dihydrophenanthrene), and five phenanthrene constituents (7-hydroxy-2,4-dimethoxy-9,10-dihydrophenanthrene(1),4,7-dihydroxy-2-methoxy-9,10-dihydrophenanthrene (2),2,7-dihydroxy-4-methoxy-9,10-dihydrophenanthrene (3),7-hydroxy-2-methoxyphenanthrene-1,4-dione (4),7-hydroxy-2-methoxy-9,10-dihydrophenanthrene-1,4-dione (5) have also been isolated from the rhizomes of this plant [45–47]. It is believed, this genus is effective for detoxification, is antiarthritis, and used as an antidote and demulcent. Liu et al. [48] isolated six bibenzyl derivatives, namely, 2,7-dihydroxy-1-(*p*-hydroxybenzyl)-4-methoxy-9,10-dihydrophenanthrene (1); 4,7-dihydroxy-1- (*p*-hydroxybenzyl)-2-methoxy-9,10-dihydrophenanthrene (2); 3, 3′-dihydroxy-5-methoxybibenzyl (3); (2E)-2-propenoic acid-3-(4-hydroxy-3-methoxyphenyl)-tetracosyl ester (4); (2E)-2-propenoic acid-3-(4-hydroxy-3-methoxyphenyl)-pentacosyl ester (5); and pentadecyl acid (6), from the tuber of this species and evaluated for antitumor activity. The compound 3(3,3′-dihydroxy-5-methoxybibenzyl) has exhibited stronger antitumor activity than other compounds.

3.6 *Bulbophyllum* Thouars

Largest genus, plants epiphytic or lithophytic, small to large in size; one- to two-leaved, arising from apex of pseudobulb. Inflorescence arises from base of pseudobulb or from rhizome, flowers small to large, flowers with the foot of the column is hinged attached to the column, petals shorter than the dorsal sepal.

3.6.1 *Bulbophyllum acutiflorum* A.Rich. (Syn: Bulbophyllum albidum (Wight) Hook.f.)

Found in the Eastern Himalayas at an elevation ranging from 900 to 1800 m, warm- to cool-growing epiphyte, pseudobulbs globose-ovoid; leaves lanceolate, bilobed at apex; inflorescence, umbel, six- to eight-flowered; flowers during spring to winter.

Kanikkar tribe of Agasthyamalai Hills uses leaves and pseudobulbs of *B. albidum* (*B. acutiflorum*) for strengthening weak uterus for conception. A preliminary study conducted to analyze the phytochemicals has shown that it contains saponins, steroids, and flavonoids along with other phytochemicals [49].

3.6.2 *Bulbophyllum cariniflorum* Rchb.

Found in the Northeastern Himalayas and eastern Southeast Asia at an elevation ranging from 1100 to 2100 m, cool-growing epiphyte; pseudobulbs five-angled, deeply grooved; leaves 2–3 elliptic pointed at apex; inflorescence 10- to 20-flowered, flowers with a fetid smell.

Paste made by mixing 2 g of dried root, 1 g of black pepper, and 5 ml of cow milk is taken half spoon orally with a cup of water by women for 5–10 days to induce abortion during first trimester of pregnancy in the districts of Mondanala and Sutanguni in the Niyamgiri Hill Ranges of Odisha, India [10].

3.6.3 *Bulbophyllum fusco-purpureum* Wight

Found in southern Western Ghats of India; pseudobulbs ovoid taper toward apex, one-leaved; leaves oblong, obtuse; inflorescence semi-erect, four- to five-flowered; flowers yellowish pink; flowering April–May.

Tribals living in Reserve Forest of Nilgiri use paste of pseudobulb to cure skin diseases. The pseudobulbs are washed and made into paste. The paste is applied on affected parts [50].

3.6.4 *Bulbophyllum kaitiense* Rchb.f.

B. kaitiense is an epiphytic, endemic orchid found in South India at an elevation ranging from 2000 to 2800 m in Kolli hills of Eastern Ghats, India. It grows trees or rocks, forms a dense mat-like structure; leaves 9–13 cm long, inflorescence umbel, five- to six-flowered, mentum in flowers absent, flowering winters to late spring.

B. kaitiense is used to cure several diseases and likely to have anticancer, antioxidant, anti-inflammatory, and antimicrobial activity [51]. Terpenoids, flavonoids, saponins, tannins, coumarin, and quinine carbohydrates are present in the pseudobulbs of *B. kaitiense*. The ethanolic extract of roots has antibacterial activity against 12 human pathogenic microorganisms. The ethanolic extract activity was more pronounced for fungi than bacteria [52]. Ethanolic extract of pseudobulbs also has anti-inflammatory activity [53].

3.6.5 *Bulbophyllum sterile* (Lam.) Suresh Syn: *Bulbophyllum nilgherrense* Wight

Warm- to cool-growing creeping epiphyte; found in India, Nepal, Bangladesh, and Myanmar at an elevation ranging from 1300 to 1700 m. Pseudobulbs fleshy, yellowish green, four-angled; leaves erect, oblong-elliptic, obtuse at apex; inflorescence many-flowered; flowering December to January.

The crushed whole plant is used to cure pimples and skin allergy by the tribal living in Reserve Forests of Nilgiri, Tamil Nadu, India. The whole plant is washed and crushed. The plant paste is applied externally on affected parts [50]. Pseudobulbs and leaves of *B. nilgherrense* are grounded into fine paste and mixed with cow's milk and administered orally for curing leucoderma [54]. Barthel [55] derived some active compound from the pseudobulbs of *B. nilgherrense* (=*B. sterile*) and claimed that these are effective in treating various disorders of cardiovascular system. The invention claimed that it has the only 1/6th of the cost of conventional treatment and has no side effects. Chopped pseudobulbs are cooked in coconut oil, and oil is applied to cure rheumatism, and the paste made from whole plant is applied on swellings [22].

3.7 *Coelogyne* Lindl.

Genus *Coelogyne* consists of nearly 200 species distributed in the Northeastern Himalayas and South India (Western Ghats), Nepal, Bhutan, Upper Myanmar, China, and Vietnam. Plants are epiphytic or lithophytic. Pseudobulbs ovoid, conical or cylindrical; two-leaved (rarely one), arise from the apex of pseudobulb, leathery; inflorescence racemose or panicled; flowers small to large. Thirty-four species occur in India. The two *C. cristata* and *C. stricta* (*C. elata*) are medicinally important.

3.7.1 *Coelogyne cristata* Lindl.

Found in northeastern India, West Bengal, Bhutan, and Nepal at an elevation ranging from 800 to 1800 m. Plants epiphytic, pendulous; pseudobulbs shiny, partially covered with sheath; leaves linear-lanceolate, pointed at tip; inflorescence one or two, arising from the base of the pseudobulb, up to 30 cm long; flowers lightly scented, white with a yellow spot on the lip; flowering March–April.

The fruits resin of *C. cristata* is used to heal the bone fractures of domestic animals. Some people also apply its resin externally on the injured portion of the body for decocting the blood [56]. *C. cristata* and *Pholidota imbricata* are the most frequently used plants in treating bone fractures in Kumaon Hills of Uttarakhand [57]. Ethanolic extract of *C. cristata* restored trabecular bone (both in femoral and tibial bones) without producing uterine estrogenicity changes when fed to ovariectomized mice. Coelogin, a compound isolated from *C. cristata*, enhanced the markers of osteoblastic differentiation and activity in vitro in the mice. The findings supported the claim of folk tradition of Kumaon region for using *C. cristata* in the treatment of fractured bones. They proposed that ethanolic extract of *C. cristata* and its pure compound coelogin has potential in the management of postmenopausal osteoporosis [57]. Majumder et al. [58, 59] isolated coelogin and coeloginin and two novel 9,10-dihydrophenanthrene derivatives, coeloginanthridin and coeloginanthrin, from the whole plant of *C. cristata*. The ethanol extracts of leaves and pseudobulbs of *C. cristata* exhibit potential antimicrobial properties against *Staphylococcus aureus* [60]. This bacterium commonly acts as a commensal of the human microbiota but can become an assertive pathogen and can cause skin infections including abscesses, respiratory infections such as sinusitis, and food poisoning. Pramanik [61] reported pharmacognostic characters along with physicochemical and fluorescence values for *C. cristata* pseudobulbs as a diagnostic tool for the standardization of the medicinal plant product.

3.7.2 *Coelogyne stricta* (D. Don) Schltr.

Found in the Northeastern Himalayas, Bhutan, and Nepal at an elevation ranging from 1400 to 1600 m. Plants are epiphyte or lithophytes. Pseudobulbs long, shiny, compressed, sheathed at base; leaves elliptic oblong, acute leathery; inflorescence 21–40 cm long, erect; flowers yellowish white to pinkish white, yellow to orange spot on the lip; flowering April–June.

The pseudobulbs are used to treat fever and headache [62]. In tribal and folk medicine, it is also used to cure bone fractures, fever, and headache [63]. A hot drink made from pseudobulbs of *C. cristata* relieves from constipation and also acts an aphrodisiac [64]. Pseudobulbs are applied externally to promote bone healing of fractured limbs in northeastern India [65]. The ethyl acetate extract is exhibit potential antibacterial activity against *Pseudomonas aeruginosa* and *Salmonella enteric*, *Enterococcus faecalis* and *Bacillus* subtilis. The MIC (minimum inhibitory concentration) value was 0.250 mg/ml and The MIC value for *Salmonella enterica*, *Corynebacterium* sp. and *Candia albicans* was 0.750 mg/ml, 0.500 mg/ml and 0.500 mg/ml, respectively. Ethyl acetate extract of *C. stricta* leaves inhibits human cervical cell lines (HeLA) cell growth at concentration 50 μg/ml [66]. Majumdar et al. [58] isolated 9,10dihydrophenanthropyrone (Coelonin) from *C. elata* (= *C. stricta*).

3.8 *Cymbidium* Swartz.

The widely distributed genus consists of about 52 species. Plants are epiphytic or terrestrial. Pseudobulbs short to long covered with leaf bases, older are without leaves; leaves clasping the pseudobulbs; inflorescence erect, arching, or pendulous with a few- to many-flowered; flowers showy, spreading small to large. Several species of this genus are medicinally important, but *C. aloifolium* is widely used in India.

3.8.1 *Cymbidium aloifolium* (L.) Sw.

In India, this species occurs in the Northeastern Himalayas and Southern India at an elevation ranging from sea level to 1100 m. Hot- to warm-growing; pseudobulbs small clasped with leaf bases; leaves fleshy, linear-oblong, notched at apex. Inflorescence one or two arising from the base of pseudobulb, pendulous; flowers 3–4 cm across; flowering April–July.

Leaves, roots, as well as whole plant are used to cure various body ailments. The paste made from the aerial roots is used to cure fractured bones [20, 54]. Dried powder *C. aloifolium* roots (2 g), dried ginger (2 g), and black pepper (1 g) are taken with a cup of cow's milk twice a day for 2 months to reduce paralysis [10]. The juice from the pod used against earache [22, 67]. The leaf juice alone or mixed with salt is used to treat earache [20], otitis and inflammatory conditions [68], and boils and fever [69]. The plant is emetic and purgative [54]. The salep made from the pseudobulbs is used as a tonic and in treating weakness of eyes, chronic illness, vertigo, burns, and sores [69].

Several phenanthrenes aloifol I and II, coelonin, 6-*O*-methoxycoelonin, batatasin III, and gigantol [70], cymbinodin A [71], and cymbinodin B [72] have been isolated from *C. aloifolium*. In a test on Swiss albino mice, ethanolic extract of *C. aloifolium* at 200 and 400 mg kg^{-1} body weight showed central nervous system (CNS) depressant effects [69]. Ethanolic leaf extract of *C. aloifolium* leaves has analgesic and anti-inflammatory activities in mice [73]. The leaves of *C. aloifolium* contain flavonoids, reducing sugars, cyanogenic glycosides, terpenoids, and tannins [27, 74]. The analgesic and anti-inflammatory effect may be due to the presence of tannins and flavonoids in the leaf extract.

3.9 *Dactylorhiza* Necker ex Nevski

The genus is consists of about 111 species and has pan-global distribution. Cold-growing terrestrials characterized by two to three flat tubers; leaves linear-lanceolate. In India, *Dactylorhiza hatagirea* Soo is used as medicine.

3.9.1 *Dactylorhiza hatagirea* Soo

Found in Pakistan, Afghanistan, Nepal, Bhutan, and Tibet at an elevation of 2500–5000 m. Plants are terrestrial perennials up to a height of 70 cm. Tubers are five lobbed; leaves broadly lanceolate; dense flowering inflorescence; flowers usually purple but rarely white; flowers during summer.

D. hatagirea is an important aphrodisiac in Ayurvedic and Unani system of medicine and used to increase vigor and vitality. The salep made from the tubers of *D. hatagirea* is extensively used in local medicine as a nervine tonic for its astringent and aphrodisiac properties [75, 76]. In Turkey, root powder is also used to make ice-cream and beverages [77]. The tubers are used to cure diabetes, diarrhea, dysentery, paralysis, convalescence, impotence, and malnutrition [40] and to cure kidney complaints and aphrodisiac [78]. In Kashmir, root tuber extract is commonly used for curing fever, whereas leaf extract is used for dysentery.

Administering root extract of *D. hatagirea* daily 200 mg kg^{-1} body weight for 28 days to male rats increased testosterone concentrations from 2.33 ng ml^{-1} to 9 ng ml^{-1}. Increased libido was evident by an increase in sexual parameters like mount, intromission, and ejaculatory frequency as well as latency [79]. The tuber extract has antibacterial activity against Gram-positive and Gram-negative bacteria, including *Escherichia coli* [80]. Izu et al. [81] isolated five new compounds (Dctylorhin A-E) and two natural compounds named dactyloses A and B. Due to over-exploitation for medicinal purpose and lack of cultivation, the plant has become endangered, and IUCN has categorized the plant as critically endangered.

3.10 *Dendrobium* Sw.

The genus *Dendrobium* is the largest genus of family Orchidaceae consisting about 900 species. The species is distributed throughout Asia, Australia, and Europe. Nearly 100 species found in India. There are 32–40 species of this genus used for the preparation of Shinhu, a popular tonic. In India a few species such as *D. cruentum*, *D. ovatum*, *D. nobile*, *D. moschatum*, etc. are used to treat minor ailments; however, *Dendrobium nodosum* (*Flickingeria nodosa*) and *Dendrobium plicate* (*Flickingeria fimbriata*) and *Dendrobium macraei* Lindl. (*Flickingeria macraei*) are widely used in Ayurvedic healthcare system in India.

3.10.1 *Dendrobium nodosum* Dalzell (Syn. *Flickingeria nodosa* (Dalzell) Seidenf.)

Plants are creeping epiphytes. Rhizomes branched terminating into pseudobulbs; pseudobulbs grooved compressed, fusiform; leaves sessile arise at the apex of pseudobulb, oblong-elliptic; flowers small (about 1 cm), white distributed in Southeast Asia at an elevation ranging from sea level to 2100 m. The plant is a source of Jivanti mentioned in *Charak Samhita*.

In folk medicine of Karnataka, halva is prepared from *Flickingeria nodosa* (*D. nodosa*), and it is consumed as astringent, aphrodisiac, and expectorant. Besides, it is also useful for curing asthma, bronchitis, throat infections, and dermatological infections and also acts as a blood purifier [82]. Juice of the pseudobulb and leaves is administered for asthma. The plant is also used for the treatment of asthma, bronchitis, consumption, fever, burning sensation, biliousness, and diseases of the blood [22]. The cold chloroform extract of pseudobulbs shows good antifungal activity against *Trichophyton mentagrophytes*, a fungus that causes athlete's foot, inflammation of skin, and infection of face, scalp, and body [83].

3.10.2 *Dendrobium plicatile* Lindl. (Syn. *Flickingeria fimbriata* (Blume) A.D. Hawkes; *Dendrobium macraei* Lindl.)

The species is widely distributed in the Himalayas, China, India, Sri Lanka to Southeast Asia, and Papua New Guinea at an elevation ranging from 700 to 1700 m. Plants are epiphyte with creeping and branched rhizome. The branched rhizomes terminating into yellowish green pseudobulbs; leaves oblong-lanceolate, leathery; inflorescence one- to three-flowered.

The plant cures from disorders of disorders of the bile, blood, and phlegm. It is generally used in decoctions with other plants having similar properties and sometimes used alone as a stimulant and tonic, the latter to treat debility associated with seminal loss [2]. In West Bengal, India, it is used in Rasayana therapy and sold as Jibanti [84]. It is used for treating asthma, bronchitis, sore throat, stomach ache, biliousness, and fever in Uttar Pradesh [65]. For curing skin allergies, hill tribes of

Odisha consume one spoonful of root paste of *Flickingeria macraei* (*Dendrobium plicatile*) with 1 g of black pepper powder on an empty stomach for 21 days. They also apply it directly on eczematous lesions [10].

3.11 *Eulophia* R. Br.

The genus consists of nearly 210 species distributed in the tropical and sub-tropical region of Africa, Asia, India, Australia, and America. Primarily, eulophias are terrestrial, but a few species have also adapted to epiphytic and lithophytic mode of living. About 28 species are found in India. These are distributed in the Northeastern Himalayas and Peninsular regions of the country. *Eulophia* species have gained prominence due to nutraceutical ability and rejuvenating and their effects, anti-fatigue, anticancer, and aphrodisiac properties. The eulophias are extensively utilized as tribal food and medicine. The eulophias may lead to newer sources of drugs in modern society.

3.11.1 *Eulophia dabia* (D.Don) Hochr (Syn. *Eulophia campestris* Wall. Ex Stapf.)

Terrestrial herb, hot- to cool-growing small- to medium-sized plants, height 20–40 cm; pseudobulb, cylindric, globose, leafless during flowering; leaves two, linear-lanceolate, acute; inflorescence erect, emerges, raceme; flowers 2.0–2.5 cm, pink; found in India, Bangladesh, Sri Lanka and Western Himalayas; flowering April–May.

Sudhanshu et al. [85] screened *Eulophia camprestis* (*Eulophia dabia*) for the presence phytochemicals, viz., alkaloids, flavonoids, tannins, saponins, glycosides, etc., and noticed that phytochemicals have considerable antioxidant activity against radical scavenging assay and could be used in malaria therapy. *Eulophia camprestis* (*Eulophia dabia*) rhizomes are used as a tonic and effective in curing stomach problems, cough, paralysis, and also used as aphrodisiac [86], for worm infestation, and for scrofula [40].

3.11.2 *Eulophia epidendraea* (J.Koenig ex Retz.) C.E.C.Fisch

Terrestrial medium-sized; pseudobulbs subterranean; leaves ovoid, linear, two to three; inflorescence one- to two-flowered; flowers fragrant; flowering December–January, found in India, Bangladesh, Sri Lanka, and Western Himalayas.

The leaves and tubers of *E. epidendraea* are traditionally used for the treatment of tumor, abscess, and wound healing of animals [87]. Maridass and Ramesh [88] isolated four phytochemicals (β-sitosterol, β-sitosterolglucoside, β-amyrin, and β-lupeol) from the tubers and four flavonoids (apigenin, luteolin, kaempferol, and quercetin) from the leaves. The ethanolic extract of tuber (500 mg kg^{-1}) of *E. epidendraea* has antidiarrheal activity [89]. Maridass [27] reported significant healing response in rats treated with 100 mg ml^{-1} *E. epidendraea* tuber extract. The tuber extract of *E. epidendraea* has a stimulating effect on collagen synthesis. Maridass et al. [87] observed a reduction of blood glucose levels when alloxan-induced rats were induced with 200 mg ethanolic tuber extract *E. epidendraea*. Besides antidiabetic ethanolic extract of tubers also has hypoglycemic and protective effects in the liver and kidney, a significant weight gain was observed in rats when they were administered with 100 and 300 mg kg^{-1} body weight.

3.11.3 *Eulophia herbacea* Lindl.

Terrestrial herb, small to medium in size, cool-growing; pseudobulb above the ground, ovoid; leaves two to three, linear-lanceolate; inflorescence erect six- to ten-flowered; flowers fragrant; flowering June; found in China, Bangladesh, India, Western Himalayas, Laos, Myanmar, and Thailand at an elevation ranging from 1000 to 2000 m.

E. herbacea is used in the treatment of tumor of the scrofulous gland of the neck [90]. It is also used to make salep and nutritious beverages from dried tubers of various species of orchids including *Eulophia.* It shows multiple activities such as anticancer, nutritional, antihyperlipidemic, antioxidant, antiarthritic, anti-inflammatory, antimicrobial, and immunomodulator [90]. Phytochemical analysis revealed the presence of acidic compounds, carbohydrates, amino acids, mucilage, tannins, steroids and triterpenoid [91]. Crushed bulbs are fried in mustard oil, and the residue is applied on rheumatism thrice a day till cure [92]. 10 g dried tuber of *E. herbacea*, 5 g dried leaves of *Withania somnifera*, 5 g dried leaves of *Curculigo orchioides*, and 5 g black pepper are grounded into powder taken orally with a cup of water for 20 days against aphrodisiac by Dongria Kandha tribe of Niyamgiri Hills in Odisha, India. The leaf decoction is also used against vermifuge [10].

3.11.4 *Eulophia nuda* Lindl. (Syn. Eulophia spectabilis Suresh)

Terrestrial, large-sized, hot- to warm-growing herb. Pseudobulbs subterranean, round, leafless at flowering; leaves 3–4 elliptic-lanceolate, acuminate appear after flowering; inflorescence erect, 20–30 long, many-flowered (12–20); flowers greenish purple.

The tubers of *E. nuda* are used to treat a tumor, scrofulous infection, anthelmintic, bronchitis, and juice from the roots is used to treat the snakebite [92]. Salep or Salam Mishri prepared from the roots of *E. nuda* and *Orchids latifolia* and said to be aphrodisiac addition to other medicinal properties [24]. Jagdale et al. [93] tested aphrodisiac claims of these two orchids and found that found crude drugs are nontoxic up to a level of 2 g kg^{-1} body weight in adult rats. Administering the crude extract of *E. nuda* and *O. latifolia* to adult rats increased mounting behavior but *Orchis* fed rats had higher mounts than the *Eulophia* fed. The remarkable increase in the organ weights, as well as sperm counts, a significant increase in the protein, hemoglobin, and testosterone content, was observed over the control. *E. nuda* contains phytochemically active compounds such as alkaloids, flavonoids, saponins, cardiac glycosides, terpenoids, and steroids having significance in medicine [94].

3.11.5 *Eulophia ochreata* Lindl.

Terrestrial plant 20–30 cm tall, warm- to cool-growing herb; pseudobulb conical ovoid; leaves two to five, oblong-lanceolate to ovate, plicate; inflorescence erect, many-flowered; flowers golden yellow; flowering in June; endemic, found in the Western Ghats, India, at an elevation of 1000 m.

Tubers of *E. ochreata* are rich source of plant fiber, protein, and carbohydrates which can be consumed as nutrition for healthy growth and adequate protection against diseases arising from malnutrition; however, a study on anti-nutritional/ toxicological factors and biological evaluation of nutrient contents is necessary before recommending them as source of food. *E. ochracea* tubers contain anti-nutrients (alkaloids, saponins, and steroids) and need adequate processing before they are consumed or fed to animals [95]. Tubers of *E. ochreata* are used for the treatment of stomachache and as an astringent, anti-fatigue, aphrodisiac, and anthelminthic and act as a blood purifier. The tubers are also used in cough, cold, and heart troubles [96]. In Rajasthan, *E. ochreata* is locally known as Mishri, and the crushed bulbs are used to cure diarrhea [97]. Tubers of *E. ochreata* are a potential source of natural antioxidant and can be exploited for its nutraceutical and medicinal properties [98]. The tubers contain a large quantity of white mucilage; they are astringent, used as a nutritive tonic, aphrodisiac, anthelmintic, and blood purifier [96, 99, 100]. They are also used to cure cough, cold, and heart troubles [26, 101, 102]. It is mainly used as a tonic by the local tribes in the study area.

3.12 *Habenaria* R. Br.

Terrestrial herbaceous plants bear ellipsoid, fusiform, or ovoid underground tubers; the leaves fall off during winters, and new leaves emerge during early summer or monsoon. The genus consists of about 600 species distributed in America, Asia, and Africa. In India, the genus is represented by 61 species, occurring in Himalayan, Central, and Peninsular India. *Riddhi* (*Habenaria intermedia*) and *Vriddhi* (*Habenaria edgeworthii*) are the critical ingredients of *Ashtaverga* in *Ayurvedic* system of medicine.

3.12.1 *Habenaria commelinifolia* (Roxb.) Wall. ex Lindl.

Terrestrial, warm- to cool-growing herb found in the Himalayas, India, Nepal, Myanmar, Thailand, and Vietnam at an elevation ranging from 375 to 2000 m. Plants are 90 cm tall; tubers ellipsoidal; leaves oblong, lanceolate; inflorescence raceme many-flowered (10–15), species flowers during the end of the monsoon.

The tubers of this species have long been utilized to cure spermatorrhea among tribal of Orissa. To cure spermatorrhea, equal quantity of dried tubers of *Habenaria commelinifolia* and roots of *Saraca indica* are boiled in one liter of water till the volume is reduced to 100 ml. The decoction (6–8 drops) is administered to a patient orally on an empty stomach for 10 days [10]. The tubers of this plant can cure snakebite and snakebite wounds, roots cure fever, and the whole plant can cure nose bleeding [62].

3.12.2 *Habenaria edgeworthii* (Hook f. ex Colett) R.K.Gupta (Syn: *Habenaria acuminata* Lindl. syn *Platanthera edgeworthii* (Hook. f. ex Collett) R.K. Gupta; *Herminium edgeworthii* (Hook.f. ex Collett) X.H.Jin.)

The species has been placed under genus *Herminium* as *Herminium edgeworthii* (Hook.f. ex Collett) X.H.Jin. In India, this species is popularly known as *Habenaria edgeworthii*, hence described under *Habenaria*. The species is found in Western Himalayas on open grassy slopes in association with *Satyrium nepalense* at altitude ranging from 2500 to 3000 m. The plants are 40–80 cm tall and bear many small flowers with 15–25-cm-long inflorescence; it flowers during the month of September.

The dried tubers are commonly known as *Vriddhi* and important constituent of Ashtavarga. The tubers of this species are used to prepare *Chyavanprash* and also used as health tonic, blood purifier, and rejuvenator [103].

3.12.3 *Habenaria intermedia* D.Don.

The species is found in moist open grasslands of the Western and Eastern Himalayan regions of the country at an elevation ranging from 2000 to 2500 m. Plants are up to 50 cm tall and bear 10–15-cm-long plicate ovate-oblong, acuminate leaves. Inflorescence four- to six-flowered; flowers are large 5–6 cm across. The species flowers from July to September.

The underground tubers of this plant are steamed or boiled and dried (*Ridhi*) before marketing. Tubers are used in *Asokaghrta*, *Amrtaprasa Ghrta*, *Dashmularishta*, and *Chagaladya Ghrita*, and it cures fever, e.g., *Ksaya*, *Raktavikara*, and *Murchha* (http://www.ayurveda.hu/api/API-Vol-5.pdf). It is a constituent of *Chyavanprash* and also used as a tonic, expectorant, rejuvenator, and lifespan promoter [30, 102]. The tubers are known to promote intellect, aphrodisiac, depurative, anthelmintic, rejuvenating, and tonic. Habu et al. [104] reported that tubers of *H. intermedia* have anti-stress/adaptogenic activity due to presence of scopoletin and gallic acid or their synergistic properties and the activity might be mediating through an antioxidant mechanism.

3.12.4 *Habenaria longicorniculata* Graham

Terrestrial, medium-sized warm- to cool-growing herb distributed in India and Sri Lanka at an elevation of 800–1000 m. Plants up to 80 cm tall; tubers unequal ellipsoid; leaves five to seven clustering at base, lanceolate, acute at tip; inflorescence with one to four scented flowers. The species is distributed in central and peninsular regions of the country.

The tuber paste is mixed with an equal amount of turmeric powder and applied externally on affected parts for 2 weeks to cure leucoderma [50].

3.12.5 *Habenaria marginata* Colebr.

Terrestrial, hot- to cool-growing herb found in China, India, Nepal, Bangladesh, Myanmar, and Thailand at an elevation of 100–1200 m. Tubers cylindrical; leaves sessile linear-oblong, margined with yellow; inflorescence many-flowered; blooms during summer. The plant can be easily identified by the presence of a cluster of marginated leaves at the base, 6–8 small greenish yellow flowers on 10–25-cm-tall plants during monsoon.

The tubers (250 g) are boiled in 1 l of water until the volume is reduced to 250 ml, and then decoction is mixed with 5 ml of honey and taken daily on an empty stomach for 14 days to cure malignant ulcer. According to Joshi et al. [30], thoroughly boiled plant extract is useful in curing flatulence.

3.12.6 *Habenaria roxburghii* Nicolson

Terrestrial, hot- to warm-growing herb with two to three orbicular, elliptic leaves lying on the ground. Plants are up to 30 cm tall; inflorescence 20–25-cm-long bearing 10–20 white fragrant flowers. The species is endemic to the Western Ghats and found in shady locations of forest floors. It flowers from July to September.

The tubers' extract of this species eaten with sugar gives a cooling effect to the body [54]. Eating tubers before breakfast controls burning micturition. Tuber also cures from diabetes [105]. The other species of *Habenaria* also used in herbal medicine. The crushed leaves of *Habenaria pectinata* can cure snakebite and snakebite wounds, whereas tubers relieve arthritis [20]. The mixture of *Habenaria plantaginea* tubers, black pepper (*Piper nigrum*), and garlic (*Allium sativum*) is useful for curing chest and stomach pain [54]; *Habenaria susannae* are used to cure blebs or bullae occurring on the palm [20].

3.13 *Malaxis* Sol ex. Sw.

The genus *Malaxis* consists of nearly 300 species distributed from tropical to temperate regions of the world. Nineteen species are found in India. They are distributed in temperate to tropical regions of the country. Two species, namely, *M. acuminata* (Rishabhaka) (Now known as *Crepidium acuminatum*) and *M. muscifera* (Jeevika), are components of Ayurvedic formulation Ashtaverga. The dried pseudobulbs are used for the preparation of herbal tonic (Chyawanprash) and to cure tuberculosis. Lack of proper production supply system and increasing demands of herbal drugs are promoting the practice of adulteration and substitution, causing the degradation of the desired therapeutic effect of plant species used in Ayurveda [106].

3.13.1 *Malaxis acuminata* D. Don (Syn Microstylis wallichii Lindl.) (Currently known as *Crepidium acuminatum* (D. Don) Szlach.)

Terrestrial rarely lithophytic herbaceous perennial found under shady locations of forest floors. It is distributed in India (Himachal Pradesh to Arunachal Pradesh, Madhya Pradesh), Nepal, Bhutan, China, Thailand, and Myanmar. *M. acuminate* (*Crepidium acuminatum*) plant up to 25 cm height, distributed all over India, 2000–3000 m altitude. Pseudobulbs are conical fleshy smooth, shiny 2–9 cm long, and mucilaginous.

Powder made from dried pseudobulbs is used in Chyavanprash, a nutritive tonic. It is reported to cure tuberculosis, bleeding diathesis, fever, phthisis, and bronchitis, soothe burning sensation, and enhance sperm production (spermatogenesis) [30, 56, 103]. *M. auminata* pseudobulbs contain piperitone, citronellal, eugenol, Limonene, 1,8-cineole, *p*-cymene, *O*-Methylbatatasin, and cetyl alcohol and two sugars, namely, glucose and rhamnose [107]. Recently, one sterol, β-sitosterol, has also been isolated from the ethyl acetate fraction of pseudobulbs. The plant has become vulnerable in its natural habitat. The loss of forest cover and an extensive collection of rhizomes from the wild are major causes of its threats in natural habitat.

3.13.2 *Malaxis muscifera* (Lindl.) Ktze.

Terrestrial, cool-growing herb, pseudobulbs long-sheathed and many-nerved; leaves two not equal, ovate-lanceolate; inflorescence ribbed, erect, many-flowered; flowers small; flowering time June to August. The species is distributed in India, Bhutan, and Nepal at an elevation ranging from 2600 to 4300 m.

The herb is an important ingredient of *Chyavanprash*. Additionally, it is used to cure disorders related to blood, inflammation, male sterility, fever, dysentery, external and internal hemorrhage, and general weakness. It is an aphrodisiac. It is also used for insect bite and rheumatism [102, 108].

3.14 *Pholidota* Lindl. ex Hook.

Plants are lithophyte or epiphyte with creeping rhizome; often the leaves are in pairs and sometimes solitary; inflorescence are drooping and densely flowered. It consists of 27 species distributed from India to South China, Malaysia, Indonesia, and New Guinea. Seven species are found in India. *Pholidota chinensis* and *Pholidota pallida* are medicinally important.

Pseudobulbs are crushed to expel the spines and applied on the body [22]. For curing inflammation, freshly collected pseudobulbs are made into a paste including the juice from the coconut kernel, and the paste is applied on the inflamed areas until the swelling or inflammation subsides [22]. Rhizome paste is applied for finger abscess [54]. Finely macerated pseudobulbs are made into a paste with mustard oil and applied to joints to remove rheumatic pains [109]. Pseudobulbs of *P. chinensis* are used to cure duodenal ulcer, scrofulous glands of neck, and toothache, and tincture made from pseudobulb are used to treat bronchitis [110]. Pseudobulb of *P. imbricata* extract can cure abdominal pain, and rheumatism and leaf and root paste when applied externally heals fractures [105]. *P. pallida* pseudobulbs are used for getting rid of intestinal worms and abdominal pain, and roots are used for rheumatism [20]. Bi et al. [111] isolated *n*-nonacosane, cyclopholidone, *n*-dotriacontanoic acid, *n*-octacostyl ferulate, cyclopholidonol, cycloneolitsol, and beta-sitosterol, respectively, from *P. chinensis*. Guo et al. [112] isolated six stilbenoids, (bibenzyldihydrophenanthrene) ether, from *P. yunnanensis*, and all the isolated compound were found to inhibit nitric oxide production in a murine macrophage-like cell line activated by lipopolysaccharide and interferon gamma.

3.15 *Rhynchostylis* Blume

The genus consists of three to four species distributed from India to the Philippines. Two species are found in India. Only *Rhynchostylis retusa* (Linn.) Bl. is used for medicinal purpose. A single leaf is made into a paste without using water and applied externally to cure throat inflammation [22]. Fresh roots 3–4 g and 2 g leaf bud of *Pisum sativum* are made into a paste; about 1 g paste is administered to the patient orally with water on an empty stomach two times a day for 7 days to cure blood dysentery. The plant is also used for softening the skin, and the leaf paste is applied externally for healing wounds [10]. The whole plant preparation is used to treat asthma, tuberculosis, infantile epilepsy, kidney stone, and menstrual disorders [105]. Methanol leaf extract of *R. retusa* has analgesic and anti-inflammatory activity. Methanolic leaf extract inhibited acetic acid-induced writhing and carrageenan-induced paw edema in mice when they were administered with 200 and 400 mg kg^{-1} and 100 and 200 mg kg^{-1} extract, respectively [113]. For treating malarial fever, decoction of the fresh roots is made and stored. 5 g paste of young shoot of *Andrographis paniculata* (Burm. f.) Wall. ex Nees along with 100 ml of this decoction is taken orally twice a day for 5 days day till it is cured [21].

3.16 *Vanda* Jones ex R Br.

Vanda is the most important genus known for its horticultural as well as therapeutic value. The name *Vanda* owes its origin from *Sanskrit* name for *Vanda tessellata*. This genus consists of 70 plus species distributed in India, Indonesia, Malaysia, Thailand, China, the Philippines, China, and Australia. There are 13 species known to occur within the political boundaries of India, among which *V. spathulata* and *V. tessellata* are medicinally important. Most of the species under this genus are epiphytic or lithophytic with monopodial growth habit and strap-shaped leaves arranged in a fishbone pattern. Flowers are large, and color ranges from green to blue.

3.16.1 *Vanda spathulata* (L.) Spreng.

The golden yellow flowers of *V. spathulata* are being eaten raw to reduce asthma or powdered and 0.5–1 g powder mixed with honey and taken twice a day orally to get rid of asthma. Flowers are also consumed against asthma and mania [26]. Flowers are taken to cure tuberculosis, asthma, and mania. Plant juice is given to temper bile and to abate frenzy [22].

3.16.2 *Vanda tessellata* (Lindl.) Rchb. f. (Syn. *Vanda roxburghii* R. Br.)

This perennial herb is generally called a "Rasna" in Ayurvedic system of medicine. Roots, leaves, or whole plant are used to treat various kinds of ailments. The roots are used to treat rheumatism and similar disorders [54]. The aerial roots and leaves are grounded with a tender bud of *Phoenix loureirii*, and the paste is plastered over fractured bone, and the extract of the same five spoonful is served twice a day orally till cure by Konda Reddis and Koyas tribe in Andhra Pradesh [54]. The roots of this plant found to have alexiteric and antipyretic properties. Root enters into the composition of various medicated oils for external application in nervous disorders and rheumatism [26]. The paste made from the leaves when applied on body alleviates fever. Shanavasakhan et al. [22] carried out an extensive survey for the medicinal use of this plant in Kerala and noticed that purpose and method of its use varied with the region. In Palakkad, leaf juice is applied for earache; in Kollam, leaf poultice is applied to relieve sprains, lumbago, and back pain, and Waynad juice from the leaves and aerial roots mixed with neem oil and garlic is used for treating earache. It has also been used to treat bronchitis inflammation, hiccup, piles, and boils on the scalp [114]. Usman et al. [115] studied pharmacognostical properties of "Rasna" and found dried roots are brown, fragrant, bitter, and longitudinally wrinkled, whereas the powder is muddy brown in color. Kumar et al. [116] reported aphrodisiac property of alcoholic extract of *V. tessellata* flowers. Alcoholic extract of flowers of *V. tessellata* at doses of 50 and 200 mg kg^{-1} increased mating performance and also increased male and female ratio of resulting offspring without any toxicity. Methanol and acetate extract of *V. tessellata* is found to have antibacterial activity against a number of pathogenic and fungi. Melanin (VR1), a compound that was isolated, has a very strong activity against bacteria [117].

Some other Vanda species like *V. coerulea*, *V. teres* (now known as *Vandopsis undulata*), and *V. cristata* (Syn. *Trudellia cristata*) also used for medicinal purpose. The flower juice of *V. coerulea* is used as an eye drop solution to cure glaucoma and blindness [118]. Leaves of *V. cristata* are used to cure cough, whereas the extract from leaves is used to inhibit a number of foodborne pathogens like *Klebsiella pneumoniae*, *E. coli*, and *Salmonella typhi* [119, 120].

4 Conclusion

India has a long tradition of using orchids and other medicinal plants in traditional medicine system such as Ayurveda, Siddha, Unani, and Homeopathy. However, overexploitation and apathy for use modern techniques for commercial use have pushed many medicinal orchids to rarity. A very few orchids in India have been subjected to search for active principle compound and clinical evaluation despite the fact that a plethora of information is available on their use in traditional as well as tribal medicine system. Further, more emphasis has to be laid down on mass propagation, cultivation practices, and post-harvest handling practices along with breeding of medicinal orchids for a better quality of products.

References

1. Jeffrey D, White JD, O'Keefe BR, Sharma J, Javed G, Nukala V, Ganguly A, Khan IA, Kumar NB, Mukhtar HF, Pauli GF, Walker L, Sivaram Rajaraman SP, Trimble EL (2016) India-United States dialogue on traditional medicine: toward collaborative research and generation of an evidence base. J Glob Oncol. https://doi.org/10.1200/JGO.17.00099. Published online November 16, 2017

2. Teoh ES (2016) Medicinal orchids of Asia. Springer International Publishing, Cham
3. Virk JK, Gupta V, Kumar S, Singh R, Bansal P (2017) Ashtawarga plants – suffering a triple standardization syndrome. J Tradit Complement Med 7:392–399
4. Kong JM, Goh NK, Chia LS, Chia TF (2003) Recent advances in traditional plant drugs and orchids. Acta Pharmacol Sin 24:7–21
5. Prasad LV. (2002) In: Chaudhury RR, Rafei UM (eds) Indian system of medicine and homoeopathy traditional medicine in Asia. WHO-Regional Office for South East Asia, New Delhi, pp 283–286
6. Ravishankar B, Shukla VJ (2007) Indian system of medicine: a brief profile. Afr J Tradit Complement Altern Med 4:319–337
7. Kant R, Verma J, Thakur K (2012) Distribution pattern, survival threats and conservation of Ashtavarga orchids in Himachal Pradesh Northwest Himalaya. Plant Arch 12:165–168
8. Sharma P, Mahajan N, Garg P, Singh G, Dadhwal S, Sharma S (2011) *Malaxis acuminata*: a review. Int J Res Ayurveda Pharm 2:422–425
9. Middha A, Purohit S (2011) Determination of free radical scavenging activity in herbal supplement: Chyawanprash. Int J Drug Dev Res 3:328–333
10. Dash PK, Sahoo S, Subhasisa B (2008) Ethnobotanical studies on orchids of Niyamgiri Hill ranges, Orissa, India. Ethnobot Leafl 12:70–78
11. Panda SP, Sharief MU, Hameed SS, Pramanik A (2015) Traditional phytotheraphic record of orchids of Odisha and their conservation strategies. Ann Plant Sci 4:1204–1207
12. Ray S, Sainkhediya J, Chouhan DS (2015) Rare and threatened flowering plants of Harda. IJSR 6:2119–2123
13. Jabin F (2011) A guiding tool in Unani Tibb for maintenance and preservation of health: a review study. Afr J Tradit Complement Altern Med 8:140–143
14. Develi IN, Dönmez MN, Kozan N, Karababa E (2015) Rheological properties of salep powder-milk mixture. J Food Sci Technol 52:6556–6564
15. Pragada PM, Rao GMN (2012) Ethnoveterinary medicinal practices in tribal region of Andhra Pradesh, India. Bangladesh J Plant Taxon 19:7–16
16. Ekka A (2013) Some plants used by Hill-Korwa in their health care from Chhattisgarh. Int J Life Sci Biotechnol Pharma Res 2:198–203
17. Deb CR, Deb MS, Jamir NS, Imchen T (2009) Orchids in indigenous system of medicine in Nagaland, India. Pleione 3:209–211
18. Perme N, Choudhury SN, Choudhury R, Natung T, De B (2015) Medicinal plants in traditional use at Arunachal Pradesh, India. Int J Phytopharm 5:86–98
19. Shapoo GA, Kaloo ZA, Ganie AH, Singh S (2013) Ethnobotanical survey and documentation of some orchid species of Kashmir Himalaya, J&K-India. Int J Pharm Biol Res 4:32–40
20. Hossain MM (2011) Therapeutic orchids: traditional uses and recent advances: an overview. Fitoterapia 82:102–140
21. Tiwari AP, Joshi B, Ansari AA (2012) Less known ethnomedicinal uses of some orchids by the tribal inhabitants of Amarkantak Plateau, Madhya Pradesh, India. Nat Sci 10:33–37
22. Shanavaskhan AE, Sivadasan M, Alfarhan AH, Thomas J (2012) Ethnomedicinal aspects of angiospermic epiphytes and parasites of Kerala, India. Indian J Tradit Knowl 11:250–258
23. Kuvar SD, Bapat UC (2010) Medicinal plants used by kokani tribals of Nasik district Maharashtra to cure cut and wounds. Indian J Tradit Knowl 9:114–115
24. Anonymous (2006) The wealth of India, revised edn. Council of Scientific and Industrial Research, New Delhi
25. Mishra SB, Dwivedi S, Shashi A, Prajapati K (2008) Ethnomedicinal uses of some plant species by ethnic and rural peoples of the Salem district of Tamil Nadu with special reference to the conservation of vanishing species. Ethnobot Leafl 12:873–887
26. Chopra RN, Chopra IC, Nayar SL (1956) Glossary of Indian medicinal plants. Council of Scientific and Industrial Research, New Delhi
27. Maridass M, Raju G, Ghanthikumar S (2008) Tissue – regenerative responses on tuber extracts of *Eulophia epidendraea* (Retz.) Fischer, in: Wistar Rat. Pharmacologyonline 3:631–636

28. Anuradha V, Rao P (1994) Praemorsin, a new phenanthropyran from *Acampae praemorsa*. Phytochemistry 37(3):909–910
29. Kala S, Senthilkumar S (2010) Antimicrobial activity of Acanthephippium bicolor, Lindley, Malaysian Journal of Microbiology 6(2):140–148
30. Joshi GC, Tewari LM, Lohani N, Upreti K, Jalal JS, Tewari G (2009) Diversity of orchids in Uttarakhand and their conservation strategy with special reference to their medicinal importance. Rep Opin 1:47–52
31. Anuradha V, Prakash NS (1998) A phenanthropyran from *Aerides crispum*. Phytochemistry 48:185–186
32. Linthoingambi L, Das AK, Singh PK, Ghosh SK (2013) Medicinal uses of orchids by tribes in India: a review. Int J Curr Res 5:2796–2798
33. Ghanaksh A, Kaushik P (1999) Antibacterial effect of *Aerides multiflora* Roxb. A study in vitro. J Orchid Soc India 13:65–68
34. Majumder PL, Banerjee S, Sen S (1995) Stilbenoids from the orchids *Agrostophyllum callosum* and *Coelogyne flaccida*. Phytochemistry 39:649–653
35. Majumder PL, Banerjee S, Sen S (1996) Three stilbenoids from the orchid *Agrostophyllum callosum*. Phytochemistry 42:847–853
36. Majumder PL, Banderjee S, Lahari S, Mukhoti N, Sen S (1998) Dimeric phenanthrenes from two *Agrostophyllum* species. Phytochemistry 47:855–860
37. Majumder PL, Sen S, Banerjee S (1999) Agrostophyllol and isoagrostophyllol, two novel diastereomeric 9,10-dihydrophenanthropyran derivatives from the orchid *Agrostophyllum callosum*. Tetrahedron 55:6691–6702
38. Majumdar PL, Majumder S, Sen S (2003) Triterpenoids from the orchids *Agrostophyllum breirps* and *Agrostophyllum callosum*. Phytochemistry 62:591–596
39. Biswas A, Bari MA, Roy M, Bhadra SK (2010) Inherited folk pharmaceutical knowledge of tribal people in the Chittagong Hill tracts Bangladesh. Indian J Tradit Knowl 9:77–89
40. Singh A, Duggal S (2009) Medicinal orchids: an overview. Ethnobot Leafl 13:351–363
41. Majumder PL, Ghosal S (1991) Arundinol, a new triterpene from the orchid *Arundina bambusifolia*. J Indian Chem Soc 68:88–91
42. Majumder PL, Ghosal S (1993) Two stilbenoids from the orchid *Arundina bambusifolia*. Phytochemistry 32:439–444
43. Majumder PL, Ghosal S (1994) Two stilbenoids from the orchid *Arundina bambusifolia*. Phytochemistry 35:205–208
44. Ahmad N (1997) Wild flowers of Bangladesh. UPL, Dhaka
45. Liu MF, Ding Y, Zhang DM (2005) Phenanthrene constituents from rhizome of *Arundina graminifolia*. Zhongguo Zhong Yao Za Zhi 30:353–356
46. Liu MF, Han Y, Xing DM, Shi Y, Xu LZ, Du LJ, Ding YI (2004) A new stilbenoid from *Arundina graminifolia*. Asian Nat Prod Res 6:229–232
47. Liu MF, Han Y, Xing DM, Wang W, Xu LZ, Du LJ, Ding Y (2005) One new benzyldihydrophenanthrene from *Arundina graminifolia*. Asian Nat Prod Res 7:767–770
48. Liu M, Lv H, Ding Y (2012) Antitumoral bibenzyl derivatives from tuber of *Arundina graminifolia*. Zhongguo Zhong Yao Za Zhi 37(1):66–70
49. Lalitharani S, Mohan VR, Maruthupandian A (2011) Pharmacognostic investigations on *Bulbophyllum albidum* (Wight) Hook. f. Int J PharmTech Res 3:556–562
50. Saranya B, Gopalan R, Narmatha bai V, Mahendran G (2012) Rare and threatened orchid species used for various diseases by Irulars, Vellingiri hills, Coimbatore. Int J Sci Res Publ 2:1–2
51. Kalaiarasan A, Kumar P, Ahmed John S (2012) Bio-chemical investigation of *Bulbophyllum kaitense* rechib root by GC-MS. Eastern Ghats of India. Nat Sci 10:29–31
52. Kalaiarasan PK, Ahmed John S (2012) Antimicrobial activity of *Bulbophyllum kaitense*. Rechib. Stem eastern penisular flora in India. J Nat Sci 10:41–44
53. Kalaiarasan A, Ahmed John S (2012) In-vitro screening for anti-inflammatory activity of *Bulbophyllum kaitense* Rechib. Pseudobulb extract by HRBC method, eastern peninsular flora in South India. Int J Sci Res Publ 2:1–5

54. Rajendran AN, Rao R, Ravi Kumar K, Henry AN (1997) Some medicinal orchids of Southern India. Anc Sci Life 17:10–14
55. Barthel M (2008) Pharmaceutical compositions containing bulbophyllum and their use for treating illnesses. Patent grant No. 7939112; publication date Sep 4 2008
56. Kumari P, Joshi GC, Tewari LM (2011) Diversity and status of ethno-medicinal plants of Almora district in Uttarakhand, India. Int J Biodivers Conserv 3:298–326
57. Sharma C, Kumari T, Arya KR (2014) Ethnopharmacological survey on bone healing plants with special references to *Pholidota articulata* and *Coelogyne cristata* (Orchidaceae) used in folk tradition of Kumaon, Uttarakhand, India. Int J Pharma Res Health Sci 2:185–190
58. Majumder PL, Laha S, Datta N (1982) Coelonin, a 9.10-dihydrophenanthrene from the orchids *Coelogyne ochracea* and *Coelogyne elata*. Phytochemistry 21:478–480
59. Majumder PL, Sen S, Majumder S (2001) Phenanthrene derivatives from the orchid *Coelogyne cristata*. Phytochemistry 58(4):581–586
60. Lyudmyla B, Halyna T, Zbigniew O, Lyudmyla K, Oleksand G (2016) Antimicrobial activity screening of extrants from leaves and pseudobulbs of Coelogyne cristata Lindl. (Orchidaceae). In: The scientific proceedings of the international network AgroBioNet, agrobiodiversity for improving nutrition, health and life quality, pp 40–44
61. Pramanick DD (2016) Pharmacognostic studies on the pseudobulb of Coelogynecristata Lindl. (Orchidaceae)-an epiphytic orchid of ethno-medicinal importance. J Pharmacogn Phytochem 5:120–123
62. Jain SK (1991) Dictionary of Indian folk medicine and ethnobotany. Deep Publications, Lucknow
63. Agrawal DP (2005) Himalayan medicine system and its materia medica. http://www.indianscience.org/essays.pdf
64. Dhiman AK (2004) Medicinal plants of Uttaranchal state. Chaukhambha Sanskrit Series Office, Varanasi
65. Trivedi VP, Dixit RS, Lal VK (1980) Orchids in the drug markets of Bareilly, Kanpur and nearby districts. Nagarjun (Calcutta) 23(8):157–163
66. Moin S, Shibu BS, Wesley PS, Devi BC (2012) Bioactive potential of *Coelogyne stricta* (D. Don) Schltr: an ornamental and medicinally important orchid. J Pharm Res 5:2191–2196
67. Joseph J (1977) Quest of medicinal plants and re-establishment of their medicinal virtues. In: Atal C, Kapur BM (eds) Cultivation and utilization of medicinal and aromatic plants. Regional Research Laboratory, Jammu Tawi
68. Kovacs A, Vasas A, Hohmann J (2007) Natural phenanthrenes and their biological activity. Phytochemistry 69:1084–1110
69. Howlader MA, Alam M (2011) Central nervous system depressant effects of the ethanolic extract of *Cymbidium aloifolium* (L.). J Appl Pharm Sci 1:60–62
70. Juneja RK, Sharma SC, Tandon JS (1987) Two substituted bibenzyls and a dihydrophenanthrene from *Cymbidium aloifolium*. Phytochemistry 26:1123–1125
71. Barua AK, Ghosh BB, Ray S, Patra A (1990) Cymbinobin-A, a phenanthraquinone from *Cymbidium aloifolium*. Phytochemistry 29:3046–3047
72. Ghosh BB, Ray S, Bhattacharyya P et al (1992) Cymbinodin B, a phenanthraquinone from *Cymbidium aloifolium*. Indian J Chem 31B:557–558
73. Howlader MA, Alam M, Ahmad KT, Khatun F, Apu AS (2011) Antinociceptive and anti-inflammatory activity of the ethanolic extract of *Cymbidium aloifolium* (L.). Pak J Biol Sci 14:909–911
74. Radhika B, Murthy JVVSN, Grace DN (2013) Preliminary phytochemical analysis & anti-bacterial activity against clinical pathogen of medicinally important orchid *Cymbidium aloifolium* (L.) Sw. Int J Pharm Sci Res 4:3925–3931
75. Vij SP, Srivastav RC, Mainra AK (1992) On the occurrence of *Dactylorhiza hatagirea* (D.Don) Soo in Sikkim. Orchid News 8–9:14–15
76. Lal B, Negi HR, Singh RD, Ahuja PS (2004) Medicinal uses of *Dactylorhiza hatagirea* among the natives of higher altitudes in Western Himalaya. J Orchid Soc India 18:97–100
77. Bulpitt CJ (2005) Occasional paper: the uses and misuses of orchids in medicine. Q J Med 98:625–631

78. Kala CP (2005) Indigenous uses, population density, and conservation of threatened medicinal plants in protected areas of the Indian Himalayas. Conserv Biol 19:368–378
79. Thakur M, Dixit VK (2007) Effect of some vajikaran herbs on pendiculation activities and in vitro sperm count in male. Sex Disabil 25:203–207
80. Ranpal S (2009) An assessment of status and antibacterial properties of *Dactylorhiza hatagirea* in Annapurna Conservation Area (A case study of Paplekharka, Lete VDC, Mustang). B. Sc. Forestry Research thesis submitted to Tribhuvan University, Institute of Forestry, Pokhara
81. Izu HK, Aneko EK, Omimori TT (1999) Studies on Nepalese crude drugs, chemical constituents of panch aunle, the roots of *Dactylorhiza hatagirea* D. Chem Pharm Bull 47:1618–1625
82. Rao TA, Sridhar S (2007) Wild orchids in Karnataka. A pictorial compendium, Institute of Natural Resources Conservation, Education, Research and Training (INCERT), Bangalore
83. Nagananda GS, Satishchandra N (2013) Antimicrobial activity of cold and hot successive pseudobulb extracts of *Flickingeria nodosa* (Dalz.) Seidenf. Pak J Biol Sci 16:1189–1193
84. Chakrabarty M, Datta GK, Ghosh S, Debnath PK (2001) Induction of antioxidative enzyme by the ayurvedic herb *Desmotrichum fimbriatum* Bl. in mice. Indian J Exp Biol 39:485–486
85. Sudhanshu, Rao N, Mittal S, Vishal, Menghani E (2012) Antioxidant agents alternative source for malaria disease. Indian J Appl Pharm 4:14–16
86. Medhi RP, Chakrabarti S (2009) Traditional knowledge of NE people on conservation of wild orchids. Indian J Tradit Knowl 8:11–16
87. Maridass M, Thangavel K, Raju G (2008) Antidiabetic activity of tuber extract of *Eulophia epidendraea* (Retz.) Fisher (Orchidaceae) in alloxan diabetic rats. Pharmacologyonline 3:606–617
88. Maridass M, Ramesh U (2010) Investigation of phytochemical constituents from *Eulophia epidendraea*. Int J Biol Technol 1:1–7
89. Maridass M (2011) Anti diarrhoeal activity of rare orchid *Eulophia epidendraea* (Retz.) Fisher. Nat Pharm Technol 1:5–10
90. Anonymous (2003) The wealth of India: a dictionary of India raw material and industrial products. Council of Scientific and Industrial Research, New Delhi
91. Tatiya A, Surana S, Bhavsar S, Patil D, Patil Y (2012) Pharmacognostic and preliminary phytochemical investigation of *Eulophia herbacea* Lindl. tubers (Orchidaceae). Asian Pac J Trop Dis 2:S50–S55
92. Sikarwar RLS, Phathak B, Jaiswal A (2008) Some unique ethnomedical properties of tribal communities of Chitrakoot, Madya Pradesh. Indian J Tradit Knowl 7:613–617
93. Jagdale SP, Shimpi SD, Chachad D (2009) Pharmacological studies of salep. J Herb Med Toxicol 3:153–156
94. Ruchi KS, Upadhyaya R, Trivedi US, Tiwari S (2012) Qualitative phytochemical analysis of *Eulophia nuda* Lindl. an endangered orchid. Int J Pharm Res Bio-Sci 1:456–462
95. Aberoumand A (2010) Evaluating potential values of phytate, trypsin inhibitors and polyphenols in plants based diets. Pak J Food Sci 20:50–53
96. Nadkarni N (1976) Indian natural medical, vol 1. Popular Prakashan, Bombay
97. Meena AK, Rao MM (2009) Folk herbal medicines used by the Meena community in Ali Aberoumand and S.S. Deokule proximate and mineral composition of wild coco (*Eulophia ochreata* L.) tubers in Iran. J Food Agro-Ind 2:203–209
98. Jagtap S, Gilda S, Bhondave P, Paradkar A, Pawar P, Harsulkar A (2009) Validation of the potential of *Eulophia ochreata* L. tubers for its anti-inflammatory and antioxidant activity. Pharmacologyonline 2:307–316
99. Maridass M, Victor B, Ramesh U (2005) Ethnobotanical information of *Eulophia epidendraea* (Retz) Fischer (Orchidaceae) in the Kambli Malaikovil Forest, Tirunelveli district, Tamil Nadu. Bombay Nat Hist Soc 102:255–257
100. Jain A, Katewa S, Galav P, Ambika N (2007) Unrecorded ethnomedicinal uses of biodiversity from Tadgarh-Raoli wild life sanctuary, Rajasthan, India. Acta Bot Yunnanic 29:337–344

101. Aswal BS (1992) Less known medicinal uses of three plants from Kumaun Himalaya (India) Indian. J For 15:76–77
102. Chauhan NS (1990) Medicinal orchids of Himachal Pradesh. J Orchid Soc India 4:99–105
103. Dey AC (1982) Indian medicinal plants used in Ayurvedic preparations. Bishan Singh Mahender Pal Singh, Dehradun
104. Habbu PV, Smita DM, Mahadevan KM, Shastry RA, Biradar SM (2012) Protective effect of *Habenaria intermedia* tubers against acute and chronic physical and psychological stress paradigs in rats. Braz J Pharmacogn 22:568–579
105. Pérez Gutiérrez RM (2010) Orchids: a review of uses in traditional medicine, its phytochemistry and pharmacology. J Med Plants Res 4:592–638
106. Chinmay R, Kumari S, Bishnupriya D, Mohanty RC, Dixit R, Padhi MM, Babu R (2011) Phyto-pharmacognostical studies of two endangered species of Malaxis (Jeevak and Rishibhak). Pharmacogn J 3:77–85
107. Bhatnagar JK, Handa SS, Duggal SC (1970) Chemical investigation on *Microstylis wallichii*. Planta Med 20:157–161
108. Chauhan RS, Nautiyal MO, Prasad P, Purohit H (2008) Ecological features of an endangered medicinal orchid – *Malaxis muscifera* (Lindley) Kuntze in Western Himalaya. McAllen Int Orchid Soc J 9:8–12
109. Sarkar PK, Agrawal VV (1978) Notes on *Pholidota pallida* Lindl. and its use in Ranchi district Bihar. Bull Bot Surv India 20:182–183
110. Nadkarni AK (1954) Dr. K.M. Nadkarni's Indian materia medica, vol 2, 3rd edn. Popular Book Depot, Bombay
111. Bi ZM, Wang ZT, Xu LS, Xu GJ (2004) Studies on chemical constituents of *Pholidota yunnanensis*. Zhongguo Zhong Yao Za Zhi 29:47–49
112. Guo XY, Wang J, Wang NL, Kitanaka S, Liu HW, Yao XS (2006) New stilbenoids from *Pholidota yunnanensis* and their inhibitory effects on nitric oxide production. Chem Pharm Bull (Tokyo) 54:21–25
113. Al-Amin M, Sultana GNN, Hossain CF (2012) Analgesic and anti-inflammatory activities of *Rhynchostylis retusa*. Biol Med 3:55–59
114. Kirtikar PK, Basu BD (1975) Indian medicinal plants, 2nd edn. (Reprint Ed. 1975). M/s Bishen Singh MahendraPal Singh, Dehradun
115. Usman MRB, Chhaya G, Surekha CG, Lalit S, Samadhan B (2012) Pharmacognostical Evaluations of Some Herbal Plant. IJARPB 2(3):302–310
116. Kumar PKS, Subramoniam A, Pushpangadan P (2000) Aphrodisiac activity of *Vanda tessellata* (Roxb.) Hook.Ex Don extract in male mice Indian. J Pharmacol 32:300–304
117. Ahmad F, Sayeed A, Islam A, Abdus Salam SM, Sadik G, Sattar MA, Astaq Mohal Khan GRM (2002) Antimicrobial activity of extracts a glycoside from *Vanda roxburghii* R Br. Pak J Biol Sci 5:189–191
118. Kumar S (2002) The medicinal plants of North-East India. Scientific Publishers, Jodhpur
119. Singh MP, Dey S (2005) Indian medicinal plants. Satish Serial Publishing House, Delhi
120. Reddy CS, Pattanaik C, Murthy MSR, Reddy KN (2005) Orchids of Eastern Ghats, India. EPTRI-ENVIS Newslett 11:6–12

Traditionally Used Medicinal *Dendrobium*: A Promising Source of Active Anticancer Constituents

16

Mukti Ram Paudel, Hari Datta Bhattarai, and Bijaya Pant

Contents

Abstract

Dendrobium represents one of the most important genera of the Orchidaceae family, having medicinal and ornamental value. *Dendrobium* species have been traditionally used as first-rate medicinal herbs in the treatment of a variety of disorders, such as nourishing the stomach and enhancing the production of body fluids. Many species of this genus are the sources of tonic for astringent, analgesic, anti-pyretic, antioxidant, antimicrobial, antidiabetic, anticancer, anti-inflammatory, anti-metastasis, and antiangiogenesis because they have alkaloids, aromatic compounds, sesquiterpenoids, and polysaccharides as main components. This chapter includes the active constituents, extract and pure isolate, from 23 *Dendrobium* species and their effect in the anticancer, anti-metastasis, and antiangiogenesis.

M. R. Paudel (✉) · H. D. Bhattarai · B. Pant
Central Department of Botany, Tribhuvan University, Kathmandu, Nepal
e-mail: mr.paudel@cdbtu.edu.np; muktiram.paudel@cdb.tu.edu.np

J.-M. Mérillon, H. Kodja (eds.), *Orchids Phytochemistry, Biology and Horticulture*, Reference Series in Phytochemistry, https://doi.org/10.1007/978-3-030-38392-3_16

Keywords

Antiangiogenesis · Anticancer · Anti-metastasis · Bibenzyl · *Dendrobium* · Fluorenone · Phenanthrene

1 Introduction

Dendrobium species represent the most important plants, ornamentally and medicinally. They have been used for a thousand years as first-rate herbs and prized folk medicine in China, India, Australia, and other countries of the world [1]. In traditional Chinese medicine (TCM), they are the source of tonic, nourishing the stomach and enhancing the production of body fluids, astringent, analgesic, and anti-inflammatory substances [2]. A common name "Shi-Hu" is used in TCM for 30 species of *Dendrobium* under two monographs, Dendrobii Caulis (Shi-Hu) and Dendrobii Officinalis Caulis (Tie-Pi Shi-Hu) [1, 2]. The Chinese consider *Dendrobium* as one of the 50 fundamental herbs used to treat all kinds of ailments and use *Dendrobium* tonic for longevity [3]. In Indian Ayurvedic system of medicine, the commonly used orchids are "Salem" (*Orchis latifolia* and *Eulophia latifolia*), "Jewanti" (*Dendrobium alpestre*), "Shwethuli" (*Acampe papillosa*), and "Rasna" (*Vanda tessellata*) [4]. *Dendrobium teretifolium* and *D. discolor* have been used for treating different ailments as dysentery, relieving pain, and controlling ringworm, and *Dendrobium speciosum* has been employed as emergency bush food in Australian traditional medicine [1].

In light of traditional importance as a medicinal plant, knowledge on the constituents of various *Dendrobium* species and their pharmacological activities has been growing, and methodologies have been developed for effective propagation [5, 6]. Indeed, a large number of pharmacological activities have been assigned to different *Dendrobium* species, such as anti-inflammatory, anti-platelet aggregation, hepatoprotective, anti-fibrotic, antiviral, antifungal, antimicrobial, antioxidant, antidiabetic, neuroprotective, immunomodulatory, and anticancer [7, 8].

Cancer is the second biggest health problem after cardiovascular diseases accounting for an estimated 9.56 million deaths worldwide in 2017. The number of cancer deaths increased between 1990 and 2017 by 66% [9, 10]. It is caused by dysfunction of gene coding for proteins such as growth factors, growth factor receptors, anti-apoptotic proteins, transcription factors, and tumor suppressors [11]. Treatment of cancer currently includes the surgical removal of cancerous tissue, radiotherapy, chemotherapy, and a combination of chemotherapy and radiotherapy. The use of anticancer drugs (chemotherapy), while often more beneficial when used in conjugation with radiation therapy or surgery, is nonetheless a key line of treatment [10, 12, 13]. Because anticancer drugs are the mainstay of chemotherapy, it is important to discover novel anticancer drugs with diverse activity, a novel mechanism of action, and minimal issues of toxicity [14–16]. In parallel, there is increasing evidence for the potential compounds from different *Dendrobium* species

Fig. 1 Some *Dendrobium* species rich in anticancer compounds: *D. chrysanthum* (**a**), *D. fimbriatum* (**b**), *D. moniliforme* (**c**), and *D. polyanthum* (**d**). (Photo credit: B.B. Raskoti for *D. chrysanthum* and *D. polyanthum*)

(Fig. 1) as inhibitors of various stages of tumorigenesis and associated processes, underlining the importance of these products in cancer prevention and therapy [7, 17]. Cancer chemoprevention by anticancer drugs derived from *Dendrobium* species has shown promising results against various malignancies [18]. Therefore, they potentially represent an inexhaustible source of chemicals for the discovery of new anticancer drugs.

Knowing that *Dendrobium* species have been traditionally used for the preparation of natural remedies, they are worth investigating by scientists who are looking for new natural ingredients, which could become the main feature of a novel mechanism of action in cancer. This book chapter is mainly focused on the constituents of *Dendrobium* species which have been subjected to investigation of cancer.

2 Anticancer Compounds Isolated from *Dendrobium* Species

As the *Dendrobium* species have been used in traditional medicines [2, 6], their misidentification and adulteration led to a loss of therapeutic potency and potential intoxication [5]. For decades, fast-developing molecular techniques using DNA fingerprinting, DNA sequencing, and DNA microarray have been applied extensively to authenticate medicinal materials, including various *Dendrobium* species [5]. Chemical studies on *Dendrobium* species have been conducted since the 1930s, while alkaloids, aromatic compounds, sesquiterpenoids, and polysaccharides have been identified as the main components [19–22]. To date, various *Dendrobium* species are known to produce a variety of secondary metabolites. The biological activities and pharmacological actions of all the isolated compounds were investigated [23]. More than 100 compounds from 42 *Dendrobium* species including 32 alkaloids, 6 coumarins, 30 bibenzyls, 4 fluorenones, 22 phenanthrenes, and 7 sesquiterpenoids have been identified and discussed [6, 8, 18, 24, 25]. The active anticancer bibenzyls, 3,4,3′-trimethoxy-5,4′-dihydroxybibenzyl (**1**), 3,4-dihydroxy-3′,4′-dimethoxybibenzyl (**2**), 4-(3-hydroxy-4-methoxyphenethyl)-2,6-dimethoxylphenol (**3**), 4,4′-dihydroxy-3,5-dimethoxybibenzyl (**4**), 4,5,4′-trihydroxy-3,3′-dimethoxybibenzyl (**5**), 5,3′-dihydroxy-3,4-dimethoxy-bibenzyl (**6**), aloifoll (**7**), chrysotoxine (**8**), dendrocandin B (**9**), dendrocandin I (**10**), dendrocandin U (**11**), dendrofalconerol A (**12**), dendrosignatol (**13**), dengraols A and B (**14**) and (**15**), denofficin (**16**), erianin (**17**), fimbriadimerbibenzyl A, B, E, F, and G (**18–22**), gigantol (**23**), longicornuol A (**24**), moscatilin (**25**), and tristin (**26**) (Fig. 2); phenanthrenes, chrysotoxol B (**27**), confusarin (**28**), cypripedin (**29**), denbinobin (**30**), epheranthol B (**31**), lusianthrindin (**32**), moniliformediquinone (**33**), and moscatin (**34**) (Fig. 3); and fluorenones, 1,4,5-trihydroxy-7-methoxy-9*H*-fluoren-9-one (**35**) and dendroflorin (**36**) (Fig. 4), have been isolated from different *Dendrobium* species.

3 Methods of Screening Anticancer Effect

Cell proliferation, inducing apoptosis, growth suppressors, activating invasion, metastasis, and angiogenesis are the major hallmarks of cancer [26]. They acquired during the development of human cancers. Several in vitro and in vivo assays have been developed to evaluate each hallmark feature of cancer, and the selection of particular assay mainly depends on the specific research question(s) to be examined. Nowadays, a wide range of in vitro and in vivo assays (some of them are described here) are available for cytotoxicity, apoptosis, cell migration, and invasion; angiogenesis in cancer cells is commonly used in cancer drug discovery.

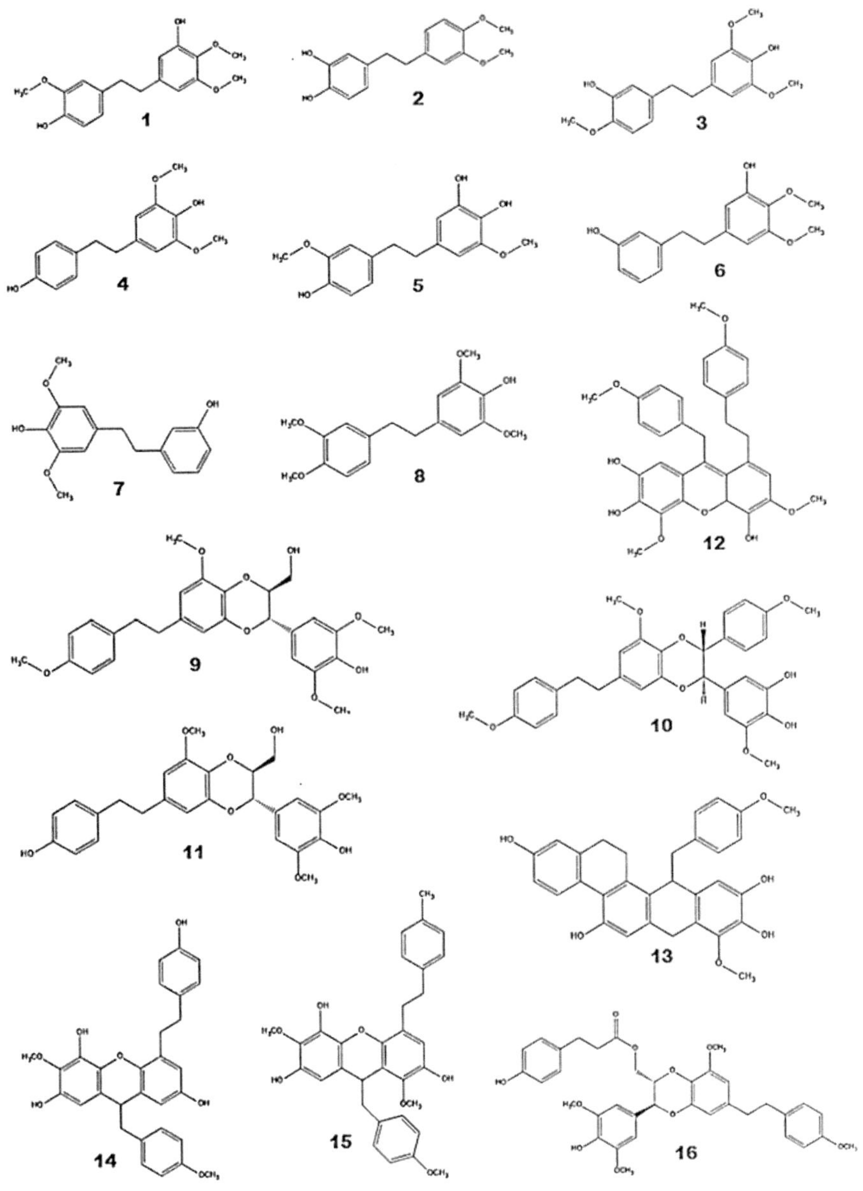

Fig. 2 (continued)

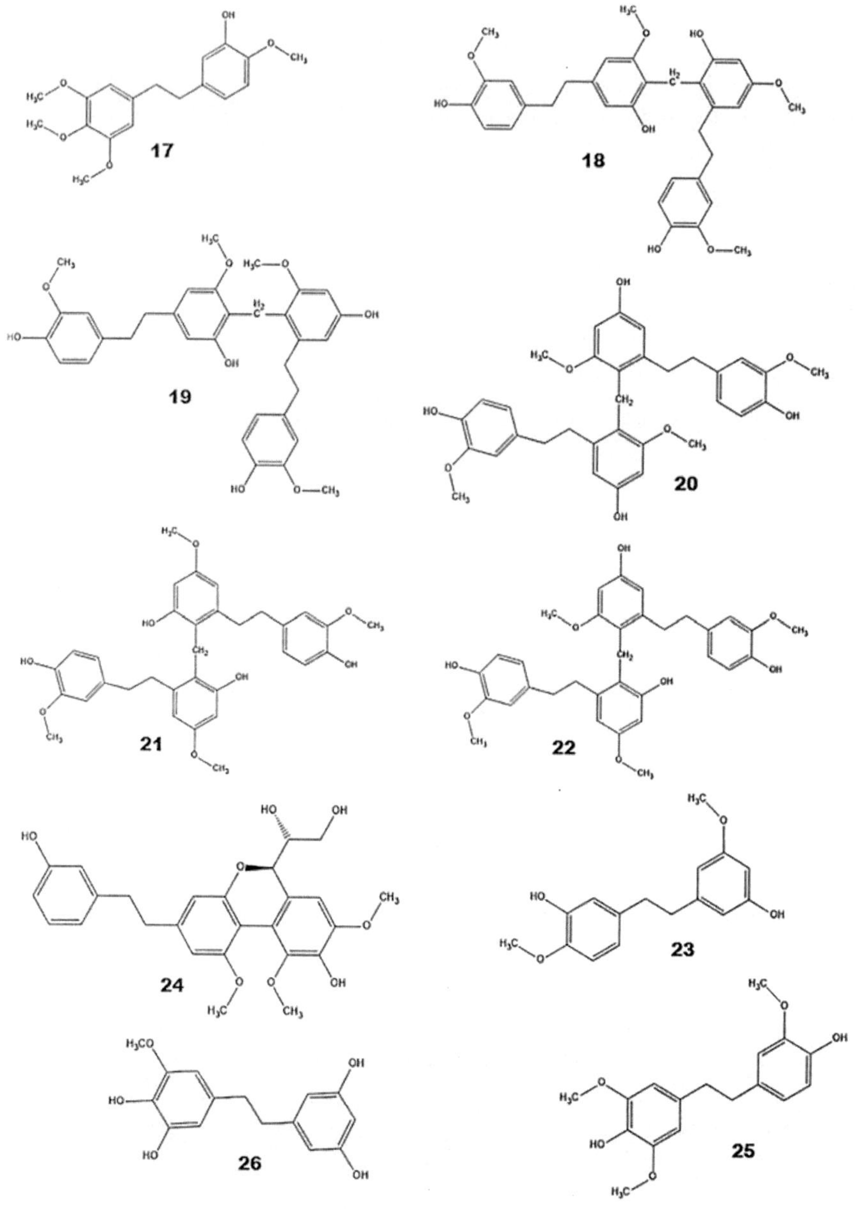

Fig. 2 Structure of anticancer bibenzyl derivatives (**1**–**26**)

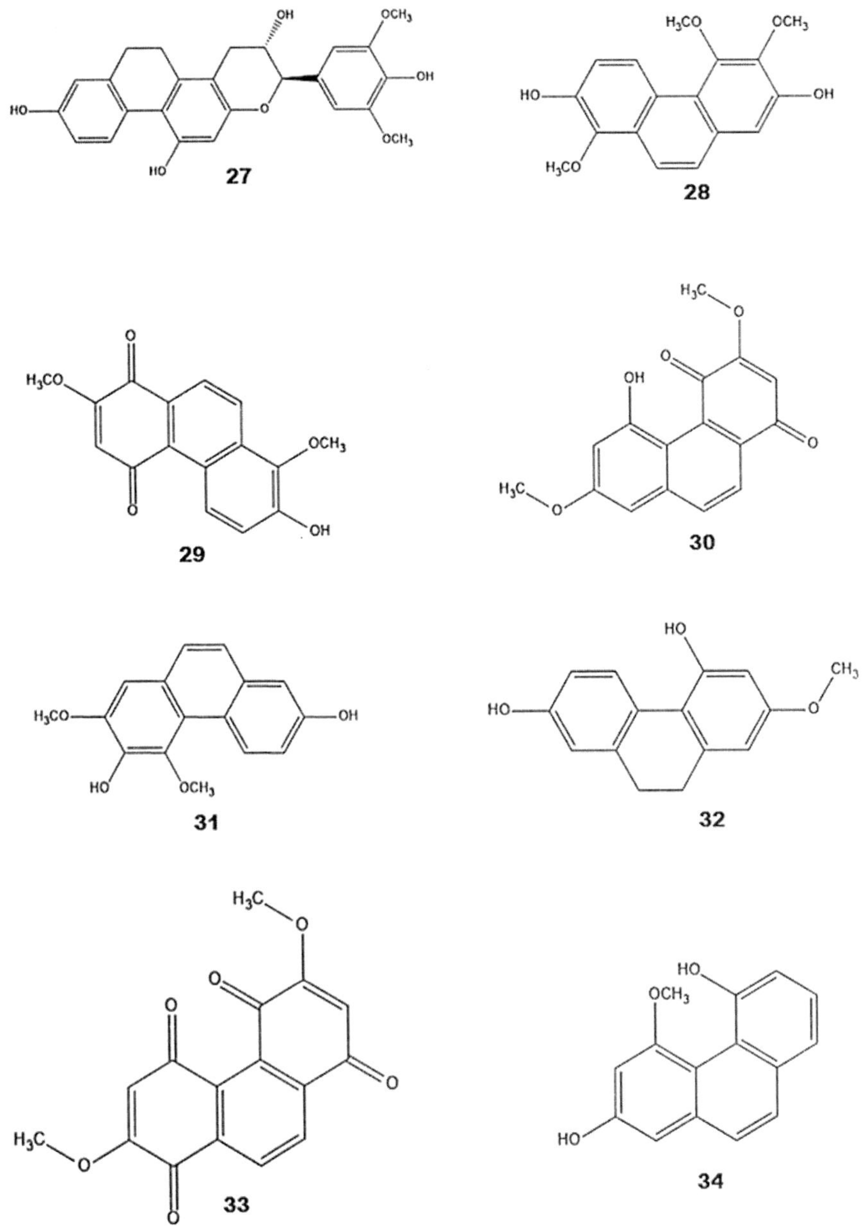

Fig. 3 Structure of anticancer phenanthrene derivatives (**27**–**34**)

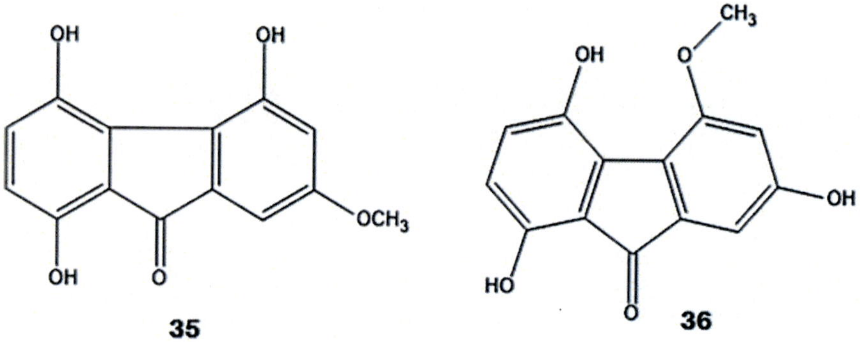

Fig. 4 Structure of anticancer fluorenone derivatives (**35** and **36**)

3.1 In Vitro Assay

A number of in vitro assays are available to investigate the anticancer potential of plant-derived extract or drug such as trypan blue dye exclusion assay, resazurin cell growth inhibition assay, LDH (lactic acid dehydrogenase) assay, SRB (sulforhodamine B) assay, MTT (3-[4,5-dimethylthiazole-2-yl]-2,5-diphenyltetrazolium bromide) assay, MTS (3-(4,5-dimethylthiazol-2-yl)-5-(3-carboxymethoxyphenyl)-2-(4-sulfophenyl)-2H-tetrazolium) assay, and XTT (2,3-bis[2-methoxy-4-nitro-5-sulfophenyl]-2*H* tetrazolium-5-carboxyanilide inner salt) assay [12, 26].

3.1.1 Trypan Blue Dye Exclusion (TBDE) Assay

TBDE assay is the most commonly performed test to assess the viability of cells. In this assay, the cells are washed with HBSS (Hank's buffered salt solution) and centrifuged for 10–15 min at 10,000 rpm, and this entire step is repeated thrice. The cells are suspended in a known quantity of HBSS, and the cell count is adjusted to 2×10^6 cells mL^{-1}. The cell suspension is distributed into Eppendorf tubes. The cells are exposed to drug dilutions and incubated at 37 °C for 3 h. Thereafter, in the dye exclusion test, i.e., the equal quantity of the drug-treated cells are mixed with trypan blue (0.4%) and incubated for 1 min. It is then loaded in a hemacytometer, and viable and nonviable cells are counted. Viable cells do not take up color, whereas dead cells take up color. However, if taken longer, live cells also generate and take up color. The percentage of growth inhibition is calculated using the following formula:

$$\text{Growth Inhibition }(\%) = 100 - (\text{Total cells} - \text{Dead cells})/\text{Total cells} \times 100$$

3.1.2 Resazurin Cell Growth Inhibition (RCGI) Assay

RCGI (also called as alamar blue) assay measures the cellular viability as well as the function of the mitochondria. In this assay, cells are treated with trypsin (0.025%) and EDTA (0.25 mM) for 5 min. Subsequently, cells are washed with PBS, counted and seeded in 96-well plate containing 5 × 10^3 cells per well, and incubated for overnight growth. Then, cells are treated with test samples and kept for 48 h incubation, followed by the addition of 20 μL of resazurin (0.01%) and further incubation for 1–2 h at 37 °C. Fluorescence of 96-well plate is measured by a multi-well plate reader at excitation and an emission wavelength of 540 and 590 nm, respectively, and inhibitory concentration (IC_{50}) values are calculated. The IC_{50} value is the amount of anticancer drug needed to inhibit 50% of cell proliferation.

3.1.3 Lactic Dehydrogenase (LDH) Assay

Lactic dehydrogenase activity spectrophotometrically measured the cellular lysates at 340 nm by analyzing NADH reduction during the pyruvate-lactate transformation. Cells are lysed with 50 mM Tris-HCl buffer with 20 mM EDTA and sodium dodecyl sulfate (SDS, 0.5%) at pH 7.4. Further cells are disrupted by sonication and centrifuged at 13,000 rpm for 15 min. The assay mixture for the enzymatic analysis consists of 33 μL of a sample in 48 mM PBS with 1 mM pyruvate and 0.2 mM NADH at pH 7.5. The percentage of LDH released is calculated as a percentage of the total amount, considered as the sum of the enzymatic activity present in the cellular lysate and that in the culture medium.

3.1.4 Sulforhodamine B (SRB) Assay

SRB is a bright pink aminoxanthene dye that binds to the basic amino acids in mild acidic condition and dissociates under basic condition. Cells are seeded in 96-well plate at 5000–10000 cells per well. The difference in cell number seeded adjusts for differences in the growth rate of the various cell lines. Cells are allowed to adhere to the wells overnight, and then the test samples are added in wells. Water is added to the control well at 1:10 dilution in a medium. The plate is incubated under 5% CO_2 at 37 °C for 3 days. The cells are fixed by the addition of cold 50% trichloroacetic acid to a final concentration of 10%. After 1 h incubation at 4 °C, the cells are washed five times with deionized water. The cells are then stained with 0.4% SRB dissolved in 1% acetic acid for 15–30 min and subsequently washed five times with 1% acetic acid to remove an unbound stain. After the plate is air-dried at room temperature, the bound dye is solubilized with 10 mM Tris base, and the plate is read on a multi-well plate reader at 595 nm.

3.1.5 3-(4,5-Dimethylthiazole-2-yl)-2,5-Diphenyltetrazolium Bromide (MTT) Assay

MTT assay, based on the conversion of the purple tetrazolium salt to yellow tetrazolium crystals of formazan of metabolically active cells, provides a quantitative determination of viable cells. Cells are seeded in the 96-well plate at a cell density of 2 × 10^5 cells per well in 100 μL of cell-cultured medium and allow to grow in a CO_2 incubator for 24 h at 37 °C. The medium is then replaced by fresh medium containing different concentrations of the test sample and allow for 48-h incubation. Thereafter, 20 μL MTT (5 mg mL^{-1} in PBS) is added to each well and incubate for 4 h. The medium is removed, and 200 μL DMSO is added to each well to dissolve the MTT metabolic product. The plate is shaken at 150 rpm for 5 min, and optical density is measured at 560 nm.

3.1.6 3-(4,5-Dimethylthiazol-2-yl)-5-(3-Carboxymethoxyphenyl)-2-(4-Sulfophenyl)-2H-Tetrazolium (MTS) Assay

Like MTT, it is also a reliable, convenient, and economical method. This method is performed in a similar way like MTT, but MTS is used to perform this assay.

3.1.7 2,3-Bis(2-Methoxy-4-Nitro-5-Sulfophenyl)-2H Tetrazolium-5-Carboxyanilide Inner Salt (XTT) Assay

In order to measure the proliferation response, the XTT assay is used. The tetrazolium salt, XTT, is especially useful in quantifying viable cells. This assay is designed for the spectrophotometric quantification of cell growth and viability without the use of radioactive isotopes and is based on the cleavage of yellow tetrazolium salt, XTT, to form an orange formazan dye by metabolically active cells. XTT cleavages into an orange formazan dye by the mitochondrial enzyme, dehydrogenase, occur exclusively in living cells. Cells are grown in a growth medium with 10% FBS in 96-well plate. They are treated with appropriate test sample for 24 h. An XTT assay is performed at the end of the incubation. Briefly, 50 mL of XTT labeling mixture solution is added to each well, and the cells are incubated at 37 °C for 4 h. The formazan crystals formed is soluble in aqueous solution, and optical density at 450 nm is compared with that of control wells with a screening multi-well plate reader. The reference wavelength is 650 nm.

3.2 In Vivo Assay

3.2.1 Induction of Ehrlich Ascites Carcinoma

Antitumor activity of the test compounds is determined using Ehrlich ascites carcinoma (EAC) tumor model in mice. The ascitic carcinoma-bearing mice (donor) are used for the study, 15 days after tumor transplantation. The animals are divided into groups of normal mice (a), tumor-bearing mice (b), tumor-bearing mice treated with the standard drug (c), and tumor-bearing mice treated with test drug (d). The ascitic fluid is drawn using an 18-gauge needle with a sterile syringe. A small amount is tested for microbial contamination. Tumor viability is determined by trypan blue exclusion test, and cells are counted by hemacytometer. The ascitic fluid is suitably diluted in normal saline to get a concentration of 10^6 cells mL^{-1} of tumor cell suspension. This is injected intraperitoneally to obtain ascitic tumor. The mice are weighed on the day of tumor inoculation and then once in 3 days thereafter. Treatment is started on the tenth day of tumor inoculation. Standard (one dose) is injected on the tenth day intraperitoneally. The drug is administered from the tenth day for 5 days intraperitoneally, after the administration of the last dose followed by 18-h fasting. The animals in each group are kept to check the mean survival time (MST) of the tumor-bearing hosts. Antitumor effects of a drug are assessed by observation of following parameters: (i) percentage increase in weight as compared to day first – 0 weight, (ii) median survival time and increase in lifespan (%ILS), and (iii) hematological parameters [12, 26].

4 Anticancer Effects of *Dendrobium* Species

The death of the cells can be either accidental or regulated. Regulated cell death is genetically strictly operated and is an essential biological phenomenon for multicellular organisms [27]. In healthy cells, it is involved in several biological processes, such as development, homeostasis, embryogenesis, maturation of the immune system, and protecting an organism from microbial attacks. The activity of regulated cell death is not limited to functions in healthy cells – it plays a significant role in several pathological conditions, such as cancer, chronic inflammation, and neurodegenerative diseases [28]. In such conditions, the mechanisms of death are impaired by increasing or decreasing cells' ability to self-destruct. Accidental or necrotic cell death is not controlled, and it occurs as a result of physical (e.g., temperature), chemical (e.g., pH), or mechanical (e.g., shearing) stimuli [29, 30].

Cancer is the uncontrolled virtually autonomous growth of abnormal cells that can arise in any organ or tissue of the body. It is a heterogeneous disease that develops through the transformation of healthy cells to rapidly and uncontrollably dividing cancer cells as a result of, for example, oxidative stress, gene oncogenes, or viruses [30]. The immune system usually isolates and destroys the abnormal cells before they proliferate enough to be noticeable as a tumor. Free radicals can compromise immune cell function reducing immune responses which can allow the abnormal cells to continue growing [31].

Extracts of many *Dendrobium* species as well as their isolate of pure compounds have been studied as a valuable natural source of promising anticancer agents (Table 1). Indeed, cell proliferation is the most important characteristic of cancer/tumor cells and could be indicated by the viability of the cells. Due to the high diversity of the types of cancer, specific mechanisms of action have been explored to assign to cytotoxic, antitumor, antiproliferative, anti-metastatic, antimigratory, or apoptotic properties in vitro or in vivo and suggest their potential anticancer effects.

4.1 Cytotoxicity Effect

The extracts and isolate of pure compounds of different *Dendrobium* species have capacities to kill the growth of various types of cancer cells. Even if some researches have shown preliminary anticancer effects of *Dendrobium*, the specific mechanisms have not been fully clarified. The various solvent extracts of *D. amoenum*, *D. chrysanthum*, *D. crepidatum*, *D. formosum*, *D. longicornu*, *D. moniliforme* (synonym, *D. candidum*), and *D. signatum* have in vitro anticancer effect on different types of human cancer cell lines, cervical cancer (HeLa) and brain tumor (U-251) [32, 35, 62, 63], colon cancer (HCT-116) [66], breast cancer (MCF-7) [65],

Table 1 Isolates of different *Dendrobium* species and their anticancer effect on different cancer cell lines

Dendrobium species	Extract/active compound isolate	Biological target	IC_{50} value	Activity	Ref.
D. amoenum	Methanol extract	HeLa cell line	110.22 μgmL^{-1}	Anticancer	[32]
		U-251 cell line	550.55 μgmL^{-1}		
D. brymerianum	Compound **25**	H460 cell line	196.7 μgmL^{-1}	Cytotoxic, antimigratory	[33]
	Compound **23**		23.4 μgmL^{-1}		
	Compound **32**		65.0 μgmL^{-1}		
	Compound **36**		125.8 μgmL^{-1}		
D. catenatum	Protein extract	HepG2, SGC-7901, MCF-7 cell lines	73.38%, 78.91%, 86.8%	Cytotoxic, antiproliferative	[34]
	Peptides		21.72–33.42% 27.81–33.94% 30.02–41.80%		
D. crepidatum	Methanol extract	HeLa cell line	194.14 μgmL^{-1}	Anticancer	[35, 36]
		U-251 cell line	301.99 μgmL^{-1}		
	Ethanol extract	T-cell lymphoma	325 μgmL^{-1}		
D. chrysanthum	Ethanol extract	T-cell lymphoma	400 μgmL^{-1}	Anticancer	[23, 36]
	Compound **25**	FaDu cell line	2.55 μM		
D. chrysotoxum	Compound **17**	T47D, 143B, HUVECs, HeLa, MG63.2 cell lines	80–160 nM, 58.19 nM, 34.2 nM, 8.3 μM, 88.69 nM	Antiproliferative, apoptosis, cytotoxic, antimigratory	[37–41]
	Compound **35**	K562, HL-60, A549, BEL-7402, SGC-7901 cell lines	32.18, 10.39, 18.40, 1.49, 15.48 μgmL^{-1}		
	Compound **36**	K562, HL-60, A549, BEL-7402, SGC-7901 cell lines	26.65, 10, 9.03, 0.97, 5.53 μgmL^{-1}		
	Compound **27** (10^{-4} mol L^{-1})	A549, HL-60 cell lines	81.1%, 65.0%		
	Compound **28** (10^{-4} mol L^{-1})	A549, HL-60 cell lines	97.3%, 55.7%		

	Compound **34** (10^{-4} mol L^{-1})	A549, HL-60 cell lines	93.4%, 47.2%		
	Compound **31** (10^{-4} mol L^{-1})	A549, HL-60 cell lines	62.4%, 76.3%		
	Compound **17** (10^{-4} mol L^{-1})	A549, HL-60 cell lines	73.2%, 68.3%		
	Compound **23** (10^{-4} mol L^{-1})	A549, HL-60 cell lines	61.5%, 32.4%		
	Compound **26** (10^{-4} mol L^{-1})	A549, HL-60 cell lines	38.2%, 32.6%		
D. denneanum	Polysaccharides	S180 induced mice	72.04%	Antitumor	[42]
D. densiflorum	Compound **29**	H460 cell line	>50 μM	Anticancer, apoptosis	[43]
D. draconis	Compound **23**	H460 cell line	>20 μM	Antimigratory, apoptosis	[44–47]
D. ellipsophyllum	Compound **5**	H460, H292, H23 cell lines	100, 100, 188.89 μM	Apoptosis, antimigratory	[48–50]
D. falconeri	Compound **12**	H460 cell line	>10 μM	Antimigratory	[51, 52]
D. fimbriatum	Compound **3**, **18**–**22**, and **25**	HL-60, SMMC-7721, A-549, SW480, MCF-7 cell lines	2–21.23 μM	Cytotoxic	[53]
D. formosum	Ethanolic extract	T-cell lymphoma	350 μgmL^{-1}	Antitumor	[54]
D. gratiosissimum	Compound **14**, **15**, **23**, **25**	HL-60	2.1, 6.4, 0.082, 10.6 μM	Cytotoxic	[55]
D. loddigesii	Compound **25**	HUVECs, A-549, H23, MDA-MB-231, HCT-116, SCC, A375 cell lines	4.5–20 μM	Antiproliferative, antiangiogenesis, antimigratory, apoptosis	[56–61]
D. longicornu	Acetone extract	U-251 cell line	620.56 μgmL^{-1}	Anticancer	[62]
	Ethanol extract	HeLa cell line	294.70 μgmL^{-1}		

(continued)

Table 1 (continued)

Dendrobium species	Extract/active compound isolate	Biological target	IC_{50} value	Activity	Ref.
D. moniliforme	Methanol extract	HeLa, HCT-116, MCF-7 cell lines, 26-M3.1 induced mice	155.80 μgmL^{-1}, 1–2 mgmL^{-1}	Anticancer, cytotoxic, anti-metastasis, antiproliferative, Apoptosis	[63–71]
	Ethanol extract	U-251 cell line	772.50 μgmL^{-1}		
	Compound **30**	K562 cell line	1.84 μM		
	Compound **33**	HSC-T6, BNL CL.2, PC-3 and DU-145 cell lines	0.5–3.0 μM		
D. nobile	Compound **30**	K562, PC-3, SNU-484, SK-Hep-1, HeLa cell lines	1.84, 7.5, 7.9, 16.4, 22.3 μM	Cytotoxic, antimigratory	[69, 72–77]
	Denobilone A	HeLa, MCF-7, A549 cell lines	9.8, 9.4, 9.9 μM		
	Lactone derivatives	HeLa, MCF-7, A549 cell lines	18.2–25.3, 15.3–30.0, 19.2–28.2 μM		
	Polysaccharides	S180 induced mice, HL-60, HepG2 cell lines	2.5 mgmL^{-1}, 200 μgmL^{-1}		
D. officinale	Aqueous extract	MNNG-induced gastric tumorigenesis in rats	2.4–4.8 g	Apoptosis, anticancer	[22, 78, 79]
	Compound **4**, **9**, **11**, **16**, **23**, **25**	HeLa cell line	8.0–92.4 μM		
	Polysaccharides	SGC-7901 xenograft mice	<1.0 g		
D. polyanthum	Compound **25** (10^{-4} mol L^{-1})	A549, HL-60 cell lines	67.5%, 53.0%	Cytotoxic	[80]

D. pulchellum	Compound **8**	H460, H23 cell lines	127.34, 145.47 μM	Anti-metastasis	[59, 81, 82]
	Compound **25**	H23 cell line	33.41 μM		
D. signatum	Ethanol extract	MCF-7, NCI-H187 cell lines	–	Cytotoxic	[83, 84]
	Compound **13**	MDA-231, HepG2, HT-29 cell lines	25.2, 51.3, 30.4 μM		
	Compound **2**		80.5, 137.8, 88.2 μM		
	Compound **9**		25.9, 49.6, 29.3 μM		
	Compound **10**		42.5, 89.7, 52.0 μM		
	Compound **12**		30.3, 50.3, 35.7 μM		
D. sinense	Compound **1**	SGC-7901, BEL-7402, K562 cell lines	16.7, >30, >30 μM	Cytotoxic	[85]
	Compound **7**		12.8, >30, >30 μM		
	Compound **6**		7.8, 11.7, 15.7 μM		
	Compound **24**		>30, 10.0, 10.3 μM		

T-cell (Dalton's) lymphoma [36, 54], and MCF-7 and lung cancer (NCI-H187) [84]. The extract has induced the apoptosis on HCT-116 cell line by chromatin condensation and formation of apoptotic bodies. During the apoptosis, the expression of caspase 3, caspase 9, and Bax was made elevated, whereas the expression of Bcl-2 and pro-inflammatory COX-2, iNOS, and NF-κB was made low [66]. The growth inhibition mechanism of the MCF-7 cell by the extract was induced by enhancing the cell cycle arrest in G_2/M phase and regulating the key biomarkers (tumor growth-associated biomarkers including Erα, IGPBP2, IGFBP4, and GATA3 and apoptosis-associated biomarkers including ELF5, p53, p21, p18, CDH1, CDH2, and p12) [65]. Also, treatment with *D. candidum* at any concentration and any time point caused no inhibitory effect on cell proliferation of normal breast epithelial (MCF10A) cell line [65].

The majority of the anticancer effect of extracts deal with in vitro evaluations. Only a few of them were deepened in vivo. *D. candidum* extract is effective in the prevention of chemically induced colon carcinogenesis in C57BL/6 mice by increasing the serum SOD level and decreasing the levels of pro-inflammatory cytokines IL-6, IL-12, TNF-α, and IFN-γ [67]. The ethanolic extract of *D. formosum* has cytotoxic activity on T-cell (Dalton's) lymphoma bearing mice. It has significant increase in apoptotic cell death in a dose- and time-dependent manner by arresting the cells in the G_2/M phase of the cell cycle with a treatment of 150 mg/kg body weight [54]. Similarly, the gastric carcinogenesis in rats was inhibited by oral administration of *D. officinale* extract (4.8 and 2.4 g/kg). The extract could downregulate the expression of malondialdehyde (MDA) and 8-hydroxy-2-deoxyguanosine and upregulate the activity of glutathione peroxidase as well as IL-2 during *N*-methyl-N′-nitro-N-nitrosoguanidine (MNNG)-induced gastric tumorigenesis in rats. It also reduced the level of inflammatory cytokines including activin A, Agrin, IL-1α, ICAM-1 (intracellular adhesion molecule-1), and TIMP-1 (tissue inhibitor of matrix metalloproteinase-1) and increased the level of IL-10. Additionally, it increases the protein level of Bax and caspase-3 and decreases the expression of Bcl-2 [79]. *D. officinale* polysaccharide can inhibit the growth of human gastric cancer cell (SGC-7901) xenograft in nude mice [78].

Mainly, the bibenzyl (see Fig. 1), phenanthrene (see Fig. 2), and fluorenone (see Fig. 3) derivatives isolated from different *Dendrobium* species seem to be very active in promising anticancer compounds. There are more than 25 bibenzyl derivatives have been isolated from *Dendrobium* species which have a potential anticancer effect on different types of cancer cell lines. Compound **17**, a promising anticancer bibenzyl compound, has been isolated from the stems of *D. chrysotoxum*. It has an antitumor effect in estrogen receptor (ER) positive breast cancer cells (T47D), and its effect has been evaluated on multiple cancer-associated pathways. Besides, compound **17** induced apoptosis in T47D cells by reducing Bcl-2 expression and activating caspase signaling and also suppressed the expression of CDKs and caused cell cycle arrest. Meanwhile, compound **17** did not affect the proliferation of MCF10A cell line [41]. Compound **17** inhibited the growth of HeLa cells and induced apoptosis in a dose- and time-dependent manner, inducing cell cycle arrest at the G_2/M phase. Its treatment increased the expression of Bax and caspase-3 but decreased levels of Bcl-2 and phosphorylated-ERK1/2 [39]. The effect of compound **17** on osteosarcoma (OS) cell lines (143B and MG63.2) was also explored and further elucidated the underlying molecular mechanisms of action. It inhibits cell proliferation and induces cell cycle G_2/M arrest by regulating cell cycle-related proteins. Furthermore, it induced cell apoptosis by activating both extrinsic (expression of downstream apoptotic protein caspase-8) and intrinsic (caspase-9) pathways [40]. Bibenzyl derivatives like compounds **4**, **9**, **11**, **14**, **15**, **16**, **23**, and **25** obtained from *D. brymerianum, D. gratiosissimum*, and *D. officinale* possessed significant cytotoxicity against human lung cancer (H460), HL-60 cells, and HeLa cell lines [22, 33, 55]. Similarly, compounds **1**, **6**, **7**, and **24**, isolated from *D. sinense*, have possessed different degrees of cytotoxicity on human gastric carcinoma (SGC-7901), human hepatocellular carcinoma (BEL-7402), and chronic myelogenous leukemia (K562) cell lines [85]. Bibenzyls derivatives from *D. fimbriatum* such as compound **18–22** exhibited broad spectrum and cytotoxicity with five human cancer cell lines: promyelocytic leukemia (HL-60), hepatocellular carcinoma (SMMC-7721), lung carcinoma (A-549), breast cancer (MCF-7), and colorectal adenocarcinoma (SW480) [53]. Furthermore, compounds **3** and **25** also have the most cytotoxic effect from the same plant [53]. Additionally, bibenzyl derivatives, compounds **2**, **9**, **10**, **12**, and **13**, isolated from the whole plant of *D. signatum*, have appreciable cytotoxic activity against three human cancer cell lines, including breast cancer (MDA-23), liver hepatocellular carcinoma (HepG2), and colorectal adenocarcinoma (HT-29) cells [83]. Compound **5** from *D. ellipsophyllum* had demonstrated high cytotoxicity on lung cancer cells (H23, H460 and H292) and revealed a significant increase of early and late apoptosis with an absence of necrosis cell death. The upregulation of a tumor repressor protein p53 was elucidated in lung cancer cells [48].

Compound **30**, promising phenanthrene, isolated from *D. nobile* and *D. moniliforme*, has anticancer effect in many cancer cell lines, including pancreatic adenocarcinoma, leukemia, and glioblastoma [64, 73]. It induces human glioblastoma multiforme (GBM) cell apoptosis through IκB kinase inactivation, followed by Akt and fork head in rhabdomyosarcoma dephosphorylation and caspase-3 activation signaling cascade [86]. Furthermore, it induces apoptosis in lung and colorectal cancer cell via Akt inactivation, Bad activation, mitochondrial dysfunction, apoptosis-inducing factor releasing, and DNA damage [87, 88]. The combination of compound **30** with Fas ligand reduces the concentration of Fas ligand needed to activate caspases and cell apoptosis. This compound inhibits nuclear factor-κB and induces apoptosis via ROS generation, and this effect takes place in a MAPK-independent pathway [64]. Also, compound **30** has been reported to increase the levels of tubulin polymerization and deregulation of Bcr-Abl signaling to inhibit human leukemia (K562) cell proliferation. Furthermore, it significantly suppressed the expression of Bcr-Abl and phosphorylation of CrkL, a crucial tyrosinase kinase and an adaptor protein in chronic myeloid leukemia, respectively [69]. Similarly, phenanthrene compound **27–29**, **31–34** obtained from *D. chrysotoxum*, *D. densiflorum*, *D. moniliforme*, and *D. brymerianum* possessed significant cytotoxicity against several human cancer cell lines [33, 38, 43, 71].

Compound **35** and **36**, fluorenone derivatives obtained from *D. chrysotoxum* and *D. brymerianum*, possessed significant cytotoxicity against different human cancer cell lines: lung (H460), leukemia (K562 and HL-60), lung (A549), hepatoma (BEL-7402), and stomach (SGC-7901) [33, 37].

4.2 Anti-metastasis Effect

Metastasis is dissemination of cancer cells from primary tumors to secondary sites through the blood and lymphatic system and forms new tumors on other parts. It consists of several steps involving detachment of malignant transformed cells from the primary tumor site, attachment to the extracellular matrix (ECM), degradation of the ECM and basement membrane (invasion), cell migration, and establishment of secondary tumors [89]. It is the leading cause of mortality among cancer patients. Epithelial to mesenchymal transition (EMT) is the hallmark of cancer metastasis [46]. EMT is a process during which epithelial cells acquire migratory and invasion ability [59]. The change of cancer cells from epithelial to mesenchymal phenotypes facilitates the aggressiveness of cancer [90]. Several signaling proteins are known to be responsible for metastasis, such as a protein tyrosine kinase called focal adhesion kinase (FAK), ATP-dependent tyrosine kinase (Akt), cell division cycle 42 (Cdc-42), metalloproteinases (MMPs), Ras-related C3 botulinum toxin substrate 1 (Rac-1), E-cadherin, N-cadherin, vimentin, slug, and caveolin-1 (Cav-1) [12, 26]. EMT was suppressed by compound **12** in expression from E-cadherin to N-cadherin and a decrease in the protein expression level of slug and vimentin [51]. It also suppressed Cav-1, which is a protein implicated in aggressiveness, and downregulated Akt in lung cancer (H460) cells. The expression of migration-related integrin, including integrin β-1 and integrin α-4, was significantly reduced in response to compound **12** [52]. The migration and invasion of lung cancer (H292) cells by compound **5** were reduced by decreasing migration-regulating proteins, including integrins αv, α4, β1, β3, and β5, as well as FAK, Rac-1, and Cdc-42 [49, 50]. Compound **25** can attenuate migration and invasion in human lung cancer cells (H23) associated with an attenuation of endogenous reactive oxygen species (ROS), in which hydroxyl radical (OH•) was identified as a dominant species [59]. It also suppresses the migration and metastasis of human breast cancer (MDA-MB-231) and lung cancer (H460) cells by inhibiting mRNA and protein expression of Twist, Akt phosphorylation, N-cadherin, and Cav-1 [33, 91]. Compound **23** has the inhibitory effect on lung cancer (H292 and H460) cell migration, downregulation of Cav-1, and activation of Akt and Cdc-42, thereby suppressing filopodia formation. Besides, it greatly decreases EMT markers, including N-cadherin, vimentin, and slug, leading to significant suppression of protein kinase B, extracellular signal-regulated kinase, and Cav-1 survival pathways during the detached condition [33, 44–47]. Compound **30** suppresses invasion and metastasis in human gastric cancer (SNU-484) and prostate cancer (PC-3) cells. Tumor invasion and metastasis are often associated with the enhanced synthesis and/or activation of matrix-degrading enzymes Rac-1 and matrix metalloproteinases (MMPs), among which MMP-2 and MMP-9 are of central importance [72, 73,

86]. Increased FAK activation is tightly associated with enhanced migratory behavior and cancer metastasis, and FAK signaling regulates the formation and turnover of focal adhesions in cells in a moving state [44]. Likewise, increased Akt phosphorylation is associated with metastasis behavior in some cancer cells and has been shown, in certain cases, to be the downstream effector of FAK [44]. Recent evidence suggests that Cav-1 plays an essential role in cancer aggressiveness and metastasis, and its overexpression is closely associated with increased cancer migration. High level of Cav-1 is found in many cancer types, and overexpression of Cav-1 results in increased motility of human lung cancer (H460) cells, while knockdown of the protein causes the opposite effect [47, 51].

Only two isolated compounds from *Dendrobium* species were evaluated in vivo: the antimigratory and anti-metastatic effects of compound **25** on human breast cancer in an MDA-MB-231 metastatic model. Compound **25** (100 mg/kg) significantly suppresses breast cancer metastasis to the lungs and reduced the number of metastatic lung nodules and lung weight without causing any toxicity [91]. An in vivo orthotopic osteosarcoma (OS) model was established by intra-tibial injection of OS 143B cells to confirm the antitumor effect of compound **17**. The mice were injected with 5% DMSO intraperitoneally every other day for seven times in total. Compound **17** markedly inhibited the growth of OS with no major organ-related toxicity [40]. Besides the in vitro evaluation of *D. candidum* extract, it has anti-metastatic effect in BALB/c mice with tumors propagated by the injection of colon carcinoma cells (26-M3.1). It has reduced the serum cytokine levels of IL-6, IL-12, TNF-alpha, and IFN-gamma to a greater extent. The most prominent anti-metastatic effect of extract has been associated with the marked decrease in expression of MMP-2 and MMP-9, together with a marked increase in expression of TIMP-1 and TIMP-2 [66, 68].

4.3 Antiangiogenesis Effect

Angiogenesis is a process that is known as the formation of new blood vessels with the help of existing blood vessels and to play a major role in cancer growth and metastasis [92]. Therefore, angiogenesis is an important factor in the progression of cancer. Angiogenesis is stimulated when tumor tissues require nutrients and oxygen. New blood vessels can feed growing tumors with nutrients and oxygen, allowing cancer cells to spread (metastasis). Angiogenesis has a four-step process: (i) the basement membrane in tissue is injured locally, (ii) endothelial cells activated by angiogenic factors migrate, (iii) endothelial cells proliferate and stabilize, and (iv) angiogenic factors continue to influence the angiogenic process [26]. It is regulated by both activator and inhibitor molecules. However, upregulation of the activity of angiogenic factors is itself not sufficient for angiogenesis. Many different proteins have been identified as an angiogenic factor, including vascular endothelial growth factor (VEGF), basic fibroblast growth factor (bFGF), platelet-derived growth factor (PGF), angiogenin, transforming growth factor-α (TGF-α), TGF-β, and tumor necrosis factor-α (TNF-α) [12, 26, 92]. Among these, VEGF and its receptors play an important role in angiogenesis. For in vitro angiogenesis assays, human umbilical vein endothelial cells (HUVECs) and human dermal microvascular endothelial cells have commonly been used as a model system. The effects of compound **25** were assessed on VEGF- and bFGF-induced angiogenesis in cultured HUVECs in vitro and in vivo [56]. It significantly inhibited the growth of lung cancer cell line (A549) and suppressed growth factor-induced neovascularization. Moreover, VEGF- and bFGF-induced cell proliferation, migration, and tube formation of HUVECs were markedly inhibited by compound **25** [57]. Compound **25** impeded angiogenesis by suppressing the activation of VEGF receptor 2 (Flk-1/KDR) and c-Raf-MEK1/2-ERK1/2 signals [93]. Compound **17** downregulates the expression of inflammation factors through the regulation of IDO-induced tumor cell angiogenesis mimicry and endothelial cell-dependent angiogenesis by targeting JAK2/STAT3 pathway [94]. Compound **17** induced depolymerization of F-actin and β-tubulin more prominently in proliferating endothelial cells. It also inhibits high glucose-induced retinal angiogenesis by blocking ERK1/2-mediated HIF-1α and suppressing VEFG-induced activation of VEGFR2 [95, 96].

5 Production of Anticancer Compounds Through In vitro Culture of *Dendrobium* Species

Many *Dendrobium* species are highly demanded in traditional herbal medicine, and they are the sources of modern anticancer compounds as described earlier. The wild resources of them have been depleted by overexploitation to meet their demand for medicinal use and health products [97, 98]. In general, plant cell and tissue culture techniques provide an alternative way to produce clonal plants for mass production as well as to conserve germplasm for future uses [99, 100]. Also, this is an excellent way to produce quality plant materials for agriculture, forestry and horticulture, and bioactive secondary metabolites for pharmaceutical industries [101]. Recently, this technique has received greater attention for producing plant-specific bioactive compounds which have applications in pharmaceutical, cosmetic, and nutraceutical industries [101, 102]. Efforts have been made for the establishment of cell, callus, protocorm-like bodies (PLBs), and organ cultures of some *Dendrobium* species for the production of value-added compounds [103–107]. The use of bioreactors for large-scale cultivation of cells, calluses, and PLBs of some *Dendrobium* species has become feasible for the production of bioactive secondary metabolites [108, 109]. By contrast, biomass production for harvesting bioactive compounds is aimed at maximizing the growth of callus, PLBs, and tissues containing high amounts of bioactive metabolites [108]. Some stress signaling substances such as salicylic acid, methyl jasmonate, and thidiazuron is added to the culture medium for increasing the content of useful secondary metabolites like alkaloids, polysaccharides, and aromatic compounds [103, 108, 109]. However, pure compounds have not been isolated so far from the in vitro culture of *Dendrobium* species. Preliminary research showed anticancer activity of in vitro culture of *D. amoenum*, *D. crepidatum*, and *D. longicornu* toward cervical (HeLa) and brain tumor (U251) cell lines [110]. This paves the pathways for the future potential of promising anticancer compounds from in vitro raised tissue of *Dendrobium* species. Thus, plant cell and tissue culture could be an alternative and suitable tool for improvement, enhancement, and production of desirable bioactive compounds which also help to minimize the pressure on the natural population of *Dendrobium* species and result in their sustainable utilization.

6 Conclusion

Organic extracts and isolated compounds, with various chemical structures from different *Dendrobium* species, explored to have an anticancer effect toward different cancer cells in vitro and/or in tumor-bearing mice in vivo. Altogether 36 anticancer compounds under three groups, bibenzyl, phenanthrene, and fluorenone, have been isolated from *Dendrobium* species, among them, 26 compounds of bibenzyl, 8 compounds of phenanthrene, and 2 compounds of fluorenone derivatives. Inhibition of the cancer cell proliferation, induction of apoptosis, suppression of metastasis, and angiogenesis have been discussed in detail of the aforementioned compounds. Out the 36 compounds, **5**, **17**, **23**, **25**, and **30** have been largely discussed in vitro and in vivo anticancer effect on the different cancer cell lines. The application of tissue culture technique has proven as an alternative strategy for the production of anticancer compounds, especially when the plant resources are overexploited. Importantly, in some cases, the anticancer principles are devoid of cytotoxicity toward normal cells, and their potencies relative to anticancer drugs currently in use may not all be known, but hopefully, some of these anticancer principles can be developed into chemopreventive therapeutics.

References

1. Hossain MM (2011) Therapeutic orchids: traditional uses and recent advances – an overview. Fitoterapia 82:102–140
2. Bulpitt CJ, Li Y, Bulpitt PF, Wang J (2007) The use of orchids in Chinese medicine. J R Soc Med 100:558–563
3. Gutiérrez RMP (2010) Orchids: a review of uses in traditional medicine, its phytochemistry and pharmacology. J Med Plants Res 4:592–638
4. Bulpitt CJ (2005) The uses and misuses of orchids in medicine. Q J Med 98:625–631
5. Ng TB, Liu J, Wong JH, Ye X, Sze SCW, Tong Y, Zhang KY (2012) Review of research on Dendrobium, a prized folk medicine. Appl Microbiol Biotechnol 93:1795–1803
6. Xu J, Han QB, Li SL, Chen XJ, Wang XN, Zhao ZZ, Chen HB (2013) Chemistry, bioactivity and quality control of Dendrobium, a commonly used tonic herb in traditional Chinese medicine. Phytochem Rev 12:341–367
7. da Silva JAT, Ng TB (2017) The medicinal and pharmaceutical importance of Dendrobium species. Appl Microbiol Biotechnol 101:2227–2239
8. Lam Y, Ng TB, Yao RM, Shi J, Xu K, Sze SCW, Zhang KY (2015) Evaluation of chemical constituents and important mechanism of pharmacological biology in *Dendrobium* plants. Evid Based Complement Alternat Med 2015:841752

9. Stanaway JD, Afshin A, Gakidou E, Lim SS, Abate D, Abate KH, Abbafati C, Abbasi N, Abbastabar H, Abd-Allah F (2018) Global, regional, and national comparative risk assessment of 84 behavioural, environmental and occupational, and metabolic risks or clusters of risks for 195 countries and territories, 1990–2017: a systematic analysis for the Global Burden of Disease Study 2017. Lancet 392:1923–1994
10. Siegel RL, Miller KD, Jemal A (2019) Cancer statistics, 2019. CA Cancer J Clin 69:7–34
11. Millimouno FM, Dong J, Yang L, Li J, Li X (2014) Targeting apoptosis pathways in cancer and perspectives with natural compounds from mother nature. Cancer Prev Res 7:1081–1107
12. Gordon JL, Brown MA, Reynolds MM (2018) Cell-based methods for determination of efficacy for candidate therapeutics in the clinical Management of Cancer. Diseases 6:85
13. Bray F, Ferlay J, Soerjomataram I, Siegel RL, Torre LA, Jemal A (2018) Global cancer statistics 2018: GLOBOCAN estimates of incidence and mortality worldwide for 36 cancers in 185 countries. CA Cancer J Clin 68:394–424
14. Guilbaud N, Kraus-Berthier L, Meyer-Losic F, Malivet V, Chacun C, Jan M, Tillequin F, Michel S, Koch M, Pfeiffer B (2001) Marked antitumor activity of a new potent acronycine derivative in orthotopic models of human solid tumors. Clin Cancer Res 7:2573–2580
15. Greenwell M, Rahman P (2015) Medicinal plants: their use in anticancer treatment. Int J Pharm Sci Res 6:4103–4112
16. Moraes DFC, de Mesquita LSS, do Amaral FMM, de Sousa Ribeiro MN, Malik S (2017) Anticancer drugs from plants. In: Malik S (ed) Biotechnology and production of anti-cancer compounds. Springer, Cham, pp 121–142
17. Hossain M, Akter S, Uddin S (2020) Screening of bioactive phytochemicals in some indigenous epiphytic orchids of Bangladesh. In: Khasim S, Hegde S, Gonzalez-Arnao M, Thammasiri K (eds) Orchid biology: recent trends & challenges. Springer, Singapore, pp 481–506
18. Cakova V, Bonte F, Lobstein A (2017) Dendrobium: sources of active ingredients to treat age-related pathologies. Aging Dis 8:827–849
19. Kierkegaard P, Pilotto AM, Leander K (1970) The constitution and relative configuration of crepidine, an alkaloid from *Dendrobium crepidatum* Lindl. Acta Chem Scand 24:3757–3759
20. Majumder PL, Guha S, Sen S (1999) Bibenzyl derivatives from the orchid *Dendrobium amoenum*. Phytochemistry 52:1365–1369
21. Li Y, Wang CL, Wang YJ, Wang FF, Guo SX, Yang JS, Xiao PG (2009) Four new bibenzyl derivatives from *Dendrobium candidum*. Chem Pharm Bull 57:997–999
22. Ren G, Deng WZ, Xie YF, Wu CH, Li WY, Xiao CY, Chen YL (2020) Bibenzyl derivatives from leaves of *Dendrobium officinale*. Nat Prod Commun 15:1–5
23. Nam B, Ryu SM, Lee D, Jung CH, Jin CH, Kim JB, Lee IS, Han AR (2019) Identification of two new phenanthrenes from Dendrobii Herba and their cytotoxicity towards human hypopharynx squamous carcinoma cell (FaDu). Molecules 24:2339
24. Tang H, Zhao T, Sheng Y, Zheng T, Fu L, Zhang Y (2017) *Dendrobium officinale* Kimura et Migo: a review on its ethnopharmacology, phytochemistry, pharmacology, and industrialization. Evid Based Complement Alternat Med 2017:7436259
25. Sut S, Maggi F, Dall'Acqua S (2017) Bioactive secondary metabolites from orchids (Orchidaceae). Chem Biodivers 14:e1700172
26. Ediriweera MK, Tennekoon KH, Samarakoon SR (2019) In vitro assays and techniques utilized in anticancer drug discovery. J Appl Toxicol 39:38–71
27. Galluzzi L, Bravo-San Pedro JM, Vitale I, Aaronson SA, Abrams JM, Adam D, Alnemri ES, Altucci L, Andrews D, Annicchiarico-Petruzzelli M et al (2015) Essential versus accessory aspects of cell death: recommendations of the NCCD 2015. Cell Death Differ 22:58–73
28. Gali-Muhtasib H, Hmadi R, Kareh M, Tohme R, Darwiche N (2015) Cell death mechanisms of plant-derived anticancer drugs: beyond apoptosis. Apoptosis 20:1531–1562
29. Montero J, Sarosiek KA, DeAngelo JD, Maertens O, Ryan J, Ercan D, Piao H, Horowitz NS, Berkowitz RS, Matulonis U et al (2015) Drug-induced death signaling strategy rapidly redicts cancer response to chemotherapy. Cell 160:977–989

30. Liu Y, Levine B (2015) Autosis and autophagic cell death: the dark side of autophagy. Cell Death Differ 22:367–376
31. Nimse SB, Pal D (2015) Free radicals, natural antioxidants, and their reaction mechanisms. RSC Adv 5:27986
32. Paudel MR, Pant B (2017) Cytotoxic activity of crude extracts of *Dendrobium amoenum* and detection of bioactive compounds by GC-MS. Botanica Orientalis: J Plant Sci 11:38–42
33. Klongkumnuankarn P, Busaranon K, Chanvorachote P, Sritularak B, Jongbunprasert V, Likhitwitayawuid K (2015) Cytotoxic and antimigratory activities of phenolic compounds from *Dendrobium brymerianum*. Evid Based Complement Alternat Med 2015:350410
34. Zheng Q, Qiu D, Liu X, Zhang L, Cai S, Zhang X (2015) Antiproliferative effect of *Dendrobium catenatum* Lindley polypeptides against human liver, gastric and breast cancer cell lines. Food Funct 6:1489–1495
35. Paudel MR, Chand MB, Pant B, Pant B (2019) Assessment of antioxidant and cytotoxic activities of extracts of *Dendrobium crepidatum*. Biomol 9:478
36. Prasad R, Koch B (2016) In vitro anticancer activities of ethanolic extracts of *Dendrobium crepidatum* and *Dendrobium chrysanthum* against T-cell lymphoma. J Cytol Histol 7:432
37. Chen Y, Li Y, Qing C, Zhang Y, Wang L, Liu Y (2008) 1,4,5-Trihydroxy-7-methoxy-9H-fluoren-9-one, a new cytotoxic compound from *Dendrobium chrysotoxum*. Food Chem 108:973–976
38. Hu J, Fan W, Dong F, Miao Z, Zhou J (2012) Chemical components of *Dendrobium chrysotoxum*. Chin J Chem 30:1327–1330
39. Li M, He Y, Peng C, Xie X, Hu G (2018) Erianin inhibits human cervical cancer cell through regulation of tumor protein p53 via the extracellular signal-regulated kinase signaling pathway. Oncol Lett 16:5006–5012
40. Wang H, Zhang T, Sun W, Wang Z, Zuo D, Zhou Z, Li S, Xu J, Yin F, Hua Y (2016) Erianin induces G2/M-phase arrest, apoptosis, and autophagy via the ROS/JNK signaling pathway in human osteosarcoma cells in vitro and in vivo. Cell Death Dis 7:e2247
41. Sun J, Fu X, Wang Y, Liu Y, Zhang Y, Hao T, Hu X (2016) Erianin inhibits the proliferation of T47D cells by inhibiting cell cycles, inducing apoptosis and suppressing migration. Am J Transl Res 8:3077–3086
42. Fan YJ, Luo AX (2011) Evaluation of anti-tumor activity of water-soluble polysaccharides from *Dendrobium denneanum*. Afr J Pharm Pharmacol 5:415–420
43. Wattanathamsan O, Treesuwan S, Sritularak B, Pongrakhananon V (2018) Cypripedin, a phenanthrenequinone from *Dendrobium densiflorum*, sensitizes non-small cell lung cancer H460 cells to cisplatin-mediated apoptosis. J Nat Med 72:503–513
44. Charoenrungruang S, Chanvorachote P, Sritularak B, Pongrakhananon V (2014) Gigantol, a bibenzyl from *Dendrobium draconis*, inhibits the migratory behavior of non-small cell lung cancer cells. J Nat Prod 77:1359–1366
45. Charoenrungruang S, Chanvorachote P, Sritularak B, Pongrakhananon V (2014) Gigantol-induced apoptosis in lung cancer cell through mitochondrial-dependent pathway. Thai J Pharm Sci 38:67–73
46. Unahabhokha T, Chanvorachote P, Sritularak B, Kitsongsermthon J, Pongrakhananon V (2016) Gigantol inhibits epithelial to mesenchymal process in human lung cancer cells. Evid Based Complement Alternat Med 2016:4561674
47. Unahabhokha T, Chanvorachote P, Pongrakhananon V (2016) The attenuation of epithelial to mesenchymal transition and induction of anoikis by gigantol in human lung cancer H460 cells. Tumor Biol 37:8633–8641
48. Hlosrichok A, Sumkhemthong S, Sritularak B, Chanvorachote P, Chaotham C (2018) A bibenzyl from *Dendrobium ellipsophyllum* induces apoptosis in human lung cancer cells. J Nat Med 72:615–625
49. Chaotham C, Pongrakhananon V, Sritularak B, Chanvorachote P (2014) A bibenzyl from *Dendrobium ellipsophyllum* inhibits epithelial-to-mesenchymal transition and sensitizes lung cancer cells to anoikis. Anticancer Res 34:1931–1938

50. Chaotham C, Chanvorachote P (2015) A bibenzyl from *Dendrobium ellipsophyllum* inhibits migration in lung cancer cells. J Nat Med 69:565–574
51. Pengpaeng P, Sritularak B, Chanvorachote P (2014) Dendrofalconerol A sensitizes anoikis and inhibits migration in lung cancer cells. J Nat Med 69:178–190
52. Pengpaeng P, Sritularak B, Chanvorachote P (2015) Dendrofalconerol A suppresses migrating cancer cells via EMT and integrin proteins. Anticancer Res 35:201–206
53. Xu FQ, Xu FC, Hou B, Fan WW, Zi CT, Li Y, Dong FW, Liu YQ, Sheng J, Zuo ZL (2014) Cytotoxic bibenzyl dimers from the stems of *Dendrobium fimbriatum* Hook. Bioorg Med Chem Lett 24:5268–5273
54. Prasad R, Koch B (2014) Antitumor activity of ethanolic extract of *Dendrobium formosum* in T-cell lymphoma: an in vitro and in vivo study. Biomed Res Int 2014:753451
55. Zhang CF, Wang M, Wang L, Iinuma M, Zhang M, Xu LS, Wang ZT (2008) Chemical constituents of *Dendrobium gratiosissimum* and their cytotoxic activities. Indian J Chem 47:952–956
56. Tsai AC, Pan SL, Liao CH, Guh JH, Wang SW, Sun HL, Liu YN, Chen CC, Shen CC, Chang YL (2010) Moscatilin, a bibenzyl derivative from the India orchid *Dendrobrium loddigesii*, suppresses tumor angiogenesis and growth in vitro and in vivo. Cancer Lett 292:163–170
57. Chen TH, Pan SL, Guh JH, Liao CH, Huang DY, Chen CC, Teng CM (2008) Moscatilin induces apoptosis in human colorectal cancer cells: a crucial role of c-Jun NH2-terminal protein kinase activation caused by tubulin depolymerization and DNA damage. Clin Cancer Res 14:4250–4258
58. Chen CA, Chen CC, Shen CC, Chang HH, Chen YJ (2013) Moscatilin induces apoptosis and mitotic catastrophe in human esophageal cancer cells. J Med Food 16:869–877
59. Kowitdamrong A, Chanvorachote P, Sritularak B, Pongrakhananon V (2013) Moscatilin inhibits lung cancer cell motility and invasion via suppression of endogenous reactive oxygen species. Biomed Res Int 2013:765894
60. Ho CK, Chen CC (2003) Moscatilin from the orchid *Dendrobrium loddigesii* is a potential anticancer agent. Cancer Investig 21:729–736
61. Cardile V, Avola R, Graziano AC, Russo A (2020) Moscatilin, a bibenzyl derivative from the orchid *Dendrobium loddigesii*, induces apoptosis in melanoma cells. Chem Biol Interact 323:109075
62. Paudel MR, Chand MB, Pant B, Pant B (2017) Cytotoxic activity of antioxidant-riched *Dendrobium longicornu*. Pharm J 9:499–503
63. Paudel MR, Chand MB, Pant B, Pant B (2018) Antioxidant and cytotoxic activities of *Dendrobium moniliforme* extracts and the detection of related compounds by GC-MS. BMC Complement Altern Med 18:134
64. Sánchez-Duffhues G, Calzado MA, De Vinuesa AG, Appendino G, Fiebich BL, Loock U, Lefarth-Risse A, Krohn K, Muñoz E (2009) Denbinobin inhibits nuclear factor-κB and induces apoptosis via reactive oxygen species generation in human leukemic cells. Biochem Pharmacol 77:1401–1409
65. Sun J, Guo Y, Fu X, Wang Y, Liu Y, Huo B, Sheng J, Hu X (2016) *Dendrobium candidum* inhibits MCF-7 cells proliferation by inducing cell cycle arrest at G2/M phase and regulating key biomarkers. Onco Targets Ther 9:21–30
66. Zhao X, Sun P, Qian Y, Suo H (2014) *D. candidum* has in vitro anticancer effects in HCT-116 cancer cells and exerts in vivo anti-metastatic effects in mice. Nutr Res Pract 8:487–493
67. Wang Q, Sun P, Li G, Zhu K, Wang C, Zhao X (2014) Inhibitory effects of *Dendrobium candidum* Wall ex Lindl. on azoxymethane- and dextran sulfate sodium-induced colon carcinogenesis in C57BL/6 mice. Oncol Lett 7:493–498
68. Li G, Sun P, Zhou Y, Zhao X, Chen F (2014) Preventive effects of *Dendrobium candidum* Wall ex Lindl. on the formation of lung metastases in BALB/c mice injected with 26-M3.1 colon carcinoma cells. Oncol Lett 8:1879–1885
69. Huang YC, Guh JH, Teng CM (2005) Denbinobin-mediated anticancer effect in human K562 leukemia cells: role in tubulin polymerization and Bcr-Abl activity. J Biomed Sci 12:113–121

70. Hsu JL, Lee YJ, Leu WJ, Dong YS, Pan SL, Uang BJ, Guh JH (2014) Moniliformediquinone induces in vitro and in vivo antitumor activity through glutathione involved DNA damage response and mitochondrial stress in human hormone refractory prostate cancer. J Urol 191:1429–1438
71. Tseng TH, Lin WL, Chen ZH, Lee YJ, Shie MS, Lee KF, Shen CH, Kuo HC (2016) Moniliformediquinone as a potential therapeutic agent, inactivation of hepatic stellate cell and inhibition of liver fibrosis in vivo. J Transl Med 14:263
72. Lu TL, Han CK, Chang YS, Lu TJ, Huang HC, Bao BY, Wu HY, Huang CH, Li CY, Wu TS (2014) Denbinobin, a phenanthrene from *Dendrobium nobile*, impairs prostate cancer migration by inhibiting Rac1 activity. Am J Chin Med 42:1539–1554
73. Song JI, Kang YJ, Yong HY, Kim YC, Moon A (2012) Denbinobin, a phenanthrene from *Dendrobium nobile*, inhibits invasion and induces apoptosis in SNU-484 human gastric cancer cells. Oncol Rep 27:813–818
74. Luo AX, Fan YJ (2011) Immune stimulating activity of water-soluble polysaccharide fractions from *Dendrobium nobile* Lindl. Afr J Pharm Pharmacol 5:625–631
75. Wang JH, Luo JP, Zha XQ, Feng BJ (2010) Comparison of antitumor activities of different polysaccharide fractions from the stems of *Dendrobium nobile* Lindl. Carbohydr Polym 79:114–118
76. Zhou XM, Zheng CJ, Wu JT, Chen GY, Chen J, Sun CG (2016) Five new lactone derivatives from the stems of *Dendrobium nobile*. Fitoterapia 115:96–100
77. Zhou XM, Zheng CJ, Gan LS, Chen GY, Zhang XP, Song XP, Li GN, Sun CG (2016) Bioactive phenanthrene and bibenzyl derivatives from the stems of *Dendrobium nobile*. J Nat Prod 79:1791–1797
78. Zhang L, Wang F, Ren X (2018) Inhibitory effect of *Dendrobium officinale* polysaccharide on human gastric cancer cell xenografts in nude mice. Food Sci Technol 38:78–83
79. Zhao Y, Liu Y, Lan XM, Xu GL, Sun YZ, Li F, Liu HN (2016) Effect of *Dendrobium officinale* extraction on gastric carcinogenesis in rats. Evid Based Complement Alternat Med 2016:1213090
80. Hu JM, Zhao YX, Miao ZH, Zhou J (2009) Chemical components of *Dendrobium polyanthum*. Bull Kor Chem Soc 30:2098–2100
81. Bhummaphan N, Pongrakhananon V, Sritularak B, Chanvorachote P (2018) Cancer stem cell-suppressing activity of chrysotoxine, a bibenzyl from *Dendrobium pulchellum*. J Pharmacol Exp Ther 364:332–346
82. Chanvorachote P, Kowitdamrong A, Ruanghirun T, Sritularak B, Mungmee C, Likhitwitayawuid K (2013) Anti-metastatic activities of bibenzyls from *Dendrobium pulchellum*. Nat Prod Commun 8:115–118
83. Mittraphab A, Muangnoi C, Likhitwitayawuid K, Rojsitthisak P, Sritularak B (2016) A new bibenzyl-phenanthrene derivative from *Dendrobium signatum* and its cytotoxic activity. Nat Prod Commun 11:657–659
84. Chimsook T (2016) Phytochemical screening, total phenolic content, antioxidant activities and cytotoxicity of *Dendrobium signatum* Leaves. MATEC Web Conf 62:03005
85. Chen XJ, Mei WL, Cai CH, Guo ZK, Song XQ, Dai HF (2014) Four new bibenzyl derivatives from *Dendrobium sinense*. Phytochem Lett 9:107–112
86. Weng HY, Hsu MJ, Chen CC, Chen BC, Hong CY, Teng CM, Pan SL, Chiu WT, Lin CH (2013) Denbinobin induces human glioblastoma multiforme cell apoptosis through the IKKα–Akt–FKHR signaling cascade. Eur J Pharmacol 698:103–109
87. Kuo CT, Hsu MJ, Chen BC, Chen CC, Teng CM, Pan SL, Lin CH (2008) Denbinobin induces apoptosis in human lung adenocarcinoma cells via Akt inactivation, Bad activation, and mitochondrial dysfunction. Toxicol Lett 177:48–58
88. Chen TH, Pan SL, Guh JH, Chen CC, Huang YT, Pai HC, Teng CM (2008) Denbinobin induces apoptosis by apoptosis-inducing factor releasing and DNA damage in human colorectal cancer HCT-116 cells. Naunyn Schmiedeberg's Arch Pharmacol 378:447–457
89. Chanvorachote P, Chamni S, Ninsontia C, Phiboonchaiyanan PP (2016) Potential anti-metastasis natural compounds for lung cancer. Anticancer Res 36:5707–5717

90. Hassan M, Watari H, AbuAlmaaty A, Obha Y, Sakuragi N (2014) Apoptosis and molecular targeting therapy in cancer. Biomed Res Int 2014:150845
91. Pai HC, Chang LH, Peng CY, Chang YL, Chen CC, Shen CC, Teng CM, Pan SL (2013) Moscatilin inhibits migration and metastasis of human breast cancer MDA-MB-231 cells through inhibition of Akt and Twist signaling pathway. J Mol Med 91:347–356
92. Nishida N, Yano H, Nishida T, Kamura T, Kojiro M (2006) Angiogenesis in cancer. Vasc Health Risk Manag 2:213–219
93. Guan L, Zhou J, Lin Q, Zhu H, Liu W, Liu B, Zhang Y, Zhang J, Gao J, Feng F (2019) Design, synthesis and antitumour and anti-angiogenesis evaluation of 22 moscatilin derivatives. Bioorg Med Chem 27:2657–2665
94. Su C, Zhang P, Liu J, Cao Y (2017) Erianin inhibits indoleamine 2, 3-dioxygenase–induced tumor angiogenesis. Biomed Pharmacother 88:521–528
95. Gong YQ, Fan Y, Wu DZ, Yang H, Hu ZB, Wang ZT (2004) In vivo and in vitro evaluation of erianin, a novel anti-angiogenic agent. Eur J Cancer 40:1554–1565
96. Yu Z, Zhang T, Gong C, Sheng Y, Lu B, Zhou L, Ji L, Wang Z (2016) Erianin inhibits high glucose-induced retinal angiogenesis via blocking ERK1/2-regulated HIF-1α-VEGF/VEGFR2 signaling pathway. Sci Rep 6:34306
97. da Silva JAT, Cardoso JC, Dobránszki J, Zeng SJ (2015) Dendrobium micropropagation: a review. Plant Cell Rep 34:671–704
98. Pant B, Paudel MR, Chand MB, Pradhan S, Malla BB, Raskoti BB (2018) Orchid diversity in two community forests of Makawanpur District, Central Nepal. J Threat Taxa 10:12523–12530
99. Pant B (2013) Medicinal orchids and their uses: tissue culture a potential alternative for conservation. African J Plant Sci 7:448–467
100. Pant B, Pradhan S, Paudel MR, Shah S, Pandey S, Joshi PR (2019) Various culture techniques for the mass propagation of medicinal orchids from Nepal. Acta Hortic 1262:109–124
101. Pant B (2014) Application of plant cell and tissue culture for the production of phytochemicals in medicinal plants. In: Adhikari R, Thapa S (eds) Infectious diseases and nanomedicine II. Springer, India, pp 25–39
102. Espinosa-Leal CA, Puente-Garza CA, García-Lara S (2018) In vitro plant tissue culture: means for production of biological active compounds. Planta 248:1–18
103. Wang HQ, Jin MY, Paek KY, Piao XC, Lian ML (2016) An efficient strategy for enhancement of bioactive compounds by protocorm-like body culture of *Dendrobium candidum*. Ind Crop Prod 84:121–130
104. Cui HY, Murthy HN, Moh SH, Cui YY, Paek KY (2015) Establishment of protocorm suspension cultures of *Dendrobium candidum* for the production of bioactive compounds. Hortic Environ Biotechnol 56:114–122
105. Cui HY, Murthy HN, Moh SH, Cui YY, Lee EJ, Paek KY (2014) Production of biomass and bioactive compounds in protocorm cultures of *Dendrobium candidum* Wall ex Lindl. using balloon type bubble bioreactors. Ind Crop Prod 53:28–33
106. Bhattacharyya P, Kumaria S, Diengdoh R, Tandon P (2014) Genetic stability and phytochemical analysis of the in vitro regenerated plants of *Dendrobium nobile* Lindl., an endangered medicinal orchid. Meta Gene 2:489–504
107. Zheng YP, Jiang W, Liao FL, Lu HF (2012) Optimization of light quality for production of alkaloid and polysaccharide in *Dendrobium candidum* Wall. ex Lindl. J Med Plants Res 6:560–565
108. Park SY, Ho TT, Paek KY (2020) Medicinal orchids: production of bioactive compounds and biomass. In: Khasim S, Hegde S, Gonzalez-Arnao M, Thammasiri K (eds) Orchid biology: recent trends & challenges. Springer, Singapore, pp 439–450
109. Murthy HN, Paek KY, Park SY (2018) Micropropagation of orchids by using bioreactor technology. In: Lee YI, Yeung ECT (eds) Orchid propagation: from laboratories to greenhouses – methods and protocols. Springer, Germany, pp 195–208
110. Paudel MR (2019) Biological activities of in vivo and in vitro plant materials of some *Dendrobium* species. PhD thesis, Tribhuvan University, Nepal

A Review on Phytochemistry, Nutritional Potential, Pharmacology, and Conservation of *Malaxis acuminata*: An Orchid with Rejuvenating and Vitality Strengthening Properties

17

Renu Suyal, Sandeep Rawat, R. S. Rawal, and Indra D. Bhatt

Contents

Abstract

Malaxis acuminata is one among the 300 species of the genus with medicinal properties and hence used in traditional Indian medicine system. The species is a perennial, monopodial, threatened terrestrial orchid distributed in moist ground and in rocks laden with mosses in south Asia including, Himalaya and southern

R. Suyal · R. S. Rawal · I. D. Bhatt
G.B. Pant National Institute of Himalayan Environment, Almora, Uttarakhand, India
e-mail: idbhatt@gbpihed.nic.in

S. Rawat (✉)
G.B. Pant National Institute of Himalayan Environment, Sikkim Regional Centre, Gangtok, Sikkim, India
e-mail: sandeep_rawat15@rediffmail.com

J.-M. Mérillon, H. Kodja (eds.), *Orchids Phytochemistry, Biology and Horticulture*, Reference Series in Phytochemistry, https://doi.org/10.1007/978-3-030-38392-3_15

Indian hills, Australia, and western region of South America. Medicinally, the species is used in Ayurvedic formulations in the preparation of energetic tonic with adaptogenic, immunomodulating, rejuvenating, and other health benefits. Various essential nutrients and pharmacological compounds are identified and detected in the pseudobulb of the species. The species has been successfully validated for antioxidant, antiaging, UV-A blocking, anti-inflammatory, anti-proliferative, and antimicrobial activities, which supported its traditional use also. Propagation methods for large-scale multiplication of the species are available but need further refining for robustness for farming purposes. Various research gap areas and possible research areas for harnessing the potential of the species have been highlighted in the end of the chapter.

Keywords

Astavarga · Anti-aging · *Malaxis* · Himalaya · Medicinal plant · Orchid · Vitality strengthening

Abbreviations

A/F	Abundance and frequency ration
AAE	Ascorbic acid equivalent
ABTS	2,2-Azinobis (3-ethylbenzoline-6-sulfonic acid) radical scavenging assay
BA	6-Benzylaminopurine
CITES	The Convention on International Trade in Endangered Species of Wild Fauna and Flora
DPPH	1,1-Diphenyl-2-picrylhydrazyl radical scavenging assay
IBA	Indole-3-butyric acid
IC_{50}%	Inhibitory concentration.
LOX	Lipoxygenase
NAA	1-Naphthaleneacetic acid
NCBI	National Center for Biotechnological Information
ppm	Parts per million
ROS	Reactive oxygen species
SPF	Sun Protection Factor
TDZ	Thidiazuron

1 Introduction

Malaxis is a genus of orchids that prefer terrestrial, epiphytic, or occasionally holomycotrophic habitats in mountainous region. The genus is distributed worldwide and consists of more than 300 species, but its diversity is highly concentrated in southeast Asia. Among the 19 species found in India, *Malaxis acuminata* D. Don. (Syn. *Crepidium acuminatum* D. Don. Szlach.) is a perennial, monopodial, medicinally important threatened terrestrial orchid [1]. The herb is found in temperate and subtropical region on moist, shady places in the pine, oak forests, moist ground, and in rocks laden with mosses within an altitude between 1200 and 2200 m asl [2]. The species is found in mountain grasslands of south Asia, Australia, and western region of South America [3–5]. Pseudobulbs of the species are important ingredients of the traditional Indian Ayurvedic formulation "Chyavanprash" [4], which is considered as a polyherbal energetic tonic with adaptogenic, immunomodulating, rejuvenating, and other health benefits [6].

Traditionally, in India, pseudobulbs *of M. acuminata* are used in the treatment of several diseases such as seminal debility, internal and external hemorrhages, dysentery, fever, emaciation and weakness, forcing its collection from wild habitats to be utilized in pharmaceutical industries. Generally, due to rapid loss of specific habitats and overexploitation, several orchids are listed in Appendix II of the Convention on International Trade in Endangered Species of Wild Fauna and Flora (CITES) [7]. The objective of this chapter is to summarize all the research findings available on various aspects, such as botanical description and distribution, ethnopharmacology, phytochemistry, propagation, and conservation measures of *M. acuminata*. Also, a systematic approach for conservation and sustainable utilization of the species has been suggested. Based on all the compiled information, research gap has also been discussed. This chapter provides the basis for further studies on conservation and development of identifying better therapeutic agents and health products from the species.

2 Botanical Description of the Species

M. acuminata is a perennial, medium-sized terrestrial orchid with a height of 30 cm and consists of pseudobulbs and fibrous roots at the base. The plant is erect and has a small stem composed of aerial flowering axis and basal swollen stem (pseudobulbs), bearing nodes and internodes arising from the base of mother rhizome. Several adventitious roots arise below this rhizomatous structure, firmly anchoring the plant into the soil. Rhizomatous structure and pseudobulbs are covered by sheathing leaf bases, which appear marginally on nodes. The leaves are simple, three to four in number, alternate, ovate to lanceolate, membranous, 5–15-cm long, and has an acute apex with a sheathing leaf base [8]. Leaf constants recorded as, palisade ratio as ~0.8, stomatal index ~8.4, stomata number ~70 nos., stomata size ~0.03 × 0.03 mm, and anomocytic type of stomata [9]. The flowers are terminal racemes, yellowish green in color with a purple tinge, and 3 mm in diameter [10, 11]. Fruit is capsule, seeds minute and powdery, and ovoid in shape. Pseudobulbs are 3–9 cm long, fleshy, smooth, shining, greenish, covered with membranous sheath, and slightly mucilaginous [4, 8]. The plant flowers from July to August and fruiting takes place from September to October [10]. Anatomical and histochemical studies revealed the presence of endophytic mycorrhizal fungus in the root and protocorm, and the endophytic bacterium (probably *Corynebacterium*) detected in the leaves and root

Fig. 1 Plant of *Malaxis acuminata* along with inflorescence and pseudobulbs

hairs [9]. Anatomical similarity between rhizomes and pseudobulbs indicates that species can be propagated from its rhizomes as well as pseudobulbs (Fig. 1).

3 Habitat, Distribution, and Ecology

M. acuminata grows in group that contains 5–25 individuals in moist, shady, and humus-rich forest floors and forms symbiotic relationship with mycorrhizal fungi that nourishes this species [7]. In India, this species is distributed in Himalaya from Jammu Kashmir to Arunachal Pradesh, including states such as Himachal Pradesh, Uttarakhand, Assam, Nagaland, Manipur, Mizoram, and Tripura [4]. In India, the species also been reported in Andaman Islands, Kerala, Anaimalai Hills, east Godavari district of Andhra Pradesh, and Madhya Pradesh [3, 4, 12, 13]. Outside India, the species is found in Myanmar, Thailand, Malaysia, Laos, Cambodia, Southern China, the Philippines, Australia, Peru, and Ecuador in mountain grasslands [3, 5].

Studies on quantitative assessment and ecologocal studies of the threatened medicinal plant along with the assessment of availability, growth preferences, distribution pattern and habitat preference are vital steps for development of conservation measures [14]. However, only few studies are available on ecological aspect confined to Indian Himalayan region (Table 1), which shows low population density across the surveyed populations and indicates poor availability of the species in the wild. Loss of habitat and illegal collection puts this species at higher risk of endangerment [15]. Dominant associates reported for *M. acuminata* at most sites are *Roscoea procera, Thalictrum foliolosum, Valeriana jatamansi, Rumex*

Table 1 Quantitative assessment of *M. acuminata* in different regions

Region	Habitat types	Density (plant/ m^2)	Frequency (%)	A/F	References
Uttarakhand	Banj Oak, mixed forest and oak-pine	1.9–2.4	13–23	–	[16]
Kumaun Himalaya	*Quercus leucotrichophora* and *Cedrus deodara*	1.70–15.30	50–90	0.03–0.24	[7]
Arakot-Khadikal forest, Tehri	Moist, shady, and humus-rich forest floor	0.2	20	–	[17]
Chail Wildlife Sanctuary, Himachal Pradesh	Different type	2.0	10	0.2	[18]

nepalensis, Oxalis corniculata, Asparagus curillus, Potentilla fulgense, and *Polygonum* species [7, 16–18]. However, other regions of the world remain unexplored in this context, which is needed to formulate a global strategy for conservation of the species.

4 Medicinal and Other Uses

M. acuminata has been used in traditional Indian medicine system in the preparation of various polyherbal formulations and tonics and in folklore medicines. Pseudobulbs of the species are used in the preparation of Chyavanprash and other Ayurvedic formulations like Astavarga, Astavarga-churna, Chitrakadi-taila, Vachadi-taila, Mahakalyan-taila, Jivaniya-ghrita, Mahamayura-ghrita, Vajikarma-taila, Brahini-gutika, and Himvana agada [4]. Among these, Chyavanprash is a well-recognized formulation that is used as immunomodulator, health promoter, rejuvenator, and brain tonic due to its antiaging, antioxidant, cardioprotective, and adaptogenic properties [6, 19]. Ayurvedic properties of the plant species has been described as, sweetness in taste, cold in potency, pacifies vata and aggravates kapha [20]. Pseudobulbs of *M. acuminata* are considered as sweet, refrigerant, and febrifuge [4].

Traditionally, paste of pseudobulbs is applied topically for insect bites and is used in the treatment of rheumatism along with other herbal plants [21]. Its swollen stem is considered as refrigerant, aphrodisiac, styptic, anti-dysenteric, febrifuge, and tonic. Decoction of bulb is used to increase the quantity of semen or to stimulate the production of semen [22]. Pseudobulbs are used in the treatment of skin diseases, piles, and burning sensation [7]. Pseudobulbs and rhizomes are used for bronchitis and is used as a tonic by local people of Uttarakhand [23]. It is used to treat sterility, vitiated condition of pitta and vata, seminal weakness, internal and external hemorrhages, dysentery, fever, emaciation, and general debility [24]. Rhizome and pseudobulbs of the species are edible and are consumed in northeastern India [9, 25]. Various medicinal uses of the species described in the literature are given in Table 2.

Table 2 Medicinal uses of *M. acuminata* reported in different parts of the world

Medicinal uses	Parts used	Utilized form	Reference
Burning sensation, bronchitis, fever, tuberculosis, and weakness	Rhizome, root, pseudobulb	Dry powder	[26–28]
Tonic and bleeding diathesis	Pseudobulbs	Fresh	[5]
Bronchitis and tonic	Pseudobulbs	Fresh	[23]
Refrigerant and febrifuge	Pseudobulb	–	[29]
Cooling, spermopiotic, analgesic, internal and external hemorrhages	Pseudobulb		[30]
Rheumatism	Pseudobulbs	Mixture with other plants	[21]
Aphrodisiac	Pseudobulbs	Powder	[31]
Styptic, anti-dysenteric, emaciation	Swollen stem	–	[24]

5 Nutritional Composition

The edible pseudobulbs of the species are rich in essential nutrients, minerals, vitamins, and other metabolites (Table 3). Analysis of pseudobulbs shows the following: total ash content between 1.49 and 6.9 (% w/w), moisture content 6.8%, total fat content 1.45%, and carbohydrate content 112 μg/ml [10, 32]. Pseudobulbs are also rich in valuable minerals such as copper (6.48 ppm), zinc (43 ppm), manganese (35 ppm), iron (331 ppm), potassium (21,600 ppm), calcium (9000 ppm), magnesium (2800 ppm), and aluminum (198 ppm) [33]. High levels of important vitamins are also reported in the pseudobulbs, specifically α-tocopherol (12.00–9.80 mg/100 dw) and γ-tocopherol (695.00–786.7 mg/100 g dw) [33]. Likewise, valuable classes of secondary metabolites also reported in the species. Total phenolic contents (1.72 mg/g), total tannins (1.69 mg/g), total flavonoid (1.71 mg/g), and total flavonol (1.81 mg/g) have also been reported in significant quantity in the pseudobulbs [34], and these metabolites are known for their antioxidant, anticancer, antidiabetic, and various medicinal properties [35–38] (Table 3).

Among these important metabolites, some are thoroughly analyzed. Among the different fatty acids, linoleic acid (18:2ω6: 61.20–65.23%), α-linolenic acid (18:3ω3, 15.50–18.10%), and oleic acid (18:1ω9, 12.00–14.87) were the major constituents; however, palmitic acid (16:0, 6.00–5.90%) stearic acid (18:0, 2.10–2.50%), γ-linolenic acid (18:3ω6, 2.20–1.87%), eicosanoic acid (20:0, 0.81–0.69%), eicosenoic acid (20:1, 0.42–0.52%) and eicosadienoic acid (20:2, 0.04–0.07%) were also present in trace amount [33]. Among these, α-linolenic acid and γ-linolenic acid are considered as essential fatty acids, which cannot be synthesized by the human body [40]. Besides their structural role, these essential fatty acids play a significant role in cellular signaling and activating or inhibiting transcription factors such as NF-κB, which is linked to pro-inflammatory cytokine production [41].

A significant amount of organic acids were also found in the pseudobulbs, and among them. Acetic acid, propenoic acid, malonic acid, succinic acid, propanoic acid, fumaric acid, itaconic acid, and pipecolic acid were majorly found [2]. These

Table 3 Chemical composition of pseudobulbs and leaf and stem extract of *M. acuminata*

S. no.	Nutritional content	Available content	References
Nutrients			
1	Total ash content (%w/w)	1.49–6.90	[9, 32]
2	pH (10% solution)	6.80	[32]
3	Moisture content (%)	53.00	[32]
4	Total fat content (%)	1.45	[39]
5	Alkaloid content (%)	5.00	[39]
6	Resin content (%)	0.90	[39]
7	Crude fiber content (%)	5.10	[39]
8	Saponin content (%)	2.00	[39]
9	Carbohydrate content(μg/g)	112.00	[39]
Minerals (ppm)			
10	Cu	6.48	[33]
11	Zn	43.00	[33]
12	Mn	35.00	[33]
13	Fe	331.00	[33]
14	K	21600.00	[33]
15	Ca	9000.00	[33]
16	Mg	2800.00	[33]
17	Al	198.00	[33]
18	Ba	26.70	[33]
19	B	55.60	[33]
20	Mo	0.30	[33]
21	Cl	156.00	[33]
22	Co	1.41	[25]
Vitamins and other secondary metabolites			
23	α-Tocopherol (mg/100 g)	9.80–12.00	[33]
24	γ-Tocopherol (mg/100 g)	695.00–786.70	[33]
25	Total phenolic content (mg/g)	1.72	[34]
26	Total tannins (mg/g)	1.69	[34]
27	Total flavonoid content (mg/g)	1.71	[34]
28	Total flavonol content (mg/g)	1.81	[34]

metabolites not only act as a reserve energy resource for the plants but also regulate physiological activities of the species by adjusting pH level of the cell, modulate cellular transport through membrane and secondary messenger of cell signaling, and play a vital role during cold stress [42–45]. Similarly, various essential (L-valine, L-leucine, L-threonine, L-phenylalanine, and L-methionine) and non-essential amino acids were also recorded in pseudobulbs [2, 46]. Thus, consumption of pseudobulbs fulfills various basic dietary requirements. However, compositional variation noted in different studies emphasized the requirement of agro-climatic, agronomic, post-harvesting management techniques for obtaining quality produce from the species. In addition, variability in such qualitative and quantitative traits is always desirable for varietal improvement through breeding programs (Table 4).

Table 4 Essential nutrients detected in the rhizome of *M. acuminata*

Class	Compound	References
Fatty acid	Linoleic acid, γ-linolenic acid, oleic acid, stearic acid, hexadecanoic acid, heptadecanoic acid, palmitic acid; eicosanoic acid, and eicosadienoic acid	[33]
Organic acids	Acetic acid, propenoic acid, malonic acid, succinic acid, propanoic acid, fumaric acid, itaconic acid, pipecolic acid, hydroxybutyric acid, malic acid, ribonic acid, hydrocinnamic acid, shikimic acid, D-xylonolactone, and gluconolactone	[2]
Sugars and glycosides	Galactose, D-glucose, D(+)-sucrose, D-mannose, fructose, β-maltose, α-D-glucopyranose, levoglucosan, methyl beta-glucofuranoside, 15,β-galacto-pyranose, and α-D-(+)-mannopyranose	[2]
Amino acids and amines	L-alanine, L-valine, L-leucine, L-glycine, L-serine, L-threonine, L-proline, L-aspartic acid, L-phenylalanine, L glutamic acid, L-asparagine, ornithine, ethanolamine, and L-valine	[2, 46]

6 Bioactive Compounds Isolated from *M. acuminata*

Phytochemical analysis of methanolic extract of pseudobulbs, leaf, and stem showed that they are rich in flavonoids, phenolic acid, sterols, and alcohols (Table 5). Among the different flavonoids and phenolic acids, catechin, phloridzin, rutin, caffeic acid, chlorogenic acid, 3-hydroxy benzoic acid, 4-hydroxy benzoic acid, protocatechuic acid, 3-hydroxy cinnamic acid, and p-coumaric acid are found in the species [2, 34]. Also, GC-MS analysis showed the presence of different volatile compounds such as sibutramine, limonene, diethylene glycol, p-cymene, eugenol, benzene, piperitone, and uridine the methanolic extracts of pseudobulbs. More recently, Singh et al. (2017) identified and characterized isorhamnetin O-glycoside, bulbophythrin A, gigantol, batatasin III, lusianthrin, 2,3-dimethoxy-9,10-dihydrophenanthrene-4,7-diol, lipar acid C, and 3′-O-methylbatatasin from ethyl acetate fraction obtained from methanolic extract of the rhizomes of *M. acuminate* using HPLC-ESI-QTOF-MS/MS analysis. This fraction showed strong antiproliferative activity in comparison with standard doxorubicin against cancer cell lines, such as A549 (70.29%), DLD1 (73.12%), MCF-7 (79.10%), and DU145 (68.65%) [47].

Similarly, dry root powder of *M. acuminata* extracted with 9:1 methanol/water solution yielded several fractions and 67 compounds including diethyl phthalate, heptadecanoic acid, methyl ester, cyclohexadecanolide, octadecanoic, diethyl phthalate, heneicosane, hexadecane, docosane, pentadecanoic acid, and others were detected by gas chromatography/mass spectrometry (GC-MS) analysis. Few of them have been reported as pharmacologically active molecules [48]. Although various compounds have been identified in the species, their quantitative assessment and other biological activity remain unexplored. However, quantitative information on these compounds and their hereditability, variability, and bioactivity are mandatory for commercialization of this product.

Table 5 Bioactive compounds isolated and identified in *M. acuminata*

Class	Compound	References
Flavonoid	Catechin, phloridzin, and rutin	[34]
Phenolic acid	Caffeic acid, chlorogenic acid, ellagic acid, 3-hydroxy benzoic, 4-hydroxy benzoic, protocatechuic acid, 3-hydroxy cinnamic acid, and p-coumaric acid	[2, 34]
Sterols	Stigmasterol and β-sitosterol	[2]
Volatile compounds	Sibutramine, limonene, diethylene glycol, p-cymene, eugenol, benzene, and piperitone	[2]
Other compounds	1,2-Benzenedicarboxylic acid, 11-hexadecenoic acid, 13-octadecenal, 1-butanol, 3-methyl-, acetate, 2,3-dimethoxy-9,10-dihydrophenanthrene-4,7-diol, 2,6-diisopropylnaphthalene, 3′-o-methylbatatasin, 6-octadecenoic acid, 8-octadecenoic acid, 9-octadecenal, batatasin III, bulbophythrin A, butyl oleate, cerasynt, cis-oleic acid, cyclopentadecanolide, diethyl phthalate, cyclopentanetridecanoic acid, e-8-hexadecen-1-ol acetate, gigantol, heneicosane, heneicosanoic acid, hydrofol acid, Isorhamnetin o-glycoside, lignoceric acid, lipar acid C, lusianthrin, margaric acid, methyl (6e)-6-octadecenoate, methyl margarate, methyl ricinoleate, methyl tetradecanoate, n-octadecanoic acid, octadec-9-enoic acid, octadecanoic acid, octyl-thiirane, oxacyclododecan-2-one, tert-hexadecanethiol, tetradecenal, tetradecyl-oxirane, thiirane, and triarachine	[47, 48]
Alcohols	Glycerol, ribitol, and myo-inositol	[2]

7 Biological and Pharmacological Activities

Biological and pharmacological properties of isolated compounds, solvent extracts or solvent fractions reported in the pseudobulb of *M. acuminata* has been described under following sections and a summary of these properties is given in Table 6.

7.1 Antioxidant Activity

In the human body, various physiological processes are continuously mediated by production and exchange of free radicals. During stress, higher production of these oxidative agents is escaped by antioxidant system of the body and these oxidants create oxidative stress through degeneration of macromolecules, resulting in aging process along with chronic degenerative diseases such as diabetes, neurodegenerative disease, cancer, Parkinson's disease, and Alzheimer's disease. Diet supplemented with adequate amount of antioxidants can prevent degradation of macromolecules and related chronic disease [49, 50]. Giri et al. [33] reported that the pseudobulbs of the species have strong antioxidant potential, as evaluated by various in vitro free radical scavanging and reducing assays such as, 2,2-azinobis(3-ethylbenzoline-6-sulfonic acid) radical assay (ABTS: 4.02 mM AAE/100 g dry weight) and 1,1-diphenyl-2-picrylhydrazyl radical assay (DPPH: 1.10 mM AAE/

Table 6 Biological and pharmacological activity of *M. acuminata*

S. No.	Pharmacological/biological properties	Plant part used	Extract/ experimental model	References
1	Antioxidant activity	Pseudobulbs, leaf and stem	Methanolic extract	[32, 34, 39]
2	Antiaging activity	Leaf and stem	Methanolic extract	[52]
3	Anti-inflammatory activity	Leaf and stem	Methanolic extract	[52]
4	Sun protection factor (SPF) and UV-A blocking activity	Leaf and stem	Methanolic extract	[52]
5	Antiproliferative activity	Pseudobulbs	Ethanolic extract/human cancer cell lines	[47]
6	Antimicrobial activity (against *E. coli, S. aureus, P. aeruginosa, B. subtilis, K. aerogenes, P. mirabilis, C. albicans*	Pseudobulb	Solvent extracts	[58, 59]

100 g dry weight), ferric reducing anti-oxidant properties (FRAPS: 1.18 mM AAE/ 100 g dry weight) and superoxide scavenging assays (0.16 unit/mg dry weight).

Similarly, Garg et al. [51] reported DPPH scavenging properties and ferric ion reducing properties of butanol extract of pseudobulbs of the species in dose-dependent manner. Also, Bose et al. [52] reported that methanolic leaf extract of in vitro-derived plants showed significant antioxidant activity (IC_{50}: 42.66 μg/mg) compared to standard ascorbic acid (IC_{50}: 38.24 μg/mg), and aqueous stem extracts of wild plant showed moderate DPPH activity(178.56 μg/mg).

7.2 Antiaging Activity

Pseudobulbs of *M. acuminata* are used in the preparation of vitality strengthening and rejuvenating formulation due to its antiaging properties. Antioxidants have ability to maintain the structural integrity of skin and thus, its use has been adopted as an important strategy for skin glowing mechanism in order to develop anti-aging products. Bose et al. [52] reported that methanolic leaf extract showed anti-collagenase activity with IC_{50} of 32.52 μg/ml [standard EDTA (IC_{50}: 35.45 μg/ml)] and very strong elastase inhibitory activity with IC_{50} of 32.24 μg/ml [standard oleanolic acid (IC_{50}: 30.56 μg/ml)], respectively. Hence, collagen and elastin are known to maintain skin structural integrity and elasticity of skin [53]; therefore, *M. acuminata* could be used as one of the potential rejuvenators and antiaging agents.

7.3 Sun Protection Factor (SPF) and UV-A Blocking Activity

Various environmental exposures such as harmful radiations (e.g., sun-light and ultraviolet radiation) lead to oxidation of the lipids, proteins, and DNA of the outer surface of the body. Skin as a protective covering of the body also gets damaged by these reactive radicals or reactive oxygen species (ROS). UV-A (400–315 nm) and UV-B (315–280 nm) radiations contribute predominantly to extrinsic premature photoaging [54]. Plant-based formulations have been reported to absorb ultraviolet radiations (UV-A and UV-B) and act as potential solar filters in developing new sunscreen formulations. Methanolic extracts of both wild and in vitro-derived plants of *M. acuminata* showed promising UV-A blocking potentials, whereas leaf and stem extracts at a concentration of 250 μg/ml showed 21.68 and 27.64 μg/ml in vitro SPF values. These values can be attributed to the presence of various polyphenolic and flavonoid compounds present in the plant extracts [52].

7.4 Anti-Inflammatory Activity

In aging process, immune response with the influence of oxidative stress is chronically started to degenerate, leads to chronic systemic inflammation. Various pro-inflammatory mediators such as, cytokines and chemokines are involved in the development of chronic inflammation and the immune-senescence process [55]. Regulation of such inflammation can be prevented at a certain level by consumption of plant-based formulations. Among the enzymatic machineries involved in the prevention of chronic systemic inflammation process, hyaluronidase is the key enzyme responsible for increased inflammation, angiogenesis, fibrosis, and collagen deposition in wound healing [56]. Bose et al. [52] showed that methanolic leaf (IC_{50}: 14.32 μg/ml) and stem extracts (IC_{50}: 16.2 μg/ml) of tissue culture-derived plants exhibited promising anti-5-LOX activity [known inhibitor of 5-LOX (IC_{50}: 2.56 ± 0.4 μg/ml)] and strongest anti-hyaluronidase activity (IC_{50}: 60.36 μg/ml) as compared to oleanolic acid (IC_{50}: 32.45 μg/ml).

7.5 Antiproliferative Activity

Sulphorhodamine B assay showed that ethanol extract and its fractions exhibited antiproliferative activity against four human cancer cell lines, i.e., A549 (non-small cell lung cancer cells), DU145 (human prostate carcinoma), DLD1 (human colorectal adenocarcinoma), and MCF-7 (human breast adenocarcinoma). The ethyl acetate fraction obtained from methanolic extract showed a potent antiproliferative activity (A549: 70.29%), DLD1: 73.12%, MCF-7: 79.10%, and DU145: 68.65% inhibition) in comparison with standard doxorubicin against cancer cell lines (A549: 80.13%), DLD1: 64.45%, MCF-7: 79.82%, and DU145: 89.26% inhibition). However, ethanol extract and its n-butanol fraction produced a moderate antiproliferative activity [47].

7.6 Antimicrobial Activity

Minimal microbial static concentration (MIC) assay showed that the ethanol and methanol extracts of *M. acuminata* were found to be highly active against both *P.*

aeruginosa and *S. aureus* strains. The plants demonstrated antimicrobial properties against Gram-negative bacterial strains [57]. Also, different extracts (hexane, chloroform, ethyl acetate, ethanol, and residual aqueous) of pseudobulbs of *M. acuminata* exhibited antibacterial activity against four bacterial strains: two Gram-positive strains (*Staphylococcus aureus* MTCC 87 and *Bacillus subtilis* MTCC 121) and two Gram-negative strains (*Escherichia coli* MTCC 40 and *Pseudomonas aeruginosa* MTCC 424). Among these strains, *E. coli* (20 mm) *and B. subtilis* (15.33 mm) showed maximum zone of inhibition (ZOI) in chloroform extract [58]. Similarly, Sharma et al. [59] analyzed antifungal (against *Candida albicans*) and antibacterial activities of butanol extracts against *Escherichia coli, Klebsiella aerogenes, Pseudomonas aeruginosa*, *Proteus mirabilis, Staphylococcus aureus* and reported strong antifungal activity against *Candida albicans* strain with 32 mm inhibition zone at a concentration of 50 mg/mg as compared to 28 mm inhibition zone of standard antifungal compound fluconazole in 50 μg/mg concentration [59]. This antimicrobial property of the species can be utilized in the preparation of preservative agent formulations and food products.

8 Propagation and Cultivation Effort

Propagation and cultivation efforts of *M. acuminata* are at preliminary level. A summary of the multiplication and propagation techniques is given in Table 7.

8.1 Vegetative Propagation Techniques

In order to conserve and multiplicate *M. acuminata* in its natural habitat, Tamta et al. [60] propagated *M. acuminata* by vegetative methods through nodal cuttings,

Table 7 Various research and results of in vitro propagation conducted in *M. acuminata*

Explant	Culture media	Response	References
Adventitious shoot buds	MS+ 3 mg/L TDZ+ 0.5 mg/L NAA; MS+ 3 mg/LTDZ +0.5 mg/L NAA+ 0.4 mM spermidine; MS+ 4 mg/L IBA + 1.5 mg/L activated charcoal (AC)	96% organogenesis; 100% shoot induction with 14.6 shoots per explant; 96% rooting with 3.3 roots per shoots	[62]
Pseudobulbs segment	MS + 1 mg/L BAP+ 1 mg/L NAA+ 2 g/L activated charcoal (AC)	65% explant responded with proliferation of protocorm-like body	[63]
Seeds	MS+ sucrose (3%, w/v) + 4 μM NAA	85% germination after 135 days	[64]
Nodal segment	MS+ sucrose (3% w/v) + 3 μM NAA+ 3 μM BA	75% survival rate with maximum plant height, and number of leaves, shoots, and roots	[58]

i.e. using tip, middle, and bottom parts and whole pseudobulb (control) in experimental plots at Deodar and Oak Forests at Chakrata, Dhanolti and Mussoorie Forest areas of Uttarakhand Himalaya. Maximum survival (up to 85%) and optimum growth were observed in whole pseudobulb as compared to other propagules.

8.2 In Vitro Propagation Techniques

Seed germination and micropropagation studies are available for the species for mass propagation. Arenmongla and Deb [61] used immature seeds of 7–8 weeks after pollination to study the effect of culture condition on asymbiotic seed germination. Immature seeds cultured on MS medium containing sucrose (3%) and 4 μM α- naphthalene acetic acid under diffused light condition (20 μ mol/m^2/s) showed 85% seed germination after 135 days as compared to full light condition (40 μ mol/m^2/s). Germinated seeds were converted into protocorm-like bodies, which further differentiated into young plantlets. Rooted plantlets with well-expanded leaves and distinct bulbs were obtained in medium supplemented with sucrose (3%), activated charcoal (0.3%), and NAA and BA (3 μM, each), and up to 15 shoots and protocorm-like body were achieved. After hardening, about 75% survival was achieved after 2 months of transfer into the field.

In a study conducted by Cheruvathur et al. (2010) [62], adventitious shoot buds were induced from internodal explants grown on Murashige and Skoog (MS) medium supplemented with different concentrations of 6-benzyladenine (BA), kinetin (Kn), and thidiazuron (TDZ). TDZ at 3 mg/l induced the highest frequency (82%) of organogenic explants. In the presence of 3 mg/l TDZ and 0.5 mg/l NAA, the frequency of organogenesis was 96% with a mean number of 6.1 shoots per explant. Highest frequency of shoot induction (100%) and mean shoot number per explant (14.6) were observed on MS medium with 3 mg/l TDZ, 0.5 mg/l NAA, and 0.4 mM spermidine, and highest frequency of rooting (96%) and mean number of roots per shoot (3.3) were observed on MS medium with 4 mg/l indole-3-butyric acid (IBA) and 1.5 mg/l activated charcoal (AC). In another study, using pseudobulb segments of *M. acuminata* as explant on MS media, supplemented with 1 mg/L BAP, 1 mg/L NAA, and 2 g/L activated charcoal (AC), 65% explant response was obtained with proliferation of protocorm-like bodies [63]. Similarly, in vitro propagation of *M. acuminata* using nodal segment on MS medium fortified with sucrose (3% w/v), 3 μM NAA, and 3 μM BA showed bud formation after 3 weeks of inoculation with the following: development of 18 shoot buds; plant height, 2.4 cm; number of leaves, 4.5; and roots, 4.0 [58]. The hardened plants that were transferred to community potting mix containing mixture of charcoal pieces, chopped forest litter, coconut husk, sand and black soil showed 75% survival rate.

Propagation protocols using seed germination as well as micropropagation techniques are well developed for *M. acuminata*. However, genetic fidelity and quality of plants raised by tissue culture have not been analyzed. During micropropagation protocol development studies, small pseudobulbs were obtained, but the accumulation of medicinally important ingredients was not analyzed; however, such in vitro raised tissues may be very important for direct production of secondary metabolites [65]. Many different in vitro approaches have been used in plant system for increased biosynthesis and the accumulation of bioactive metabolites. These methods includes, suspension culture, biotransformation, and *Agrobacterium*-mediated transformation [65–67]. These new technologies served to enhance the continued produtivity of phytochemicals, especially medicinal compounds from higher plants as a renewable source. It is expected that continuous and intensified efforts in this field will lead to controllable and successful biotechnological production of specific, valuable, and as yet unknown plant chemicals.

9 Future Prospective and Conclusion

M. acuminata is well recognized for its vitality strengthening and rejuvenating properties and thus used in traditional Ayurvedic formulations. Pharmacologically, this species has been evaluated for antiaging, antioxidant, antiinflammatory, sun protection and UV-A blocking, antiproliferative, and anti-microbial activities [2, 32, 34, 47, 52, 58–59,]. All these assays were based on in vitro experiments; however, to reach any conclusive remark, these activities need further evidences at molecular and clinical levels. Further, before development of any pharmacological products, clinical efficacy needs to be analyzed. Also, various molecules extracted from the species using various solvents and fractions need to be examined for these pharmacological activities in deeper manner for supporting its ethno-pharmacological significance. Generally, a particular compound-rich fraction or extract of the species is considered to be responsible for a particular specific biological activity, but actual and reasonable mechanism of action at physiological level needs exact information on active molecule. Similarly, statistical difference among phytochemicals with any activity is required for efficacy testing during the drug development.

Market success of medicinal plant as a drug depends upon the content of active ingredient present in the species. However, drug derived from plant required consistent and continuous supply of the active ingredient overcomming the issue of seasonality. Content of secondary metabolites is highly influenced by different factors such as growing conditions, seasons, climatic conditions, sun light exposure, altitude, along with genetic makeup [36, 38, 68–70]. Therefore, selection of elite genotype and climatic conditions are essentially required for obtaining higher phytochemical content and better pharmacological activity. Identification of suitable growing conditions, agronomic practices, and inheritance of the active metabolite need more scientific exploration.

Genetic information of bioactive compound can improve the quality traits. In NCBI database, only 38 nucleotide sequences are available (retrieved on July 12, 2020), and most of them belong to genomic region of DNA. Thus, gene related markers, such as, ESTs, or ant other functional markers are scared in the species. In the recent past, modern biotechnology tools such as high throughput sequencing technology has become a reliable tool for the generation of large set data of expression part at sequence level in shorter time and cost effective manner.

However, in mordern arena it bacome a very useful tool for trait-specific marker development for molecular breeding, genetic diversity analysis or population genetics studies [71, 72]. Currently, whole-genome and transcriptome sequencing are also preferred method to identify novel genome-wide polymorphic SSR markers, and is cost-effective irrespective of model and non-model plants [73]. Such information can be useful for rapid genotyping, molecular breeding, and functional marker identification for trait improvement.

Propagation protocols such as tissue culture, seed germination, and clonal propagation for rapid multiplication are available in the species [58, 62–64], but a more robust technique needs to be developed for the benefit of farmers. With the product development, demand of the species will increase consequently, which will create more pressure in the species on its wild stock. Further, robust and rapid propagation method for quality-related traits can facilitate farmers to enhance yield and productivity.

Overall, on the basis of available information on *M. acuminata*, it can be concluded it has proven vitality strengthening and rejuvenating properties. Various pharmacological activities has been carried out in support of its traditional uses. However, most of these phytochemical and pharmacological studies are in its preliminary stage and need comprehensive scientific evidences. Today, *M. acuminata* is collected from wild populations for commercial supply, and therefore robust propagation protocols and quality planting material are needed to farmers to cultivate this species for sustainability and conservation.

Acknowledgments The authors thank the Director of G.B. Pant National Institute of Himalayan Environment for the support and encouragement.

Funding information:

This study was partially funded by the Botanical Garden Scheme of Ministry of Environment, Forest & Climate Change (MoEF&CC), Government of India, New Delhi (F.N. BSI-290/6/2013-Tech; September 29, 2013).

References

1. Shukla PK, Chaubey OP (2008) Threatened wild medicine plants, assessment, conservation and managemet. Anmol Publications Pvt Ltd, New Delhi
2. Bose B, Choudhury H, Tandon P, Kumaria S (2017) Studies on secondary metabolite profiling, anti-inflammatory potential, *in vitro* photoprotective and skin-aging related enzyme inhibitory activities of *Malaxis acuminata*, a threatened orchid of nutraceutical importance. J Photochem Photobiol B Biol 173:686–695
3. Polunin O, Stainton A (1977) Flowers of the Himalaya. Oxford University Press, Calcutta
4. Balkrishna A, Srivastava A, Mishra RK, Patel SP, Vashistha RK, Singh A, Jadon V, Saxena P (2012) Astavarga plants-threatened medicinal herbs of the North-West Himalaya. Int J Med Arom Plants 2:661–676
5. Yonzone R, Lama D, Bhujel RB, Rai S (2013) Diversity, distribution and present availability status of *Malaxis* Soland. ex Sw. (Orchidaceae) in Darjeeling Himalaya of WB, India. Lifesci Leaflet 9:18–28
6. Govindarajan R, Singh DP, Rawat AKS (2007) High performance liquid chromatographic method for the quantification of phenolics in 'Chyavanprash', a potent Ayurvedic drug. J Pharma Biomed Anal 43:527–532

7. Lohani N, Tewari LM, Joshi GC, Kumar R, Kishor K, Upreti BM (2013) Population assessment and threat categorization of endangered medicinal orchid *Malaxis acuminata* D. Don. from North-West Himalaya. Int J Conserv Sci 4:493–502
8. Chowdhary HJ, Wadhwa BM (1984) Flora of Himachal Pradesh. Botanical Survey of India, Calcutta
9. Sharma YP, Rani J, Raina R, Bandana K (2014) New insights into the morphology of *Malaxis acuminata* D Don. Int J Farm Sci 4:136–146
10. Chinmay R, Suman K, Bishnupriya D, Mohanty RC, Dixit R, Padhi MM, Babu R (2011) Phyto-Pharmacognostical studies of two endangered species of *Malaxis* (Jeevak and Rishibhak). Pharmacog J 3:77–85
11. Uma E, Rajendran R, Muthukumar T (2015) Morphology, anatomy and mycotrophy of pseudobulb and subterranean organs in *Eulophia epidendraea* and *Malaxis acuminata* (Epidendroideae, Orchidaceae). Flora 217:14–23
12. Bose TK, SK Bhattacharjee, P Das, Basak UC (1980) Orchids of India. Naya Prokash, Calcutta
13. Reddy CS, Pattanaik C, Murthy MSR, Reddy KN (2006) Floristic census of orchids of Eastern Ghats, India. The Botanica 56:79–76
14. Suyal R, Bhatt ID, Rawal RS, Tewari LM (2019) Status of two threatened astavarga herbs, *Polygonatum cirrhifolium* and *Malaxis muscifera*, in West Himalaya: conservation implications. Proc Natl Acad Sci, India Sect B: Biol Sci 2019. https://doi.org/10.1007/s40011-019-01144-3
15. Kant R, Verma J, Thakur K (2012) Distribution pattern, survival threats and conservation of 'Astavarga' orchids in Himachal Pradesh, Northwest Himalaya. Plant Arch 12:165–168
16. Jalal JS, Rawat GS (2009) Habitat studies for conservation of medicinal orchids of Uttarakhand, Western Himalaya. African J Plant Sci 3:200–204
17. Dobhal A (2016) Diversity of medicinal plants in Arakot-Khadikal forest area of district Tehri Garhwal, Uttarakhand, India. Environ Conserv J 17:47–66
18. Bhardwaj A, Verma RK, Rana JC (2017) Phytosociology of terrestrial orchid species in *Cedrusdeodara* (Roxb. ex D. Don) G. Don forest of western Himalaya, India. Environ Ecol 35:3373–3377
19. Sharma R, Martins N, Kuca K, Chaudhary A, Kabra A, Rao MM, Prajapati PK (2019) Chyawanprash: a traditional Indian bioactive health supplement. Biomol Ther 9:161
20. Singh AP, Sandhu AS (2005) A dictionary of medicinal plants. Singhal S. Sundeep Publishers, New Delhi
21. Chakarvarty HL (1976) Plant wealth of Iraq. A dictionary of economic plants, vol 1. Botany Directorate, Ministry of Agriculture and Agrarian Reform (Iraq), Baghdad
22. Jalal JS, Kumar P, Pangtey YPS (2008) Ethnomedicinal orchids of Uttarakhand, western Himalaya. Ethnobot Leaflets 12:1227–1230
23. Gaur RD (1999) Flora of district Garhwal North West Himalaya. Transmedia, Srinagar
24. Jadhav D (2008) Medicinal plants of Madhya Pradesh and Chhattisgarh. Daya Publishing House, Delhi
25. Rai V, Agarwal M, Khatoon S, Rawat AKS, Mehrotra S (2001) Estimation of Co and Mn in some medicinal plants. Bull Environ Contaminat Toxicol 66:427–432
26. Aggarwal S, Zettler LW (2010) Reintroduction of an endangered terrestrial orchid, *Dactylorhiza hatagirea* (D. Don) Soo, assisted by symbiotic seed germination: first report from the Indian subcontinent. Nat Sci 8:139–145
27. Arditti J, Ernst R (1984) Physiology of germinating orchid seeds. In: Arditti J (ed) Orchid biology: reviews and perspectives III. Cornell University Press, New York, pp 177–222
28. Subedi A, Kunwar B, Choi Y, Dai Y, van Andel T, Chaudhary RP, de Boer HJ, Gravendeel B (2013) Collection and trade of wild-harvested orchids in Nepal. J Ethnobiol Ethnomed 9:64
29. Warrier PK, Nambiar VPK, Ramankutty C (1997) Indian medicinal plants: a compendium of 500 medicinal plants. Arya Vaidya Sala, Oriental Longman, Chennai
30. Khajuria AK, Kumar G, Bisht NS (2017) Diversity with ethnomedicinal notes on orchids: a case study of Nagdev forest range, Pauri Garhwal, Uttarakhand, India. J Med Plant Stud 5:171–174
31. Chauhan NS (1990) Medicinal orchids of Himachal Pradesh. J Orchid Soc India 4:99–105

32. Arora M, Singh S, Mahajan A, Sembi JK (2017) Propagation and phytochemical analysis of *Crepidium acuminatum* (D. Don) Szlach. IOSR J Pharm Biol Sci 12:14–20
33. Lohani N, Tewari LM, Joshi GC, Kishor K, Kumar S, Tewari G, Joshi N (2013) Chemical composition of *Microstylis wallichii* Lindl. from Western Himalaya. J Med Plant Plants Res 7:2289–2292
34. Giri L, Belwal T, Bahukhandi A, Suyal R, Bhatt ID, Rawal RS, Nandi SK (2017) Oxidative DNA damage protective activity and anti-oxidant potential of Ashtvarga species growing in the Indian Himalayan region. Ind Crop Prod 102:173–179
35. Rawat S, Jugran A, Giri L, Bhatt ID, Rawal RS (2011) Assessment of antioxidants properties in fruits of *Myrica esculenta*: a popular wild edible species in Indian Himalayan region. Evid Based Compl Alt Med 2011:512787
36. Bhatt ID, Dauthal P, Rawat S, Gaira K, Jugran A, Rawal RS, Dhar U (2012) Characterization of essential oil composition, phenolic content and antioxidant properties in wild and planted individuals of *Valeriana jatamansi* Jones. Sci Hort 136:61–68
37. Bahukhandi A, Rawat S, Bhatt ID, Rawal RS (2013) Influence of solvent types and source of collection on total phenolic content and antioxidant activities of *Acorus calamus* L. Natl Acad Sci Lett 36:93–99
38. Suyal R, Rawat S, Rawal RS, Bhatt ID (2019) Variability in morphology, phytochemicals and antioxidants in *Polygonatum verticillatum* (L.) All. populations under different ecological habitats in Himalaya. Environ Monit Assess 191-S:783
39. Arora M, Kaur G, Singh S, Mahajan A, Sembi JK (2019) Quantification of phytochemicals in the pseudobulbs of *Crepidium acuminatum* (D. Don) Szlach-a critically endangered medicinal plant. Curr Trend Biotech Pharm 13:366–375
40. Whitney E, Rolfes SR (2008) Understanding nutrition, 11th edn. Thomson Wadsworth, California
41. Calder PC (2004) n− 3 Fatty acids, inflammation, and immunity—relevance to postsurgical and critically III patients. Lipids 39:1147–1161
42. Finkemeier I, König AC, Heard W, Nunes-Nesi A, Pham PA, Leister D et al (2013) Transcriptomic analysis of the role of carboxylic acids in metabolite signaling in Arabidopsis leaves. Plant Physiol 162:239–253
43. De Angeli A, Zhang J, Meyer S, Martinoia E (2013) AtALMT9 is a malate-activated vacuolar chloride channel required for stomatal opening in Arabidopsis. Nat Commun 4:1804
44. Dyson BC, Miller MA, Feil R, Rattray N, Bowsher C, Goodacre R et al (2016) FUM2, a cytosolic fumarase, is essential for acclimation to low temperature in Arabidopsis thaliana. Plant Physiol 172:118–127
45. Drincovich MF, Voll LM, Maurino VG (2016) On the diversity of roles of organic acids. Front Plant Sci 7:1592
46. Rajurkar NS, Gaikwad KN (2014) Identification and quantification of amino acids from medicinally important plants by using high-performance thin-layer chromatogramphy. J Liq Chromat Relat Technol 37:2197–2205
47. Singh D, Kumar S, Pandey R, Hasanain M, Sarkar J, Kumar B (2017) Bioguided chemical characterization of the antiproliferative fraction of edible pseudo bulbs of *Malaxis acuminata* D. Don by HPLC-ESI-QTOF-MS. Med Chem Res 26:3307–3314
48. Raval SS, Vaghela PG, Mandavia MK, Golakiya BA (2016) Phytochemical analysis of *Malaxis acuminata* D. Don (Jeevak), an ingredient of Jeevaniya Group of Ashtavarga. Indian J Agric Biochem 29:155–160
49. Rawat S, Bhatt ID, Rawal RS (2011) Total phenolic compounds and antioxidant potential of *Hedychium spicatum* Buch. Ham. ex D. Don in west Himalaya, India. J Food Compos Anal 24:574–579
50. Bhatt ID, Rawat S, Rawal RS (2013) Antioxidants in medicinal plants. In: Chandra S, Lata H, Verma A (eds) Biotechnology for medicinal plants. Springer-Verlag, Berlin/Heidelberg, pp 296–326
51. Garg P, Aggarwal P, Sharma P, Sharma S (2012) Antioxidant activity of the butanol extract of *Malaxis acuminata* (Jeevak). J Pharm Res 5:2888–2288

52. Bose B, Kumaria S, Choudhury H, Tandon P (2017) Insights into nuclear DNA content, hydrogen peroxide and antioxidative enzyme activities during transverse thin cell layer organogenesis and ex vitro acclimatization of *Malaxis wallichii*, a threatened medicinal orchid. Physiol Mol Biol Plants 23:955–968
53. Oikarinen A (2004) Connective tissue and aging. Int J Cosm Sci 26:107–107
54. Amaro-Ortiz A, Yan BD, Orazio JA (2014) Ultraviolet radiation, aging and the skin: prevention of damage by topical cAMP manipulation. Molecules 19:6202–6219
55. Chung HY, Kim DH, Lee EK, Chung KW, Chung S, Lee B et al (2019) Redefining chronic inflammation in aging and age-related diseases: proposal of the Seno-inflammation concept. Aging Dis 10:367–382
56. Sunitha K, Suresh P, Santhosh MS, Hemshekhar M, Thushara RM, Marathe GK, Thirunavukkarasu C, Kemparaju K, Kumar SM, Girish KS (2013) Inhibition of hyaluronidase by N-acetyl cysteine and glutathione: role of thiol group in hyaluronan protection. Int J Biol Macromol 55:39–46
57. Bharal A, Kashyap M, Sohpal VK, Sembi JK (2014) Evaluation of antimicrobial properties of terrestrial orchids (collected from northern Himalayas) against certain human pathogens. Int J Bioassays 3:3036–3039
58. Arora M, Kaur G, Kahlon PS, Mahajan A, Sembi JK (2017) Pharmacognostic evaluation and antimicrobial activity of endangered ethnomedicinal plant *Crepidium acuminatum* (D. Don) Szlach. Pharmacog J 9:S56–S63
59. Sharma P, Nipun M, Pankaj G, Gurukirpal S, Sumit D, Sakshi S (2011) *Malaxis acuminata*: a review. Int J Res Ayurveda Pharm 2:422–425
60. Tamta BP, Sharma AK, Puni L, Singh A (2015) Propagation and conservation of endangered orchid *Microstylis wallichii* Syn *Malaxis acuminata* (Jeevak) in its natural habitats of Uttarakhand Himalayas. Int J Sci Technol 4:424–432
61. Arenmongla T, Deb CR (2012) Germination of immature embryos and multiplication of *Malaxis acuminata* D. Don, an endangered therapeutically important orchid, by asymbiotic culture in vitro. Indian J Biotechnol 11:464–469
62. Cheruvathur MK, Abraham J, Mani B, Thomas TD (2010) Adventitious shoot induction from cultured internodal explants of *Malaxis acuminata* D. Don, a valuable terrestrial medicinal orchid. Plant Cell Tissue Org Cult 101:163–170
63. Kaur S, Bhutani KK (2010) Micropropagation of *Malaxis acuminata* D. Don: a rare orchid of high therapeutic value. J Med Arom Plants 1:29
64. Arenmongla T, Deb CR (2012) Studies on *in vitro* morphogenetic potential of nodal segments of *Malaxis acuminata* D. Don. Appl Biol Res 14:156–163
65. Giri L, Jugran A, Rawat S, Dhyani P, Andola H, Bhatt ID, Rawal RS, Dhar U (2012) *In vitro* propagation, genetic and phytochemical assessment of *Habenaria edgeworthii*: an important Astavarga plant. Acta Physiol Plant 34:869–875
66. Mishra J, Bhandari H, Singh M, Rawat S, Agnihotri RK, Mishra S, Purohit S (2011) Hairy root culture of *Picrorhiza kurroa* Royle ex Benth.: a promising approach for the production of picrotin and picrotoxinin. Acta Physiol Plant 33:1841–1846
67. Giri L, Dhyani P, Rawat S, Bhatt ID, Nandi SK, Rawal RS, Pande V (2012) *In vitro* production of phenolic compounds and antioxidant activity in callus suspension cultures of *Habenaria edgeworthii*: a rare Himalayan medicinal orchid. Ind Crop Prod 39:1–6
68. Badhani A, Rawat S, Bhatt ID, Rawal RS (2015) Variation in chemical constituents and antioxidant activity in yellow Himalayan (*Rubus ellipticus* smith) and hill raspberry (*Rubus niveus* Thunb.). J Food Biochem 39:663–372
69. Rawat S, Bhatt ID, Rawal RS (2017) Geographical and environmental variation in chemical constituents and antioxidant properties in *Roscoea procera* wall. J Food Biochem 41:e12302
70. Jugran AK, Bhatt ID, Rawal RS, Rawat S (2020) Essential oil composition, phenolics and antioxidant activities of *Valeriana jatamasi* Jones (Caprifoliaceae) at different phenological stages. Plant Biosyst. (in press)

71. Sharma H, Kumar P, Singh A, Aggarwal K, Roy J, Sharma V, Rawat S (2020) Development of polymorphic EST-SSR markers and their applicability in genetic diversity evaluation in *Rhododendron arboreum*. Mol Biol Rep 47:2447–2457
72. Sharma H, Bhandawat A, Rawat S (2020) Cross-transferability of SSR markers developed in *Rhododendron* species of Himalaya. Mol Biol Rep 47.: (in press). https://doi.org/10.1007/s11033-020-05606-0
73. Dhyani P, Sharma B, Singh P, Masand M, Seth R, Sharma RK (2020) Genome-wide discovery of microsatellite markers and, population genetic diversity inferences revealed high anthropogenic pressure on endemic populations of *Trillium govanianum*. Ind Crop Prod 154:112698

Phytochemistry, Pharmacology, and Conservation of *Ansellia africana*: A Vulnerable Medicinal Orchid of Africa

18

Paromik Bhattacharyya, Shubhpriya Gupta, and Johannes Van Staden

Contents

Abstract

Medicinal plants are natural reserves of various therapeutic biomolecules. Orchids occupy a significant position among these. *Ansellia africana* is one of the most important orchids used in various pharmacopeias worldwide, especially in Traditional African Pharmacopeia (TAP). South Africa, in particular, houses approximately 494 species of orchids with a 75% rate of endemism. *A. africana* is a wonder orchid having large reserves of prized biomolecules that provide remedies to chronic ailments such as Alzheimer's disease. Apart from its

P. Bhattacharyya · S. Gupta · J. Van Staden (✉)
Research Centre for Plant Growth and Development, School of Life Sciences, University of KwaZulu-Natal Pietermaritzburg, Scottsville, South Africa
e-mail: rcpgd@ukzn.ac.za

J.-M. Mérillon, H. Kodja (eds.), *Orchids Phytochemistry, Biology and Horticulture*, Reference Series in Phytochemistry, https://doi.org/10.1007/978-3-030-38392-3_17

pharmaceutical importance, the plant is of high importance in the horticultural industry, primarily because of its flowers. However, as with many orchid species, the wild populations of this fascinating orchid species are under severe threat. Presently *A. africana* has been categorized as "vulnerable" and mandates systematic as well as scientific approaches to conserve as well as sustainably utilize its economic potentials. Keeping into consideration the huge potentiality of *A. africana* as a medicinal herb, the present chapter documents all the major research aspects with a focused aim to provide holistic information about the pharmaco-horticultural importance of this prized orchid species of Africa.

Keywords

Medicinal orchids · Alzheimer's disease · Traditional African pharmacopeia (TAP) · Phytomedicine

1 Introduction

Global climate change, along with an increased rate of deforestation, has adversely affected the distribution of flora and fauna on a rapid scale stressing the need to conserve natural habitats on a priority basis. Africa as a continent is one of the main nuclei of the few remaining biodiversity reserves of the world. Among the various parts, the southern part of Africa is a rich reserve for various flora and fauna [1, 2]. Being a part of the African mega biodiversity hotspot, South Africa holds a rich reserve of various herbaceous plant species including rare, endangered, and threatened (RET) – medicinal aromatic plants (MAP) [2, 3, 4]. In Africa, a large population of people depends a lot on traditional healers for the treatment of various chronic ailments. Orchids are a vital part of African traditional medicine as in other traditional pharmacopoeias of the world. However, the exact time of inclusion of orchids (for medicinal purposes) in African traditional medicine is not known [2, 5]. In South African traditional pharmacopeias, approximately 49 orchid species are being used [3, 6]) of which *Ansellia africana* Lindl stands out prominently. Traditionally various parts of *A. africana* have been used in the Traditional African Pharmacopeia (TAP) for centuries [2, 3]. It is reported that the smoke generated after burning the stem and roots of *A. africana* has exhibited a vital role in the treatment of ailments involving the central nervous system (CNS) [1]. The stem and root infusions obtained by *A. africana* have been reported to possess aphrodisiac properties. Along with its medicinal usage, *A. africana* is highly valued for its beautiful flowers which are reported to have a high shelf life in comparison to other orchid species. The medicinal orchids of Southern Africa have not been extensively studied for taxonomical aspects and very little research has been done on the phenotypic and genotypic diversity of wild populations of *A. africana* and its associated species. These lacunae can be filled by performing research focused on the systematic studies along with their conservation and ethnopharmacological aspects.

2 *Ansellia africana* as a Species

Orchids are one of the most diversified forms of plant taxa. Being one of the mega biodiversity hotspots in the world, the orchid biodiversity in southern Africa is tremendous of which *A. africana* deserves special mention [1]. *A. africana* is an epiphyte and generally grows in clusters on trees with a predominant presence in the semitropical areas (Fig. 1a, b). The roots are specially adapted to provide better attachment to the substratum on which it grows along with the absorption of nutrients and moisture for its survival in harsh and rugged climatic conditions. Unlike other epiphytes, the roots of *A. africana* are of needle shape pointing upward which later on multiply exponentially and form a dense clump around the pseudobulb which gathers the senescing leaves and tissue debris upon which the orchid survives (Fig. 1c). It blooms generally in the dry winters and produces a surplus of

Fig. 1 (**a**) *Ansellia africana* plants in the wild (**b**) Plants growing in the wild with attached capsules after pollination (**c**) Clustered needle like pointed roots of *A. africana*

yellow or greenish-yellow flowers that are marked with light or deep brown spots from which it has got its popular name "leopard orchid." It has a reported distribution in tropical Africa along with Namibia, Botswana, and Swaziland, along with the Northern Cape in South Africa. The orchid has a specific habitat specificity and is found to grow in hot and dry river valleys [1, 7]. The natural populations of *A. africana* are under tremendous anthropogenic pressure and are facing the risk of extinction. Keeping into consideration its present threatened status, IUCN has categorized *A. africana* as "vulnerable." Like most orchid species *A. africana* is also enlisted within Appendix −1 of CITES.

3 Ethnomedicinal, Horticultural, and Traditional Uses

A. africana is one of the most prized medicinal orchid species, not only in Africa but also globally. In TAP, it is one of the most important ingredients primarily because of its huge reserves of prized biomolecules [1–3]. Traditionally, stem infusions and smoke from *A. africana* are used by Zulu traditional healers as antidotes for bad dreams [3], whereas leaf and stem extracts were used by the Mpika tribes of Zambia in controlling madness and related mental disorders [8]. The orchid extract has also found important usage as an aphrodisiac along with imparting various protective charms [8]. Apart from its ethnomedicinal usage, the showy flowers *of A. africana* make it an important candidate taxa for the horticultural industry along with a high shelf life. In short, *A. africana* is one of the few orchid species which has both horticultural and medicinal usage, making it a prized species for collectors and export.

4 *Ex Situ* Conservation Using *In Vitro* Technologies

The rapid depletion of orchid bio-resources from their natural habitats demands urgent strategies for their conservation. Due to various developmental activities and other anthropogenic pressures in the African mega biodiversity hotspot, the wild populations of various orchids including *A. africana* are facing the risk of fragmentation as well as extinction. The attractive flowers and medicinal properties of *A. africana* have made it a highly traded orchid species leading to indiscriminate collection from the wild, thus rendering the species to become threatened in the wild [1, 2, 9]. Lack of comprehensive annual data on annual trade of *A. africana* is also a major concern.

Habitat destruction has various far-reaching impacts, which are not only the loss of precious gene pools useful in plant development or biosynthesis of new compounds but also the loss of a pharmaceutically important source of various vital compounds. The modern tools of biotechnology can be utilized for the propagation and conservation of plant genetic resources [10]. In general, these could be accomplished both by in situ and ex situ methods. These techniques were initially introduced for plant species having agricultural and horticultural importance, but are now rapidly being applied to the collection, propagation, preservation, and evolution of rare and endangered plant germplasm. In situ conservation, which is an ideal and dynamic approach that allows plants to interact and coevolve, includes the protection of genetic resources in the natural environment through the protection of the environment itself [11]. It is the most suitable option to ensure the natural growth, proliferation, and perpetuation of the species. Globally, to promote the cause of in situ conservation, many forested areas have been categorized as national parks, wildlife sanctuaries, and biosphere reserves. However, it is costly to maintain and is highly susceptible to natural calamities. On the other hand, ex situ conservation programs have played an important role in acclimatization, rehabilitation, multiplication, and judicious exploitation. Biotechnological approaches of conservation are complementary to conventional methods. These can directly assist plant conservation programs through molecular marker technology, molecular diagnostics, *in vitro* technologies, and cryopreservation [10, 12].

One of the major reasons behind the depletion of natural reserves of orchids is their extremely low rates of seed germination in nature, along with dependence on symbiotic fungal strains which augments the process of seed germination in nature [13]. Like other orchids, *A. africana* also has a very low rate of seed germination in nature which is less than 3%, and efforts are being made to supplement the process using asymbiotic seed germination techniques [9]. It was observed that the growth medium on which the seeds were germinated impacts the process rather than the sterilization process. In addition, dark preconditioning after inoculation significantly increased the rate of seedling growth particularly the rhizoids in the protocorms [9]. However, recently Papenfus et al. [14], reported a positive synergy of smoke-water on the germination of *A. africana*. Smoke is generally reported to have a conducive effect in enhancing plant vigor and influences pollen growth [15, 16]. The research done by Papenfus et al. [14] provided the first insights on the impact of the two smoke-derived chemical compounds, i.e., karrikinolide (KAR_1) and tri-methylbutenolide (TMB) on the *in vitro* seed germination and seedling growth of *A. africana*.

It was found that the half-strength MS medium supplemented with 1:250 (v:v) smoke-water (SW) significantly improved the germination rate index (GRI) and the development rate index (DRI) of *A. africana* seeds. The SW-treated seeds in MS medium significantly enhanced the production of large protocorm-like bodies (PLBs). However, the KAR_1-treated seeds showed no significant effect on the germination and development of seeds, whereas, TMB-treated seeds significantly reduced the GRI and DRI of *A. africana* seeds [14].

Along with the development of symbiotic seed germination modules, the improvement of micropropagation procedures can assure sufficient supply of plant material along with the probability of producing elite genotypes with a subsequent reduction in over-collection of the wild germplasm (Fig. 2a, b). There exist a few reports on the high frequency *in vitro* propagation, acclimatization, and clonal fidelity assessment of the microclones of *A. africana* [7, 14]. The media supplemented with *meta*-Topolin Riboside (*m*TR) enhanced the shoot and root length, leaf number, frequency of root organogenesis, and fresh weight of *A. africana* [17].

Fig. 2 (**a**) Asymbiotic germination of seeds (**b**) Stages of germinating seed (**c**) Initiation of plantlet from nodal explant (**d**) Proliferation of multiple shoots (**e**) Synthetic seed encapsulated in sodium alginate beads (**f**) Proliferated roots with prominent white needle-like structures (**g**) Hardened micropropagated plants of *A. africana*

The report also demonstrated the efficacy of *m*TR which is a derivative of *meta*-Topolin over the conventional cytokinin-based PGRs like BA and TDZ. Working in similar lines, Bhattacharyya et al. [7] demonstrated the comparative efficacy of aromatic cytokinin *m*T in terms of ascertaining genetic stability and regeneration potential over conventional cytokinin BA. The research also provided vital insights into the role of phenolic elicitors like phloroglucinol singly and also in combination with conventional auxins like indole Butyric Acid (IBA) in induction, proliferation, and branching of *A. africana* roots which are highly valued for their medicinal properties [17] (Fig. 1b and 2c, d).

Being a recalcitrant species, the development of artificial seed technology in *A. africana* deserves special mention. Synthetic seed technology has played a pivotal role in orchid biotechnology as it offers tremendous potential for easy handling, micropropagation, and long- and short-term storage of plant germplasm through cryopreservation techniques [18–20]. Furthermore, the successful development of synthetic seed technology is largely dependent on the formation and production of viable artificial seeds that would be able to transform into complete plantlets [18]. Recently, Bhattacharyya et al. [20] reported the synthetic seed production and short-term storage of *A. africana* using protocorm-like body (PLB) segments. The developed propagule provides vital insights into the role of aromatic cytokinin *meta*-Topolin and its derivatives (*m*TR, *m*TTHP, and *mem*TTHP) in the short-term storage of *A. africana* synthetic seeds. The MS media supplemented with 7.5 μM *mem*-TTHP showed the highest response percentage of encapsulated PLBs which were successfully stored for 75 days at 8 °C. The *in vitro* regeneration of *A. africana* will be helpful in maintaining its population in the wild and can be utilized in the conservation of various other orchids that are endangered (Fig. 2e).

Apart from the formulation of conservation strategies, *in vitro* propagation techniques have contributed significantly to the growth of the pharmaceutical industry over the past several decades in a multidisciplinary manner, including varietal improvement and development of elite cultivars and production of secondary metabolites [21–25]. PGRs such as auxins and cytokinins (CK) regulates the activities of phenylalanine ammonia-lyase and chalcone synthase enzymes, thereby affecting the biochemical synthesis of phenolic acids [26, 27]. Amoo et al. [28] reported that CK supplemented MS medium significantly enhanced the various bioactive metabolites including phenolics and iridoids in *Aloe arborescens*. The orchids have distinctive somatic embryos which are known as protocorm-like bodies or PLBs and are reported to contain crucial bioactive metabolites that are present in wild plants [29–31]. Therefore, these PLBs can be propagated on a large scale as they are highly differentiated and can be used as a substitute for wild medicinal sources for exploiting the therapeutic potential. Bhattacharyya et al. [32] reported that *A. africana* PLBs are a potential source of biologically important phenolic acids such as hydroxybenzoic and hydroxycinnamic acid derivatives. They discovered that the treatment of topolins (*m*T, *m*TR, *Mem* T, and *Mem*TR) and TDZ significantly influenced the biosynthesis and accumulation of various phenolic acids (including hydroxybenzoic and hydroxycinnamic acid derivatives) and exponentially enhanced the biomass, FW, and DW of *A. africana* PLBs (Fig. 2f, g).

Also, natural antioxidants, which are present in fruits, vegetables, and medicinal plants, have received much attention and have been studied extensively since they are effective free radical scavengers and are assumed to be less toxic than synthetic antioxidants. Rapid multiplication rate, higher genetic stability, and significantly higher antioxidant activity reported in the study ensure the utility of this micropropagation protocol developed for further utilization in the ex-situ conservation and commercial utilization of *A. africana* [17, 32]. In short, micropropagation is an important biotechnological tool, which largely assists in combating the various biodiversity conservation issues arising primarily due to unplanned over-collection and habitat destruction. Proper use of plant tissue culture modules can further facilitate the production of elite cultivars with significantly higher yields of secondary metabolites on a mass scale, which will cater to both medicinal as well as horticultural industries.

5 Phytoconstituents and Biological Activity

5.1 Antimicrobial and Membrane Damaging Activity

The ethnomedicinal applications of medicinal herbs are useful not only in conservation but also in cultural tradition and biodiversity at community levels [33]. The whole plant of *A. africana* has been reported to be used in the treatment of respiratory disorders such as asthma [34, 35], while its shoots are being used in the treatment of lice [36]. However, studies on the biological activities of *A. africana* are rare. Penduka et al. [37] provided some vital insights into the antimicrobial activities of *A. africana* by testing its extract against *Moraxella catarrhalis* (clinical isolate), *Klebsiella pneumoniae* (ATCC 4352), *Staphylococcus aureus* (ATCC 25925), and *Mycobacterium smegmatis* (ATCC 14468). The exhibited minimum inhibitory concentration (MIC) value of *A. africana* extracts was significantly low ranging from 2.5 to 10 mg/l. One of the most significant findings of the research is the efficacy of *A. africana* extracts against the *M. tuberculosis* strains. The African subcontinent is severely engulfed by the phenomenon of malnutrition which further magnifies the risk of tuberculosis in the region. The situation gets more aggravated due to the lack of modern healthcare systems. The potential activity of *A. africana* extracts against *M. tuberculosis* strains provides a new prospect in the development of indigenous medicines and drugs at an affordable rate [37]. Comparative antimicrobial activity of root and stem extracts of *A. africana* was also estimated, which closely supports the traditional use of the whole plant in treating respiratory problems [34, 35]. Furthermore, the studies revealed no antagonistic influence of the plant extract with antibiotic ciprofloxacin, which provides baseline information on the utilization of *A. africana* in multiple drug therapy (MDT) after *in vivo* confirmation [37]. Interestingly, plant extracts exhibited a low membrane-damaging activity against *M. smegmatis* strains which can be due to the thick and waxy cell wall of Mycobacteria which makes it have a highly impermeable outer surface, enabling Mycobacteria strains to withstand extreme environmental conditions also in the presence of antibacterial agents [38]. The findings of Penduka et al. [37] validates the traditional use of *A. africana* by traditional healers in the treatment of skin, respiratory, and soft tissue infections. The research findings also envisage the fact that the strategies for the treatment of infectious diseases are multidimensional and more dedicated investigations are required using *in vivo* models along with compound isolation (Table 1).

Table 1 Biological activities attributed to *Ansellia africana*

Name of the biological activity	Plant parts used	Salient findings	References
COX-1 and COX-2 assay	Leaves, stems, and roots	1. Roots of *A. africana* showed high COX-1 and COX-2 inhibition activity	Chinsamy et al. [4]
		2. The dicholoromethane (DCM) root extract of *A. africana* exhibited the highest EC_{50} activity among seven south African medicinal orchid species tested, i.e., 0.25 ± 0.10 mg/ml	
Acetylcholinisterase inhibitory activity	Leaves, stems, and roots	Highest acetylcholinesterase inhibitory activity exhibited among the seven most potential medicinal orchid species tested. The most potent extract was the ethanolic root extract	
β-Carotene bleaching activity	Leaves, stems, and roots	Exhibited moderate β-carotene bleaching activity	
Mutagenic activity	Leaves, stems, and roots	DCM root extract and ethanolic leaf, stem, and root extract exhibited mutagenic effects	
Antioxidant power	Leaves, stems, and roots	High antioxidant levels determined by FRAP (Fluorescence Recovery After Photobleaching) assay	
Antimicrobial activity	Leaves, stem, and roots	1. Demonstrated high efficacy against both gram-positive and gram-negative strains of bacteria.	Penduka et al. [37]
		2. In extremely low concentration (2.5-10 mg/l) root and stem extracts showed high efficacy.	
		3. Potentiality to combat pathogenicity of *Mycobacterium tuberculosis* and *Staphylococcus aureus* causing chronic respiratory track ailments in humans.	
Phenolic acid activity	Protocorm-like bodies (PLB)	Modulating the production of biopharmaceutically important phenolic acids, i.e., hydroxybenzoic acid and hydroxycinnamic acid derivatives in protocorm-like bodies (PLB)	Bhattacharyya et al. [32]

5.2 Acetylcholinesterase Inhibitory, Antiinflammatory, and Antioxidant Activity

In the recent past, a lot of medicinal plant bioprospection has been carried out based on their activities related to the CNS [39]. Reports on the traditional use of *A. africana* in the treatment of mental disorders like insanity, hyper fits, nervous disorders have been confirmed to have prominent acetylcholinesterase (AChE) inhibitory activity [2, 4]. A total of 64 African plant species are reported to exhibit potent AChE activity among which *A. africana* figures prominently [2, 39]. The findings of Chinsamy et al. [4] have revealed that the ethanolic root extracts of *A. africana* have a promising AChE activity which is in close synchrony to its traditional usages [1].

Traditionally, *A. africana* is also reported to possess potent anti-inflammatory activity and is being used by traditional healers as an aphrodisiac [1, 3, 8]. Recent studies on the phytochemical constitution of *A. africana* revealed the presence of high levels of gallotanin in *A. africana* plant parts [4]. Gallotanin is a prized molecule and is reported to possess various biological activities such as anti-inflammatory activity [37]. In recent times, Alzheimer's disease has become a major concern for society. Inflammatory responses, cholinergic system, and oxidative stress often collectively account for the various symptoms prevalent in aged persons and Alzheimer-affected patients [37].

One of the primary enzymes which are involved in anti-inflammatory responses is cyclooxygenase (COX). In general, COX enzymes are being classified into COX-1 and COX-2. According to Bohlin et al. [40], flavonoids, naphthoquinones, alkylamides, phenolic phenyl-propane derivatives are some of the compounds involved in COX enzyme inhibition. Chinsamy et al. [4] reported that the extracts from the roots of *A. africana* showed high COX-1 and COX-2 inhibition activity. The highest EC50 activity (0.25 ± 0.10 mg/ml) was shown by dichloromethane (DCM) root extract of *A. africana*. Also, the aqueous concoctions of *A. africana* plant parts exhibited high levels of COX-1 and COX-2 activity which explains the practice of Mpika tribes and other tribes of Africa to administer warm aqueous concoctions of *A. africana* plants in the treatment of bad dreams or madness [1]. Apart from AChE activity and anti-inflammatory activity the *A. africana* extracts exhibited a potent antioxidant and antimutagenic activity justifying the importance and significance of *A. africana* in TAP.

6 Molecular Biology Approaches

The use of DNA fingerprinting has a wide range of applications. It has been used in forensic science to solve criminal cases and settle parental disputes. It has a wide spectrum of applications in the field of plant sciences; it is used to identify genetic diversity within breeding populations, to positively identify and differentiate accessions, cultivars, and species that might be challenging to illustrate due to related phenotypic characters or indistinct traits, and to identify plants encompassing genes of interest (such as the confirmation of transformation events). As the natural populations are exposed to several genetic factors or forces affecting the amount and kind of genetic variation such as mutation, chance events like genetic drift and founder events, selection, migration, and a species mating system [41], a careful assessment of these factors and their role in the formulation of conservation strategies is essential.

The sustainability of micropropagation techniques depends upon the production of true-to-type plants and maintenance of the genetic integrity of the *in vitro*-raised plants so that the advantage in the use of elite genotypes over natural seedlings is maintained. However, *in vitro* techniques are known to induce clonal variations. Since the first observation and report of clonal variations [42], it remains an issue of great concern for tissue culture-raised plants as there are reports of genetic disparities in micropropagated plants [43, 44]. The occurrence of these variations depends on several factors such as the source of the explants, media composition, and cultural conditions [45]. Thus, monitoring the degree of genetic variability within the *in vitro*-raised plants is of prime consideration for the effective commercial utilization of the technique and also for large-scale production of true-to-type plants of the desired genotype [46]. In the recent past, various high-frequency regeneration protocols have been developed in orchids including *A. africana* with high rates of genetic stability [17, 20, 47].

7 Future Research Prospects

7.1 Bioassays and *In Vivo* Model-Based Studies

A. africana offers various promising leads both in terms of floriculture and biopharmaceuticals. This orchid is highly prized for its showy flowers along with its therapeutic properties, primarily due to the reserves of various bio entities in its aerial parts [1]. This chapter summarized the existing ethnobotanical and horticultural uses, phytochemistry, pharmacological activities along with conservation and molecular insights on *A. africana* [1, 32].

To make this prized ethnomedicinal herb more acceptable, systematic clinical trials evaluating its in-depth biological activities using *in vivo* models is needed [1]. Also, more systematic studies on various toxicological as well as mutagenic properties must be taken up. In short, there is an urgent need for systematic clinical trials to establish the efficacy of *A. africana* in medicine. Along with clinical trials, attempts should also be made to mass propagate elite cultivars of *A. africana* using various biotechnological tools [1, 17, 19]. The PLBs can serve as reserves of important medicinal compounds like phenolic acids which can be further enhanced by the use of suitable elicitors [32]. Fast regeneration protocols, which will be cheap and reproducible, must be developed to further facilitate the sustainable utilization of *A. africana* germplasm. To facilitate such endeavors, interventions of innovative biotechnological tools such as cryopreservation methods and bioreactors are required [1, 17].

7.2 Drug Discovery

The current trend of research in drug discovery from traditionally reported medicinal herbs includes a multidirectional approach comprising of botanical, biological, phytochemical, and molecular approaches. Discovery of new biological molecules provides significantly important research leads against chronic disorders such as cancer, malaria, and various neurodegenerative disorders [4, 5]. In recent years, various plant-based drugs of immense medicinal importance like arteether, dendrobine, gigantol, moscatillin, galantamine, nitisinone, etc. have made a major impact in the areas of drug discovery from traditional ethnomedicinal herbs [48]. Various research studies on African medicinal herbs with special reference to *A. africana* have revealed that it houses various phytochemical entities that are largely unexplored, requiring thorough clinical evaluation. The results exhibited by various fractions of *A. africana* plant extract in the treatment of neurodegenerative disorders provides evidence that it houses molecules that may play a substantial role in the treatment of CNS disorders [1, 4].

7.3 Next-Generation Sequencing and Transcriptome Data Mining

Development of next-generation sequencing (NGS) has opened new gateways for exploring the genomes of various nonmodel plants which might play a pivotal role in deciphering the medicinal properties in various traditional medicinal herbs [49]. To date, several molecular and analytical methods have been applied in the identification and genotyping of various medicinally important orchid species, mostly dendrobiums. However, none to date have taken genome sequencing of the orchids into account. So far, only the genome of *D. officinale* has been reported [50]. Development of NGS-based approach attempting to sequence the selected medicinal orchid transcriptomes like *A. africana* to identify and characterize transcripts potentially contributing to their observed medicinal properties will largely help in answering various unanswered questions in plant therapeutics [1]. In addition to it, understanding the medicinal potential of *A. africana* might illuminate the basic understanding of the genes involved in the biosynthesis and channelization of bioactive secondary metabolites such as phenylpropanoids, alkaloids, and terpenoids. In addition to that, knowledge of NGS provides a better understanding of the various biological interactions such as stress tolerance and mycorrhizal association. Utilizing this advanced biotechnological tool, the important role played by symbiotic association with fungi in nature and how it controls the germination pathways can be deciphered for *A. africana.*

7.4 Endophyte Mapping and Metabolite Production

Microorganisms are known to play an important role in most ecosystems including soil and plant habitats. The microorganisms associated with plants maintain their biological diversity in terrestrial ecosystems through various biological processes [51]. The endophytic microorganisms (including bacteria, fungi, virus, and protozoa) reside in intercellular spaces of stems, petioles, roots, and leaves of plants without causing any disease [16, 52]. Endophytes are well known to promote plant growth, plant productivity, and assist their hosts to withstand biotic and abiotic stresses by the production of countless biologically active metabolites [53, 54]. Besides this, endophytes are known to produce a larger number of metabolites and enzymes as validated by a number of endophyte culture studies [16, 55–57]. Many of these metabolites are the same as produced by their respective hosts and may have potential benefits in the pharmaceutical industry [58, 59]. Various orchids (i.e., *Bletilla ochracea*, *Dendrobium nobile*, *Dendrobium loddigesii*, *Dendrobium aqueum*, *Pecteilis Susannae*, *Vanda coerulea*, etc.) have been explored for their endophytes and their secondary metabolites [60–64]. However, the naturally occurring endophytic microorganisms present in *A. africana* are not yet known. There is a need to investigate and map the endophytic diversity (particularly bacterial and fungal endophytes) of *A. africana* which might be a source of vital metabolites including phytohormones which may have a positive effect on the development of this endangered species. The associated endophytic flora of *A. africana* may serve as beneficial microorganisms that could be used in its propagation and improving acclimatization and vigor [65].

7.5 Phylogeography and DNA Barcoding

Scientific analysis of the phytogeography that is mainly influenced by the occurrence of concerted evolution which plays a significant role in conservation schemes of RET plants particularly in case of orchids. In concerted evolution, hundreds to thousands of tandemly repeated copies of DNA such as ribosomal DNA (rDNA) evolve in a concerted way. Therefore, the copies of related genes have more similarities within the species as compared to between the species. The homogenization of these multiple-copy genes through unequal crossing-over and high frequency of gene conversion is the key factor in influencing the phenomenon of concerted evolution. On the other hand, under distinct circumstances, the vital nature of the tandem repeats of such genes gets disturbed and incomplete intra-genomic deviation appears under such conditions. Incomplete concerted evolution for r-DNA has been reported in several plant species including orchids. *A. africana* is one of the species from South Africa that is highly endangered due to high rates of deforestation and habitat destruction, therefore, studies on phylogeography and existing genetic variations using molecular markers can be of great implication. The studies related to the genetic variability of several orchid species have demonstrated an irregular distribution of genetic diversity among different geographic zones. Furthermore, being an endangered orchid, DNA barcoding of *A. africana* and other related South African orchid species is of relevance. DNA barcodes such as rbcL (ribulose-bisphosphate carboxylase), matK (Megakaryocyte-Associated Tyrosine Kinase), psbA-trnH, rpoC1, and ITS2 (internal transcribed spacer 2) are popular worldwide. Few studies have employed psbA-trnH barcode for identifying species of medicinal pteridophytes and within the genus *Dendrobium*. Chen et al. [66] have shown that ITS2 is a universal barcode in the identification of plants, as 92.7% cases from 6600 samples in seven phyla (Angiosperms, Gymnosperms, Ferns, Mosses, Liverworts, Algae, and Fungi) has been correctly identified by ITS2. Consequently, ITS2 region has been useful in differentiating plants from various families including Orchidaceae. Therefore, these potential barcodes can be used for authentication of orchids including *A. africana*.

8 Conclusions

The orchid *A. africana* holds intriguing prospects, not only in the field of medicinal plant research but also in the horticultural industry because of its showy flowers and it can be one of the most desired orchid species for commercial growers. Moreover, research insights providing its potent activity on the CNS make it an important plant species for bioprospecting of drugs against Alzheimer's disease. It also possesses potent antibacterial activity which can further strengthen Africa's indigenous medicine production. Efforts are required for further studies, especially evaluating its *in vivo* biological activities along with toxicological and mutagenic properties to better validate the safety of these different plant-derived compounds. Also, to establish its efficacy, there is a need for preclinical and clinical trials. However, the high demand for *A. africana* in national and international markets has led to its overexploitation and habitat destruction. This has resulted in the loss of the wild population of *A. africana*. Therefore, for the successful commercialization of this threatened taxon wide research is needed that includes conservation practices and a sustainable supply of plants. This can be achieved by utilizing biotechnological techniques such as micropropagation, cryopreservation, and bioreactors. Detailed research on synthetic seed technology (artificially encapsulated somatic embryos) is required for the improvement in germination frequency of *A. africana* synthetic seeds and subsequent plantlet growth in the soil such that it can be used on a commercial scale. Furthermore, hairy root culture can be used as a model system to improve the valuable phytochemicals of *A. africana*. To prevent misidentification and possible adulteration of *A. africana*, quality control protocols are needed. In the future, new research findings may increase the present therapeutic importance of *A. africana* and its future use in modern medicine. In short, *A. africana* is one of the most prized orchid species occurring in Africa, which has the potential to boost the African bio-economy by promoting phyto-horticultural ventures.

Acknowledgments PB and SG thank the University of KwaZulu-Natal, South Africa for financial support in the form of postdoctoral fellowships. The authors are grateful to the Microscopy and Microanalysis Unit (MMU), UKZN, Pietermaritzburg for microscopic assistance. We are thankful to Dr. Heino B. Papenfus, Kelp Products International (Pty) Ltd., Simon's Town, South Africa, Mrs. Louise Van Staden, Pietermaritzburg, South Africa and Mrs. Lee Warren, Senior Administrative Assistant, Research Centre for Plant Growth and Development, University of KwaZulu-Natal, Pietermaritzburg, South Africa for providing the photographs of wild plants as well as seed germination photographs of *A. africana*.

References

1. Bhattacharyya P, Van Staden J (2016) *Ansellia africana* (Leopard orchid): a medicinal orchid species with untapped reserves of important biomolecules—a mini review. S Afr J Bot 106:181–185
2. Chinsamy M, Finnie JF, Van Staden J (2011) The ethnobotany of South African medicinal orchids. S Afr J Bot 77:2–9
3. Hutchings A, Scott AH, Lewis G, Cunningham A (1996) Zulu medicinal plants: an inventory. University of Natal Press, Scottsville
4. Chinsamy M, Finnie JF, Van Staden J (2014) Anti-inflammatory, antioxidant, anti-cholinesterase activity and mutagenicity of South African medicinal orchids. S Afr J Bot 91:88–98
5. Hossain MM (2011) Therapeutic orchids: traditional uses and recent advances—an overview. Fitoterapia 82:102–140
6. Germishuizen G, Meyer NL (2003) Plants of southern Africa: an annotated checklist. National Botanical Institute, Pretoria
7. Vasudevan R, Van Staden J (2011) Cytokinin and explant types influence *in vitro* plant regeneration of leopard orchid (*Ansellia africana* Lindl.). Plant Cell Tissue Organ Cult 107:123–129
8. Gelfand M (1985) The traditional medical practitioner in Zimbabwe: his principles of practice and pharmacopoeia. Mambo Press, Gweru
9. Vasudevan R, Van Staden J (2010) *In vitro* asymbiotic seed germination and seedling growth of *Ansellia africana* Lindl. Sci Hortic (Amsterdam) 123:496–504
10. Tandon P, Sharma M, Swarup R (2005) Biodiversity: status and prospects. Narosa Publishing House, New Delhi
11. Deb CR, Tandon P (2004) Establishment of an embryogenic suspension culture of *Pinus kesiya* (Khasi pine) from various explants. Indian J Biotechnol 3:445–448
12. Tandon P, Kumaria S (1998) Threats to plant diversity in high altitude of North-East India and conservation of rare and endangered plants using biotechnological approaches. In: Science at High Altitude. Proceedings of the National Symposium, Allied Publishers Ltd, India, pp 140–147
13. Paul S, Kumaria S, Tandon P (2012) An effective nutrient medium for asymbiotic seed germination and large-scale in vitro regeneration of *Dendrobium hookerianum*, a threatened orchid of northeast India:1–7. https://doi.org/10.1093/aobpla/plr032
14. Papenfus HB, Naidoo D, Pošta M et al (2015) The effects of smoke derivatives on *in vitro* seed germination and development of the leopard orchid *Ansellia africana*. Plant Biol 18:289–294
15. Ngoroyemoto N, Gupta S, Kulkarni MG et al (2019) Effect of organic biostimulants on the growth and biochemical composition of *Amaranthus hybridus* L. S Afr J Bot 124:87–93
16. Gupta S, Chaturvedi P, Kulkarni MG, Van Staden J (2020) A critical review on exploiting the pharmaceutical potential of plant endophytic fungi. Biotechnol Adv 39:107462
17. Bhattacharyya P, Kumar V, Van Staden J (2017) Assessment of genetic stability amongst micropropagated *Ansellia africana*, a vulnerable medicinal orchid species of Africa using SCoT markers. S Afr J Bot 108:294–302
18. Ara H, Jaiswal U, Jaiswal VS (2000) Synthetic seed: prospects and limitations. 78:1438–1444
19. Gantait S, Bustam S, Sinniah UR (2012) Alginate-encapsulation, short-term storage and plant regeneration from protocorm-like bodies of *Aranda Wan Chark Kuan* 'Blue' ?? *Vanda coerulea* Grifft. ex. Lindl. (Orchidaceae). Plant Growth Regul 68:303–311
20. Bhattacharyya P, Kumar V, Van Staden J (2018) *In vitro* encapsulation based short term storage and assessment of genetic homogeneity in regenerated *Ansellia africana* (leopard orchid) using gene targeted molecular markers. Plant Cell Tissue Organ Cult 133:299–310
21. Nadeem M, Kumar A, Nandi SK, Palni LMS (1998) Tissue culture of medicinal plants with particular reference to Kumaun Himalaya. In: Proceedings of the workshop on Himalayan medicinal plants-potential and prospects. Kosi-Katarmal, Almora, pp 5–7
22. Rout GR, Samantaray S, Das P (2000) *In vitro* manipulation and propagation of medicinal plants. Biotechnol Adv 18:91–120

23. Verpoorte R, Contin A, Memelink J (2002) Biotechnology for the production of plant secondary metabolites. Phytochem Rev 1:13–25
24. Yesil-Celiktas O, Nartop P, Gurel A et al (2007) Determination of phenolic content and antioxidant activity of extracts obtained from *Rosmarinus officinalis* 'calli. J Plant Physiol 164:1536–1542
25. Shinde AN, Malpathak N, Fulzele DP (2010) Determination of isoflavone content and antioxidant activity in *Psoralea corylifolia* L. callus cultures. Food Chem 118:128–132
26. Deikman J, Hammer PE (1995) Induction of anthocyanin accumulation by cytokinins in Arabidopsis thaliana. Plant Physiol 108:47–57
27. Szopa A, Ekiert H (2014) *Schisandra chinensis* (Turcz.) Baill. (Chinese magnolia vine) *in vitro* cultures. In: Govil JN (ed) Recent progress in medicinal plants. Biotechnology and genetic engineering II, vol 39. Studium Press LLC, Houston, pp 405–434
28. Amoo SO, Aremu AO, Van Staden J (2013) Shoot proliferation and rooting treatments influence secondary metabolite production and antioxidant activity in tissue culture-derived *Aloe arborescens* grown *ex vitro*. Plant Growth Regul 70:115–122
29. Tsay HS, Gau TG, Chen CC (1989) Rapid clonal propagation of *Pinellia ternata* by tissue culture. Plant Cell Rep 8:450–454
30. Roy J, Banerjee N (2003) Induction of callus and plant regeneration from shoot-tip explants of *Dendrobium fimbriatum* Lindl. var. *oculatum* Hk. f. Sci Hortic (Amsterdam) 97:333–340
31. Martin KP, Madassery J (2006) Rapid *in vitro* propagation of *Dendrobium* hybrids through direct shoot formation from foliar explants, and protocorm-like bodies. Sci Hortic (Amsterdam) 108:95–99
32. Bhattacharyya P, Kumar V, Grúz J, Doležal K, Van Staden J (2019) Deciphering the phenolic acid reserves and antioxidant activity within the protocorm like bodies of *Ansellia africana*: a vulnerable medicinal orchid. Ind Crop Prod 135:21–29
33. Mesfin YM, Hailemariam D, Biadglign S, Kibret KT (2014) Association between HIV/AIDS and multi-drug resistance tuberculosis: a systematic review and meta-analysis. PLoS One 9: e82235
34. Bandeira SO, Gaspar F, Pagula FP (2001) African ethnobotany and healthcare: emphasis on Mozambique. Pharm Biol 39:70–73
35. Luo X, Pires D, Aínsa JA et al (2011) Antimycobacterial evaluation and preliminary phytochemical investigation of selected medicinal plants traditionally used in Mozambique. J Ethnopharmacol 137:114–120
36. Ndawonde BG, Zobolo AM, Dlamini ET, Siebert SJ (2007) A survey of plants sold by traders at Zululand muthi markets, with a view to selecting popular plant species for propagation in communal gardens. Afr J Range Forage Sci 24:103–107
37. Penduka D, Mthembu W, Cele KH et al (2018) Extracts of *Ansellia africana* and *Platycarpha glomerata* exhibit antibacterial activities against some respiratory tract, skin and soft tissue infections implicated bacteria. S Afr J Bot 116:116–122
38. He Z, De Buck J (2010) Cell wall proteome analysis of *Mycobacterium smegmatis* strain MC2 155. BMC Microbiol 10:121
39. Masondo NA, Stafford GI, Aremu AO, Makunga NP (2019) Acetylcholinesterase inhibitors from southern African plants: an overview of ethnobotanical, pharmacological potential and phytochemical research including and beyond Alzheimer's disease treatment. S Afr J Bot 120:39–64
40. Bohlin L, Göransson U, Alsmark C, Weden C, Backlund A (2010) Natural products in modern life science. Phytochem Rev 9:279–301
41. Hedrick PW (2001) Conservation genetics: where are we now? Trends Ecol Evol 16:629–636
42. Braun AC (1959) A demonstration of the recovery of the crown-gall tumor cell with the use of complex tumors of single-cell origin. Proc Natl Acad Sci U S A 45:932
43. Feyissa T, Welander M, Negash L (2007) Genetic stability, *ex vitro* rooting and gene expression studies in *Hagenia abyssinica*. Biol Plant 51:15–21
44. Peyvandi M, Noormohammadi Z, Banihashemi O et al (2009) Molecular analysis of genetic stability in long-term micropropagated shoots of *Olea europaea* L.(cv. Dezful). Asian J Plant Sci 8:146–152

45. Salvi ND, George L, Eapen S (2001) Plant regeneration from leaf base callus of turmeric and random amplified polymorphic DNA analysis of regenerated plants. Plant Cell Tissue Organ Cult 66:113–119
46. Larkin PJ, Scowcroft WR (1981) Somaclonal variation—a novel source of variability from cell cultures for plant improvement. Theor Appl Genet 60:197–214
47. Gantait S, Kundu S, Ali N, Sahu NC (2015) Synthetic seed production of medicinal plants: a review on influence of explants, encapsulation agent and matrix. Acta Physiol Plant 37:1–12
48. Balunas MJ, Kinghorn AD (2005) Drug discovery from medicinal plants. Life Sci 78:431–441
49. Egan AN, Schlueter J, Spooner DM (2012) Applications of next-generation sequencing in plant biology. Am. J. Bot. 99:175–185
50. Yan L, Wang X, Liu H, Tian Y, Lian J, Yang R et al (2015) The genome of *Dendrobium officinale* illuminates the biology of the important traditional Chinese orchid herb. Mol Plant 8:922–934
51. Hardoim PR, Van Overbeek LS, Berg G et al (2015) The hidden world within plants: ecological and evolutionary considerations for defining functioning of microbial endophytes. Microbiol Mol Biol Rev 79:293–320
52. Bacon CW, White J (2000) Microbial endophytes. CRC Press
53. Larriba E, Jaime MDLA, Nislow C et al (2015) Endophytic colonization of barley (*Hordeum vulgare*) roots by the nematophagous fungus *Pochonia chlamydosporia* reveals plant growth promotion and a general defense and stress transcriptomic response. J Plant Res 128:665–678
54. Nassimi Z, Taheri P (2017) Endophytic fungus *Piriformospora indica* induced systemic resistance against rice sheath blight via affecting hydrogen peroxide and antioxidants. Biocontrol Sci Tech 27:252–267
55. Tan RX, Zou WX (2001) Endophytes: a rich source of functional metabolites. Nat Prod Rep 18:448–459
56. Gupta S, Chaturvedi P (2015) Phytochemical screening and extracellular enzymatic enumeration of foliar endophytic fungal isolates of *Centella asiatica* (L.) urban. Int J Pharm Sci Rev Res 35:21–24
57. Gupta S, Bhatt P, Chaturvedi P (2018) Determination and quantification of asiaticoside in endophytic fungus from *Centella asiatica* (L.) urban. World J Microbiol Biotechnol 34:111
58. Kumara PM, Zuehlke S, Priti V et al (2012) *Fusarium proliferatum*, an endophytic fungus from *Dysoxylum binectariferum* Hook. f, produces rohitukine, a chromane alkaloid possessing anti-cancer activity. Antonie Van Leeuwenhoek 101:323–329
59. Mousa WK, Raizada MN (2013) The diversity of anti-microbial secondary metabolites produced by fungal endophytes: an interdisciplinary perspective. Front Microbiol 4:65
60. Tao G, Liu ZY, Hyde KD et al (2008) Whole rDNA analysis reveals novel and endophytic fungi in *Bletilla ochracea* (Orchidaceae). Fungal Divers 33:101–112
61. Chen XM, Dong HL, Hu KX, Sun ZR, Chen J, Guo SX (2010) Diversity and antimicrobial and plant-growth-promoting activities of endophytic fungi in *Dendrobium loddigesii* Rolfe. J Plant Growth Regul 29:328–337
62. Chutima R, Dell B, Vessabutr S, Bussaban B, Lumyong S (2011) Endophytic fungi from *Pecteilis susannae* (L.) Rafin (Orchidaceae), a threatened terrestrial orchid in Thailand. Mycorrhiza 21:221–229
63. Aggarwal S, Nirmala C, Beri S et al (2012) *In vitro* symbiotic seed germination and molecular characterization of associated endophytic fungi in a commercially important and endangered Indian orchid *Vanda coerulea* Griff. Ex Lindl. Eur J Environ Sci 2:33–42
64. Parthibhan S, Rao MV, Kumar TS (2017) Culturable fungal endophytes in shoots of *Dendrobium aqueum* Lindley–an imperiled orchid. Ecol Genet Genom 3:18–24
65. Yuan Z, Chen Y, Yang Y (2009) Diverse non-mycorrhizal fungal endophytes inhabiting an epiphytic, medicinal orchid (*Dendrobium nobile*): estimation and characterization. World J Microbiol Biotechnol 25:295
66. Chen S, Yao H, Han J, Liu C, Song J, Shi L, Zhu Y, Ma X, Gao T, Pang X (2010) Validation of the ITS2 region as a novel DNA barcode for identifying medicinal plant species. PloS One. https://doi.org/10.1371/journal.pone.0008613

Dendrobium sp.: In vitro Propagation of Genetically Stable Plants and Ethnomedicinal Uses

19

Leimapokpam Tikendra, Nandeibam Apana, Angamba Meetei Potshangbam, Thoungamba Amom, Ravish Choudhary, Rajkumari Sanayaima, Abhijit Dey, and Potshangbam Nongdam

Contents

Abstract

Plants belonging to the *Dendrobium* genus occupy a dominant position among the orchids because of their high ornamental and therapeutic values. They are widely popular in the international floriculture trade as they bear stunning flowers with diverse coloration, varied forms, and patterns. The ethnomedicinal uses of these orchids are also prominently found due to the possession of immense medicinal properties. Excessive exploitation through rampant unregulated

L. Tikendra · N. Apana · A. M. Potshangbam · T. Amom · P. Nongdam (✉)
Department of Biotechnology, Manipur University, Canchipur, Manipur, India
e-mail: nongpuren@gmail.com

R. Choudhary
DSST, Indian Agricultural Research Institute, New Delhi, India

R. Sanayaima
DDU College, University of Delhi, New Delhi, India

A. Dey
Department of Life Sciences, Presidency University, Kolkata, India
e-mail: abhijit.dbs@presiuniv.ac.in

J.-M. Mérillon, H. Kodja (eds.), *Orchids Phytochemistry, Biology and Horticulture*,
Reference Series in Phytochemistry, https://doi.org/10.1007/978-3-030-38392-3_30

collection and widespread habitat destruction have dwindled the *Dendrobium* natural populations at an alarming rate. The fast and reliable micropropagation techniques offer an alternative to the slow conventional methods of *Dendrobium* propagation. The rapid in vitro regeneration of genetically stable *Dendrobiums* is essential for effective germplasm conservation and large-scale orchid commercialization. Several genetic stable *Dendrobiums* have been successfully propagated by ascertaining the clonal fidelity of the regenerants using different molecular markers. This chapter focuses on the uses of *Dendrobiums* as important ethnomedicine, their in vitro propagation, and clonal assessment for producing genetically stable orchids using DNA markers.

Keywords

Dendrobiums · Phytochemicals · Alkaloids · Micropropagation · In vitro propagation · Genetic variation · Somaclonal variation · Genetic stability · DNA markers

1 Introduction

Orchids, the incredible flowering plants which belong to the Orchidaceae family, comprise about 1000 genera and 35,000 species [1]. They have captivating floristic characters and color patterns with diverse forms and growth habits and are distributed worldwide from tropics to high alpine [2, 3]. They are considered as luxurious plants because of their exquisite beauty and fragrance of flowers, brilliance in coloration, and remarkable range in sizes and manifold shapes [4]. The genus *Dendrobium*, composed of about 1400 species, is significant among the orchids because of their high ornamental and medicinal values [5, 6]. They exhibit tremendous diversity with numerous interspecific hybrids and are widely distributed geographically in most parts of Asia, Australia, and Europe [7, 8]. Many *Dendrobiums* possess extraordinarily beautiful flowers with varied floral patterns and forms, making them one of the most sought after ornamental plants in the International floricultural market. The large part of contemporary orchid trade is mostly dominated by artificially propagated plants and hybrids of *Dendrobium*, *Cymbidium*, and *Phalaenopsis* orchids [9, 10]. *Dendrobiums* are also in huge demand for the pharmaceutical industry due to the rich content of diverse useful phytochemicals.

The natural populations of these multiutility orchids have witnessed drastic reduction due to excessive unregulated collection for illegal trade and rampant habitat destruction [11]. The whole Orchidaceae is listed in the Red Data Book of International Union of Conservation of Nature (IUCN), and the entire *Dendrobiums* are included in Appendix II of threatened species of plants and animals under CITES [12, 13]. Rapid large-scale propagation of the orchids is the need of the hour to meet the increasing commercial demands and for effective germplasm conservation. But conventional propagation methods are slow, labor-intensive, and extremely time-consuming. Plant tissue culture techniques may substitute the conventional approaches for effective conservation and commercialization of *Dendrobiums*. Many workers have successfully micropropagated several *Dendrobiums* using different explants [11, 14–19]. But there is a possibility of the emergence of somaclonal variation among the regenerated orchids as plant tissues are routinely exposed to different stress conditions during in vitro culture. Incorporation of plant growth regulators, long culture cycle, and differential explant regeneration patterns might change the genetic makeup of the regenerants leading to somaclonal variation [20, 21]. Genetic variation among the in vitro propagated plants is unwanted if the primary regenerants are the desired end products. Production of genetically stable plants similar to the elite mother plants is beneficial to orchid cut-flower industry as it assists in uniform blooming during predictable periods fulfilling the market demands. The genetically stable *Dendrobiums* can be propagated by ascertaining the clonal fidelity of the regenerants using different DNA molecular markers. Many investigators have successfully used several DNA markers for genetic homogeneity assessment of different micropropagated *Dendrobiums* [17, 22–25]. In this chapter, we highlight the ethnomedicinal uses of *Dendrobiums* for therapeutic control of ailments and recent works on the in vitro propagation of genetically stable plants using different DNA markers.

2 Ethnomedicinal Uses

The study of how people of a particular culture and area use indigenous plants in their lives for daily health management and other requirements is called ethnobotany [26]. Schultes [27] had termed it as "The study of the relationship which existed between people of ancient societies and their environment." Every society harbors a specific medical culture or "Ethnomedicine," which is concerned with the cultural interpretations of health, illness, disease prevention, and local healing practices [28]. Since time immemorial, the traditional healers in any ethnic community use wild plants for making indigenous medicines. These local medicine men also used *Dendrobiums* for folk medicine preparation in the form of herbal paste or medicinal concoctions to treat different ailments [29]. The ethnomedicinal uses of *Dendrobiums* date back to twenty-eighth century B.C when their therapeutic application was mentioned in "Material medica" during the time of Emperor "Shen-Nung" [1, 30]. They are prominently employed in the traditional Chinese medicine (TCM) by utilizing about 30 different *Dendrobium* species under *Dendrobii Caulis* (Shi-Hu) and *Dendrobii officinalis* Caulis (Tie-Pi Shi-Hu) [6, 31]. They are used in Chinese folk medicines as a tonic, analgesic, fluid body enhancer, and anti-inflammatory substances [32]. *Dendrobiums* are also applied in Indian Ayurveda medicine with *D. alpestre* as a source of "Jewanti" and *D. teretifolium*, *D. macraei*, *D. densiflorum*, *D. fimbriatum*, and *D. discolor* in the management of dysentery, pain, pimples, skin eruption, liver upset, nervous debility, asthma, bronchitis, throat trouble, and fever and is used as an aphrodisiac [33–41]. Many *Dendrobiums* have rich contents of phytochemical compounds such as gigantol, moscatilin, dendrobinae, mucilage, dendrobine, denbinobine, dendroside derivatives, nobilin D, nobilin E and nobilone in *D. nobile* [31, 42, 43]; dendromoniliside derivatives, moniliformin, 4-phenanthrenequinone, and daucosterol in *D. monoliforme* [44, 45]; dibutyl phthalate, ethyl haematommate, methyl B-orcinol carboxylate, N-docosyl trans-ferulate, and ferulaldehyde in *D. longicornu* [1]; jebantine and jibantic acid in *D. macraei* [1, 46]; flavanthrin, coelonin, iusianthridin, moscatin, gigantol, dibutyl phthalate, and P-hydroxyphenylpropionic methyl ester in *D. aphyllum* [1]; dendrocandin derivatives, amotin, amoenin, flaccidin, and 3,4-dihydroxy-5,4 dimethoxybibenzyl in *D. candidum* [1, 47]; crepidine, crepidamine, and dendrocrepine in *D. crepidatum* [43]; homoeridictyol, scoparone, bibenzyl, densiflorol, cypripedin, gigantol, moscatilin, tristin, naringenin, homoeriodictyol, moscatin, and scoparone in *D. densiflorum* [1, 48]; isoamoenylin, amoenylin and moscatilin in *D. amoenum* [1, 49]; rotundatin, moscatin, moscatilin, and scopoletin in *D. moschatum* [50]. The presence of the diverse alkaloids, flavonoids, and glycosides in *Dendrobiums* makes them highly important medicinal herbs with antimicrobial, anti-inflammatory, anticancer, antioxidative, and antiviral properties [51]. The ethnomedicinal uses of some important *Dendrobiums* are described below.

D. amoenum Wall. Ex. Lindl.: It is an important epiphytic orchid with high floricultural and medicinal values. The dried stems are powdered to prepare a decoction, which can be used as a tonic [40]. The fresh paste obtained from grounded pseudobulbs may be applied to treat burnt skin and dislocated bones [52].

D. aphyllum (Roxb.) C.E.C. Fisch.: The leaves can be grounded after drying completely and made into a fine paste with water. The paste obtained from the leaves can be put on abnormal or deformed parts of the head of a newly born baby to get into normal shape [53]. The leaves poultice is also used in the treatment of boils and pimples in the skin [54].

D. aurantiacum Rchb. F: The dried stems of the orchid are utilized as traditional or folk medicines in China for their antipyretic, eyesight improving, immune-modulatory, anti-oxidant, and antiaging properties [45]. The infusion or decoction obtained from the leaves is used in diabetic treatment [55].

D. candidum Wall ex. Lindl.: It is considered a crucial herbal medicine in South and Southeast Asia [56]. The decoction obtained from the leaves is used to treat diabetes in China as it showed stimulation of insulin secretion from the beta cells while inhibiting glucagon production [57]. They may also be used in maintaining the tonicity of the stomach and promoting the secretion of body fluid [58]. There are reports of employing them in relieving the symptoms of throat inflammation, gastritis, dehydration, and blurred vision [59].

D. chrysotoxum Lindl.: The medicinal properties of the plant are bestowed by the presence of diverse polysaccharides and phenanthrenes derivatives in leaves and pseudobulbs. The anti-inflammatory activities of the plant are due to the presence of erianthridin [55]. The liquid extract, which is prepared by boiling the leaves, can be used as tonic and antipyretic [41].

D.chrysanthum Wall: The dried and grounded leaves are used for anti-pyretic, eyes-benefitting, and immune-regulatory purposes. They are also applied for skin improvement and treatment of some skin diseases [60].

D. crepidatum Griff: Stems and pseudobulbs are plant parts used for making herbal medicines. The dried stems of the plant are employed for managing cancer and diabetes and in the treatment of cataracts and fever [1, 61]. The grounded pseudobulbs are made into a paste with some water used to treat fractured and dislocated bones [52, 62].

D. densiflorum Lindl.: The plant is used in traditional or folk medicines as a yin tonic in China. The dried and grounded pseudobulbs are made into a paste which is then used to promote body fluid secretion and strengthen the stomach and relieve fatigue and pain [48]. They are also employed as body immune booster and in the treatment of boils, pimples, and other skin rashes [40, 43].

D. fimbriatum Hook: The infusion or decoction of the leaves is consumed as a tonic as it promotes body fluid secretion. The paste prepared from the dried leaves can be put on the surface of the fractured part of the body for setting the cracked bones [63]. The whole plant can be used to treat liver upset and controlling debility and nervous breakdown [36].

D. loddigesii Rolfe: This is one of the most prominent medicinal orchids in Southern China [64]. The infusion or decoction made from the leaves is used as a tonic for nourishing the stomach and stimulating the secretion of body fluid. It is also utilized for lowering body temperature during fever and also acts as an anticancer agent [65].

D. macrostachyum Lindl.: Juice can be extracted from the juvenile leaves and tender shoot tips of the orchid. The extracted juice may be used as effective ear drops for curing ear pain and treating boils, pimples, and other skin rashes [31, 66, 67].

D. macraei Lindl.: The whole plant is useful for treating bronchitis, throat problem, fever, asthma, and as an aphrodisiac [68]. It can also be utilized as a tonic for general debility [69]. The dried tubers can be made into powder by grinding and used as a stimulant for lowering blood pressure and curing skin allergy [62].

D. microbulbon A.Rich: This is a small rare epiphytic orchid having useful medicinal properties [70]. The leaves are crushed to make a paste that can be applied on the stomach for curing stomach pain [71]. The bulbs are also eaten by the tribal people of Gujarat, India, as a source of food [72].

D. moschatum Lindl.: The plant is used in traditional medicines for its antimicrobial, anticancer, antiallergic, and anti-inflammatory activities due to the high content of phenanthrenes [73]. The liquid extract from the leaves is utilized as ear drops for curing earaches. The pastes prepared from the dried and powdered pseudobulbs are used to treat dislocated and fractured bones [62, 74].

D.moniliforme (L.) Sw.: It is an important medicinal orchid which is widely distributed in China [75]. The pseudobulbs, after drying, can be boiled or soaked in hot water to get infusion or decoction. The infusion from the dried stems is used as an aphrodisiac and for antipyretic, analgesic, and tonic purposes [76].

D. nobile Lindl.: The dried pseudobulbs are ground to powder and mixed with water to form liquid extract, which can be used as a tonic for nourishing the stomach, promoting the body fluid secretion, and reducing fever [77]. It is also utilized in managing tuberculosis, general debility, lowering salivation, night sweats, and anorexia [78]. Fresh dried stems can also be employed in the preparation of the drugs that work as an aphrodisiac, analgesic, and life longevity [79].

D. officinale Kimura et Migo: It is a crucial herbal medicine in many Asian countries for hundreds of years due to its high content of polysaccharides [80]. The plant can be employed as a nourishing yin tonic for promoting body fluid production [81]. They are also reportedly used in improving immunity, strengthening memory, preventing and combating cancers, and prohibiting thrombokinesis [82].

D.thyrsiflorum Rchb.f: This is the most widely used orchid in Chinese herbal preparation after *D. nobile* due to its widespread distribution and strong reproductive ability [83]. The high content of scoparone and coumarins in its system produces effects of relaxing smooth muscles, expanding vessels, and anticoagulating blood [84]. The orchid has been used as a good immune-modulator due to the rich presence of diverse polysaccharides [85]. Apart from managing many chronic disorders, it is also employed as a stimulant and commercial dye [86].

D. tosaense Makino: This is a medicinal orchid that locals consume as a quality health food in Taiwan [87]. The infusion or decoction can be prepared by either soaking the leaves in hot water or boiling them. The decoction made from the leaves is used to treat anxiety and panic attacks in China [88].

Other *Dendrobiums*: The whole plant of *D. longicornu* is used to treat fever and coughs [89]. The dried pseudobulbs of *D. primulinum* are grounded to make into a pulp, which is used as an immune enhancer [90]. The paste made from the pseudobulbs of *D. transparens* is applied for curing fractured and dislocated bones [62]. The stem parts of *D. huoshanense* are also utilized for promoting body immunity and fluid secretion and also to treat throat inflammation, stomach pain, and ophthalmic disorders [91]. The paste prepared from the dried pseudobulb of *D. heterocarpum* is used to treat fractured and dislocated bones [52]. The juice extracted from the fresh plant of *D. ovatum* can be given internally for controlling stomach pain. It also stimulates the bile and acts as a laxative to the intestines [54, 92]. The pulp made from the pseudobulbs of *D. monticola* is used in the treatment of boils, pimples, and other skin rashes [54]. The pseudobulbs of *D. tokai* on the other hand, are used as oral contraceptives in India [93].

3 In Vitro Propagation of *Dendrobiums*

3.1 Culture Media and Plant Growth Regulators

The success of orchid in vitro propagation depends mainly on the choice of culture media and plant growth regulators (PGRs) used. This is because varied quantities of inorganic and organic nutrients are provided to the growing tissues depending on the culture media type employed to propagate orchids [94]. A defined culture medium contains minor and major inorganic salts, carbon source, amino acids, and several vitamins. However, as per requirement, the culture media can be incorporated with organic acids, organic nitrogenous compounds, and other plant extracts [95, 96]. Normal media consist of few mineral salts with 30 mM each of inorganic nitrogen and potassium, and ammonium salts at the range of 2–20 mM. Low concentration (1–3 mM) of calcium, sulfate, phosphate, and magnesium salts is sufficient to impart in vitro tissue growth. Sucrose is integrated into the medium as only carbon source, but glucose, fructose, and other sugars like mannose and galactose may also be included as the carbon source to assist cell growth [97]. Vitamins (thiamine HCl, nicotinic acid, pyridoxine HCl, riboflavin, biotin, folic acid), an amino acid (glycine) may be incorporated at varying concentrations depending on the kind of medium and specific culture requirement. Nitrogen in the culture medium was known to affect in vitro seed germination in several orchids [98, 99]. The nitrogen requirement for seed germination is provided by NH_4^+ and NO_3^- [100].

Ammonium nitrate is the source of nitrogen in Murashige and Skoog (MS) medium [101], while ammonium sulfate provides nitrogen in Mitra, Knudson C (KC), Vacin, and Went (VW), and B5 media [102–105]. The high seed germination and subsequent development for many *Dendrobiums* in MS medium may be attributed to the presence of ammonium nitrate and rich content of macro- and micronutrients [92, 106]. The seed germination rate and culture growth on nutrient media vary for different orchids as the nutritional requirement is species-specific [9, 107]. The in vitro propagation of *Dendrobiums* is generally performed on MS, Mitra, VW, KC, and B5 media. The media are appended with different growth hormones to enhance culture growth and differentiation. Cytokinins like benzyl amino purine (BAP), kinetic (KN), isopentenyl adenine (2ip), and thidiazuron (TDZ) are used for shoot initiation, multiplication, and plant regeneration. Auxins such as indole-3-acetic acid (IAA), indole-3- butyric acid (IBA), 1- naphthalene acetic acid (NAA), and 2,4-Dichlorophenoxyacetic acid (2,4-D) are essential for rooting induction and multiplication apart from inducing cell division and cell expansion. The cytokinin and auxin are often employed in combination at different concentrations to promote shoot and root development leading to complete orchid regeneration.

3.2 Explants (Selection and Surface Sterilization)

The selection of the right explant is one of the key steps for effective micropropagation of *Dendrobiums*. The wrong choice of explant seriously undermines the success of in vitro orchid propagation. The explant culture response is affected by several factors such as genotypes, physiological stage of mother plants and explant source, age, size, density, and its portion in donor/mother plant [108]. The choice of the explant is dependent on plant material availability, seasonal abundance, medium type and culture environment, age of the tissue explant, and other physiological factors [109]. Several workers gave explant preference to immature seeds [110–112], nodal part [113–116], shoot tips [90, 117], pseudobulb segment, and axillary buds [118–120] for in vitro propagation of *Dendrobiums*. Juvenile tissues must be chosen as they have more regeneration capability compared to differentiated ones. Explants from the in vitro derived plants are favorable due to their high regeneration potential, less exudation of phenolic compounds, and non-requirement of disinfection before starting a tissue culture process.

The surface sterilization of explants is crucial as it removes the microbial contaminants from its surface before being inoculated to the nutrient medium. The failure to eliminate the infectious agents during the process leads to unsuccessful orchid micropropagation. Culture contamination with microbes is undesirable as it compromises the normal in vitro culture development [121]. Active competition exists between the microbes and growing tissues for nutrients, and the presence of contaminants often leads to increase culture mortality, variable growth, tissue necrosis, and failure in shoot and root multiplication [122]. Different chemical agents are employed for explant sterilization of *Dendrobiums*. Choosing proper concentration and treatment duration is vital for minimizing explant tissue injury due to the toxic nature of the disinfectants. Complete decontamination of explants with sterile distilled water is required after every chemical treatment. Mostly mercuric chloride (1–2%) and sodium hypochlorite (4–8%) have been used for explant surface sterilization to initiate culture in several *Dendrobiums* [11, 17–19, 123, 124].

3.3 In Vitro Propagation of *Dendrobiums* Using Different Explants

3.3.1 Seed Culture

Dendrobiums are epiphytic orchids whose seeds do not have thick seed coat. Compared to the heavily lignified seed coat of terrestrial orchids, lignin and cuticular materials are absent in the seed coats of *Dendrobiums*. The seeds are simpler and easier to germinate, unlike terrestrial orchid seeds, as they can easily absorb water and nutrient from the culture environment [109]. The mature seeds as explants are preferred as they produce a higher germination percentage than the immature seeds [107]. The disinfectant concentration for surface sterilization of the mature seeds in dehisced capsules should be low as the thin seed coat of *Dendrobiums* may not tolerate the harsh chemical effect of the sterilizing agent [125]. The seeds, when inoculated on to appropriate nutrient medium, germinate by enlarging the embryos and transforming into a highly meristematic spherical shape structure called protocorms [126]. The protocorms, under appropriate growth conditions, subsequently develop into complete seedlings with well-formed leaves and roots.

Parthibhan et al. [127] cultured seeds of *D. aqueum* on half-strength MS basal medium to produce protocorms. The protocorms were used to initiate in vitro propagation of *D. aqueum* on half-strength MS medium augmented with different cytokinins (BAP, 2ip, KN, and TDZ) and auxins (IBA, NAA, 2, 4-D) at varying concentrations (1.0, 3.0, 5.0, 7.0, and 10.0 mgL^{-1}) along with other natural additives of banana extract (BE) and coconut water (CW) (1%, 3%, 5%, 7%, and 10%). The highest shoot number (9.3) per explants was obtained on 3.0 mgL^{-1} NAA enriched medium followed by the production of seven shoots per explants on medium incorporated with 3% of BE. Rooting of the shoots was best observed with 8.75 roots per shoot on half MS medium supplemented with 5.0 mgL^{-1} IBA. Root elongation was maximum with root length measuring at 1.48 cm in 7.0 mgL^{-1} NAA incorporated medium. Well-developed plantlets were transferred to small pots with brick pieces and charcoal mixture (1:1) along with layers of mosses for better acclimatization before they were hardened and transferred to a greenhouse. Nontachaiyapoon et al. [128] studied the effectiveness of eight putative orchid mycorrhizal fungi from three orchid genera in promoting in vitro seed germination and protocorm development of *D. draconis*. The seed and protocorm developmental stages were examined weekly by culturing on MS medium and Oat meal agar (OMA), which was inoculated with one of the eight fungal isolates (PV-QS-0-2, C1-Dt-TC-1, CS-QS-0.1, Da-KP-0-1, PV-PC-1-1, C3-DT-TC-2, Pch-Qs-0-3, PV-QS-0-1, and C1-QS-0-1). The seeds of *D. draconis* germinated successfully (100%) in all the treatments tested in 2 weeks, but further seed development varied considerably with different treatment types. Three fungal isolates of different anamorphic species of *Tulasnella* (C1-DT-TC-1, PV-PC-1-1, and 3-DT-TC-2) produced significant protocorm development, but none of the fungal isolates performed better than the normal MS medium in regard to seed germination, protocorm formation, and seedling development.

Paul et al. [129] studied the seed germination response and in vitro propagation of *D. hookerianum* on MS, Mitra, KC, and B5 media. The seed germination was fastest in MS medium with the highest seed germination percentage (95.27%) achieved in 3–4 weeks. The longest time in seed germination (7–8 weeks) with the lowest germination percentage (51.38%) was observed in B5 medium. The protocorm development, leaf, and root organogenesis and subsequent seedling development were superior in MS medium as compared to the other three media. The study suggested that growth hormone incorporation into the media was not essential for the stimulation of orchid growth. Nongdam and Tikendra [112] adopted seed culture to in vitro propagate *D. chrysotoxum* on Mitra medium enriched with different combinations and concentrations of auxins and cytokinins. Maximum seed germination was attained in 2 weeks when seeds were grown on medium supplemented with 2.0 mgL^{-1} BAP, 2.0 mgL^{-1} IAA, and 0.4% activated charcoal (AC). A higher concentration of cytokinins (BAP or KN) with a low level of auxin (NAA or IBA) promoted shoot and leaf multiplication. But the reduction in shoot response was noticed when the medium was incorporated with the only cytokinin, thereby suggesting the synergetic effect of auxin and cytokinin in shoot and leaf development. IBA was better than NAA in inducing rooting, but root formation was more pronounced when auxin was incorporated with a low concentration of cytokinin.

Utami et al. [130] germinated seeds of *D. lasianthera* on VW medium fortified with varying concentrations of peptones (1.0, 2.0, 3.0 gL^{-1}). Maximum seed germination (100%) and shoot formation (84%) were accomplished with medium enriched with 2.0 mgL^{-1} peptone. The role of organic additives on subsequent shoot development was also examined by inoculating the nascent seedlings with 1–2 leaves on a medium incorporated with different organic nutrients. The shoot and root growth in the seedlings were much improved in medium containing 15% coconut water. Tikendra et al. [11] reported the in vitro propagation of *D. thyrsiflorum* using seeds obtained from unripe capsules. Multiple shoot formation was observed when either BAP or KN was present in the Mitra medium. But raising the concentration of cytokinin increased shoot production and shoot length in the culture. Medium containing 2.0 mgL^{-1} KN generated more shoots (3.16 ± 0.47)

compared to BAP (2.18 ± 0.54) at a similar concentration. The presence of both cytokinin (BAP or KN) and auxin (IAA, IBA, or NAA) in the medium gave more shoots than the medium supplemented with only cytokinin. Shoot formation was the highest (3.83 ± 0.48) when the medium was appended with 1.0 mgL^{-1} KN and 2.5 mgL^{-1} IAA. Among the auxins tested, IAA produced the best root initiation with maximum root formation (6.32 ± 0.37) attained in 2.0 mgL^{-1} IAA-enriched medium. Tikendra et al. [24] also accomplished effective seed germination and subsequent protocorm formation of *D. moschatum* in Mitra medium supplemented with different concentrations of BAP (Figs. 1a & b). Shooting initiation was

Fig. 1 In vitro propagation of *D. moschatum* by seed culture. (**a**) Swelling of seeds indication successful germination in M + 2.4 mgL^{-1} BAP. (**b**) Protocorm formation witnessed in M + 2.4 mgL^{-1}BAP. (**c**) In vitro shoot initiation after protocorm development in M + 0.6 mgL^{-1}TDZ. (**d**) Shoot multiplication and leaf formation in M + 1.2 mgL^{-1}TDZ + 1.2 mgL^{-1} NAA. (**e**) High root multiplication observed in M + 1.2 mgL^{-1}IBA + 1.2 mgL^{-1} TDZ. (**f**) Hardening of healthy and well-developed plantlets

prominent when TDZ and BAP were incorporated singly or in combination with auxin in the medium (Fig. 1c). High shoot multiplication and leaf formation were noticed in medium supplemented either with 2.4 mgL^{-1} BAP or1.2 mgL^{-1} TDZ + 1.2 mgL^{-1} NAA (Fig. 1d). IAA was the least responsive to in vitro rooting as compared to IBA and NAA. Root formation was high in the medium supplemented either with 1.2 mgL^{-1}IBA + 1.2 mgL^{-1} TDZ or 1.2 mgL^{-1} NAA +1.2 mgL^{-1} TDZ (Fig. 1e). The well-developed seedlings were hardened after acclimatizing them in the greenhouse condition using bricks, charcoal, and coconut husk (1:1:1) as a potting mixture with 95% survival rate (Fig. 1F).

Lin et al. [19] germinated the seeds of *D. cariniferum* on half-strength MS and MS basic medium. The seeds showed faster germination in half-strength MS compared to full- strength MS medium after 30 days of culture. They noticed the influence of pH of the medium on the protocorm multiplication rate with the highest proliferation recorded in the medium with pH 5.7 and the lowest observed under pH 5.9. Different concentrations of NAA and BAP were incorporated into the medium to test their influence on protocorm differentiation. Protocorm proliferation and differentiation were best noticed in medium with 0.1 mgL^{-1} NAA and 1.0 mgL^{-1} BAP compared to other hormonal combinations. Protocorm differentiation to seedlings was influenced by the level of peptone concentration. Medium integrated with 1.5 mgL^{-1} peptone produced the fast-growing thickest seedlings with the best root development.

3.3.2 Shoot Tips

Different explants like shoot tip, nodal and pseudobulb segment, floral stalk, and inflorescence were employed apart from the seeds to in vitro propagate several *Dendrobiums*. While the seeds differentiated into protocorms, other explants gave rise to protocorm-like bodies (PLBs) before developing into plantlets [131]. Morel [132] first coined the term "Protocorm like bodies," which were the structures similar to the seed-derived meristematic protocorms obtained directly from tissue explants and or/callus under in vitro condition. The PLBs differentiate into multiple shoot buds, but growth hormone concentration and combination are the keys to successful PLB formation and multiplication and subsequent development into seedlings [133].

Sharma and Tandon [134] employed excised shoot apices to grow *D. wardianum* on MS medium incorporated with various inorganic and organic sources like calcium nitrate, ammonium sulfate, urea, and amino acids. Direct multiple shoot formation was observed in the medium supplemented with 2.5 mgL^{-1}calcium nitrate and 0.5 mgL^{-1} urea, apart from generating PLBs, which subsequently developed into complete plantlets. Kanjilal et al. [135] used a transverse section of shoots obtained from 8 weeks old in vitro grown seedlings to propagate *D. moschatum* on KC medium enriched with 15% coconut water (CW) and varied concentration of growth hormones. Medium supplemented with 15% CW and 1.0 mgL^{-1} 2,4-D prompted PLB formation but PLB proliferation was more prominent in medium augmented with 15% CW, 3.0 mgL^{-1} IBA, and 2.0 mgL^{-1} NAA. The explant survival rate was 93%, with the formation of 7.6 PLBs per explant. Rooting development was best with 1.0 mgL^{-1} NAA supplemented in the potting mixture consisting of bricks, charcoal chips, sand, and soil in 1:1:1:1 ratio. Roy and Banerjee [136] used shoot tips for in vitro propagation of *D. fimbriatum* on KC medium supplemented with different concentrations of NAA and BAP. Callus formation was observed in NAA and BAP supplemented medium with 66.67% of the shoot tip explants responding to callus formation. However, the hormone-free medium produced PLBs instead of callus formation, which subsequently developed into shoots. The culture upon transferred to medium enriched with 4.0 mgL^{-1} BA and 0.5 mgL^{-1} NAA induced shoot proliferation through axillary branching. Malabadi et al. [137] inoculated a thin transverse section (1 mm) from the shoot tips of in vitro cultured *D. nobile* on Mitra medium appended with different concentrations of triacontanol (TRIA). Medium supplemented with 4.0 mgL^{-1} TRIA proved effective for PLB initiation from shoot tips and further shoot bud proliferation with 93% of the explants producing many PLBs and shot buds in the culture. Roy et al. [66] tested the growth potential of shoot tips of *D. chrysotoxum* on KC medium augmented with different concentrations of TDZ, BAP, and NAA. Medium incorporated with either 2.0 μM TDZ or BAP recorded the highest callusing, while the presence of 0.5 μM NAA in the medium produced 69-fold increase in callus weight in 3 months. High PLBs formation through the intervening callus phase was noticed with 1.0 μM NAA-enriched medium. The direct PLB formation from the explant was also witnessed depending upon the type of cytokinin used and its dosage. Medium augmented either with 1.0 μM TDZ or 8.0 μM BAP produced similar PLBs yield in the culture.

Pornpienpakdee et al. [138] found the development of PLBS from shoot tips of *Dendrobium* Eiskul hybrid on VW medium appended with six different chitosan molecules. Medium with either 10 mgL^{-1} P-70 or 20 mgL^{-1} P-90 chitosan produced optimal PLB initiation and multiplication. But the presence of 20 mgL^{-1} O-80 chitosan in the medium was essential for shoot induction in PLBs. Medium incorporated with 10 mgL^{-1} O-80 chitosan promoted plantlet regeneration after shoot induction of the PLBs. The increased concentration of P-70 chitosan from 10 mgL^{-1} to 16 mgL^{-1} in the medium enhanced the in vitro to ex vitro transplanting efficiency from 95 to 100%. Asghar et al. [139] employed lateral shoots (8.0 cm) as explants to propagate *D. nobile* in phytotechnology medium (O753) supplemented with BAP and KN along with CW as additives. Medium with 2.0 mgL^{-1} BAP produced a maximum number of shoots (4.33), while the most extended shoots (4.18 cm) were obtained in medium augmented with 1.5 mgL^{-1} KN. Incorporation of higher concentrations of BAP, KN (3.0 mgL^{-1}), and CW (300 mlL^{-1}) led to necrosis, low growth, and yellowing of shoots, thus hampering shoot growth and development. High rooting percentage (97.5%), root number (4.70), and root length (3.4 cm) were noticed in IBA supplemented medium. Rooting development was more pronounced with IBA than with NAA though the increased concentration of both auxins (3.0 mgL^{-1}) resulted in an inadequate rooting response.

Pant and Thapa [90] initiated shooting of shoot tips (0.3–0.5 mm) derived from in vitro propagated seedlings of *D. primulinum*, on MS medium fortified with different growth hormones. Maximum shoots (4.5 shoots per explant) was recorded in MS medium enriched with 1.5 mgL^{-1} of IBA. The root formation (three roots per shoot) was best observed 3 weeks after the in vitro regenerated shoots were transplanted to a rooting medium with 0.5 mgL^{-1} IBA. Pradhan et al. [117] grew *D. densiflorum* using shoot tips on MS medium appended with varied combinations of cytokinins and auxins. Shooting initiation was noticed in medium supplemented with 0.5 mgL^{-1} NAA and BAP (0.5–2.0 mgL^{-1}). Maximum shooting (four shoots per explants) was witnessed in MS medium incorporated with 2.0 mgL^{-1} BAP and 0.5 mgL^{-1} NAA. Rooting of the in vitro propagated shoots was accomplished on a medium integrated with auxins (NAA, IAA, or IBA) at varying concentrations (0.5--2.0 mgL^{-1}). IBA proved more effective in rooting compared to IAA and NAA. The best rooting response (4.5 roots per shoot) was produced in MS medium with 1.5 mgL^{-1} IBA. Winarto et al. [16] initiated shoot tip culture of *Dendrobium* 'Zahra FR 62' leading to PLB formation and multiplication in half-strength MS medium augmented with 1.0 mgL^{-1} TDZ and 0.5 mgL^{-1} IBA. The initial PLB production was high, with 5–10 new PLBs generated from 3–5 PLBs in 4–5 months. PLB proliferation and plantlet conversion took place with a medium incorporated with 2% sucrose and 0.05 mgL^{-1} BA.

3.3.3 Pseudobulb Segments

Vij and Pathak [140] employed a pseudobulb segment (0.5–1.0 cm) from a 40-week old axenic culture to grow *D. chrysanthum* in MS medium supplemented with different PGRs. Among the growth hormone tested, IAA, gibberellic acid (GA_3), or BAP did not elicit any culture response. Medium enriched with 1.0 mgL^{-1} NAA and 1.0 mgL^{-1} KN along with 2.0 gmL^{-1} yeast extract and 25 ml urea evoked organogenic response. Thirty-seven percent of the explant responded, leading to the development of complete plants with well-developed leaves and roots in 18–25 weeks.

Yasugi and Shinto [118] also observed positive shooting responses from the pseudobulb segment on MS medium supplemented with 0.1 mgL^{-1} NAA and 0.1 mgL^{-1} BAP. Shoot generation was the highest in BAP- and NAA-enriched medium with the formation of 2.3–2.5 shoots per pseudobulb segment in 8 weeks of culture. The robust rooting response was also noticed in the same hormonal combinations with the production of 5.8–9.0 roots per shoot and corresponding shoot length ranging from 5.3–7.3 mm. Sharma et al. [119] successfully induced axillary bud development from pseudobulb explants on MS medium augmented with 7.5 mgL^{-1}IAA and 20 mgL^{-1} BAP. Medium supplemented with 2.0 mgL^{-1} BAP generated maximum shoot number (39) with increased shoot length and pronounced bulblet formation. Incorporation of IAA in the medium induced rooting of in vitro shoots with 90% of them successfully rooted. Hossain et al. [123] utilized pseudobulb sections derived from in vitro grown seedlings of *D. aggregatum* to assess culture morphogenetic response. MS medium augmented with 1.0 mgL^{-1} BAP and 0.5 mgL^{-1} picloram produced multiple shootings with the highest shoot formation of 7.75 shoots per explant in about 35 days. The presence of 0.5 mgL^{-1} IAA in the medium induced rooting, and well-rooted plantlets were successfully hardened with 80% survival rate by transferring them to community pots filled with sterilized bricks pieces, charcoal, and peat moss (1:1:1).

3.3.4 Nodal Segment

Nodal segments were used by Gao et al. [141] to grow *D. nobile* in different culture media viz., MS, B5, and KC supplemented with different concentrations of BAP and NAA. The best shoot induction was observed in MS + 0.5 mgL^{-1} BAP + 0.2 mgL^{-1} NAA, while maximum shoot multiplication was noticed in MS + 3.0 mgL^{-1} BAP + 0.5 mgL^{-1} NAA. IBA promoted root initiation, but the highest rooting response was found in MS medium fortified with 0.1 mgL^{-1} NAA. Bai et al. [142] tested in vitro culture response of axenic nodal segments on either half or full-strength MS medium appended with BAP, NAA, and IBA. Induction of adventitious bud was witnessed in either hormone-free half or full-strength MS medium. BAP promoted shooting and in vitro shoots were rooted completely on half-strength MS medium enriched with IBA at a concentration range of 0.2–0.4 mgL^{-1} and 0.1% activated charcoal (AC). Bhattacharyya et al. [143] propagated *D. aphyllum* by implanting a thin cell layer obtained from nodal segments on MS medium augmented with varied combinations of growth regulators and meta-topolin. Shoot proliferation was best with an average production of 39.4% shoots per explant in medium with 15 mM meta-topolin, 10 μM TDZ, and 10 μM $AgNO_3$. The rooting frequency was maximum (82.34%) in half-strength MS medium incorporated with 15 μM IBA. Dohling et al. [100] employed the nodal segment to propagate *D. longicornu* on MS medium with 3% sucrose, 0.8% agar, and growth hormones (NAA, 2,4-D, and BAP) having a concentration in the range 1–50 μM. NAA produced the highest shoot bud number without PLBs formation, and the maximum shoot number was generated in 30 μM NAA-enriched medium. The presence of 2, 4-D, on the other hand, gave a variable response by producing both shoots and PLBs. When 15 or 20 μM of 2, 4-D was associated with 15 μM BAP, only PLB formation took place with the occurrence of the highest PLB conversion (41.48%). Bhattacharyya et al. [17] tested the efficiency of nodal segments of *D. crepidatum* for PLB and shoot formation on MS medium containing diverse concentrations of either TDZ or KN with 0.5 mgL^{-1} NAA. Incorporation of TDZ singly at 3 mgL^{-1} produced a shooting response frequency of 55%, which was enhanced to 97% by adding 0.5 mgL^{-1}NAA in the medium. Influence of different polyamine concentration on shoot multiplication was also examined for explants subcultured in MS+ 2.0 mgL^{-1} TDZ + 0.5 mgL^{-1} NAA. Medium supplemented with 0.8 mM putrescine, 2.0 mgL^{-1} TDZ, and 0.5 mgL^{-1} NAA generated maximum shoots (11.8 per explant), while the highest rooting response was achieved in 2.0 mgL^{-1} IBA incorporated medium.

Pant et al. [144] documented in vitro organogenesis from the nodal segment in *D. fimbriatum*. Medium supplemented either with BAP or KN singly did not produce any effective shoot organogenesis. But medium with BAP and NAA at a similar concentration of 17.76 μM showed shoot bud initiation, proliferation, and shoot multiplication after 10 weeks of culture. However, enhancement of hormone concentration level in the medium lowered the shoot formation. The regenerated shoots were best rooted in a rooting medium with 8.88 μM of NAA, producing maximum root number (9.20 ± 0.24) and longer root length (2.68 ± 0.01 cm). The well-developed seedlings were initially acclimatized in small pots with vermiculite for about 2 weeks. Axenic stem nodal segments were used by Zhang and Gao [124] to induce PLB formation in *D. officinale* on solid half-strength MS medium containing 2.0 mgL^{-1} BAP, 0.5 mgL^{-1} NAA, and 30 gL^{-1} sugar. The axillary buds developed from the stem nodes in 14 days gave rise to PLBs, which differentiated into shoot tips. The induced PLBs with emerging shoot tips, when shifted to H16 media, differentiated into cluster buds, which on further subculturing subsequently developed into well-rooted seedlings. The final hardening was performed by transferring the partially acclimatized seedlings to thermocol pots filled with brick pieces, charcoal, and mosses (1:1:1) with 84% survival rate.

3.3.5 Flower Stalk Node

Flower stalk nodes were utilized to in vitro propagate *Dendrobium* hybrid Sonia 17 and 28 by Martin et al. [145]. Bud break was initiated when the flower stalk node was cultured on half-strength MS medium enriched with either 6.97 μM KN and 15% coconut water (CW) or 13 μM BAP alone. KN in the medium improved shoot bud formation as five shoots per shoot bud were formed when the excised shoots were transferred to KN-enriched medium. BAP, on the other hand, promoted PLB formation with shoot buds developing into multiple PLBs on 44.4 μM BAP incorporated medium. Further conversion of PLBs into shoots was accomplished when they were moved to half-strength MS medium augmented with 67 μM KN. The rooting of the shoots was best on half-strength MS medium incorporated with 2 gL^{-1} AC, and the complete seedlings with leaves and roots were acclimatized and hardened with 80% survival rate.

The recent in vitro propagation works on *Dendrobiums* performed by several researchers using different explants in the last 5 years are listed in Table 1.

½ MS = half strength Murashige and Skoog medium; 2,4-D = 2,4-dichlorophenoxy acetic acid; *CW* = coconut water; *BAP* = 6-benzylamino purine; *FT* = foliar fertilizer; GA_3 = gibberellic acid; *IAA* = indole-3- acetic acid; *IBA* = indole-3- butyric acid; *KC* = Knudson C medium; *KN* = kinetin; *MS* = Murashige and Skoog medium; *M* = Mitra medium; *NAA* = 1- napthaleneacetic acid; *Pic* = picloram; *PLB(s)* = protocorm-like bodies; *PM* = phytamax medium; *TDZ* = thidiazuron; *VW* = Vacin and Went; *ZN* = zeatin

4 Genetic Stability of In Vitro Propagated *Dendrobiums*

4.1 Somaclonal Variation

The primary objective of micropropagation is the production of identical clones genetically similar to the elite mother plants. Genetic differences may appear among the propagated plants as the cells and tissues are continuously confronted with various stress conditions during in vitro culture. The genetic variation detected in the plants derived from any in vitro cultured cells or tissues was termed "Somaclonal variation" by Larkin and Scowkraft [167]. Somaclonal variation in the regenerated plants is not only detrimental for elite genotype conservation but also undesirable for

Table 1 In vitro propagation of different *Dendrobiums* from various explants in the last 5 years

Species	Explant used	Culture medium having optimal growth response	References
D. antennatum Lindl	Seeds	• 100% seed germination was observed in three different combinations: MS + 10% CW; MS+ growmore (1.0 mgL^{-1}) + 10% CW; MS+ growmore (1.0 mgL^{-1}) + 50 mgL^{-1} spring onion • Plantlet height (58.0 ± 5.0 mm) was maximum in MS + growmore (1.0 mgL^{-1}) + 10% CW, while the root number (5.9 ± 2.1) and root length (33.3 ± 7.8 mm) were superior in MS+ growmore (1.0 mgL^{-1}) + 50 mgL^{-1} spring onion • Shoot number was maximum (1.4 ± 0.5) in MS + 10% CW	[146]
D. aurantiacum Rchb.f	Shoot tips	• Callusing was 100% successful in all the combinations tested but MS + 10.0 mgL^{-1} 2,4-D took the shortest period (3 days) along with the highest cell concentration (43.73 x 10^6)	[147]
D. aqueum Lindl	Seeds-derived protocorms	• ½ MS + 3.0 mgL^{-1} NAA induced the maximum shoot number (9.4 ± 1.81) per explant while the shoot length was highest (1.52 ± 0.20 cm) in ½ MS + 7.0 mgL^{-1} NAA • ½ MS + 5.0 mgL^{-1} IBA recorded the highest root number (8.75 ± 1.18) per explant while the longest root length (1.48 ± 0.13 cm) was witnessed in ½ MS + 7.0 mgL^{-1} NAA	[148]
	Stem transverse thin cell layers (tTCLs)	• Highest (42.67 ± 0.58) globular somatic embryogenesis (SE) per tTCL was observed in ½ MS + 1.5 $mgdm^{-3}$ 2iP • Indirect SE callus response was best (53.33%) in ½ MS + 1.5 $mgdm^{-3}$ 2iP + 1.0 $mgdm^{-3}$ IBA • Direct SE response of 25.0% was observed in ½ MS + 0.5 $mgdm^{-3}$ BAP + 1.5 $mgdm^{-3}$ 2iP	[149]
D. bensoniae Rchb.f	Shoot nodes	• Shoot induction (80%), shoot multiplication (4.66 ± 0.57 per explants), and leaves per explant (9.33 ± 1.15) were highest in	[150]

(continued)

Table 1 (continued)

Species	Explant used	Culture medium having optimal growth response	References
		MS + 2.0 mgL^{-1} BAP • Root induction (90%) was maximum in MS + 1.0 mgL^{-1} BAP + 1.5 mgL^{-1} IBA • Root multiplication was highest (10.35 ± 0.07 per explant) in MS + 0.5 mgL^{-1} BAP + 1.0 mgL^{-1} IBA	
D. candidum wall. Ex Lindl	Shoot derived PLBs	• Earliest (7 days) callus formation with 100% rate of callus induction in MS + 1.0 mgL^{-1} 2,4-D + 0.5 mgL^{-1} KN, but the re-differentiation rate was lower than 30% • Re-differentiation rates were superior (about 50%) in MS + 10.0 mgL^{-1} NAA + 0.25 mgL^{-1} BAP; MS + 10.0 mgL^{-1} NAA + 0.5 mgL^{-1} BAP; MS + 10.0 mgL^{-1} NAA + 0.25 mgL^{-1} KN; MS + 10.0 mgL^{-1} NAA + 0.5 mgL^{-1} KN	[151]
D. cariniferum Rchb. f	Seeds	• Seed germination was faster in ½ MS basal medium • Medium at pH 5.7 witnessed high rate (5.8 ± 0.92) of protocorm proliferation, with temperature (23 ± 2 °C) and light intensity (1000 lux) optimal for protocorm multiplication • Protocorm proliferation was highest (10.27 ± 0.52) in MS + 0.1 mgL^{-1} NAA + 1.0 mgL^{-1} BAP • Seedling growth with best (10.01 ± 0.28) rooting response was observed in MS + 1.5 mgL^{-1} peptone • Culture period (60 days), temperature (23 ± 2 °C), and light intensity (1500 ~ 2000 lux) were the optimal conditions for bioactive compounds accumulation	[19]
D. chryseum Rolfe	Protocorms	• Shoot number (18.75 ± 0.48 shoots per culture) was highest in ½ MS + 2.0 mgL^{-1} KN + 10% CW, while the shoot length was longest (2.0 ± 0.20 cm) in ½ MS + 1.0 mgL^{-1} GA_3 + 10% CW	[152]

(continued)

Table 1 (continued)

Species	Explant used	Culture medium having optimal growth response	References
		• Rooting (5.0 roots; 1.7 ± 0.17 cm) was best achieved in ½ MS + 1.5 mgL^{-1} IAA	
D. chrysotoxum Lindl	In vitro-derived nodal segments	• Highest regeneration response (100%) with 20.70 ± 0.0 PLBs observed in MS (liquid) + 5.37 μM NAA. The plantlet development was also earliest (9 weeks) in this hormonal combination	[153]
D. fimbriatum hook	Nodal segments	• Shoot multiplication (14.00 ± 0.47) and shoot length (1.50 ± 0.02) were superior in MS + 17.76 μM BAP + 17.76 μM NAA • Maximum root number (9.20 ± 0.24) and root length (2.68 ± 0.01 cm) were witnessed in MS + 8.88 μM NAA	[18]
D. Hybrid	In vitro-derived seedlings	• FT NPK (32:10:10) had significantly higher growth response than ½ MS medium • FT + tomato extract produced the highest fresh weight (1.8 g), and height (6.78 cm) of the seedlings • Leaf number (5.86), and root number (9.58) were highest in FT + mungbean sprout extract, while the root length (3.72 cm) was maximum in the medium containing potato extract	[154]
D. lasianthera J.J.Sm	Seeds	• MVW + 3.0 gL^{-1} peptone recorded the best seed germination (84%) • MS + 15% CW regenerated the highest length of plantlet (3.4 ± 1.7 cm), leaves (1.9 ± 0.5 cm), roots (1.8 ± 0.4 cm), and maximum number of leaves (5.2 ± 2.1) and roots (6.0 ± 3.6) after 16 weeks of culture	[130]
D. moschatum Sw.	Seeds	• Overall shoot multiplication was superior in MT medium appended with TDZ • Root formation with high (7.46 ± 0.64 roots) in M + 1.2 mgL^{-1} TDZ + 1.2 mgL^{-1} IBA • Maximum shoot length (6.41 ± 0.54 cm) was noticed in MT + 0.6 mgL^{-1} TDZ, while the	[24]

(continued)

Table 1 (continued)

Species	Explant used	Culture medium having optimal growth response	References
		highest root length (6.92 ± 0.60 cm) was witnessed in MT + 1.2 mgL^{-1}TDZ + 1.2 mgL^{-1} IBA	
D. nobile Lindl	Seeds, PLBs	• M + 1.0 mgL^{-1} NAA recorded the earliest seed germination (2.05 ± 0.05 weeks), protocorm development (4.40 ± 0.14 weeks), and seedlings formation (13.25 ± 0.50 weeks) along with the highest frequency of seed germination (98.50 ± 1.3%) • The highest PLB formation per explant (10.0 ± 0.4) and maximum shoot per explant (14.0 ± 0.0) were observed in M + 20% CW along with the earliest development of plantlets (14.0 ± 0.40 weeks)	[155]
	Nodal segments	• Shoot proliferation was highest (21.8 ± 0.5) in MS+ 1.0 mgL^{-1} meta-topolin +0.8 mg^{L-1} putrescine • Rooting frequency was maximum (10.1 ± 0.4) in ½ MS+ 2.0 mg^{L-1} IBA + 0.5 mg^{L-1} phloroglucinol	[17]
		• The highest shoot induction (80%) and shoot multiplication (4.32 ± 1.05 per explant) were achieved in MS + 2.5 mgL^{-1} BAP • Rooting was best (90%) in MS + 1.0 mgL^{-1} IBA + AC	[156]
D. ovatum (Willd.) Kranzl	PLBs	• Spherule formation from micro seeds was highest (94.7 ± 1.01%; 95.5 ± 0.97%) in modified ½ MS incorporated separately with 0.1 mgL^{-1} ZN or 10% CW • Maximum (348.6 ± 6.90) PLB development was observed in ½ MS + 0.1 mgL^{-1} ZN • PLB conversion to plantlets was most efficient (93.0 ± 1.65%) in ½ MS + 10% CW	[157]
D. palpebrae Lindl	Seeds	• Seed germination (80%) was best in PM + 2.0% sucrose than other media (KC, MS, MVW) experimented • In vitro flowering occurred only in MS + 0.8% agar +3.0% sucrose +0.5 mgL^{-1} pic +1.0 mgL^{-1} BAP, and MS+ PM + 2.0% sucrose	[158]

(continued)

Table 1 (continued)

Species	Explant used	Culture medium having optimal growth response	References
		+0.5 mgL^{-1} NAA + 1.0 mgL^{-1} BAP • Root length (4.63 ± 0.22 cm) and root number (2.53 ± 0.14) were highest in MS + 0.5 mgL^{-1} IBA	
D. "pompadour" (hybrid of *D.* Louis Bleriot X *D. Phalaenopsis*)	Meristem tips	• Highest survival rate (64%) and occurrence of the green and visible development (3.2) from the meristem-tips were noticed in VW (liquid) + 0.1 mgL^{-1} NAA • A frequency of 87.7% clones or 49.5% of the total explants were virus free	[159]
D. primulinum Lindl	Seeds	• Seed germination started in 2 weeks in MS + BAP and the maximum capability of seedling growth (4.5 ± 1.29) was obtained in MS + 1.5 mgL^{-1} BAP • Protocorm formation was earliest in basal MS medium but enhanced when the medium was incorporated with 0.5 mgL^{-1} each of BAP and NAA	[160]
D. Red bull	Shoot tips	• After 150 days of culture, maximum number of shoots (7.66) was observed in MS + 3.0 mgL^{-1} BAP + 1.0 mgL^{-1} NAA • Shoot length was highest (21.19 cm) in MS +3.0 mgL^{-1} BAP + 1.5 mgL^{-1} NAA • Root multiplication was maximum (4.67) in MS + 4.0 mgL^{-1} BAP + 1.5 mgL^{-1} NAA	[161]
D. signatum Rchb.f	Seeds	• MS + 10% potato extract; ½ MS + 10% potato extract; MS + 5.0% mashed banana recorded 100% seed germination • ½ MS basal and ½ MS + 2.0 mgL^{-1} TDZ +0.5 mgL^{-1} NAA produced the highest shoot proliferation (67.0 ± 8.33%), while the root multiplication was best (5.3 ± 2.02) in ½ MS + 2.0 mgL^{-1} BAP + 0.5 mgL^{-1} NAA	[162]
D. Sonia	Rhizome buds	• Shoot induction was best in MS + 11 μM BAP • Shoot multiplication, PLB formation, and rooting was maximum in MS + 11 μM	[163]

(continued)

Table 1 (continued)

Species	Explant used	Culture medium having optimal growth response	References
		BAP + 11.42 μM IAA • The monochromatic light spectra best suited for shoot initiation and proliferation was yellow light (at an intensity of 24.6 μmol/m2/s and 590 nm wavelength). High shoot proliferation rate (98%) with 290 shoots per ten explants was recorded under yellow light influence • Although the leaf area (32.1 ± 1.22 mm^2), and fresh weight (5.5 ± 0.40gm) were highest in yellow monochromatic light, the shoot length (5.4 ± 0.39 cm) and root length (2.1 ± 0.33 cm) were superior under blue light (at intensity of 22.5 μmol/m2/s and 470 nm wavelength)	
D. Sonia "Earsakul"	Petals and sepals	• ½ MS + 1.0 mgL^{-1} BAP + 0.5 mgL^{-1} NAA produced maximum (75.0 ± 11.2%) meristemoid induction of the petal tissues transiently transformed by *Agrobacterium tumefaciens* strain EHA105 harboring pCAMBIA-1301 that infiltrated the petal tissues • Larger (0.5–2.0 mm) and highest (4.8 ± 1.8) mean meristemoids tissue per explant was observed in ½ MS (liquid) +1.0 mgL^{-1} BAP + 1.0 mgL^{-1} NAA • The maximum (90 ± 10.0%) survival rate with 0.0% bacterial contamination was observed in the agroinfiltrated petal tissues treated with 20 mgL^{-1} meropenem	[164]
D. Sonia "red jo"	PLBs	• Seedlings grew better in VW + NaCl (5.0–40 mM) than medium without NaCl or with $CaSiO_3$ or proline • Maximum shoot (0.61 ± 0.2 cm) and root length (0.81 ± 0.36 cm) were evident in VW + 5.0 mM NaCl • Fresh weight (0.11 ± 0.04) was maximum in VW + 40 mM NaCl	[165]
D. thyrsiflorum Rchb.f	Seeds	• Successful seed germination occurred in 2–3 weeks of culture • Highest (3.49 ± 0.96) shoot multiplication was observed in	[11]

(continued)

Table 1 (continued)

Species	Explant used	Culture medium having optimal growth response	References
		M + 1.0 mgL^{-1} KN + 1.0 mgL^{-1} IAA • Maximum root multiplication (6.52 ± 0.37) was recorded in M + 2.0 mgL^{-1} IAA • Transferring of plantlets from M + 1.0 mgL^{-1} KN + 1.0 mgL^{-1} IAA to new combination of M + 1.0 mgL^{-1} KN + 2.5 mgL^{-1} IAA further increased the shoot number (3.83 ± 0.48)	
D. trankimianum T Yukawa	Latent buds	• With 92.06% of PLBs formation, MS (modified) + 1.5 mgL^{-1} TDZ + 0.5 mgL^{-1} NAA produced the highest (14.11) PLB regeneration per explant • Maximum shoot number (22.35 shoots/explant) and shoot length (1.96 cm) were recorded in MS + 1.5 mgL^{-1} BAP • Among the various effects of mash, medium incorporated with 60 g ripe banana per liter of medium regenerated the highest shoots/explant (25.11), and shoot length (2.12 cm) • With 98.51% rooting, ½ MS + 0.5 mgL^{-1} NAA produced the highest root number (7.91) and root length (4.01 cm)	[166]

economic profits if the primary regenerants are the desired end products [21, 168]. Therefore, it is imperative to ascertain clonal fidelity to propagate only the genetically stable plants.

Choosing proper explant for culture initiation is critical for preserving the genetic stability of the regenerated plants. Genetic variation is anticipated among in vitro clones propagated through seed explants as they are derived from the fusion product of gametes from genetically different parents. Utilization of meristematic tissues like shoot tips, axillary,, and stem nodes as starting materials for micropropagation is likely to generate genetically identical plants by reducing variation [169]. The mature and differentiated tissues like roots, leaves, and stems produce more variation than juvenile explants with preexisting meristems due to the intervening callus phase [119, 170]. The callus-mediated plant propagation is involved with the risk of genetic aberration as callus formation is associated with the dedifferentiation phase, which is generally followed by abnormal cell divisions [171]. As explant preparation from the same donor plant escalates the chance of variant production, it is important to select a donor plant with inherent genetic composition and genome uniformity [172]. Plants generated through somatic embryogenesis are more genetically stable than those propagated through organogenic differentiation, as DNA methylation is less in early embryogenesis [171].

Factors like longer explant disinfection in sterilants, high hormone concentration, long culture duration/ subculture cycle, the callus transition phase, explant cell/tissue heterogeneity, and other spontaneous mutations maybe the reasons for the occurrence of a genetic aberration among the in vitro clones [20, 21]. The chances of somaclone production in prolong culture increase as there is a possibility of greater accumulation of genetic variation during successive subcultures [173]. The rise in subculture frequency may also intensify the rate of somaclonal variation produced by nucleotide sequence alternation rather than quantitative changes as shown by constant C-value even after the seventh subculture cycle of olive genotypes [174]. Some growth hormones at specific concentrations/in different combinations may incite mutation producing genetic variation among the regenerants [175, 176]. Carvalho et al. [177] observed the increased generation of somaclonal variants with 2,4-D in prolonged culture. DNA methylation rate was raised in the presence of 2, 4-D, which changed the DNA ploidy level resulting in the generation of variant clones. Arnhold-Schmitt [178] observed changes in chromosome arrangement and DNA methylation in carrot callus culture when IAA was present with inositol in the medium. The plant growth regulator ratio also affected the in vitro genetic changes as low and high incidence of the variant "mantled" flowering was witnessed in high auxin/cytokinins and high cytokinins/ auxin ratio, respectively, by Eeuwens et al. [179] in oil palm.

4.2 Genetic Stability Assessment Using DNA Markers

The long-term benefit of *Dendrobium* micropropagation lies in the production of genetically stable plants identical to elite mother plants. Genetically uniform plant production confers a great advantage to the cut-flower industry as it helps in the production of uniform blooming during predictable periods to meet the market demands. If the genetic uniformity can be retained for a longer duration, the overall production cost will be reduced significantly, and the whole process would be highly profitable. The genetic variation is generally due to alteration in chromosome numbers (aneuploidy and polyploidy), chromosome structure (due to deletion, duplication, insertion, or translocation), DNA base mutations, and gene amplification or methylation, as in the case of epigenetics [180–185]. Different strategies that may be employed to assess the genetic stability include morphological characters based phenotypic identification, cytological analysis of chromosomal alteration, and isozyme-based variation study [186–188]. Flow cytometry analysis can also be performed to complement the traditional cytological studies to give a correct assessment of any change in ploidy [189, 190]. Since these methods are associated with limitations, more effective PCR-based DNA markers have been used for genetic stability testing of the micropropagated orchids.

DNA markers have become a versatile tool in plant genotyping in the last three decades. The two-classes of DNA markers, viz., the codominant markers such as simple sequence repeat (SSR), restriction fragment length polymorphism (RFLP), amplified fragment length polymorphism (AFLP), and single nucleotide polymorphism (SNP); and the dominant markers like random amplified polymorphic DNA (RAPD), inter-simple sequence repeat (ISSR), inter-retrotransposon targeted amplified region (IRAP) and start codon targeted polymorphism (SCoT), have allowed the researchers to study the somaclonal variation present either in a single known locus or at different loci in the plant genome, respectively [191–197]. RAPD, the simplest and inexpensive DNA marker, is extensively used to detect somaclonal variation but is coupled with drawbacks of low reliability and reproducibility [198]. ISSR markers are more effective and reliable with longer primers (15–30 mers) and higher annealing temperature, giving greater stringency than RAPD [199]. But both the markers are simple, fast, and cost-effective and use only a single primer to amplify the genomic DNA [200, 201]. They are also easier technically than RFLP, SSR, and AFLP markers as no prior sequence information is required for generating DNA amplification products [201, 202]. With the introduction of gene-targeted markers such as SCoT and TRAP (targeted region amplified polymorphism), the polymorphism can be detected with reproducible bands and accurate results [22]. SCoT is an emerging marker that gains popularity over other dominant markers such as RAPD and ISSR. This marker system requires a single primer during amplification and is based on the short-ranged conserved region that flanked the "ATG" start codon in the plant genes [203, 204]. Arbitrary markers like RAPD and ISSR may fail to detect variations as their genetic information is based on the noncoding regions of DNA not linked to functional traits. SCoT markers are derived either from the gene itself or its immediate flanking regions and are associated with functional genes and their corresponding traits [205]. TRAP technique employs a public express sequence tag (EST) database to design primers against the annotated EST sequences for the detection of polymorphic markers to link the EST sequences with its respective phenotypes [206]. TRAP markers may yield more accurate estimates of genetic variation than other markers [207].

The epigenetic nature of somaclonal variation is caused by methylation that does not change the DNA sequence, affecting the cell's ability to read the genes, thereby the gene activity [208]. In such cases, specific markers like methylation-sensitive amplified polymorphism (MSAP), RFLP, or identification via gene expression can be employed for the detection of somaclonal variants [209–213]. The use of a single marker system for genetic stability testing of the micropropagated plants does not always guarantee accurate results [214]. So, utilization of two or more marker types is crucial for genetic clonal assessment as this will help validate the outcomes of variability analysis by different markers [196, 215, 216].

Ferreira et al. [120] carried out RAPD analysis to check for possible genetic alterations in *Dendrobium* Second Love plants which were originated from six consecutive subcultures. Twenty RAPD primers produced 172 bands with band number ranging from 5 to 12 per primer. Two phenotypic alterations of fasciated and elongated leaves were detected during the culture, but the RAPD analysis did not reveal any genetic polymorphism among the phenotypically variant plants. This showed that variation might not necessarily correspond to alteration in the DNA sequence. The plants with fasciated or elongated leaves generated normal shoots on further subculturing under similar culture conditions, which affirmed the notion that phenotypical variations did not ensure changes in the DNA sequence. RAPD analysis did not divulge any genetic polymorphism in the micropropagated *Dendrobium* Second Love in six consecutive subcultures (540 days). The study presented the possibility of direct in vitro propagation of genetically stable plants under the influence of TDZ as a sole promoter for multiple shoot initiation and rapid proliferation. Song et al. [85] tested the genetic stability of callus-derived plants of *D.nobile* using ISSR markers. Thirty eight ISSR primers were used, which generated 141 scorable bands with an average of 3.7 bands per primer. Monomorphic banding patterns were observed for shoots obtained from the callus subcultured for less than 15 cycles, but a low degree of polymorphism was detected for those derived from callus with more than 16 culture cycles. The genetic uniformity was maintained among the 20 randomly selected plants generated from shoots obtained from 15 cycles of callus subculture. The finding illustrated the genetic stability of *D. nobile* regenerants obtained from callus-derived shoots, which had been subcultured for 15 cycles on MS medium enriched with 4.0 mM NAA. Bhattacharyya et al. [217] used two markers, viz., RAPD and SCoT, to evaluate the genetic stability of micropropagated *D. nobile*. Seven RAPD and 15 SCoT primers, which gave reproducible and scorable bands, were selected after screening 80 RAPD and 35 SCoT primers. The band number for RAPD ranged from 2 to 6, while it varied from 4 to12 bands for SCoT markers. The PIC values recorded for RAPD and ISSR markers were 0.92 and 0.76, respectively, while resolving power (R_P) ranged from 3.66 to 10 for RAPD and 4 to 12 for SCoT markers. The cumulative RAPD and SCoT data showed only five polymorphic bands, indicating a high level of genetic monomorphism (97%) between the in vitro clones and mother plant. The observation of high Rp and PIC values for both the marker systems indicated greater marker efficiency in detecting the genetic stability of the in vitro propagated orchids. The Mantel test also showed a high correlation between RAPD and SCoT markers ($r = 0.51$), portraying similar efficacy of the two markers in identifying variability between the regenerants.

Investigation on the genetic stability of regenerated *Dendrobium* Bobby Messina (DBM) following cryopreservation procedure was performed by Antony et al. [22] using RAPD markers. Many RAPD primers were screened, but only ten primers generated distinct and reproducible bands were chosen for the study. The percentage of polymorphic bands found for ten primers examined in the three cryopreserved DBM viz., DBVG2P1, DBVG2P2, and DBVG2P3 was between 20 and 39.9%. Variation in the RAPD banding profiles indicated the existence of somaclonal variation within the regenerated plants. The clonal variability may appear due to DMSO toxicity- (PVS2), freezing-, or thawing-induced injury or the regeneration phase [218]. The study disclosed a high polymorphism in cryopreserved plantlets 18 months post-cryopreservation compared to the earlier report of only 10%

polymorphism in DBM cryopreserved for 3 months. Antony et al. [22] further performed a somaclonal variation detection study of cryopreserved PLBs of *Dendrobium* Bobby Messina using targeted region amplification polymorphism (TRAP) and SCoT markers for producing genetically stable plants. PLBs, non-cryopreserved PLBs, and thawed cryopreserved PLBs were examined for genetic uniformity using the molecular markers. Eight pairs of TRAP primers yielded distinct, scorable, and reproducible bands with size varying from 100 to 2000 bp. The TRAP primers disclosed absolute polymorphism in the cryopreserved PLBs. In non-cryopreserved PLBs, only primer TRAP 20-6B generated monomorphism, while the remaining seven TRAP primers displayed both complete and partial polymorphism. The four SCoT primers produced specific and reproducible amplification fragments with band size ranging from 500 to 3000 bp. four SCoT primers (S26, S32, S33, and S36) generated polymorphism for both the cryopreserved and non-cryopreserved PLBs. The investigation revealed that both TRAP and SCoT markers could be effectively used to detect somaclonal variation in regenerated cryopreserved PLBs, which will assist in the in vitro propagation of genetically stable *Dendrobium* Bobby Messina.

Bhattacharyya et al. [219] assessed the genetic stability of ISO (indirect shoot organogenesis) and DSO (direct shoot organogenesis) regenerated *D. thrysiflorum* using SCoT and ISSR markers. Thirty six SCoT and Twenty five ISSR primers were screened to select 8 SCoT and 5 ISSR primers, which yielded reproducible and unambiguous bands. SCoT markers identified a small degree of polymorphism with 3.22 and 8% clonal variability among the DSO-and ISO-derived plants, respectively. However, ISSR markers detected low polymorphism of 4.76% in the ISO-derived plants, while no genetic variation was witnessed between the DSO-derived plants. Pooled SCoT and ISSR data approach showed a very low variability of 1.88% in the DSO generated plants and 6.52% polymorphism within ISO-derived plants. The study showed the effectiveness of SCoT to reveal more variability among the regenerated plants than the ISSR markers. Molecular analysis disclosed the DSO regenerated plants to be more genetically stable than ISO propagated plants of *D. thyrsiflorum*, which might be due to hormone-induced stress during PLB formation, shoot induction, and long culture cycle. Wannajindaporn et al. [220] executed a genetic variation assessment of 25 *Dendrobium* "Earsakul" mutants obtained from NaN_3 treatment along with ten untreated control plants using ISSR markers. Marker analysis of the putative mutants using 11 ISSR primers produced 173 fragments across all genotypes with 39 polymorphic fragments generating 22.5% polymorphism. The ISSR fragments size varied from 140 bp (ISSR 835) to 5000 bp (ISSR 835) with ISSR 827 giving the highest polymorphism (47.8%), and the lowest of 39.1% by ISSR 811. But the untreated ten controls exhibited similar DNA banding profiles with no existence of polymorphism among them. The study demonstrated the effectiveness of ISSR markers to identify *Dendrobium* mutants derived from NaN_3-induced PLBs, allowing earlier selection and producing greater genetic stability among the clones by reducing mutant population size. Bhattacharyya et al. [17] assessed the genetic stability of acclimatized plants of *D. crepidatum* using SCoT and ISSR markers. Eight SCoT yielded 30 amplified bands with 3 polymorphic bands generating 10% polymorphism, while 5 ISSR primers produced 18 reproducible bands with 100% monomorphism among the regenerants. Combined RAPD and ISSR data revealed low clonal variability of 6.38% with a Jaccard's coefficient varying from 0.94 to 1.00. The genetically stable *D. crepidatum* was successfully propagated with the two markers detecting a very low level of genetic polymorphism among the regenerants.

To produce genetically stable plants of *Dendrobium* Sabin Blue, Ching et al. [23] examined the presence of somaclonal variation in PLBs treated at different concentrations of NAA, kinetin, TDZ, and AC using ISSR and direct amplification of minisatellite DNA region (DAMD) markers. Nine ISSR and eleven DAMD primers were utilized to evaluate the genetic differentiation of PLBs subcultured under the influence of different additives for 2 years. PLBs under the treatment of 1.5 mgL^{-1} kinetin harbored the highest genetic variation while the protocorms grown on medium supplemented with either 4 mgL^{-1} TDZ or 0.5gL^{-1} AC exhibited the maximum genetic stability. The study observed that genetic stability assessment of PLBs for long-term culture maintenance should be conducted with molecular markers so that only the genetically uniform plants were propagated. Bhattacharyya et al. [221] employed IRAP and ISSR markers to check the genetic consistency of *D. aphyllum* micropropagated through a transverse thin cell layer approach. Five IRAP and nine ISSR primers, which produced scorable reproducible bands, were chosen for the genetic stability test of in vitro propagated orchids. The IRAP produced 26 scorable bands with an average of 5.20 bands per primer, while ISSR generated 50 amplifiable bands with 5.55 average bands per primer. Both the markers produced two polymorphic bands each because of which IRAP and ISSR detected 7.69 and 4% polymorphism, respectively. The cumulative IRAP and ISSR data analysis revealed a polymorphism of 5.26% among the regenerants and mother plants. IRAP markers were more effective than ISSR in polymorphism detection as ISSR targeted a particular region of the genome, which might not display the genetic variation among the clones. But IRAP, which was a retrotransposon-targeted molecular marker, affirmed the results of ISSR marker analysis. IRAP markers are appropriate for somaclonal variability detection as the movement of transposable elements in the genome is an important contributing factor for the emergence of somaclonal variation [167, 222]. The investigation revealed high genetic stability within the plantlets of *D. aphyllum* propagated through the use of *t*-TCL as an explant source.

Tikendra et al. [24, 25] successfully in vitro propagated genetically stable *D. moschatum* and *D. chrysotoxum* by assessing the clonal fidelity using RAPD and ISSR markers. The two markers are dominant markers that require no prior information of the targeted sequence of the plant genome. Genetic stability assessment conducted in these *Dendrobiums* using DNA markers revealed a high degree of monomorphism among the regenerants and mother plants. Experimentally, the leaf samples from the mother plants and randomly selected micropropagated plants were used for genomic DNA extraction using CTAB (Cetyltrimethylammonium bromide) method [223]. The quantity and purity of the extracted genomic DNA were checked using a UV spectrophotometer and 0.8% agarose gel electrophoresis, respectively. The genomic DNA was subjected to PCR amplification using different RAPD and

ISSR primers. The amplified DNA fragments were separated on 2.0% agarose gel electrophoresis in 1X Tris-acetate EDTA buffer and stained with 0.5μgL^{-1} ethidium bromide (EtBr). A 1 kb DNA ladder was used to ascertain the size of the separated unknown DNA fragments. The agarose gel containing the separated bands was visualized and photographed using a gel documentation system. The experimental steps involved in the molecular genetic stability assessment of in vitro propagated *D. chrysotoxum* and *D. moschatum* are illustrated in Fig. 2.

In the genetic stability assessment of the micropropagated *D. chrysotoxum*, 12 RAPD and 11 ISSR primers were selected after screening based on the production of reproducible and scorable bands. Twelve RAPD primers generated 74 scorable bands, out of which 73 bands were monomorphic, ensuing 98.81% of monomorphism. Similarly, 11 ISSR primers produced 76 reproducible bands, from which 73 bands were monomorphic, producing 97.47% of monomorphism. In *D. moschatum*, genetic homogeneity analysis using 10 primers each of RAPD and ISSR markers yielded 48 and 54 scorable bands out of which 45 and 53 bands were monomorphic, resulting in 95.2% and 98.0% of monomorphism, respectively. The combined use of two marker systems ensured that the outcome of RAPD analysis was validated by the follow-up results of the ISSR markers. The ISSR generated a higher average number of bands per primer than RAPD markers indicating its effectiveness in analyzing genetic polymorphism. The banding profiles of OPF-14 (Fig. 3A) and UBC-814 (Fig. 3B) in *D. chrysotoxum*, and so also the amplification profiles of OPC-08 (Fig. 3C) and UBC-868 (Fig. 3D) in *D. moschatum*, showed total similarity in their respective banding pattern between the regenerants and mother plants.

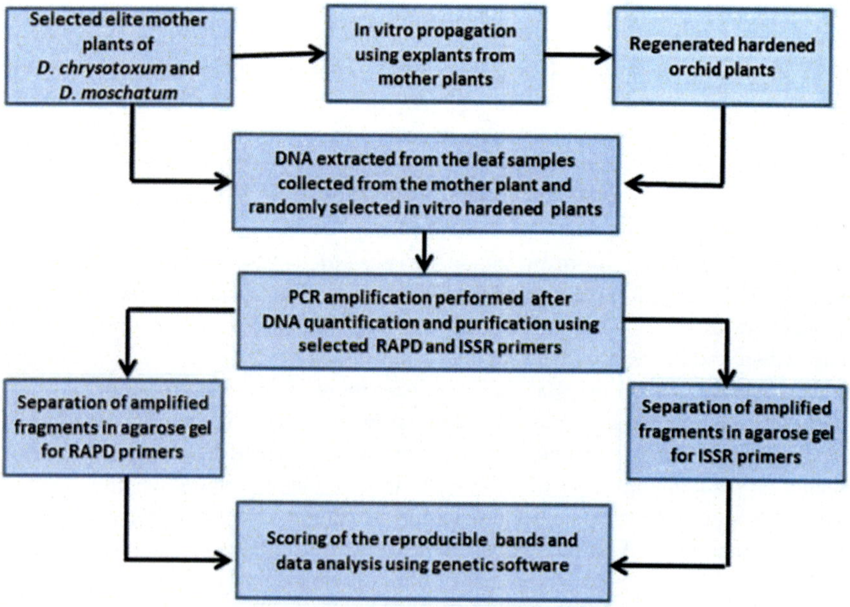

Fig. 2 Experimental steps involved in the genetic stability assessment of in vitro propagated *Dendrobiums* using RAPD and ISSR molecular markers

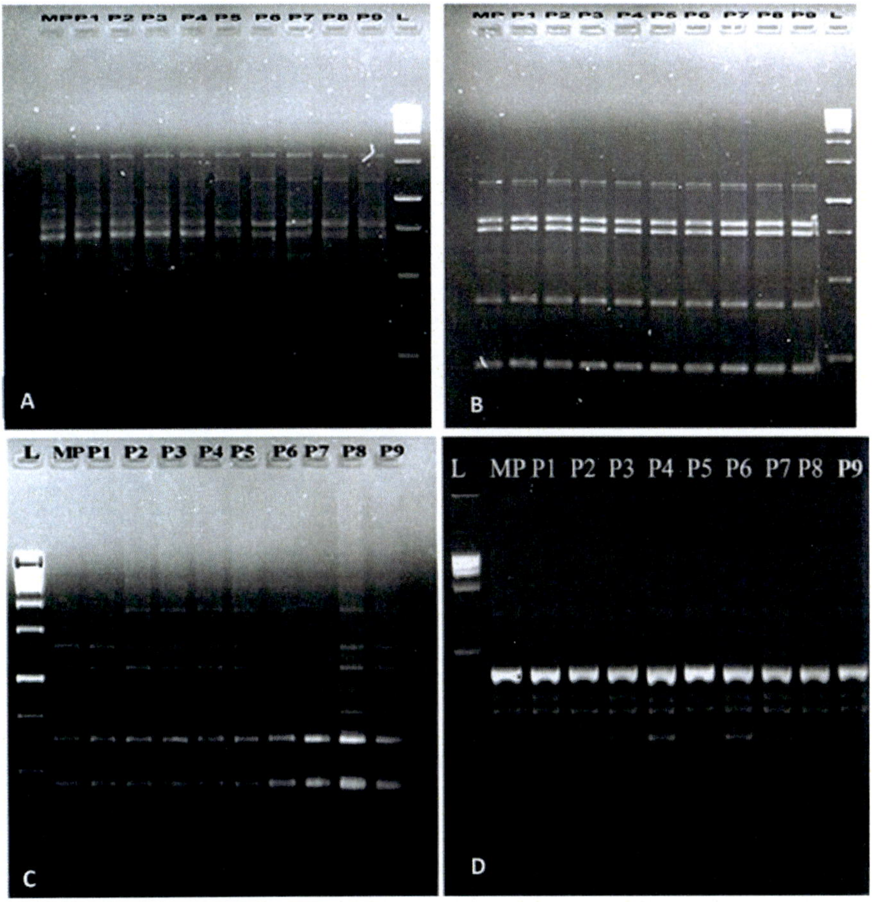

Fig. 3 Banding profiles of OPF-14 RAPD primer (**a**) and UBC-814 ISSR primer (**b**) in *D. chrysotoxum;* OPC-08 RAPD primer (**c**) and UBC-868 ISSR primer (**d**) in *D. moschatum*

The genetic association and closeness of mother plant (MP) and in vitro clones (P1-P9) can be portrayed by constructing UPGMA dendrograms. The dendrograms showed the clustering of MP with P1, P2, P3, and P4 in *D. chrysotoxum*, while MP was associated with P1, P2, P3, P4, P5, P6, P7, and P8 in a major cluster in *D. moschatum* (Fig. 4).

The consistency of genetic similarity defined by the cluster analysis was checked by performing PCoA (Principal coordinate analysis). PCoA revealed the grouping of the regenerants and the mother plants similar to the clustering patterns exhibited by UPGMA dendrograms. The correlation between the different markers employed in the analysis was checked by performing the Mantel test. The genetic matrices of RAPD and pooled RAPD+ISSR showed no significant correlation for *D. chrysotoxum* ($r = 0.620$; $p = 0.02$) and *D. moschatum* ($r = 0.966$; $p = 0.10$). But, significant correlation was observed between ISSR and RAPD-ISSR pooled matrices in *D. chrysotoxum* ($r = 0.936$, $p = 0.002$), and *D. moschatum* ($r = 0.753$,

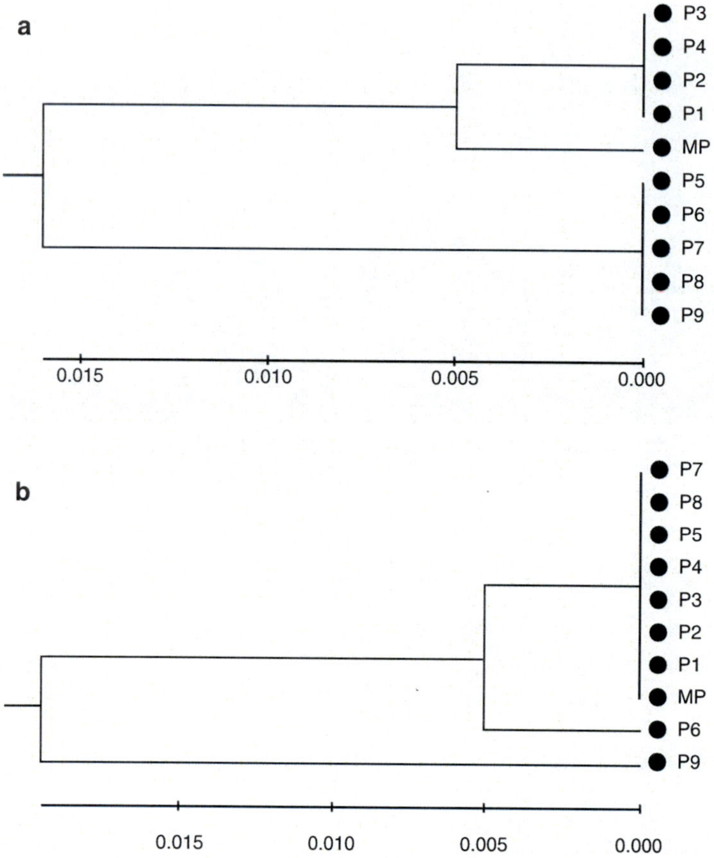

Fig. 4 Dendrograms depicting the clustering pattern between the mother plants and in vitro regenerants of (**a**) *D. chrysotoxum and* (**b**) *D. moschatum*

$p = 0.010$). In both the investigations, the ISSR markers were more effective than RAPD in determining the genetic stability of in vitro propagated *Dendrobiums*. Galvan et al. [224] and Ajibade et al. [225] also demonstrated the effectiveness of ISSR over RAPD markers in clonal fidelity assessment of regenerants. The ISSR markers are distributed in the entire genome allowing amplification of genomic DNA in a greater number of fragments per primer than RAPD markers.

5 Conclusions

Dendrobiums are extensively used in traditional medicines to treat diverse ailments due to the inherent therapeutic properties they acquired from numerous phytochemical contents. The in vitro propagation of different *Dendrobiums* is successfully accomplished utilizing mainly seeds, shoot tips, pseudobulb, and nodal segments as reliable explants. Clonal fidelity assessment using DNA markers is performed to in vitro propagate genetically stable *Dendrobiums*. Of the several DNA markers employed for clonal evaluation, mostly ISSR, IRAP, and SCoT are preferred to the less reliable and inconsistent RAPD markers. The application of two markers system is more appropriate than using a single marker in genetic stability assessment as the result of variability analysis of one marker type can be validated by that of the others. The rapid and mass propagation of genetically stable orchids will ensure the effective conservation and commercialization of *Dendrobiums*, narrowing down the existing gap between demand and supply.

Acknowledgments LT and PN are thankful to UGC (University Grant Commission), New Delhi, India, for financial support.

References

1. Lam Y, Ng TB, Yao RM, Shi J, Xu K, Sze SCW, Zhang KY (2015) Evaluation of chemical constituent and important mechanism of pharmacological biology in *Dendrobium* plants. Evid Based Complementary Altern Med 2015:841752
2. Nongdam P, Nirmala C (2011) In vitro rapid propagation of *Cymbidium aloifolium* (L.) Sw.: a medicinally important orchid via seed culture. J Biol Sci 11:254–260
3. Wraith J, Norman P, Pickering C (2020) Orchid conservation and research: an analysis of gaps and priorities for globally red listed species. Ambio 49:1601–1611
4. Nongdam P, Nirmala C (2012) In vitro seed germination and mass propagation of *Cymbidium dayanum* Reichb: an important ornamental orchid of North-East India. Trends Hortic Res 2 (2):28–37
5. Xiaohua J, Singchi C, Yibo L (2008) Taxonomic revision of *Dendrobium monifolium* complex (orchidaceae). Sci Hort 120:143–145
6. Paudel MR, Bhattarai HD, Pant B (2020) Traditionally used medicinal *Dendrobium*: a promising source of active anticancer constituents. In: Orchids phytochemistry, Biology horticulture: fundamentals and applications. Springer, Cham. pp 1–26. https://doi.org/10.1007/978-3-030-11257-8_16-1.
7. Wood HP (2006) The dendrobiums. Timber Press, Portland
8. Moudi M, Go R, Yien CYS, Saleh MN (2013) A review on molecular systematic of the Genus *Dendrobium* Sw. Acta Biol Malays 2:71–78
9. Nongdam P, Nirmala C, Tiwari R (2006) In vitro multiplication of *Cymbidium pendulum* orchids via embryo culture. Plant Cell Biotech Mol Biol 7:145–150
10. Hinsley A, Boer HJD, Fay MF, Gale SW, Gardiner LM, Gunasekara RS, Kumar P, Masters S, Metusala D, Roberts D, Veldman S, Wong S, Phelps J (2018) A review of the trade in orchids and implications for conservation. Bot J Linn Soc 186:435–455
11. Tikendra L, Amom T, Nongdam P (2018) Effect of phytohormones on rapid in vitro propagation of *Dendrobium thyrsiflorum* Rchb.f.: an endangered medicinal orchid. Pharmacogn Mag 14:495–500
12. Senthilkumar S (2001) Problems and prospects of orchid mycorrhizal research. J Orchid Soc India 15:23–32
13. Wraith J, Pickering C (2018) Tourism and recreation of global threat to orchids. Biodivers Conserv 26:3407–3420
14. Tee CS, Wong CQ, Lam XL, Maziah M (2010) A preliminary study of protocorm-like bodies (PLBs) induction using leaf explants of *Vanda* and *Dendrobium* orchids. Asia Pac J Mol Biol Biotechnol 18:189–191

15. Zhao XB, Wu WJ, Pang L, Cai GX (2012) Study on key technology of rapid culture and propagation for *Dendrobium candidum* as rare and endangered medicinal materials. J TCM Univ Hunan 32(3):27–30. (in Chinese with English abstract)
16. Winarto B, Rachmawati F (2013) In vitro propagation protocol of *Dendrobium* 'Gradita 31' via protocorm like bodies. Thammasat Int J Sci Technol 18(2):54–68
17. Bhattacharyya P, Kumaria S, Tandon P (2016) High frequency regeneration protocol for *Dendrobium nobile*: a model tissue culture approach for propagation of medicinally important orchid species. S Afr J Bot 104:232–243
18. Paul P, Joshi M, Gurjar D, Shailajan S, Kumaria S (2017) In vitro organogenesis and estimation of β-sitosterol in *Dendrobium fimbriatum* Hook.: an orchid of biopharmaceutical importance. S Afr J Bot 113:248–252
19. Lin W, Wang J, Xu X, Wu Y, Qiu D, He B, Sarsaiya S, Ma X, Chen J (2020) Rapid propagation in vitro and accumulation of active substances of endangered *Dendrobium cariniferum* Rchb. F. Bioengineered 11:386–396
20. Rawat JM, Rawat B, Agnihotri RK, Chandra A, Nautiyal S (2013) In vitro propagation, genetic and secondary metabolite analysis of *Aconitum violaceum* Jacq.: a threatened medicinal herb. Acta Physiol Plant 35:2589–2599
21. Krishna H, Alizadeh M, Singh D, Singh U, Chauhan N, Eftekhari M, Sadh RK (2016) Somaclonal variations and their applications in horticultural crops improvement. 3. Biotech 6:54
22. Antony JJJ, Shamshir RA, Poobathy R, Chew BL, Subramaniam S (2015) Somaclonal variations were not induced by the cryopreservation: levels of somaclonal variations of in vitro and thawed protocorms of *Dendrobium bobby* Messina analysed by SCoT and TRAP DNA markers. S Afr J Bot 100:148–157
23. Chin CK, Lee ZH, Mubbarakh SA, Antony JJJ, Chew BL, Subramaniam S (2019) Effects of plant growth regulators and activated charcoal on somaclonal variations of protocorm-like bodies (PLBs) of *Dendrobium sabin* blue orchid. Biocatal Agric Biotechnol 22:101426
24. Tikendra L, Amom T, Nongdam P (2019) Molecular genetic homogeneity assessment of micropropagated *Dendrobium moschatum* Sw. - a rare medicinal orchid, using RAPD and ISSR markers. Plant Gene 19:100196
25. Tikendra L, Koijam AS, Nongdam P (2019) Molecular markers based genetic fidelity assessment of micropropagated *Dendrobium chrysotoxum* Lindl. Meta Gene. https://doi.org/10.1016/j.mgene.2019.100562
26. Chandra S, Rawat DS (2015) Medicinal plants of the family Caryophyllaceae: a review of ethno medicinal uses and pharmacological properties. Integr Med Res 4:123–131
27. Schultes RE (1962) The role of ethnobotanist in search for new medicinal plants. Lloydia 25 (4):257–266
28. Quinlan MB (2011) Ethnomedicine. Wiley online library https://doi.org/10.1002/9781444395303.ch19
29. Pohle P (1990) Useful plants of Manang district: a contribution to the ethnobotany of the Nepal. Himalaya, Franz Steiner Verlag Wiesbaden GmbH, Stuttgart
30. Withner CL (1959) The orchid: a scientific survey. Ronald Press Company, New York
31. Hossain MM (2011) Therapeutic orchids: tradicinal uses and recent advances - an overview. Fitoterapia 82:102–104
32. Bulpitt CJ (2007) The use of orchids in Chinese medicine. J R Soc Med 100:558–563
33. Vaidya B, Shrestha M, Joshee N (2000) Report on Nepalese orchid species with medicinal properties. In: Watanabe T, Takano A, Bista MS, Bista, Siju HK (eds) The Himalayas plants, can they save us? Proceeding of Nepal-Japan joint symposium on conservation and utilization of Himalayan medicinal resources, Society for the conservation and development of Himalayan Medicinal Resources (SCDHMR), Japan, pp 146–152
34. Joshi KK, Joshi SD (2000) Genetic heritage of medicinal and aromatic plants of Nepal Himalayas. Buddha Academy Publisher and Distributors, Pvt, Kathmandu
35. Bulpitt CJ (2005) The uses and misuses of orchids in medicine. QJM 98:625–631

36. Baral SR, Kurmi PP (2006) A compendium of medicinal plants of Nepal. Publisher Rachana Baral, Mass Printing Press, Kathmandu
37. Subedi A (2011) New species, pollinator interactions and pharmaceutical potential of Himalayan orchids. Ph. D. Thesis, Leiden University, The Netherlands
38. Joshi G, Tewari LM, Lohani N, Upreti K, Jalal JS, Tewari G (2009) Diversity of orchids on Uttarakhand and their conservation strategy with special reference to their medicinal importance. Rep Opin 1:47–52
39. Meitei A, Pamarthi RK, Kumar R, Bhutia NT, Rai D, Kiran BP, Singh AK, Gazmer R, Singh DR (2019) *Dendrobium nobile* Lindl. in Indian traditional medicine, a phytochemical approach. Indian J Horticul 76(3):557–560
40. Pant B, Raskoti BB (2013) Medicinal orchids of Nepal. Himalayan Map House, Pvt Ltd, Kathmandu
41. Nongdam P (2014) Ethno-medicinal uses of some orchids of Nagaland, North-East India. Res J Med Plant 8(3):126–139
42. Matsuda H, Morikawa T, Xie H, Yoshikawa M (2004) Antiallergic phenanthrenes and stilbenes from the tubers of *Gymnadenia conopsea*. Planta Med 70:847–855
43. Arditti J (1992) Fundamentals of orchid biology. Wiley, New York, pp 243–260
44. Arora M, Mahajan A, Sembi JK (2007) A review on phytochemical and pharmacological potential of family orchidaceae. Int Res J Pharm 8(10):9–24
45. Guteirrez RMP (2010) Orchids: a review of uses in traditional medicine, its phytochemistry and pharmacology. J Med Plant Res 4(8):592–638
46. Khasim S, Rao PRM (1999) Medicinal importance of orchids. The Botanica 49:86–91
47. Li Y, Wang CL, Guo SX, Yang JS (2008) Two new compounds from *Dendrobium candidum*. Chem Pharm Bull 56:1477–1479
48. Fan C, Wang W, Wang Y, Qin G, Zhao W (2001) Chemical constituents from *Dendrobium densiflorum*. Phytochemistry 57:1255–1258
49. Venkateswarlu S, Raju MS, Subbarahu GV (2002) Synthesis and biological activity of isamoenylin, a metabolite of *Dendrobium amoenum*. Biosci Biotechnol Biochem 66:2236–2238
50. Miyazawa M, Shimamura H, Nakamura S, Kameoka H (1997) Antimutagenic activity of gigantol from *Dendrobium nobile*. J Agric Food Chem 45:2849–2853
51. Vaidya BN (2019) Nepal global hotspot for medicinal orchids. In: Medicinal plants. pp 35–80. https://doi.org/10.1007/978-3-030-31269-5_3
52. Arditti J, Ernst R (1984) Physiology of orchid seed germination. In: Arditi J (ed) Orchid biology reviews and perspectives. Cornell University Press, New York
53. Hossain MM (2009) Traditional therapeutic uses of some indigenous orchids of Bangladesh. Med Aromat Plant Sci Biotechnol 42:101–106
54. Singh DK (2001) Morphological diversity of the orchids of Orissa/Sarat Misra. In: Pathak P, Sehgal RN, Shkhar N, Sharma M, Sood ABSMPS (eds) Orchids: science and commerce. New Delhi, Bishen Singh Mahendra Pal Singh, pp 35
55. Yang L, Wang Z, Xu L (2006) Simultaneous determination of phenols (bibenzyl phenanthrene and fluorenone) in *Dendrobium* species by high performance liquid chromatography with diode array detection. J Chromatograp 1104:230–237
56. Zhao P, Wang W, Feng FS, Wu F, Yang ZQ, Wang WJ (2007) High frequency shoot regeneration through transverse thin cell layer culture in *Dendrobium candidum* Wall Ex Lindl. Plant Cell Tissue Organ Cult 90:131–139
57. Wu HS, Xu JH, Chen LZ, Sun JJ (2004) Studies on antihyperglycemic effect and its mechanism of *Dendrobium candidum*. Zhongguo Zhong Yao Za Zhi 29:160–163
58. Zheng Y (2005) Shihuin China pharmacopoeia. Chemical Technology Press, Beijing, pp 61–63
59. Ding YP, Wu LS, Yu LW (1998) Optimized harvesting period for *Dendrobium candidum*. China J Chinese Materia Medica 23:458–460
60. Bensky D, Gemble A (1993) A Chinese herbal medicine Materia Medica. Eastland Press, Seattle

61. Xu J, Han BQ, Li SL, Chen XJ, Wang XN, Zhao ZZ et al (2013) Chemistry, bioactivity and quality control of *Dendrobium*, a commonly used tonic herbin traditional Chinese medicine. Phytochem Rev 12:341–367
62. Pant B (2013) Medicinal orchids and their uses: tissue culture a potential alternative for conservation. Afric J Plant Sci 7(10):448–467
63. Bi ZM, Wang ZT, Xu LS, XU GJ (2003) Studies on the chemical constituents of *Dendrobium fimbriatum*. Yao Xue Xue Bao 38: 526–529
64. Jin H, Xu ZX, Chen JH, Han SF, Ge S, Luo YB (2009) Interaction between tissue-cultured seedlings of *Dendrobium officinale* and mycorrhizal fungus (*Epulorhiza sp.*) during symbiotic culture. Chin J Plant Ecol 33:433–441
65. Ho CK, Chen CC (2003) Moscatilin from the orchid *Dendrobium loddigesii* is a potential anticancer agent. Cancer Investig 21:729–736
66. Roy J, Naha S, Majumdar M, Banerjee N (2007) Direct and callus mediated protocorm-like body induction from shoot-tips of *Dendrobium chrysotoxum* Lindl. (Orchidaceae). Plant Cell Tissue Organ Cult 90:31–39
67. Nimisha PS, Hiranmai Yadav R (2012) Proximate and phytochemical analysis of *Dendrobium macrostachyum* Lindl. Int J Pharmaceutic Sci 4(1):385–386
68. Arditti J, Clementsm MA, Fast G, Handley G, Nishimura (1982) Orchid seed germination and seedling culture -a manual. In: Arditti J (ed) Orchid biology -reviews and perspectives, vol II. Cornell University Press, Ithaca, pp 243–370
69. Dressler RL (1993) Phylogeny and classification of the orchid family. Cambridge University Press, New York
70. Rama Rao V (2002) Distributional status survey of threatened plants of Gujarat. PhD Thesis, Sardar Patel University, Vallabh Vidyanagar
71. Gavali D, Sharma D (2004) Traditional knowledge and biodiversity conservation in Gujarat. Indian J Trad Knowledge 3:51–58
72. Kirtikar KR, Bashu BD (2001) Orchidaceae *Dendrobium*. In: Indian medicinal plants, 2nd edn. Oriental Press, India, pp 3316–3317
73. Kovacs A, Vasas A, Hohmann J (2008) Natural phenanthrenes and their biological activity. Phytochemistry 69:1084–1110
74. Balzarini J, Neyts J, Schols D, Hosoya M, Van Damme E, Peumans W, De Clercq E (1992) The mannose-specific plant lectins from *Cymbidium* hybrid and *Epipactis helleborine* and the (N-acetylglucosamine) N-specific plant lectin from *Urtica dioca* are potent and selective inhibitors of human immunodeficiency virus and cytomegalovirus replication in vitro. Antivir Res 18:191–207
75. Wu ZG, Jiang W, Chen SL, Mantri N, Tao ZM, Jiang CX (2016) Insights from the cold transcriptome and metabolome of *Dendrobium officinale*: global reprogramming of metabolic and gene regulation networks during cold acclimation. Front Plant Sci 7:1653
76. Chen KK, Chen AL (1935) The alkaloid of Chin-Shih-Hu. J Biol Chem 111:653–658
77. Liu QF, Zhao WM (2009) A new dedonbrine type alkaloid from *Dendrobium nobile*. Chin Chem Lett 14:278–279
78. Chauhan NS (1999) Medicinal and aromatic plants of Himachal Pradesh. Indus Publishing, New Delhi, p 632
79. Devi PU, Selvi S, Devipriya D, Murugan S, Suja S (2009) Antitumor and antimicrobial activities and inhibition of in-vitro lipid peroxidation by *Dendrobium nobile*. Afr J Biotechnol. https://doi.org/10.5897/AJB2009.000-9292
80. Zhu S, Niu Z, Xue Q, Hui W, Xie X, Ding X (2018) Accurate authentication of *Dendrobium officinale* and its closely related species by comparative analysis of complete plastomes. Acta Pharm Sin B 8(6):969–980
81. Lin Y, Li J, Li B, He T, Chun Z (2010) Effects of light quality on growth and development of protocorm-like bodies of *Dendrobium officinale* in vitro. Plant Cell Tissue Organ Cult 105 (3):329–335

82. Zhou Y, Honda M, Zhu H, Zhang Z, Guo X, Li T, Li Z, Peng X, Nakajima K, Duan L (2015) Spatiotemporal sequestration of miR165/166 by *Arabidopsis argonaute* 10 promotes shoot apical meristem maintenance. Cell Rep 10:1819–1827
83. Zhang GN, Zhang CF, Wang ZT, Xu LS (2004) Chemical constituents of *Dendrobium thyrsiflorum* Rchb f (II). Chinese J Natural Med 2(2):78–81
84. Zhang D, Zhang JJ (2005) Effect of *Coeloglossum viride* var. bracteatum extract on oxidation injury in sub-acute senescent model mice. Zhongguo Yi Xue Ke Xue Bao 27:729–733
85. Song C, Kang J, Ji X, Fu C, Wen X, Yu L, Qiao G (2007) Efficient plant regeneration and genetic fidelity assessment of in vitro-derived plants of *Dendrobium nobile*- an endangered medicinal and ornamental herb. Orchid Sci Biotechnol 1:51–55
86. Yan L, Wang CL, Wang YJ, Gua SX, Yang JS, Chen XM, Xiao PG (2009) Three new bibenzyl derivatives from *Dendrobium candidum*. Chem Pharm Bull 57:218–219
87. Wu JW, Hu CY, Shahid MQ et al (2013) Analysis on genetic diversification and heterosis in autotetraploid rice. Springer Plus 2:439
88. You HL, Park JD, Baek NI, Kim S, Ahn BZ (1995) In vitro and in vivo antimural phenanthrenes from the aerial parts of *Dendrobium nobile*. Planta Med 61:178–180
89. Manandhar NP (1995) A survey of medicinal plants of Jajarkot district, Nepal. J Ethnopharmacol 48(1):1–6
90. Pant B, Thapa D (2012) In vitro mass propagation of an epiphytic orchid, *Dendrobium primulinum* Lindl. through shoot tip culture. Afr J Biotech 11(42):9970–9974
91. Yuan Y, Yu M, Jia Z, Song X, Liang Y, Zhang J (2018) Analysis of *Dendrobium huoshanense* transcriptome unveils putative gene associated with active ingredients synthesis. BMC Genomics 19:978
92. Kartikar KR, Basu BD (1981) Indian medicinal plants, vol 2. International Book Distribution, Dehradun
93. Mk A, Rashid MH, Hossain MS, Salam MA, Rouf MA (2002) In vitro seed propagation of Dendrobium (*Dendrobium transparens*) orchid as influenced by different media. Biotechnology 1:111–115
94. Teixeira da Silva JA (2013) The role of thin cell layers in regeneration and transformation in orchids. Plant Cell, Tissue and Organ Cult 113:149–161
95. Zahara M, Datta A, Boonkorkaew P (2016) Effects of sucrose, carrot juice and culture mediaon growth and net CO2 exchange rate in *Phalaenopsis* hybrid 'pink'. Sci Hortic 205:17–24
96. Utami ESW, Hriyanto S (2020) Organic compounds: contents and their role in improving seed germination and protocorm development in orchids. Int J Agronomy 2020:2795108
97. Zahara M, Datta A, Boonkorkaew P, Mishra A (2017) The effects of different media, sucrose concentrations and natural additives on plantlet growth of *Phalaenopsis* hybrid 'pink'. Braz Arch Biol Technol 60:e160149
98. Anderson AB (1996) The reintroduction of *Plantanthera ciliaris* in Canada. In: Allen C (ed) North American native terrestrial orchids propagation and production. North American Native Terrestrail Orchid Conference, Germantown, pp 73–76
99. Stewart SL, Kane ME (2006) Asymbiotic seed germination and in vitro seedling development of *Habenaria macroceratitis* (Orchidaceae), a rare Florida terrestrial orchid. Plant Cell Tissue Organ Cult 86:147–158
100. Dohling S, Kumaria S, Tandon P (2012) Multiple shoot induction from axillary bud cultures of the medicinal orchid, *Dendrobium longicornu*. AoB Plants pls032. https://doi.org/10.1093/aobpla/pls032
101. Murashige T, Skoog F (1962) A revised medium for rapid growth and bio assays with tobacco tissue culture. Physiologia Planturam 15(3):473–497
102. Mitra GC, Prasad RN, Roychowdhury A (1976) Inorganic salts and differentiation of protocorms in seed callus of orchid and correlative changes in its free amino acid content. Indian J Exp Biol 14:350–351

103. Knudson L (1946) A new nutrient solution for germination of orchid seed. Amer Orchid Soc Bull 15:214–217
104. Vacin E, Went FW (1949) Some pH changes in nutrient solutions. Bot Gaz 110:605–613
105. Gamborg OL, Miller RA, Ojima K (1968) Nutrient requirements of suspension culture of soybean root cells. Exp Cell Res 50(1):151–158
106. Kramer PJ, Kozlowski TT (1979) Physiology of woody plants. Academic Press, New York, p 811
107. Suzuki RM, Moreira VC, Pescador R, Melo FW (2012) Asymbiotic seed germination and in vitro seedling development of the threatened orchid *Hoffmennseggella cinnaberina*. In Vitro Cell Dev Biol 48:500–511
108. Yildiz M (2012) The prerequisite of the success in plant tissue culture: high frequency shoot regeneration. In: Leva A, Rinaldi LMR (eds) Recent advances in plant in vitro culture, 1st edn. InTech Rijeka, London, UK pp 63–90. https://doi.org/10.5772/51097
109. Teixeira da Silva JA, Dobranszki J, Cardoso JC, Zeng SJ (2015) Mocropropagation of *Dendrobium*: a review. Plant Cell Rep 34:671–704
110. Kong Q, Yuan SY, Végvári GY (2007) Micropropogation of an orchid *Dendrobium strongylanthum* Rchb.f. Int J Hortic Sci 13:61–64
111. Kaewduangta W, Reamkatog P (2011) Effect of modification medium on growth development of *Dendrobium parishii* in vitro. Am Eurasian J Agric Environ Sci 11:117–121
112. Nongdam P, Tikendra L (2014) Establishment of an efficient in vitro regeneration protocol for rapid and mass propagation of *Dendrobium chrysotoxum* lindl. using seed culture. Sci World J. https://doi.org/10.1155/2014/740150
113. Bhadra SK, Barua AK, Bhattacharjee B, Hossain MM (2002) In vitro micropropagation of *Dendrobium aphyllum* (Roxb.) G. E. C. Fisher. Bangladesh J Genet Biotechnol 3:47–50
114. Bai MF, Wu TL, Huang M, Zhang TG (2004) Rapid propagation of *Dendrobium loddigesii* Rolfe by tissue culture. Seed 23:44–46. (in Chinese with English abstract)
115. Huang WC, Yin LQ, Hu YH, Wang XQ, Zhao XF, Li XF (2008) In vitro rapid propagation of *Dendrobium fimbriatum*. J Shanghai Jiaotong Univ (Agric Sci) 26:584–587. (in Chinese with English abstract)
116. Guan P, Shi JM (2009) Tissue culture of stem segment and induction of floral buds of *Dendrobium denndanum*. Lishizhen Med Materia Med Res 20:205–206. (in Chinese with English abstract)
117. Pradhan S, Paudel YP, Pant B (2013) Efficient regeneration of plants from shoot tip explants of *Dendrobium densiflorum* Lindl., a medicinal orchid. Afr J Biotechnol 12:1378–1383
118. Yasugi S, Shinto H (1994) Formation of multiple shoots and regenerated plantlets by culture of pseudobulb segment in *nobile* type *Dendrobium*. Plant Tissue Cult Lett 11(2):153–156
119. Sharma U, Rao VR, Nohan JSS, Reddy AS (2007) In vitro propagation of *Dendrobium microbulbon* A. Rich—a rare ethnomedicinal herb. Indian J Biotechnol 6:381–384
120. Ferreira WDM, Kerbauy GB, Costa APP (2006) Micropropagation and genetic stability of a *Dendrobium* hybrid (Orchidaceae). In Vitr Cell Dev Biol Plant 42:568–571
121. Webster S, Mitchell SA, Ahmad MH (2003) A novel surface sterilization method for reducing fungal and bacterial contamination of field grown medicinal explants intended for in vitro culture. In Proceedings of 17th SRC conference entitled "Science and technology for economic development: Technology driven agriculture and agro processing SRC, Jamaica
122. Oyebanji OBO, Nweke O, Odebunmi NB, Galadima MS, Idris UN, Nnodi A, Afolabi S, Ogbadu GH (2009) Simple, effective and economical explant-surface sterilization protocol for cowpea, rice and sorghum seeds. Afri J Biotechnol 8(20):5395–5399
123. Hossain MM, Sharma M, Pathak P (2013) In vitro propagation of *Dendrobium aphyllum* (Orchidaceae)—seed germination to flowering. J Plant Biochem Biotechnol 22:157–167
124. Zhang X, Gao J (2020) In vitro tetraploid induction from multigenotype protocorms and tetraploid regeneration in *Dendrobium officinale*. Plant Cell Tissue Organ Cult 141:289–298
125. Tomita M, Tomita M (1997) Plant regeneration from immature seed -derived callus of *Cypripedium macranthos* Swartz var. taiwanianum (Masamune) F. Maekawa. Breed Sci 47:279281

126. Hynson NA, Madesen TP, Selosse MA, Adam IKU, Ogura-Tsujita Y, Roy M et al (2013) The physiological ecology of mycoheterotropy. In: Merckx VSFT (ed) Mycoheterotropy: the biology of plants living on fungi. Springer, Berlin, pp 297–342
127. Parthibhan S, Senthil Kumar T, Rao MV (2015) Phenology and reintroduction strategies for *Dendrobium aqueum* Lindley – an endemic, near threatened orchid. J Nat Conserv 24:68–71
128. Nontachaiyapoon S, Sasirat S, Manoch L (2011) Symbiotic seed germination of *Grammatophyllum speciosum* Blume and *Dendrobium draconis* Rchb. f., native orchids of Thailand. Sci Hortic 130:303–308
129. Paul S, Kumaria S, Tandon P (2012) An effective nutrient medium for asymbiotic seed germination and large-scale in vitro regeneration of *Dendrobium hookerianum*, a threatened orchid of northeast India. AoB Plants plr032. https://doi.org/10.1093/aobpla/plr032
130. Utami ESW, Hariyanto S, Manuhara YSW (2017) In vitro propagation of the endangered medicinal orchid, *Dendrobium lasianthera* JJ Sm through mature seed culture. Asian Pac J Trop Biomed 7:406–410
131. Phillip VJ, Nainar SAZ (1986) Clonal propagation of *Vanilla planifolia* (Salisb.) Ames using tissue culture. J Plant Physiol 122:211–215
132. Morel G (1966) Meristem culture clonal propagation of orchids. Orchid Dig 30:45–49
133. Dohling S, Das MC, Kumaria S, Tandon P (2007) Conservation of splendid orchids of North-East India. Biodivers. Its significance. IK International Publishers, New Delhi, pp 354–365
134. Sharma A, Tandon P (1992) In vitro culture of *Dendrobium wardianum* Warner: morphogenetic effects of some nitrogenous adjuvants. Indian J Plant Physiol 35(1):80–85
135. Kanjilal B, DeSarkar D, Mitra J, Dutta BK (1999) Stem disc culture: development of a rapid mass propagation method of *Dendrobium moschatum* (Buch-Ham) Swartz—an endangered orchid. Curr Sci 77:497–500
136. Roy J, Banerjee N (2003) Induction of callus and plant regeneration from shoot-tip explant of *Dendrobium fimbriatum* Lindl. var. oculatum Hk. f. Sci Hortic 97:333–340
137. Malabadi RB, Mulgund GS, Kallappa N (2005) Micropropagation of *Dendrobium nobile* from shoot tip sections. J Plant Physiol 162:473–478
138. Pornpienpakdee P, Singhasurasak R, Chaiyasap P, Pichyangkura R, Bunjongrat R, Chadchawan S, Limpanavech P (2010) Improving the micropropagation efficiency of hybrid *Dendrobium* orchids with chitosan. Sci Hortic 124:490–499
139. Asghar S, Ahmad T, Ahmad I, Yaseen M (2011) In vitro propagation of orchid (*Dendrobium nobile*) var. Emma white. Afr J Biotechnol 10:3097–3103
140. Vij SP, Pathak P (1989) Micropropagation of *Dendrobium chrysanthum* Wall. through pseudobulb segments. J Orchid Soc India 3:25–28
141. Gao PY, Chen JM, Gan Q (2002) Tissue culture of stem segment and plantlet regeneration of *Dendrobium nobile*. Chin Tradit Herbal Drug 33:1031–1033. (in Chinese with English abstract)
142. Bai Y, Bao YH, Wang WQ, Yan YN (2006) Tissue culture and rapid propagation of *Dendrobium crepidatum* Lindl. ex Paxt. Plant Physiol Commun 42:R28–S68. (in Chinese)
143. Bhattacharyya P, Paul P, Kumaria S, Tandon P (2018) Transverse thin cell layer (t-TCL) mediated improvised micropropagation protocol for endangered medicinal orchid *Dendrobium aphyllum* Roxb: an integrated phytomolecular approach. Acta Physiol Plant 40:137
144. Pant B, Sha HS, Shrestha R, Pandey S, Joshi PR (2017) An overview on orchid endophytes. In: Varma A, Prasad R, Tuteia N (eds) Mycorrhiza-nutrient uptake biocontrol ecorestoration. Springer, Berlin, pp 503–524
145. Martin KP, Geevarghese J, Jodheph D, Madassery J (2005) In vitro propagation of *Debdrobium* hybrids using flower stalk node explants. Indian J Exp Biol 43:280–285
146. Nugroho JD, Arobaya AYS, Tanur EA (2019) Propagation of *Dendrobium antennatum* Lindl via seed culture in vitro using simple medium: fertilizer and complex organic based medium. HIYATI J Biosci 26(3):133–138
147. Ma NL, Khoo SK, Lee JX, Soon CF, Shukor NAAB (2020) Efficient micropropagation of *Dendrobium aurantiacum* from shoot explant. Plant Sci Today 7(3):476–482

148. Parthibhan S, Kumar TS, Rao MV (2015) In vitro regeneration from protocorms in *Dendrobium aqueum* Lindley- an imperilled orchid. J Nat Conserv 24:68–71
149. Partibhan S, Rao MV, da Silva T, Kumar TS (2018) Somatic embryogenesis from stem thin cell layers of *Dendrobium aqueum*. Biol Plant 62(3):439–450
150. Riva SS, Islam A, Hoque ME (2016) In vitro regeneration and rapid multiplication of *Dendrobium bensoniae*, an indigenous ornamental orchid. The Agriculturists 14(2):24–31
151. Refish NMR, Wang L, Fu C, Xu X, Jin W, Li M, Yu L (2016) Establishment and optimization of high efficiency embryogenic callus induction system in *Dendrobium candidum*. African J Plant Sci 10(4):77–83
152. Maharajan S, Thakur LS, Thapal BB, Pradhan S, Pant KK, Joshi GP, Pant B (2020) In vitro propagation of the endangered orchid *Dendrobium chryseum* Rolfe from protocorms culture. Nepal J Sci Technol 19(1):39–47
153. Kaur S (2017) In vitro regeneration of shoots from nodal explants of *Dendrobium chrysotoxum* LINDL. J Hortic Res 25(1):27–34
154. Hapsoro D, Septiana VT, Ramadiana S, Yusnita Y (2018) A medium containing commercial foliar fertilizer and some organic additives could substitute MS medium for in vitro growth of *Dendrobium* hybrid seedlings. J Floratek 13(1):11–22
155. Kaur S, Bhandari P, Bhutani KK (2015) Chracterization of bioactive compounds at seedling stage and optimization of seed germination, culture multiplication and *Dendrobium nobile* Lindl. - a study in vitro. Int J Adv Res 3(4):1041–1052
156. Priyanka S, Verma LS, Satyanarayana E, Subhankar (2018) In vitro regeneration and rapid multiplication of *Dendrobium nobile*. Int J Chem Studies 6(6):1286–1288
157. Shetty V, Thomas A, Pujari I, Babu VS (2015) Asymbiotic hypergeneration of protocorm like bodies -an efficient and simple micropropagation strategy for conserving the therapeutic ornamental *Dendrobium ovatum*. Int J Recent Sci Res 6(12):8009–8015
158. Bhowmik TK, Rahman MM (2020) Micropropagation of commercially important orchid *Dendrobium palpebrae* Lindl. through in vitro developed psudobulb culture. J Adv Biotechnol Exp Ther 3(3):225–232
159. Kunagorn N, Roopkam C, Aumroong P, Anukul N (2017) Meristem tip culture of *Dendrobium* orchid for boosting efficiency of hygienic large scale micropropagation. Acta Hortic 1155: 419–424
160. Adhikari H, Pant B (2019) In vitro seed germination and seedling growth of the orchid *Dendrobium primulinum* Lindl. African J Plant Sci 13(12):324–331
161. Mamun AA, Islam MM, Ahmed M, Ghose GC (2018) In vitro mass propagation from shoot tip of *Dendrobium Red Bull*-an endangered epiphytic orchid. Plant Tissue Cult Biotech 28 (2):161–169
162. Rattana K, Sangchanjiradet S (2017) Micropropagation of *Dendrobium signatum* Rchb.F. Pertanika J Trop Agric Sci 40(4):577–586
163. Billore V, Jain M, Suprasanna P (2017) Monochromic radiation through light-emitting diode (LED) positively augments in vitro shoot regeneration in orchid (*Dendrobium Sonia*). Can J Biotech 1(2):50–58
164. Sahagun J, Kongbangkerd A, Ratannasut K (2018) Organogenic potential of *Dendrobium* floral tissues for stable transformation applications. Philippine J Sci 147(4):667–676
165. Obsuwan K, Serayheap K, Thepsithar C (2019) Effects of calcium silicate and proline-induced salt tolerance on the in vitro propagation of *Dendrobium Sonia* 'Red jo'. Acta Hortic 1262:87–91
166. Bing HYN, Tham DT, Vinh TT, Hoi QV, Cong VK, Duy NV (2018) In vitro propagation of the new orchid *Dendrobium trankimianum* T. Yukawa. J Biotechnol 16(4):649–657
167. Larkin P, Scowcroft W (1981) Somaclonal variation—a novel source of variability from cell cultures for plant improvement. Theor Appl Genet 60:197–214
168. Lakshmanan V, Venkataramareddy SR, Neelwarne B (2007) Molecular analysis of genetic stability in long-term micropropagated shoots of banana using RAPD and ISSR markers. Electron J Biotechnol. https://doi.org/10.2225/vol10-issue1-fulltext-12

169. Sahijram L, Soneji JR, Bollamma KT (2003) Analyzing somaclonal variation in micropropagated bananas (*Musa spp*). In Vitro Cell Dev Biol 39:551–556
170. Saravanan S, Sarvesan R, Vinod MS (2011) Identification of DNA elements involved in somaclonal variants of *Rauvolfia serpentine* (L.) arising from indirect organogenesis as evaluated by ISSR analysis. Indian J Sci Technol 4:1241–1245
171. Vázquez AM (2001) Insight into somaclonal variation. Plant Biosyst 135:57–62
172. Kunitake H, Nakashima T, Mori K, Tanaka M, Mii M (1995) Plant regeneration from mesophyll protoplasts of Lisianthus (*Eustoma grandiflorum*) by adding activated charcoal into protoplast culture medium. Plant Cell Tissue Organ Cult 43:59–65
173. Zayova E, Vassilevska IR, Kraptchev B, Stoeva D (2010) Somaclonal variations through indirect organogenesis in eggplant (*Solanum melongena* L.). Biol Divers Conserv 3:1–5
174. Farahani F, Yari R, Masoud S (2011) Somaclonal variation in Dezful cultivar of olive (*Olea europaea* subsp. europaea). Gene Conserve 10:216–221
175. Siragusa M, Carra A, Salvia L, Puglia A, De Pasquale F, Carimi F (2007) Genetic instability in calamondin (*Citrus madurensis* Lour.) plants derived from somatic embryogenesis induced by diphenylurea derivatives. Plant Cell Rep 26:1289–1296
176. Sun S, Zhong J, Li S, Wang X (2013) Tissue culture-induced somaclonal variation of decreased pollen viability in torenia (*Torenia fournieri* Lind.). Bot Stud 54(1):36
177. De Carvalho SR, Luis ZG, Scherwinski-Pereira JE (2014) The histodifferentiation events involved during the acquisition and development of somatic embryogenesis in oil palm (*Elesis guineensis* Jacq.). Plant Growth Regul 72:67–80
178. Arnhold-Schmitt B (1993) Rapid changes in amplification and methylation pattern of genomic DNA in cultured carrot root explants (*Daucus carota* L.). Theor Appl Genet 85:793–800
179. Eeuwens CJ, Lord S, Donough CR, Rao V, Vallejo G, Nelson S (2002) Effects of tissue culture conditions during embryoid multiplication on the incidence of "mantled" flowering in clonally propagated oil palm. Plant Cell Tissue Organ Cult 70:311–323
180. Kaeppler SM, Kaeppler HF, Rhee Y (2000) Epigenetic aspects of somaclonal variation in plants. Plant Mol Biol 43:179–188
181. Ngezahayo F, Dong Y, Liu B (2007) Somaclonal variation at the nucleotide sequence level in rice (*Oryza sativa*) as revealed by RAPD and ISSR markers, and by pairwise sequence analysis. J Appl Genet 48:329–336
182. Leva AR, Petruccelli R, Rinaldi LMR (2012) Somaclonal variation in tissue culture: a case study with olive. In: Leva AR, Rinaldi LMR (eds) Recent advances in plant in vitro culture. INTECH Open Access Publisher, Croatia, pp 123–150
183. Neelankandan AK, Wang K (2012) Recent progress in the understanding of tissue culture-induced genome level changes in plants and potential applications. Plant Cell Rep 31:597–620
184. Tiwari JK, Chandel P, Gupta S, Gopal J, Singh BP, Bhardwaj V (2013) Analysis of genetic stability of in vitro propagated potato microtubers using DNA markers. Physiol Mol Biol Plants 19:587–595
185. Azizi P, Hanafi MM, Sahebi M, Harikrishna JA, Taheri S, Yassoralipour A, Nasehi A (2020) Epigenetic changes and their relationship to somaclonal variation: a need to monitor the micropropagation of plantation crops. Funct Plant Biol. https://doi.org/10.1071/fp19077
186. Sabir A, Newbury HJ, Todd G, Catty J, Ford-Lloyd BV (1992) Determination of genetic stability using isozymes and RFLPs in beet plants generated in vitro. Theor Appl Genet 84:113–117
187. Dixit S, Mandal BB, Sangeeta A, Srivastava PS (2003) Genetic stability assessment of plants regenerated from cryopreserved embryogenic tissues of *Dioscorea bulbifera* l. using RAPD, biochemical and morphological analysis. Cryo Letters 24:77–84
188. Mallon R, Rodriguez-Oubina J, Gonzalez M (2010) In vitro propagation of the endangered plant *Centaurea ultreiae*: assessment of genetic stability by cytological studies, flow cytometry and RAPD analysis. Plant Cell Tissue Organ Cult 101:31–39
189. Loureiro J, Rodriguez E, Dolezel J, Santos C (2006) Comparison of four nuclear isolation buffers for plant DNA cytometry. Ann Bot 98:679–689

190. Bairu MW, Aremu AO, van Staden J (2011) Somaclonal variation in plants: cause and detection methods. Plant Growth Regul 63:147–173
191. Butiuc-Keu A, Farkas A, Cristea V (2016) Genetic stability assessment of in vitro plants by molecular markers. Studia Univeritatis Babes - Bolyai Biologia 61(1):107–114
192. Singh SR, Dalal S, Singh R, Dhawan AK, Kalia RK (2013) Ascertaining clonal fidelity of micropropagated plants of *Dendrocalamus hamiltonii* Nees et Arn. ex Munro using molecular markers. In Vitro Cell Dev Biol 49:572–583
193. Bandupriya HDD, Iroshini WWMA, Perera SACN, Vidhanaarachchi VRM, Fernando SC, Santha ES, Gunathilake TR (2017) Genetic fidelity testing using SSR marker assay confirms trueness to type of micropropagated coconut (*Cocos nucifera* L) plantlets derived from unfertilized ovaries. The Open Plant Sci J 10:46–54
194. Raynalta E, Elina J, Sudarsono SD (2018) Clonal fidelity of micro propagated *Phalaenopsis* plantlets based on assessment using eighteen Ph-Pto SNAP marker loci. Agrivita J Agri Sci 40 (3):390–402
195. Rohela GK, Jogam P, Bylla P, Reuben C (2019) Indirect regeneration and assessment of genetic fidelity of acclimated plantlets by SCoT, ISSR and RAPD markers in *Rauwolfia tetraphylla* L.: an endangered medicinal plant. Biomed Res Int 2019:3698742
196. Dey A, Nongdam P, Nandy S, Mukherjee S, Tikendra L, Hazra AK, Pandey DK (2020) Polyamine elicited aristolochic acid production in in vitro clonally fidel *Aristolochia indica* L.: an ISSR and RAPD markers and HPTLC based study. S Afr J Bot https://doi.org/10.1016/j.sajb.2020.06.018
197. Dey A, Nandy S, Nongdam P, Tikendra L, Mukherjee A, Mukherjee S, Pandey DK (2020) Methyl jasmonate and salicylic acid elicit indole alkaloid production and modulate antioxidant defence and biocidal properties in *Rauvolfia serpentina* Benth. ex Kurz. in vitro culture. S Afri J Bot 135:1–17
198. Muralidharan K, Wakeland EK (1993) Concentration of primer and template qualitatively affects products in random- amplified polymorphic DNA PCR. BioTechniques 14:362–364
199. Amom T, Tikendra L, Rahaman H, Potshangbam A, Nongdam P (2018) Evaluation of genetic relationship between 15 bamboo species of North-East India based on ISSR marker analysis. Mol Biol Res Commun 7:7–15
200. Zerini HN, Jafari H, Ramandi HD, Bolandi AR, Karimishahri (2019) A comparative assessment of DNA fingerprinting assays of ISSR and RAPD markers for molecular diversity of Saffron and other *Crocus* spp. in Iran. Nucleus 62:39–50
201. Gianchino RRA (2020) Investigation of the genetic variation of anise (*Pimpinella anisum* L.) using RAPD and ISSR markers. Genet Resour Crop Evol 67:763–780
202. Zietkiewicz E, Rafalski A, Labuda D (1994) Genome fingerprinting by simple sequence repeat (SSR) anchored polymerase chain reaction amplification. Genomics 20:176–183
203. Collard BCY, Mackill DJ (2009) Start codon targeted (SCoT) polymorphism: a simple, novel DNA marker technique for generating gene-targeted markers in plants. Plant Mol Biol Rep 27:86–93
204. Amom T, Tikendra L, Apana N, Goutam M, Sonia P, Koijam AS, Potshangbam AM, Rahaman H, Nongdam P (2020) Efficiency of RAPD, ISSR, iPBS, SCoT and phytochemical markers in the genetic relationship study of five native and economical important bamboos of North-East India. Phytochemistry 174:112330
205. Amom T, Nongdam P (2017) The use of molecular marker methods in plants: a review. Int J Cur Res Rev 9:1–9
206. Hu J, Vick BA (2003) Target region amplification polymorphism: a novel marker technique for plant genotyping. Plant Mol Biol Rep 21:289–294
207. Liu Q, Chen J, Corlett RT, Fan X, Yu D, Yang H, Gao J (2015) Orchid conservation in the biodiversity hotspot of Southwestern China. Conserv Biol 29:1563–1572
208. Smulders M, de Klerk G (2011) Epigenetics in plant tissue culture. Plant Growth Regul 63:137–146

209. Hao YJ, Wen XP, Deng XX (2004) Genetic and epigenetic evaluations of citrus calluses recovered from slow-growth culture. J Plant Physiol 161:479–484
210. Baurens FC, Causse S, Lagavre T (2008) Methylation-sensitive amplification polymorphism (MSAP) protocol to assess CpG and CpNpG methylation in the banana genome. Fruits 63:117–123
211. Morcello F, Gagneur C, Adam H, Richaud F, Singh R, Cheah SC, Rival A, Duval Y, Trebear JW (2006) Somaclonal variation in micropropagated oil palm. Characterization of two novel genes with enhanced expression in epigenetically abnormal cell lines and in response to auxin. Tree Physiol 26:585–594
212. Wang D, Zhao J, Bai Y, Ao Y, Guo C (2017) The variation analysis of DNA methylation in wheat carrying gametocidal chromosome 3c from *Aegilops triuncialis*. Int J Mol Sci 18:1738
213. Nadeem MA, Nawaz MA, Shahid MQ, Dogan Y, Comertpay G, Yildiz M, Hatipoglu R, Ahmed F, Alsaleh A, Labhane N et al (2018) DNA molecular markers in plant breeding: current status and recent advancement in genomic selection and genome editing. Biotechnol Biotechnol Equip 32:261–285
214. Dey A, Hazra AK, Nongdam P, Nandy S, Tikendra L, Mukharjee A, Banerjee S, Mukherjee S, Pandey DK (2019) Enhanced bacoside content in polyamine treated in vitro raised *Bacopa monnieri* (L.) Wettst. S Afr J Bot 123:259–269
215. Gantait S, Mandai N, Bhattacharyya S, Kanti Das P (2010) Determination of genetic integrity in long-term micropropagated plantlets of *Allium ampeloprasum* L. using ISSR markers. Biotechnol 9:218–223
216. Rathore NNS, Rai MK, Phulwaria M, Rathore NNS, Shekhawat NS (2014) Genetic stability in micropogated *Cleome gynandra* revealed by SCoT analysis. Acta Physiol Plant 36:555–559
217. Bhattacharyya P, Kumaria S, Diengdoh R, Tandon P (2014) Genetic stability and phytochemical analysis of the in vitro regenerated plants of *Dendrobium nobile* Lindl., an endangered medicinal orchid. Meta Gene 2:489–504
218. Rani V, Raina SN (2000) Genetic fidelity of organized meristem-derived micropropagated plants: a critical reappraisal. In Vitro Cell Dev Biol Plant 36:319–330
219. Bhattacharyya P, Kumaria S, Job N, Tandon P (2015) Phyto-molecular profiling and assessment of antioxidant activity within micropropagated plants of *Dendrobium thyrsiflorum*: a threatened, medicinal orchid. Plant Cell Tissue Organ Cult 122:535–550
220. Wannajindaporn A, Poolsawat O, Chaowiset W, Tantasawat PA (2015) Evaluation of genetic variability in *in vitro* sodium azide-induced *Dendrobium* 'Earsakul' mutants. Genet Mol Res 13(3):5333–5342
221. Bhattachryya P, van Staden J (2018) Molecular insights into genetic diversity and population dynamics of five medicinal *Eulophia* species: a threatened orchid texa of Africa. Physiol Mol Biol Plants 24:631–641
222. Poczai P, Varga I, Laos M et al (2013) Advances in plant gene targeted and functional markers: a review. Plant Methods 9:6
223. Doyle JJ, Doyle JL (1990) Isolation of plant DNA from fresh tissue. Focus 12:13–15
224. Galvan MZ, Bornet B, Balatti PA, Balachandra M (2003) Inter simple sequence repeat (ISSR) markers as a tool for the assessment of both genetic diversity and gene pool origin in common bean (*Phaseolus vulgaris* L.). Euphytica 132:297–301
225. Ajibade SR, Weeden NF, Chite SM (2000) Inter simple sequence repeat analysis of genetic relationships in the genus *Vigna*. Euphytica 111:47–55

Eulophia spp.: In Vitro Generation, Chemical Constituents, and Pharmacological Activities

20

Varsha Shriram and Vinay Kumar

Contents

Abstract

Eulophia R.Br. ex Lindl represents one of the largest, wide-spread, and important genera of the family *Orchidaceae*. *Eulophia* shows an extraordinary kind of morphological diversity and occupies a wide variety of habitats. Current records testify that it encompasses around 230 species, out of which 203 are the accepted ones. This genus is of prime importance because of its distinctive ecology and broad-spectrum ornamental and therapeutic properties. Crude solvent extracts and phytochemicals have been assessed from the members of this genus and found to possess potent pharmacological activities including anticancer, anti-diabetic, anti-inflammatory, and DNA protection among others. Keeping this in view, we are presenting herein an account on the distribution and botanical description of the genus *Eulophia*, chemical constituents reported from the

V. Shriram (✉)
Department of Botany, Prof. Ramkrishna More Arts, Commerce and Science College, Savitribai Phule Pune University, Pune, India
e-mail: varshashriram76@gmail.com

V. Kumar
Department of Biotechnology, Modern College of Arts, Science and Commerce, Savitribai Phule Pune University, Pune, India

J.-M. Mérillon, H. Kodja (eds.), *Orchids Phytochemistry, Biology and Horticulture*, Reference Series in Phytochemistry, https://doi.org/10.1007/978-3-030-38392-3_31

Eulophia species along with their ethno/pharmacological activities/claims besides their evidenced-based therapeutic activities. Since many species of *Eulophia* are reportedly threatened, their conservation is required on urgent basis. However, low germination and survival rate in natural conditions along with requirement for specific fungal associations for germination and growth, *in situ* conservation approaches have met with limited success. Plant tissue culture technologies as emerging platforms for *in vitro* regeneration, phytochemical production and *ex situ* conservation for the members of *Eulophia* are discussed with a perspective approach.

Keywords

Anticancer · Asymbiotic germination · Eulophia · *In vitro* regeneration · Orchidaceae · Phenanthrene · Phytomolecules · Protocorms · Protocorm-like bodies

1 Introduction

The family *Orchidaceae* comprises highly evolved monocot plants, representing second largest and most diversified group, which boasts a worldwide distribution of an estimated 28,000 species and 736 genera [13, 21]. Family *Orchidaceae* belongs to the Asparagales (order) of class monocotyledonae of Angiospermae (Sub-division) in the kingdom Plantae. The word orchid was coined by Theophrastus, originated from Greek word *Orchis* as the tubers' shape in many species from the genus resembling to the testicles. Orchids were the symbol of virility for the Greeks, whereas for Chinese people these were the plants of the kings' fragrance [26]. Orchids hold a peak position in plant evolution, often used for ornamental purpose because of their incredible diversity and beautiful flowers besides their great value in horticulture and plant-based medicines [4]. They exhibit broad diversity in morphology, growth form, life history, and habitat. These are herbaceous, perennial plants, either terrestrial or epiphytic. The epiphytes can absorb water from the surrounding air. Most of the members are autotrophs while some are saprophytic, being helped to obtain nourishment by their rhizospheric fungi. Orchid flowers show tremendous variations in size (2 mm diameter in genus *Pleurothallism* and more than 30 cm in *Brassia*). Further, the diversity is observed in the methods of pollination as well as types of pollinators. Orchid species have evolved to imitate several other organisms like bee, fly, etc. among other creatures they try to attract. Some have developed elaborate systems of water traps and tunnels, hinged petals, and sticky packets of pollen devices often described as *devious* or *deceitful*.

Orchids are distributed through wide ecological conditions. They occur in all the terrestrial ecosystems except pole and desert regions. This family primarily found in tropical regions, whereas many species are spread in the northern and southern temperate zones. Utmost diversity is observed in tropical regions where rainforest, temperature lush, and high humidity conditions prevail. Many species in the northern temperate zones are found in bogs, prairies, grasslands, and hardwood forests.

Orchids often blossom in high altitude regions (4500 m above mean sea level). A large number of orchids are found in cloud forests in tropical regions on mountain-sides, often covered with mosses and lichens. This is a favorable habitat for epiphytic orchids. Few orchid species flourish in desert conditions. A few species like *Brassavola* are known to grow on mangrove roots, whereas some species grow on bare rocks [20]. Some orchid species are widespread (e.g., *Ionopsis utricularioides*); however, some are restricted to single mountain [20].

Orchids are historically used as food, medicines, adhesives, perfumes, and flavoring agents. For instance, Vanilla (a product obtained from the orchid *Vanilla planifolia*) is commercially important flavoring agent derived from orchid. They comprise one of the topmost horticultural plant families and their cultivation is highly significant for global nursery industry [3]. The orchids have taken an important place in the flower industry because of their charm, shelf-life, productivity, and ease with packing and/or shipping. Orchids account for a large share of global floriculture trade both as cut flowers and as potted plants. They have been used across the world in various traditional medicinal systems [67].

The record of orchids might begin with their uses in the therapeutic purpose. Chinese were the first to describe and cultivate various orchids [35]. These plants first received recognition in the traditional herbal writings of China and Japan, and they were the first to describe medicinal use of orchids [10]. In recent times, more species belonging to different genera have been reported for their medicinal properties and this list is expected to expand in future [25, 64]. As reported in the literature, various species of orchids are used as antimicrobial, anti-inflammatory, antioxidant, anticancer, anti-pyretic, anti-mutagenic, anti-convulsive, anti-helmintic, anti-hepatotoxic, wound healing, anti-platelet, anti-diabetic, anti-allergic, immunomodulatory, anti-aging, pain reliever, antiviral, and herbicidal agents [39, 42, 83, 89]. Phytomolecules of orchids have been investigated and several constituents have been reported [44]. The phytochemicals from these plants are mainly alkaloids, stilbenoids and phenanthrenes, flavonoids, terpenoids, essential oils, glycosides, and coumarins. Phenanthrene derivatives are the common phyto-constituent in orchids [27]. Hydroxylated and 9,10-dihydroxy phenanthrenes are most commonly found in rhizome/ bulbs/ tubers, roots of these plants besides other derivatives. Phenanthrene derivatives are synthesized from stilbenes that originate from cinnamic acid derivatives. Biosynthesis of 9, 10 dihydroxy phenanthrenes is via phenylpropionic acid derivatives.

Eulophia R.Br. ex Lindl is an important and largest terrestrial genus of the *Orchidaceae*. According to current records, it encompasses around 230 species [90] and out of that 203 are the accepted ones. It belongs to the subtribe Eulophiinae (subfamily Epidendroideae; tribe Cymbidieae) [74]. The center of diversity of the genus is Africa, mostly found in the palaeotropics, although six species extend into the neotropics, and *Eulophia* shows an extraordinary morphological diversity and occupies a surprisingly wide variety of habitats. This genus is of prime importance due to its distinctive ecology and broad-spectrum ornamental and therapeutic properties. Keeping this in view, we are presenting herein an account on the distribution and botanical description of *Eulophia*, chemical constituents reported from the

Eulophia species along with their ethno/pharmacological activities/claims besides their evidenced-based therapeutic activities. Further, many species of *Eulophia* are under IUCN threatened categories and hence conservation of the genus is the focus for the conservation biologist. However, low germination and survival rate in natural conditions along with requirement for specific fungal associations for germination and growth, *in situ* conservation approaches have met with limited success. Plant tissue culture technologies as emerging platforms for *in vitro* regeneration, phytochemical production, and *ex situ* conservation for the members of *Eulophia* are discussed with a perspective approach.

2 Distribution and Botanical Description

Eulophia belongs to the family Orchidaceae and its members are known as "corduroy orchids" [38], while in India they are known as "Amarkand." The name *Eulophia* is of Greek origin (eu-lopos), describing the "beautiful crest," denoting lip-crests [23]. This genus name was coined by Robert Brown, and was established by Lindley in 1821. *Eulophia* are predominantly terrestrial, rarely lithophytic and sympodial herbs whereas 2 are epiphytic as found on Madagascar. It shows wide distribution across the world and frequently found in Africa and Asia; some species are deciduous while few are evergreen. Africa is considered as a center of diversity of this genus with an extensive diversity of habitats comprising the desert-margins, marshes, and dry savanna woodlands besides wet tropical forests. It is large pan-tropical genus distributed in shady rainforests with grass or shrubs or in open scrub or woodland in the tropics and subtropics of Africa, India, Asia, Queensland, and the Americas, while some of its species (such as *E. petersii*) are habituated to arid climates (desert species). Some species have shown broad latitudinal spread as in *E. cucullata,* which can be correlated to the polyploidy.

Phylogenetic studies reveal that genus *Eulophia* is paraphyletic. It is either autotrophic or at times chlorophyll deficient. Root system is adventitious, slender to stout, produced from the tubers'/ pseudobulbs' base often with a well-defined white velamen. Pseudo-bulbs/ tubers are the perennating organ either above the ground and rhizomatous or tuberous if subterranean. They are cylindrical/ fusiform/ conical/ ovoid/ irregular, with several nodes, homoblastic, and grow into a chain of tubers. Leaves appear at or after anthesis. Green Leaves usually present, may be highly reduced in some species, scale-like and brown or even buff. Green leaves thin-textured or fleshy or coriaceous, petiolate, long and narrow, linear/ lanceolate/ ovate/ elliptic/ sometimes pleated, with or without prominent longitudinal veins, acute to acuminate, articulate or not to a sheathing or petiole-like leaf base, sometimes overlapping and forming a false stem. Flowering stalk (scape) produced with leaves from the current year's growth that are enclosed with subscarious tubular sheaths.

Inflorescence is erect, terminal/ lateral/ basal laxly to sub-densely, many flowered or rarely reduced to a solitary flower, usually racemose or rarely paniculate, simple or rarely branching; bracts are persistent. Flowers are small to large, green/ brown/ showy and brightly colored at times bicolored, resupinate/ non-resupinate. Sepals and petals either similar or petals are much broader. Dorsal sepal free, oblong/ elliptic/ lanceolate/ oblanceolate, reflexed, erect/ porrect; lateral sepals oblique at base and decurrent on column foot. Labellum / Lip flat, sessile, 3-lobed, free to base or fused to base of the column, lacking a claw, usually with a callus 2 or 3 ridged or papillose on upper surface, either spurred at the base or without spurred or saccate. If spur is present, it may be obscure and sac-like. Labellum disk-like/ entirely tuberculate/ nerves tuberculate/ verrucose/ keeled; mid-vein present on side lobes often raised into a ridge at the base, lateral lobes free / fused to base of column, mid-lobe flat or convex. Columns are short to long, with or without a column-foot. Ovary is cylindrical, grooved. This genus includes 203 accepted species and 2 are the cross (hybrid species), for the complete list of species and the description refer to http://www.plantsoftheworldonline.org/taxon/urn:lsid:ipni.org:names:325750-2).

3 Phytochemical Constituents

Owing to the wide-spectrum uses of *Eulophia* genus, several researchers have focused on investigation of its phytomolecules and their bioactivities in the recent past, refer Table 1. Key attention was to isolate and to identify their respective phyto-constituents and then exploring them for broad-spectrum bioactivities. Several important biomolecules with tremendous therapeutic values have been reported from a number of *Eulophia* species. Kovács et al. [45] investigated the phytochemical constituents of *E. ochreata* and *E. nuda* and confirmed the presence of essential minerals, polyphenols, saponins, alkaloids, phytic acid among others in these species. Several phenanthrenes have been reported from *E. nuda* till date. Diverse Phenanthrene derivatives are most commonly occurring phytomolecules in this genus, for instance Bhandari and Kapadi [6] isolated crystalline molecule and characterized it as 9,10-Dihydrophenanthrene from ethanol extract of tubers of *E. nuda* and named it as eulophiol with molecular formula $C_6H_60_4$. Later in 1985, Nudol, a phenanthrene from *E. nuda* with molecular formula $C_{16}H_{14}0_4$ and mol. Mass 270) was reported by Bhandari et al. [7]. A research group from Thailand and Australia took the interest and described six phenanthrene derivatives from *E. nuda* 9,10-Dihydro-2,5-dimethoxyphenanthrene-1,7-diol (mol. Formula $C_{16}H_{16}0_4$; mol. Mass 272), 9,10-dihydro-4-methoxyphenanthrene-2,7-diol (mol. Formula $C_{15}H_{14}O_3$; mol. Mass 242.0941), l,5-Dimethoxyphenanthrene-2,7-diol (mol. Formula $C_{16}H_{14}O_{4;}$ mol. Mass 270), 1,5,7,-trimethoryphenanthrene-2,6-diol (mol. Formula $C_{17}H_{16},O_{5;}$ mol. Mass 300.0098), 5,7-dimethoxyphenanthrene-2,6-diol (mol. Formula $C_{16}H_{14}O_{4;}$ mol. Mass 270.28), and 4,4′,8,8′-tetramethoxy [1,1′-biphenanthrenel-2,2′,7,7′-tetrol (mol. Formula $C_{32}H_{26}O_{8;}$ mol. Mass 538)]. Besides, 4-Hydroxybenzaldehyde and 4-hydroxybenzyl alcohol were reported by this group [91]. Further, by the same research group in 1989 two more phenanthrene derivatives, viz., 9,10-dihydro-l-(4′-hydroxybenzyl) -4,7-dimethoxyphenanthrene-2,8-diol ($C_{23}H_{22}O_5$) with the molecular mass 378, 1-(4′-hydroxybenzyl)-4,8 dimethoxyphenanthrene- 2,7-diol ($C_{23}H_{20}O_5$) with molecular mass 376 were

Table 1 List of phytochemicals (pure compounds) reported from *Eulophia* species

Eulophia species	Plant part	Pure Compound reported	References
Eulophia epidendrea (JKoen) Schltr.	Leaves	Apigenin, luteolin, kaempferol, and quercetin Tuber fractions β-sitosterol, β-sitosterolglucoside, β-amyrin and lupeol	Maridass and Ramesh [51]
Eulophia nuda Lindl.	Tuber	n-hexacosyl alcohol and lupeol	Merchant et al. [56]
		1,5-dihydroxy-2,7-dimethoxy-9,10-dihydrophenanthrene (Eulophiol)	Bhandari and Kapadi [6]
		2,7-dihydroxy-3,4-dimethoxyphenanthrene (Nudol)	Bhandari et al. [7]
		9,10-dihydro-2,5-dimethoxyphenanthrene-1,7-diol	Tuchinda et al. [91]
		9,10-dihydro-4-methoxyphenanthrene-2,7-diol	Tuchinda et al. [91]
		1,5-dimethoxyphenanthrene-2,7-diol	Tuchinda et al. [91]
		1,5,7,-trimethoxyphenanthrene-2,6-diol	Tuchinda et al. [91]
		5,7-dimethoxyphenanthrene-2,6-diol	Tuchinda et al. [91]
		4,4′,8,8′-tetramethoxy [1,1′-biphenanthrene]-2,2′,7,7′ tetrol.	Tuchinda et al. [91]
		4-hydroxybenzaldehyde	Tuchinda et al. [91]
		4-hydroxybenzyl alcohol	Tuchinda et al. [91]
		9,10-dihydro-1-(4′-hydroxybenzyl)-4,7 dimethoxyphenanthrene-2,8-diol	Tuchinda et al. [92]
		1-(4′-hydroxybenzyl)-4,8-dimethoxyphenanthrene-2,7-diol	Tuchinda et al. [92]
		3,4′-dihydroxy-3′,5,5′-trimethoxybibenzyl bis(4-hydroxybenzyl) ether	Tuchinda et al. [92]
	Tubers: water extract	Alkaloids, flavonoids, saponins, carbohydrates, steroids, triterpenoids,	Bhatt et al. [8]

(continued)

Table 1 (continued)

Eulophia species	Plant part	Pure Compound reported	References
		tannins, phenolics, coumarins and anthraquinones	
Eulophia ochreata Lindl.	Fresh tubers	9, 10-Dihydro-2, 5-Dimethoxyphenanthrene-1, 7-diol 5, 7-Dimethoxyphenanthrene-2, 6-diol	Datla et al. [15]

described from same species [92]. Moreover, 3,4′-dihydroxy-3′,5,5′-trimethoxybibenzyl with molecular weight 304 and bis(4-hydroxybenzyl) ether with molecular weight 230 were found [92].

Various important classes of secondary metabolites including flavonoids, reducing sugars, cyanogenic glycosides terpenoids, and tannins were reported from methanolic extracts of *E. epidendrea* of tubers [53]. The same group further reported the presence of flavonoids, sterols, and terpenoids in the fractions collected from leaf and root extracts of *E. epidendrea* [51]. Major phytochemicals reported from species of *Eulophia* are presented in Table 1.

4 Pharmacological Activities

4.1 Ethnopharmacological Uses

Rhizomes or tubers of many *Eulophia* species are regularly consumed by the tribes, especially in India, both as food and for therapeutic materials for maintenance of health and longevity. In traditional medical practices like the Ayurveda, it is frequently recommended as an expectorant, tonic, diuretic, astringent, and a soft purgative [64]. The tubers of several *Eulophia* species have traditional medicinal usages against scrofulous glands of the neck, bronchitis, and rheumatoid arthritis, besides being used as a vermifuge, appetizer, anti-fatigue, anti-helminthic, anti-bellyache, anti-scrofulous blood purifier, anti-helminthic, anti-tumor, wound healing, and an aphrodisiac agent. For instance, fresh juice of *E. campestris* tubers is used in traditional systems for treating diarrhea, and dysentery, besides its use as a laxative and an appetizer. The rhizomes of *E. campestris* have tonic properties and are also used as aphrodisiac [11]. The tubers are useful against worm infestation and scrofula [83]. *E. epidendrea* tubers paste is useful against boils and is used as breast pain reliever for feeding mothers [75]. The tubers of *E. epidendrea* are used traditionally for curing diarrhea and tumors [69]. The tubers of this orchid are also used as appetizer and as blood purifier and are particularly helpful during heart problems [52]. Similarly, *E. herbacea* tubers are used for reducing liver swelling [2], for increasing the sperm-count [2], for treating pimples [70], while fried tubers in mustard oil are externally applied for rheumatism [82].

E. nuda Lindl., is an important and heavily explored species of this genus for its tremendous therapeutic potentials. Its tubers are used for treating worm infestation and scrofula [83], skin rashs, acidity, piles, anorexia, anthrax, and stomach complaints [79], rheumatoid arthritis [49], cancer, asthma, and bronchitis [34]. The whole tubers of *E. nuda* are used to get relief from abdominal pain [14], while the root juices are used against snakebites [82]. The "Salep" of *E. ochreata* tubers is used for the treatment of sexual impotency and male sterility [27, 43], the tubers paste is used against asthma and acute bronchitis [32, 33], whereas the tuber-decoction works as antidote for snakebite and to cure leukemia [49].

4.2 Evidence-Based Pharmacological Activities

Investigations have been carried out to scientifically validate the ethnopharmacological activities and therapeutic claims of some important species of *Eulophia,* including *E. nuda*, *E. campestris*, *E. herbacea*, and *E. epidendrea*, among others. Most of the investigated species have wide-spectrum and strong therapeutic activities, thus confirming their ethnopharmacological importance. The *E. campestris* tubers have tremendous binding properties [97], and thus their mucilage is heavily used for their binding attributes in drug tablet formulations, which ultimately regulate the drug-release rate [22]. Yadav et al. [96] evaluated the effects of "Salep" of *E. campestris* on glycation inhibitory activities, and the authors reported that the application of "Salep" reduced the formation of glycated products/ advanced glycation end-products. Similarly, mucilage of the tubers of *E. herbacea* has been reported as an excellent suspending agent, besides its properties like less sedimentation rates, high viscosity, acidic pH, and abilities to re-disperse make it a good pharmaceutical adjuvant [9]. The methanolic extracts from the tubers of *E. epidendrea* exhibited anti-diarrheal potencies in experimental rats, as confirmed by the abilities of the extract in inhibiting the intestinal fluid accumulation and the propulsive movements of the intestinal contents [50]. These investigations hold significance for validating the folk medicines claims. Jagdale et al. [30] validated the traditional uses of *E. nuda* tubers as an aphrodisiac on mice models and reported significant aphrodisiac potentials of this plant. Jagtap et al. [31] reported anti-inflammatory and antioxidant activities from the methanol extracts of *E. ochreata*. Besides, solvent extracts of *E. ochreata* showed noteworthy antibacterial activities against *Bacillus subtilis*, *Staphylococcus Aureus*, and *Escherichia coli* [18]. The tubers of *E. ochreata* have also been reported to possess antioxidant, antiglycan, and amylase inhibitory activities and thus might prove a potential source for identifying anti-diabetic molecules [63]. *E. ochreata* have also shown high antioxidative activities [18, 62, 63]. In a comprehensive *in vitro* assessment, Kumar et al. [47] studied phytochemical profiling, and antioxidant and free-radical scavenging activities of different solvent extracts of *E. nuda* and the authors attributed these potencies to high amounts of phenols, flavonoids, vitamins, and carotenoids. All the extracts showed a broad-spectrum antioxidant properties including DNA protection from hydroxyl-radical-induced damage [47]. The results provided the scientific basis for the traditional uses of this orchid as a natural antioxidant and phytotherapeutic agent. Our group further reported antioxidant, free-radical scavenging, and cytotoxicity activities of *E. nuda* tuber extracts [61]. The Reversed Phase High Performance Liquid Chromatography-Electrospray Ionization/Mass Spectrometry-based phytochemical profiling of ethyl acetate extract confirmed the presence of a total of 37 compounds including some known antioxidants like catechin, tocopherol, and trigallic among others, and the antioxidants were reported to act via induction of nuclear transcription factor-erythroid-2 related factor (Nrf2) and hemeoxygenase-1 (HO-1) pathways [61]. Further, Preedapirom et al. [73] reported aphrodisiac activities of *E. macrobulbon* crude extract on erectile dysfunction in male aged rats. Jansakul [36] studied potential applications of ethanolic extract of *E. macrobulbon* tubers, and an isolated constituent, 1-(4′-hydroxybenzyl)-4,8-dimethoxyphenanthrene-2,7-diol on human erectile dysfunction. Authors in an attempt to study the underlying mechanism of action, they reported the relaxant mechanism of on human corpus cavernosum.

In similar vein, several therapeutically potent, pure molecules have been reported from the members of *Eulophia*. For instance, our group isolated a phenanthrene derivative 9,10-dihydro-2,5-dimethoxyphenanthrene-1,7-diol from *E. nuda* which showed excellent cytotoxic activity (better efficacy than the standard anticancer molecule carboplatin) against human cancer cells [79]. This compound was also isolated from *E. ochreata* and it effectively inhibited lipopolysaccharide-induced and Toll-like receptors-mediated, nuclear factor κB-activated inflammatory genes; however, it reduced both lipopolysaccharide-induced tumor necrosis factor (TNF-α) release and carrageenan-induced paw edema in rats [15]. Overall, these results confirmed the anti-inflammatory potencies of this plant, besides highlighting the underlying mechanism targeted by this compound. Tatiya et al. [86] successfully attempted the bioassay-guided isolation of 1-phenanthrenecarboxylic acid 1, 2, 3, 4, 4a, 9, 10, 10a-octahydro-1, 4a-dimethyl-, methyl ester from E. herbacea and this compound showed strong anti-proliferative activities against human cancer cell lines, thus confirming the traditional anticancer uses of the tuber in the folklores. Upadhyay et al. [93] isolated phenanthrene derivative Eulophiol from *Eulophia* species, and examined its potential application in inhibition of immune stimulation involving Toll-like receptor ligands, especially TLR-4, the authors labeled it as a potent Toll-like receptor signaling antagonist. Temkitthawon et al. [88] isolated a new phenanthrene, 9,10-dihydro-4-(4′-hydroxybenzyl)-2,5-dimethoxyphenanthrene-1,7-diol, besides three known phenanthrenes, 1-(4′-hydroxybenzyl)-4,8-dimethoxyphenanthrene-2,7-diol, 9,10-dihydro-2,5-dimethoxyphenanthrene-1,7-diol, and 1,5,7-trimethoxyphenanthrene-2,6-diol from the tubers of *E. macrobulbon*, and the phytomolecules were found to be potent phosphodiesterase inhibitors. Wisutthathum et al. [95] undertook pharmacological characterization of the vascular actions of the *E. macrobulbon* ethanolic extract and its active compound, 1-(4′-hydroxybenzyl)-4,8-dimethoxyphenanthrene-2,7-diol using isolated pulmonary arteries from rats having pulmonary arterial hypertension induced by monocrotaline. The pulmonary arterial hypertension was reported to be improved by the tuber extract via pulmonary arteries relaxation mediated through endothelial nitric oxide, reduced Ca^{2+}-mobilization, besides reduced arteries wall thickness and the right ventricular hypertrophy [95]. Ethnopharmacological uses and evidenced-based pharmacological activities of different *Eulophia* species are listed in Table 2.

5 In vitro Regeneration, Phytochemical Production, and Conservation

A large number of orchid species including *Eulophia* are under threat due to their over-exploitation, beside loss of habitats and the pollinators [3, 84]. This genus mostly in marginal situations is exposed to several challenges like unsuitable environmental conditions besides predation, parasitism, or competition are also major threats to the population of its species. Physical alterations, habitat destruction and degradation, climatic changes and challenges coupled with the over-exploitation, and the introduction of non-native species have worsen the situation and consequently contributed to the decline in numbers and distribution of *Eulophia* genus [65], it has further resulted into an increasing number of its species in IUCN Red Lists.

Eulophia shows vegetative propagation by tubers and sexual reproduction by seeds. They can also be propagated by stem and rhizome cuttings. These vegetative methods of propagation are helpful because they produce exact clones unlike sexual reproduction. However, asexual reproduction is exceptionally slow and usually produces 2–4 plants in a year. This difficulty in natural population drives many medicinal as well as horticultural orchids including few *Eulophia* species to be endangered and some are even reached the extinction [67]. Seed germination is another method of propagation but this is not genetically identical to the parent. Besides, *Eulophia* flowers only after they are 4–5 years old. In nature, only around 5% flowers get pollinated, leaving the ovules of 95% flowers as unfertilized, a major hindrance in capsule formation. Further, though each capsule bears many seeds but only 5% seeds germinate in their natural environment besides their slow growth characteristics [57]. Suppressed endosperm and lack of nutrients also hamper the germination and early vegetative growth, and thus these plants require highly specialized symbiotic fungal associations towards fulfilling their nutritional requirements [57, 72].

In situ conservation of orchids under threat is thus a difficult task and necessitate viable *ex situ* alternatives for their multiplication and conservation [48]. Plant tissue culture-based technologies have emerged in recent years as sound platforms for mass multiplication and germplasm conservation of orchids, besides *in vitro* production platforms for high-value secondary metabolites. Plant tissue culture techniques have made significant contributions in multiplication of many threatened orchid species. As per the recent reports, seeds can be germinated asymbiotically without the association of any fungal partner in a nutrient medium for best results [67].

In initial attempts, the micropropagation of *E. dabia* was successfully achieved described by Sharma and Vij [78] followed by the micropropagation of *E. hormusjii*

Table 2 List of plant species from *Eulophia* genus, their ethnopharmacological uses, and evidence-based pharmacological activities of their crude extracts and/or pure molecules

Eulophia species	Plant part	Extract/Pure molecule	Ethnopharmacological/Pharmacological activities	References
A. Ethnopharmacological Uses				
Eulophia campestris Wall.	Tuber	Fresh Juice	Gastro-intestinal disorders (diarrhea, dysentery, stomach pain, laxative), appetizer	Chanda et al. [11]
Eulophia campestris Wall.	Rhizome	–	Tonic, stomach problem, aphrodisiac, cough, cold	Medhi and Chakrabarti [55]
Eulophia campestris Wall.	Tuber	Mucilage	Worm infestation, scrofula	Singh and Duggal [83]
Eulophia dabia (D. Don.) Hochr.]	Tuber	–	Cough, cold	Joshi et al. [39]
Eulophia epidendrea (JKoen) Schltr.	Tuber	Paste	Ease the pain due to milk clotting	Rajendran et al. [75]
Eulophia epidendrea (JKoen) Schltr.	Tuber	–	Tumor, diarrhea	Patil and Mahajan [69]
Eulophia epidendrea (JKoen) Schltr.	Tuber	–	Appetizer, anthelmintic, aphrodisiac, stomachic, worm infestation, blood purifier, heart troubles	Maridass et al. [52]
Eulophia epidendrea (JKoen) Schltr.	Tuber			Narkhede et al. [64]
Eulophia graminae Lindl.	Tuber	Extract	Earache	Karuppusamy [41]
Eulophia herbacea Lindl	Tuber	Extract	Reduces the liver swelling	Ahirrao et al. [2]
Eulophia herbacea Lindl	Tuber	Roasted	Increases the sperm count	Ahirrao et al. [2]
Eulophia herbacea Lindl	Tuber	Crushed; fried in mustered oil	Rheumatism	Sikarwar et al. [82]

(continued)

Table 2 (continued)

Eulophia species	Plant part	Extract/Pure molecule	Ethnopharmacological/Pharmacological activities	References
Eulophia herbacea Lindl	Tuber	–	Bellyache	Dey and Nath [19]
Eulophia herbacea Lindl	Tuber	Paste	Treatment for pimples	Patil and Patil [70]
Eulophia herbacea Lindl	Seed	Powder	Weakness (Fatigue)	Tayade and Patil [87]
Eulophia herbacea Lindl			Hypolipidemic, antidiabetic and anti-oxidant activity	Tatiya et al. [85]
Eulophia herbacea Lindl			Anabolic and Reproductive Activity	Patil et al. [71]
Eulophia macrobulbon	Root	Extract	inflammatory and antioxidant effect and anticancerogenic potential exerted	Schuster et al. [77]
Eulophia mannii (Rchb. f.) Hook. f. (EM)	Tuber	Extract	Antioxidant activity	Narkhede et al. [64]
Eulophia nuda Lindl.	Tuber	Extract	Worm infestation and scrofula	Singh and Duggal [83]
Eulophia nuda Lindl.	Tuber	Raw material without processing	Skin rash, acidity, piles, anorexia, anthrax, stomach complaints	Shriram et al. [79]
Eulophia nuda Lindl.	Raw tuber	Raw material without processing	Rheumatoid arthritis	Mali and Bhadane [49]
Eulophia nuda Lindl.	Tubers	Extract	Anticancer, anti-asthmatic, anti-bronchitis	Jain et al. [34]
Eulophia nuda Lindl.	Whole plant	Paste	Boils, abscesses	Hossain [27]
Eulophia nuda Lindl.	Root	Juice	Treatment for snakebite	Sikarwar et al. [82]
Eulophia nuda Lindl.	Tubers	Extract	Anti-inflammatory activity	Abhyankar and Upadhyay [1]

Eulophia nuda Lindl.	Whole tuber	Raw material without processing	Abdominal pain due to non-menstruation, Spermatorrhea, Leucorrhea	Das et al. [14]
Eulophia nuda Lindl.	Tuber	Extract	Vermifuge, blood purifier	Sastri [76]
Eulophia nuda Lindl.	–	–	Appetizer, tonic	Patil and Mahajan [69]
Eulophia nuda Lindl.	–	–	Rheumatoid arthritis	Mali and Bhadane [49]
Eulophia nuda Lindl.	–	–	Anthelmintic, bronchitis	Merchant et al. [56]
Eulophia nuda Lindl.	–	–	Snake bite	Sikarwar et al. [82], Abhyankar and Upadhyay [1]
Eulophia nuda Lindl.	–	–	Worm infestation and scrofula	Singh and Duggal [83]
Eulophia nuda Lindl.	–	–	Gastric problems	Kapale [40]
Eulophia nuda Lindl.	Tuber	Aquous methanol extract	DNA damage protecting activity	Kumar et al. [47]
Eulophia nuda Lindl.	–	–	Anti-glycation effect	Yadav et al. [96]
Eulophia nuda Lindl.	–	–	Aphrodisiac activities of Salep in adult Swiss male mice	Jagdale et al. [30]
Eulophia nuda Lindl.	–	–	Anti-inflammatory activity	Tuchinda et al. [92]
Eulophia nuda Lindl.	–	Acetone extract	Antibacterial activity against *Escherichia coli, Pseudomonas aeruginosa, Staphylococcus aureus*	Nagulwar et al. [59]
Eulophia nuda Lindl.	–	–	Antifungal Activity against *Aspergillus niger, A. flavus, Candida albicans*	Nagulwar et al. [59]
Eulophia nuda Lindl.	–	–	Hepatoprotective activity (Acute ccl4 induced hepatotoxicity in rats)	Nagulwar et al. [59]
Eulophia nuda Lindl.	Tuber	Ethanol extract	Anti-mycobacterial activity	Gupta et al. [24]
Eulophia ochreata Lindl	Tuber	Salep	Treatment of sexual impotency and male sterility	Hossain [27], Das et al. [14]
Eulophia ochreata Lindl	Tuber	Paste	Asthma, acute bronchitis	Jain et al. [32], Jain et al. [33]
Eulophia ochreata Lindl	Tuber	Powder	Increase the stamina for physical activities	Jagtap et al. [29]

(continued)

Table 2 (continued)

Eulophia species	Plant part	Extract/Pure molecule	Ethnopharmacological/Pharmacological activities	References
Eulophia ochreata Lindl	Tuber	Extract	For restoring general health, strength, vigor	Jagtap et al. [31]
Eulophia ochreata Lindl	Tuber	Decoction	Antinode in snakebite, cure leukemia	Mali and Bhadane [49]
Eulophia pratensis Lindl.	Tuber	Paste	To remove scrofulous gland in the neck	Hossain [27]
Eulophia ramentaceae Lindl Ex. Wight	Tuber	–	Impotency related problems	Bhagaonkar and Kadam [5]
B. Pure Molecules and/or their Pharmacological Activities				
Eulophia ochreata Lindl	Tuber	9, 10-Dihydro-2, 5-Dimethoxyphenanthrene-1, 7-diol	Inhibits inflammatory signaling mediated by Toll-like receptors in human THP1 cells	Upadhyay et al. [93] Datla et al. [15]
Eulophia ochreata Lindl	Tuber	9, 10-Dihydro-2, 5-Dimethoxyphenanthrene-1, 7-diol 5, 7-Dimethoxyphenanthrene-2, 6-diol	Antioxidant activity	Kshirsagar et al. [46]
Eulophia nuda Lindl.	Tuber	9, 10-Dihydro-2, 5-Dimethoxyphenanthrene-1, 7-diol	Anti-proliferative activity against Human cancer cells (MCF 7)	Shriram et al. [79]

using rhizome segments [94]. McAlister and Van Staden [54] reported *in vitro* seed germination of *E. cucullata*, *E. streptopetala*, and *E. petersii* on MS medium [58], medium supplemented with 3% sucrose and 0.01% myoinositol. Chang et al. [12] reported *in vitro* seed germination of the orchid *E. graminea*, a species native to Taiwan, and was successfully developed into rhizomes. The *in vitro* plants could flower and produce the fruits (capsule) through the *autogamous* mating system. Obtained seeds were sown *in vitro* and in this manner four generations were cultured over a period of 4 year [12].

Panwar et al. [68] reported a standardized method for *in vitro* propagation and tuberization of *E. nuda*, using tuber explants on MS medium supplemented with 6-benzyladenine (BA) and additives (ascorbic acid, adenine sulfate, arginine, and citric acid). Multiplication of shoots was achieved successfully up to three subcultures [68]. The rate of *in vitro* tuber formation was comparably higher than in the natural conditions. Rooting of shoots was achieved and the *in vitro* generated plantlets were acclimatized to the greenhouse conditions [68]. It was thus an efficient method for regeneration, multiplication, and production of large number of tuberous plantlets of E. nuda. Similarly, an optimized protocol for micropropagation of *E. nuda* from axillary bud segments was reported by Shroti and Upadhyay [81] using MS medium supplemented 2,4-D. The explants developed protocorm like bodies (PLBs) within 6–8 weeks of inoculation on the growth medium, and the subculture PLBs on basal MS medium differentiated plantlets into tubers.

A method for asymbiotic seed germination, seedling development, and establishment of *in vitro* generated plantlets of *E. nuda* was reported by our group [60]. The authors optimized several parameters such as most suitable seed age, culture medium compositions, type and concentrations of phytohormones, additives and/or supplements, and reported a standardized and reproducible method for asymbiotic seed germination and plantlet regeneration of *E. nuda* (Figs. 1 and 2; [60]). These findings hold significance and may help in mass multiplication and conservation of this and other closely related species. Further, ploidy analysis of *in vitro* raised plants, first work of this kind in this plant, revealed that these plantlets kept their genome stable and true-to-typeness with mother plants (Fig. 3), a key consideration for conservation of any species without genetic alteration.

We conducted a study on another species of the genus, *E. ochreata* [80], and described a first report on direct as well as indirect *in vitro* plant regeneration of this important orchid. A number of parameters comprising type of explant, artificial growth medium types and compositions, and phytohormones were standardized for optimal plant regeneration. Among the explants, the PLBs were found to be the best choice for induction, proliferation of shoots, as well as for callus production [80]. The MS medium supplemented with 2.5 mg L^{-1} BAP and 1.0 mg L^{-1} Kin proved best for shoot multiplication with synchronized growth. The number of shoots was further enhanced with subcultures on same media composition; achieving up to 40 shoots per explant after 3 such cycles of 30 days each. The shoots were successfully rooted *in vitro* on ¼ strength MS fortified with activated charcoal and additives; rooted plantlets were acclimatized greenhouse conditions. The similar

Fig. 1 Asymbiotic germination of *Eulophia nuda* seeds obtained from capsules. (**a**) Capsules of plants growing in wild conditions, (**b**) Micrograph of seed with embryo (arrow) at pre-germination stage, (**c**) Scanning electron micrographs of *E. nuda* seed (arrow), (**d**) Scanning electron micrograph showing testa-breaking leading to protocorm formation (arrows), (**e**) Shoot emergence from PLBs (arrow). Reprinted with permission from Springer, Nanekar et al. [60]. https://doi.org/10.1007/s40011-014-0353-4; https://link.springer.com/article/10.1007/s40011-014-0353-4

ploidy levels of *in vitro* regenerated plants in comparison with the field-grown mother plants confirmed the true-to-typeness of tissue culture generated plantlets.

Similarly, *in vitro* germination of seeds, differentiation of embryos, as well as *ex vitro* seedling production was reported from *in vitro* rhizome-like bodies of *E. promensis* by Hossain [28]. Through the subcultures, profuse proliferation of protocorms was observed which later developed rhizome-like bodies (RLBs). These RLBs further produced in between one to three seedlings per RLB. Authors thus developed a fast and cost-effective method for the micropropagation of *E. promensis* which can potentially be used for other orchids as well for their propagation and conservation. Symbiotic germination of *Eulophia* species was established successfully by Ochora et al. [66] using oats medium and germination medium supplemented with banana homogenate and activated charcoal. Further, *in vitro* asymbiotic germination and clonal propagation of *E. cullenii* was reported by

Fig. 2 Asymbiotic seed germination and plant developmental stages of *Eulophia nuda*, (**a**) Seed germination and PLB formation after 60 days of inoculation on BM1 + CW, (**b**) Protocorm-derived seedling formation on BM1 + CW containing 1 mg/L each of NAA and BAP after 120 days of seed inoculation, (**c** and **d**) Stages of shoot multiplication on BM1 + CW containing 2.5 mg/L BAP and 1.5 mg/L Kin, 60 days after seedling inoculation and further shoot proliferation after two subcultures of 30 days each, (**e**) *In vitro* rooting of microshoots on MS medium supplemented with 200 mg/L activated charcoal and 2 mg/L IBA, (**f**) Acclimatized plants of *E. nuda* in greenhouse, inset: tuber of *E. nuda*. Bar = 1 cm. Reprinted with permission from Springer, Nanekar et al. [60]. https://doi.org/10.1007/s40011-014-0353-4; https://link.springer.com/article/10.1007/s40011-014-0353-4

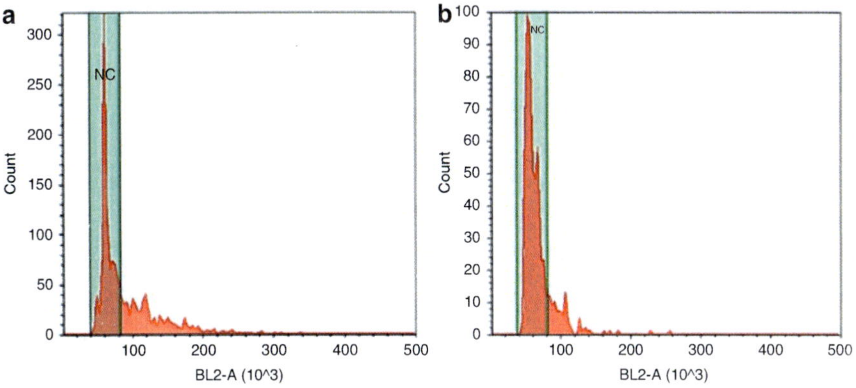

Fig. 3 Histograms of flow cytometric analysis of field-grown (**a**) and *in vitro* raised plants (**b**) of *E. nuda*. Reprinted with permission from Springer, Nanekar et al. [60] https://doi.org/10.1007/s40011-014-0353-4; https://link.springer.com/article/10.1007/s40011-014-0353-4

Decruse et al. [17]. These authors like previous reports used organic supplements like coconut water, peptone, yeast extract, and casein hydrolysate for enhancing the protocorm growth followed by shoot development and then linear mini-rhizomes formation. These findings thus hold significance from eco-restoration perspectives of orchids as well.

In vitro asymbiotic seed germination and plantlet development in *E. nuda* on Knudson-C medium was also reported by Dawande and Gurav [16]. Symbiotic seed culture of *E. alta* was well attempted by Johnson [37] using fungal isolates collected from its roots, and authors found that seedlings co-cultured with fungal isolate (Ealt-396) grew more rapidly than the asymbiotic seedlings. Authors advocated that the symbiotically grown seedlings to be more appropriate for reintroduction to natural areas than their asymbiotic counterparts.

6 Conclusion

Orchids represent highly evolved and valuable plants, extensively utilized for ornamental and therapeutic purposes. They have remarkable ethnopharmacological applications in traditional medicinal systems and many of these claims have been scientifically validated through pharmacological assessments of crude extracts as well as pure molecules from these plants. *Eulophia* represents a diverse group of orchids with tremendous potentials, commercially, ecologically, and therapeutically; however, owing to these potentials many of the species of this genus are over-exploited and thus belong to threatened taxa. Considering the low germination and survival rate in natural conditions along with requirement for specific fungal associations for germination and growth, *in situ* conservation approaches have met with limited success. Plant tissue culture technologies have emerged as potent means for large-scale production and conservation of important members of *Eulophia* genus. There are some important reports in recent years describing micropropagation of a number of species of *Eulophia*; however, most the species are remained to be explored, and thus more of such investigations are required. Besides, though many of the *Eulophia* species are known to biosynthesize potent bioactive phytomolecules, there are very few or no reports on *in vitro* production of these phytomolecules using plant cell and tissue cultures of these plants, future investigations need to be conducted to fill this big gap.

Acknowledgments VS wish to acknowledge the financial assistance from the University Grants Commission (UGC), Government of India [No.: F 39-426/2010 (SR)].

References

1. Abhyankar RK, Upadhyay R (2011) Ethnomedicinal studies of tubers of Hoshangabad M.P. Bull Environ Pharmcol Life Sci 1:57–59
2. Ahirrao YA, Patil MV, Patil DA (2008) On identities and ethnomedicinal claims of some common botanicals sold by vendors in North Maharashtra, India. Indian J Tradit Knowl 2:9–13

3. Alam MK, Rashid MH, Hossain MS, Salam MA, Rouf MA (2002) In vitro seed propagation of Dendrobium (Dendrobium transparens) orchid as influenced by different Media. Biotechnol 1:111–115
4. Arenmongla T, Deb CR (2012) Germination of immature embryos and multiplication of Malaxis acuminate D. Don, an endangered therapeutically important orchid, by asymbiotic culture in vitro. Indian J Biotechnol 11:464–469
5. Bhagaonkar PY, Kadam VN (2006) Ethnopharmacology of Bajara tribe of Umarkhed taluka, district Yavatmal, Maharashtra for reproductive disorders. Indian J Tradit Knowl 5:336–341
6. Bhandari SR, Kapadi AH (1983) A 9, 10-dihydrophenanthrene from tubers of *Eulophia nuda*. Phytochemistry 22:747–748
7. Bhandari SR, Kapadi AH, Majumder PL, Joardar M, Shoolery JN (1985) Nudol, a phenanthrene of the orchids *Eulophia nuda*, *Eria carinata* and *Eria stricta*. Phytochemistry 24:801–804
8. Bhatt D, Jethva K, Zaveri M (2020) Phytopharmacognostical study of tubers of Eulophia nuda Lindl. Int J Pharm Sci Res 49:3483–3488
9. Bhurat MR, Kawatikwar PS, Sanghavi RS, Patil PP, Salunke PA, Kapure SV (2012) Evaluation of Eulophia herbacea tubers mucilage as an innovative suspending agent. J Pharm Res 5:321–323
10. Bulpitt CJ (2005) The uses and misuses of orchids in medicine. QJM: An Int J Med 98:625–631
11. Chanda R, Mohanty JP, Bhuyan NR, Kar PK, Nath LK (2007) Medicinal plants used against gastrointestinal tract disorders by the traditional healers of Sikkim Himalayas. Indian J Tradit Knowl 6:606–610
12. Chang C, Hu WH, Chen YC, Su YL, Chiu YT (2010) *In vitro* flowering and mating system of *Eulophia graminea* Lindl. Bot Stud 51:357–362
13. Chase MW, Cameron KM, Freudenstein JV, Pridgeon AM, Salazar G, Van den Berg C, Schuiteman A (2008) An updated classification of Orchidaceae. Bot J Linn Soc 177(2):151–174
14. Das S, Dash SK, Padhy SN (2003) Ethno-medicinal informations from Orissa State, India, A review. J Hum Ecol 14:165–227
15. Datla P, Kalluri MD, Basha K, Bellary A, Kshirsagar R, Kanekar Y et al (2010) 9,10-Dihydro-2,5- dimethoxyphenanthrene-1,7-diol, from Eulophia ochreata, inhibits inflammatory signalling mediated by Toll like receptors. Br J Pharmacol 160:1158–1170
16. Dawande V, Gurav R (2015) *In-vitro* seed germination and effect of growth regulators on subsequent development of protocorms of *Eulophia nuda* Lindl. Int J Appl Sci Biotechnol 3: 243–247
17. Decruse SW, Reny N, Shylajakumari S, Krishnan PN (2013) *In vitro* propagation and field establishment of *Eulophia cullenii* (Wight) Bl., a critically endangered orchid of Western Ghats, India through culture of seeds and axenic seedling-derived rhizomes. In Vitro Cell Dev Biol-Plant 49:520–528
18. Devkar S, Jagtap SD, Kale Y, Kasote D (2009) Antibacterial activity of Eulophia ochreata L. tubers. J Herb Med Toxicol 3:31–33
19. Dey A, Nath De J (2010) A survey of ethnomedicinal plants used by the tribals of Ajoydha hill region, Purulia District, India. Am-Eurasian J Sustain Agric 4:280–290
20. Dodson CH (2020) Orchid. Encyclopædia Britannica. Available at https://www.britannica.com/plant/orchid. Accessed 20 Aug 2020
21. Fay MF (2018) Orchid conservation: how can we meet the challenges in the twenty-first century? Bot Stud 59:16
22. Ghule BV, Darwhekar GD, Jain DK, Yeole PG (2006) Evaluation of binding properties of Eulophia campestris Wall. Mucilage. Indian J Pharm Sci 68:566–569
23. Gledhill D (2002) The names of plants, 3rd edn. Cambridge University Press. https://doi.org/10.1017/CBO9780511754951
24. Gupta VK, Kaushik A, Chauhan DS, Ahirwar RK, Sharma S, Bisht D (2018) Anti-mycobacterial activity of some medicinal plants used traditionally by tribes from Madhya Pradesh, India for treating tuberculosis related symptoms. J Ethnopharmacol 227:113–120
25. Gutierrez RMP (2010) Orchids: A review of uses in traditional medicine, its phytochemistry and pharmacology. J Med Plants Res 4:592–638

26. Hew CS (2001) Ancient Chinese orchid cultivation: a fresh look at an age-old practice. Scientia Horticulturae 87(1–2):1–10
27. Hossain MM (2011) Therapeutic orchids: traditional uses and recent advances – an overview. Fitoterapia 82(2):102–140
28. Hossain MM (2015) *Ex vitro* seedling development from *in vitro* rhizome-like bodies in *Eulophia promensis* Lindl.: a new technique for orchid propagation. J Bot:207694. https://doi.org/10.1155/2015/207694
29. Jagtap SD, Deokule SS, Bhosle SV (2008) Ethanobotanical uses of endemic and RET plants by Pawra tribe of Nandurbar district, Maharashtra. Indian J Tradit Knowl 7:311–315
30. Jagdale SP, Shimpi S, Chachad D (2009) Pharmacological studies of Salep. Herb Med Toxicol 3:153–156
31. Jagtap SD, Deokule SS, Pawar PK, Harsulkar AM (2009) Traditional ethnomedicinal knowledge confined to the Pawra tribe of Satpura Hills, Maharashtra, India. Ethnobot Lealf 13:98–115
32. Jain A, Katewa SS, Galav PK, Sharma P (2005a) Medicinal plant diversity of wildlife sanctuary, Rajasthan, India. J Ethnopharmacol 102:143–157
33. Jain A, Katewa SS, Galav PK (2005b) Some phytotherapeutic claims by tribals of southern Rajasthan. Indian J Tradit Knowl 4:291–297
34. Jain B, Kumane SC, Bhattacharya S (2006) Medicinal flora of Madhya Pradesh and Chhattisgarh – a review. Indian J Tradit Knowl 5:237–242
35. Jalal JS, Kumar P, Pangtey YPS (2008) Ethnomedicinal Orchids of Uttarakhand, Western Himalaya. Ethnobot Leafl 1:164
36. Jansakul C, Yorsin S, Naphatthalung J, Tachanaparugse K, Changwichai K, Chootip K, Ingkaninan K (2019) Relaxant mechanism of Eulophia macrobulbon ethanolic extract and 1-(4'-hydroxybenzyl)-4, 8-dimethoxyphenanthrene-2, 7-diol on human corpus cavernosum. Funct Foods Health Dis 9:328–340
37. Johnson TR, Stewart SL, Dutra D, Kane ME, Richardson L (2007) Asymbiotic and symbiotic seed germination of *Eulophia alta* (Orchidaceae)-preliminary evidence for the symbiotic culture advantage. Plant Cell Tissue Organ Cult 90:313–323
38. Jones DL (2006) A complete guide to native orchids of Australia including the island territories. Frenchs Forest, N.S.W, New Holland, p 358
39. Joshi GC, Tewari LM, Lohani N, Upreti K, Jalal JS, Tewari G (2009) Diversity of orchids in Uttarakhand and their conservation strategy with special reference to their medicinal importance. Rep Opin 1:47–52
40. Kapale R (2012) Ethnobotany of Baiga tribals with reference to utilization of forest resources in Achanakmar biosphere reserve (India). Bull Env Pharm Life Sci 1(6):73–76
41. Karuppusamy S (2007) Medicinal plants used by Paliyan tribes of Sirumalai hills of southern India. Nat Prod Radiance 6:436–442
42. Kayalvizhi K, Divya K, Sankari A (2020) Medicinal orchids – an overview. Biotica Res Today 2:1084–1087
43. Kirtikar KR, Basu BD (1918) Indian medicinal plants. Springer, Allahabad, pp 1242–1243
44. Kong JM, Goh NK, Chia LS, Chia TF (2003) Recent advances in traditional plant drugs and orchids. Acta Pharmacol Sin 24:7–21
45. Kovács A, Vasas A, Hohmann J (2008) Natural phenanthrenes and their biological activity. Phytochemistry 69:1084–1110
46. Kshirsagar RD, Kanekar YB, Jagtap SD, Upadhyay SN, Rao R, Bhujbal SP et al (2010) Phenanthrenes of Eulophia ochreata Lindl. Int J Green Pharm 4:147–152
47. Kumar V, Lemos M, Sharma M, Shriram V (2013) Antioxidant and DNA damage protecting activities of *Eulophia nuda* Lindl. Free Radicals Antioxid 3:55–60
48. Mahendran G, Muniappan V, Ashwini M, Muthukumar T, NarmathaBai V (2013) Asymbiotic seed germination of *Cymbidium bicolor* Lindl. (Orchidaceae) and the influence of mycorrhizal fungus on seedling development. Acta Physiol Plant 35:829–840
49. Mali PY, Bhadane VV (2008) Some rare plants of ethnomedicinal properties from Jalgaon district of Maharashtra. Int J Green Pharm 2:76–78
50. Maridass M (2011) Anti diarrhoeal activity of rare orchid Eulophia epidendraea (Retz.) Fisher. Nat Pharm Technol 1:5–10

51. Maridass M, Ramesh U (2010) Investigation of phytochemical constituents from *Eulophia epidendraea*. Int J Biol Technol 1:1–23
52. Maridass M, Victor B, Ramesh U (2005) Ethnobotanical information of Eulophia epidendraea (Retz) Fischer (Orchidaceae) in the Kambli Malaikovil Forest, Tirunelveli district, Tamil Nadu. J Bombay Nat Hist Soc 102:255
53. Maridass M, Hussain MI, Raju G (2008) Phytochemical survey of orchids in the Tirunelveli hills of South India. Ethnobot Leafl 12:705–712
54. McAlister BG, Van Staden J (1998) *In vitro* culture of *Eulophia* species. S Afr J Bot 64:264–266
55. Medhi RP, Chakrabarti S (2009) Traditional knowledge of NE people on conservation of wild orchids. J Tradit Knowl 8:11–16
56. Merchant JR, Shah RS, Hirwe SN (1962) Chemical investigation of Eulophia nuda Lindl. Curr Sci 31:95
57. Mohanty P, Paul S, Das MC, Kumaria S, Tandon P (2012) A simple and efficient protocol for the mass propagation of *Cymbidium mastersii*: an ornamental orchid of Northeast India. AoB PLANTS pls023. https://doi.org/10.1093/aobpla/pls023
58. Murashige T, Skoog F (1962) A revised medium for rapid growth and bio assays with tobacco tissue cultures. Physiol Plant 15:473–497
59. Nagulwar VP, Nandgave M, Mahajan MS, Deshpande SA (2017) Phytochemical screening and evaluation of pharmacological activities of *Eulophia nuda* Lind. Tuber extracts. Int J Pharm Sci Res 8:3516–3523
60. Nanekar V, Shriram V, Kumar V, Kishor PK (2014) Asymbiotic in vitro seed germination and seedling development of Eulophia nuda Lindl., an endangered medicinal orchid. Proc Natl Acad Sci India Sect B: Biol Sci 84:837–846
61. Nanekar V, Shriram V, Khare T, Kumar V (2020) Nrf2/HO-1 mediated antioxidant activities, cytotoxicity analysis and LCESI/MS profiling of Eulophia nuda L. Nat Prod J 10:69–79
62. Narkhede AN, Jagtap SD (2015) Screening of Amarkand species with respect to their poly-phenolic content and free radical quenching potential. Int J Pharm Bio Sci 6:1123–1133
63. Narkhede AN, Nirmal PS, Tupe R, Kulkarni OP, Harsulkar AM, Jagtap SD (2012) In vitro antioxidant, antiglycation and alpha amylase inhibitory potential of Eulophia ochreata L. J Pharm Res 5:2532–2537
64. Narkhede AN, Jagtap SD, Nirmal PS, Giramkar SA, Nagarkar BE, Kulkarni OP, Harsulkar AM (2016) Anti-fatigue effect of *Amarkand* on endurance exercise capacity in rats. BMC Complement Altern Med 16:23–27
65. Norah MM (2014) Conservation status of *Eulophia* species (orchidaceae) in Uganda, East Africa. Asian J Conserv Biol 3:94–100
66. Ochora J, Stock WD, Linder HP, Newton LE (2001) Symbiotic seed germination in twelve Kenyan orchid species. Syst Geogr Plants 71:585–596
67. Pant B (2013) Medicinal orchids and their uses: tissue culture a potential alternative for conservation. African J Plant Sci 7:448–467
68. Panwar D, Ram K, Shekhawat NS (2012) *In vitro* propagation of *Eulophia nuda* Lindl., an endangered orchid. Sci Hortic 139:46–52
69. Patil MC, Mahajan RT (2013) Ethnobotanical potential of Eulophia species for their possible biological activity. Int J Pharm Sci Rev Res 21:297–307
70. Patil SL, Patil DL (2007) Ethnomedicinal plants of Dhule District, Maharashtra. Nat Prod Radiance 6:148–151
71. Patil MC, Patil CJ, Mahajan RT (2017) Evaluation of the anabolic and reproductive activity of *Eulophia herbacea* Lindl. in male rats. Br J Med Health Res 4:13–28
72. Paul S, Kumaria S, Tandon P (2012) An effective nutrient medium for asymbiotic seed germination and large-scale in vitro regeneration of *Dendrobium hookerianum*, a threatened orchid of northeast India. AoB Plants pls032. https://doi.org/10.1093/aobpla/plr032
73. Preedapirom W, Changwichit K, Srisawang P, Ingkaninan K, Taepavarapruk P (2018) Aphrodisiac activity of *Eulophia macrobulbon* extract on erectile dysfunction in male aged rats. Biomed Res Int:6217029. https://doi.org/10.1155/2018/6217029
74. Pridgeon AM, Cribb PJ, Chase MW, Rasmussen F (eds) (2009) Genera Orchidacearum. Epidendroideae (part 2). Oxford University Press, Oxford

75. Rajendran A, Rao NR, Kumar KR, Henry AN (1997) Some medicinal orchids of Southern India. Anc Sci Life 17:1–4
76. Sastri BN (1952) The wealth of India, A dictionary of Indian raw materials and industrial products: raw materials, vol 3. Council of Scientific Industrial Research, New Delhi, p 221
77. Schuster R, Zeindl L, Holzer W, Khumpirapang N, Okonogi S, Viernstein H, Mueller M (2017) Eulophia macrobulbon – an orchid with significant anti-inflammatory and antioxidant effect and anticancerogenic potential exerted by its root extract. Phytomedicine 24:157–165
78. Sharma M, Vij SR (1986) Micropropagation of Eulophia dabia (D. Don) Hochr. Orchid News, 2(11)
79. Shriram V, Kumar V, Kishor PB, Suryawanshi SB, Upadhyay AK, Bhat MK (2010) Cytotoxic activity of 9,10-dihydro-2,5-dimethoxyphenanthrene-1,7-diol from *Eulophia nuda* against human cancer cells. J Ethnopharmacol 128:251–253
80. Shriram V, Nanekar V, Kumar V, Kishor PB (2014) *In vitro* regeneration and ploidy level analysis of *Eulophia ochreata* Lindl. Indian J Exp Biol 52:1112–1121
81. Shroti RK, Upadhyay R (2014) In vitro micropropagation of Eulophia nuda Lind an endangered terrestrial orchid through PLB (protocorm like bodies). Int J Plant Animal Environ Sci 4:172–176
82. Sikarwar RL, Pathak B, Jaiswal A (2008) Some unique ethnomedicinal perceptions of tribal communities of Chitrakoot, Madhya Pradesh. Indian J Tradit Knowl 7:613–617
83. Singh A, Duggal S (2009) Medicinal orchids: an overview. Ethnobot Leaflets 13:351–363
84. Swarts ND, Dixon KW (2009) Terrestrial orchid conservation in the age of extinction. Ann Bot 104:543–556
85. Tatiya AU, Puranik PM, Surana SJ, Patil YS, Mutha RE (2013) Evaluation of hypolipidemic, antidiabetic and anti-oxidant activity of *Eulophia herbacea* tubers. Bangladesh J Pharmacol 8: 269–275
86. Tatiya AU, Bari N, Surana SJ, Kalaskar MG (2014) Effect of bioassay guided isolation of 1-phenanthrene carboxylic acid from Eulophia herbacea Lindl. Tubers on human cancer cell lines. Res J Phytochem 8:155–161
87. Tayade SK, Patil DA (2005) Hitherto untapped plantlore from Nandurbar district (Maharashtra). Nat Prod Radiance 4:46–50
88. Temkitthawon P, Changwichit K, Khorana N, Viyoch J, Suwanborirux K, Ingkaninan K (2017) Phenanthrenes from *Eulophia macrobulbon* as novel phosphodiesterase-5 inhibitors. Nat Prod Commun 12:1934578X1701200121
89. Teoh ES (2016) Medicinal orchids of Asia. Springer International Publishers, Basel
90. Thomas S (1998) Note on nomenclature and typification of Eulophia hians Spreng. (Orchidaceae). Kew Bull 1005–1007
91. Tuchinda P, Udchachon J, Khumtaveeporn K, Taylor WC, Engelhardt LM, White AH (1988) Phenanthrenes of Eulophia nuda. Phytochemistry 27:3267–3271
92. Tuchinda P, Udchachon J, Khumtayeeporn K, Taylor WC (1989) Benzylated phenanthrenes from Eulophi nuda. Phytochemistry 28:2463–2466
93. Upadhyay S, Kanetkar Y, Kshirsagar R, Rajgopal V, Datla P, Bellary A, Singh SP (2009) Toll like receptor (TLR) signaling antagonist. US Patent 20090215908
94. Vij SP, Sood A, Pathak P (1989) On the utility of rhizome segments in micropropagating *Eulophia hormusjii* Duth. J Orchid Soc India 3:41–45
95. Wisutthathum S, Demougeot C, Totoson P, Adthapanyawanich K, Ingkaninan K, Temkitthawon P, Chootip K (2018) Eulophia macrobulbon extract relaxes rat isolated pulmonary artery and protects against monocrotaline-induced pulmonary arterial hypertension. Phytomedicine 50:157–165
96. Yadav DP, Chhipa RC, Singh B (2012) Anti-glycation effect of whitton root (eulophia nuda) in-vitro condition. Int J Pharm Sci Res 3:3502
97. Yan-Qiu Z, Rong-Yi Z, Rui M, Zhi-Qiong T, Yu-Jie W (2009) Endophytic fungi of roots of Eulophia flava. Southwest China J Agric Sci 22:675–680

Cyrtopodium glutiniferum, an Example of Orchid Used in Folk Medicine: Phytochemical and Biological Aspects

21

Carlos Fernando Araujo-Lima, Israel Felzenszwalb, and Andrea Furtado Macedo

Contents

Abstract

The center of *Cyrtopodium* spp. diversity is in a typical Brazilian savanna-like formation. *C. glutiniferum* has thick big exposed pseudobulbs and yellow high paniculate inflorescences. This species has been taxonomically confused with many other correlates, which also have yellow flowers. *C. glutiniferum* is located in hotspots in Brazil and is relevant ethnopharmacologically. The plant's bulb traditional uses include the treatment of abscesses and wound healing. In previous work, phenanthrene was the most abundant subclass in metabolomic analysis, however the most abundant molecules were dihydroformononetin, caffeic acid

Carlos Fernando Araujo-Lima and Israel Felzenszwalb contributed equally with all other contributors.

C. F. Araujo-Lima · I. Felzenszwalb
Laboratory of Environmental Mutagenesis, Department of Biophysics and Biometry, Rio de Janeiro State University, Rio de Janeiro, Brazil

Roberto Alcantara Gomes Institute of Biology, Universidade do Estado do Rio de Janeiro, UERJ, Rio de Janeiro, Brazil

A. F. Macedo (✉)
Integrated Laboratory of Plant Biology, Department of Botany, Institute of Biosciences, Federal University of Rio de Janeiro State, UNIRIO, Rio de Janeiro, Brazil
e-mail: andrea.macedo@unirio.br

J.-M. Mérillon, H. Kodja (eds.), *Orchids Phytochemistry, Biology and Horticulture*, Reference Series in Phytochemistry, https://doi.org/10.1007/978-3-030-38392-3_33

4-O-glucoside (glucocaffeic acid), and arbutin. As *C. glutiniferum* is traditionally used to treat skin lesions, our research group has verified the positive anti-inflammatory and antiproliferative effects of the extract against activated mononuclear cells. However, probably the therapeutic effect of this orchid is not directly related to the elimination of the pathogen but to the control of tissue lesions mediated by acute inflammation, acting against the cells of the immune response, involved in the occurrence of abscesses.

Keywords

Ethnopharmacology · Ethnobotany · *Cyrtopodium* · Antimicrobial · Antioxidant · Anti-inflammatory · Plant tissue culture

Abbreviations

ATCC	American Type Culture Collection
CFU	Colony-forming unit
FDA	United States Food and Drug Administration Agency
IAA	Indole-3-acetic acid
OA	Oatmeal agar
PLBs	Protocorm-like bodies
UHPLC-MS/MS	Ultra-high-performance liquid chromatography tandem mass spectrometry
UV	Ultraviolet radiation

1 Introduction

The *Cyrtopodium* genus (Epidendroideae: Cymbidieae: Cyrtopodiinae) has approximately 50 species of neotropical scope, from southern Florida to northern Argentina, mostly terrestrial, with few epiphytes [1–6]. Most species are found in Brazil, about 40, followed by Bolivia and Venezuela with 9 species each [7]. Only two species have been registered in the USA (*C. polyphyllum* and *C. punctatum*), however, *C. polyphyllum* has been naturalized [8]. In fact, the center of genus diversity is in the "cerrado," a typical Brazilian savanna-like formation, whose physiognomies include grassy field ("campo limpo"), grass-herb-sub-shrub field ("campo sujo"), and semideciduous xeromorphic medium tall forest ("cerrado" sensustricto) [9]. Despite the great importance of the genus object of this work, it is relevant to state that the classification and, therefore, the identification of *Cyrtopodium* species in Brazil have been revised due to innumerable problems concerning the material deposited in the herbariums. Many holotypes that were deposited consisted of incomplete plant material, mainly without a flower. Due to these problems, epitypes have been used for the correct identification of *Cyrtopodium* species. Terrestrial species with large pseudobulbs and yellow flowers, mainly, have been the subject of a major revision due to their taxonomic complexity.

Species of this group, which were collected: (a) in the northern part of South America, in the Amazon basin, and, probably, along the north and northeast coast of

Brazil are *C. andersonii* (Lamb. Ex Andrews) R. Br. and (b) in southeastern Brazil are *C. glutiniferum* Raddi or *C. cardiochilum* Lindl [9]. The last two species are considered co-specific [10]. Consequently, older articles published, with plant material collected in Brazil, may present identification errors. For this reason, we have chosen to make a more general overview of the genus *Cyrtopodium*, with special attention to *Cyrtopodium glutiniferum*.

2 The Genus *Cyrtopodium* R. Br. (Orchidaceae) and *Cyrtopodium glutiniferum* Raddi: Biological, Cultivation, and Phytochemical Aspects

Brazilian *Cyrtopodium* species occur according to the soil and the surrounding vegetation, most of which are terrestrial and bloom at the end of the dry season and at the beginning of the rainy season [1]. Few studies on pollination have been carried out on this taxon which can occur through optional self-pollination, with rain-assisted autogamy, as well as cross-pollination, by the action of large bees by food deceit or not [6, 11–15]. There are reports that some species produce fewer fruits and seeds due to the low pollination rate [3, 6].

The most common Brazilian biomes occurrence of these species are: (a) "cerrado"; (b) rocky field vegetation ("campo rupestre"), a vegetation with more or less continuous herbaceous stratum intermixed by small shrubs or subshrubs on sandy soil with pebbles or graves; (c) "restinga" vegetation, a lowland coastal plain vegetation on lacustrine and marine sands that occur discontinuously along the Brazilian coast within the Atlantic Forest – which is considered one of the global biodiversity hotspots; and (d) along the coast on rocks [9, 11, 16, 17].

Some *Cyrtopodium* species are endemic to Brazil and are classified as "vulnerable" (*Cyrtopodiumtriste* Rchb. f. and Warm.; *Cyrtopodium palmifrons* Rchb. f. and Warm.), "endangered" (*Cyrtopodium poecilum var. roseum* Bianch. and J. A. N. Bat.; *Cyrtopodium lissochiloides* Hoehne and Schltr.), "critically endangered" (*Cyrtopodium lamellaticallosum* J. A. N. Bat. and Bianch.; *Cyrtopodium latifolium* Bianch. and J. A. N. Bat.; *Cyrtopodium linearifolium* J. A. N. Batista and Bianchetti), and "data deficient" (*Eulophia ruwenzoriensis* Rendle) [12]. Notably, *Cyrtopodium* species from southeastern Brazil, including *C. glutiniferum,* are found growing in the remnants of the Atlantic rain forest and in sandy coastal plain habitats [18–21]. Precisely, populations in lowland areas where land is desirable for real estate development and habitat exploitation are vanishing [22]. Despite the considerable economic importance of their showy flowers and the medicinal value of the pseudobulbs, most species of the genus *Cyrtopodium* are unknown to horticulturists and are rarely seen in cultivation. Large-scale propagation is a prerequisite to meet future pharmaceutical and ornamental requirements, and to prevent eradication of this highly valuable plant [7].

Therefore, due to issues of endemism, ethnopharmacological importance, ornamental value, original habitat destruction, modification and fragmentation, low seed production, and pressure on the preservation of the group, since many species have been overcollected and only exist in areas of environmental preservation [3, 23], studies have been developed to establish cultivation protocols. Among these publications, some stand out and will be commented below.

In vitro asymbiotic seed propagation was carried out for *C. punctatum,* where a higher germination rate was registered in the P723 medium, in the dark and greater seedling development in the same medium, but with a photoperiod of 16 h [3]. Guo et al. [24] established a protocol for the production of *C. paranaense* seedlings, from in vitro root tips cultivation, with the formation of protocorm-like body (PLB). In this work, it was verified that the PLBs were formed from the tips of the roots or the stele of the root of the mother plant, maintaining an initial dependence on its vascular tissue [24]. Picolotto et al. also established a *C. paludicolum* cultivation protocol from root tips [25]. Vogel and Macedo [7] produced the only work on in vitro production of *C. glutiniferum.* In this work it was verified that germination was faster under white and blue light and highest under green light. The three light conditions also induced the development of protocorms. Indole-3-acetic acid (IAA) positively affected protocorm-like bodies (PLBs), shoots, and roots multiplication from protocorms (Fig. 1). Later, Guimarães et al. [26] published work on the symbiotic propagation of *C. glutiniferum* with positive results for germination and growth on oatmeal agar (OA) medium inoculated with the mycorrhizal fungus *Epulorhiza* sp. *C. glutiniferum* seems to have a preference for strains of *Epulorhiza* and that fungus digestion is essential to protocorm development [27]. Similarly, in studies carried out with *C. paludicolum* and *C. saintlegerianum* it was observed that symbiotic seed germination is more beneficial than asymbiotic germination [23, 28]. Environmental restoration models may require basic biotechnology techniques such as in vitro germination, since this technique has an advantage of not producing individuals genetically identical to the matrix, generating variability. This is the exact opposite of micropropagation, another tissue culture technique. As for *Cyrtopodium* species, which represent a valuable germplasm, in vitro cultivation techniques for germplasm propagation and conservation are essential [29].

Cyrtopodium glutiniferum Raddi occurs only in Brazil (endemic), in the Atlantic forest, in the states of Espírito Santo, Rio de Janeiro, and Minas Gerais, common, especially, in rock outcrops, granite (inselbergs) but also possible to be observed developing in sandy soil. There are three worldwide hotspots of inselberg plant diversity. Southeastern Brazil, where *C. glutiniferum* is located, is one of them [17].

As described previously, *C. glutiniferum* is subjected to conditions of exposure to full sun, high UV incidence, lack of soil, constant winds, water and nutrient scarcity, difficulty in affixing roots, and high temperatures (Fig. 2a), including the group of succulent plants, which are tolerant to dissection [17, 30].

C. glutiniferum has thick big exposed (not buried) fusiform pseudobulbs (30–90 cm high) (Fig. 2a) and yellow high (90–240 cm long) paniculate axillary inflorescences, with simple or 1–4 ramifications that rise from the developing shoot (Fig. 2b). Its large flowers (expanded lip with 1.6–2.3 cm long), completely or predominantly yellow, appear at the end of the dry season and beginning of the rainy season (August–October) (Fig. 2c). The ovate sepals may be greenish-yellow or yellow-brown in color and have brown spots at the apex, while the callus may have orange spots that occasionally extend to the surrounding area and the base of the lateral lobes (Fig. 2c) [7, 31]. In *C. glutiniferum* pollination occurs through sexual propagation (i.e., gene recombination) [7].

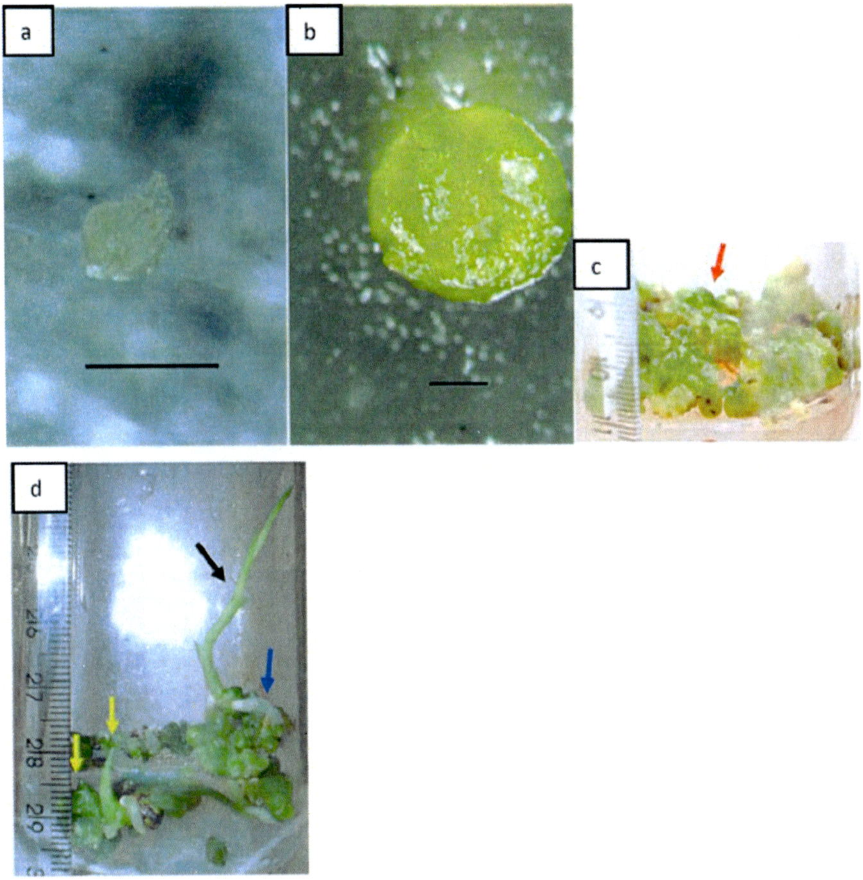

Fig. 1 Representative photographs of the sequence of events and stages of in vitro germination of *C. glutiniferum* seed: (**a**) seed germination after 50 days, (**b**) protocorm formation after 50 days, (**c**) 4-weeks-old PLBs (arrow – PLB mass), and (**d**) plantlets and PLBs after 4 weeks of culture on IAA medium (black arrow – plantlet; blue arrow – root with velamen; yellow arrows – bud protuberance). Scale bars = 1 mm. (Source: Vogel and Macedo [7])

As previously explained, many species of the genus have already been confused with each other because of their close resemblance [5]. *C. glutiniferum* has already been mistakenly identified as *C. cardiochilum*, *C. andersonii*, or *C. paranaenses Schltr*. Due to its similarity with *C withneri* L. C. Menezes has not yet enabled a taxonomic solution [7, 31].

3 Ethnopharmacological Aspects of *Cyrtopodium glutiniferum*

Popularly known as "*Sumaré*," the genus *Cyrtopodium* is used on the production of ointments in Brazil, with no or poor distinction of the species utilized [32]. The genus has ethnopharmacological relevance, being used for treatment of chest colds,

Fig. 2 *Cyrtopodium glutiniferum* in inselbergh at Morro da Urca, Rio de Janeiro. (**a**) Whole plant with pseudobulbs, leaves, and inflorescence; (**b**) detail of flower buds and flowers [7]; (**c**) flower detail [31]

tuberculosis, and hemoptysis [33], as a topic antibiotic [34] and to treat perforation wounds [24]. Species of the genus are also used in the form of ointments for the treatment of lesions on the eyelids; in the form of juice to treat abscesses, folliculitis, or in the form of syrups to treat cough and pertussis [10, 32]. Researchers reported that extracts from pseudobulbs of *C. paranaenses* Schltr. and *C. andersonii* R. Br.

are useful for healing wounds [7, 35–37]. *C. punctatum* is part of some phytopharmaceutical formulations for treating abscesses and burn [38]. The *C. andersonii* pseudobulb is used in traditional medicine to treat inflammatory symptoms, with healing properties and antihemorrhagic activity [38]. Previous studies have already reported the presence of glucomannan with possible immunomodulatory, anti-inflammatory, and gastroprotective activity [2, 38]. In an article published with *C. macrobulbon*, a species that is traditionally used to treat urinary infections, it was found that its extract and stilbenoids have antinociceptive, anti-inflammatory properties [39, 40]. Subsequently, it was also found that extracts of *C. paniculatum*, which are rich in stilbenoids, had a moderate toxic effect on cancer cells [41].

Stilbenoids appear to be a chemical characteristic of several species of Orchidaceae, some of which are more frequent in the *Cyrtopodium* genus [41] (Table 1). Stilbenoids are phytoalexins that protect the plant against herbivory. This chemical group has a great structural diversity that goes through stilbenes, bibenzyls, 9,10-dihydrophenanthrenes, phenanthrenes, (dihydro)-phenanthrenequinones, *p*-hydroxybenzyl-phenanthrenes, 9,10-dihydrophenanthrofurans, and phenanthrene dimers. These molecules have anti-inflammatory, antifibrotic, and antimicrobial properties, which may justify the traditional use of several species of the genus [41].

The genus *Cyrtopodium* has an ethnopharmacological basis, such as the folk use of *C. cardiochilum* Lindl. and *C. andersonni* R. Br. on gastric inflammatory diseases, mainly associated to polysaccharidic contents [2, 38, 39]. *C. macrobulbon* (La Llave and Lex.) G. A. Romero and Carnevali and *C. paniculatum* (Ruiz and Pav.) Garay are traditionally used for treating urinary infections [39] and their extracts are rich in stilbenoids and phenanthrenes derivatives, as cyrtopodinone, cyrtopodinol, and cyrtopodin. Some of these molecules showed moderate cytotoxicity activity towards a cancer cell line [40, 41]. *C. macrobulbon* is also employed in the treatment of abscesses, and as a balsamic agent [39]. *C. punctatum* (L.) Lindl. is traditionally used as an expectorant in the recovery of dry cough, and amelioration of the inflammatory symptoms of bronchitis and asthma as syrup [5, 32]. *C. punctatum* is also used as emetic, on blood pressure control [42], to treat rheumatism [43] and also for treating boils and abscesses [44].

Recently, we described a phenolic content emphasized metabolomic analysis of *C. glutiniferums* by UHPLC-MS/MS, since, as previously reported, some stilbenoids appear to be characteristic of the genus. This was the first work on metabolomics of the genus *Cyrtopodium* [45]. In a survey made from publications on the phytochemistry of the species of *Cyrtopodium* genus, it was found that some molecules, mostly stilbenoids, were reported in more than one species: glucomnann, confusarin, gigantol, potatoesin III, denthyrsinin, and shancidin (Table 1).

In our work it was found that phenanthrenes was the most abundant subclass, despite the fact that the most abundant molecules were dihydroformononetin, caffeic acid 4-O-glucoside (glucocaffeic acid), and arbutin. In this study, we also proceed an in vitro genotoxicity assessment, recommended by FDA for any chemical used in human therapeutics and also some aspects of antiproliferative, anti-inflammatory, and antioxidant activity of *C. glutiniferum* aqueous extract [45]. The genotoxicity assessment is fundamental for any natural product intended by human use. Despite

Table 1 Survey of common published molecules of *Cyrtopodium* spp.

	Confusarin	Gigantol	Batatasin III	Denthyrsinin	Shancidin	Glucomannan
C. paniculatum **Pseudobulbs** [40]	+	+	+	+	+	
C. macrobulbon **Pseudobulbs** [39]	+	+	+			
C. andersonii **Pseudobulbs** [38]						+
C. cardiochilum **Pseudobulbs** [2]						+
C. paniculatum **Roots** [41]	+	+	+	+		
C. glutiniferum **Pseudobulbs** [45]				+	+	

the popular belief that the use of plants as an alternative for the treatment of several diseases is free of hazards, the chemical composition of the vegetal extracts can imply in the occurrence of the most diverse lesions to living matter and, above all, to DNA [46]. In our investigation, *C. glutiniferum* extract induced mutagenicity only on TA100 strain at the highest tested dose (5.0 mg) on Ames Test and did not increase micronucleated RAW264.7 macrophages. Because of the facts presented, the extract cannot be considered genotoxic [45].

Our recent findings about *C. glutiniferum* extract efficacy suggest extract can be considered a good direct antioxidant, reducing the $DPPH^+$ radicals, acting as a free radical scavenger, presenting an EC_{50} of 132.6 ± 6.2 μg/mL. We also observed a dose-dependent and time-dependent anti-inflammatory response in LPS-activated macrophages (Fig. 3) by the decrease of both stimulation index (calculated by the variation in mitochondrial function using WST-1 reagent – Fig. 3a) and viable cells counts (through trypan blue method – Fig. 3b), suggesting an immunomodulatory effect. In fact, the secondary metabolites detected on *C. glutiniferum* extract are described as antiproliferative and anti-inflammatory agents [47–49].

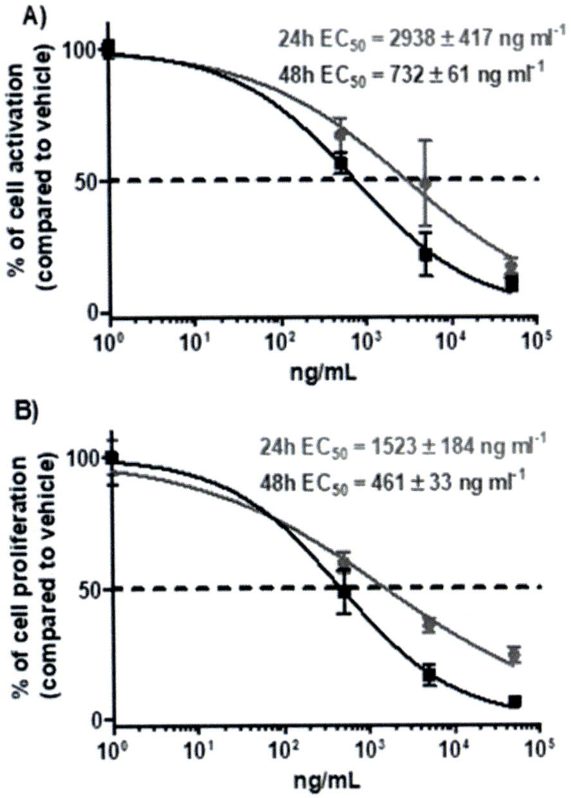

Fig. 3 Effects of *Cyrtopodium glutiniferum* extract on RAW 264.7 macrophages activation and proliferation status. After offering 1 μg/mL of *E. coli* LPS as a phlogistic agent to RAW 264.7 cells and treat them with *C. glutiniferum* aqueous extract (from 0 to 50,000 ng mL^{-1}) for 24 or 48 h, WST-1 (**a**) and trypan blue (**b**) cell counting were performed to evaluate activation and proliferation. (Source: Araújo-Lima et al. [45])

Stilbenoids and other polyphenols are described as regulators of transcriptional factors, as NF-κB, IKK and mTOR, acting directly on cell activation responses [50, 51]. In comparison to other orchids from the same genus, the pseudobulbs of *C. paranaense* are used for wound healing, but no biological activity was established scientifically. The pseudobulbs of *C. flavum*, which are popularly used to heal skin lesions in Brazilian folk medicine, have been studied as an analgesic and anti-inflammatory agent. Its 20% pseudobulb aqueous extract and a polysaccharide extracted from it, cyrtopodine, were described as anti-inflammatories, besides the absence of analgesic activity [32].

4 Novel Evidences of *C. glutiniferum* Efficacy on Skin Lesions Treatment

Regarding the ethnotherapeutical perspective, *Cyrtopodium* ointments are widely used in Brazilian traditional medicine to treat boils and abscesses, as mentioned above. Besides that, they can be considered as the most common treatment to skin and soft tissue diseases [52], furuncles, and abscesses since they have a complex pathophysiology and are mediated by both infectious and inflammatory processes [53]. A skin abscess is essentially a suppurative sequela of folliculitis, and when infection involves several adjacent follicles, producing a coalescent inflammatory mass with pus draining from multiple orifices, the larger nodule is then termed a carbuncle [52]. The tissular homeostasis misbalance caused by bacterial infection can produce pyogenic abscesses, which can culminate in hemodynamic alterations, as ischemia and subcutaneous thrombosis, resulting in tissue necrosis, decurrent of inflammation [54].

Skin abscesses are commonly caused by infectious pathogens, and the bacterial milieu can consist both in Gram-positive and Gram-negative bacteria, such as *Staphylococcus aureus* infection in boils [55] and *Escherichia coli* infection in Fournier's syndrome [56].

Trying to demystify some aspects concerning the mechanism of action of *C. glutiniferum* as therapeutic alternative for treatment of skin and soft tissue diseases, we decided to investigate the efficacy of the aqueous extract against *S. aureus* (ATCC 25923 strain) and *E. coli* (ATCC 25922 strain). The minimal inhibitory concentration (MIC) tests in flat bottom microtiter 96 wells plate were performed according to Antimicrobial Susceptibility Testing guidelines from the Institute of Clinical and Laboratory Standards Institute [57].

For antimicrobial activity experiments, the tested concentrations of *C. glutiniferum* extract were 3.00, 1.00, 0.33, 0.11, 0.04, and 0.01 mg/mL. The maximum concentration of *C. glutiniferum* extract was defined as 1/1000 of the mass found in topical ointments (3%). So, an aliquot of each strain was thawed and diluted in Mueller Hinton Broth – MHB in the proportion of 1:500, generating an inoculum of ATCC 25923 from *S. aureus* and another of ATCC 25922 from *E. coli*, both with concentration of approximately 2.0×10^5 CFU/mL. From these suspensions, 100 μL were pipetted into each well of the microtiter plates, one strain on each plate. In the first row of each, another 100 μL of MH broth without inoculum was pipetted to control the test sterility (white). The antimicrobial tetracycline (Sigma Aldrich; St. Louis, MO, USA) was used in all experiments as a positive control, in concentrations from 1.8 to 445 μg/mL.

With the addition of the inoculum, the concentration of the tested compound/ extract was adjusted to the desired value (dilution 1:2) and the final concentrations of the inocula were approximately 1.0×10^5 CFU.

Microtiter plates were incubated for 24 h at 35 °C and the growth of the samples was evaluated both visually and spectrophotometrically using the SpectraMax Plus 384 Microplate Reader spectrophotometer, measuring the absorbance of the cultures (Optical Growth Dosage) at the wavelength of 620 nm. The experiments were carried out in triplicate and repeated three times.

According to Table 2, the results of the above-mentioned extract's antimicrobial activity were not encouraging. The extract was only able to cause partial inhibition of *E. coli* 25922 in the highest concentration tested. At the other concentrations of the assay, both *E. coli* 25922 and *S. aureus* 25923 did not suffer any inhibition effect after being exposed to *C. glutiniferum* extract. These findings interpose the results of the bacterial mutagenesis model, in which death of *Salmonella enterica* serovar Typhimurium was detected in concentrations greater than 50 μg.

Considering the complexity of pathophysiological aspects on skin lesions, and correlating these data to our recent findings about the anti-inflammatory and anti-proliferative effects of *C. glutiniferum* against activated mononuclear cells, probably the therapeutic effect of this orchid is not directly related to eliminate the pathogen but in controlling the tissue injuries mediated by the acute inflammation, acting against immune response cells, involved in abscess occurrence [58–60].

Despite our negative results about antimicrobial efficacy, one of the most abundant compounds detected by metabolomics in our *C. glutiniferum* extract, arbutin, is associated to the bactericidal effect of other plant extracts from *Bergenia* genus [61], and also in essential oils [62]. The chemical nature of arbutin (derivative of hydroquinone) and its tyrosinase inhibitor effect can be related to its efficacy against different bacteria, interfering in protein scaffolds and resulting in cell wall damages, even when these effects being observed in high doses of this polyphenol [62, 63]. The other two compounds (glucocaffeic acid and dihydroformononetin) have no evidences on literature about its microbicidalefficacy.

Table 2 Antimicrobial activity of *Cyrtopodium glutiniferum* extract (in mg/mL) against *Escherichia coli* (25922 strain) and *Staphylococcus aureus* (25923 strain)

Strain	Blank	Negative control	3.00	1.00	0.33	0.11	0.04	0.01
E. coli (25922)	–	+	/	+	+	+	+	+
S. aureus (25923)	–	+	+	+	+	+	+	+

Categories: growth without inhibition: + (≥80% of growth, in comparison to negative control); partial inhibition: / (between 10% and 80% of growth, in comparison to negative control; total inhibition: – (<10% of growth, in comparison to negative control)

5 Conclusions

Cyrtopodium species from southeastern Brazil, including *Cyrtopodium glutiniferum* Raddi, are found growing in biodiversity hotspots, remnants of the Atlantic rain forest and in sandy coastal plain habitats and, besides that, *C. glutiniferum* is relevant for biotechnological and pharmacognosy purposes. The orchid is widely prescribed in Brazilian folk medicine and some of its medicinal properties had to be scientifically validated. In our previous metabolomic analysis, the most abundant molecules detected on aqueous extract were dihydroformononetin, caffeic acid 4-O-glucoside (glucocaffeic acid), and arbutin. As *C. glutiniferum* is traditionally used to treat skin lesions, as abscesses, furunculosis, and wound healing, our research group has verified the positive anti-inflammatory and antiproliferative effects of the extract against activated mononuclear cells. However, the therapeutic effect of this orchid probably is most directly related to the modulation of host immune responses and that about the elimination of the bacteria on inflammation site. The effects of *C. glutiniferum* on inflammation-mediated tissue damage can favor host cellular and tissue repair and resolutive response phase in cutaneous infections.

Acknowledgments Funding: This work was supported by the National Council of P&D of Brazil (CNPq) and Coordination of Superior Level Staff Improvement (CAPES), which granted fellowships; and the Foundation for Research of Rio de Janeiro State (FAPERJ), UERJ, UNIRIO, and CNPq, which gave financial support.

Technical support: The authors thank AykeAdnet de Lima and Renato Geraldo da Silva Filho, from the Department of Microbiology and Parasitology of Federal University of Rio de Janeiro State (UNIRIO) by technical support on antimicrobial experiments.

References

1. Batista JAN, Bianchetti LB (2004) Three new taxa in *Cyrtopodium* (Orchidaceae) from central and southeastern Brazil. Brittonia 56(3):260–274. https://doi.org/10.1663/0007-196X(2004)056[0260:TNTICO]2.0.CO;2
2. Barreto DW, Parente JP (2006) Chemical properties and biological activity of a polysaccharide from *Cyrtopodium cardiochilum*. Carbohydr Polym 64(2):287–291. https://doi.org/10.1016/j.carbpol.2005.11.038
3. Dutra D, Kane ME, Richardson L (2009) Asymbiotic seed germination and in vitro seedling development of *Cyrtopodium punctatum*: a propagation protocol for an endangered Florida native orchid. Plant Cell Tissue Organ Cult 96(3):235–243. https://doi.org/10.1007/s11240-008-9480-z
4. Chase MW, Cameron KM, Freudenstein JV et al (2015) An updated classification of Orchidaceae. Bot J Linn Soc 177(2):151–174. https://doi.org/10.1111/boj.12234
5. Romero-González GA, Carnevali G, Duno R et al (2019) A new species of *Cyrtopodium* (Orchidaceae: Cyrtopodiinae) from Belize. Phytotaxa 424(2):71–86. https://doi.org/10.11646/phytotaxa.424.2.1
6. Maciel AA, Cardoso JCF, Oliveira PE (2020) On the low reproductive success of two *Cyrtopodium* species (Orchidaceae: Cyrtopodiinae): the relative roles of biotic and abiotic pollination. Plant Species Biol 35(1):49–58. https://doi.org/10.1111/1442-1984.12260
7. Vogel IN, Macedo AF (2011) Influence of IAA, TDZ, and light quality on asymbiotic germination, protocorm formation, and plantlet development of *Cyrtopodium glutiniferum* Raddi., a

medicinal orchid. Plant Cell Tissue Organ Cult 104(2):147–155. https://doi.org/10.1007/s11240-010-9810-9
8. Brown PM (2005) Additions and emendations to the wild orchids of North America, north of Mexico. SIDA Contrib Bot 21(4):2297–2319. http://www.jstor.org/stable/41968528
9. De Barros F, Nogueira Batista JA, De Bem Bianchetti L (2003) Epitypification and taxonomic elucidation of some Brazilian taxa of *Cyrtopodium* R. Br. (Orchidaceae). Taxon 52(4):841–849. https://doi.org/10.2307/3647358
10. Ventura Lima B, Ferreira Santos A, de Fátima Fernandes A, Augusto Guimarães Fuga C, Weingart Barreto R, Vieira Silva R (2019) Novas opções de substratos para o cultivo de *Cyrtopodium cardiochilum* (Orchidaceae). Colloq Agrar 15(4):100–106. https://doi.org/10.5747/ca.2019.v15.n4.a316
11. Pansarin LM, Pansarin ER, Sazima M (2008) Facultative autogamy in *Cyrtopodium polyphyllum* (Orchidaceae) through a rain-assisted pollination mechanism. Aust J Bot 56(4):363–367. https://doi.org/10.1071/BT07135
12. Pemberton RW, Liu H (2008) Potential of invasive and native solitary specialist bee pollinators to help restore the rare cowhorn orchid (*Cyrtopodium punctatum*) in Florida. Biol Conserv 141(7):1758–1764. https://doi.org/10.1016/j.biocon.2008.04.016
13. Dutra D, Kane ME, Adams CR, Richardson L (2009) Reproductive biology of *Cyrtopodium punctatum* in situ: implications for conservation of an endangered Florida orchid. Plant Species Biol 24(2):92–103. https://doi.org/10.1111/j.1442-1984.2009.00242.x
14. Liu H, Pemberton R (2010) Pollination of an invasive orchid, *Cyrtopodium polyphyllum* (Orchidaceae), by an invasive oil-collecting bee, *Centris nitida*, in southern Florida. Botany 88(3):290–295. https://doi.org/10.1139/B10-017
15. Vinaixa M, Schymanski EL, Neumann S, Navarro M, Salek RM, Yanes O (2016) Mass spectral databases for LC/MS- and GC/MS-based metabolomics: state of the field and future prospects. TrAC Trends Anal Chem 78:23–35. https://doi.org/10.1016/j.trac.2015.09.005
16. Moreira MM, Barberena FFVA, Lopes RC (2014) Orchidaceae of the *Grumari restinga*: floristic and similarity among restingas in Rio de Janeiro state, Brazil. Acta Bot Bras 28(3):321–326. https://doi.org/10.1590/0102-33062014abb3173
17. de Paula LFA, Mota NFO, Viana PL, Stehmann JR (2017) Floristic and ecological characterization of habitat types on an inselberg in Minas Gerais, southeastern Brazil. Acta Bot Bras 31(2):199–211. https://doi.org/10.1590/0102-33062016abb0409
18. Cirne P, Zaluar LT, Scarano FR (2003) Plant diversity, interspecific associations, and postfire resprouting an a sandy spit in a Brazilian coastal plain. Ecotropica 9:33–38
19. De Fraga CN, Peixoto AL (2004) Florística e ecologia das Orchidaceae das restingas do estado do Espírito Santo. Rodriguesia 55(84):5–20. https://doi.org/10.1590/2175-78602004558401
20. Zamith LR, Scarano FR (2000) Produção de mudas de espécies das Restingas do. Acta Bot Bras 18(1):161–176. http://www.scielo.br/pdf/abb/v18n1/v18n1a14
21. Cunha MFB, Forzza RC (2007) Orchidaceae no Parque Natural Municipal da Prainha, RJ, Brasil. Acta Bot Bras 21(2):383–400. https://doi.org/10.1590/s0102-33062007000200013
22. Salimon CI, Negrelle RRB (2001) Natural regeneration in a quaternary coastal plain in southern Brazilian Atlantic Rain Forest. Braz Arch Biol Technol 44(2):155–163. https://doi.org/10.1590/S1516-89132001000200008
23. de Carvalho OC, de Paiva Neto VB, Padilha DRC et al (2018) *Cyrtopodium paludicolum* germination with two *Tulasnella* isolates. Acta Bot Bras 32(1):107–112. https://doi.org/10.1590/0102-33062017abb0181
24. Guo WL, Chang YCA, Kao CY (2010) Protocorm-like bodies initiation from root tips of *Cyrtopodium paranaense* (Orchidaceae). HortScience 45(9):1365–1368. https://doi.org/10.21273/hortsci.45.9.1365
25. Rotili Nunes Picolotto D, Borges de Paiva Neto V, de Barros F, Roxana Castro Padilha D, Cláudia Ferreira da Cruz A, Campos Otoni W (2017) Micropropagation of *Cyrtopodium paludicolum* (Orchidaceae) from root tip explants. Crop Breed Appl Biotechnol 17:191–197. https://doi.org/10.1590/1984-70332017v17n3a30

26. Guimarães FAR, Pereira MC, Felício C d S et al (2013) Symbiotic propagation of seedlings of *Cyrtopodium glutiniferum* Raddi (Orchidaceae). Acta Bot Bras 27(3):590–596. https://doi.org/10.1590/S0102-33062013000300016
27. Pereira MC, Rocha DI, Veloso TGR et al (2015) Characterization of seed germination and protocorm development of *Cyrtopodium glutiniferum* (Orchidaceae) promoted by mycorrhizal fungi *Epulorhiza* spp. Acta Bot Bras 29(4):567–574. https://doi.org/10.1590/0102-33062015abb0078
28. Sousa KCI, De Araújo LG, De Sousa Silva C et al (2019) Seed germination and development of orchid seedlings (*Cyrtopodium saintlegerianum*) with fungi. Rodriguesia 70. https://doi.org/10.1590/2175-7860201970004
29. Rodrigues LA, de Paiva Neto VB, Boaretto AG et al (2015) In vitro propagation of *Cyrtopodium saintlegerianum* Rchb. f. (Orchidaceae), a native orchid of the Brazilian savannah. Crop Breed Appl Biotechnol 15(1):10–17. https://doi.org/10.1590/1984-70332015v15n1a2
30. Couto DR, Francisco TM, Manhães VC, Dias HM., Pereira MCA (2017) Floristic composition of a Neotropical inselberg from Espírito Santo state, Brazil: an important area for conservation. Check List 13(1):2043–2054
31. do Brasil F. 2020 Flora do Brasil. Jardim Botânico do Rio de Janeiro. http://floradobrasil.jbrj.gov.br/. Acessed 24 June 2020
32. Silva AG, Boldrini RF, Kuster RM (2013) Os sumarés cicatrizantes da medicina tradicional brasileira, ou, as surpresas químicas ativas do desconhecido gênero *Cyrtopodium* (Orchidaceae). Nat Line 11:152–154
33. Corrêa MP, de Azeredo Penna L (1984) Dicionário Das Plantas Úteis Do Brasil e Das Exóticas Cultivadas: HL (Vol. 4). Ministério da Agricultura, Instituto Brasileiro de Desenvolvimento Florestal
34. Gonçalves Brasileiro B, Ramos Pizziolo V, Soares Raslan D, Mashrouah Jamal C, Silveira D (2006) Antimicrobial and cytotoxic activities screening of some Brazilian medicinal plants used in Governador Valadares district. Rev Bras Cienc Farm 42(2):195–202. https://doi.org/10.1590/s1516-93322006000200004
35. Pereira CEDB, Felcman J (1998) Correlation between five minerals and the healing effect of Brazilian medicinal plants. Biol Trace Elem Res 65(3):251–259. https://doi.org/10.1007/bf02789100
36. Boscolo OH, De Senna Valle L (2008) Plantas de uso medicinal em Quissamã, Rio de Janeiro, Brasil. Iheringia Ser Bot 63(2):263–272
37. Boscolo OH, Lucia RRMV, Fernandes, De L (2010) An ethnobotanical survey as subsidy for the generation of researches related to biotechnology. Int Res J Biotechnol 1:1–6
38. Parente JP, Adão CR, Da Silva BP, Tinoco LW (2014) Structural characterization of an acetylated glucomannan with antiinflammatory activity and gastroprotective property from *Cyrtopodium andersonii*. Carbohydr Res 391(1):16–21. https://doi.org/10.1016/j.carres.2014.03.021
39. Morales-Sánchez V, Rivero-Cruz I, Laguna-Hernández G, Salazar-Chávez G, Mata R (2014) Chemical composition, potential toxicity, and quality control procedures of the crude drug of *Cyrtopodium macrobulbon*. J Ethnopharmacol 154(3):790–797. https://doi.org/10.1016/j.jep.2014.05.006
40. Auberon F, Olatunji OJ, Herbette G et al (2016) Chemical constituents from the aerial parts of *Cyrtopodium paniculatum*. Molecules 21(10):1–18. https://doi.org/10.3390/molecules21101418
41. Auberon F, Olatunji OJ, Raminoson D et al (2017) Isolation of novel stilbenoids from the roots of *Cyrtopodium paniculatum* (Orchidaceae). Fitoterapia 116:99–105. https://doi.org/10.1016/j.fitote.2016.11.015
42. Teoh ES (2019) Orchids as aphrodisiac, medicine or food. Springer, Singapore
43. Magalhães K do N, Guarniz WAS, Sá KM et al (2019) Medicinal plants of the Caatinga, northeastern Brazil: ethnopharmacopeia (1980–1990) of the late professor Francisco José de Abreu Matos. J Ethnopharmacol 237:314–353. https://doi.org/10.1016/j.jep.2019.03.032

44. Martínez GJ, Barboza GE (2010) Natural pharmacopoeia used in traditional Toba medicine for the treatment of parasitosis and skin disorders (Central Chaco, Argentina). J Ethnopharmacol 132(1):86–100. https://doi.org/10.1016/j.jep.2010.07.049
45. Araújo-Lima CF, Paula da Silva Oliveira J, Coscarella IL et al (2020) Metabolomic analysis of *Cyrtopodium glutiniferum* extract by UHPLC-MS/MS and in vitro antiproliferative and genotoxicity assessment. J Ethnopharmacol 253:112607. https://doi.org/10.1016/j.jep.2020.112607
46. Sponchiado G, Adam ML, Silva CD et al (2016) Quantitative genotoxicity assays for analysis of medicinal plants: a systematic review. J Ethnopharmacol 178:289–296. https://doi.org/10.1016/j.jep.2015.10.026
47. Dvorakova M, Landa P (2017) Anti-inflammatory activity of natural stilbenoids: a review. Pharmacol Res 124:126–145. https://doi.org/10.1016/j.phrs.2017.08.002
48. Oliviero F, Scanu A, Zamudio-Cuevas Y, Punzi L, Spinella P (2018) Anti-inflammatory effects of polyphenols in arthritis. J Sci Food Agric 98(5):1653–1659. https://doi.org/10.1002/jsfa.8664
49. Russo GL, Tedesco I, Spagnuolo C, Russo M (2017) Antioxidant polyphenols in cancer treatment: friend, foe or foil? Semin Cancer Biol 46:1–13. https://doi.org/10.1016/j.semcancer.2017.05.005
50. Malaguarnera L (2019) Influence of resveratrol on the immune response. Nutrients 11(5):1–24. https://doi.org/10.3390/nu11050946
51. Yahfoufi N, Alsadi N, Jambi M, Matar C (2018) The immunomodulatory and anti-inflammatory role of polyphenols. Nutrients 10(11):1–23. https://doi.org/10.3390/nu10111618
52. Atanaskova N, Tomecki KJ (2010) Innovative management of recurrent furunculosis. Dermatol Clin 28(3):479–487. https://doi.org/10.1016/j.det.2010.03.013
53. Pinkus H (2009) Short communication. Clin Chem Lab Med 26(9):517–518. https://doi.org/10.1515/cclm.1988.26.9.579
54. Roje Z, Roje Ž, Matić D, Librenjak D, Dokuzović S, Varvodić J (2011) Necrotizing fasciitis: literature review of contemporary strategies for diagnosing and management with three case reports: torso, abdominal wall, upper and lower limbs. World J Emerg Surg 6(1):1–17. https://doi.org/10.1186/1749-7922-6-46
55. Chambers HF (2008) Clinical decisions management of skin and soft-tissue infection incision and drainage plus anti-MSSA therapy:1063–1067
56. Dornelas MT, Correa M d PD, Barra FML et al (2012) Fournier's syndrome: a 10-year evaluation study. Rev Bras Cir Plást 27(4):600–604
57. Clinical and Laboratory Standards Institute (CLSI) (2015) M07-A10: methods for dilution antimicrobial susceptibility tests for bacteria that grow aerobically; approved standard – tenth edition. Clin Infect Dis 35(2). www.clsi.org
58. Lei Z, Zhang D, Lu B, Zhou W, Wang D (2018) Activation of mast cells in skin abscess induced by *Staphylococcus aureus* (*S. aureus*) infection in mice. Res Vet Sci 118(8):66–71
59. Kobayashi SD, Malachowa N, Deleo FR (2015) Pathogenesis of *Staphylococcus aureus* abscesses. Am J Pathol 185(6):1518–1527. https://doi.org/10.1016/j.ajpath.2014.11.030
60. Natsis NE, Cohen PR (2018) Coagulase-negative *Staphylococcus* skin and soft tissue infections. Am J Clin Dermatol 19(5):671–677. https://doi.org/10.1007/s40257-018-0362-9
61. Żbikowska B, Franiczek R, Sowa A, Połukord G, Krzyzanowska B, Sroka Z (2017) Antimicrobial and antiradical activity of extracts obtained from leaves of five species of the genus *Bergenia*: identification of antimicrobial compounds. Microb Drug Resist 23(6):771–780. https://doi.org/10.1089/mdr.2016.0251
62. Moghrovyan A, Sahakyan N, Babayan A, Chichoyan N, Petrosyan M, Trchounian A (2019) Essential oil and ethanol extract of oregano (*Origanum vulgare* L.) from Armenian flora as a natural source of terpenes, flavonoids and other phytochemicals with antiradical, antioxidant, metal chelating, tyrosinase inhibitory and antibacterial activity. Curr Pharm Des 25(16):1809–1816. https://doi.org/10.2174/1381612825666190702095612
63. Ma C, He N, Zhao Y, Xia D, Wei J, Kang W (2019) Antimicrobial mechanism of hydroquinone. Appl Biochem Biotechnol 189(4):1291–1303. https://doi.org/10.1007/s12010-019-03067-1

Phenanthrenes from Orchidaceae and Their Biological Activities

22

Andrea Vasas

Contents

Abstract

Orchidaceae is one of the largest plant family. Its members are distributed worldwide, and many of them are used in traditional medicines for the treatment of different diseases. In the last few decades, phytoconstituents with great chemical diversity were isolated from several orchids, and a wide range of biological activities were detected either for crude extracts or for pure compounds. Among the isolated compounds, phenolic derivatives, especially stilbenoids and phenanthrenes, are the most important ones. In the plant kingdom, orchids are the best sources of phenanthrenes – around 400 components were isolated from 100 species. Investigation of the phenanthrene content of orchids can be contributed to place taxonomically questionable or rather controversial species into the proper genera. Pharmacological investigation of Orchidaceae phenanthrenes revealed that several compounds possess promising activities – mainly antiproliferative, antimicrobial, anti-inflammatory, and antioxidant effects. This chapter summarizes the isolated

A. Vasas (✉)
Department of Pharmacognosy, University of Szeged, Szeged, Hungary
e-mail: vasasa@pharmacognosy.hu

J.-M. Mérillon, H. Kodja (eds.), *Orchids Phytochemistry, Biology and Horticulture*, Reference Series in Phytochemistry, https://doi.org/10.1007/978-3-030-38392-3_34

phenanthrenoids, their occurrence in Orchidaceae species, chemotaxonomical significances, and biological activities of the most promising representatives.

Keywords

Orchidaceae · *Bletilla* · *Dendrobium* · Phenanthrenes · Phenanthroquinones · Chemotaxonomy · Antiproliferative

1 Introduction

Orchidaceae, one of the largest and most evolved family of the flowering plants, comprises approximately 27,000 species in 900 genera [1]. Orchids are widely distributed, and the diversity of the plants increases toward the tropic and especially tropical mountains. Approximately 73% of orchids are epiphytic. The most represented genera are *Bulbophyllum* ($n = 1000$), *Dendrobium* ($n = 900$), *Eria* ($n = 500$), *Maxillaria* ($n = 420$), *Oncidium* ($n = 420$), *Liparis* ($n = 350$), *Eulophia* ($n = 200$), and *Calanthe* ($n = 150$) [2, 3]. Many of the orchids are used in traditional medicines for the treatment of different diseases, e.g., tumors, infectious diseases, pain and fever, diabetes, skin diseases, problems concerning circulation, digestive, respiratory, and reproduction organs [4, 5]. In the last decades, several studies have been performed to justify the medicinal use of various plants in the treatment of diseases. As a result of these investigations, a wide range of chemical compounds, including alkaloids, bibenzyl derivatives (stilbenes and phenanthrenes), flavonoids, and terpenoids, were identified from orchids, and diuretic, anti-inflammatory, antipyretic, antirheumatic, myorelaxant, antitumor, antimicrobial, hypoglycemic, and neuroprotective activities of the extracts and pure compounds were detected [6, 7]. Moreover, phenanthrenes are important phytoalexins and act as defense against diverse biotic and abiotic factors [8, 9].

Phenanthrenes and stilbenoids are the most relevant components present in Orchidaceae species. Phenanthrenes are aromatic secondary metabolites, occurring relatively rare in nature. To date, only a few phenanthrene-containing plant families ($n = 17$) have been identified, and less than 500 phenanthrenes were isolated [5]. Most of these metabolites were described from Orchidaceae, Juncaceae, Combretaceae, and Dioscoreaceae species [5]. However, the best sources of phenanthrenes are orchids; approximately 75% of the compounds were isolated from Orchidaceae species. As phenanthrenes are derived from stilbenes by oxidative coupling of the aromatic rings of stilbene precursors, it is not surprising that they occur together in orchids. The phenanthrene core could also be generated from stilbenes by UV irradiation in the presence of oxidants [10, 11], and fungal infection also increases the formation of phenanthrenes [9]. After the biosynthesis of the phenanthrene scaffold, further modifications are catalyzed by different oxidases. Most of the phenanthrenes are stable, as it was confirmed by an investigation proving that heat processing did not affect significantly the anti-inflammatory activity of bibenzyls and phenanthrenes isolated from *Bletilla striata* tuber [12]. In case of phenanthrenes, certain substituents seem to be specific to certain families, e.g., *p*-hydroxybenzyl- and stilbene-substituted compounds are almost exclusively found in Orchidaceae species. Therefore, these components are considered by some as chemotaxonomic markers.

Although phenanthrenes comprise a relatively small group of natural products, discovering new phenanthrene derivatives and evaluating their prospective biological activities have become of great interest to many research groups worldwide. Several phenanthrenoids possess promising antitumor, antimicrobial, and anti-inflammatory activities. For some of them, studies have been conducted to elucidate their possible mode of action and to deduce structure–activity relationships. Based on 213 references, this review covers the 376 phenanthrenes that have been isolated from orchids, their occurrence in Orchidaceae species, chemotaxonomical significances, and biological activities of the most promising compounds.

2 Occurrence of Phenanthrenes in Orchidaceae Species

Phenanthrenes isolated so far from natural sources may be classified into three major groups: mono-, di-, and triphenanthrenes. The diversity of monophenanthrenes is due to the number and type of the connecting substituents and the saturation of the bond between C-9 and C-10. Compounds with saturated C-9–C-10 bond are named as dihydrophenanthrenes, while 9,10-dehydro derivatives are the phenanthrenes. 9,10-Dehydro and dihydro derivatives occur almost equally in monophenanthrenes; about 60% of the compounds are dihydrophenanthrenes [7]. Another group of monophenanthrenes is the phenanthrenequinones. Di- and triphenanthrenes can be classified by the type of connecting monomers, and the location of the linkage. Up to now, only a few compounds of the triphenanthrene group have been described [5].

According to the literature data, the best sources of phenanthrenes are Orchidaceae species, approximately 75% of the identified compounds isolated from Orchidaceae species (Table 1). Mono-, di-, and triphenanthrenes were equally isolated from orchids. Interestingly, up to now, phenanthrene trimers were isolated only from orchids.

In many cases, depictions of the compounds are not correct, as rings A and C are usually replaced. During biosynthesis, incorporation of dihydro-*m*-coumaric acid into a 9,10-dihydrophenanthrene (hircinol), through a bibenzyl, was confirmed by Fritzemeier and Kindl [8]; therefore, most of the Orchidaceae phenanthrenes, substituted with hydroxy and methoxy groups at C-2 and C-4 on one of their rings, have to be depicted as ring A. In many cases, the hyroxy group is changed to methoxy group at C-2. In most cases, ring C is substituted with a hydroxy group at C-7.

Table 1 Phenanthrenes in orchids

	Species	Compound	Ref.
1	*Aerides rosea*	**28, 191**	[168]
2	*Agrostophyllum callosum*	**4, 17, 18, 20, 187, 342, 343**	[38, 66, 169]
3	*A. khasiyanum*	**342, 343**	[169]
4	*Appendicula reflexa*	**1, 17, 174, 177, 193, 339**	[41]
5	*Arundina graminifolia*	**1, 4, 5, 15, 83, 88, 258, 272, 275, 289, 290**	[42, 67, 70]
6	*Bletilla formosana*	**1, 24, 30, 33, 72–79, 107, 110, 170, 174, 177, 226, 228, 229, 241, 244, 258, 290, 293–295, 297, 313, 315, 324, 342, 349–351, 353**	[43, 104]
7	*Bletilla ochracea*	**229, 231, 242, 243, 245**	[170]
8	*Bletilla striata*	**1, 3, 5, 8, 9, 11, 16, 24, 25, 29, 69, 72–75, 81, 82, 88, 93–95, 110, 122, 129, 134, 140–144, 147, 148, 165, 170–175, 179, 180, 192, 226, 227, 229–231, 244, 253, 284, 285, 289, 290, 294, 297, 300, 306, 307, 313, 318, 319, 322, 324–326, 331–335, 344–348, 350, 352, 353, 356, 373**	[12, 37, 44, 71, 106, 107, 138, 139, 117, 120, 135, 154, 164, 171, 172]
9	*B. yunnanensis*	**81, 291, 313, 353**	[158]
10	*Brasiliorchis prophyrostele*	**17, 181**	[173]
11	*Bulbophyllum gymnopus*	**35, 208, 211, 216**	[174]
12	*B. leapardium*	**185**	[175]
13	*B. odoratissimum*	**1, 43, 166, 220, 272, 301, 302**	[45, 176, 162, 176]
14	*B. reptans*	**1, 35, 172, 208, 211, 216, 289, 353, 364**	[24]
15	*B. retusiusculum*	**103, 224**	[134]
16	*B. rigidum*	**358**	[177]
17	*B. vaginatum*	**1, 17, 19, 24, 35–37, 136, 172, 174, 176, 178, 184, 186, 190, 208–210, 357, 359**	[46, 123, 178]
18	*Calanthe arisanensis*	**45–48, 268, 278, 279**	[30, 31]
19	*Cephalantheropsis gracilis*	**49, 50**	[32]
20	*Cremastra appendiculata*	**1, 4, 9, 12–14, 92, 120, 127, 169, 172, 194, 195, 221, 222, 232–239, 286, 287, 290, 321, 324, 337, 340, 341, 353–355, 360–363, 376**	[26, 26, 26, 48, 90, 92, 108, 153, 163, 179]
21	*Cirrhopetalum andersonii*	**102, 223**	[132, 133]

(continued)

Table 1 (continued)

	Species	Compound	Ref.
22	*C. maculosum*	**353**	[155]
23	*Coelogyne cristata*	**41, 160, 162, 215**	[129, 130, 180]
24	*C. elata*	**1**	[47]
25	*C. flaccida*	**20**	[66]
26	*C. flavida*	**157**	[128]
27	*C. longipes*	**157, 158, 160, 166**	[89]
28	*C. ochracea*	**1, 252, 256, 271**	[22]
29	*Cymbidium aloifolium*	**1, 17, 276, 277**	[49, 141, 142]
30	*C. finlaysonianum*	**1, 5, 17, 25, 80, 170, 177, 269, 276**	[50]
31	*Cymbidium* Great Flower Marie Laurencin	**51, 52, 53, 269**	[34]
32	*Cymbidium* Great Flower Marie Laurencin	**54, 281**	[34]
33	*C. pendulum*	**212, 218**	[181]
34	*Cypripedium tibeticum*	**272, 274**	[143]
35	*Cyrtopodium paniculatum*	**1, 5, 24, 33, 39, 55, 121, 123, 132, 170, 174, 196, 211, 212, 217, 240, 258, 267, 290, 297, 313, 316, 323, 324, 327, 328, 336, 353**	[17, 51]
36	*Dendrobium amoenum*	**159**	[131]
37	*D. aphyllum*	**283**	[148]
38	*D. brymerianum*	**5, 172**	[72]
39	*D. chrysanthum*	**166, 190, 248, 254**	[85, 93]
40	*D. chrysotoxum*	**122, 145, 166, 170, 190, 272**	[114, 182]
41	*D. crystallinum*	**247**	[124]
42	*D. denneanum*	**2, 8, 56–60, 166, 197–199**	[13]
43	*D. densiflorum*	**5, 166, 212, 259, 272**	[73, 183]
44	*D. draconis*	**8, 32, 261**	[145]
45	*D. fimbriatum*	**62**	[15]
46	*D. formosum*	**1, 5, 8, 24, 32, 174, 211, 261**	[52]
47	*D. hainanense*	**43, 44**	[184]
48	*D. loddigesii*	**5, 31, 166, 168, 200, 246**	[14, 88]
49	*D. longicornu*	**32, 260**	[144]
50	*D. moniliforme*	**8, 42, 63, 136, 270, 282**	[35, 147]
51	*D. moscatum*	**166**	[87]
52	*D. nobile*	**1, 5, 8, 10, 23–26, 64, 66, 170, 172, 176, 184, 190, 201–204, 211, 270, 292**	[18, 74, 74, 101, 185–187]
53	*D. palpebrae*	**5, 190, 212, 338**	[76]
54	*D. plicatile*	**1, 5, 8, 23–25, 31, 46, 48, 69, 166, 170, 174, 212, 225, 258, 288**	[23, 53, 77, 159]

(continued)

Table 1 (continued)

	Species	Compound	Ref.
55	*D. primulinum*	**67**	[16]
56	*D. rotundatum*	**31, 35, 166, 174, 208**	[86, 87]
57	*D. scabrilingue*	**1, 5, 38**	[54]
58	*D. sinense*	**262**	[146]
59	*D. thyrsiflorum*	**5, 8, 9, 23–25, 100, 101, 166, 167, 174, 211, 212, 255, 263, 288, 303, 310, 329, 330, 369, 370, 372**	[75, 188]
60	*D. venustum*	**5, 308**	[78, 166]
61	*Ephemerantha fimbriata*	**26, 184, 214**	[20]
62	*E. lonchophylla*	**23–25, 176, 182, 258, 270**	[149, 189]
63	*Epidendrum rigidum*	**23, 214**	[190]
64	*Eria carinata*	**174**	[191]
65	*E. confusa*	**211, 217**	[192]
66	*E. flava*	**1, 25, 172, 176, 289**	[40, 55]
67	*E. stricta*	**174**	[191]
68	*Eulophia herbacea*	**164**	[193]
69	*E. macrobulbon*	**1, 16, 71, 172, 180, 188, 212, 229**	[56, 194, 195]
70	*E. nuda*	**1, 21, 22, 68, 82, 174, 176, 188, 189, 212, 356**	[57, 109, 191, 196]
71	*E. ochreata*	**16, 170**	[197]
72	*E. petersii*	**1, 5, 21, 167, 188**	[58]
73	*Flickingeria fimbrata*	**1, 5, 8, 23–25, 30, 34, 170, 174, 211, 212, 266, 273**	[39, 150]
74	*Gavilea lutea*	**9**	[198]
75	*Gomesa recurva*	**185**	[199]
76	*Gymnadenia conopsea*	**1, 9, 75, 90, 91, 226, 289, 297**	[59, 60]
77	*Liparis nakaharai*	**205**	[36]
78	*L. nervosa*	**353–355, 367, 368, 374**	[157]
79	*L. regnieri*	**206, 207**	[91]
80	*Loroglossum hircinum*	**7, 8**	[69]
81	*Luisia indivisa*	**5, 167**	[79]
82	*L. volucris*	**371**	[200]
83	*Maxillaria densa*	**24, 27, 174, 183, 214, 216**	[201–203]
84	*Monomeria barbata*	**69, 81, 96, 106, 245, 288, 314, 323, 346, 375**	[101, 110, 161]
85	*Nidema boothii*	**5, 25, 40, 212, 258**	[80]
86	*Odontioda* Marie Noel "Velano"	**269, 280**	[152]
87	*Odontoglossum harvengtense* "Tutu"	**190**	[204]

(continued)

Table 1 (continued)

	Species	Compound	Ref.
88	*Oncidium isthmi*	**264**	[151]
89	*Orchis militaris*	**4**	[68, 69]
90	*Otochilus fusca*	**157**	[128]
91	*O. porrectus*	**312**	[159]
92	*Pholidota articulata*	**157, 159, 161**	[126]
93	*P. cantonensis*	**5, 155, 265**	[81]
94	*P. chinensis*	**1, 2, 4–6, 8, 9, 24, 33, 65, 108–110, 113, 124, 177, 296, 299, 303, 309–311, 317**	[61, 62, 97, 115]
95	*P. imbricata*	**219, 246**	[124]
96	*P. yunnanensis*	**105, 114, 115, 156, 298, 308**	[100, 102, 125]
97	*Pleione bulbocodioides*	**1, 5, 88, 104, 111, 116–119, 125, 126, 128, 132, 140–144, 146–148, 226, 294, 320, 365, 366**	[21, 25, 63, 64, 99, 116, 118, 119, 121]
98	*P. yunnanensis*	**1, 5, 11, 75, 81, 133–135, 226, 353**	[111, 122
99	*Scaphyglottis livida*	**1, 5, 176, 213**	[82, 205]
100	*Spiranthes sinensis*	**70, 83–86, 89, 98, 99, 131, 137–139, 149–153, 252, 257, 258, 304**	[95, 112, 113, 156]
101	*S. sinensis var. amoena*	**4, 87, 98, 99, 130, 154, 257, 305**	[65, 94]
102	*Sunipia scariosa*	**5, 157, 158**	[83]
103	*Thunia alba*	**5, 176, 212, 289, 353**	[84]
104	*Vanda coerulea*	**1, 17, 156, 157**	[127]

2.1 Monophenanthrenes

Around 80% of the phenanthrenes reported from orchids are monophenanthrenes. All of them are substituted, most frequently at C-2, C-5, and C-7, but substitution is freasible at every other place. Substitution at positions C-9 and C-10 are quite rare. Five dihydrophenanthrenes, substituted with hydroxy (**56–60**) group at C-9, were isolated from *Dendrobium denneanum* [13]. Loddigesiinol A (**200**), identified from *Dendrobium loddigesii* [14], **62** from *D. fimbriatum* [15], **67** from *D. primulinum* [16], **55** from *Cyrtopodium paniculatum* [17], **202** from *D. nobile* [18], and phoimbrtol (**219**) from *Pholidota imbricata* are also substituted with a hydroxy group at C-9 [19]. Fimbriol A (**214**) from *E. fimbriata* [20], **190** from several orchids, and **202** from *D. nobile* are substituted with a methoxy group at C-9 [18]. There are some examples, e.g., bulbocodioidins A–D (**116–119**) [21], ochrolone (**163**) [22], plicatol A (**225**) [23], reptanthrin (**364**) [24], bulbocodioidins E (**365**), and F (**366**) of 9,10-disubstituted phenanthrenes [25].

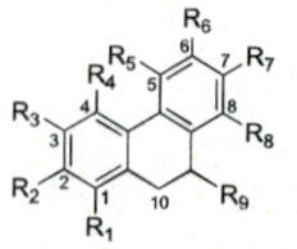

	R1	R2	R3	R4	R5	R6	R7	R8	R9
1	H	OH	H	OMe	H	H	OH	H	H
2	H	OH	H	OH	H	H	OH	H	H
3	H	OMe	H	OMe	H	H	OMe	H	H
4	H	OMe	H	OMe	H	H	OH	H	H
5	H	OMe	H	OH	H	H	OH	H	H
6	H	OMe	H	OH	H	OH	H	H	H
7	H	OMe	H	OMe	OH	H	H	H	H
8	H	OH	H	OMe	OH	H	H	H	H
9	H	OMe	H	OH	OH	H	H	H	H
10	H	OH	H	OMe	H	H	OMe	H	H
11	H	OMe	H	O-Glc	H	H	OH	H	H
12	H	O-Glc	H	OMe	H	H	OH	H	H
13	H	O-Glc	H	H	OMe	H	OH	H	H
14	H	O-Glc	H	H	OMe	H	O-Glc	H	H
15	H	OMe	H	OMe	H	H	OH	OH	H
16	H	OH	H	OMe	H	H	OMe	OH	H
17	H	OH	H	OMe	H	OMe	OH	H	H
18	H	OMe	OMe	H	OMe	H	OMe	H	H
19	H	OH	H	OMe	H	OH	OH	H	H
20	H	OMe	OH	H	OMe	H	OH	H	H
21	H	OMe	H	OH	H	H	OMe	OH	H
22	H	OH	H	OMe	H	H	OMe	OH	H
23	H	OMe	OMe	OH	H	H	OH	H	H
24	H	OH	OMe	OMe	H	H	OH	H	H
25	H	OMe	OH	OMe	H	H	OH	H	H
26	H	OH	OH	OMe	OH	H	H	H	H
27	H	OH	OMe	OMe	OH	H	H	H	H
28	H	OH	H	OH	H	OMe	OH	H	H
29	H	OH	H	OMe	H	H	OH	OMe	H
30	OH	OH	H	OMe	H	H	OH	H	H
31	H	OH	H	OMe	OH	H	H	H	OH
32	H	OH	H	OMe	OH	H	OH	H	H
33	H	OH	OMe	OMe	H	OMe	OH	H	H
34	OMe	OMe	H	H	OMe	OMe	OMe	H	H
35	H	OH	OMe	H	OMe	OMe	OH	H	H
36	H	OH	OMe	H	OMe	OH	OH	H	H
37	H	OH	OH	H	OMe	OH	OH	H	H
38	H	OH	OMe	OMe	H	H	OMe	H	H
39	H	OH	OMe	OMe	H	OH	OMe	OMe	H
40	OMe	OH	H	H	OMe	OH	OMe	H	H
41	OMe	OMe	OH	H	OH	OH	H	H	H
42	H	OMe	H	OH	OMe	OMe	H	OH	H
43	H	OMe	OMe	OMe	H	H	OH	H	H
44	CH_2OMe	OH	OMe	OMe	H	H	OH	H	H
45	H	OMe	H	OH	OMe	OH	H	H	H
46	OH	H	H	OH	OMe	OH	OMe	H	H

	R1	R2	R3	R4	R5	R6	R7	R8	R9
47	OH	H	H	OH	OMe	OMe	OH	H	H
48	OH	H	H	OH	OMe	OMe	OMe	H	H
49	H	OH	OMe	OMe	H	H	OH	OMe	H
50	H	OH	OMe	OMe	H	OMe	OH	OMe	H
51	H	OMe	OH	O-Glc	OH	H	H	H	H
52	H	OMe	OH	OMe	OH	H	H	H	H
53	OH	OMe	OH	OMe	OH	H	H	H	H
54	H	OMe	OH	OMe	OH	OH	H	H	H
55	H	OH	OMe	OMe	H	H	OH	H	OH
56	H	OH	H	OH	O-Glc	H	H	H	OH
57	H	O-Glc	H	OMe	OH	H	H	H	OH
58	H	H	H	OH	OMe	OH	H	H	OH
59	H	OH	H	OMe	OH	H	OH	H	OH
60	H	OH	H	OH	OMe	H	OH	H	OH
61	H	OH	H	OMe	OH	H	H	H	OH
62	H	OH	H	OH	OH	H	H	H	OH
63	OH	H	OMe	OMe	OH	H	OMe	H	H
64	H	OMe	OH	OMe	H	H	OMe	H	H
65	H	OMe	H	OH	H	H	OMe	OH	H
66	H	OH	OMe	OMe	H	H	OMe	OH	H
67	H	OH	H	O-Glc	OH	H	H	H	OH
68	H	OH	H	OMe	H	H	OMe	OH	H
69	OH	OMe	H	OH	H	H	OH	H	H
70	H	OMe	H	OMe	H	H	OH	CHO	H
71	H	OH	H	OMe	p-HB	H	OMe	OH	H
72	H	OH	p-HB	OMe	H	H	OH	H	H
73	H	OH	p-HB	H	OMe	H	OH	p-HB	H
74	p-HB	OH	p-HB	OMe	H	H	OH	H	H
75	p-HB	OH	H	OMe	H	H	OH	H	H
76	H	OH	p-HB	H	OMe	p-HB	OH	p-HB	H
77	p-HB	OH	H	OMe	H	H	OMe	H	H
78	p-HB	OH	H	OMe	H	H	OH	OMe	H
79	p-HB	OH	H	OMe	H	H	OH	p-HB	H
80	p-HB	OH	H	OMe	H	OMe	OH	H	H
81	p-HB	OMe	H	OH	H	H	OH	H	H
82	p-HB	OH	H	OMe	H	H	OMe	OH	H
83	H	OH	H	H	OMe	H	OMe	p-HB	H
84	prenyl	OH	H	H	OMe	H	OMe	p-HB	H
85	prenyl	OH	H	H	OMe	H	H	p-HB	H
86	p-HB	OH	prenyl	OH	OH	H	OMe	p-HB	H
87	p-HB	OMe	H	OH	H	H	OH	prenyl	H
88	H	OH	H	OH	OH	H	OMe	p-HB	H
89	p-HB	OH	p-HB	OH	OH	H	OMe	p-HB	H
90	p-HB	OMe	H	OH	OH	H	H	H	H
91	H	OMe	p-HB	OH	H	H	OH	H	H

Almost all of the monophenanthrenes are substituted with hydroxy group(s); those *O*-glycosides [e.g., cremaphenanthrene M (**236**), **250**, **360**, and **361**] that do not bear this functional group are suggesting to be biogenetically also hydroxylated [5, 26–29]. The hydroxy group is usually linked at C-2 and/or C-7. There are only few compounds (e.g., **2**, **62**) substituted with only hydroxy groups [13, 15]. The methoxy group is the second most common substituent, linking mainly at C-2, C-4, C-5, and C-6. Most of the Orchidaceae phenanthrenes are methoxylated, e.g., calanphenanthrene A (**45**), calanhydroquinones A–C (**46–48**), calanquinones A–C (**268**, **278**, **279**) [30, 31], cephanthrenes A (**49**) and B (**50**) [32], marylaurencinols A–C (**52–54**) [33], and ephemeranthoquinone C (**281**) [34], **42**, **63**, **64** [18, 35] and **205** [36], identified from *Calanthe arisanensis*, *Cephalantheropsis gracilis*, *Cymbidium* Great Flower Marie Laurencin, *Dendrobium moniliforme*, *D. nobile*, and *Liparis nakaharai*. Compounds **3**, **171**, and **179**, isolated from *Bletilla striata* [37], the 9,10-dihydrophenanthrene callosumin (**18**) and its phenanthrene pair callosuminin (**187**) from *A. callosum* [38], and **34** from *Flickingeria fimbrata*, substituted with only methoxy groups [39].

The most common monophenanthrenes are the 9,10-dihydrophenanthrene coelonin (**1**) and its phenanthrene pair flavanthrinin (**172**) [40], hircinol (**8**), and its unsaturated pair plicatol-B (moscatin, **166**), lusianthridin (**5**) and lusianthrin (**167**), and erianthridin (**24**) and nudol (**174**). Coelonin (**1**) was isolated from almost every investigated Orchidaceae species [*Appendicula reflexa* [41], *Arundina graminifolia* [42], *B. formosana* [43], *B. striata* [12, 37, 44], *Bulbophyllum reptans* [24], *B. odoratissimum* [45], *B. vaginatum* [46], *Coelogyne elata* [47], *C. ochracea* [47], *Cremastra appendiculata* [26, 48], *Cymbidium aloifolium* [49], *C. finlaysonianum* [50], *Cyrtopodium paniculatum* [51], *D. formosum* [52], *D. plicatile* [53], *D. scabrilingue* [54], *Eria flava* [55], *Eulophia macrobulbon* [56], *E. nuda* [57], *E. petersii* [58], *Flickingeria fimbrata* [39], *G. conopsea* [59, 60], *Pholidota chinensis* [61, 62], *Pleione bulbocodioides* [63, 64], and *Spiranthes sinensis* var. *amonea* [65], and it is a common constituent of several dimers. Orchinol (**4**) is also a common 9,10-dihydrophenanthrene identified from *A. callosum* [38, 66], *A. graminifolia* [67], *C. appendiculata* [48], *O. militaris* [68, 69], *Pholidota chinensis* [61], and *S. sinensis* var. *amoena* [65]. Lusianthridin (**5**) was isolated from several orchids such as *Arundia graminifolia* [70], *B. striata* [44, 71], *Cymbidium finlaysonianum* [50], *Cyrtopodium paniculatum* [51], *D. brymerianum* [72], *D. densiflorum* [73], *D. formosum* [52], *D. nobile* [74, 75], *D. palpebrae* [76], *D. plicatile* [53, 77], *D. scabrilingue* [54], *D. venustum* [78], *E. petersii* [58], *Flickingeria fimbrata* [39], *L. indivisa* [79], *N. boothii* [80], *P. cantonensis* [81], *P. bulbocodioides* [63, 64], *S. livida* [82], *S. scariosa* [83], and *T. alba* [84]. Plicatol-B (**166**) was identified not only from *Dendrobium* species *D. chrysanthum* [85], *D. plicatile* [23], *D. rotundatum* [86, 87], *D. densiflorum* [73], *D. moscatum* [87], and *D. loddigesii* [88], but also from *B. odoratissimum* [45],and *Coelogyne longipes* [89].

In Orchidaceae phenanthrenes, mono- (glucose) or disaccharide units (evolved from glucose, apiose, and rhamnose) can also be joined to the skeleton. Monoglucosides were isolated from *Bletilla striata* (**11**, **93–95**, **249**, **251**), *Cremastra appendiculata* (**12**, **13**) [90], *Cymbidium* Great Flower Marie Laurencin (**51**) [33], *Dendrobium primulinum* (**67**) [16], and *Liparis regnieri* (**206**, **207**) [91]. Moreover, disaccharide-substituted phenanthrenes, such as phenanthrene-glucoapioside (**198**), glucorhamnoside (**199**) [13], and diglucosides (**14**, **221**, **222**, **248**, **250**, **360**, **361**), were also reported from orchids (*B. striata*, *C. appendiculata*, *Dendrobium chrysanthum*, and *Dendrobium denneanum*) [26, 90, 92, 93].

The presence of prenyl-substituted phenanthrenes is relatively uncommon in nature. However, seven prenylated compounds, namely sinensols B (**84**), C (**85**) and F (**86**), spirasineol A (**87**), sinensol G (spiranthol-A, **98**), spiranthol-B (**99**), and spiranthesol (**305**) were isolated from *Spiranthes sinensis* and *S. sinensis* var. *amonea*, one of them (**305**) is a dimer substituted with two prenyl groups [65, 94, 95, 112, 113]. A formyl group when present occurs at C-8 (spiranthesphenanthrene D, **70**), while carboxyl group at C-4 (**252**) in the case of compounds isolated from *Coelogyne ochracea* and *S. sinensis* [95, 96].

Considering the stilbene origin of almost all of the phenanthrenes, it is axiomatical that stilbene-substituted phenanthrenes also occur in orchids. Surprisingly, this type of phenanthrenes is quite rare in plants. Most of the compounds ($n = 13$) were isolated from orchids. In cases of phochinenins I (**108**), J (**109**), and K (**110**) [97], isolated from *Pholidota chinensis*, lusianthridin (**5**) is connected with batatasin III (dihydrostilbene), and in phochinenin L (**113**) [97] a dihydrophenanthrene, isolated previously from a liverwort (*Plagiochila spinulosa*), is linked to the stilbene thunalbene [98]. Other Orchidaceae phenanthrenes, e.g., bleformin C (**107**), phochinenin K (**110**), bulbocodioidin H (**111**), bletilladin D (**112**), and phoyunnanins A (**114**) and B (**115**) are also derived from 9,10-dihydrophenanthrenes, substituted with stilbenes [25, 43, 44, 97, 99, 100]. Contrary to these phenanthrenes, where stilbenes and phenanthrenes are connected via C-C coupling, monbarbatain E (**106**) is joined to stilbostemin E (a dihydrostilbene moiety) through an *O* atom [101]. Similar compounds [shancilin (**104**), and phoyunnanin (**105**)] were isolated from *Pleione bulbocodioides* and *Pholidota yunnanensis* [63, 102]. Interestingly, up to now, only stilbene-substituted mono-9,10-dihydrophenanthrenes were identified from orchids; stilbene-substituted mono-, di-, and tri-9,10-dehydrophenanthrenes are not published in the literature.

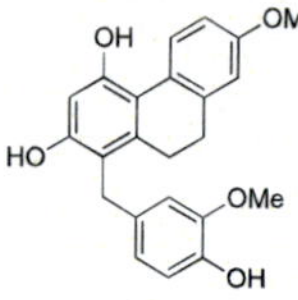

92

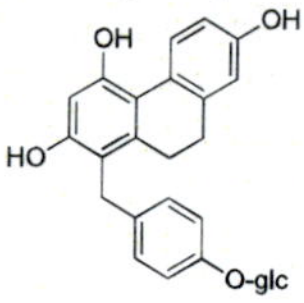

93

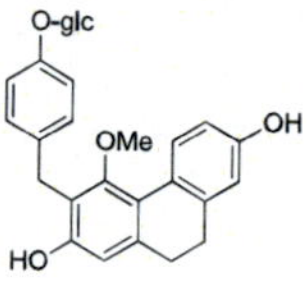

94

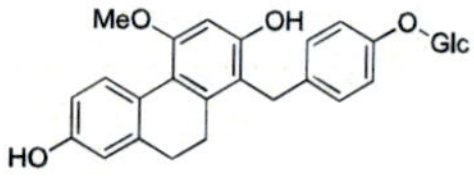

95

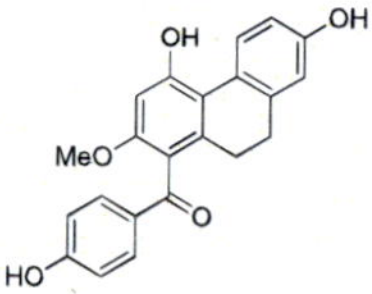

96

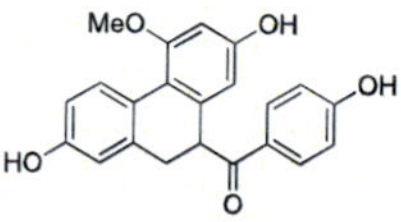

97

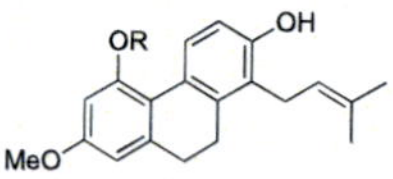

98 R = H
99 R = Me

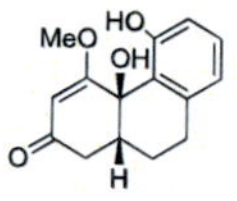

100

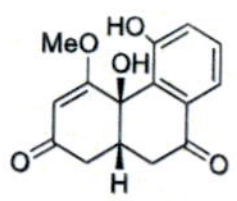

101

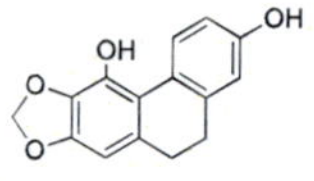

102

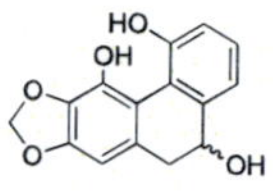

103

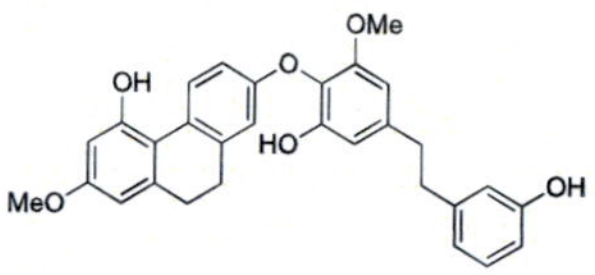

104

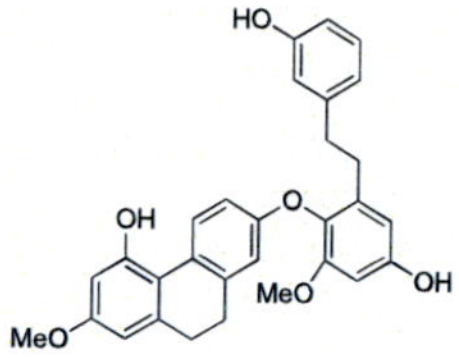

105

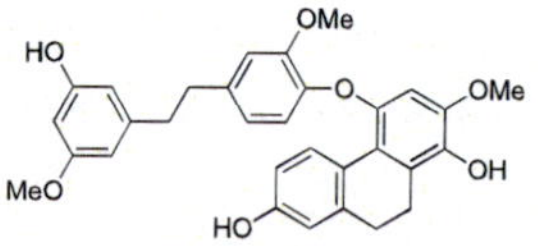

106

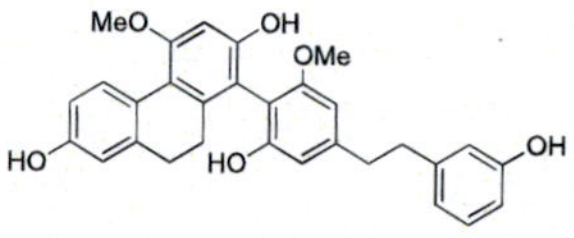

107

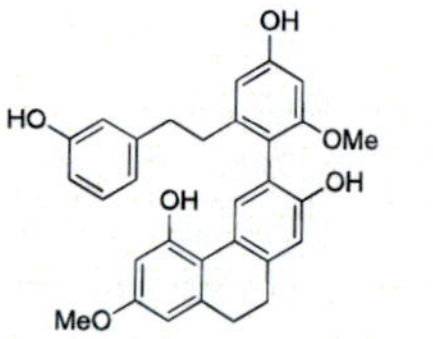

108

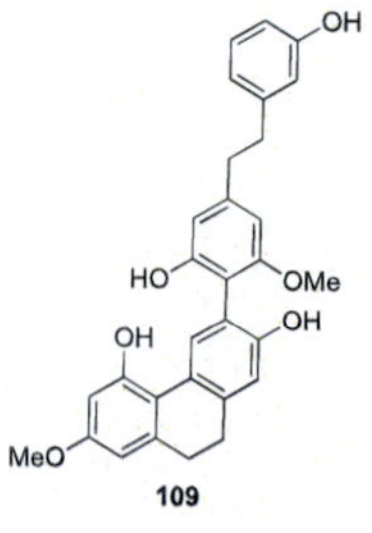

109

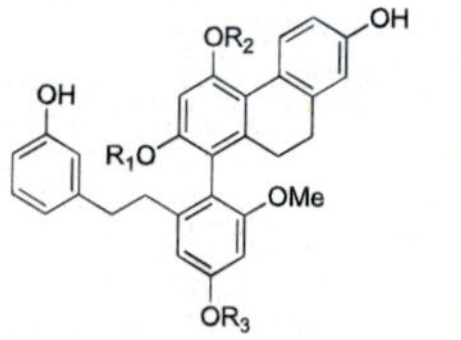

110 R_1 = Me, R_2 = H, R_3 = H
111 R_1 = H, R_2 = Me, R_3 = H
112 R_1 = H, R_2 = H, R_3 = Me

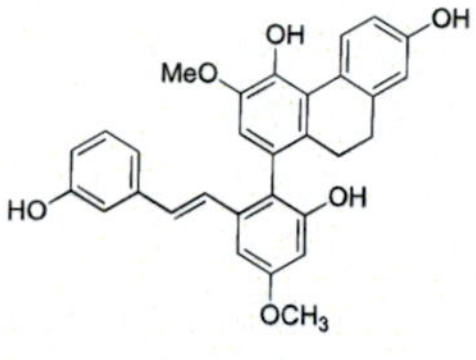

113

Hydroxybenzyl-substituted monophenantheres are quite usual in Orchidaceae species. Such compounds were isolated from the members of *Arundia* (**83**, **88**) [67, 70], *Bletilla* [e.g., **72–79**, **82**, **88**, **241–244**, bleochrin G (**245**)] [12, 43, 44, 71, 103–107, 170], *Cremastra* (e.g., **92**) [108], *Cymbidium* (**80**) [50], *Eulophia* (**82**) [109], *Gymnadena* [**75**, gymconopins A (**90**) and B (**91**)] [59, 60], *Monomeria* (**245**) [110], *Pleione* (**78**, **81**, **88**) [63, 111], and *Spiranthes* [sinensols A (**83**), B (**84**), C (**85**), F (**86**), spirasineol-A (**87**), and sinensol H (**89**)] genera [65, 112, 113].

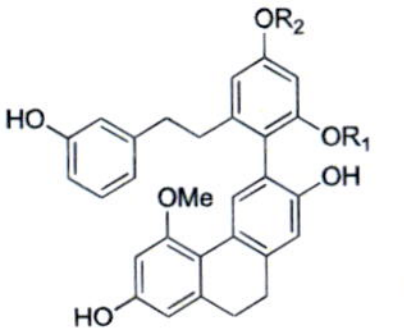

114 R_1 = H, R_2 = Me
115 R_1 = Me, R_2 = H

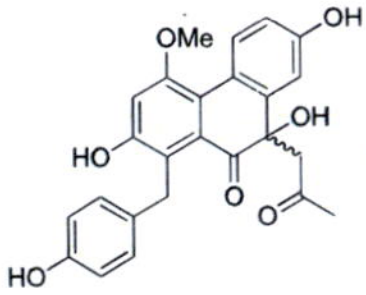

116

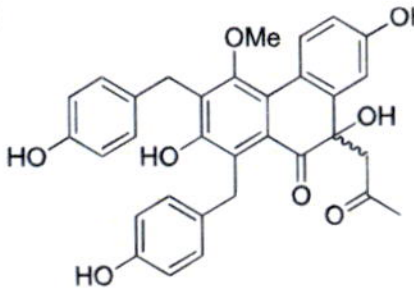

117

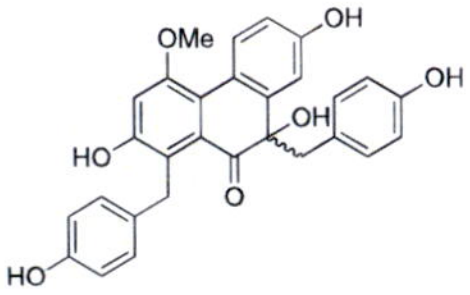

118

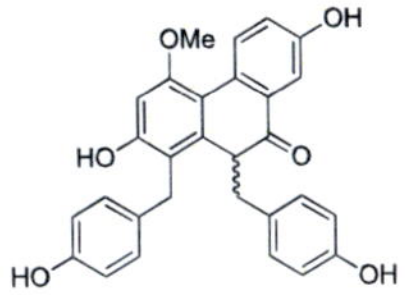

119

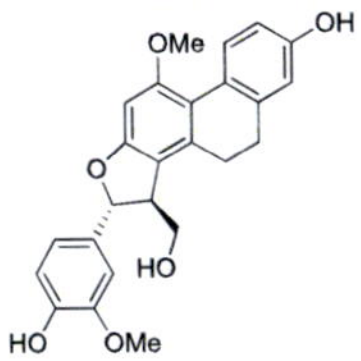

120

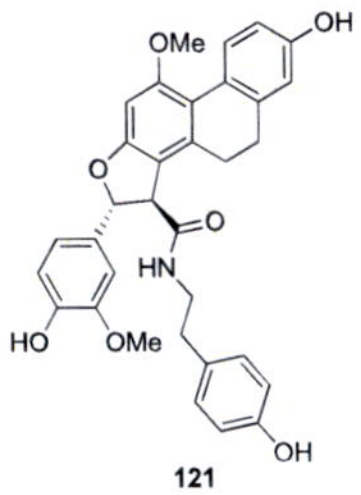

121

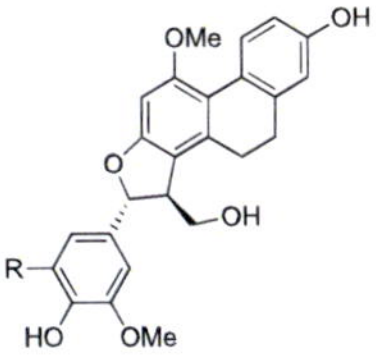

122 R = H
123 R = H

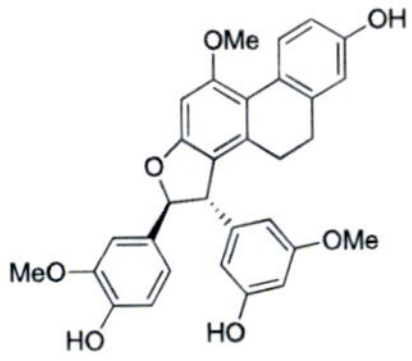

124

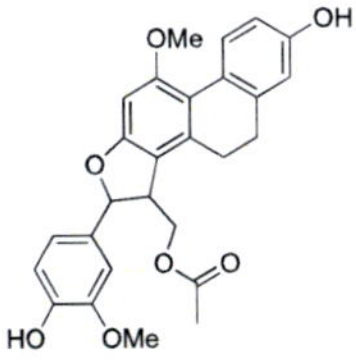

125

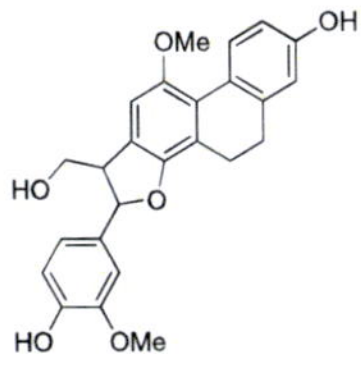

126

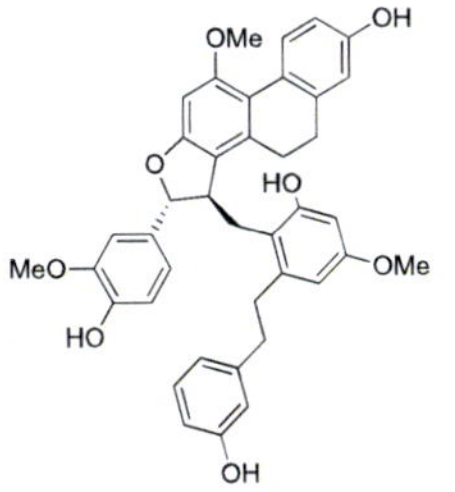

127

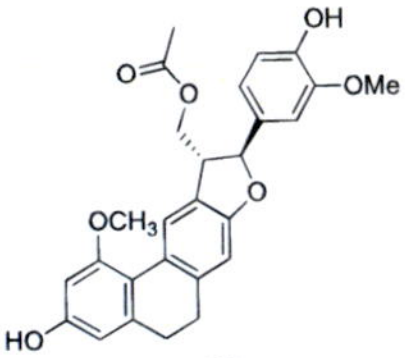

128

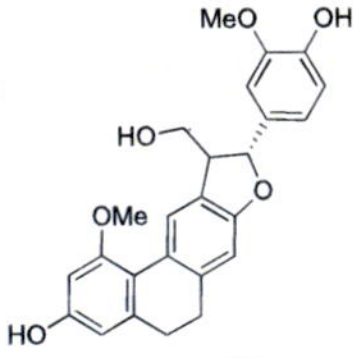

129

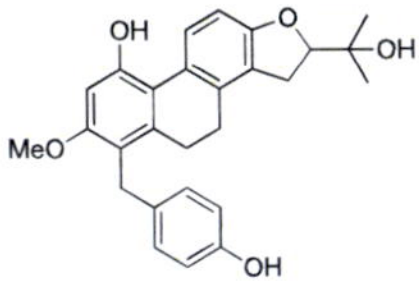

131

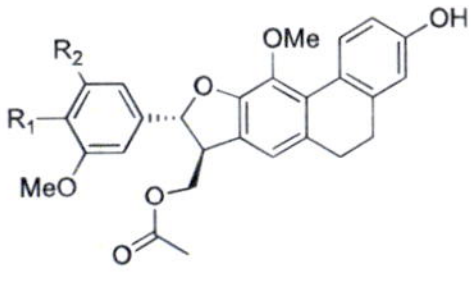

133 R_1 = OH, R_2 = OMe
134 R_1 = H, R_2 = OH

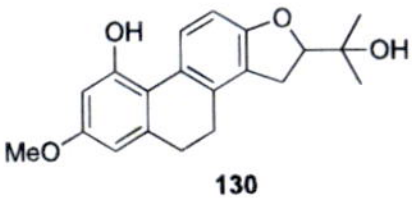

130

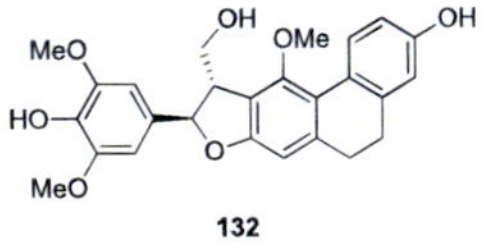

132

Another group of monophenanthrenes in orchids involves phenanthrofurans and phenanthropyrans. Furan and pyran rings are generally linked to the phenanthrene skeleton at C-1–C-2 (**120–127**, **129**, **140–146**) [17, 44, 90, 99, 114–121], but other connections [e.g., C-2–C-3 (**131–134**) [17, 112, 122], C-5–C-6 (**136**) [35, 123], C-6–C-7 (**135**, **148**, **149**) [112, 119, 120, 122], and C-7–C-8 (**138**, **139**, **150–153**)] also occur [95]. Interestingly, all of the furan ring-containing derivatives are dihydrophenanthrenes, while loddigesiinol B (**246**) is the only phenanthropyran, which has an unsaturated phenanthrene skeleton [19, 22]. Phenanthrenelactones were reported from *Dendrobium crystallinum* (**247**) [124].

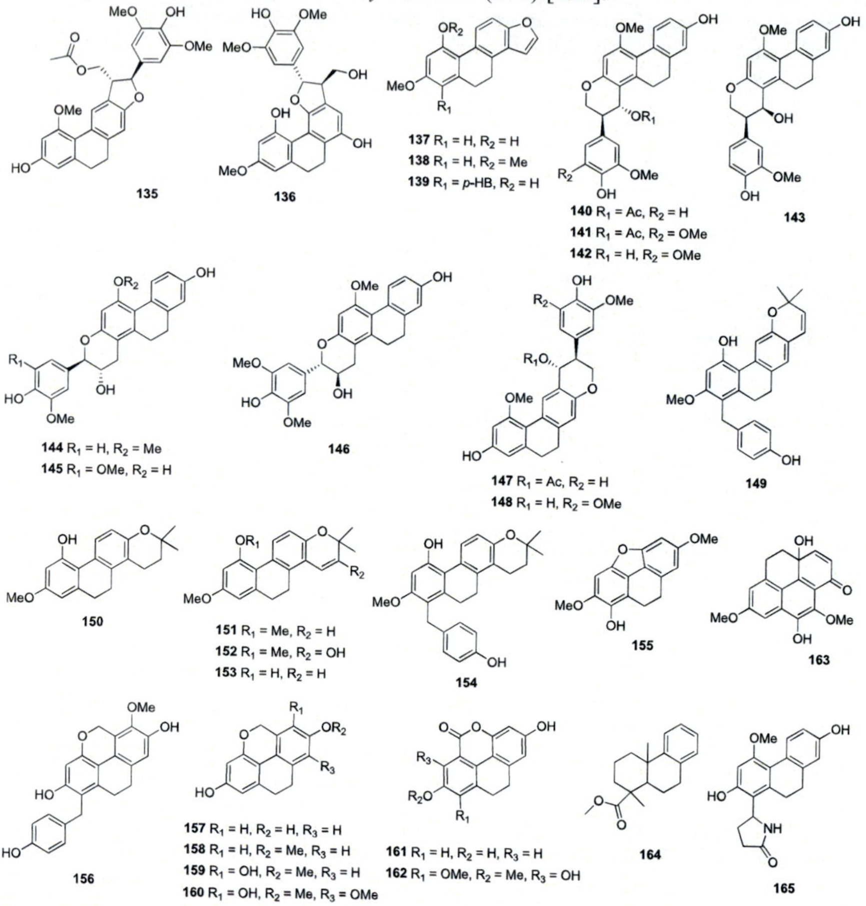

In case of **155**–**163**, a ring closure is performed between C-4 and C-5 forming a five- (**155**) or six-membered (**156**–**162**) oxygen-heterocyclic ring or a cyclohexane ring (**163**). These compounds were isolated from *Pholidota* [81, 125, 126], *Vanda* [127], *Coelogyne* [22, 89, 128–130], *Otochilus* [128], *Sunipia* [83], and *Dendrobium* species [131]. Some monophenanthrenes have additional different substituents, e.g., in case of cirrhopetalanthridin (**102**) and bobulretin B (**103**), and their 9,10-unsaturated pairs cirrhopetalin (**223**) and bobulretin A (**224**) a methylenedioxy group is connected at C-2–C-3 [132–134]. Two spirolactone-substituted phenanthrenes (**253**, **254**) were isolated from *Bletilla striata* and *D. chrysanthum* [85, 135].

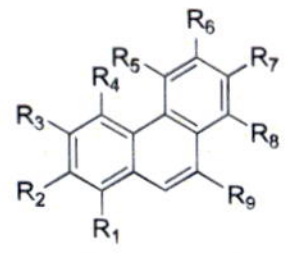

	R1	R2	R3	R4	R5	R6	R7	R8	R9
166	H	OH	H	OMe	OH	H	H	H	H
167	H	OMe	H	OH	H	H	OH	H	H
168	H	OMe	H	OMe	OH	H	H	H	H
169	H	OMe	H	OMe	H	H	OH	H	H
170	H	OMe	OH	OMe	H	H	OH	H	H
171	H	OMe	H	OMe	H	H	OMe	H	H
172	H	OH	H	OMe	H	H	OH	H	H
173	H	OH	H	OMe	H	H	OMe	H	H
174	H	OH	OMe	OMe	OH	H	OH	H	H
175	H	OH	H	OMe	H	H	OH	OMe	H
176	H	OMe	OH	OMe	H	H	OH	H	H
177	H	OH	OMe	OMe	H	OMe	OH	H	H
178	H	OH	OH	OMe	H	H	OH	H	H
179	H	OMe	OMe	OMe	H	H	OMe	H	H
180	H	OMe	OH	OMe	H	H	OH	OMe	H
181	OH	OH	H	OMe	H	OH	OH	H	H
182	H	OH	H	H	OH	OMe	OMe	H	H
183	H	OH	OMe	OMe	OH	H	H	H	H
184	H	OH	OH	OMe	OH	H	H	H	H
185	OMe	OH	OMe	H	OH	H	H	H	H
186	H	OH	H	OMe	H	OMe	OH	H	H
187	H	OMe	H	OMe	H	OMe	OMe	H	H
188	H	OH	H	OMe	H	H	OH	OMe	H
189	H	OH	H	OMe	H	H	OMe	OH	H
190	H	OH	H	OMe	OH	H	H	H	OMe
191	H	OH	H	OMe	H	OH	OH	H	H
192	H	OH	OH	OMe	H	H	OH	H	H
193	H	OH	OMe	OMe	H	OMe	OMe	OH	H
194	H	O-Glc	H	OMe	H	H	OH	O-Glc	H
195	H	O-Glc	H	OMe	H	H	O-Glc	OH	H
196	OMe	OH	OMe	H	OMe	OMe	H	H	H
197	H	O-Glc	H	OMe	OH	H	H	H	H
198	H	O-Glc-Api	H	OMe	OH	H	H	H	H
199	H	O-Glc-Rha	H	OMe	OH	H	H	H	H
200	H	OH	H	OMe	OMe	H	H	H	OH

	R1	R2	R3	R4	R5	R6	R7	R8	R9
201	H	OH	OMe	OMe	H	H	OMe	OH	H
202	H	OH	OH	OMe	OH	H	H	H	OMe
203	H	OH	OMe	OMe	OH	H	H	OMe	H
204	H	OMe	H	OMe	H	H	OH	OMe	H
205	H	OMe	OMe	OMe	H	H	OH	H	H
206	H	OH	OMe	OMe	H	H	O-Glc	H	H
207	H	O-Glc	OMe	OMe	H	H	OH	H	H
208	H	OH	OMe	OMe	H	OMe	OH	H	H
209	H	OH	OH	OMe	H	OMe	OH	H	H
210	H	OH	OH	OMe	H	OH	OH	H	H
211	H	OH	OMe	OMe	H	H	OH	OMe	H
212	H	OMe	OH	OMe	H	H	OH	OMe	H
213	H	OMe	OH	OMe	H	H	OH	H	H
214	H	OH	OMe	OMe	OH	H	H	OMe	H
215	H	OH	H	OH	H	OH	OMe	OMe	H

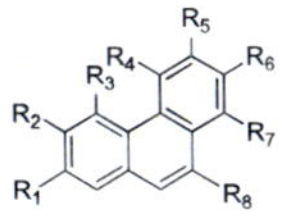

	R1	R2	R3	R4	R5	R6	R7	R8
216	OH	OMe	OMe	H	H	OH	H	H
217	OH	OMe	OMe	H	OH	OMe	OMe	H
218	OMe	OH	OMe	H	OMe	OMe	OMe	H
219	OMe	H	OH	OMe	H	H	H	OH
220	OMe	OH	OMe	H	OMe	OH	H	H
221	O-Glc	H	OMe	H	H	OH	O-Glc	H
222	O-Glc	H	OMe	H	H	O-Glc	OH	H

Phenanthroquinones are supposed to be the oxidative derivatives of phenanthrenes or dihydrophenanthrenes; however, the detailed biosynthetic pathway of these components needs further investigations. Phenanthrenequinones and alkylated 1,4-phenanthrenequinones can be formed from the analogous oxygenated phenanthrenes and abietane-type diterpenoids, respectively [5, 136, 137]. In the past two decades, several phenanthroquinones were identified from orchids belonging to different genera, e.g., *Arundia* (**258**, **272**, **275**) [42, 67], *Bletilla* (**258**) [43, 138, 139], *Bulbophyllum* (**272**) [45], *Calanthe* (**268**, **278**, **279**) [30, 31], *Coelogyne* (**256**, **271**) [140], *Cymbidium* (**276**, **277**, **269**, **281**) [33, 34, 50, 141, 142], *Cyrtopodium* (**258**, **267**) [51], *Cypripedin* (**272**, **274**) [143], *Dendrobium* (**258–262**, **270**, **272**, **282**) [52, 74, 77, 144–148], *Eulophia* (**258**) [149], *Flickingeria* (**266**, **273**) [150], *Nidema* (**258**) [80], *Oncidium* (**264**) [151], *Odontioda* (**280**) [152], *Pholidota* (**265**) [81], and *Spiranthes* (**257**, **258**) [94, 95]. These compounds were found to be diversely substituted – hydroxy, methyl, and methoxy groups being the most frequent moieties attached to the main skeleton.

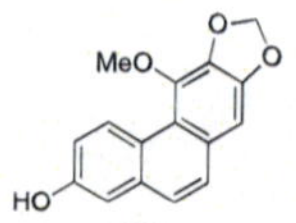

223

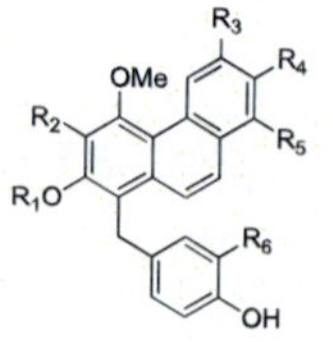

	R1	R2	R3	R4	R5	R6
226	H	H	H	OH	H	H
227	H	H	H	OMe	H	H
228	H	OMe	H	OH	H	H
229	H	H	H	OH	OMe	H
230	H	H	OH	OMe	H	H
231	H	H	H	OMe	OH	H
232	H	H	H	OH	O-Glc	H
233	H	H	H	OH	H	OMe
234	H	H	OH	OH	H	OMe
235	H	H	H	OMe	H	OMe
236	H	H	H	O-Glc	H	OMe
237	Glc	H	H	OH	H	OMe
238	H	H	H	OH	O-Glc	OMe
239	H	H	H	O-Glc	OH	OMe
240	H	OMe	H	H	OMe	H

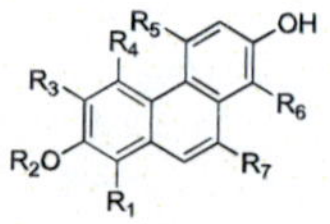

	R1	R2	R3	R4	R5	R6	R7
241	H	H	OMe	OMe	H	*p*-HB	H
242	H	Me	OH	OMe	H	*p*-HB	H
243	H	Me	OH	OMe	H	H	*p*-HB
244	*p*-HB	H	H	H	OMe	*p*-HB	H
245	*p*-HB	H	*p*-HB	OMe	H	*p*-HB	H

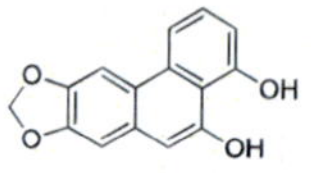

224

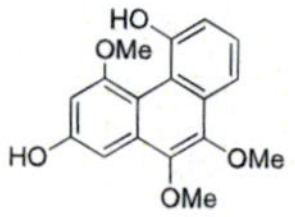

225

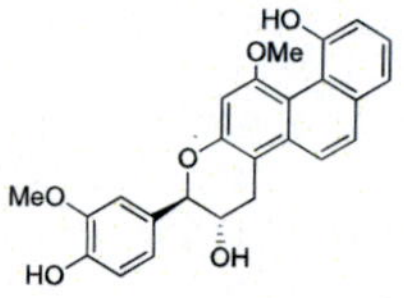

246

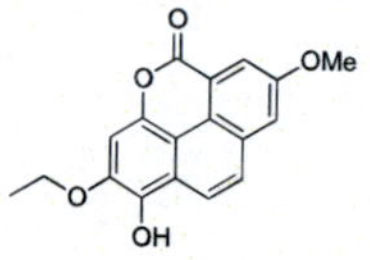

247

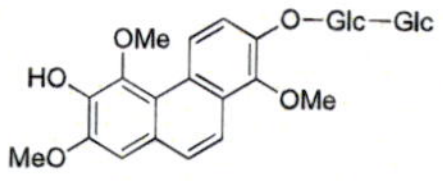

248

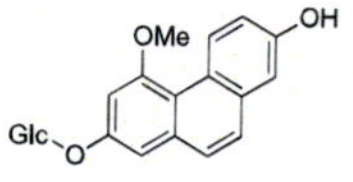

249

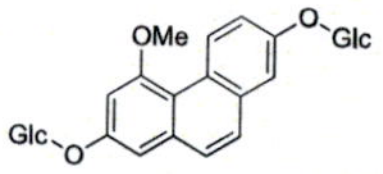

250

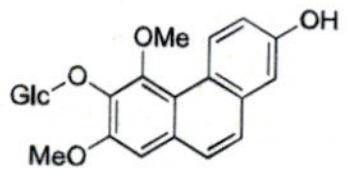

251

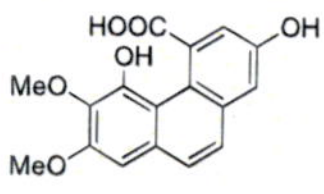

252

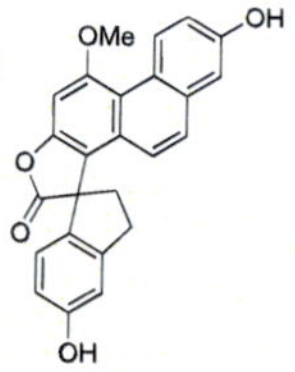

253

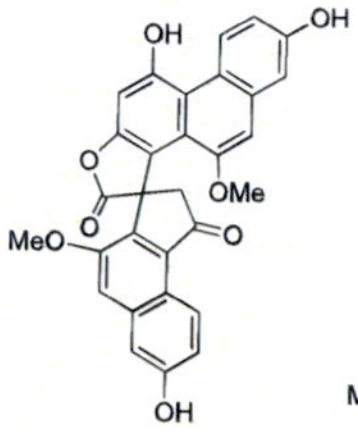

254

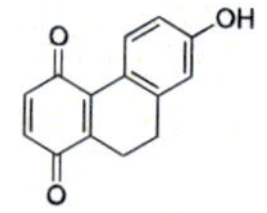

255

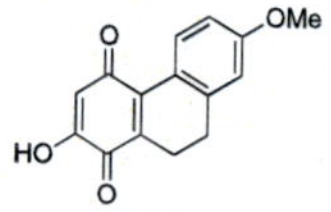

256

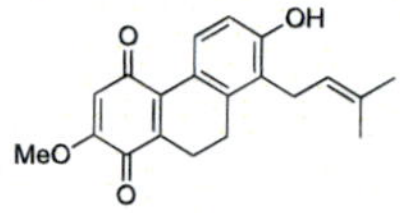

257

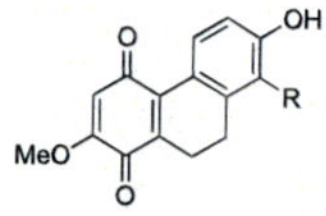

258 R = H
259 R = OMe

Although phenanthrene-1,4-quinones are the most common phenanthrenequinones, the carbonyl group can also be attached at C-2, C-3, C-5, and/or C-8 (e.g., **282–287**) [138, 139, 148, 153]. The saturation level of phenanthrenequinones can vary depending on the type of the phenanthrene unit.

Interestingly, the most common 9,10-dihydrophenanthrenes have their own phenanthrene pairs, e.g., coelonin (**1**) and flavanthrinin (**172**), orchinol (**4**) and dehydroorchinol (**169**), lusianthridin (**5**) and lusianthrin (**167**), hircinol (**8**) and plicatol-B/moscatin (**166**), **3** and **171**, and loroglossol (**7**) and **168**.

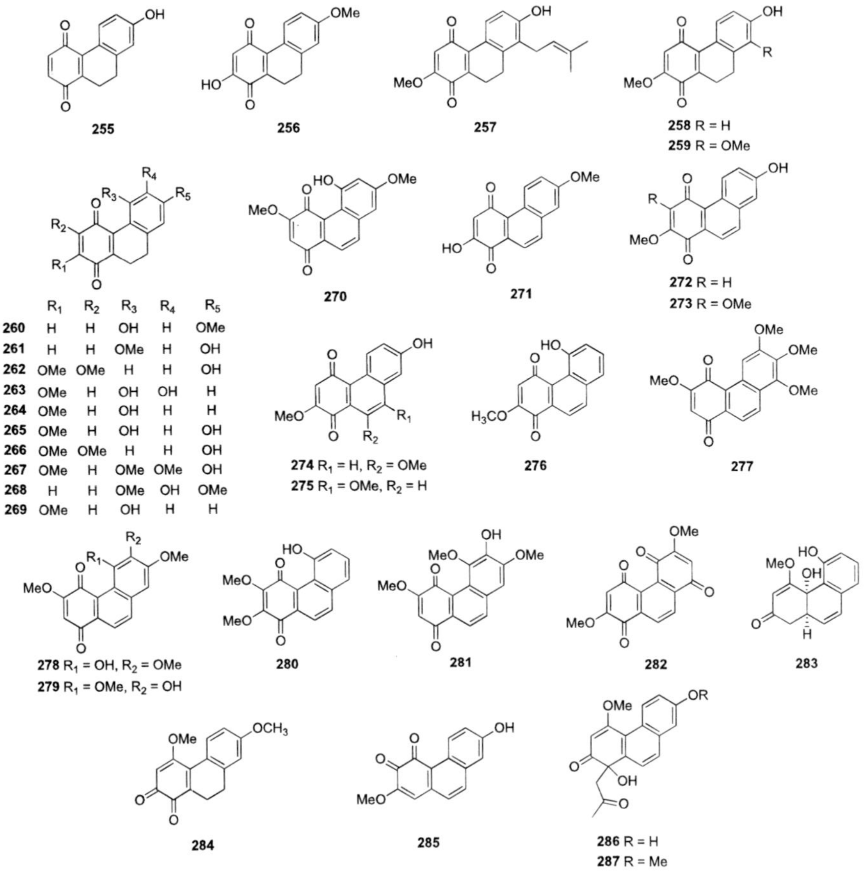

2.2 Di- and Triphenanthrenes

More than 85 dimers and 2 trimers have been described from orchids. Monophenanthrenes can connect through their functional groups or directly via C–C coupling to form di- or triphenanthrenes. The monomers can be linked together at many different positions (1-1’, 1-3’, 1-6’, 1-8’, 2-7’, 3-6’, 6-6’, 8-6’, and 8-8’); usually, C-1 is involved in the connection. The linking monomers have usually been isolated previously from the same plant, or the same genus. Di- and triphenanthrenes are even more rare than monophenanthrenes.

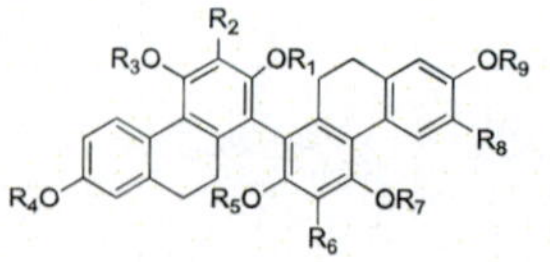

	R1	R2	R3	R4	R5	R6	R7	R8	R9
288	Me	H	H	H	Me	H	H	H	H
289	H	H	Me	H	H	H	Me	H	H
290	H	H	Me	H	H	H	Me	Me	H
291	Me	H	Me	Me	Me	H	Me	H	Me
292	H	OMe	Me	Me	H	OMe	Me	H	Me
293	H	OMe	Me	H	H	OMe	Me	H	H

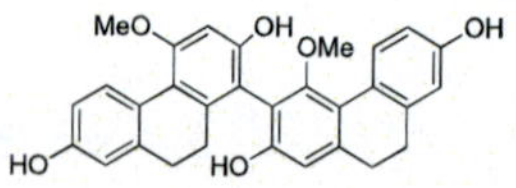

294

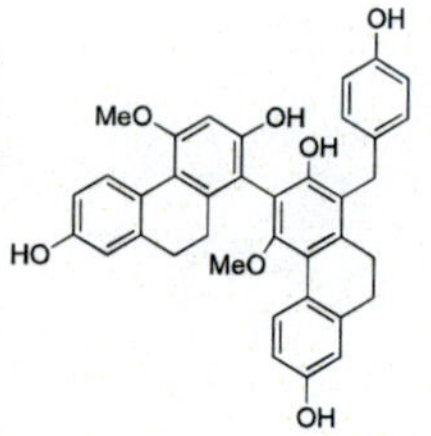

295

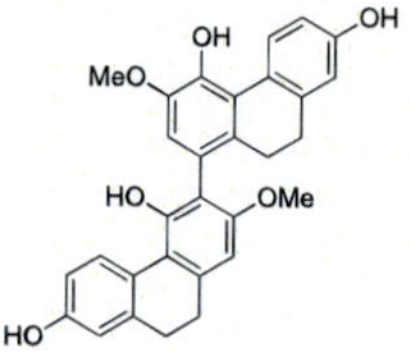

296

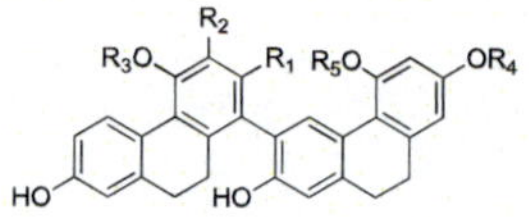

	R1	R2	R3	R4	R5
297	OH	H	Me	H	Me
298	OMe	H	H	Me	H
299	H	OMe	H	H	Me

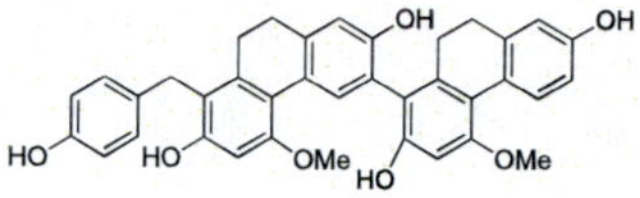

300

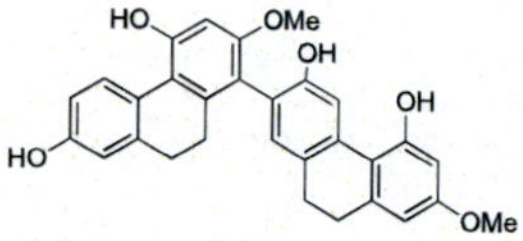

301

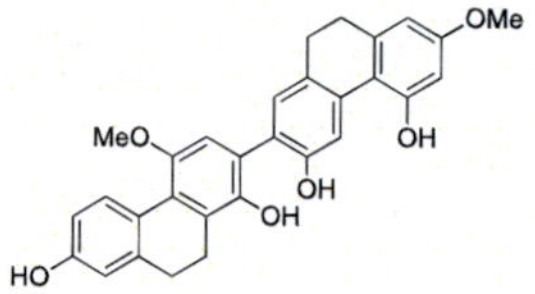

302

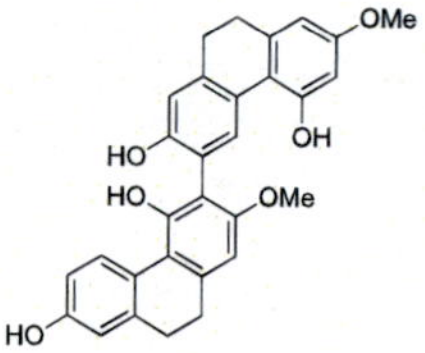

303

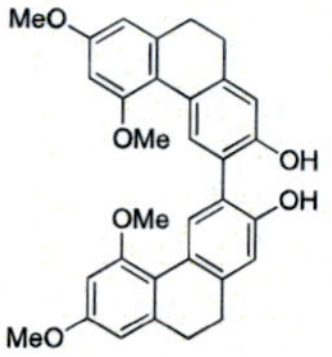

304

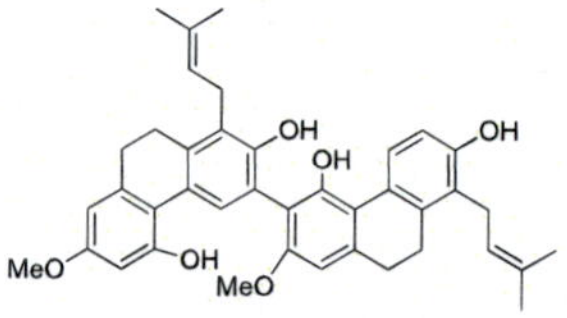

305

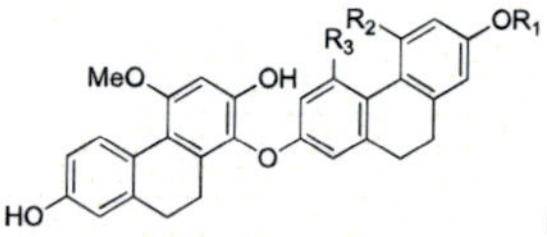

306 R_1 = H, R_2 = H, R_3 = OMe
307 R_1 = Me, R_2 = OH, R_3 = H

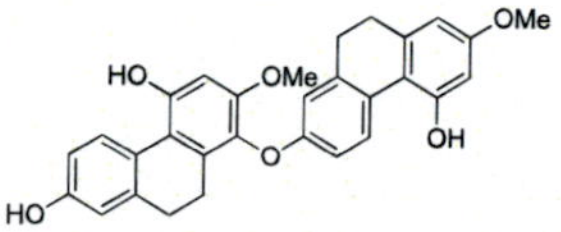

308

Most dimeric compounds isolated from orchids are supposed to be formed by the coupling of prevalent monomeric units, e.g., lusianthridin (**288**, **298**, **303**, **346**) [99, 100, 115, 154], coelonin (**289**, **294**, **295**, **297**, **300**) [43, 51, 67, 107], flavanthrinin (**353**) [43, 84, 155], orchinol (**304**) [95, 156], nudol (**368**) [157], 2,4,7-trimethoxy-9,10-dihydrophenanthrene (**291**) [158], and 2-hydroxy-4,7-dimethoxyphenathrene (**355**) [153, 157]. Interestingly, a prenylated dimer [spiranthesol (**305**)] is also identified from *Spiranthes sinesis* which is composed of two 3-6' connected sinensol G (**98**) monomers [94]. Among the identified compounds, symmetrical dimers also occur, e.g., in case of **291**, two 2,4,7-trimethoxy-9,10-dihydrophenanthrene (**3**) monomers are attached through C-1 and C-1' [158].

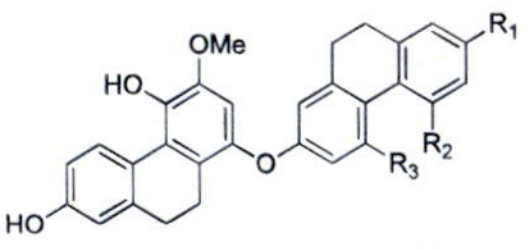

309 R_1 = OH, R_2 = OMe, R_3 = H
310 R_1 = OMe, R_2 = OH, R_3 = H
311 R_1 = OH, R_2 = H, R_3 = OMe

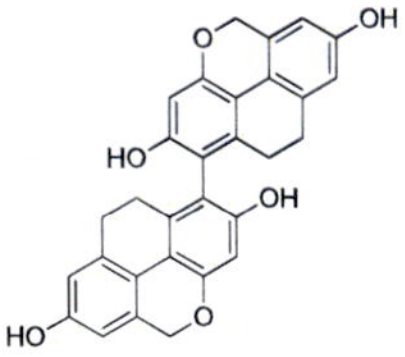

312

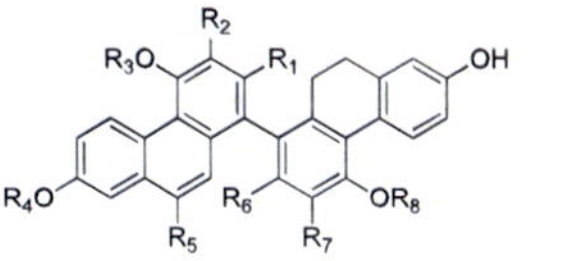

	R_1	R_2	R_3	R_4	R_5	R_6	R_7	R_8
313	OH	H	Me	H	H	OH	H	Me
314	OMe	H	H	H	H	OMe	H	H
315	OH	OMe	Me	H	H	OH	OMe	Me
316	OH	OMe	Me	H	H	OH	H	Me
317	H	OMe	H	H	H	H	OMe	H
318	OH	H	Me	Me	H	OH	H	Me
319	OH	H	Me	Me	H	OH	OMe	H
320	OH	H	Me	H	OMe	OH	H	Me

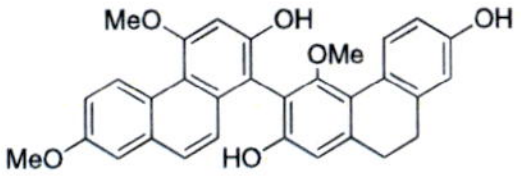

321

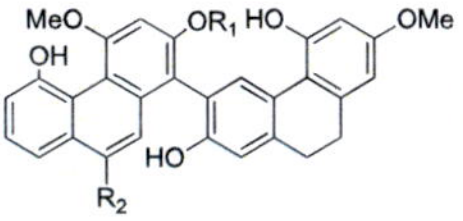

329 R_1 = Me, R_2 = OH
330 R_1 = H, R_2 = H

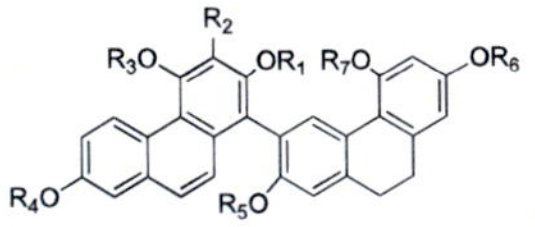

	R_1	R_2	R_3	R_4	R_5	R_6	R_7
322	H	H	Me	Me	H	Me	H
323	Me	H	H	H	H	Me	H
324	H	H	Me	H	H	H	Me
325	H	H	Me	Me	Me	H	Me
326	Me	H	H	Me	H	Me	H
327	H	OMe	Me	H	H	Me	H
328	H	OMe	Me	H	H	H	Me

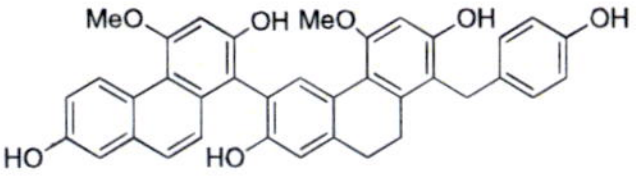

331

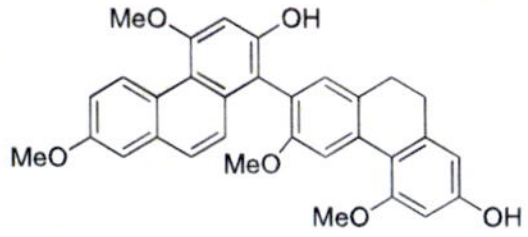

332

333 R_1 = OMe, R_2 = H
334 R_1 = H, R_2 = OMe

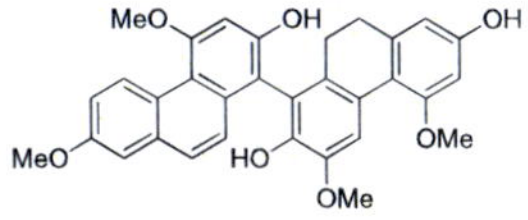

335

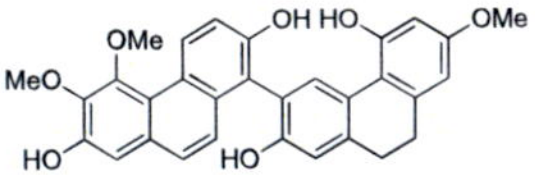

336

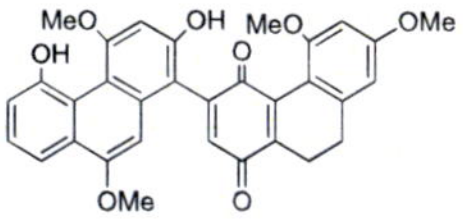

338

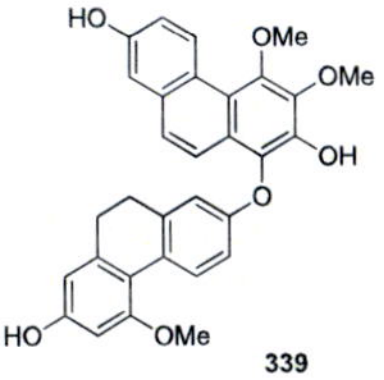

339

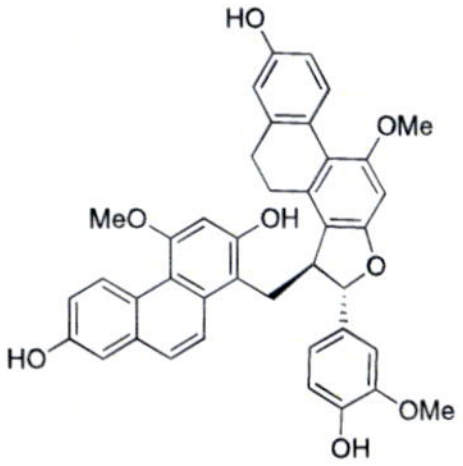

337

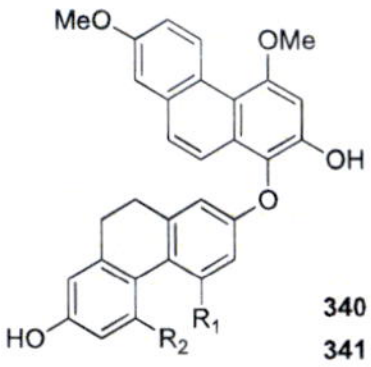

340 R_1 = H, R_2 = OMe
341 R_1 = OMe, R_2 = H

The C–C coupling of two flavidin monomers (**157**) resulted in 8,8'-biflavidin (**312**) [159]. Different monomers, 9,10-dihydro and -dehydro derivatives, can also be connected producing a great variety of dimers. In blestriarene A (**290**), two 9,10-dihydro derivatives, namely coelonin (**1**) and 6-methoxy coelonin (**17**), are joining through their 1-1' [160], while in blestriarene B (**313**) a moscatin (**166**) unit connects to coelonin (**1**) [104]. In bleformin I (**315**), a nudol (**174**) and an erianthridin (**24**), and in coeludol A (**316**), a coelonin (**1**) and a nudol (**174**) are connected [43, 51].

The coupling of lusianthridin (**5**) with lusianthrin (**167**) resulted in the formation of monbarbatains B (**314**) and C (**323**), differing only in the positions of the linkages [5, 161]. In bulbophythrin B (**301**), a lusianthridin (**5**) and a cannabidihyrophenanthrene (**6**) monomer are connecting, through their C-1 and C-7' [162]. In many cases, **173** is one of the members of dimers, e.g., cremaphenanthrene C (**321**) [163], **318**, **319**, **322**, **332** [164], **335** [154], and **354** [153]. In case of liparisphenanthrene C (**374**), two **173** monomers are joining through an ether bridge [157]. Similarly, in blestrins A (**306**) and B (**307**), the two monomers [two coelonin (**1**) in case of A, and a coelonin (**1**) and a lusianthridin (**5**) in case of B] connect forming an ether bridge [165]. Phoyunnanin E (**308**), phochinenis C–E (**309**–**311**), and blestrins C (**333**) and D (**334**) are also dimers, in which the two monomers are connecting through a hydroxy group, resulted in an ether bridge [115, 166, 167]. Dendropalpebrone (**338**), isolated from *D. palpebrae*, is the only dimer containing a phenanthrene and a phenanthroquinone moiety [76].

Until now, compounds **360** and **361**, isolated from *C. appendiculata*, are the only glycosylated phenanthrene dimers that have been reported from nature [26]. Presumably, these diphenanthrenes are formed by the coupling of a flavanthrinin (**172**) and a flavanthrinin-diglucoside. Several diphenanthrenes are formed by the coupling of 2,5-dihydroxy-6-methoxy-9,10-dihydrophenanthrene, identified from a liverwort species *Plagiochila spinulosa* with other, more common monomers, e.g., lusianthridin (**296**, **310**) and coelonin (**299**, **309**), described from orchids [115]. In case of phochinenin B (**317**), the same liverwort dihydrophenanthrene connects with its previously unknown unsaturated form [115]. Compound **292**, a constituent of *D. nobile* [18], is a homodimer of two 2-hydroxy-3,4,7-trimethoxy-9,10-dihydrophenanthrene, identified previously from *P. spinulosa* [98].

Up to now, only two triphenanthrenes are described from orchids; monbarbatain D (**375**) is most likely derived from the coupling of three 9,10-dihydrophenanthrene lusianthridin (**5**) units through C-1–C-8–C-6 [161], while in **376**, three phenanthrene [two flavanthrinin (**172**) and a 2-hydroxy-4,7-dimethoxyphenanthrene (**173**)] monomers are connected [153].

2.3 Occurrence of phenanthrenes in Orchidaceae

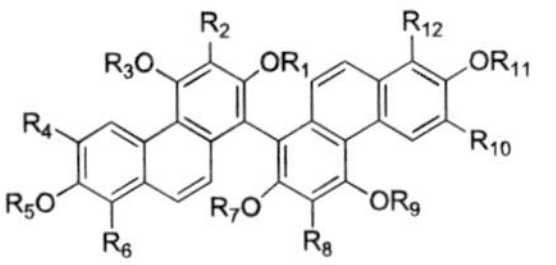

	R1	R2	R3	R4	R5	R6	R7	R8	R9	R10	R11	R12
342	H	H	Me	OMe	H	H	H	H	Me	OMe	H	H
343	H	H	Me	OH	Me	H	H	H	Me	OH	Me	H
344	H	H	Me	H	H	H	Me	H	Me	H	H	H
345	Me	H	H	H	H	OMe	Me	H	H	H	H	OMe
346	Me	H	H	H	H	H	Me	H	H	H	H	H
347	Me	H	H	H	H	H	Me	H	H	H	Me	H
348	Me	H	H	H	H	OMe	Me	H	H	H	H	H
349	H	H	Me	OMe	H	H	H	H	Me	OH	Me	H
350	H	H	Me	H	H	H	H	H	Me	OMe	H	H
351	H	H	H	H	H	H	H	H	Me	H	H	OMe
352	H	H	Me	H	H	OMe	H	H	Me	H	H	OMe
353	H	H	Me	H	H	H	H	H	Me	H	H	H
354	H	H	Me	H	H	H	H	H	Me	H	Me	H
355	H	H	Me	H	Me	H	H	H	Me	H	Me	H
356	H	H	Me	H	H	OMe	H	H	Me	H	H	OMe
357	H	OH	Me	OMe	H	H	H	OH	Me	OMe	H	H
358	H	OMe	Me	H	H	H	H	OMe	Me	H	H	H
359	H	OH	Me	OMe	H	H	H	OH	Me	OMe	H	H
360	H	H	Me	H	H	H	Glc	H	Me	H	Glc	H
361	H	H	Me	H	Me	H	Glc	H	Me	H	Glc	H
362	H	H	Me	H	H	H	H	H	Me	OH	H	H
363	H	H	Me	H	H	H	H	H	Me	H	H	OMe

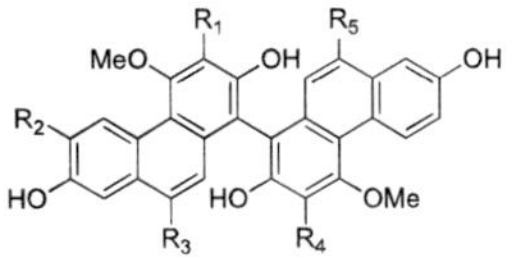

	R1	R2	R3	R4	R5
364	OMe	H	OMe	OMe	OMe
365	H	OMe	H	H	OMe
366	H	H	OMe	H	OMe

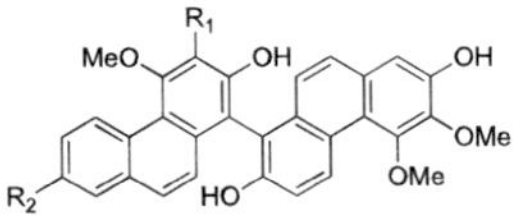

367 R_1 = H, R_2 = OMe
368 R_1 = OMe, R_2 = OH

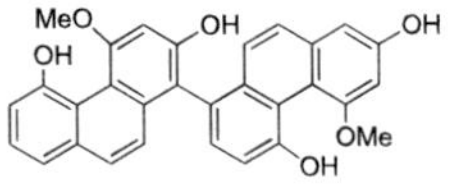

369

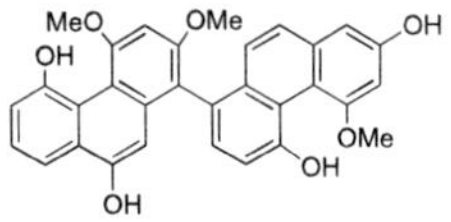

370

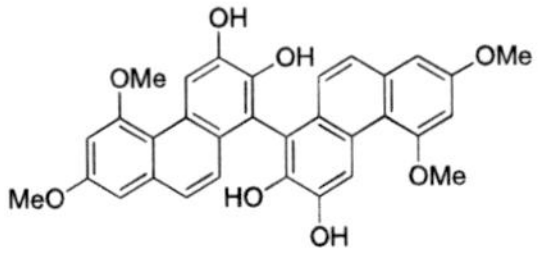

371

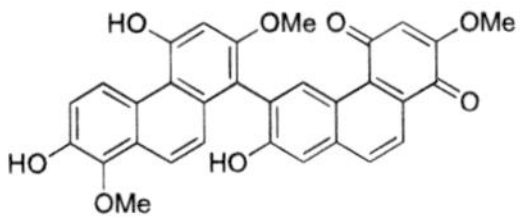

372

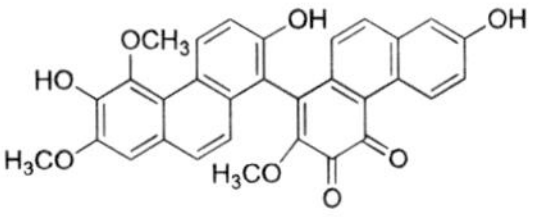

373

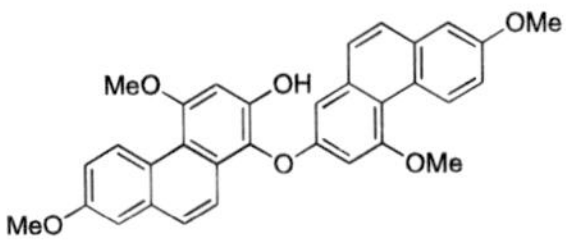

374

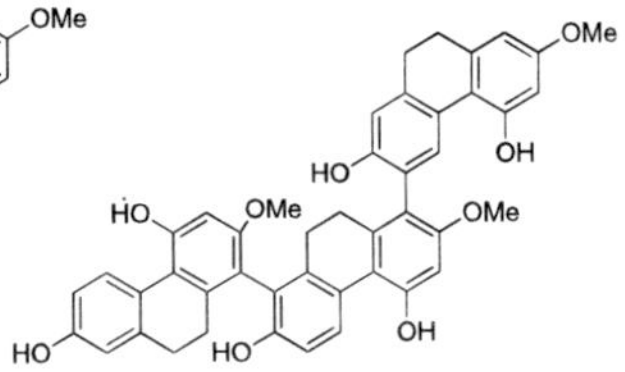

375

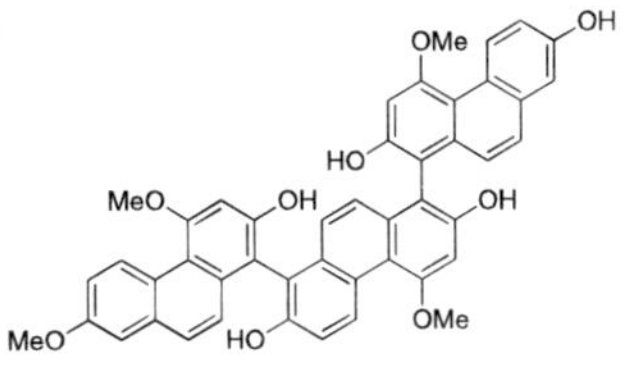

376

3 Chemotaxonomical Significance

Orchidaceae phenanthrenes have chemotaxonomical significance, as prenyl-, *p*-hydroxybenzyl-, and stilbene-substituted phenanthrenes were isolated almost exclusively from orchids. Prenyl-substituted phenanthrenes were isolated only from *Spiranthes sinensis*, and these compounds are substituted also with *p*-hydroxybenzyl group(s). Stilbene-substituted phenanthrenes were identified from *Pleione*, *Pholidota*, *Monomeria*, *Bletilla,* and *Cremastra* species. Either prenyl- or stilbene-substituted phenanthrenes belong to the 9,10-dihydrophenanthrene group, and with the exception of spiranthesol (**305**), all of them are monophenanthrenes. *p*-Hydroxybenzyl-substituted phenanthrenes were isolated from species belonging to the genera *Arundia*, *Bletilla*, *Eulophia*, *Cremastra*, *Cymbidium*, *Dendrobium*, *Gymnadena*, *Pleione*, and *Spiranthes*. Moreover, almost all of the Orchidaceae phenanthrenes are substituted with hydroxy and/or methoxy groups at C-2, C-4, and C-7.

The phenanthrene-containing species belong to the subfamilies Cypripedioideae, Epidendroideae, Higher Epidendorideae, and Orchidoideae. Among them, Epidendroideae is the best source of these compounds; members of 23 genera (e.g., *Arundia*, *Bletilla*, *Cremastra*, *Coelogyne*, *Pleione*, *Pholidota*, *Bulbophyllum*, *Monomeria*, and *Dendrobium*) have been reported to contain phenanthrenes. Several genera (n = 11) of the subfamily Higher Epidendroideae, especially members of subtribe Cyrtopodiinae (e.g., *Eulophia*, *Cyrtopodium*, and *Cymbidium*), contain phenanthrenes, and their structure is very close to that isolated from Epidendroideae species. *Spiranthes* and *Gymnadenia* species belonging to the subfamily Orchidoideae contain *p*-hydroxybenzyl- and/or prenyl-substituted compounds. *Cypripedium tibeticum* belonging to the subfamily Cypripedioideae is the only member of this subfamily from which phenanthrenes (**272**, **274**) have been reported [143].

Bletilla, *Cremastra*, *Pleione*, and *Pholidota* genera belonging to the subfamily Epidendroideae contain the widest variety of phenanthrenes (n = 80, 39, 33, and 36, respectively). From the members (*Bulbophyllum* and *Cirrhopetalum* species) of the subtribe Bulbophyllinae (tribe Podochileae), methylenedioxy-substituted phenanthrenes (**102**, **103**, **223**, **224**) were isolated. *Dendrobium* and *Flickingeria* species belonging to the subtribe Dendrobiinae, tribe Podochileae, are rich in phenanthroquinones.

4 Pharmacological Activities of Orchidaceae Phenanthrenes

4.1 Traditional Uses of Orchids

Several phenanthrene-containing plants are traditionally applied for treating different diseases mainly in Asia, Africa, and South America, e.g., the tubers of *C. appendiculata* are one of the main constituents of a popular traditional Chinese medicine (TCM) product, "Shan-Ci-Gu," which has been used for the treatment of different types of cancer in Asia [26]. Pseudobulbs of *B. striata* and *M. barbata* are used to cure pulmonary disorders in China [41, 110, 164]. The tubers of *Bletilla yunnanensis* have long been used in TCM ("Baiji") in southwest China to treat tuberculosis, pneumonorrhagia, and pneumonophthisis [158]. A traditional medicine, namely "Xiao Huang Cao Shi-Hu," is prepared from the stems of different *Dendrobium* species, e.g., *D. aphyllum*, *D. chrysotoxum*, *D. denneanum*, *D. nobile*, and *D. officinale*, and applied to alleviate stomach disorders and fever [13, 14, 114,

148]. *P. chinensis* has been used in TCM for the treatment of high blood pressure and headache [115]. *S. sinensis* is also a well-known member of TCM, and it is applied to cure inflammatory diseases and different types of cancer [156]. The whole plant or the pseudobulbs of *Pholidota yunnanensis* are applied to alleviate cough, trauma, and stomachache [125]. *Eulophia* species are used in different parts of India for their anticancer, immunomodulatory, aphrodisiac, and anthelminthic activities [196]. In Indian folk medicine, *Eulophia ochreata* is used as an antiphlogistic agent. It also used for its rejuvenating and aphrodisiac properties, and tuber sap is applied externally for curing rheumatism [197].

4.2 Biological Activities of Orchidaceae Phenanthrenes

The pharmacological activities (e.g., antimicrobial, antioxidant, anti-inflammatory, cytotoxic, and neuroprotective effects) of almost all newly isolated compounds have been investigated, and several compounds showed multiple bioactivities. Based on the accumulated data, denbinobin (**270**) can be considered one of the most promising phenanthrene, and because of its novel mechanisms of action it could be a potential anticancer lead compound for further development.

4.2.1 Antiproliferative Activity

Many Orchidaceae phenanthrenes showed significant cytotoxic activities against different cancer cell lines. Among them, only compounds with IC_{50} values lower than 10 μM are presented in this section. Ddenbinobin (**270**) has shown noteworthy activities in different test systems. Denbinobin (**270**) causes cytotoxicity in several different mechanisms of actions, involving 1) apoptosis-induction through caspase-dependent and -independent ways [206], and by the activation of proapoptotic Bax protein, and inactivation of the antiapoptotic Bcl-2 [207]; 2) blocking of NF-κB activation [208]; 3) enhancing the synthesis of reactive oxygen species (ROS) [208]; 4) downregulation of DcR3 which sensitized the cells toward programmed cell death [209, 210]; and 5) antimigratory effects through reducing Src kinase activity, inhibition of calcium-binding cell migration protein (S100A8), and by down-regulation of matrix MMP-2 and MMP-9 metalloproteinases [207]. A structure-activity relationship (SAR) study revealed that the quinone substruction present in this compound is essential for the antiproliferative effect [208].

The *D. nobile* metabolites lusianthridin (**5**) and denbinobin (**270**) exerted cytotoxic effects against A549 human lung carcinoma [ED_{50}: 7.7 μM (**5**); 1.3 μM (**270**)], SK-OV-3 human ovary adenocarcinoma [ED_{50}: 9.4 μM (**5**); 3.5 μM (**270**)], and HL-60 human promyelocytic leukemia [ED_{50}: 9.8 μM (**5**); 0.11 μM (**270**)] cell lines. Denbinobin (**270**) proved to be more effective than lusianthridin (**5**). The methylated derivatives of these compounds did not exert cytotoxic effect; therefore, it can be concluded that the presence of a free phenolic hydroxy group in the molecule is essential for the inhibitory activity. Moreover, lusianthridin (**5**) possessed antitumor activity in ICR mice, implanted intraperitoneally with sarcoma 180 at a dose of 20 mg/kg [74]. In another investigation, lusianthridin (**5**) suppressed cancer stem cells in lung cancer cells through downregulation of Src-STAT3-c-Myc pathways. The compound inhibited the formation of tumor spheres of primary lung cancer cells. Moreover, after cancer stem cell attenuation by **5**, the lung cancer cells exhibited significantly higher susceptibility to chemotherapeutic drugs. Lusianthridin (**5**) strongly suppressed CSC phenotypes of H460 and H292 cells [78].

Two dimeric phenanthrenes [denthyrsinol (**369**) and denthyrsinone (**372**)] and the monophenanthrene denthirsinin (**212**), isolated from *D. thyrsiflorum*, showed significant cytotoxicity against HeLa, K-562 and MCF-7 cells with IC_{50} values 9.3 μM (HeLa), 1.6 μM (K-562) for denthyrsinol (**369**), 9.9 μM (HeLa), 6.0 μM (K-562), and 3.5 μM (MCF-7) for denthyrsinone (**372**), and 2.7 μM (HeLa), 2.3 μM (K-562), and 4.8 μM (MCF-7) for denthirsinin (**212**), respectively, using the MTT assay. SAR investigations displayed that the dimerization of phenanthrenes is a very important factor for the inhibition of cancer cell growth [188]. The *D. chrysotoxum* metabolite 4,9-dimethoxyphenanthrene-2,5-diol possessed cytotoxic activity against human hepatoma (BEL-7402) (IC_{50} 1.8 μM) and gastric cancer (SGC-7901) (IC_{50} 2.9 μM) cell lines [182]. Densiflorol B (**272**) was proved to be active against human leukemia HL-60 (IC_{50} value 4.6 μM), human lung adenocarcinoma A549 (IC_{50} value 8.7 μM), and human hepatoma BEL-7402 (IC_{50} value 7.2 μM) tumor cell lines. Cisplatin was used as a positive control (IC_{50} values 0.7 μM on HL-60, 0.4 μM on A549, and 0.18 μM on BEL-7402, respectively) [45]. Compound **220**, isolated from *B. odoratissimum*, possessed cytotoxic activity against A549 and SGC-7901 tumor cell lines with IC_{50} values 3.4 and 1.1 μM [176].

Calanquinone A (**278**) was proved to be cytotoxic against different [MCF-7 breast (IC_{50} 0.03 μM), PC-3 prostate (IC_{50} 0.16 μM), A549 lung (IC_{50} 0.19 μM), HCT-8 colon (IC_{50} 0.20 μM), KB nasopharyngeal (IC_{50} 0.32 μM), DU145 prostate (IC_{50} 0.34 μM), and KB-VIN vincristine-resistant nasopharyngeal (IC_{50} 0.45 μM), respectively] cancer cell lines. Moreover, it showed an improved drug resistance profile compared to the positive control paclitaxel [31]. The phenanthrene dimers bulbophythrins A (**302**) and B (**301**) possessed significant cytotoxic activity against the HL-60 (IC_{50} 1.3 and 4.0 nM), K562 (IC_{50} 6.1 and 2.8 nM), SGC-7901 (IC_{50} 6.1 nM for **301**), and A549 (IC_{50} 9.2 and 1.2 nM) cell lines, respectively. These inhibition values were comparable to that of the positive control cisplatin [IC_{50} 2.3 nM (HL-60), IC_{50} 0.27 nM (K562), IC_{50} 0.7 nM (SGC-7901), and IC_{50} 1.5 nM (A549)] [162]. In a WST-8 assay, ephemeranthoquinone B (**269**), a secondary metabolite of *Cymbidium* Great Flower Marie Laurencin, possessed cytotoxic activity (IC_{50} 2.8 μM) which was comparable to that of the positive control mitomycin C (IC_{50} 0.1 μM) [33]. Monbarbatain B (**314**) inhibited the proliferation of HL-60 cells (IC_{50} 7.3 μM), and this activity was comparable with that of cisplatin (IC_{50} value 7.7 μM) [161]. Cymbinodin A (**276**) possessed significant cytotoxic activity against NCI-H187 cells with IC_{50} 3.7 μM. Doxorubicin was used as a positive control (IC_{50} 0.24 μM) [50]. By using the MTT assay, ephemeranthoquinone B (**269**) and 5-hydroxy-2,3-dimethoxy-1,4-phenanthrenequinone (**280**), isolated from *Odontioda* Marie Noel "Velano," were found to possess specific cytotoxic effect against HL-60 cells (IC_{50} 3.0 μM for **269**, and 4.7 μM for **280**). However, none of the compounds caused apoptosis according to a DNA fragmentation assay [152]. A phenanthrenequinone (**264**) of *Oncidium isthmi* inhibited the M14 (IC_{50} 1.5 μM) and NCI-H460 (IC_{50} 5.0 μM) cell lines. Moreover, it induced apoptosis through caspase 3/7 activation and in an LDH release assay [151]. Compounds **177** and **193** significantly inhibited the effect of CDK1/cyclin B kinase (IC_{50}s 0.07 and 0.2 μM, respectively) [41].

Compound **82**, isolated from *B. striata*, was proved to be cytotoxic against HCT-15 cells (IC_{50} value 2.2 μM) using the SRB assay. Doxorubicin was used as a positive control (IC_{50} value 0.02 μM) [106]. Cypripedin (**259**), a phenanthrequinone from *D. densiflourm*, attenuated typical mesenchymal phenotypes, including migratory behavior of nonsmall cell lung cancer H460 cells, with a significant reduction of actin stress fibers and focal adhesion and with weakened anchorage-independent growth. The negative activity of the compound on epithelial-to-mesenchymal transition was due to ATP-dependent tyrosine kinase (Akt) inactivation. The observation was also supported by a similar result on another lung cancer H23 cell line, indicating that cypripedin (**259**) possesses a promising pharmacological effect on lung cancer metastasis [183]. Coelonin (**1**), lusianthridin (**5**), **69**, and calanhydroquinone C (**48**) showed potent cytotoxic activity against MDA-MB231 cells with the IC_{50} values 5.4, 8,0, 7.7, and 2.0 μM, respectively. Erianthridin (**24**), **1**, **5**, **69**, **170**, and nudol (**174**) were effective against HepG2 cells with the IC_{50} values 8.4, 9.4, 8.7, 8.8, 6.1, and 4.2 μM, respectively. Finally, compounds **1**, **5**, **69**, **48**, and **174** proved to be active against A549 cells with the IC_{50} values 5.8, 8.6, 8.9, 4.9, and 9.1 μM, respectively [53].

Liparisphenanthrene A (**367**), **354**, and **355** showed cytotoxic activity on HGC-27 cell line (IC_{50} values 9.6, 8.2, and 10.0 μM, respectively) and on HT-29 cell line (IC_{50} values 9.3 μM for **367**, and 8.5 μM for **355**) [157]. *P*-configuration of bulbocodioidin H (**111**) was cytotoxic against colon cancer (HCT-116), liver cancer (HepG2), and breast cancer (MCF-7) cell lines with IC_{50} values of 7.6, 3.8, and 3.4 μM, respectively. Cyrtonesin A (**132**) showed cytotoxic activity on MCF-7 (IC_{50} value 5.4 μM). Taxol was used as a positive control (IC_{50} values 0.02 μM for HCT-116, 0.03 μM for HepG2, and 0.01 μM for MCF-7) [25]. Bulbocodioidin A (9*R*) (**116**) was cytotoxic against HepG2 (IC_{50} value 8.1 μM) and BGC-823 (IC_{50} value 8.4 μM), while bulbocodioidin D (10*S*) (**119**) against HCT-116 (IC_{50} value 8.3 μM), HepG2 (IC_{50} value 2.3 μM), and MCF-7 (IC_{50} value 2.5 μM) cells [21].

4.2.2 Antimicrobial Activity

The antimicrobial effect of numerous phenanthrenes was tested against various (resistant and nonresistant) bacteria and fungi, and several of them inhibited the growth of the microorganisms, which might be due to their phytoalexin peculiarities. Orchinol (**4**) from *O. militaris* and hircinol (**8**) from *L. hircinum* are typical phytoalexins of orchids. Orchinol (**4**) inhibited completely the growth of *Candida lipolytica* at 100 ppm for the first 6 days, whereas hircinol (**8**) or the control did so for only 3 days [69].

The tubers of *B. striata* have been used in TCM for the treatment of pneumonorrhagia and pneumonophthisis. Therefore, its constituents lusianthridin (**5**) and **81** were tested against *Staphylococcus aureus*. Both compounds inhibited the growth of this gram-positive bacterium [MIC values 50 μg/mL (**5**) and 25 μg/mL (**81**), respectively] [71]. The biphenanthrenes blestriarene A-C (**289, 313, 353**) were effective against *S. aureus* and *Streptococcus mutans*. Blestriarene B (**313**) possessed the highest activity (MIC values 12.5 μg/mL and 6.25 μg/mL, respectively) against both test organisms [160]. The antimicrobial activities of **318, 319, 322, 332, 353** (blestriarene C), and **356**, isolated also from *B. striata*, were evaluated against gram-positive and gram-negative bacterial strains using the microdilution method. Compounds **318**, **319**, **322**, **353**, and **356** showed potent antibacterial activities against methicillin-resistant *S. aureus* ATCC 43300 and ampicillin-resistant *S. aureus* ATCC 29213, which were comparable in potency to the positive control ampicillin. Among them, compound **318** possessed the highest inhibitory activities against ampicillin-resistant *S. aureus* ATCC 29213 and methicillin-resistant *S. aureus* ATCC 43300 (MIC values of 2 and 4 μg/mL, respectively) [164]. Blestriarenes B (**313**) and C (**353**) were effective against *S. aureus* (MIC = 6.25 and 25 μg/mL, respectively) and *Streptococcus epidermidis* (MIC = 25 μg/mL for both) [158]. Coelonin (**1**) (26 μg/mL), flavanthridin (**25**) (53 μg/mL), shanciol C (**142**) (26 μg/mL), shanciol F (**143**) (52 μg/mL), shanciol D (**148**) (52 μg/mL), and blestriarene A (**290**) (6 μg/mL) from *B. striata* showed potent antibacterial activity, with MICs of 6–52 μg/mL against *S. aureus* ATCC 6538 [120].

Ephemeranthoquinone B (**269**), the metabolite of *Cymbidium* Great Flower Marie Laurencin, was active against *B. subtilis* (MIC = 4.9 μg/mL) [33]. Another phenanthrene of the plant, marylaurencinol C (**54**), had an antimycotic activity against *Trichophyton rubrum* (inhibition zone of 12.7 mm at 10 μg/disk) which was comparable to that of ketoconazole (inhibition zone of 15.7 mm at 5 μg/disk) [34]. Phenanthrenes (**88**, **272**, and **290**) from *A. graminifolia* were tested for antibacterial and antihemolytic activities. Blestriarene A (**290**) and shancidin (**88**) inhibited the growth of *S. aureus* and *E. coli* (MIC value 20 μg/mL for both), and *B. subtilis* (MIC value 40 μg/mL), by breaking the integrity of membranes and the cell wall. Meanwhile, **88**, **290**, and densiflorol B (**272**) showed antihemolytic activity (50% inhibition rate at 16 μg/mL for **88**, and **290**, and 128 μg/mL for **272**) on erythrocytes hemolysis caused by bacteria [42].

4.2.3 Anti-inflammatory Activity

The inhibitory activity of compounds, isolated from *B. formosana* (**72, 110, 295, 342**), were tested on superoxide anion generation and elastase release by human neutrophils in response to N-formyl-L-methionyl-L-leucyl-L-phenylalanine/cytochalasin B (FMLP/CB) by using idelalisib (IC_{50} values of 0.1 μM in case of superoxide anion generation, and 0.3 μM in case of elastase release, respectively) as a positive control. The 3-*p*-hydroxybenzyl-substituted 9,10-dihydrophenanthene (**72**) exerted the highest inhibitory activity against superoxide anion generation (IC_{50} value 0.2 μM). In case of **110**, in which a stilbene unit (batatasin III) and a 9,10-dihydrophenanthene connect through C-1–C-6′, high inhibitory activity was detected for elastase release (IC_{50} value of 0.6 μM). Among the phenanthene dimers, **295**, evolved by 1,3′ connection of two 9,10-dihydrophenanthene units, one of them substituted with a 1′-p-hydroxybenzyl group possessed the most potent inhibitory activity against superoxide anion generation (IC_{50} value 0.2 μM). Compound **342** showed significant inhibitory activity against elastase release (IC_{50} value 0.3 μM). In this compound, two 9,10-phenanthenes, containing methoxy groups at C-6, are connected through their C-1 [43].

Compounds **59** and **195**, isolated from *D. denneanum*, significantly inhibited the phosphorylation of IκBα, which prevents the degradation of IκBα and subsequent NF-κB activation. Therefore, these compounds may decrease the expression of proinflammatory cytokines by inhibiting the activation of NF-κB [13]. Compound **16**, isolated from *Eulophia ochreata*, inhibited the production of inflammatory cytokine TNF-α from LPS-stimulated THP-1, RAW264.7 cells and PBMCs. Moreover, it blocked TLR4-induced signaling via the NF-kB pathway by inhibiting MyD88-mediated signaling mechanisms. Treatment with **16** inhibited the expression of IL-8 and IL-β in a dose-dependent manner [211].

The secondary metabolites of *D. loddigesii*, loddigesiinols A (**200**) and B (**246**), moscatin (**166**), 5-hydroxy-2,4-dimethoxyphenanthrene (**168**), and lusianthridin (**5**), inhibited the LPS-induced NO production in RAW 264.7 macrophage cells (IC_{50} values 2.6 μM for **200**, and 10.9 μM for **246**, 6.4 μM for **166**, 5.3 μM for **168**, and 4.6 μM for **5**), compared to the positive control aminoguanidine (IC_{50} 17.5 μM), while rotundatin (**31**) and hircinol (**8**) were less active [14]. In an experiment, coelonin (**1**, IC_{50} 10.2 μM), lusianthridin (**5**) (IC_{50} 9.6 μM), hircinol (**8**, IC_{50} 26.4 μM), ephemeranthol A (**23**, IC_{50} 12.0 μM), erianthridin (**24**, IC_{50} 19.5 μM), flavanthridin (**25**, IC_{50} 34.1 μM), ephemeranthol C (**26**, IC_{50} 17.6 μM), 5,7-dimethoxyphenanthrene-2,6-diol (**176**, IC_{50} 35.7 μM), fimbriol B (**184**, IC_{50} 28.9 μM), and **203** (IC_{50} 20.4 μM), the constituents of *D. nobile*, showed anti-inflammatory activity which was comparable to the positive control aminoguanidine (IC_{50} 17.5 μM) [185]. The inhibitory activity of ephemeranthol A (**23**) mediated through NF-κB and MAPK pathways. At a concentration of 25 μg/mL, compounds **23** and **204** decreased the production of TNF-α, IL-1β, and IL-6 [187].

Compound **190** inhibited RANKL-induced formation of osteoclasts, and the lipopolysaccharide (LPS)-stimulated NO production (IC_{50} 24.1 μM) in RAW 264.7 cells. Moreover, due to its inhibitory effect on the iNOS protein expression, compound **190** scavenged the NO radical production in micromolar concentrations [204]. The LPS-induced NO-production of RAW 264.7 cells was also inhibited by a metabolite of *Pholidota yunnanensis*, 1-(4′-hydroxybenzy)-imbricatin (**156**) (IC_{50} 20.8 μM) [125]. Besides the COX-2 inhibitory activity of compound **156** and the *Vanda coerulea* metabolite 6-methoxy-coelonin (**17**) (IC_{50} values 12.0 and 5.8 μM, respectively), a dose-dependent suppression of the PGE_2 production in HaCaT cells (IC_{50} values of 12.2 and 19.3 μM, respectively) was also observed. Moreover, **17** decreased COX-2 expression induced by UV_B irradiation [127].

Phenanthrenes of *B. striata* were investigated for their antineuroinflammatory activity. The dimeric phenanthrene **352** showed the most potent anti-neuroinflammatory activity, with an IC_{50} value of 1.0 μM, followed by bleformin F (**350**, IC_{50} 5.0 μM). The heterodimer phochinenin K (**110**) also showed potent activity (IC_{50} 1.9 μM). Minocycline was used as a positive control (IC_{50} 27.2 μM) [44].

Dendrochrysanene (**254**) concentration-dependently inhibited the production of proinflammatory cytokines and iNOS. At 1.1 μg/mL concentration, the compound partially suppressed the mRNA level of TNF-α and IL-8 from LPS-activated macrophages, respectively, and at 11.2 μg/mL it significantly inhibited the amount of TNF-α and IL-8 mRNAs. Moreover, it suppressed IL-10 and iNOS mRNAs at 0.11 and 1.1 μg/mL, and at 11.2 μg/mL **254** significantly inhibited IL-10 and iNOS at the mRNA level [85].

The anti-inflammatory activity of **188** and denthyrsinin (**212**) was tested by measuring the reduction of anti-inflammatory (IL-6, TNF-α) and proinflammatory (IL-10) cytokines in the LPS-stimulated macrophage model *in vitro*. The IC_{50} values in case of the reduction of IL-6 were 23 μM for **188**, and 25 μM for **212**. Cortisol (IC_{50} value 0.5 μM) was used as a positive control. TNF-α secretion was reduced by about 50% by incubating the cells with coelonin (**1**), and about 40% after treatment of **212**. Coelonin (**1**), **212**, and the positive control dexamethason completely inhibited iNOS expression [56]. Coelonin (**1**) showed potent anti-inflammatory activity on LPS-stimulated NO production in BV-2 microglial cells, with IC_{50} value of 5.4 μM, which was stronger than the positive control quercetin (IC_{50} value 10.5 μM) [63].

Nishidono et al. investigated the effect of heat processing on the chemical constituents and NO-suppressing activity of Bletilla tuber. The Chinese Pharmacopoeia describes the heat-processing methods used on raw tubers of *B. striata* to produce the herbal medicine Bletilla tuber. The comparison of the chemical profiles of the extracts indicated that the relative content of phenanthrenes had not changed. The extracts of the tuber with (BT1) or without (BT2) heat processing showed similar IC_{50} values (120 μg/mL for BT1 and 140 μg/mL for BT2) on NO production suppressing activity in IL-1β-treated hepatocytes. Furthermore, besides bibenzyls, phenanthrenes [coelonin (**1**), flavanthrinin (**172**), **75**, and blestriarenes A (**290**) and C (**353**)] were identified as the compounds, responsible for suppressing the NO activity (30.8 μM for **1**, and 2.2 μM for **75**). The results suggested that the biological activities, such as the anti-inflammatory and hemostatic effects of Bletilla tuber, are not affected by heat processing [12].

The inhibitory effect of phoimbrtol (**219**, IC_{50} value 11.6 μM) and loddigesiinol B (**246**, IC_{50} value 10.9 μM) on nitric oxide (NO) production was stronger than those of the positive controls quercetin (17.5 μM) and resveratrol (22.0 μM) [124].

4.2.4 Antioxidant Activity

The *D. draconis* metabolite, compound **32**, exerted the same activity (IC_{50} 10.2 μM) in the DPPH radical-scavenging assay as the positive control trolox (IC_{50} 11.7 μM) [145]. Loddigesiinol A (**200**, IC_{50} 26.1 μM)), moscatin (**166**, IC_{50} 59.8 μM), and lusianthridin (**5**, IC_{50} 62.2 μM) possessed antioxidant effect in the DPPH-scavenging assay which was comparable with the positive control resveratrol (IC_{50} 28.7 μM) [14]. Confusarin (**211**), flavanthrinin (**172**), and compounds **190** and **170** were investigated by the DPPH assay for their antioxidant activity using ascorbic acid (IC_{50} 18.0 μM) as a positive control. Among the investigated compounds, **211** displayed comparable activity (IC_{50} value of 12.9 μM) to that of ascorbic acid.

According to the SAR investigations, a more pronounced antioxidant activity could be detected when hydroxy and methoxy groups are presented at *ortho* position [186].

In the DPPH radical-scavenging assay, compound **190** and moscatin **166** from *D. chrysotoxum* revealed antioxidant activity (IC_{50} 6.8 μM for **190**, and 9.2 μM for **166**) of similar potencies to ascorbic acid (IC_{50} 8.3 μM) [182]. In the same test system, monbarbatains A–D (**346**, **314**, **323**, **375**) (IC_{50} values 15.1, 12.9, 9.4, and 4.8 μM, respectively), **288** (IC_{50} 7.7 μM), and **81** (IC_{50} 12.2 μM) possessed antioxidant activity. BHA and rutin were used as positive controls (IC_{50} values 9.3 and 5.2 μM, respectively) [101, 161]. The phenanthrenes of *Odontoglossum* Harvengtense "Tutu," compound **190** and flavanthrinin (**172**), exerted NO scavenging activity (IC_{50} values 19 and 2 μM) [204]. Free radical-scavenging activity of **16** and **170** was determined by DPPH method (IC_{50} values 29.5 and 40.0 μM, respectively). The IC_{50} value of the positive control curcumin was determined as 45.0 μM [197].

Several dihydrophenanthrenes from *Pholidota chinensis* [coelonin (**1**), 2,4,7-trihydroxy-9,10-dihydrophenanthrene (**2**), lusianthridin (**5**), cannabidihydrophenanthrene (**6**), hircinol (**8**), isohircinol (**9**), erianthridin (**24**), and eulophiol (**65**)] possessed antioxidant activity, with IC_{50} values of 16.7 (**1**), 16.2 μM (**2**), 22.3 (**5**), 24.7 (**6**), 28.9 (**8**), 21.4 (**9**), 14.9 (**24**), and 27.7 (**65**), respectively, in the DPPH radical-scavenging assay [62]. Among the constituents of another *Pholidota* species (*P. yunnanensis*), imbricatin (**156**) was the most active (EC_{50} 8.8 μM) in the DPPH radical-scavenging assay. The EC_{50} values of phoyunnanin C (**298**, 26.7 μM), **288** (15.6 μM), lusianthridin (**5**, 22.3 μM), eulophiol (**65**, 27.7 μM), and **2** (10.0 μM) were comparable to that of the positive control resveratrol (21.2 μM). In case of phoyunnanins A (**114**) and B (**115**), lower activities (EC_{50} values of 55.9 and 47.3 μM, respectively) were detected [100].

In vitro free radical-scavenging assays (DPPH, ·OH) were used to confirm the antioxidant effect of coelonin (**1**), methoxycoelonin (**17**), imbricatin (**156**), and flavidin (**157**) (DPPH assay: IC_{50} values of 8.5 (**1**), 9.0 (**17**), 8.4 (**156**), and 6.6 (**157**) μM; ·OH assay: IC_{50} values of 0.03 (**1**), 0.04 (**17**), 0.08 (**156**), and 0.08 (**157**) μM, respectively). Furthermore, compounds **17** and **156** neutralized dose-dependently (IC_{50} values of 9.4 and 8.8 μM, respectively) the ROS production in H_2O_2-induced HaCaT cells [127]. Coelonin (**1**) and ochinol (**4**) showed DPPH and ABTS radical-scavenging activities (EC_{50}s 10.8 μM for both on DPPH, and 7.0 and 5.0 μM on ABTS). Blestriarene A (**290**) was the most effective in radical-scavenging measurement (IC_{50} 7.7 μM for DPPH and 4.1 μM for ABTS). Gallic acid and ascorbic acid served as positive controls for antioxidant investigations (IC_{50} 5.8 and 22.5 μM for DPPH, and 2.4 and 32.5 μM for ABTS) [48]. The DPPH radical-scavenging activity (IC_{50} value of 18.1 μM) of phoimbrtol (**219**), isolated from *P. imbricata*, was comparable to that of resveratrol ($IC_{50} = 24.5$ μM), and quercetin ($IC_{50} = 21.7$ μM) [124].

4.2.5 Other Activities

In case of confusarin (**211**)-treated (10 μM) PC12 cells, an increase of the neurite outgrowth (7.1%) was detected, after 72 h of incubation, proving the neuroprotective activity of the compound. NGF was used as a positive control (11.9% at 50 ng/mL) [114].

Coelonin (**1**, IC_{50} 13.4 μM), compound **202** (IC_{50} 9.0 μM), fimbriol B (**184**, IC_{50} 11.0 μM), and denbinobin (**270**, IC_{50} 15.2 μM), yielded from *D. nobile*, showed inhibitory activity against hepatic stellate (HSC-T6) cell line, which was comparable to that of (–)-epigallocatechin-3-gallate (IC_{50} 9.9 μM) [18]. Compounds **184**, **202**, and **270** decreased selectively, and dose- and time-dependently the viability of HSC-T6 cells. At a 10 μM concentration, apoptotic cell transformations, generated by the compounds, were observed, while at a concentration of 30 μM, necrotic changes of the cells were performed. Moreover, fimbriol B (**184**) and **202** caused caspase-dependent apoptosis in the caspase-3/7 activity assay [212].

Oxoflavidin (**161**) displayed remarkable osteogenic activity in the osteoblast ALP assay on mice calvarial osteoblasts at concentrations of 1 pM, 100 pM, and 1 nM. Cell mineralization was the most pronounced after using 100 pM of **161**. At the same concentration, applying real-time quantitative PCR measurements, compound **161** enhanced the transcription of osteogenic proteins (e.g., BMP-2, OCN) after 24 and 48 h. This observation supported the folk medicinal use of *Pholidota articulata* in bone healing [126]. Coelogin (**160**) treatment to calvarial osteoblasts led to enhanced ALP activity (a marker of osteoblast differentiation) and increased calcium nodule formation compared to control untreated cells. Moreover, coelogin (**160**) also increased the transcript levels of osteogenic gene markers such as BMP-2, Type I Col, and RUNX-2 [130].

Compound **17** was shown to possess vasorelaxant activity on rat aorta rings precontracted by phenylephrine (an α_1 receptor agonist), the $Ca_V1.2$ stimulator (S)-(−)-Bay K 8644, or depolarized with high K^+ concentrations. Compound **181** was active solely on rings stimulated by 25 mM K^+. The spasmolytic activity of **17** was significantly affected by the presence of an intact endothelium. The K_{ATP} channel blocker glibenclamide and the K_V channel blocker 4-aminopyridine significantly antagonized the vasorelaxant activity of **17**. In patch-clamp experiments, **17** inhibited Ba^{2+} current through $Ca_V1.2$ channels in a concentration-dependent manner, whereas it did not affect K^+ currents through K_{ATP} and K_V channels [173]. The antihypertensive activity of EtOH extract of *E. macrobulbon* and its constituent **229** was investigated on pulmonary arteries. It is in accordance to specific expression of PDE5 in pulmonary vasculature. Pulmonary arterial hypertension is improved by the extract acting through pulmonary artery relaxation mediated through endothelial NO, reduced Ca^{2+}-mobilization, and reduced pulmonary artery wall thickness and right ventricular hypertrophy [194]. In another investigation, compound **229** possessed potent PDE5 and PDE6 inhibitory activities (IC_{50} = 1.7 and 1.8 μM, respectively) evaluated by the [3H]cGMP radioassay method. Sildenafil was used as a positive control (IC_{50} values 0.03 and 0.05 μM, respectively) [195].

Selective BChE (butyrylcholinesterase) inhibitory effects were observed for coelonin (**1**, IC_{50} value 19.7 μM), orchinol (**4**, IC_{50} value 32.8 μM), blestriarene A (**290**, IC_{50} value 88.5 μM), and dehydroorchinol (**169**, IC_{50} value 37.8 μM). The most potent anti-BChE activity was exhibited by coelonin (**1**), which was comparable to that of galantamine (IC_{50} = 13.2 μM). Dimerization of monomeric 9,10-

dihydrophenanthrene significantly decreased the BChE activity [48]. Ochinol (**4**) and blestriarene A (**290**) inhibited the β-amyloid peptide aggregation (64.5 and 29.5% at 20 μM, respectively), suggesting that they could serve as potential agents for Alzheimer's disease drug developments [48].

Compound **261** showed significant α-glucosidase and pancreatic lipase inhibitory effects with IC_{50} values of 126.9 and 69.5 μM, respectively. A kinetic study, conducted by the Lineweaver-Burk plot method, revealed that **32** was a noncompetitive inhibitor of α-glucosidase and pancreatic lipase enzymes. Moreover, lusianthridin (**5**) showed glucose uptake stimulatory effect without toxicity on L6 myotubes at concentrations of 1 and 10 μg/mL [52].

5 Conclusions

Phenanthrenes are formed a group of natural products. To date, approx. 500 phenanthrene-type plant secondary metabolites have been identified from various sources. Phenanthrenes have received much attention in the literature over the past two decades, and a variety of potential beneficial effects have been elucidated. Discovery of new phenanthrene derivatives, either from natural sources or by synthetic routes, with promising pharmacological activities is a great challenge to many research groups worldwide.

The occurrence of phenanthrenes is not restricted to one organ, they are found in different plant parts (stem, rhizome, root, tuber bulb, and leaf); mainly, the whole plants were investigated. Whereas the biosynthetic pathways of hircinol (**8**) and orchinol (**4**) have been very well characterized [8], those of other phenanthrene derivatives are unknown.

It can be concluded that orchids are the most abundant sources of such compounds. Previously, more than 370 phenanthrenes have been isolated from approximately 100 species of 40 genera of the Orchidaceae. *Bletilla*, *Bulbophyllum*, *Cremastra*, *Cymbidium*, *Cyrtopodium*, *Dendrobium*, *Pholidota*, *Pleione*, and *Spiranthes* species are the best sources of phenanthrenes. These species are belonging to the subfamilies Epidendroideae, higher Epidendroideae, and Orchidoideae. The most investigated genus in Orchidaceae is *Dendrobium*; 26 species of this genus were studied for phenanthrene content, and 136 compounds were isolated from them. However, *Bletilla striata* is the richest source of phenanthrenes as to date 80 compounds were isolated from this plant. The most common monophenanthrenes are coelonin (**1**), lusianthridin (**5**), hircinol (**8**), erianthridin (**24**), plicatol-B (moscatin, **166**), lusianthrin (**167**), flavanthrinin (**172**), and nudol (**174**).

Because of their limited occurrence, phenanthrenes can be used as chemotaxonomic markers. Besides the plant-specific enzymes requiring for the biosynthesis of such compounds, certain phenanthrene substituents (e.g., hydroxybenzyl, prenyl, vinyl, stilbene/dihydrostilbene, and mono-/disaccharide units) are also restricted to specific families. Prenylated derivatives and stilbene/dihydrostilbene-connected phenanthrenes have been identified solely from orchids, especially from the members of the genus *Pholidota*. *p*-Hydroxybenzyl-substituted and glycosylated compounds seem to be also characteristic to orchids. These species are rich in phenanthroquinones, too. To date, only two trimeric phenanthrenes are known from natural source; both were reported from Orchidaceae species. It is also unique in orchid phenanthrenes that, because of the close coexistence of liverwort species, some of the dimers contain a monophenanthrene unit that has been isolated earlier only from *Plagiochila* species.

The identified phenanthrenes have been studied in almost all cases for their pharmacological effects, and many of them possessed diverse activities. In several cases, detailed mechanism of action investigations has also been performed. The accumulated data in case of denbionobin (**270**) implied that it could be a potential anticancer lead compound. Compound **318** showed promising inhibitory activities against ampicillin-resistant *S. aureus* and methicillin-resistant *S. aureus* (MIC values 2 and 4 μg/mL, respectively). Anti-inflammatory activity of lusianthridin (**5**), moscatin (**166**), **168**, and loddigesiinols A (**200**) and B (**246**) (IC_{50} values of 4.6, 6.4, 5.3, 2.6, and 10.9 μM, respectively), mediated through inhibition of LPS-induced NO production, could also be promising for further investigations. Compound **352** showed strong antineuroinflammatory activity (IC_{50} value of 1.0 μM). However, despite the promising *in vitro* results, *in vivo* effects of the phenanthrenes are rarely investigated. Semisynthetic modifications or synthesis of the compounds could allow to widening their biological investigations. Finally, although orchids are used widespread in traditional medicines, the toxicity and mutagenic potential of these species are seldom verified.

In conclusion, phytochemical investigation of taxonomically questionable or rather controversial orchids, focusing on their phenanthrene content, can contribute to clarify the taxonomic position of these species, and the great diversity of phenanthrenes provided by Orchidaceae species offers unique opportunities either for medicinal chemists or pharmacologists to discover new classes of therapeutics.

References

1. The Plant List:Orchidaceae http://www.theplantlist.org/1.1/browse/A/Orchidaceae/ Retrieved 18 November 2020
2. Dressler RL (1993) Phylogeny and classification of the orchid family. Dioscorides Press, Portland
3. Dressler RL (1981) The orchids: natual history and classification. Harvard University Press
4. Gutiérrez RMP (2010) Orchids: a review of uses in traditional medicine, its phytochemistry and pharmacology. J Med Plant Res 4:592–638
5. Tóth B, Hohmann J, Vasas A (2018) Phenanthrenes: a promising group of plant secondary metabolites. J Nat Prod 81:661–678
6. Chinsamy M, Finnie JF, Van Staden J (2014) Anti-inflammatory, antioxidant, anti-cholinesterase activity and mutagenicity of South African medicinal orchids. South Afr J Bot 91:88–98
7. Kovács A, Vasas A, Hohmann J (2008) Natural phenanthrenes and their biological activity. Phytochemistry 69:1084–1110

8. Fritzemeier KH, Kindl H (1981) 9,10-Dihydrophenanthrenes as phytoalexins of Orchidaceae. Biosynthetic studies *in vitro* and *in vivo* proving the route from L-phenylalanine to dihydro-*m*-coumaric acid, dihydrostilbene and dihydrophenanthrenes. Eur J Biochem 133:545–550
9. Reinecke T, Kindl H (1994) Inducible enzymes of the 9,10-dihydro-phenanthrene pathway. Sterile orchid plants responding to fungal infection. Mol Plant Microbe Interact 7:449–454
10. Mallory FB, Wood CS, Gordon JT (1964) Photochemistry of stilbenes. III. Some aspects of the mechanism of photocyclization to phenanthrenes. J Am Chem Soc 86:3094–3102
11. Morgan DD, Horgan SW, Orchin M (1970) Photocyclization of stilbene analogs I. The oxidative photocyclization of 1,3-distyrylbenzene. Tetrahedron Lett 49:4347–4350
12. Nishidono Y, Tanaka K, Ishii T, Okada R, Okuyama T, Nishizawa M, Norimoto H, Murayama C, He D (2020) Effect of heat processing on the chemical constituents and NO-suppressing activity of Bletilla tuber. J Nat Med 74:219–228
13. Lin Y, Wang F, Yang LJ, Chun Z, Bao JK, Zhang GL (2013) Anti-inflammatory phenanthrene derivatives from stems of *Dendrobium denneanum*. Phytochemistry 95:242–251
14. Ito M, Matsuzaki K, Wang J, Daikonya A, Wang NL, Yao XS, Kitanaka S (2010) New phenanthrenes and stilbenes from *Dendrobium loddigesii*. Chem Pharm Bull 58:628–633
15. Xu FQ, Xu FC, Hou B, Fan WW, Zi CT, Li Y, Dong FW, Liu YQ, Sheng J, Zuo ZL, Hu JM (2014) Cytotoxic bibenzyl dimers from the stems of *Dendrobium fimbriatum* Hook. Bioorg Med Chem Lett 24:5268–5273
16. Ye Q, Mei Y, Yang P, Cheng L, Kong D (2016) A new 9,10-dihydrophenanthrene glycoside from *Dendrobium primulinum*. Chem Nat Compd 52:381–383
17. Auberon F, Olatunji OJ, Herbette G, Raminoson D, Antheaume C, Soengas B, Bonté F, Lobstein A (2016) Chemical constituents from the aerial parts of *Cyrtopodium paniculatum*. Molecules 21:1–18
18. Yang H, Sung SH, Kim YC (2007) Antifibrotic phenanthrenes of *Dendrobium nobile* stems. J Nat Prod 70:1925–1929
19. Wang J, Wang T, Xie P, Yin G, Li X (2014) New phenanthrene derivatives with nitric oxide inhibitory and radical-scavenging activities from *Pholidota imbricata* Hook. Nat Prod Res 28: 251–256
20. Tezuka Y, Yoshida Y, Kikuchi T, Xu GJ (1993) Constituents of *Ephemerantha fimbriata*. Isolation and structure elucidation of two new phenanthrenes, fimbriol-A and fimbriol-B, and a new dihydrophenanthrene, ephemeranthol-C. Chem Pharm Bull 41:1346–1349
21. Shao SY, Wang C, Han SW, Sun MH, Li S (2019) Phenanthrenequinone enantiomers with cytotoxic activities from the tubers of *Pleione bulbocodioides*. Org Biomol Chem 17:567–572
22. Anuradha V, Rao NSP, Udayabhaskar M (1995) Ochrolone, a pyrene from *Coelogyne ochracea*. Phytochemistry 39:1429–1431
23. Honda C, Yamaki M (2000) Phenanthrenes from *Dendrobium plicatile*. Phytochemistry 53: 987–990
24. Majumder PL, Pal S, Majumder S (1999) Dimeric phenanthrenes from the orchid *Bulbophyllum reptans*. Phytochemistry 50:891–897
25. Wang C, Shao SY, Han SW, Li S (2019) Atropisomeric bi(9,10-dihydro)phenanthrene and phenanthrene/bibenzyl dimers with cytotoxic activity from the pseudobulbs of *Pleione bulbocodioides*. Fitoterapia 138:104313
26. Liu XQ, Li XP, Yuan WK, Yuan QY, Qin BH (2016) Two new phenanthrene glucosides from *Cremastra appendiculata* and their cytotoxic activities. Nat Prod Commun 11:477–479
27. Liu XQ, Li XP, Yuan QY (2015) A new biphenanthrene glucoside from *Cremastra appendiculata*. Chem Nat Compd 51:1035–1037
28. Yamaki M, Kato T, Bai L, Inoue K, Takagi S (1993) Phenanthrene glucosides from *Bletilla striata*. Phytochemistry 34:535–537
29. Liu L, Li J, Zeng KW, Jiang Y, Tu PF (2015) Five new benzylphenanthrenes from *Cremastra appendiculata*. Fitoterapia 103:27–32

30. Lee CL, Chang FR, Yen MH, Yu D, Liu YN, Bastow KF, Morris-Natschke SL, Wu YC, Lee KH (2009) Cytotoxic phenanthrenequinones and 9,10- dihydrophenanthrenes from *Calanthe arisanensis*. J Nat Prod 72:210–213
31. Lee CL, Nakagawa-Goto K, Yu D, Liu YN, Bastow KF, Morris-Natschke SL, Chang FR, Wu YC, Lee KH (2008) Cytotoxic calanquinone A from *Calanthe arisanensis* and its first total synthesis. Bioorg Med Chem Lett 18:4275–4277
32. Chang CF, Hsu YL, Lee CY, Wu CH, Wu YC, Chuang TH (2015) Isolation and cytotoxicity evaluation of the chemical constituents from *Cephalantheropsis gracilis*. Int J Mol Sci 16: 3980–3989
33. Yoshikawa K, Ito T, Iseki K, Baba C, Imagawa H, Yagi Y, Morita H, Asakawa Y, Kawano S, Hashimoto T (2012) Phenanthrene derivatives from *Cymbidium* great flower Marie Laurencin and their biological activities. J Nat Prod 75:605–609
34. Yoshikawa K, Baba C, Iseki K, Ito T, Asakawa Y, Kawano S, Hashimoto T (2014) Phenanthrene and phenylpropanoid constituents from the roots of *Cymbidium* great flower 'Marylaurencin' and their antimicrobial activity. J Nat Med 68:743–747
35. Zhao N, Yang G, Zhang Y, Chen L, Chen Y (2016) A new 9,10-dihydrophenanthrene from *Dendrobium moniliforme*. Nat Prod Res 30:174–179
36. Kuo W, Huang YL, Shen CC, Shieh BJ, Chen CC (2007) Prenylated benzoic acids and phenanthrenes from *Liparis nakaharai*. J Chin Chem Soc 54:1359–1362
37. Yamaki M, Kato T, Bai L, Inoue K, Takagi S (1991) Methylated stilbenoids from *Bletilla striata*. Phytochemistry 30:2759–2760
38. Majumder PL, Banerjee S, Sen S (1996) Three stilbenoids from the orchid *Agrostophyllum callosum*. Phytochemistry 42:847–852
39. Wu YP, Liu WJ, Zhong WJ, Chen YJ, Chen DN, He F, Jiang L (2017) Phenolic compounds from the stems of *Flickingeria fimbrata*. Nat Prod Res 31:1518–1522
40. Majumder PL, Banerjee S (1988) Structure of flavanthrin, the first dimeric 9,10-dihydrophenanthrene derivative from the orchid *Eria flava*. Tetrahedron 44:7303–7308
41. Apel C, Dumontet V, Lozach O, Meijer L, Guéritte F, Litaudon M (2012) Phenanthrene derivatives from *Appendicula reflexa* as new CDK1/cyclin B inhibitors. Phytochem Lett 5: 814–818
42. Yan X, Tang B, Liu M (2018) Phenanthrenes from *Arundina graminifolia* and *in vitro* evaluation of their antibacterial and anti-haemolytic properties. Nat Prod Res 32:707–710
43. Lin CW, Hwang TL, Chen FA, Wu TS, Huang CH, Hung HY, Wu TS (2016) Chemical constituents of the rhizomes of *Bletilla formosana* and their potential anti-inflammatory activity. J Nat Prod 79:1911–1921
44. Zhou D, Chen G, Ma YP, Wang CG, Lin B, Yang YQ, Li W, Koike K, Hou Y, Li N (2019) Isolation, structure elucidation, optical resolution, and antineuroinflammatory activity of phenanthrene and 9,10-dihydrophenanthrene derivatives from *Bletilla striata*. J Nat Prod 82: 2238–2245
45. Chen Y, Xu J, Yu H, Qing C, Zhang Y, Wang L, Liu Y, Wang J (2008) Cytotoxic phenolics from *Bulbophyllum odoratissimum*. Food Chem 107:169–173
46. Leong YW, Kang CC, Harrison LJ, Powell AD (1997) Phenanthrenes, dihydrophenanthrenes and bibenzyls from the orchid *Bulbophyllum vaginatum*. Phytochemistry 44:157–165
47. Majumder P, Laha S, Datta N (1982) Coelonin, a 9,10-dihydrophenanthrene from the orchids *Coelogyne ochracea* and *Coelogyne elata*. Phytochemistry 21:478–480
48. Tu Y, Huang J, Li Y (2018) Anticholinesterase, antioxidant, and beta-amyloid aggregation inhibitory constituents from *Cremastra appendiculata*. Med Chem Res 27:857–863
49. Juneja RK, Sharma SC, Tandon JS (1987) Two substituted bibenzyls and a dihydrophenanthrene from *Cymbidium aloifolium*. Phytochemistry 26:1123–1125
50. Lertnitikul N, Pattamadilok C, Chansriniyom C, Suttisri R (2020) A new dihydrophenanthrene from *Cymbidium finlaysonianum* and structure revision of cymbinodin-A. J Asian Nat Prod Res 22:83–90

51. Auberon F, Olatunji OJ, Raminoson D, Muller CD, Soengas B, Bonte F, Lobstein A (2017) Isolation of novel stilbenoids from the roots of *Cyrtopodium paniculatum* (Orchidaceae). Fitoterapia 116:99–105
52. Inthongkaew P, Chatsumpun N, Supasuteekul C, Kitisripany T, Putalun W, Likhitwitayawuid K, Sritularak B (2017) α-Glucosidase and pancreatic lipase inhibitory activities and glucose uptake stimulatory effect of phenolic compounds from *Dendrobium formosum*. Rev Bras Farmacogn 27:480–487
53. Chen DN, Wang YY, Liu WJ, Chen YJ, Wu YP, Wang JX, He F, Jiang L (2020) Stilbenoids from aerial parts of *Dendrobium plicatile*. Nat Prod Res 34:323–328
54. Sarakulwattana C, Mekboonsonglarp W, Likhitwitayawuid K, Rojsitthisak P, Sritularak B (2020) New bisbibenzyl and phenanthrene derivatives from *Dendrobium scabrilingue* and their α-glucosidase inhibitory activity. Nat Prod Res 34:1694–1701
55. Majumder PL, Banerjee S (1990) Two stilbenoids from the orchid *Eria flava*. Phytochemistry 29:3052–3055
56. Schuster R, Zeindl L, Holzer W, Khumpirapang N, Okongi S, Viernstein H, Mueller M (2017) *Eulophia macrobulbon* – an orchid with significant anti-inflammatory and antioxidant effect and anticancerogenic potential exerted by its root extract. Phytomedicine 24:157–165
57. Tuchinda P, Udchachon J, Khumtaveeporn K, Taylor WC, Engelhardt LM, White AH (1988) Phenanthrenes of *Eulophia nuda*. Phytochemistry 27:3267–3271
58. Blitzke T, Masaoud M, Schmidt J (2000) Constituents of *Eulophia petersii*. Fitoterapia 71: 593–594
59. Matsuda H, Morikawa T, Xie H, Yoshikawa M (2004) Antiallergic phenanthrenes and stilbenes from the tubers of *Gymnadenia conopsea*. Planta Med 70:847–855
60. Morikawa T, Xie H, Matsuda H, Wang T, Yoshikawa M (2006) Bioactive constituents from Chinese natural medicines. XVII. Constituents with radical scavenging effect and new glucosyloxybenzyl 2-isobutylmalates from *Gymnadenia conopsea*. Chem Pharm Bull 54:506–513
61. Chen Y, Cai S, Deng L, Xia Q, Du LF, Cui GZ, Li J, Zhou XM, Ye Q, Zhou Y et al (2015) Separation and purification of 9,10-dihydrophenanthrenes and bibenzyls from *Pholidota chinenesis* by high-speed countercurrent chromatography. J Sep Sci 38:453–459
62. Wang J, Wang L, Kitanaka S (2007) Stilbene and dihydrophenanthrene derivatives from *Pholidota chinensis* and their nitric oxide inhibitory and radical-scavenging activities. J Nat Med 61:381–386
63. Bai L, Yamaki M, Takagi S (1996) Stilbenoids from *Pleione bulbocodioides*. Phytochemistry 42:853–856
64. Li Y, Zeng KW, Zhao MB, Jiang Y, Li J, Tu PF, Wu ZH (2017) A new prenylated flavone from *Pleione bulbocodioides*. J Asian Nat Prod Res 19:738–743
65. Tezuka Y, Ueda M, Kikuchi T (1989) Studies on the constituents of Orchidaceae plants. VIII. Constituents of *Spiranthes sinensis* (Pers.) Ames var. *amoena* (M. Biebersin) Hara. (1). Isolation and structure elucidation of spiranthol-A, spiranthol-B, and spirasineol-A, new isopentenyldihydrophenanthrenes. Chem Pharm Bull 37:3195–3199
66. Majumder PL, Banerjee S, Maiti DC, Sen S (1995) Stilbenoids from the orchids *Agrostophyllum callosum* and *Coelogyne flaccida*. Phytochemistry 39:649–653
67. Auberon F, Olatunji OJ, Krisa S, Antheaume C, Herbette G, Bonte F, Merillon JM, Lobstein A (2016) Two new stilbenoids from the aerial parts of *Arundina graminifolia* (Orchidaceae). Molecules 21:1430
68. Baller A, Corrodi H, Gaumann E, Hardegger E, Kern H, Winterhalter-Wild N (1957) Wilting agents and antibiotics. XX. Induced defensive substances in the Orchidaceae. 1. Helv Chim Acta 40:1062–1066
69. Fisch MH, Flick BH, Arditti J (1973) Structure and antifungal activity of hircinol, loroglossol and orchinol. Phytochemistry 12:437–441
70. Liu MF, Han Y, Xing DM, Wang W, Xu LZ, Du LJ, Ding Y (2005) One new benzyldihydrophenanthrene from *Arundina graminifolia*. J Asian Nat Prod Re.:767–770

71. Takagi S, Yamaki M, Inoue K (1983) Antimicrobial agents from *Bletilla striata*. Phytochemistry 22:1011–1015
72. Klongkumnuankarn P, Busaranon K, Chanvorachote P, Sritularak B, Jongbunprasert V, Likhitwitayawuid K (2015) Cytotoxic and antimigratory activities of phenolic compounds from *Dendrobium brymerianum*. Evid Based Complement Alternat Med 350410
73. Fan C, Wang W, Wang Y, Qin G, Zhao W (2001) Chemical constituents from *Dendrobium densiflorum*. Phytochemistry 57:1255–1258
74. Lee YH, Park JD, Baek NI, Kim SI, Ahn BZ (1995) *In vitro* and *in vivo* antitumoral phenanthrenes from the aerial parts of *Dendrobium nobile*. Planta Med 61:178–180
75. Zhou XM, Zheng CJ, Gan LS, Chen LS, Chen GY, Zhang XP, Song XP, Li GN, Sun CG (2016) Bioactive phenanthrene and bibenzyl derivatives from the stems of *Dendrobium nobile*. J Nat Prod 79:1791–1797
76. Kyokong N, Muangnoi C, Thaweesest W, Kongkatitham V, Likhitwitayawuid K, Rojsitthisak P, Sritularak B (2019) A new phenanthrene dimer from *Dendrobium palpebrae*. J Asian Nat Prod Res 21:391–397
77. Yamaki M, Honda C (1996) The stilbenoids from *Dendrobium plicatile*. Phytochemistry 43: 207–208
78. Bhummaphan N, Petpiroon N, Prakhongeheep O, Sritularak B, Chanvorachote P (2019) Lusianthridin targeting of lung cancer stem cells via Src-STAT3 suppression. Phytomedicine 62:152932
79. Majumder PL, Lahiri S (1990) Lusianthrin and lusianthridin, two stilbenoids from the orchid *Lusia indivisa*. Phytochemistry 29:621–624
80. Hernandez-Romero Y, Rojas JI, Castillo R, Rojas A, Mata R (2004) Spasmolytic effects, mode of action, and structure-activity relationships of stilbenoids from *Nidema boothii*. J Nat Prod 67:160–167
81. Li B, Ali Z, Chan M, Li J, Wang M, Abe N, Wu CR, Khan IA, Wang W, Li SX (2017) Chemical constituents of *Pholidota cantonensis*. Phytochemistry 137:132–138
82. Estrada S, Rojas A, Mathison Y, Israel A, Mata R (1999) Nitric oxide/cGMP mediates the spasmolytic action of 3,4'-dihydroxy-5,5'-dimethoxybibenzyl from *Scaphyglottis livida*. Planta Med 65:109–114
83. Yang J, Jiang J, Huang G, Liu B, Liu Y, Zhan R, Chen Y (2014) Two new flavanone glycosides from *Sunipia scariosa*. Biochem Syst Ecol 57:317–321
84. Majumder PL, Roychowdhury M, Chakraborty S (1998) Thunalbene, a stilbene derivative from the orchid *Thunia alba*. Phytochemistry 49:2375–2378
85. Yang L, Qin LH, Bligh SWA, Bashall A, Zhang CF, Zhang M, Wang ZT, Xu LS (2006) A new phenanthrene with a spirolactone from *Dendrobium chrysanthum* and its anti-inflammatory activities. Bioorg Med Chem 14:3496–3501
86. Majumder PL, Pal S (1992) Rotundatin, a new 9,10-dihydrophenanthrene derivative from *Dendrobium rotundatum*. Phytochemistry 31:3225–3228
87. Majumder PL, Sen RC (1987) Structure of moscatin – a new phenanthrene derivate from *Dendrobium rotundatum*. Indian J Chem 26B:18–20
88. Chen CC, Wu LG, Ko FN, Teng CM (1994) Antiplatelet aggregation principles of *Dendrobium loddigesii*. J Nat Prod 57:1271–1274
89. Chen Y, Shi X, Liu Y, Li Y, Zhang Y (2013) Aromatic compounds from *Coelogyne longipes*. Biochem Syst Ecol 50:72–74
90. Wang Y, Guan SH, Meng YH, Zhang YB, Cheng CR, Shi YY, Feng RH, Zeng F, Wu ZY, Zhang JX, Yang M, Liu X, Li Q, Chen XH, Bi KS, Guo DA (2013) Phenanthrenes, 9,10-dihydrophenanthrenes, bibenzyls with their derivatives, and malate or tartrate benzyl ester glucosides from tubers of *Cremastra appendiculata*. Phytochemistry 94:268–276
91. Ren J, Qian XP, Guo YG, Li T, Yan SK, Jin HZ, Zhang WD (2016) Two new phenanthrene glycosides from *Liparis regnieri* Finet and their antibacterial activities. Phytochem Lett 18:64–67
92. Liu XQ, Li XP, Yuan QY (2016) Two new phenanthrene glucosides from *Cremastra appendiculata*. Chem Nat Compd 52:23–25

93. Ye QH, Zhao WM, Qin GW (2003) New fluorenone and phenanthrene derivatives from *Dendrobium chrysanthum*. Nat Prod. Res 17:201–205
94. Tezuka Y, Ji L, Hirano H, Ueda M, Nagashima K, Kikuchi T (1990) Studies on the constituents of orchidaceous plants. IX. Constituents of *Spiranthes sinensis* (Pers.) Ames var. amoena (M. Bieberson) Hara. (2). Structures of spiranthesol, spiranthoquinone, spiranthol-C, and spirasineol-B, new isopentenyldihydrophenanthrenes. Chem Pharm Bull 38:629–635
95. Liu L, Yin QM, Yan X, Hu C, Wang W, Wang RK, Luo X, Zhang XW (2019) Bioactivity-guided isolation of cytotoxic phenanthrenes from *Spiranthes sinensis*. J Agric Food Chem 67: 7274–7280
96. Anuradha V, Rao NSP, Bhaskar MU (1994) Ochrolic acid, a precursor to phenanthropyrones, from *Coelogyne ochracea*. Phytochemistry 36:1515–1517
97. Yao S, Tang CP, Ye Y, Kurtán T, Kiss-Szikszai A, Antus S, Pescitelli G, Salvadori P, Krohn K (2008) Stereochemistry of atropisomeric 9,10-dihydrophenanthrene dimers from *Pholidota chinensis*. Tetrahedron: Asymmetry 19:2007–2014
98. Connolly JD, Rycroft DS, Srivastava DL, Cole WJ, Ifeadike P, Kimbu SF, Singh J, Hughes M, Thom C, Gerhard U, Organ AJ, Smith RJ, Harrison LJ (1999) Aromatic compounds from the liverwort *Plagiochila spinulosa*. Phytochemistry 50:1159–1165
99. Liu XQ, Guo YQ, Gao WY, Zhang TJ, Yan LL (2008) Two new phenanthrofurans from *Pleione bulbocodioides*. J Asian Nat Prod Res 10:453–457
100. Guo XY, Wang J, Wang NL, Kitanaka S, Yao XS (2007) 9, 10-Dihydrophenanthrene derivatives from *Pholidota yunnanensis* and scavenging activity on DPPH free radical. J Asian Nat Prod Res 9:165–174
101. Yang M, Cai L, Fang H, Yang S, Fang Y, Ding Z (2014) Phenolic compounds from *Monomeria barbata*. Chem Nat Compd 50:88–92
102. Guo XY, Wang J, Wang NL, Kitanaka S, Liu HW, Yao XS (2006) New stilbenoids from *Pholidota yunnanensis* and their inhibitory effects on nitric oxide production. Chem Pharm Bull 54:21–25
103. Yamaki M, Bai L, Inoue K, Takagi S (1990) Benzylphenanthrenes from *Bletilla striata*. Phytochemistry 29:2285–2287
104. Lin YL, Chen WP, Macabalang AD (2005) Dihydrophenanthrenes from *Bletilla formosana*. Chem Pharm Bull 53:1111–1113
105. Bai L, Kato T, Inoue K, Yamaki M, Takagi S (1993) Stilbenoids from *Bletilla striata*. Phytochemistry 33:1481–1483
106. Woo KW, Park JE, Choi SU, Kim KH, Lee KR (2014) Phytochemical constituents of *Bletilla striata* and their cytotoxic activity. Nat Prod Sci 20:91–94
107. Bai L, Kato T, Inoue K, Yamaki M, Takagi S (1991) Blestrianol A, B and C, biphenanthrenes from *Bletilla striata*. Phytochemistry 30:2733–2735
108. Liu L, Li J, Zeng KW, Li P, Tu PF (2013) Three new phenanthrenes from *Cremastra appendiculata* (D. Don) Makino. Chin Chem Lett 24:737–739
109. Tuchinda P, Udchachon J, Khumtaveeporn K, Taylor WC (1989) Benzylated phenanthrenes from *Eulophia nuda*. Phytochemistry 28:2463–2466
110. Yang MH, Cai L, Li MH, Zeng XH, Yang YB, Ding ZT (2010) Three new phenanthrenes from *Monomeria barbata* Lindl. Chin Chem Lett 21:325–328
111. Dong HL, Liang HQ, Wang CL, Guo SX, Yang JS (2013) Shancigusins E-I, five new glucosides from the tubers of *Pleione yunnanensis*. Magn Reson Chem 51:371–377
112. Lin YL, Huang RL, Don MJ, Kuo YH (2000) Dihydrophenanthrenes from *Spiranthes sinensis*. J Nat Prod 63:1608–1610
113. Lin YL, Wang WY, Kuo YH, Liu YH (2001) Homocyclotirucallane and two dihydrophenanthrenes from *Spiranthes sinensis*. Chem Pharm Bull 49:1098–1101
114. Hu J, Fan W, Dong F, Miao Z, Zhou J (2012) Chemical components of *Dendrobium crepidatum* and their neurite outgrowth enhancing activities. Chin J Chem 30:1327–1330
115. Yao S, Tang CP, Li XQ, Ye Y (2008) Phochinenins A-F, Dimeric 9,10-dihydrophenanthrene derivatives, from *Pholidota chinensis*. Helv Chim Acta 91:2122–2129

116. Liu XQ, Yuan QY, Guo YQ (2009) Two new stilbenoids from *Pleione bulbocodioides*. J Asian Nat Prod Res 11:116–121
117. Yamaki M, Bai L, Kato T, Inoue K, Takagi S (1993) Three dihydrophenanthropyrans from *Bletilla striata*. Phytochemistry 32:427–430
118. Bai L, Yamaki M, Yamagata Y, Takagi S (1996) Shanciol, a dihydrophenanthropyran from *Pleione bulbocodioides*. Phytochemistry 41:625–628
119. Bai L, Yamaki M, Takagi S (1998) Flavan-3-ols and dihydrophenanthropyrans from *Pleione bulbocodioides*. Phytochemistry 47:1125–1129
120. Jiang S, Chen CF, Ma XP, Wang MY, Wang W, Xia Y, Zhang N, Pan WD, Wu MK (2019) Antibacterial stilbenes from the tubers of *Bletilla striata*. Fitoterapia 138104350
121. Bai L, Masukawa N, Yamaki M, Takagi S (1982) Four stilbenoids from *Pleione bulbocodioides*. Phytochemistry 48:327–331
122. Dong H, Wang C, Li Y, Guo S, Yang J (2010) Complete assignments ^{1}H and ^{13}C NMR data of three new dihydrophenanthrofurans from *Pleione yunnanensis*. Magn Reson Chem 48:256–260
123. Leong YW, Harrison LJ (2004) A biphenanthrene and a phenanthro[4,3-b]furan from the orchid *Bulbophyllum vaginatum*. J Nat Prod 67:1601–1603
124. Wang L, Zhang CF, Wang ZT, Zhang M, Xu LS (2009) Five new compounds from *Dendrobium crystallinum*. J Asian Nat Prod Res 11:903–911
125. Dong FW, Fan WW, Xu FQ, Wan QL, Su J, Li Y, Zhou L, Zhou J, Hu JM (2013) Inhibitory activities on nitric oxide production of stilbenoids from *Pholidota yunnanensis*. J Asian Nat Prod Res 15:1256–1264
126. Sharma C, Dixit M, Singh R, Agrawal M, Mansoori MN, Kureel J, Singh D, Narender T, Arya KR (2015) Potential osteogenic activity of ethanolic extract and oxoflavidin isolated from *Pholidota articulata* Lindley. J Ethnopharmacol 170:57–65
127. Simmler C, Antheaume C, Lobstein A (2010) Antioxidant biomarkers from *Vanda coerulea* stems reduce irradiated HaCaT PGE-2 production as a result of COX-2 inhibition. PLoS One 5:1–9
128. Majumder PL, Datta N, Sarkar AK, Chakraborti J (1982) Flavidin. A novel 9,10-dihydrophenanthrene derivative of the orchids *Coelogyne flavida*, *Pholidota articulata* and *Otochilus fusca*. J Nat Prod 45:730–732
129. Majumder P, Bandyopadhyay D, Joardar S (1982) Coelogin and Coeloginin: Two Novel 9,l0-dihydrophenanthrene derivatives from the orchid *Coelogyne cristata*. J Chem Soc Perkin Trans I:1131–1136
130. Sharma C, Kumari T, Mansoori MN, Dixit M, Shukla P, Singh D, Bhandari SPS, Narender T, Arya KR (2014) Ethanolic extract of *Coelogyne cristata* Lindley (Orchidaceae) and its compound coelogin promote osteoprotective activity in overiectomized estrogen deficient mice. Phytomedicine 21:1702–1707
131. Veerraju P, Rao NSP, Rao LJ, Rao KVJ Rao PRMn(1989) Amoenumin, a 9,10-dihydro-5H-phenanthro-(4,5-b,c,d)-pyran from *Dendrobium amoenum*. Phytochemistry 28:950951
132. Majumder PL, Basak M (1991) Two stilbenoids from the orchid *Cirrhopetalum andersonii*. Phytochemistry 30:3429–3432
133. Majumder PL, Basak M (1990) Cirrhopetalin, a phenanthrene derivative from *Cirrhopetalum andersonii*. Phytockemistry 29:1002
134. Sun J, Zhan R, Chen Y, Zhang Y, Chen L (2018) A new phenanthrene and a new 9,1dihydrophenanthrene from *Bulbophyllum retusiusculum*. Nat Prod Res 32:2447–2451
135. Yamaki M, Bai L, Kato T, Inoue K, Takagi S, Yamagata Y, Tomita KI (1993) Blespirol, a phenanthrene with a spirolactone ring from *Bletilla striata*. Phytochemistry 33:1497–1498
136. Alder AC, Rüedi P, Eugster CH (1984) Plectranthone A, B, C und D. Diterpenoide Phenanthren-1,4-dione aus Blattdrüsen einer *Plectranthuus* sp. (Labiatae). Helv Chim Acta 67:1003–1011
137. Krohn K, Loock U, Paavilainen K, Hausen BM, Schmalle HW, Kiesele H (2001) Synthesis and electrochemistry of annoquinone-A, cypripedin methyl ether, denbinobin and related 1,4-phenanthrenequinones. Arkivoc 1:88–130

138. Xiao S, Xu D, Zhang M, Linhu L, Ding L, Zhou S, Zhou Y (2016) A novel phenanthrene-1,2-dione from *Bletilla striata*. Chin J Org Chem 36:638–641
139. Sun A, Pang S, Liu J, Lin J, Xu R (2016) Two novel phenanthrenequinones with anti-cancer activity isolated from *Bletilla striata*. Bioorg Med Chem Lett 26:2375–2379
140. Bhaskar MU, Rao LJM, Rao NSP, Rao PRM (1991) Ochrone A, a novel 9,10-dihydro-1,4-phenanthraquinone from *Coelogyne ochracea*. J Nat Prod 54:386–389
141. Barua AK, Ghosh BB, Ray S, Patra A (1990) Cymbinodin-A, a phenanthraquinone from *Cymbidium aloifolium*. Phytochemistry 29:3046–3047
142. Ghosh BB, Ray S, Bhattacharyya P, Datta PK, Mukherjee BB, Patra A, Banerjee AK, Barua AK (1992) Cymbinodin B, a phenanthraquinone from *Cymbidium aloifolium*. Indian J Chem 31B:557–558
143. Liu D, Ju J, Zou Z, Lin G, Yang J (2005) Isolation and structure determination of cypritibetquinone A and B, two new phenanthraquinones from *Cypripedium tibeticum*. Acta Pharm Sin 40:255–257
144. Hu JM, Chen JJ, Yu H, Zhao YX, Zhou J (2008) Five new compounds from *Dendrobium longicornu*. Planta Med 74:535–539
145. Sritularak B, Anuwat M, Likhitwitayawuid K (2011) A new phenanthrenequinone from *Dendrobium draconis*. J Asian Nat Prod Res 13:251–255
146. Chen XJ, Mei WL, Zuo WJ, Zeng YB, Guo ZK, Song XQ, Dai HF (2013) A new antibacterial phenanthrenequinone from *Dendrobium sinense*. J Asian Nat Prod Res 15:67–70
147. Lin TH, Chang SJ, Chen CC, Wang JP, Tsao LT (2001) Two phenanthraquinones from *Dendrobium moniliforme*. J Nat Prod 64:1084–1086
148. Yang D, Liu LY, Cheng ZQ, Xu FQ, Fan WW, Zi CT, Dong FW, Zhou J, Ding ZT, Hu JM (2015) Five new phenolic compounds from *Dendrobium aphyllum*. Fitoterapia 100:11–18
149. Tezuka Y, Hirano H, Kikuchi T, Xu GJ (1991) Constituents of *Ephemerantha lonchophylla*; isolation and structure elucidation of new phenolic compounds, ephemeranthol-A, ephemeranthol-B, and ephemeranthoquinone, and of a new diterpene glucoside, ephemeranthoside. Chem Pharm Bull 39:593–598
150. Chen DN, Wu YP, Chen YJ, Liu WJ, Wang JX, He F, Jiang L (2017) Two new stilbenoids from aerial parts of *Flickingeria fimbriata*. J Asian Nat Prod Res 212:117–122
151. Williams RB, Martin SM, Hu JF, Garo E, Rice SM, Norman VL, Lawrence JA, Hough GW, Goering MG, O'Neil-Johnson M, Eldridge GR, Starks CM (2012) Isolation of apoptosis-inducing stilbenoids from four members of the Orchidaceae family. Planta Med 78:160–165
152. Masuda Y, Suzuki R, Sakagami H, Umemura N, Shirataki Y (2012) Novel cytotoxic phenanthrenequinone from *Odontioda* Marie Noel 'Velano'. Chem Pharm Bull 60:1216–1219
153. Xue Z, Li S, Wang S, Wang Y, Yang Y, Shi J, He L (2006) Mono-, bi-, and triphenanthrenes from the tubers of *Cremastra appendiculata*. J Nat Prod 69:907–913
154. Shi Y, Zhang B, Lu Y, Qian C, Feng Y, Fang L, Ding Z, Cheng D (2017) Antiviral activity of phenanthrenes from the medicinal plant *Bletilla striata* against influenza A virus. BMC Compl Altern Med 17:273
155. Majumder PL, Pal A, Joardar M (1990) Cirrhopetalanthrin, a dimeric phenanthrene derivative from the orchid *Cirrhopetalum maculosum*. Phytochemistry 29:271–274
156. Li CY, Liu J, Su XH, Yuan ZP, Zhong YJ, Li YF, Liang B (2013) New dimeric phenanthrene and flavone from *Spiranthes sinensis*. J Asian Nat Prod Res 15:417–421
157. Liu L, Yin QM, Zhang XW, Wang W, Dong XY, Yan X, Hu R (2016) Bioactivity-guided isolation of biphenanthrenes from *Liparis nervosa*. Fitoterapia 115:15–18
158. Yang X, Yang J, Xu C, Lv J, Wang C, Song P (2016) Antimicrobial stilbenoids from *Bletilla yunnanensis*. Chem Nat Compd 52:19–22
159. Shi X, Li Y, Liu Y, Jiang J, Wang L, Zhang Y, Chen Y (2010) A new 9,10-dihydrophenanthropyran dimer and a new natural 9,10-dihydrophenanthropyran from *Otochilus porrectus*. Biochem Syst Ecol 38:842–845
160. Yamaki M, Bai L, Inoue K, Takagi S (1989) Biphenanthrenes from *Bletilla striata*. Phytochemistry 28:3503–3505

161. Yang M, Cai L, Tai Z, Zeng X, Ding Z (2010) Four new phenanthrenes from *Monomeria barbata* Lindl. Fitoterapia 81:992–997
162. Xu J, Yu H, Qing C, Zhang Y, Liu Y, Chen Y (2009) Two new biphenanthrenes with cytotoxic activity from *Bulbophyllum odoratissimum*. Fitoterapia 80:381–384
163. Liu L, Li J, Zeng KW, Jiang Y, Tu PF (2016) Five new biphenanthrenes from *Cremastra appendiculata*. Molecules 21:1089/1–10.
164. Qian CD, Jiang FS, Yu HS, Shen Y, Fu YH, Cheng DQ, Ding ZS, Gan LS (2015) Antibacterial biphenanthrenes from the fibrous roots of *Bletilla striata*. J Nat Prod 78:939–943
165. Bai L, Yamaki M, Inoue K, Takagi S (1990) Blestrin A and B, bis(dihydrophenanthrene)ethers from *Bletilla striata*. Phytochemistry 29:1259–1260
166. Phiboonchaiyanan PP, Petpiroon N, Sritularak B, Chanvorachote P (2018) Phoyunnanin E induces apoptosis of non-small cell lung cancer cells via p53 activation and down-regulation of survivin. Anticancer Res 38:6281–6290
167. Yamaki M, Bai L, Kato T, Inoue K, Takagi S, Yamagata Y, Tomita KI (1992) Bisphenanthrene ethers from *Bletilla striata*. Phytochemistry 31:3985–3987
168. Cakova V, Urbain A, Antheaume C, Rimlinger N, Wehrung P, Bonté F, Lobstein A (2015) Identification of phenanthrene derivatives in *Aerides rosea* (Orchidaceae) using the combined systems HPLC-ESI-HRMS/MS and HPLC-DAD-MS-SPE-UV-NMR. Phytochem Anal 26: 34–39
169. Majumder PL, Banerjee S, Lahiri S, Mukhoti N, Sen S (1998) Dimeric phenanthrenes from two *Agrostophyllum* species. Phytochemistry 47:855–860
170. Li JY, Hou B, Ren FC, Yang XB, Lv YF, Kuang MT, Yang L, Zhou J, Hu JM (2018) Poly *p*-hydroxybenzyl substituted bibenzyls and phenanthrenes from *Bletilla ochracea* Schltr with anti-inflammatory and cytotoxic activity. Fitoterapia 129:214–248
171. Xiao SJ, Yuan FM, Zhang MS, Yu SY, Li JD, Yi XD, Xu DL (2017) Three new 1-(p-hydroxybenzyl)phenanthrenes from *Bletilla striata*. J Asian Nat Prod Res 19:140–144
172. Xiao SJ, Yuan FM, Zhang MS, Yu SY, Li JD, Yi XD, Xu DL (2016) Three new 1-(p-hydroxybenzyl)phenanthrenes from *Bletilla striata*. J Asian Nat Prod Res 31:1–5
173. Waratchareeyakul W, Fusi F, Durante M, Ahmed A, Knirsch W, Mas-Claret E (2020) Vasorelaxing activity of stilbenoid and phenanthrene derivatives from *Brasiliorchis porphyrostele*: involvment of smooth muscle $Ca_V1.2$ channels. Planta Med 86:631–642
174. Majumder PL, Banerjee S (1988) A ring-B oxygenated phenanthrene derivative from the orchid *Bulbophyllum gymnopus*. Phytochemistry 27:245–248
175. Majumder PL, Kar A, Shoolery JN (1985) Bulbophyllanthrin, a phenanthrene of the orchid *Bulbophyllum leopardium*. Phytochemistry 24:2083–2087
176. Chen YG, Xu JJ, Yu H, Qing C, Zhang YL, Liu Y, Wang JH (2007) 3,7-Dihydroxy-2,4,6-trimethoxyphenanthrene, a new phenanthrene from *Bulbophyllum odoratissimum*. J Korean Chem Soc 51:352–355
177. Majumder PL, Bandyopadhyay S, Pal S (2008) Rigidanthrin, a new dimeric phenanthrene derivative of the orchid *Bulbophyllum rigidum*. J Indian Chem Soc 85:1116–1123
178. Leong YW, Harrison LJ, Powell AD (1999) Phenanthrene and other aromatic constituents of *Bulbophyllum vaginatum*. Phytochemistry 50:1237–1241
179. Xue Z, Li S, Wang SJ, Yang YC, He DX, Ran GL, Kong LZ, Shi JG (2005) Studies on chemical constituents from the corm of *Cremastra appendiculata*. China Journal of Chinese Materia Medica 30:511–513
180. Majumder PL, Sen S, Majumder S (2001) Phenanthrene derivatives from the orchid *Coelogyne cristata*. Phytochemistry 58:581–586
181. Majumder PL, Sen RC (1991) Pendulin, a polyoxygenated phenanthrene derivative from the orchid *Cymbidium pendulum*. Phytochemistry 30:2432–2434
182. Li YP, Wang YJ, Chen LL (2016) Chemical constituents of *Dendrobium chrysotoxum*. Chem Nat Compd 52:90–92
183. Treesuwan S, Chanvorachote P, Pongrakhananon V, Treesuwan S, Chanvorachote P, Pongrakhananon V, Sritularak B (2018) Cypripedin diminishes an epithelial-to mesenchymal

transition in non-small cell lung cancer cells through suppression of Akt/GSK-3β signalling. Sci Rep 8:8009

184. Zhang YY, Wang P, Wang H, Chen LL, Mei WL, Dai HF, Zhang YY, Song XQ, Zuo WJ (2019) Chemical constituents from *Dendrobium hainanense*. J Asian Nat Prod Res 21:873–880
185. Hwang JS, Lee SA, Hong SS, Han XH, Lee C, Kang SJ, Lee D, Kim Y, Hong JT, Lee MK, Hwang BY (2010) Phenanthrenes from *Dendrobium nobile* and their inhibition of the LPS-induced production of nitric oxide in macrophage RAW 264.7 cells. Bioorg Med Chem Lett 20:3785–3787
186. Zhang X, Xu JK, Wang NL, Kurihara H, Yao XS (2008) Antioxidant phenanthrenes and lignans from *Dendrobium nobile*. J Chin Pharm Sci 17:314–318
187. Kim JH, Oh SY, Han SB, Uddin GM, Kim CY, Lee JK (2015) Anti-inflammatory effects of *Dendrobium nobile* derived phenanthrenes in LPS-stimulated murine macrophages. Arch Pharm Res 38:1117–1126
188. Zhang GN, Zhong LY, Bligh SWA, Guo YL, Zhang CF, Zhang M, Wang ZT, Xu LS (2005) Bi-bicyclic and bi-tricyclic compounds from *Dendrobium thyrsiflorum*. Phytochemistry 66: 1113–1120
189. Chen CC, Huang YL, Teng CM (2000) Antiplatelet aggregation principles from *Ephemerantha lonchophylla*. Planta Med 66:372–374
190. Hernandez-Romero Y, Acevedo L, de Sanchez M, Shier WT, Abbas HK, Mata R (2005) Phytotoxic activity of bibenzyl derivatives from the orchid *Epidendrum rigidum*. J Agric Food Chem 53:6276–6280
191. Bhandari SR, Kapadi AH, Majumder PL, Joardar M, Shoolery JN (1985) Nudol, a phenanthrene of the orchids *Eulophia nuda*, *Eria carinata* and *Eria stricta*. Phytochemistry 24:801–804
192. Majumder PL, Kar A (1987) Confusarin and confusaridin, two phenanthrene derivatives of the orchid *Eria confusa*. Phytochemistry 26:1127–1129
193. Tatiya AU, Bari N, Surana SJ, Kalaskar MG (2014) Effect of bioassay guided isolation of 1-phenanthrene carboxylic acid from *Eulophia herbacea* Lindi. tubers on human cancer cell lines. Res J Phytochem 8:155–161
194. Wisutthathum S, Demougeot C, Totoson P, Adthapanyawanich K, Ingkaninan K, Temkitthawon P, Chootip K (2018) *Eulophia macrobulbon* extract relaxes rat isolated pulmonary artery and protects against monocrotaline-induced pulmonary arterial hypertension. Phytomedicine 50:157–165
195. Temkifthawon P, Changwichit K, Khorana N, Viyoch J, Suwanborirux K, Ingkaninan K (2017) Phenanthrenes from *Eulophia macrobulbon* as novel phosphodiesterase-5 inhibitors. Nat Prod Commun 12:79–82
196. Shriram V, Kumar V, Kishor PB, Suryawanshi SB, Upadhyay AK, Bhat MK (2010) Cytotoxic activity of 9,10-dihydro-2,5-dimethoxyphenanthrene-1,7-diol from *Eulophia nuda* against human cancer cells. J Ethnopharmacol 128:251–253
197. Kshirsagar RD, Kanekar YB, Jagtap SD, Upadhyay SN, Rao R, Bhujbal SP, Kedia JN (2010) Phenanthrenes of *Eulophia ochreata* Lindl. Int J Green Pharm:147–152
198. Cretton S, Righi D, Sahib L, Christen P, Oyarzun A, Fajardo V, Kaiser M (2018) A new antifungal and antiprotozoal bibenzyl derivative from *Gavilea lutea*. Nat Prod Res 32:695–701
199. Savaris CR, Ghiraldi DMB, Chiavelli LUR, de Oliveira Santin SM, Milaneze-Gutierre MA, Kischkel B, Negri M, Scariot DB, Garcia FP, Nakamura CV, Pomini AM (2018) Phytochemical and biological studies of *Gomesa recurva* R. Br. (Orchidaceae): Chemotaxonomic significance of the presence of phenanthrenoids. Biochem Syst Ecol 80:11–13
200. Majumder PL, Lahiri S (1990) Volucrin, a new dimeric phenanthrene derivative from the orchid *Luisia volucris*. Tetrahedron 46:3621–3626
201. Estrada S, Toscano RA, Mata R (1999b) New phenanthrene derivatives from *Maxillaria densa*. J Nat Prod 62:1175–1178
202. Estrada S, López-Guerrero JJ, Villalobos-Molina R, Mata R (2004) Spasmolytic stilbenoids from *Maxillaria densa*. Fitoterapia 75:690–695

203. Valencia-Islas NA, Paul RN, Shier WT, Mata R, Abbas HK (2002) Phytotoxicity and ultrastructural effects of gymnopusin from the orchid *Maxillaria densa* on duckweed (*Lemna pausicostata*) frond and root tissues. Phytochemistry 61:141–148
204. Suzuki R, Tanaka T, Yamamoto M, Sakagami H, Tomomura M, Tomomura A, Satoh K, Shirataki Y (2012) In search of new biological activities of isolates from *Odontoglossum* Harvengtense 'Tutu'. In Vivo 26:993–1000
205. Estrada-Soto S, López-Guerrero JJ, Villalobos-Molina R, Mata R (2006) Endothelium-independent relaxation of aorta rings by two stilbenoids from the orchids *Scaphyglottis livida*. Fitoterapia 77:236–239
206. Yang KC, Uen YH, Suk FM, Liang YC, Wang YJ, Ho YS, Li IH, Lin SY (2005) Molecular mechanisms of denbinobin-induced anti-tumorigenesis effect in colon cancer cells. World J Gastroenterol 11:3040–3045
207. Song JI, Kang YJ, Yong HY, Kim YC, Moon A (2012) Denbinobin, a phenanthrene from *Dendrobium nobile*, inhibits invasion and induces apoptosis in SNU-484 human gastric cancer cells. Oncol Rep 27:813–818
208. Sánchez-Duffhues G, Calzado MA, de Vinuesa AG, Appendino G, Fiebich BL, Loock U, Lefarth-Risse A, Krohn K, Muñoz E (2009) Denbinobin inhibits nuclear factor-κB and induces apoptosis via reactive oxygen species generation in human leukemic cells. Biochem Pharmacol 77:1401–1409
209. Huang YC, Guh JH, Teng CM (2005) Denbinobin-mediated anticancer effect in human K562 leukemia cells: role in tubulin polymerization and Bcr-Abl activity. J Biomed Sci 12:113–121
210. Lu TL, Han CK, Chang YS, Lu TJ, Huang HC, Bao BY, Wu HY, Huang CH, Li CY, Wu TS (2014) Denbinobin, a phenanthrene from *Dendrobium nobile*, impairs prostate cancer migration by inhibiting Rac1 activity. Am J Chin Med 42:1539–1554
211. Datla P, Kalluri MD, Basha K, Bellary A, Kshirsagar R, Kanekar Y, Upadhyay S, Singh S, Rajagopal V (2010) 9,10-Dihydro-2,5-dimethoxyphenanthrene-1,7-diol, from *Eulophia ochreata*, inhibits inflammatory signalling mediated by Toll-like receptors. Br J Pharmacol 160:1158–1170
212. Yang H, Lee PJ, Jeong EJ, Kim HP, Kim YC (2012) Selective apoptosis in hepatic stellate cells mediates the antifibrotic effect of phenanthrenes from *Dendrobium nobile*. Phytother Res 26: 974–980

Orchids of Genus *Bletilla*: Traditional Uses, Phytochemistry, Bioactivities, and Commercial Importance

23

Hari Prasad Devkota, Rajan Logesh, Anjana Adhikari-Devkota, and Mukti Ram Paudel

Contents

H. P. Devkota (✉) · A. Adhikari-Devkota
Graduate School of Pharmaceutical Sciences, Kumamoto University, Kumamoto, Japan
e-mail: devkotah@kumamoto-u.ac.jp; anjana@kumamoto-u.ac.jp

R. Logesh
TIFAC-CORE in Herbal Drugs, Department of Pharmacognosy and Phytopharmacy, JSS College of Pharmacy (JSS Academy of Higher Education and Research), Udhagamandalam, Tamil Nadu, India

M. R. Paudel
Central Department of Botany, Tribhuvan University, Kathmandu, Nepal
e-mail: muktiram.paudel@cdb.tu.edu.np

J.-M. Mérillon, H. Kodja (eds.), *Orchids Phytochemistry, Biology and Horticulture*, Reference Series in Phytochemistry, https://doi.org/10.1007/978-3-030-38392-3_36

Abstract

The genus *Bletilla* comprises of six species, that is, *B. chartacea*, *B. foliosa*, *B. formosana*, *B. guizhouensis*, *B. ochracea*, and *B. striata*, mainly distributed in China, Japan, Taiwan, Vietnam, Thailand, and Myanmar. Few species of this genus are used for medicinal purposes and they have great importance for their ornamental values. *Bletilla* species are used traditionally for the treatment of respiratory diseases and as hemostatic agents. Chemical analysis has revealed the presence of various compounds such as phenanthrenes, bibenzyl derivatives, terpenoids, and saponins, among others. Similarly, pharmacological activity and evaluation of the extracts and isolated compounds have reported potent anti-inflammatory, antioxidant, and hemostatic activities. However, extensive harvesting in recent years has resulted in destruction of natural habitats and needs urgent focus for preservation. This book chapter focuses on the traditional uses, phytochemistry, and biological activities of *Bletilla* species along with their conservation practices and commercialization potential.

Keywords

Bletilla · *Bletilla striata* · Orchids · Orchidaceae · Traditional uses · Phytochemistry · Conservation · Commercialization

1 Introduction

Orchidaceae is one of the largest families of flowering plants which consists of more than 800 genera and 30,000 species [1]. Orchid flowers are well-known for their beautiful color and fragrances and cultivated worldwide for ornamental purposes [2, 3]. Many orchid species are also used in traditional medicines for the prevention and treatment of various diseases and also studied widely for their interesting bioactive chemical constituents and pharmacological activities along with their potential in drug discovery and development and use in food, cosmetic, and other industries [4–9].

The genus *Bletilla* consists of six species, that is, *B. chartacea* (King & Pantl.) Tang & F.T.Wang, *B. foliosa* (King & Pantl.) Tang & F.T.Wang, *B. formosana* (Hayata) Schltr., *B. guizhouensis* Jie Huang & G.Z.Chen, *B. ochracea* Schltr., and *Bletilla striata* (Thunb.) Rchb.f. (http://www.plantsoftheworldonline.org/taxon/urn:lsid:ipni.org:names:331150-2, retrieved on December 4, 2020). Species of *Bletilla* are native to China, Japan, Korea, Myanmar, Taiwan, Thailand, and Vietnam. They are widely cultivated for their ornamental values. Furthermore, *B. formosana*, *B. ochracea*, and *B. striata* are reported to be used in various traditional medicine

Fig. 1 Photographs of *Bletilla striata*: (**a**) plant with flowers, (**b**) flowers close-up, (**c**) rhizomes, (**d**) rhizomes close-up

systems in various respiratory disorders, swelling, hemorrhage, ulcer, and as hemostatic agents. Among these six species, *B. striata* (Fig. 1) is one of the most widely used species and is studied in detail for its traditional medicinal value, bioactive chemical constituents, and pharmacological activities. However, extensive harvesting in recent years has resulted in destruction of natural habitats and needs urgent focus for preservation. This book chapter focuses on the traditional uses, phytochemistry, and biological activities of *Bletilla* species along with their conservation practices and commercialization potential.

2 Botanical Description, Distribution, and Ecology

Bletilla, commonly known as urn orchid, is a temperate and terrestrial genus consisting of six species (*B. chartacea, B. foliosa, B. formosana, B. guizhouensis, B. ochracea*, and *B. striata*) distributed through China, Japan, Taiwan, Vietnam, Thailand, and Myanmar [10, 11]. They are small- to medium-sized herbs with subterranean, tuberous, and irregular shaped rhizome containing many long and fibrous roots. The erect stem bears a few plicate, narrowly oblong-lanceolate to linear-lanceolate deciduous leaves sheathing at the base. The inflorescence is terminal and loose with several flowers. The flowers are usually medium-sized, star-shaped with white, pink, purple, or yellow colored tepals [8].

Bletilla species vary on their size and the habitat; however, most of them are reported from tropical or temperate regions. *B. formosana* is a perennial herb found in sunny locations from 1200 to 3000 m, and *B. ochracea* is found in open forest with altitude of 900–2400 m. *B. striata* is the most commonly cultivated species that usually grows on grassland at 1100–3200 m [8].

3 Traditional Uses

Mainly three species of genus *Bletilla* are being used in traditional/folk medicine system of China, India, and other Asian countries. In China, the tubers of *B. formosana*, *B. ochracea*, and *B. striata* are biological sources for the crude drug "Bai-ji" used in the treatment of bleeding, burns, esophagus inflammation, gastritis, and ulcers [12]. The details of the medicinal uses are provided below:

B. formosana: In China, the stems are known for their properties to strengthen the lungs and to reduce swelling and bleeding. They are used for the treatment of cough, tuberculosis, pneumonia, bronchiectasis, peptic ulcers, and bleeding from nose. The scrapings of the stems are applied to treat cracks of the heel in India [8].

B. ochracea: It is reported to be used as substitute for *B. striata* in Southwest China and used in treatment of pulmonary tuberculosis, pneumonia, and ulcers [12–14].

B. striata: *B. striata* is well known for its stringent hemostatic property [15]. Tubers are the mainly used parts which are used for the treatment of swelling, bleeding from cuts, burns and trauma, cough, tuberculosis, etc. [8]. *B. striata* is used locally for the treatment of chapped skin, sores, ulcers, and wounds which helps in reducing the swelling and promoting the regeneration of tissues [3]. It has hemostasis and lung-supplementing effects, and is taken for hematemesis due to pulmonary tuberculosis and gastric ulcer. Powder or grated material is applied to swellings, cuts, cracks, and burns. It is also cultivated in the garden for ornamental purposes. Although Orchids are very difficult to cultivate, it show vigorous growth even if they are left alone [16]. It is also used in drinks, medicated diets, and also in wines for its beneficial purposes. Other uses include as cosmetic, insecticide, and coating agent [17].

4 Bioactive Compounds

Orchids are well known for their various bioactive constituents, and they are mainly derivatives of chemical classes such as hydroxyl-benzyl derivatives [18], fluorenones and stilbenoids [19], and flavones *C*-glycosides [20], among others. Few species of genus *Bletilla* have also been studied for their bioactive chemical constituents, and *B. striata* is the most studied among these species for the isolation/analysis of chemical constituents and biological activities. Studies so far have reported dihydrophenanthrene derivatives, phenanthrene derivatives, bibenzyl derivatives, triterpenoids, and saponins as major chemical constituents. The detailed chemical components of these species are explained below:

***Bletilla formosana*:** Lin et al. [21] performed the chemical analysis of the whole plants of *B. formosana.* Three new dihydrophenanthrene derivatives named as 4-methoxy-9,10-dihydrophenanthrene-1,2,7-triol, 1-(4-hydroxybenzyl)-4,7-dimethoxy-9,10-dihydrophenanthrene-2-ol, and 1,3,6-tri(4-hydroxylbenzyl)-4-methoxydihydrophenanthrene-2,7-diol (Fig. 2) were isolated and identified along with seven already known phenanthrene derivatives namely 1-(4-hydroxybenzyl)-4-methoxy9,10-dihydrophenanthrene-2,7-diol, 1,6-di(4-hydroxylbenzyl)-4-methoxy-9,10-dihydrophenanthrene-2,7-diol, 1,3-di(4-hydroxybenzyl)-4-methoxy-9,10-dihydrophenanthrene-2,7-diol, 1-(4-hydr oxybenzyl)-4-methoxyphenanthrene-2,7-diol, 1-(4-hydroxybenzyl)-4,8-di-methoxyphenanthrene-2,7-diol, 1,8-bi(4-hydroxybenzyl)-4-methoxyphenanthrene-2,7-diol, and blestriarene B; a known bibenzyl derivative, 2′,6′-bis(*p*-hydroxybenzyl)-3,3′-dihydroxy-5- methoxybibenzyl; six known flavonoids, apigenin, kaempferol, 6-methoxykaempferol, kaempferol 7-*O*-glucoside, 8-*C*-*p*-hydoxybenzylkaempferol and isorhamnetin; and three simplephenolic compounds, 3,4-dihydroxybenzaldehyde, protocatechuic acid, and *p*-(hydroxymethyl)phenyl-β-D-glucoside.

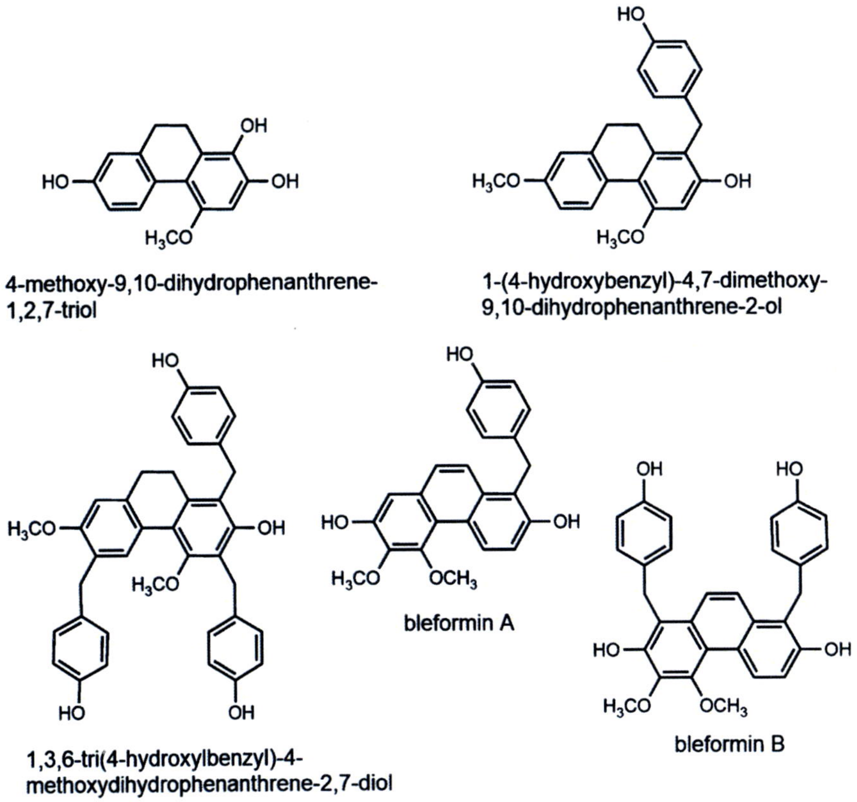

Fig. 2 Structures of some main compounds from *B. formosana*

Lin et al. [22] performed the chemical analysis of the rhizomes of *B. formosana*. Total nine new phenanthrene derivatives named as bleformins A-I (Fig. 2), a new bibenzylglucoside named as bleformin J were isolated and identified along with 45 known compounds including several phenanthrenes, bibenzyl derivatives and other chemical classes.

***B. ochracea*:** Cai et al. [13] isolated a new phenanthrene derivative, 2,7,-bis (allyloxy)-5-methoxy-3-methyl-9,10-dihydrophenanthrene along with gastrodin, gastrodigenin, and 4-hydroxybenzaldehyde (Fig. 3) and three steroids namely *β*-sitosterol, stigmasterol, and daucosterol from the tubers of *B. ochracea*. Yang et al. [23] isolated a new 2(2-methylpropyl)butanedioic acid derivative named as bletillin A and a new bibenzyl derivative named as bletillin B (Fig. 3) along with 17 known compounds such as gymnoside V, gymnoside VI, militarine, erianthridine, nudol, 1,5,7-trimethoxyphenanthrene-2,6-diol, 4,7-dihydroxy-2-methoxy-9,10-dihydrophenanthrene, 4,7-dihydroxy-1-p-hydroxybenzyl-2-methoxy-9,10-dihydro phenanthrene, 2,7-dihydroxy-1-p-hydroxybenzyl-4-methoxy-9,10-dihydrophenanthrene, blestriarene A, blestriarene B, blestriarene C, 3,3′-dihydroxy-2,6-bis(p-hydroxybenzyl)-5-methoxybibenzyl, 2′,6′-bis(p-hydroxybenzyl)-3′,5-dimethoxy-3-hydroxybibenzyl, 3, 3′-dihydroxy-4-(p-hydroxybenzyl)-5-methoxybibenzyl, 3,3′-dihydroxy-2-(p-hydroxy benzyl)-5-methoxybibenzyl, and 3′,5-dihydroxy-2-(p-hydroxybenzyl)-3-methoxy bibenzyl from the tubers. Some of the phenanthrene derivatives were reported to show antibacterial activity against Gram-positive bacteria. Li et al. [12] isolated an identified four new dihydrophenanthrenofuran derivatives named as bleochranols A-D along with 10 known phenanthrenes and 11 known bibenzylderivatives from the rhizomes. These compounds were also evaluated for their anti-inflammatory and

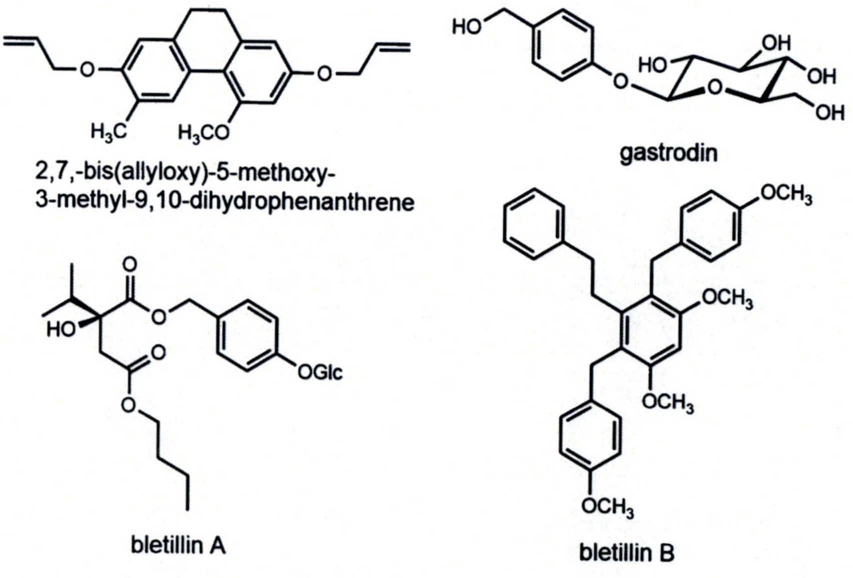

Fig. 3 Structures of some main compounds from *B. ochracea*

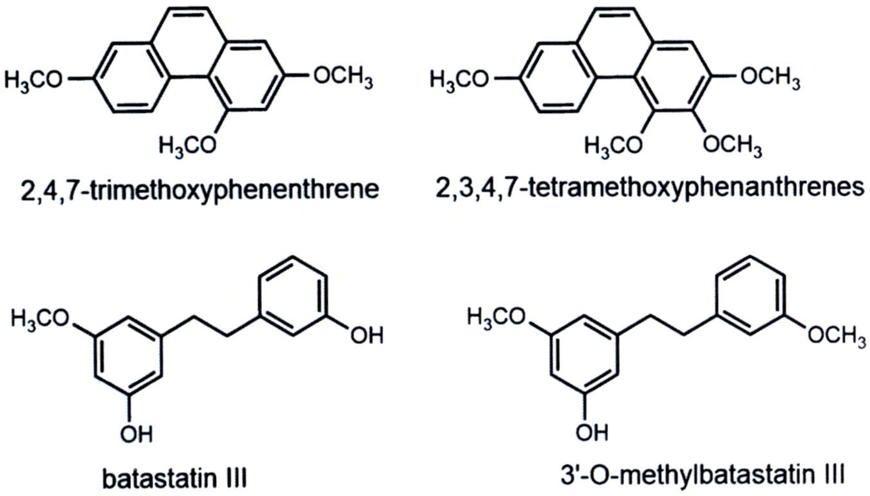

Fig. 4 Structures of some main compounds from *B. striata*

cytotoxicity activities. Li et al. [24] further isolated and characterized six new bibenzyl derivatives named as bleochrins A-F and four new phenanthrene derivatives named as bleochrins G-J along with 11 previously known compounds from the rhizomes. Recently, few studies related to the extraction, characterization and bioactive evaluation of polysaccharides from the tubers of *B. ochracea* are also reported [25].

B. striata: As mentioned before, *B. striata* is the most extensively studied species in this genus regarding chemical constituents and biological activities. In 2017, He et al. [15] have compiled the chemical constituents isolated and identified from *Bletilla striata,* where they have tabulated a total of 125 compounds belonging to different classes such as phenanthrene derivatives, biphenanthrene derivatives, dihydrophenanthrene derivatives, bibenzyls, phenanthraquinones, terpenoids, and saponins, among others. The details of these compounds can be found in the mentioned review [15], and thus they are not discussed in detail here in this chapter. Structures of some representative compounds, that is, 2,4,7-trimethoxyphenenthrene, 2,3,4,7-tetramethoxyphenanthrene, batastatin III, and 3'-O-methylbatastatin III are given in Fig. 4. In recent years, studies have also focused on the advanced techniques for the extraction, analysis, and characterization of polysaccharides from the rhizomes along with their bioactivities [26, 27].

5 Biological Activities

Owing to their extensive use in traditional medicines, and presence of interesting bioactive chemical constituents, *Bletilla* species are also studied widely for their pharmacological activities using in vitro and in vivo models. Some of these studies are discussed in detail in the following sections.

5.1 Anti-inflammatory Activity

Lin et al. have evaluated the anti-inflammatory activity of the chemical constituents of the rhizomes of *B. formosana.* The 43 compounds were isolated from the ethanolic extract and were evaluated for superoxide anion scavenging and elastase inhibitory activities. The results revealed that most of the compounds showed potent inhibitory activity against superoxide anion and elastase activity with an IC_{50} value ranging from 0.2 ± 0.1 and 0.4 ± 0.1 μM to >10 μM, which were compared with the CAL-101 used as a standard that possessed an IC_{50} value of 0.1 ± 0.1 and 0.3 ± 0.1, respectively [22].

Li et al. carried out anti-inflammatory activity evaluation of the extracts of rhizomes of *Bletilla ochracea* against murine monocytic RAW 264.7 cells and the results showed that the ethanol fraction possessed an IC_{50} value of 45.85 ± 2.21 μM, whereas the compounds 3-(4-hydroxybenzyl)-4-methoxy-9,10-dihydrophenanthrene-2,7-diol, 4-methoxy-9,10-dihydrophenanthrene-2,7-diol and 4-methoxyphenanthrene-2,7-diol showed most potent activity with an IC_{50} value of 8.17 ± 0.64, 8.81 ± 0.46, and 2.86 ± 0.17 μM, respectively [12].

Jiang et al. have carried out anti-inflammatory activity using coelonin an active compound isolated from the ethanol extract of the tubers of *B. striata*. The results showed that the compound coelonin significantly inhibited lipopolysaccharide (LPS)-induced interleukin-1β (IL-1β), interleukin-6 (IL-6), and tumor necrosis factor-α (TNF-α) expression at 2.5 μg/mL. The phosphorylation levels of the key inflammatory regulators such as nuclear factor-kappa B (NF-κB) and cyclin-dependent kinase inhibitor 1B (p27Kip1) were also significantly reduced [28].

Zu et al. carried out the pulmonary anti-inflammatory activity of the extracts from *B. striata* in RAW264.7 cells using $PM_{2.5}$. The pretreatment with the extract of *B. striata* significantly decreased the inflammatory cytokines in the macrophage. The extract also attenuated $PM_{2.5}$-induced proinflammatory protein expression and downregulated the levels of phosphorylated NF-κBp65, inhibitor of kappa B (IκB)-α, c-Jun N-terminal kinase (JNK), extracellular signal-regulated kinase (ERK), and p38 [29].

Wang et al. carried out the evaluation of anti-inflammatory activity against COX-1 and COX-2 for the compounds isolated from *B. striata*. The results showed that the compounds, (1α,3α)-1-O-[(β-D-xylopyranosyl-(1→2)-α-L-rhamnopyranosyl)]-3-O-glucopyranosyl-5α-spirostan, (1α,3α)-1-O-[(β-D-xylopyranosyl-(1→2)-α-Lrhamnopyranosyl)oxy]-3-O-D-glucopyranosyl-25(27)-ene-5αspirostan, (1α,3α)-1-O-[(β-D-xylopyranosyl-(1→2)-α-L-rhamnopyranosyl)oxy]-epiruscogenin, (1α,3α)-1-O-[(β-D-xylopyranosyl-(1→2)-α-Lrhamnopyranosyl)oxy]-epineoruscogenin, and 3-O-β-D-glucopyranosyl-3-epi-neoruscogenin exhibited significant anti-inflammatory activity with an IC_{50} value ranging from 35.5 to 96.4 μM [30].

5.2 Antioxidant Activity

Dong et al. carried out the DPPH, ABTS, and FRAP assay on the tubers of acetone fractions on the fermented and nonfermented *B. formosana*. The results revealed that the fermented *B. formosana* showed higher antioxidant activity on DPPH, ABTS, and FRAP assays with IC_{50} values of 0.417, 43.57 ± 4.72, and 120.1 ± 0.2 μg/ml, respectively [31].

Chen et al. studied the antioxidant activity of a new polysaccharide from the fibrous roots of *B. striata* against DPPH, ABTS, and superoxide anion radicals. The results showed that both the polysaccharide as well as *B. striata* extract showed potent activity. The percentage scavenging activities for polysaccharide and *B. striata* against DPPH, superoxide, and ABTS were 64.47% and 23.35%, 72.27% and 32.82%, and 29.49% and 18.90%, respectively [32].

Song et al. have carried out the antioxidant activity against DPPH, ABTS, hydroxyl, superoxide anion, and reducing power using ethanol extract of *B. striata* pure and ointment mixture. The results showed that both pure and ointment mixture showed potent antioxidant activities [33].

Zhang et al. evaluated the antioxidant activity against ABTS, and FRAP assay using BSP fraction. The results showed that *B. striata* could achieve 76% of ABTS scavenging activity, and 46.2% of FRAP ability at 10 mg/ml concentration [34].

Qu et al. have carried out the antioxidant activity against superoxide anion, hydroxyl, DPPH and chelation of ferrous ions using ethanol extract of *B. striata*. The results showed that on treating with *B. striata* it possessed significant antioxidant properties [27].

Jiang et al. have carried out the antioxidant activity against DPPH and FRAP using fractions from ethanol extract of tubers of *B. striata*. The results showed that ethanolic extracts from both FRP and PSP had strong free radical scavenging with IC_{50} values of FRP (6.2 mg/L) which was slightly lower than the positive control (2.4 mg/L) but was significantly higher than PSP (68.0 mg/L) activity, respectively [35].

Wang et al. carried out the antioxidant activity of *B. ochracea* polysaccharides (BOP) against DPPH, ABTS, hydroxyl, superoxide anion, and ferrous ions (Fe^{2+}) free radicals assay. The results showed that the BOP showed to inhibit with an EC_{50} value of 692.16, 224.09, 542.22, 600.53, and 515.70 μg/mL, respectively [25].

5.3 Cytotoxic, Antitumor, and Anticancer Activity

Li et al. have carried out cytotoxic activity against HL-60, SMMC-7721, A-549, MCF-7, and SW-480 from the ethanol extract of rhizomes of *Bletilla ochracea* and the results showed that the compounds bleochrin E, pleiobibenzynin A, pleiobibenzynin B, and 2,4-bis(p-hydroxybenzyl)-3,3′-dihydroxy-5-methoxybibenzyl were the most potent inhibitors with an IC_{50} values ranging from 0.79 to 6.57 μM [24].

Niu et al. have carried out the antitumor activity using CT26 cells from the polysaccharide fraction of tubers of *B. ochracea* (BOP). The results showed that BOP significantly decreased tumor growth in a dose-dependent manner [36].

Sun et al. have carried out the anticancer activity against MCF-7, HT-29, HUVEC, and A549 cells on the tubers of *Bletilla striata*. The compounds 7-hydroxy-2-methoxy-phenanthrene-3,4-dione and 3′,7′,7-trihydroxy-2,2′,4′-trimethoxy-[1,8′-biphenanthrene]-3,4-dione showed promising as an antiproliferative agent against cancer with an IC_{50} value ranging from 12.64 ± 2.17 to 48.35 ± 3.87 μM, respectively [37].

Zhang et al. have carried out the anti-cancer activity against GES-1 cells using BSP fraction. The results showed that BSP at higher concentrations of polysaccharides in culture medium could potentially increase permeability of cell membrane and did not significantly affect cell viability of GES1 cells after 24 h or 48 h treatment [34].

Jiang et al. have carried out the antitumor activity against HepG2 cells using fractions from ethanol extract of tubers of *B. striata*. The results showed that both FRP and PSP can dose dependently induce HepG2 cells apoptosis, which implied tumor therapeutic effect [35].

Guo et al. have carried out the cytotoxicity activity against human red blood cells using phenanthrene fraction (EF60) of *B. striata*. The results showed that the EF60 possessed at 160 μg/mL showed no cytotoxicity against human erythrocytes, and was minimally toxic to human umbilical vein endothelial cells with an IC_{50} of 75 μg/mL [38].

Wang et al. have carried out cytotoxicity activity against A-549 cells, BGC-823 cells, HepG2 cells, HL-60, MCF-7 cells, SMMC-7721, and W480 for the compounds isolated from *B. striata*. The results showed that the compounds (1α,3α)-1-O-[(β-D-xylopyranosyl-(1→2)-α-L-rhamnopyranosyl)]-3-OD-glucopyranosyl-5α-spirostan, (1α,3α)-1-O-[(β-D-xylopyranosyl-(1→2)-α-Lrhamnopyranosyl)oxy]-3-O-D-glucopyranosyl-25(27)-ene-5αspirostan, (1α,3α)-1-O-[(β-D-xylopyranosyl-(1→2)-α-L-rhamnopyranosyl)oxy]-epiruscogenin, (1α,3α)-1-O-[(β-D-xylopyranosyl-(1→2)-α-Lrhamnopyranosyl)oxy]-epineoruscogenin and 3-O-β-D-glucopyranosyl-3-epi-neoruscogenin exhibited significant cytotoxicity against all tested tumor cell lines with IC_{50} values less than 30 μM for all the cell lines used, whereas the compound bletilnoside A did not show any cytotoxicity activity [30].

5.4 Antimicrobial Activity

Yang et al. studied the antimicrobial activity against *Staphylococcus aureus*, *S. epidermidis*, *Bacillus subtilis*, *Escherichia coli*, *Candida albicans*, *C. krusei*, and *C. parapsilosis* for the compounds from the ethanol extract of the tubers of *B. ochracea*. The results showed that the compound blestriarene A possessed potent activity against three strains with an IC_{50} value ranging from 12.5 to 50 μg/mL, and the other compounds showed lesser activity against all strains [23].

Jiang et al. have performed the antibacterial activity against *S. aureus*, *B. subtilis*, and *E. coli* for the compounds isolated from the tubers of *B. striata*. The results showed the compounds 2,7-dihydroxy-3,4-dimethoxyphenanthrene and 2,7-dihydroxy4-methoxy-9,10-dihydrophenanthrene showed potent inhibition against *S. aureus* and *B. subtilis* with an MIC ranging from 26 to 53 μg/ml; and the other compounds namely 2,7-dihydroxy-3,4-dimethoxy-9,10-dihydro phenanthrene, shanciol C, shanciol D, shanciol F, and blestriarene A showed inhibition only against *S. aureus* with an MIC ranging from 6 to 53 μg/ml, whereas all the compounds did not show any inhibition against *E. coli* [39].

Jiang et al. have carried out the antibacterial activity against *S. aureus*, *B. subtilis*, and *E. coli* using ethanol extract of the dried tubers of *Bletilla striata*. The results showed that the compounds namely bletistrin G, bletistrin J, bulbocol, shancigusin C, shanciguol, and shancigusin B showed good antibacterial activities against *S. aureus* and *B. subtilis*. Among them, compounds bulbocol and shancigusin B showed potent inhibitory activities against S. aureus, with MICs of 9 and 3 μg/mL, respectively [40].

Qian et al. have carried out the antibacterial activity against Gram-positive bacteria: *S. aureus*, *S. epidermidis*, *Enterococcus faecalis*, and *B. subtilis* and Gram-negative bacteria: *E. coli* and *Proteus vulgaris* using compounds isolated from the ethanol extract of the rhizomes of *B. striata*. The results showed that the compounds 4,7,7′-trimethoxy9′,10′-dihydro(1,3′-biphenanthrene)-2,2′,5′-triol, 4,7,4′-trimethoxy-9′,10′-dihydro(1,1′-biphenanthrene)-2,2′,7′-triol, 4,7,3′,5′-tetramethoxy-9′,10′-dihydro(1,1′-biphenanthrene)-2,2′,7′-triol, 4,8,4′,8′-tetramethoxy(1,1′-biphenanthrene)-2,7,2′,7′-tetrol, and blestriarene C showed potent antibacterial activities against six Gram-positive bacteria strains, including methicillin-resistant *S. aureus* ATCC 43300 and ampicillin-resistant *S. aureus* ATCC 29213. Among them, 4,7,4′-trimethoxy-9′,10′-dihydro(1,1′-biphenanthrene)-2,2′,7′-triol showed the most potent inhibitory activities, with MICs of 2 and 4 μg/mL against ampicillin-resistant *S. aureus* ATCC 29213 and methicillin-resistant *S. aureus* ATCC 43300, respectively [41].

Guo et al. have carried out the antibacterial activity against *Staphylococcus aureus* using phenanthrene fraction (EF60) of *B. striata*. The results showed that the EF60 possessed significant minimum inhibitory concentration (MIC) values against these pathogens ranged from 8 to 64 μg/mL [38].

5.5 Hemostatic Activity

In traditional medicines, *Bletilla* species are well known for their hemostatic properties. Polysaccharides from the rhizomes of *B. striata* are reported to induce the hemostatic activity by activating adenosine diphosphate (ADP) receptor signaling pathway through P2Y1, p2y12, and PKC receptors [42].

Yang et al. have carried out the hemostatic activity in in vivo using compounds isolated from the tubers of *B. striata*. The results showed the parent compounds underwent various metabolic processes and their metabolites had various activities which could possess hemostatic activity [43].

Wang et al. have carried out hemostatic activity of the compounds isolated from *B. striata*. The results showed that the compounds (1α,3α)-1-O-[(β-D-xylopyranosyl-(1→2)-α-L-rhamnopyranosyl)]-3-OD-glucopyranosyl-5α-spirostan, (1α,3α)-1-O-[(β-D-xylopyranosyl-(1→2)-α-Lrhamnopyranosyl)oxy]-3-O-D-glucopyranosyl-25(27)-ene-5αspirostan, (1α,3α)-1-O-[(β-D-xylopyranosyl-(1→2)-α-L-rhamnopyranosyl)oxy]-epiruscogenin, (1α,3α)-1-O-[(β-D-xylopyranosyl-(1→2)-α-Lrhamnopyranosyl)oxy]-epineoruscogenin, and 3-O-β-D-glucopyranosyl-3-epi-neoruscogenin exhibited highest hemostatic activity [30].

Lu et al. have carried out the hemostatic activity using water, n-butyl alcohol, petroleum ether, and ethyl acetate fractions obtained from the ethanol extract of *B. striata*. The results showed that the water and n-butyl alcohol fractions significantly increased the platelet aggregation rate induced by ADP [44].

Wang et al. have evaluated the hemostatic activity using the extracts of tubers of *B. striata*. The results showed that *B. striata* inhibited the rat tail hemorrhage, traumatic hemorrhage of liver, and spleen in rabbits as well as traumatic hemorrhage of liver and abdominal aorta in Beagle dogs [45].

5.6 Immunological Activity

Wang et al. have carried out the immunological activity in ICR mice with the ethanol extract of *Bletilla striata* polysaccharide (BSP). The results showed on the BSP-1, BSP-2, and CBSP on thymus were −23.5%, 3.3%, and −4.1%, spleen index were −36.0%, −24.4%, and −10.4%, respectively [46].

Peng et al. have carried out the immunobiological activity using the acetone fraction of BSPF2. The results showed that BSPF2 was found to stimulate spleen cells proliferation. On the basis of this finding, it was suggested that BSPF2 may be a good source for the development of immunomodulator [47].

5.7 Anti-Fibrosis Activity

Wang et al. have carried out anti-fibrosis activity against human mesangial cells (HMCs) using aqueous extract of BSP at a concentration of 5, 10, 20, 40, 80, and 160 μg/ml. The results showed that BSPb exhibited significant anti-fibrosis activity by downregulating TGF-β RI, TGF-β RII, and α-SMA production [48].

5.8 Antiviral Activity

Shi et al. have studied the influenza A virus activity using ethanol extract of the rhizomes of *B. striata*. The results on the compounds against cytotoxicity to MDCK and reduction of CPE in MDCK cells, hemagglutination, and neuraminidase inhibition assay showed to inhibit with an IC_{50} value ranging from 0.9 ± 0.2 to 42.3 ± 3.9 μM, whereas the neuraminidase inhibition assay showed to possess IC_{50} values ranging from 16.8 ± 1.6 to 87.5 ± 10.1 μM, respectively [49].

5.9 Wound Healing Activity

Bletilla plants are also widely used in traditional medicines for wound healing activities. Song et al. have carried out the wound healing activity in vivo using ethanol extract of BSP pure and ointment mixture. The results showed that animals treated with “mixed ointment” experienced inflammatory infiltration, which was lower than that of other groups. Both "BSPG ointment" and "*Bletilla* phenolic ointment" demonstrated superior tissue repair compared to the control. Therefore, this study confirmed that the BSP and ointment mixture has excellent wound healing activities [33].

Chen et al. have carried out the wound healing activity using alcohol extract of the tubers of *B. striata* polysaccharide (BSP). The results showed that the BSP exhibited excellent healing function mainly due to its modulation of macrophages throughout inflammation and proliferation periods [50].

Diao et al. have carried out the wound healing activity using *B. striata* polysaccharide (BSP) isolated from *B. striata*. The results showed that BSP enhanced vascular endothelial cell (EC) proliferation and vascular endothelial growth factor (VEGF) expression and also changed the nitric oxide synthase (iNOS), tumor necrosis factor alpha (TNF-α), and interleukin 1 beta (IL-1β) mRNA levels and enhanced the expression of these cytokines, but has no effect on interferon gamma (IFN-γ) level, respectively [51].

5.10 Antiulcer Activity

Zhang et al. have carried out the gastric ulcer activity in vivo using BSP fraction. The results showed that the BSP treatment led to a 92% reduction in ulcer area. The histopathology reports showed that on BSP treatment the ulcer model group exhibited a remarkably gradual decrease in PAS staining intensity, compared to the normal group, respectively [34].

5.11 Anti-neuroinflammatory Activity

Zhou et al. have carried out the anti-neuroinflammatory activity against LPS-activated BV-2 microglial cells using the compounds isolated from the ethanol extract of dried tubers of *B. striata*. The results showed that the compound phochinenin K exhibited the most potent activity with an IC_{50} value of 1.9 μM [52].

5.12 Anti-Mitotic Activity

Morita et al. have carried out the antimitotic activity against K562/BCRP cells using compounds isolated from the methanol extract tubers of *B. striata*. The results showed that the stilbenoids strongly enhanced the cytotoxicity of SN-38 in K562/BCRP cells but not in K562 cells to possess antimitotic activity [53].

5.13 Anti-Tyrosinase Activity

Jiang et al. have carried out the tyrosinase inhibitory activity using fractions from ethanol extract of tubers of *B. striata*. The results showed that the ethanolic extracts from PSP showed strong tyrosinase inhibition activity in a dose-dependent manner with $IC_{50} = 751.4$ mg/L [35].

5.14 Anti-Ulcer Activity

Shi et al. have carried out the anti-ulcer activity of BSP against mice. The results showed that on BSP treatment (2, 10, and 50 mg/kg) exhibited dose-dependent inhibitory activity against the expression of Th2 type cytokines including IL-4, IL-5, and IL-13 [54]. Ke and Zhao et al. have performed the ulcerative colitis activity in mice using BSP. The results on BSP treatment showed to inhibit lymphocyte activation and secretion of related cytokines via suppressing the activation of macrophages [55]. Yu et al. have carried out the anti-ulcer activity of BSP against streptozotocin-induced diabetic ulcers in rats and the results revealed that BSP effectively stimulated inflammatory cell infiltration, promoted epithelial tissue formation and fibroblast proliferation, and increased hydroxyproline content [56].

6 Commercial Importance

B. striata and few other species of the genus are commercially important orchids due to their extensive use in traditional medicine in Asian countries and also for their ornamental purposes. Recent research regarding their bioactive chemical constituents and pharmacological activities have promoted their wide application. Many studies are also focusing on the polysaccharides from rhizomes/tubers [25–27, 57]. The mucilaginous roots of *B. striata* are also used for writing by mixing with vermilion, as insecticides and also in cosmetics [17]. However, extensive harvesting in recent years has resulted in destruction of natural habitats and need urgent focus for preservation [58]. Various new biotechnological tools have been applied in the conservation, propagation, and cultivation of commercially and medicinally important orchids in recent years [6, 59]. Cryopreservation techniques have also been utilized for *B. formosana* and *B. striata* seeds [58, 60–63]. However, there is a urgent need for proper conservation and cultivation techniques and approaches for sustainable harvesting/utilization of *Bletilla* species to meet the increasing commercial market demand.

7 Conclusions and Future Remarks

In this chapter, we compiled the available information about the traditional medicinal uses, phytochemistry and pharmacological activities of orchids belonging to the genus *Bletilla*. *B. striata* along with few other species are very important in traditional medicine systems in Asia. They are also important for their other non-medicinal uses such as for foods and cosmetics. Extensive harvesting in recent years has resulted in the decline of natural habitats, which needs immediate concern from the scientific community and other stakeholders.

References

1. Cakova V, Urbain A, Le Quéméner C et al (2015) Purification of vandaterosides from *Vanda teres* (Orchidaceae) by stepwise gradient centrifugal partition chromatography. J Sep Sci 38: 3006–3013. https://doi.org/10.1002/jssc.201500349
2. Khan H, Marya BT et al (2019) Genus Vanda: a review on traditional uses, bioactive chemical constituents and pharmacological activities. J Ethnopharmacol 229:46–53
3. De Lakshman C, Pathak P, Rao AN, Rajeevan PK (2015) Commercial orchids. De Gruyter Open, Warsaw/Berlin
4. Bulpitt CJ, Li Y, Bulpitt PF, Wang J (2007) The use of orchids in Chinese medicine. J R Soc Med 100:558–563. https://doi.org/10.1258/jrsm.100.12.558
5. Gutiérrez RMP (2010) Orchids: a review of uses in traditional medicine, its phytochemistry and pharmacology. J Med Plants Res 4:592–638. https://doi.org/10.5897/JMPR10.012
6. Pant B (2013) Medicinal orchids and their uses: tissue culture a potential alternative for conservation. African J Plant Sci 7:448–467. https://doi.org/10.5897/ajps2013.1031
7. Hadi H, Razali SNS, Awadh AI (2015) A comprehensive review of the cosmeceutical benefits of vanda species (orchidaceae). Nat Prod Commun 10:1483–1488
8. Teoh ES (2016) Medicinal orchids of Asia. Springer, Cham. https://doi.org/10.1007/978-3-319-24274-3
9. Teoh ES (2019) Orchids as aphrodisiac, medicine or food. Springer Nature Switzerland AG, Cham
10. Chen X, Gale SW, Cribb PJ (2009) BLETILLA H. G. Reichenbach, Fl. Serres Jard. Eur. 8: 246. 1853, nom. cons. In: Wu ZY, Raven PH (eds) Flora of China, Flagellariaceae through Marantaceae, vol 24. Science Press/Missouri Botanical Garden Press, Beijing/St. Louis, pp 209–210
11. Govaerts R (2020) World checklist of orchidaceae. http://wcsp.science.kew.org/. Accessed 21 Feb 2020
12. Li JY, Kuang MT, Yang L et al (2018) Stilbenes with anti-inflammatory and cytotoxic activity from the rhizomes of Bletilla ochracea Schltr. Fitoterapia 127:74–80. https://doi.org/10.1016/j.fitote.2018.02.007
13. Cai JY, Zhao L, Zhang DZ (2007) Chemical constituents from Bletilla ochracea Schltr. Chem Res Chinese Univ 23:705–707. https://doi.org/10.1016/S1005-9040(07)60153-6
14. Yu H, Dai B, Qian C et al (2016) Antibacterial activity of chemical constituents isolated from fibrous roots of *Bletilla striata*. Zhong Yao Cai 39:544–547
15. He X, Wang X, Fang J et al (2017) *Bletilla striata*: medicinal uses, phytochemistry and pharmacological activities. J Ethnopharmacol 195:20–38
16. Watanabe M, Devkota HP, Sugimura K, Watanabe T (2018) A guidebook of medicinal plant park. School of Pharmacy, Kumamoto University, Kumamoto
17. Guo Y, Zhai L, Long H et al (2018) Genetic diversity of *Bletilla striata* assessed by SCoT and IRAP markers. Hereditas 155:35. https://doi.org/10.1186/s41065-018-0074-4
18. Zeng X, Zhang S, Zhang L et al (2006) A study of the neuroprotective effect of the phenolic glucoside gastrodin during cerebral ischemia in vivo and in vitro. Planta Med 72:1359–1365. https://doi.org/10.1055/s-2006-951709
19. Yang L, Wang Z, Xu L (2006) Simultaneous determination of phenols (bibenzyl, phenanthrene, and fluorenone) in Dendrobium species by high-performance liquid chromatography with diode array detection. J Chromatogr A 1104:230–237. https://doi.org/10.1016/j.chroma.2005.12.012
20. Williams CA (1979) The leaf flavonoids of the orchidaceae. Phytochemistry 18:803–813. https://doi.org/10.1016/0031-9422(79)80019-9

21. Lin YL, Chen WP, Macabalang AD (2005) Dihydrophenanthrenes from *Bletilla formosana*. Chem Pharm Bull 53:1111–1113. https://doi.org/10.1248/cpb.53.1111
22. Lin CW, Hwang TL, Chen FA et al (2016) Chemical constituents of the rhizomes of *Bletilla formosana* and their potential anti-inflammatory activity. J Nat Prod 79:1911–1921. https://doi.org/10.1021/acs.jnatprod.6b00118
23. Yang X, Tang C, Zhao P et al (2012) Antimicrobial constituents from the tubers of *Bletilla ochracea*. Planta Med 78(6):606–610. https://doi.org/10.1055/s-0031-1298264
24. Li JY, Yang L, Hou B et al (2018) Poly p-hydroxybenzyl substituted bibenzyls and phenanthrenes from Bletilla ochracea Schltr with anti-inflammatory and cytotoxic activity. Fitoterapia 129:241–248. https://doi.org/10.1016/j.fitote.2018.07.007
25. Wang B, Xu Y, Chen L et al (2020) Optimizing the extraction of polysaccharides from bletilla ochracea schltr. Using response surface methodology (RSM) and evaluating their antioxidant activity. Processes 8:341. https://doi.org/10.3390/pr8030341
26. Shena X, Jiang W, Shena X, Jiang W (2020) Polysaccharides from *Bletilla striata*: extraction, optimization and their antioxidant activities in vitro. Med Res 4:200010. https://doi.org/10.21127/yaoyimr20200010
27. Qu Y, Li C, Zhang C et al (2016) Optimization of infrared-assisted extraction of *Bletilla striata* polysaccharides based on response surface methodology and their antioxidant activities. Carbohydr Polym 148:345–353. https://doi.org/10.1016/j.carbpol.2016.04.081
28. Jiang F, Li M, Wang H et al (2019) Coelonin, an anti-inflammation active component of *Bletilla striata* and its potential mechanism. Int J Mol Sci 20(18):4422. https://doi.org/10.3390/ijms20184422
29. Zu YY, Liu QF, Tian SX et al (2019) Effective fraction of *Bletilla striata* reduces the inflammatory cytokine production induced by water and organic extracts of airborne fine particulate matter (PM2.5) in vitro. BMC Complement Altern Med 19:369. https://doi.org/10.1186/s12906-019-2790-3
30. Wang W, Meng H (2015) Cytotoxic, anti-inflammatory and hemostatic spirostane-steroidal saponins from the ethanol extract of the roots of *Bletilla striata*. Fitoterapia 101:12–18. https://doi.org/10.1016/j.fitote.2014.11.005
31. Dong J, Zhao L, Cai L et al (2014) Antioxidant activities and phenolics of fermented *Bletilla formosana* with eight plant pathogen fungi. J Biosci Bioeng 118:396–399. https://doi.org/10.1016/j.jbiosc.2014.03.003
32. Chen Z, Zhao Y, Zhang M et al (2020) Structural characterization and antioxidant activity of a new polysaccharide from *Bletilla striata* fibrous roots. Carbohydr Polym 227:115362. https://doi.org/10.1016/j.carbpol.2019.115362
33. Song Y, Zeng R, Hu L et al (2017) In vivo wound healing and in vitro antioxidant activities of *Bletilla striata* phenolic extracts. Biomed Pharmacother 93:451–461. https://doi.org/10.1016/j.biopha.2017.06.079
34. Zhang C, Gao F, Gan S et al (2019) Chemical characterization and gastroprotective effect of an isolated polysaccharide fraction from *Bletilla striata* against ethanol-induced acute gastric ulcer. Food Chem Toxicol 131:110539. https://doi.org/10.1016/j.fct.2019.05.047
35. Jiang F, Li W, Huang Y et al (2013) Antioxidant, Antityrosinase and antitumor activity comparison: the potential utilization of fibrous root part of *Bletilla striata* (Thunb.) Reichb.f. PLoS One 8(2):e58004. https://doi.org/10.1371/journal.pone.0058004
36. Niu J, Wang S, Wang B et al (2020) Structure and anti-tumor activity of a polysaccharide from Bletilla ochracea Schltr. Int J Biol Macromol 154:1548–1555. https://doi.org/10.1016/j.ijbiomac.2019.11.039
37. Sun A, Liu J, Pang S et al (2016) Two novel phenanthraquinones with anti-cancer activity isolated from *Bletilla striata*. Bioorganic Med Chem Lett. 26(9):2375–2379. https://doi.org/10.1016/j.bmcl.2016.01.076
38. Guo JJ, Dai BL, Chen NP et al (2016) The anti-*Staphylococcus aureus* activity of the phenanthrene fraction from fibrous roots of *Bletilla striata*. BMC Complement Altern Med 16(1):491. https://doi.org/10.1186/s12906-016-1488-z

39. Jiang S, Chen CF, Ma XP et al (2019) Antibacterial stilbenes from the tubers of *Bletilla striata*. Fitoterapia 138:104350. https://doi.org/10.1016/j.fitote.2019.104350
40. Jiang S, Wan K, Lou HY et al (2019) Antibacterial bibenzyl derivatives from the tubers of *Bletilla striata*. Phytochemistry 162:216–223. https://doi.org/10.1016/j.phytochem.2019.03.022
41. Qian CD, Jiang FS, Yu HS et al (2015) Antibacterial biphenanthrenes from the fibrous roots of *Bletilla striata*. J Nat Prod 78(4):939–943. https://doi.org/10.1021/np501012n
42. Dong L, Liu X-X, Wu S-X et al (2020) Rhizoma Bletillae polysaccharide elicits hemostatic effects in platelet-rich plasma by activating adenosine diphosphate receptor signaling pathway. Biomed Pharmacother 130:110537
43. Yang C, Xia T, Wang C et al (2019) Using the UPLC–ESI–Q-TOF–MSE method and intestinal bacteria for metabolite identification in the nonpolysaccharide fraction from *Bletilla striata*. Biomed Chromatogr 33:e4637. https://doi.org/10.1002/bmc.4637
44. Lu B, Xu YM, Zhang HM, Li TJQY (2005) Effects of different extracts from Bletilla colloid on rabbit platelet aggregation. Pharm J PLA 21:330–331
45. Wang W, Cheng MH, Gao JZX (2016) Study on hemostasis of *Bletilla striata* hemostatic sponge. J Pharm Pract 34:32–35
46. Wang Y, Han S, Li R et al (2019) Structural characterization and immunological activity of polysaccharides from the tuber of *Bletilla striata*. Int J Biol Macromol 122:628–635. https://doi.org/10.1016/j.ijbiomac.2018.10.201
47. Peng Q, Li M, Xue F, Liu H (2014) Structure and immunobiological activity of a new polysaccharide from *Bletilla striata*. Carbohydr Polym 107:119–123. https://doi.org/10.1016/j.carbpol.2014.02.042
48. Wang Y, Liu D, Chen S et al (2014) A new glucomannan from *Bletilla striata*: structural and anti-fibrosis effects. Fitoterapia 92:72–78. https://doi.org/10.1016/j.fitote.2013.10.008
49. Shi Y, Zhang B, Lu Y et al (2017) Antiviral activity of phenanthrenes from the medicinal plant *Bletilla striata* against influenza a virus. BMC Complement Altern Med 17:273. https://doi.org/10.1186/s12906-017-1780-6
50. Chen Z, Cheng L, He Y, Wei X (2018) Extraction, characterization, utilization as wound dressing and drug delivery of *Bletilla striata* polysaccharide: a review. Int J Biol Macromol 120:2076–2085
51. Diao H, Li X, Chen J et al (2008) *Bletilla striata* polysaccharide stimulates inducible nitric oxide synthase and Proinflammatory cytokine expression in macrophages. J Biosci Bioeng 105:85–89. https://doi.org/10.1263/jbb.105.85
52. Zhou D, Chen G, Ma YP et al (2019) Isolation, structural elucidation, optical resolution, and Antineuroinflammatory activity of Phenanthrene and 9,10-Dihydrophenanthrene derivatives from *Bletilla striata*. J Nat Prod 82(8):2238–2245. https://doi.org/10.1021/acs.jnatprod.9b00291
53. Morita H, Koyama K, Sugimoto Y, Kobayashi J (2005) Antimitotic activity and reversal of breast cancer resistance protein-mediated drug resistance by stilbenoids from *Bletilla striata*. Bioorganic Med Chem Lett 15(4):1051–1054. https://doi.org/10.1016/j.bmcl.2004.12.026
54. Shi S, Huang Z, Luo Y et al (2012) *Bletilla striata* polysaccharide for the treatment of ulcerative colitis in mice. J China Pharm Univ 43(6):535–540
55. Ke CYZC (2011) Effects of *Bletilla striata* polysaccharides on ulcerative colitis. Chin Pharm 22:2132–2134
56. Yu L, Nie X, Pan H et al (2011) Diabetes mellitus ulcers treatment with *Bletilla striata* polysaccharide. Zhongguo Zhongyao Zazhi 36(11):1487–1491. https://doi.org/10.4268/cjcmm20111118
57. Li Q, Li K, Huang SS et al (2014) Optimization of extraction process and antibacterial activity of *bletilla striata* polysaccharides. Asian J Chem 26:3574–3580. https://doi.org/10.14233/ajchem.2014.16500
58. Wu RY, Chang SY, Hsieh TF, Sen CY (2013) Cryopreservation of *Bletilla formosana* seeds (Orchidaceae) by desiccation. Sci Hortic 157:108–112. https://doi.org/10.1016/j.scienta.2013.03.033

59. Hsiao YY, Pan ZJ, Hsu CC et al (2011) Research on orchid biology and biotechnology. Plant Cell Physiol 52:1467–1486. https://doi.org/10.1093/pcp/pcr100
60. Hirano T, Godo T, Mii M, Ishikawa K (2005) Cryopreservation of immature seeds of *Bletilla striata* by vitrification. Plant Cell Rep 23:534–539. https://doi.org/10.1007/s00299-004-0893-9
61. Hu WH, Yang YH, Liaw SI, Chang C (2013) Cryopreservation the seeds of a Taiwanese terrestrial orchid, *Bletilla formosana* (Hayata) Schltr. by vitrification. Bot Stud 54:33. https://doi.org/10.1186/1999-3110-54-33
62. Wu R-Y, Chang S-Y, Hsieh T-F et al (2016) Cryopreservation of orchid genetic resources by desiccation: a case study of *Bletilla formosana*. In: Marco-Jiménez F, Akdemir H (eds) Cryopreservation in Eukaryotes. IntechOpen. https://doi.org/10.5772/65302
63. Ishikawa K, Harata K, Mii M et al (1997) Cryopreservation of zygotic embryos of a Japanese terrestrial orchid (*Bletilla striata*) by vitrification. Plant Cell Rep 16:754–757. https://doi.org/10.1007/s002990050314

Orchids of Genus *Vanda*: Traditional Uses, Phytochemistry, Bioactivities, and Commercial Importance

24

Hari Prasad Devkota, Anjana Adhikari-Devkota, Rajan Logesh, Tarun Belwal, and Bijaya Pant

Contents

H. P. Devkota (✉) · A. Adhikari-Devkota
Graduate School of Pharmaceutical Sciences, Kumamoto University, Kumamoto, Japan
e-mail: devkotah@kumamoto-u.ac.jp; anjana@kumamoto-u.ac.jp

R. Logesh
TIFAC-CORE in Herbal Drugs, Department of Pharmacognosy and Phytopharmacy, JSS College of Pharmacy (JSS Academy of Higher Education and Research), Udhagamandalam, Tamil Nadu, India

T. Belwal
College of Biosystems Engineering and Food Science, Key Laboratory for Agro-Products Postharvest Handling of Ministry of Agriculture and Rural Affairs, Zhejiang Key Laboratory for Agro-Food Processing, Zhejiang University, Hangzhou, China

B. Pant
Central Department of Botany, Tribhuvan University, Kathmandu, Nepal
e-mail: b.pant@cdbtu.edu.np

J.-M. Mérillon, H. Kodja (eds.), *Orchids Phytochemistry, Biology and Horticulture*, Reference Series in Phytochemistry, https://doi.org/10.1007/978-3-030-38392-3_37

Abstract

The genus *Vanda* comprises of about 74 species, mainly distributed in South and Southeast Asia. Many *Vanda* species are being used in traditional medicines in India, Nepal, Bangladesh, and other south Asian countries. Among these species, only very few species are studied in detail for their bioactive chemical constituents, pharmacological activities, and are considered for commercial product development as medicines and cosmetics. In this chapter, we aim to provide an overview about the traditional uses, phytochemistry, and pharmacological activities of various important *Vanda* species along with their conservation and cultivation practices and commercial importance.

Keywords

Vanda · Orchids · Orchidaceae · Traditional uses · Phytochemistry · Commercialization

1 Introduction

Orchidaceae is one of the largest families of flowering plants, and they consist of more than 880 genera and over 30,000 species [1, 2]. They are one of the most evolutionary advanced plants and inhabit almost every habitat on earth. Orchid plants are well known for their beautiful flowers, having attractive colors, shapes, fragrance, and are widely cultivated around the world for their ornamental values [2]. Although orchids are primarily cultivated and used largely in floriculture industry, many are also used in traditional medicine as herbal plants and products, in foods, and in cosmetics around the world [3–6].

The genus *Vanda* comprises of about 74 species, distributed mainly in South and Southeast Asia [7, 8]. Many species are used as crude drugs under the name of "Rasna" in Ayurvedic formulations [9]. They are also used in other traditional medicine systems in China, India, Nepal, Bangladesh, and other south Asian countries [2, 10, 11]. Among these 74 species, only very few species are studied in detail for their bioactive chemical constituents, pharmacological activities, and are considered for commercial product development as medicines and cosmetics. In this chapter, we aim to provide an overview about the traditional uses, phytochemistry, and pharmacological activities of various important *Vanda* species along with their conservation and cultivation practices and commercial importance.

2 Botanical Description, Distribution, and Ecology

The genus *Vanda* comprises approximately 74 species and is widespread throughout South and Southeast Asia – India, Nepal, Bhutan, Bangladesh, Myanmar, Sri Lanka, Thailand, and East Asia, through southern China, Taiwan to Korea and Japan, down through Indonesia, Philippines to northern Australia, and New Guinea and the Solomon Islands [7].

The genus has monopodial growth habit, mostly epiphytic but sometimes lithophytes and terrestrial too. Plants range from small to large. *Vanda* species can form large plants with long sturdy stems with extensive aerial root systems on trees with stiffly erect, premorse tipped leaves. *Vanda* species are epiphytic monopodial herbs with ascending or rarely arching stems with short internodes and many distichous leaves and thick roots in the lower part. The colorful and long-lasting flowers are arranged in few to many-flowered axillary racemes. The thick roots serve as the primary water storage organs which allow these species to withstand drought in semidormant conditions (http://www.plantsoftheworldonline.org/taxon/urn:lsid:ipni.org:names:30077641-2, accessed on December 17, 2020). They are grouped into four categories on the basis of their leaf character, e.g., strap shaped, terete, semiterete, and channeled [12]. Inflorescence is axillary, erect, and simple with often brightly colored, sometimes fragrant, flowers. There is great diversity in floral shape and color. The flowers are small to large, few to many fleshy, heavy textured, long lasting, and yellow, brown, purple, magenta, blue, and lavender in color. The flower size varies from 2.5 to 10 cm [13]. The genus is one of the five most horticulturally important orchid genera in the world [7]. A number of species of genus *Vanda* are vulnerable to extinction in the wild and are being rare and geographically restricted in distribution [14].

3 Traditional Uses

Many *Vanda* species are being used in different systems of traditional medicines in South Asian countries for the treatment of inflammation, wounds, bone fractures, nervous disorders, and rheumatism [2, 9–11]. Among them, the uses of *Vanda coerulea* Griff. ex Lindl. (Fig. 1), *Vanda cristata* Wall. ex Lindl. (Fig. 2), *Vanda parviflora* Lindl. *Vanda spathulata* (L.) Spreng, *Vanda tessellata* (Roxb.) Hook. ex. G.Don (Syn. *Vanda roxburghii* R.Br.) (Fig. 3), and *Vanda testacea* (Lindl.) Rchb. f (Fig. 4) are commonly reported. Traditional uses of individual species of *Vanda* are mentioned below:

***Vanda coerulea*:** The juice of the leaves is used to treat diarrhea and indigestion [15]. Decoction of the flowers is used as appetizer and tonic [16].

***Vanda cristata*:** The paste prepared from whole plants or roots is used to treat cuts, wounds, and boils and to treat dislocated bones [17]. The juice of the leaves is used

Fig. 1 Photographs of *Vanda coerulea*. (Photos by H.P. Devkota)

Fig. 2 Photographs of *Vanda cristata*. (Photos by B. Pant)

to treat bronchitis, cough, tonsillitis, and weakness [15]. Leaf powder is used as expectorant and the leaf paste applied to cuts and wounds [18].

***Vanda parviflora*:** It is used to treat rheumatism, disorders of nervous system, and also as anticancer and antiviral agent [19, 20].

***Vanda spathulata*:** Powder prepared from dry flowers is used to treat asthma, maniac disorders, and neurological disorder [21].

Fig. 3 Photographs of *Vanda tessellata*. (Photos by B. Pant)

Fig. 4 Photographs of *Vanda testacea*. (Photo by B. Pant)

***Vanda tessellata*:** Roots are commonly used for the treatment of inflammatory diseases and rheumatism. Rhizome paste is applied in dislocated bones [15]. Leaf juice or paste is used to treat bronchitis, earache, rheumatism, and fever [16, 22, 23]. Paste obtained from the roots and leaves is applied for the treatment of sprains, rheumatism, and also used as antidote for spider, scorpion, and snake bite [16, 18]. Root decoction is used in the treatment of cholera [16]. Plant ash with mustered oil is used to treat bone fracture [16]. Roots are used for the treatment of rheumatism and bronchitis. Paste made from the leaves is used to treat fever [18]. Leaf juice is used to treat ear infection and skin diseases [24].

***Vanda testacea*:** Leaf extract is used to treat earache as eardrop. It is also used to treat cuts, wounds, and for antiviral activities. Leaves are used in the treatment of viral diseases and cancer. Leaf drops are used during earache [11]. Powder obtained from dried flowers and leaves is used to treat rheumatism [11, 16].

4 Phytochemistry

Orchids are rich in various bioactive chemical compounds. Hydroxyl-benzyl derivatives [25], fluorenones and stilbenoids [26], and flavones *C*-glycosides [27] are some of the most common chemical classes in plants of the Orchidaceae family. Only a few species of *Vanda* genus have been studied for their chemical constituents, and they are reported to contain phenanthrene derivatives, bibenzyl derivative, and other compounds including anthocyanins, simple phenolic compounds, and volatile compounds (Table 1). Phenanthrene derivatives (Fig. 5) are reported from *V. tessellata*, *V. parviflora*, and *V. coerulea* [28–31]. Simmer et al. isolated a bibenzyl derivative, gigantol (**8**) (Fig. 6), from the stems of *Vanda coerulea* [28]. It was also isolated and identified from the roots of *V. tessellata* [32]. Few other phenolic compounds are isolated such as tetracosylferulate (**9**), parishin (**10**), 4-(*β*-D-glucopyranosyloxy) benzyl alcohol (**11**), 2,5-dimethoxy-6,8-dihydroxy isoflavone (**12**), and gallic acid (**13**) from various *Vanda* species (Table 1, Fig. 6). Anthocyanins such as delphinidin and cyanidin derivatives were reported from the flowers of various *Vanda* hybrids [33–35]. An alkaloid, laburine acetate, was isolated from *Vanda hindsii* [36]. Some other compounds are 2,7,7-tri methyl bicyclo [2.2.1], heptane [37], and steroids ([38, 39]. Joshi et al. [40] performed the GC-MS analysis of the methanol extract of whole plant of *Vanda cristata* and reported the identification of 9-methyl-octadecenoate as a major component (53.43%) followed by palmitic acid (23.51%), 15-methyl-hexadecanoic acid methyl ester (4.86%), 10-nonadecenoic acid methyl ester (3.55%), 2-methyl-Z,Z-3,13-octadecadienol (2.95%), and 11-tridecene-1-ol (2.74%). Other minor constituents were alpha-bisabolol, 14-methyl-pentadecanoic acid methyl ester, 10-octadecenoic acid methyl ester, hexadecanoic acid, linolelaidoyl chloride, etc.

Table 1 Major bioactive compounds identified from *Vanda* species

Chemical class	Compounds	Plant source	References
Phenanthrene derivatives	Tessallatin (**1**)	*V. tessellata*	[30]
	Oxo-tessallatin (**2**)	*V. tessellata*	[31]
	Parviflorin (**3**)	*V. parviflora*	[29]
	Flavidin (**4**)	*V. coerulea*	[28]
	Imbricatin (**5**)	*V. coerulea*	[28]
	Coelonin (**6**)	*V. coerulea*	[28]
	Methoxycoelonin (**7**)	*V. coerulea*	[28]
Bibenzyl derivative	Gigantol (**8**)	*V. coerulea*, *V. tessellata*	[28] [32]
Phenolic compounds	Tetracosylferulate (**9**)	*V. tessellata*	[38]
	Parishin (**10**)	*V. parishii*	[41]
	4-(*β*-D-Glucopyranosyloxy) benzyl alcohol (**11**)	*V. parishii*	[41]
	2,5-Dimethoxy-6,8-dihydroxy isoflavone (**12**)	*V. tessellata*	[42]
	Gallic acid (**13**)	*V. tessellata*	[42]

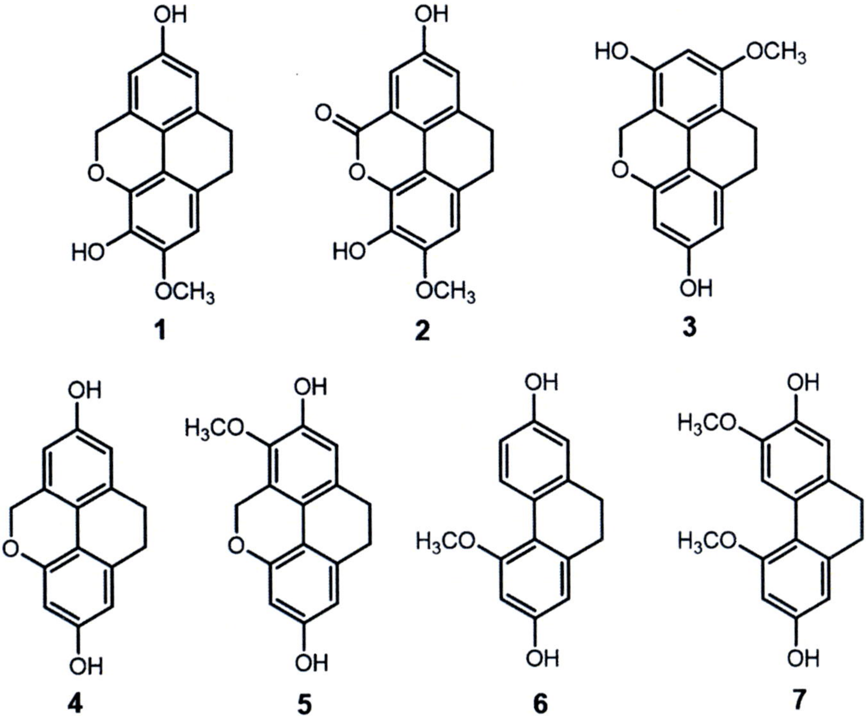

Fig. 5 Structures of phenanthrene derivatives

5 Biological Activities

Some of the medicinal species of *Vanda* genus are evaluated for their pharmacological activities such as anti-inflammatory, antioxidant, cytotoxic, antiaging, hepatoprotective, and anticonvulsant activities, among others.

5.1 Antioxidant and Anti-Inflammatory Activities

Simmler et al. [28] evaluated the antioxidant and anti-inflammatory activities of extracts obtained from *Vanda coerulea*. Preliminary experiments showed that the aqueous ethanol extract of stems exhibited strong 1–1-diphenyl-2-picrylhydrazyl (DPPH) and hydroxyl radical (OH) scavenging activities as compared to extracts of leaves and roots. Stem extract also showed potent in vitro inhibitory activity against type 2 prostaglandin (PGE-2) release from ultraviolet (UV_B) irradiated HaCaT keratinocytes. Five compounds, flavidin (**4**), imbricatin (**5**), coelonin (**6**), methoxycoelonin (**7**), and gigantol (**8**), were isolated from the extract by using bioassay-guided isolation procedures, and imbarcatin (**5**) and methoxycoelonin (**7**)

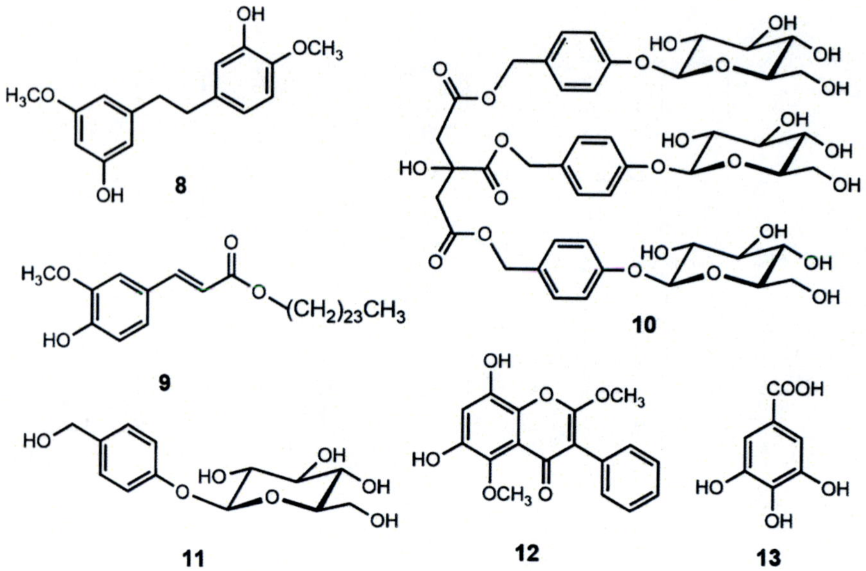

Fig. 6 Structures of eucomic acid derivatives, bibenzyl derivative, and phenolic compounds

showed potent antioxidant activities. Compounds 5, 7, and 8 also showed potent inhibitory activity against cyclooxygenase (COX)-2 expression as revealed by western blot analysis.

Chawla et al. [38] reported that for their anti-inflammatory activity, different extracts were obtained from the roots of *Vanda tessellata*. Petroleum ether, chloroform, and methanol extracts exhibited 54.2%, 42.1%, and 21.9% antiedema activity, respectively. Further chemical isolation afforded tetracosylferulate and *β*-sitosterol-D-glucoside from the petroleum ether and chloroform extracts, respectively [38].

Begum et al. reported the anti-inflammatory activity of methanol extracts of leaves and roots of *Vanda tessellata* in carragenan-induced paw edema test and also reported that the leaf extract (50 &100 mg/kg) reduced paw edema significantly at the third and fourth hours of the treatment with maximum 67.14% of inhibition, whereas root extract (100 mg/kg) also showed potential anti-inflammatory potential at third hour with 61.37% of inhibition [43].

Vijaykumar et al. [44] reported the DPPH and nitric oxide (NO) radical-scavenging activities of the petroleum ether extract of leaves of *Vanda tessellata*. Similarly, Thaakur and Pokkula [45] reported the ameliorative effects of valuated hydro-alcoholic extracts of leaves of *Vanda testacea* in axotomy-induced peripheral neuropathy in rats possibly by enhanced antioxidant activity, reduced calcium level, and inhibition of PGE2. Islam et al. reported the antioxidant activity of the methanol extract of roots of *Vanda tessellata* in DPPH and NO assays [46].

5.2 Cytotoxic Activity

Cytotoxic effects of *Vanda cristata* have been reported by Joshi et al. [40] against cervical cancer (HeLa) and glioblastoma (U251) cell lines. They reported that methanol extracts of the whole plant of *V. cristata* were effective cell growth inhibitors with significant percentage inhibition of 54.56% at the highest concentration (400 μg/ml) in Hela cell lines with IC_{50} values of 317.23 μg/ml. Similarly, in the case of glioblastoma cells (U251), *V. cristata* was significantly effective in inhibiting growth, with percentage inhibition of 61.86%, with IC_{50} of 163.66 μg/ml.

Chowdhury et al. reported on the aqueous and methanol leaf extracts of *V. tessellata* for cytotoxic activity using brine shrimp (*Artemia salina*), and the results showed very low cytotoxicity against brine shrimps [47].

Islam et al. have studied the cytotoxic activity using brine shrimp lethality bioassay for the methanolic extract of *Vanda tessellata* root, and the results showed that the extract showed significant toxicity of brine shrimp nauplii with the LC_{50} value of 25.19±0.98 μg/ml [46]. Similarly, Prakash et al. reported the potent antioxidant activities of chloroform and ethanol extracts of the leaves of *Vanda tessellata* [48].

5.3 Hepatoprotective Activity

Anwar et al. [49] reported the dose-dependent hepatoprotective activity of the petroleum ether extract of leaves of *Vanda tessellata* in rats. The activity was evaluated through the determination of serum markers such as cholesterol, triglycerides, alanine amino transferase, etc.

5.4 Antimicrobial Activity

Bhattacharjee et al. [50] evaluated the antibacterial and antifungal activities of different extracts of the whole plant of *Vanda tessellata*. Among different extracts, the chloroform extract showed potent antibacterial activity and antifungal activity.

5.5 Antiaging Activity and Cosmetic Applications

Andre et al. [51] reported the antiaging potentials of the extracts of *Vanda coerulea* for skin-hydrating properties in cosmetics composition. The extracts showed skin-hydrating property mainly by increasing the expression of aquaporin 3 and lympho-epithelial Kazal type-related inhibitor (LEKTI) protein, which resulted in the limitation of intercellular water evaporation and enhancement of transport of water in the epidermis. Similarly, Bonte et al. [52] evaluated the effects of ethanolic extracts of *Vanda coerulea* against the cutaneous aging. Various bioactive constituents were identified such as imbricatin, methoxycoelonin, and gigantol which significantly decreased the number of S phase cells and led to reduction in cyclin E and cyclin-dependent kinase 2. Similarly, Cauchard et al. [53] also evaluated the antiaging potential of *Vanda coerulea* and reported that the extracts can be used as active constituents in cosmetics to diminish the aging signs of the skin.

5.6 Antidepressant Activity

Dasari et al. [21] reported that the methanolic extract of the flowers of *Vanda spathulata* showed antidepressant activities in mice using the forced swim test and the tail suspension test. Similarly, Prakash et al. [48] reported the antidepressant activity of chloroform and ethanol extracts of the leaves of *Vanda tessellata* in rats using forced swimming test and tail suspension test.

5.7 Neuroprotective Activity

Mundugaru et al. (2020) evaluated the neuroprotective activity of the hydroalcoholic extract of *Vanda tessellata* in experimental models of ischemic hippocampal injury in rats, and the results suggested that the extract at a concentration of 200 and 400 mg/kg showed neuroprotective potentials in ischemic hippocampal injury [54].

5.8 Antinociceptive and Analgesic Activities

Chowdhury et al. reported the antinociceptive activity of the aqueous and methanol extracts of leaves of *Vanda tessellata* in antinociceptive activity in mice using acetic acid-induced writhing test, hot plate test, and tail immersion test. Extracts at 200 and 400 mg/kg showed antinociceptive activity in a dose-dependent manner [47].

Islam et al. studied the analgesic activity of the methanolic extract of the roots of *Vanda tessellata* in an acetic acid-induced writhing model of pain in mice. The extract at the dose of 200 and 400 mg/kg body weight exhibited significant reduction in acetic acid-induced writhing in mice [46]. Similarly, Begum et al. reported the analgesic activity of methanolic extract of leaves and roots in acetic acid-induced writhing and formalin-induced paw-licking models in mice. The methanolic extract of the leaves at the dose of 100 mg/kg body weight showed significant analgesic activity [43].

5.9 Other Pharmacological Activities

Apart from above-mentioned activities, various other pharmacological activities are also reported for the extracts obtained from different species of *Vanda*. For example,

Uddin et al. [55] reported the antiacetylcholinesterase and antibutyrylcholinesterase activities of the chloroform extract of *Vanda tessellata*. The ethanolic extract of roots of *Vanda tessellata* showed potent anticonvulsant activity against picrotoxin, pentylenetetrazole, and maximal electroshock-induced convulsions in mice [56]. Suresh et al. evaluated the aphrodisiac activity of the water and alcohol extracts of different plant parts (root, flower, and leaf) of *Vanda tessellata* in male mice. The evaluation was carried out by observing the mounting behavior, mating performance, and reproductive performance. The alcohol extract of the flower showed most potent activity [57]. The aqueous extract of *Vanda tessellata* showed potent wound-healing activity in rats after topical administration at a dose of 150 mg/kg/day [58].

6 Commercial Importance

Vanda orchids have very high commercial potential for their ornamental values and also for their traditional medicinal uses, bioactive constituents, and pharmacological activities. Extracts of *Vanda* plants are also gaining attentions for the potential use in cosmetics [6, 53].

Breeding technologies, basic research into orchid biology, and the application of biotechnology for their improvement and production in the orchid industry have developed everlasting interest on them, globally [5, 59]. It indicates there will be a huge demand for orchids in the future including esthetically and medicinally important *Vanda*. Because of their extraordinary floral diversification and deep color combination, *Vanda* is regarded as a commercially important group of plant in the orchids floriculture industry, both as cut flowers and as potted plants. *Vanda* has been designated as the "Queen Orchid of the East" due to its robust and large rounded flowers [60]. One of the species, *V. coerulea* or blue *Vanda,* has a magnificent flower, with purplish blue color. This species got the honor from American Orchid Society 39 times and has been used as the parental stock to develop new hybrids for more than 4000 hybrids [61]. Various verieties and hybrids are also available in India [13]. However, the proper conservation and sustainable utilization of *Vanda* species for commercial purposes is an urgent issue as the natural plant populations are being decreased due to extensive harvesting. The natural populations of *Vanda* species are decreasing, and some of them are now categorized as threatened species under IUCN Red List (http://www.iucnredlist.org/) [2]. In addition to esthetic importance, due to the high medicinal value of *Vanda,* their illegal collection for trade and consumption has resulted in more species in the threatened category [62]. Thus, various researchers are involved in their ex situ conservation by in vitro culture technique [63, 64]. Besides, the beneficial endophytes in them add on more value in them [65, 66], and this is another potential of research to investigate their role for medicinal properties of *Vanda*. Thus, advanced biotechnological tools for the lab-based culture and propagation are necessary which may in future provide solutions for large-scale production and commercial uses.

7 Conclusions and Future Recommendations

The species of *Vanda* are distributed and cultivated in various Asian countries and are widely used for ornamental and medicinal purposes. Only very few studies have been performed regarding the chemical constituents in these species and their respective biological activities. Further research is necessary in the detailed chemical isolation of active constituents and their bioactivity analysis using animal models to explore their potential for pharmacological and cosmetic applications. Extensive harvesting practices have also resulted in the decrease of natural habitat for these species; thus, sustainable harvesting practices are necessary along with the development of advanced techniques for their cultivation and propagation.

References

1. Cakova V, Urbain A, Le Quéméner C et al (2015) Purification of vandaterosides from *Vanda teres* (Orchidaceae) by stepwise gradient centrifugal partition chromatography. J Sep Sci 38: 3006–3013. https://doi.org/10.1002/jssc.201500349
2. Khan H, Marya, Belwal T et al (2019) Genus *Vanda*: a review on traditional uses, bioactive chemical constituents and pharmacological activities. J Ethnopharmacol 229:46–53
3. Bulpitt CJ, Li Y, Bulpitt PF, Wang J (2007) The use of orchids in Chinese medicine. J R Soc Med 100:558–563. https://doi.org/10.1258/jrsm.100.12.558
4. Gutiérrez RMP (2010) Orchids: a review of uses in traditional medicine, its phytochemistry and pharmacology. J Med Plants Res 4:592–638. https://doi.org/10.5897/JMPR10.012
5. Pant B (2013) Medicinal orchids and their uses: tissue culture a potential alternative for conservation. African J Plant Sci 7:448–467. https://doi.org/10.5897/ajps2013.1031
6. Hadi H, Razali SNS, Awadh AI (2015) A comprehensive review of the cosmeceutical benefits of vanda species (orchidaceae). Nat Prod Commun 10:1483–1488
7. Gardiner LM, Kocyan A, Motes M et al (2013) Molecular phylogenetics of *Vanda* and related genera (Orchidaceae). Bot J Linn Soc 173:549–572. https://doi.org/10.1111/boj.12102
8. The Plant List (2013) The Plant List. Version 1.1. In: Internet. http://www.theplantlist.org/1.1/browse/A/Orchidaceae/Vanda/. Retrived on 15 Dec 2020
9. Musharof Hossain M (2011) Therapeutic orchids: traditional uses and recent advances – an overview. Fitoterapia 82:102–140
10. Hossain MM (2009) Traditional therapeutic uses of some indigenous orchids of Bangladesh. Med Aromat Plant Sci Biotechnol 3:100–106
11. Pant B, Raskoti BB (2013) Medicinal orchids of Nepal. Himalayan Map House, Kathmandu
12. De LC, Rao AN, Rajeevan PK et al (2015) Morphological characterization in *Vanda* spp. Int J Sci Res 4:26–32
13. De Lakshman C, Pathak P, Rao AN, Rajeevan PK (2015) Commercial orchids. De Gruyter Open Ltd, Warsaw/Berlin
14. Motes MR (1997) Vandas: their botany, history and culture. Timber Press, Portland/Cambridge
15. Panda A, Mandal D (2013) The folklore medicinal orchids of Sikkim. Anc Sci Life 33:90. https://doi.org/10.4103/0257-7941.139043
16. Deb CR, Deb MS, Jamir NS, Imchen T (2009) Orchids in indigenous system of medicine in Nagaland, India. Pleione 3:209–211
17. Manandhar NP (2002) Plants and people of Nepal. Timber Press, Inc., Portland
18. Subedi A, Kunwar B, Choi Y et al (2013) Collection and trade of wild-harvested orchids in Nepal. J Ethnobiol Ethnomed:9. https://doi.org/10.1186/1746-4269-9-64
19. Singh MP, Dey S (2005) Indian medicinal plants. Satish Serial Publishing House, New Delhi

20. Rao TA (1998) Conservation of wild orchids of Kodagu in the Western Ghats. Agricultural Technologies and Services Pvt., Bangalore
21. Dasari R, Sathyavati D, Belide SK et al (2013) Evaluation of antioxidant activity of two important memory enhancing medicinal plants *Celtis timorensis* and *Vanda spathulata*. Asian J Pharm Clin Res 6:153–155
22. Vaidya B, Shrestha M, Joshee N (2000) Report on Nepalese orchid species with medicincal properties. In: Watanabe T, Takano A, Bista MS, Saiju HK (eds) Proceeding of Nepal-Japan joint symposium on conservation and utilization of Himalayan medicinal resources
23. Shengji P, Zhiwei Y (2018) Orchids and its uses in chinese Chinese medicine and health care products. Med Res Innov 2. https://doi.org/10.15761/mri.1000133
24. Rajendran A, Rao NR, Kumar KR, Henry AN (1997) Some medicinal orchids of southern India. Anc Sci Life 17:10–104
25. Zeng X, Zhang S, Zhang L et al (2006) A study of the neuroprotective effect of the phenolic glucoside gastrodin during cerebral ischemia in vivo and in vitro. Planta Med 72:1359–1365. https://doi.org/10.1055/s-2006-951709
26. Yang L, Wang Z, Xu L (2006) Simultaneous determination of phenols (bibenzyl, phenanthrene, and fluorenone) in Dendrobium species by high-performance liquid chromatography with diode array detection. J Chromatogr A 1104:230–237. https://doi.org/10.1016/j.chroma.2005.12.012
27. Williams CA (1979) The leaf flavonoids of the orchidaceae. Phytochemistry 18:803–813. https://doi.org/10.1016/0031-9422(79)80019-9
28. Simmler C, Antheaume C, Lobstein A (2010) Antioxidant biomarkers from *Vanda coerulea* stems reduce irradiated HaCaT PGE-2 production as a result of COX-2 inhibition. PLoS One 5. https://doi.org/10.1371/journal.pone.0013713
29. Anuradha V, Prakasa Rao NS (1998) Parviflorin a phenanthropyran from *Vanda parviflora*. Phytochemistry 48:181–182. https://doi.org/10.1016/S0031-9422(97)00610-9
30. Anuradha V, Prakasa Rao NS (1998) Tessallatin a phenanthropyran from *Vanda tessalata*. Phytochemistry 48:183–184. https://doi.org/10.1016/S0031-9422(97)00611-0
31. Anuradha V, Basaveswara Rao MV, Aswar AS (2008) Oxo-tessallatin, a novel phenanthrapyrone isolated from *Vanda tessalata*. Orient J Chem 24:1119–1122
32. Uddin MN, Afrin R, Uddin MJ et al (2015) *Vanda roxburghii* chloroform extract as a potential source of polyphenols with antioxidant and cholinesterase inhibitory activities: identification of a strong phenolic antioxidant. BMC Complement Altern Med 15. https://doi.org/10.1186/s12906-015-0728-y
33. Junka N, Kanlayanarat S, Buanong M, Wongs-Aree C (2012) Characterisation of floral anthocyanins and their antioxidant activity in *Vanda hybrid* (*V. teres* x *V. hookeriana*). J Food Agric Environ 10:221–226
34. Junka N, Kanlayanarat S, Buanong M et al (2011) Analysis of anthocyanins and the expression patterns of genes involved in biosynthesis in two *Vanda* hybrids. Int J Agric Biol 13:873–880
35. Tatsuzawa F, Saito N, Seki H et al (2004) Acylated anthocyanins in the flowers of *Vanda* (Orchidaceae). Biochem Syst Ecol 32:651–664. https://doi.org/10.1016/j.bse.2004.02.004
36. Brandänge S, Granelli I, Johanson R et al (1973) Studies on Orchidaceae alkaloids. XXXVI. Alkaloids from some Vanda and Vandopsis species. Acta Chem Scand 27:1096–1097. https://doi.org/10.3891/acta.chem.scand.27-1096
37. Subramoniam A, Gangaprasad A, Sureshkumar PK et al (2013) A novel aphrodisiac compound from an orchid that activates nitric oxide synthases. Int J Impot Res 25:212–216. https://doi.org/10.1038/ijir.2013.18
38. Chawla AS, Sharma AK, Handa SS, Dhar KL (1992) Chemical studies and antiinflammatory activity of *Vanda roxburghii* roots. Indian J Pharm Sci 54(4):159–161
39. Das S, Bhattacharya A, Bhattacharya AK (1967) Active constituents of *Vanda roxburghi*. J Indian Chem Soc 44:804–805
40. Joshi PR, Paudel MR, Chand MB et al (2020) Cytotoxic effect of selected wild orchids on two different human cancer cell lines. Heliyon 6:e03991. https://doi.org/10.1016/j.heliyon.2020.e03991

41. Dahmén J, Leander K (1976) The structure of parishin, a glucoside from *Vanda parishii*. Phytochemistry 15:1986–1987. https://doi.org/10.1016/S0031-9422(00)88865-2
42. Prakash B, Bias RT (2016) A novel chemical from the leaf extract of *Vanda tesselata* (Roxb.) hook. ex. G. Don. World J Pharceutical Res 5:1049–1058
43. Begum Y, Sen P, Bulbul I, Nasrin F (2018) Evaluation of analgesic and anti-inflammatory potentials of the leaf and root extracts of *Vanda roxburghii* (Roxb). J Complement Altern Med Res. https://doi.org/10.9734/jocamr/2018/40002
44. Vijaykumar K (2013) In vitro anti-oxidant activity of pet-ether extract of *Vanda tessellata* Roxb. Int Ayur Med J 1:1–4
45. Thaakur SR, Pokkula S (2013) Ameliorative effects of *Vanda testacea* in sciatic nerve transection-induced neuropathy in rats. Int J Pharma Bio Sci 4(1):271–284
46. Islam SS, Sayeed H, Shahriyar SA, Ferdous AIA (2016) Antioxident, analgesic and cytotoxic activity of methanolic extract of *Vanda roxburghii* root. Aging (Albany NY) 1:3
47. Chowdhury MA, Rahman MM, Chowdhury HRH et al (2014) Antinociceptive and cytotoxic activities of an epiphytic medicinal orchid: *Vanda tessellata* Roxb. BMC Complement Altern Med 14. https://doi.org/10.1186/1472-6882-14-464
48. Prakash B, Bais RT, Sahu RK (2018) Screening of antioxidant and antidepressant activity of *Vanda tessellata* leaves extract. Int J Green Pharm 12(4):S829–S834. https://doi.org/10.22377/ijgp.v12i04.2262
49. Anwar M, Indro S, Wisnu RW et al (2013) Hepatoprotective activity of pet-ether extract of *Vanda tessellata* Roxb. Proc 2013 J Int Conf Rural Inf Commun Technol Electr Technol rICT ICEV-T 2013:3–6
50. Bhattacharjee B, Islam T, Rahman MZ, Shahinul Islam SM (2015) Antimicrobial activity and phytochemical screening of whole plant extracts of *Vanda tessellata* (Roxb.) Hook.ex.D.Don. World J Pharm Pharm Sci 4(1):72–83
51. Andre P, Archambault JC, Renimel I (2011) Use of an extract of the orchid Vanda coerulea as a skin hydrating agent. US Patent US8039028B2
52. Bonté F, Simmler C, Lobstein A et al (2011) Action d'un extrait de Vanda coerulea sur la sénescence de fibroblastes cutanés. Ann Pharm Fr 69:177–181. https://doi.org/10.1016/j.pharma.2011.02.001
53. Cauchard JH, Pellicier F, Archambault JC, et al (2011) Orchid *vanda coerulea* as a cosmetic active agent. US Patent 0091588 A1
54. Mundugaru R, Sharma S, Sivanesan S et al (2020) Neuroprotective effect of vanda roxburghii extract in endothelin-1 (Et-1) induced hippocampal ischemic damage and ameliorate cognitive deficit. Indian J Pharm Educ Res 54(3):740–749. https://doi.org/10.5530/ijper.54.3.125
55. Uddin MJ, Rahman MM, Abdullah-Al-Mamun M, Sadik G (2015) *Vanda roxburghii*: an experimental evaluation of antinociceptive properties of a traditional epiphytic medicinal orchid in animal models. BMC Complement Altern Med 15:305. https://doi.org/10.1186/s12906-015-0833-y
56. Pathan D, Ambavade S (2014) Investigation of anticonvulsant activity of *Vanda roxburghii*. J Pharmacog Phytochem 2:95–99
57. Suresh KPK, Subramoniam A, Pushpangadan P (2000) Aphrodisiac activity of *Vanda tessellata* (Roxb.) Hook. ex Don extract in male mice. Indian J Pharmacol 32:300–304
58. Nayak BS, Suresh R, Rao AVC et al (2005) Evaluation of wound healing activity of Vanda roxburghii R.Br (Orchidacea): a preclinical study in a rat model. Int J Low Extrem Wounds 4:200–204. https://doi.org/10.1177/1534734605282994
59. Hsiao YY, Pan ZJ, Hsu CC et al (2011) Research on orchid biology and biotechnology. Plant Cell Physiol 52:1467–1486. https://doi.org/10.1093/pcp/pcr100
60. Teoh ES (2016) Medicinal orchids of Asia. Springer, Cham. https://doi.org/10.1007/978-3-319-24274-3
61. Sripotar D (2008) Non-detrimental finding of *Vanda coerulea*. NDF WORKSHOP WG 4 – Geophytes and Epiphytes case study 4 summary

62. Hinsley A, De Boer HJ, Fay MF et al (2018) A review of the trade in orchids and its implications for conservation. Bot J Linn Soc 186:435–455. https://doi.org/10.1093/botlinnean/box083
63. Pant B, Pradhan S, Paudel MR et al (2019) Various culture techniques for the mass propagation of medicinal orchids from Nepal. Acta Horticulturae 1262:109–124
64. Kaur S, Bhutani KK (2009) In vitro propagation of *Vanda testacea* (Lindl.) Reichb. f. – a rare orchid of high medicinal value. Plant Tissue Cult Biotechnol 19:1–7. https://doi.org/10.3329/ptcb.v19i1.4077
65. Chand K, Shah S, Sharma J et al (2020) Isolation, characterization, and plant growth-promoting activities of endophytic fungi from a wild orchid *Vanda cristata*. Plant Signal Behav 15. https://doi.org/10.1080/15592324.2020.1744294
66. Aggarwal S, Nirmala C, Beri S et al (2012) In vitro symbiotic seed germination and molecular characterization of associated endophytic fungi in a commercially important and endangered Indian orchid *Vanda coerulea* Griff. Ex Lindl. Eur J Environ Sci 2:33–42. https://doi.org/10.14712/23361964.2015.36

Part VI

Cosmetic Applications

Orchid Extracts and Cosmetic Benefits 25

Mayuree Kanlayavattanakul and Nattaya Lourith

Contents

M. Kanlayavattanakul (✉) · N. Lourith
School of Cosmetic Science, Mae Fah Luang University, Chiang Rai, Thailand

Phytocosmetics and Cosmeceuticals Research Group, Mae Fah Luang University, Chiang Rai, Thailand
e-mail: mayuree@mfu.ac.th; nattayal@mfu.ac.th

J.-M. Mérillon, H. Kodja (eds.), *Orchids Phytochemistry, Biology and Horticulture*, Reference Series in Phytochemistry, https://doi.org/10.1007/978-3-030-38392-3_22

Abstract

Orchid has long been used in several traditional medicines all around the world. This medicinal herb is evidenced for immunomodulatory activity and functions as longevity recipe. The most commonly used orchids in complementary medicine, Ayurvedic and traditional Chinese recipes, are *Vanda* and *Dendrobium*. In addition to these genera, different orchids are potentially to be implied for health promotion aspects including their cosmetic benefits. Orchids with scientific supports for cosmetic properties relevant for skin dryness, skin wrinkle, and aging of skin are therefore summarized in this chapter. Furthermore, traditional uses relevant to cosmetic benefits are disclosed as well as those commercialized orchid extracts in cosmetic industry. Thus, the beautiful floriculture orchids and full of availability are appreciable to be used for skin aging protection and treatment products, and flow in the stream of the consumers' awareness and preference on natural or bio-based products are presented in this context.

Keywords

Orchid · Cosmetics · Hydration · Moisturizer · Antiaging · Anti-wrinkle

1 Introduction

Orchid is evidenced as a therapeutic herb that positively affects human health. This flowering family has variety of species according to its beautiful flower and could be cultivated in all continents except Antarctica and deserts. The economic important of orchid is therefore unlimited for therapeutic uses but included floriculture proposes. That drives orchid into a huge business as a second most cut flowers. Regarding its importance in potted floriculture, orchid bleedings have been continuously taken worldwide resulting of more than 25,000 species majorly being developed in tropical and subtropical regions.

Pharmacological activities of orchids liberating variety applications of the herbs in different recipes [1–5] and the specific phytochemistry and biological activities contributing on diseases would be addressed in different chapters of this book. In this chapter, appraisal of orchids for skin treatments is objectively to be focused. The adverse effects of oxidants, radical, inflammatory mediators, and enzymes causing dryness, wrinkle, and aging of skin as well as hyperpigmentation are firstly summarized to figure out on these correlations exacerbating aging.

2 Causes and Treatment Strategies of Dryness, Greasiness, Wrinkle, and Aging of Skin

When an individual ages, the skin barrier is impaired, and this is known as chronologic aging. The turnover rate of epidermal cells slows down, and the vascular network between epidermal cells, which consists of keratinocytes, fibroblasts,

Langerhans cells, and melanocytes, and the skin elastic fibers and fluids are disrupted. In addition, these cells are decreased resulting in skin thickness reduction. Consequently, skin absorption, sensory perception, protection, secretion, and excretion are reduced including thermoregulation. Epidermis, particularly the stratum corneum, is thinner leading to skin dryness due to a reduction in water holding capacity resulting in severe skin damage. These cutaneous impairments are caused by a reduction in collagen, elastin, and hyaluronan which are synthesized by epidermal cells [6].

3 Impacts of Radical, UV, and Extracellular Matrix in Firmness, Wrinkle, and Aging of Skin

Skin aging is caused by several factors which damage cell membranes and components including lipids, proteins, and DNA. Reactive molecules with unpaired electrons or free radicals initiate cellular damage known as intrinsic, chronologic, and extrinsic aging. Natural cellular metabolism generates free radicals in a self-defense mechanism and efficiently scavenges these species and neutralizes the radicals; however, these are decreased with age. Dermal damage is also induced by UV exposure at the shorter wavelengths (UVB), which are absorbed by the epidermis prior to irradiation of keratinocytes. Meanwhile, longer wavelengths (UVA) penetrate the skin and interact with epidermal and dermal cells. Proteolytic enzyme activities are propagated resulting in degradation of collagen and elastin fibers including glycosaminoglycan (GAG), hyaluronan, chondroitin, keratin, dermatan, and heparin. They are linked to proteins such as collagen (28 types) and elastin and act as lubricants associated with the elasticity and tensile strength of skin. In addition, the matrix metalloproteinase (MMP) is a degradation enzyme of the extracellular matrix (ECM), including collagen, elastin, and GAG. These enzymes with 28 members (MMP-1 to MMP-28) are function and accelerated with age and radicals including inflammatory mediators as well. Therefore, deactivation, inhibition, and suppression of MMP, especially collagenase, elastase, and hyaluronidase, in addition to stimulation of hyaluronan synthase are regarded as the leading strategy in the management of skin aging. In addition, cellular damage results in inflammatory mediators generating free radicals and worsens intrinsic aging in turn as well as an induction of MMPs activation [6, 7].

Therefore, antioxidative molecules (e.g., superoxide dismutase, catalase, glutathione peroxidase) and nonenzymatic antioxidants (for instance, vitamin E, vitamin C, ubiquinone) to prevent free radical damage which terminate the radicals, protect against radical generation, increase self-defense mechanisms, and act as topical sun protectors limiting radical generation are contributing to antiaging products and have been extensively commercialized as over-the-counter (OTC) products. Antioxidants are therefore accepted as major therapeutic ingredients which decelerate skin aging. Consequently, commercial interest in the incorporation of antioxidants in cosmetic products is increasing, particularly in naturally derived products as they are thought to be milder, safer, and healthier. Topical OTC products are the main source of interest in treating skin disorders, including wrinkles, and protecting against aging, particularly those containing botanical ingredients.

Oxidative stresses induce inflammatory responses and activate MAPK pathway as well as NF-κB, AP-1, and pro-inflammatory cytokines and other inflammatory mediators that upregulate MMPs activities that severely propagate aging process of skin, which later generate radicals in the systems, accumulating or worsening aging of skin [8] including dryness of skin. Accordingly, anti-inflammatory and immunomodulatory agents are used in dermatology [9] not only for combating skin aging but also for allergic skin treatment and suppression of skin dryness. The treatment of excessive skin dryness is the subject of many cosmetic formulations, as this ailment can impact personal appearance and self-confidence and over time can result in reductions in elasticity and promote the generation of wrinkles. The application of skin-hydrating products thus allows for skin hydration and enhanced aesthetic appearance. Moisturizers considered safe can cause allergic skin reactions in some users, and public perceptions are shifting from synthetic products toward the use of non-irritating, natural moisturizers. Of these, plant-derived polysaccharides are actives gaining interest among consumers and researchers in the cosmetic field [10].

Skin hyperpigmentation is caused by several factors, i.e., UV radiation, radicals, inflammatory mediators, and hormones. Briefly, UV radiation causes skin hyperpigmentation by stimulating keratinocytes to secrete α-melanocyte-stimulating hormone (α-MSH), a small peptide hormone derived from proopiomelanocortin (POMC). Consequently, α-MSH binds to melanocortin 1 receptor (MC1R) expressed on melanocyte surfaces and thereafter induces melanogenesis via multiple signaling pathways resulting from cAMP, protein kinase A (PKA), cAMP response element-binding protein (CREB), and microphthalmia-associated transcription factor (MITF) activity. MITF is a key transcription factor regulating the transcription of melanogenic enzymes, i.e., tyrosinase, tyrosinase-related protein (TRP)-1, and TRP-2. In addition, UV radiation modulates nuclear factor E2-related factor 2 (Nrf2) and further activates mitogen-activated protein kinases (MAPKs). MAPKs consist of three subtypes: stress-activated protein kinases (SAPKs)/c-Jun NH2-terminal kinases (JNK), p38, and extracellular signal-regulated kinases (ERKs). JNK and p38 kinases are stimulated by pro-inflammatory cytokines and environmentally induced stresses such as exposure to UV irradiation, heat, and hydrogen peroxide, resulting in DNA damage. Melanogenesis is controlled by MAPKs, with MITF being activated by p38 phosphorylation. By contrast, ERK activation inhibits melanin synthesis by downregulating MITF expression [11, 12].

4 Orchids and Cosmetic Benefits

The potential of each orchid with cosmetic benefits is disclosed alphabetically on the basis of scientific evidences. The phytochemically active compounds will be included together with biological activities *in vitro*, *ex vivo*, and *in vivo* if appreciable.

4.1 *Ansellia africana*

Leopard orchid has long been regarded as the important source of pharmacologically active biomolecules beneficial for health [13]. This orchid posed anti-inflammatory effect by the inhibition against COX-1. Its crude root extract using CH_2Cl_2 inhibited COX-1 at EC_{50} of 0.25 ± 0.10 mg/ml. In addition, this extract showed acetylcholinesterase inhibition, although at a lesser degree than the ethanolic and ether extracts ($EC_{50} = 0.34 \pm 0.14$ and 0.24 ± 0.03 and 0.33 ± 0.03 mg/ml, respectively). EC_{50} of galantamine, the positive control, was 0.44 ± 0.10 μM [14].

4.2 *Bulbophyllum scaberulum*

Ethanolic extract of the orchid root inhibited COX-2 and acetylcholinesterase with EC_{50} of 0.44 ± 0.32 and 0.26 ± 0.00 mg/ml, while that of CH_2Cl_2 extract were 1.43 ± 0.86 and 0.02 ± 0.00 mg/ml, respectively [14].

4.3 *Dendrobium* spp.

Dendrobium is the second largest genus in the family Orchidaceae, and more than 1,100 species are cultivated in Thailand regarding continuous bleeding of the orchid to give glamor color and shape varieties. This world's major stakeholder cut orchids are widely in red-purple, pink, and white flowers, of which Sonia, Sonia Pink, Snow Rabbit, and Shavin White are the most common floriculture varieties. The methanolic extracts of these varieties were screened on *in vitro* antioxidant and tyrosinase inhibitory effect. The flower of Shavin White was best in DPPH radical scavenging activity, followed by Sonia Pink, while Snow Rabbit and Sonia were comparatively low ($IC_{50} = 463.08 \pm 15.68$, 492.83 ± 15.73, and > 500 μg/ml) once compared with the standard gallic acid, quercetin, and ascorbic acid ($IC_{50} = 0.64 \pm 0.01$, 0.85 ± 0.04, and 0.94 ± 0.05 μg/ml). Although anti-DPPH activities of the orchid flower extracts were weak, their inhibitory effects against mushroom tyrosinase were strong especially Sonia, Sonia Pink, and Shavin White ($IC_{50} = 57.38 \pm 9.26$, 83.21 ± 3.53, and 111.67 ± 2.88 μg/ml) that were more potent than the standard kojic acid ($IC_{50} = 151.73 \pm 2.06$ μg/ml), while the extract of Snow Rabbit was lower ($IC_{50} = 167.82 \pm 2.63$ μg/ml) than the standard. These activities would be governed by the active principles in terms of phenolics and flavonoids. Nevertheless, the actives profile of these floriculture *Dendrobium* was not carried out in the study [15]. The red-purple *Dendrobium* Sonia was assessed on biological activity and phytochemical profile in different research. The anthocyanin-rich extracts were prepared from the orchid flower. Of which, chemical quality in terms of total phenolics was additionally reported together with the biological activities beneficial for cosmetics, i.e., astringency, *in vitro* radical scavenging, and enzyme inhibitory activities. The 70% ethanolic and water extract at different times of extraction were highlighted as the interesting source of antioxidants and inhibitors against collagenase, elastase, and tyrosinase, which is promising for prevention of collagen and elastin degradation as per skin dullness. Moreover, the extracts were safe in NHF posed antioxidant and MMP-2 inhibitory effect activities. In addition to the safe and efficient activities in NHF, the extracts were safe in B16F10 melanoma and were proved to suppress cellular melanogenesis, in which the mechanism was revealed to be by tyrosinase and TRP-2 inhibition at a higher degree than the standard kojic acid. The biological activities of the *Dendrobium* Sonia flower were governed by their ten phenolics and three anthocyanins constituents. Of which, sinapic and ferulic acids were the major phenolics, and pelargonidin was the principal anthocyanin followed by cyanidin and keracyanin [16].

4.4 *Dendrobium candidum*

D. candidum stem is used in traditional Chinese medicine as yin tonic with inflammation treatment. This methanolic extract of medicinal herb (200, 400, and 800 mg/kg) was reported to increase serum superoxide dismutase (SOD level), while pro-inflammatory cytokines, i.e., interleukin (IL)-6, IL-12, tumor necrosis factor (TNF)-α, and interferon (IFN)-γ, were decreased as examined in an animal model for 2 weeks in a dose-dependent manner [17].

4.5 *Dendrobium chrysotoxum*

The 95% ethanolic extract of the stem was assessed upon its inhibitory effect against acetylcholine (AChE) and butyrylcholine (BChE) esterases. However, the activity of the isolated pure compounds was moderate to weak [18].

4.6 *Dendrobium denneanum*

The isolated pure compounds, 2,5-dihydroxy-4-methoxy-phenanthrene 2-*O*-β-D-glucopyranoside and 5-methoxy-2,4,7,9S-tetra-hydroxy-9,10-dihydrophenanthrene, from the stem exhibited potent anti-inflammation. iNOS was suppressed as per the inhibition against p38, JNK, MAPK, and IκBα, which suggested their dual mechanism inhibition in MPKs and NF-κB pathways [19].

4.7 *Dendrobium huoshanense*

Polysaccharides derived from *Dendrobium* orchids are found to have several health benefits especially against inflammation including those from *D. huoshanense*. The orchid stem rich in polysaccharides was prepared into the extract that majorly consists of mannose and glucose in a molar ratio of 1.89:1. The polysaccharides extract was intragastrically administrated at 100, 200, and 400 mg/kg/day for

4 weeks into cigarette smoke-induced mice. The orchid extract was shown to inhibit TNF-α and IL-1β secretion in serum. The anti-inflammatory effect was studied to be caused by NF-κB reduction as well as phosphorylation of IκB, p65, p38, and JNK. Thus, anti-inflammatory activity of *D. huoshanense* polysaccharides extract was by alleviating NF-κB and MAPK signaling pathways [20].

4.8 *Dendrobium nobile*

The stem methanolic extract of the orchid contains alkaloids (96.1%) and polysaccharides (1.2%) shown to prevent lipopolysaccharide (LPS)-induced elevation in tumor necrosis factor receptor 1 (TNFR1) mRNA and protein levels. LPS-induced activation of phosphorylated p38 mitogen-activated protein kinases (p38 MAPK) and nuclear factor kappa-B (NF-κB) pathway was also suppressed as per injection of 40, 80, and 160 mg/kg/day into rats for 14 days. Of which, the activity was pronounced at higher concentration [21]. The isolated pure compounds, ephemeranthol A and dehydroorchinol, were later confirmed upon these anti-inflammatory activities. They inhibited cellular NO production as per pro-inflammatory cytokines in a dose-dependent behavior at the cellular safety concentration ranging from 6.25 to 50 μg/ml. The significantly efficient dose (25 μg/ml) of the compounds was later on confirmed on their inhibitory effect against IL-1β and IL-6, while TNF-α was significantly suppressed by ephemeranthol A but not dehydroorchinol. Thereafter, the stronger anti-inflammatory active, ephemeranthol A, was examined upon its function in inflammatory signaling pathway. Ephemeranthol A reduced the level of phosphorylated p38 and inhibited NF-κB activation [22].

4.9 *Dendrobium officinale*

The stem of the orchid has long been used in traditional Chinese medicine claimed to reduce fever and to have immunological function. This *Dendrobium* species is therefore commercially cultivated not only in China mainland but also in Southeast Asian countries especially Thailand.

The 132 kDa polysaccharides (mannose and glucose of 3.8:1.0) derived from the orchid were shown to remarkably reduce cellular oxidative stress by the capability to inhibit ROS production (at the dose of 62.5–500 μg/ml). Furthermore, the polysaccharides extract significantly decreased the p-NF-κBp65/NF-κBp65 level induced by H_2O_2. In addition, the extract was also efficient as examined in an animal model [23]. Potency of the orchid polysaccharides against inflammation was confirmed by different *in vivo* studies as IL-1β, IL-6, IL-18, TNF-α, and IFN-γ were significantly decreased following 50, 100, and 200 mg/kg administrated into the induced-mice group [24]. Polysaccharides derived from the stem of the orchid that is mainly composed of mannose, glucose, and arabinose (molecular weight of 393.8 kDa) were orally administrated into female mice (70 mg/kg) for 10 weeks. The *Dendrobium* polysaccharides were evidenced to reduce pro-inflammatory cytokines

(TNF-α and IL-6) and MDA levels while estradiol, SOD, GSH-x, and total antioxidant capacity in serum. Moreover, it significantly balanced pro-inflammatory/anti-inflammatory cytokines ratio, the key mechanism maintaining body health and resisting damage, to normal level and was able to improve function of mitochondria by an inhibition of p53/Bcl-2 mediated mitochondrial apoptosis signaling pathway in natural aging-induced mice. Taken together, the orchid may alleviate cellular aging by inhibitory effect against NF-κB, and the orchid was suggested to be used for natural aging treatment in female [25]. The 72.1% polysaccharides extract (mannose and glucose – 19.51% and 14.03%) of the orchid stem by 80% EtOH was orally administrated into the diabetic cardiomyopathy-induced mice for 8 weeks. The treatment groups at the dose of 150 and 300 mg/kg were shown to be increased in SOD with the suppression of MDA activities. In addition, NF-κB, TNF-α, and IL-1β were significantly suppressed in a comparison with the diabetic cardiomyopathy-induced mice [26].

The pharmacological benefits of *D. officinale* stem are confirmed with the traditionally used by the scientific evidences that continuously explored. The polysaccharides extract, Dendronan®, has been therefore commercialized and proved to have protective effects against oxidative stress including its capability to increase CAT, SOD, and GSH-Px with the reduction of MDA in animal model [27]. The orchid polysaccharides are therefore potentially used for immunomodulatory activity enhancement [28, 29] associated with longevity or aging protection and treatment.

To widen its application, the different parts of the orchid are explored. Leaf is obviously one part of the medicinal herb that is essential to be revealed for its new potential uses. The leaves of 11 different strains of *D. officinale* were extracted by 80% MeOH. They exhibited anti-DPPH activity at an interesting capability, which corresponds with total flavonoids content, and rutin was shown to be the biologically active marker of the orchid leaf [30].

4.10 *Dendrobium tosaense*

Stem extract of *D. tosaense* containing quercetin was orally administrated into allergic modeling mice at the dose of 30, 100, and 300 mg/kg in a comparison with quercetin (1.6 mg/kg). Anti-allergic potential of the extract was evidenced by the significant reduction of serum IgG1 and IgE as per IL-4, IFN-γ, and IL-6 level except the low dose (30 mg/kg). Thus, the orchid was potentially to be used for atopic dermatitis therapy or other allergic disorder [31].

4.11 *Eulophia hereroensis*

Tuber extract of this orchid with CH_2Cl_2 showed anti-activity against COX-1, COX-2, and acetylcholinesterase activity (EC_{50} = 0.87 ± 0.28, 1.17 ± 0.15, and 0.23 ± 0.16 mg/ml). However, its ether extract posed only COX-2 and acetylcholinesterase inhibitions (EC_{50} = 1.12 ± 0.33 and 1.20 ± 0.24 mg/ml) [14].

4.12 *Eulophia macrobulbon*

This tropical orchid is a major floriculture species in Thailand and in other Southeast Asian countries. It has been used to treat insect bites in local Thai folk remedy. Its root was therefore extracted, challenged on activity against DPPH radical. The crude extract scavenged 9 ± 0% of the radical, whereas that of the standard ascorbic acid was 67 ± 1%, at the same test concentration of 100 μg/ml. Fractionation of the crude extract was undertaken to improve antioxidant activity to 51 ± 3% and 44 ± 2%, respectively. The crude extract and these two potent fractions were revealed to have cellular anti-inflammatory activities. The secretions of IL-6, IL-10, and TNF-α, inflammatory mediators, were shown to be suppressed at the same test concentration at 100 μg/ml. The suppression (%) of the crude extract were 30 ± 7, 67 ± 10, and 81 ± 9. Meanwhile, the potent fractions were 12 ± 1% and 24 ± 14%, 60 ± 10% and 77 ± 4%, and 106 ± 11% and 81 ± 14%, respectively. Of which, the capability of the orchid and orchid active fractions were noted potent against IL-6. Thus, the IC_{50} of each sample against these mediators were examined and were shown to be 54, 25, and 54 μg/ml, respectively [32].

4.13 *Eulophia petersii*

This medicinal orchid inhibited COX-1 especially its CH_2Cl_2 extracts of the stem, pseudobulb, and roots ($EC_{50} = 1.49 \pm 0.05$, 0.87 ± 0.12, and 1.41 ± 0.64 mg/ml) and posed acetylcholinesterase inhibitory effect ($EC_{50} = 0.39 \pm 0.04$, 0.51 ± 0.14, and 0.51 ± 0.05 mg/ml) [14].

4.14 *Tridactyle tridentata*

South Africans traditionally employed the orchid root in the remedy recipes, in which its CH_2Cl_2 extract was later on confirmed for its anti-inflammatory effects via the capability against COX-1 and acetylcholinesterase activities ($EC_{50} = 1.47 \pm 0.89$ and 0.46 ± 0.01 mg/ml) [14].

4.15 *Vanda coerulea*

This orchid is commonly called blue orchid with regard to its anthocyanin constituents [33]. The hydroalcoholic stem extract displayed *in vitro* radical scavenging activity. In addition, the isolated pure compounds posed inhibitory effect against PGE-2 production in irradiated HaCaT cell line and UVB-induced COX-2 expression as well [34].

The ethanolic extract of the stem was isolated to give active pure compounds. The orchid-derived stilbenoid, imbricatin, methoxycoelonin, and gigantol replicated senescence of normal human skin fibroblasts and were able to restore the percentage at a rate equivalent to that of young cells together with the recovery of the cyclin E and cyclin-dependent kinase 2 (cdk2). These results highlight the potential of the orchid as raw material to fight against the visible signs of skin aging [35].

4.16 *Vanda roxburghii*

The orchid leaf extract (aqueous) was evidenced to have wound healing properties as examined in an excision wound animal model. Topical application of the extract at a dose of 150 mg/kg/day consecutively for 10 days was shown to reduce the wound diameter by 60%, while the control group has 48% reduction. The wounds were fully healed in 13 days, whereas those of the control group were 20 days. Moreover, the significant increment ($p < 0.005$) in wet and dry granulation tissue weights, hydroxyproline, and hexosamine contents were detected. Of which, the pro-healing action was suggested to be by an attribution to increase collagen deposition or to better alignment and maturation or both [36]. Moreover, its root extracts were prepared and screened for antioxidant activity, in which the chloroform extract was exhibited as the most potent fraction. IC_{50} against DPPH and OH radicals were 5.76 ± 0.42 and 7.96 ± 0.61 μg/ml, while those of the standard catechin were 4.55 ± 0.33 and 9.45 ± 0.57 μg/ml, respectively. In addition, antioxidant activity of the extract assessed by FRAP assay was comparable to catechin (1.34 ± 0.04 and 1.48 ± 0.03 at 100 μg/ml). The active principles in terms of total phenolics and flavonoids contents of the extract were noted to be significantly higher than the other extracts (85.9 ± 1.03 mg GAE/g dried extract and 300.1 ± 0.61 mg CE/g dried extract). The extract additionally posed inhibitory effects against AChE and BChE at IC_{50} of 221.13 and 82.51 μg/ml, while those of the standard donepezil and galantamine were 5.21 and 8.91 μg/ml, respectively. The active compound responsible for these activities was found to be gigantol [37].

4.17 *Vanda teres*

The methanolic extract of *V. teres* stem was isolated to yield the active compounds. The isolated eucomic acid and vandateroside II were shown to increase cytochrome c oxidase activity and/or expression, without enhancing cellular mitochondrial content. In addition, they contributed to stimulate respiratory functions in keratinocytes. Thus, these orchid isolated compounds were suggested to become new natural ingredients for antiaging preparations to remedy age-related disorders such as skin aging [38].

The orchids with cosmetic benefits presented above are therefore summarized in Table 1. The therapeutic uses of orchids relevant to cosmetic benefits are shown in Table 2. Furthermore, orchid extracts commercially available for cosmetic preparation are disclosed in Table 3 together with INCI (International Nomenclature of Cosmetic Ingredients) name and CAS (Chemical Abstracts Service) number for further reference.

Table 1 Candidate orchid with cosmetic properties (anti-inflam., anti-inflammatory)

Botanical name	Part uses	Biological activity			Phytochemical active	References
		In vitro	Ex vivo	In vivo		
Ansellia africana Lindl.	Root	Anti-inflam.				[13]
Bulbophyllum scaberulum (Rolfe) Bolus	Root	Anti-inflam.				[14]
Dendrobium spp.	Flower	Antioxidant, anti-tyrosinase			Phenolics and flavonoids	[15]
		Astringency, antioxidant, anti-collagenase, anti-elastase, anti-tyrosinase	NHF; antioxidant and anti-MMP-2 B16F10 melanoma; anti-melanogenesis, anti-tyrosinase, anti-TRP-2		10 phenolics; majorly sinapic and ferulic acid 3 anthocyanins; pelargonidin, cyanidin, keracyanin	[16]
D. candidum Wall. ex Lindl.	Stem			Anti-inflam. in animal model		[17]
D. chrysotoxum Lindl.	Stem	Anti-inflam.				[18]
D. denneanum Kerr	Stem		Anti-inflam.		2,5-dihydroxy-4-methoxy-phenanthrene 2-*O*-β-D-glucopyranoside, 5-methoxy-2,4,7,9S-tetra-hydroxy-9,10-dihydrophenanthrene	[19]
D. huoshanense	Stem			Anti-inflam. in animal model	Polysaccharides	[20]
D. nobile Lindl.	Stem	Anti-inflam.	Anti-inflam.	Anti-inflam. in animal model	Alkaloids, polysaccharides, ephemeranthol A, dehydroorchinol	[21, 22]

(continued)

Table 1 (continued)

Botanical name	Part uses	Biological activity			Phytochemical active	References
		In vitro	Ex vivo	In vivo		
D. officinale Kimura & Migo	Stem		Antioxidant, anti-inflam.	Anti-inflam., immunomodulatory in animal models	Polysaccharides	[23–29]
	Leaf	Antioxidant			Flavonoids	[30]
D. tosaense Makino	Stem			Anti-allergy, anti-inflam. in animal model	Quercetin	[31]
Eulophia hereroensis Schltr.	Tuber	Anti-inflam.				[14]
E. macrobulbon (E.C.Parish & Rchb.f.) Hook. f.	Root	Antioxidant	Anti-inflam.			[32]
E. petersii (Rchb. f.) Rchb. f.	Stem, pseudobulb, root	Anti-inflam.				[14]
Tridactyle tridentata (Harv.) Schltr.	Root	Anti-inflam.				[14]
Vanda coerulea Griff. ex Lindl.	Stem	Antioxidant	Anti-inflam.		Anthocyanins, stilbenoid, imbricatin, methoxycoelonin, gigantol	[33–35]
V. roxburghii R.Br.	Leaf			Wound healing in animal model	Phenolics, flavonoids, gigantol	[36, 37]
V. teres	Stem		Antioxidant		Eucomic acid, vandateroside II	[38]

Table 2 Therapeutic uses of orchids relevant to cosmetic benefits

Botanical name	Part uses	Uses	References
Acampe papillosa (Lindl.) Lindl.	Root	Cooling agent, astringent, asthma	[39]
Anocetochilus formosanus Hayata	Tuber	Antioxidant	[40, 41]
Arundina graminifolia (D. Don) Hochr.	Rhizome	Antibacterial, body ache treatment	[2, 42]
Bletilla striata (Thunb.) Rchb. f.	Whole plant	For cracked feet and hands, anti-inflam., emollient for burn and skin diseases, tonic	[3]
Cleisostoma williamsonii (Rchb. f.) Garay	Whole plant	Astringent	[39]
Coelogyne corymbosa Lindl.	Pseudobulb	Wound healing	[42]
C. punctulata Lindl.			[36]
Cymbidium aloifolium (L.) Sw.	Seed	Wound healing	[43]
Dendrobium alpestre Royle	Pseudobulb	Tonic, pimples and skin problems	[41]
D. crumenatum Sw.	Leaf	Pimples	[36]
D. monticola Hunt & Summerh.	Pseudobulb	Pimples and skin eruptions	
D. nobile Lindl.	Pseudobulb	Burn soothing	
	Seed	Wound healing	
	Stem	Longevity	
D. ovatum (Wild.) Kranzl.	Whole plant	Emollient	[39]
Eulophia dabia (D. Don) Hochr.	Tuber	Astringent	[2, 39]
Flickingeria macraei (Lindl.) Seidenf.	Pseudobulb	Tonic, skin allergy and eczema	[42]
Luisia tenuifolia Blume	Rhizome, leaf	Emollient	[39, 42]
Rhynchostylis retusa Blume	Whole plant	Emollient, wound, asthma and skin diseases curing, inhibitory effect against *Bacillus subtilis* and *Escherichia coli*	[39, 42, 44]
Vanda tessellate (Roxb.) Hook. Ex G. Don	Whole plant	Inflammation, tonic, scalp boils	[2, 39, 44]

5 Conclusions

Orchid has long served as the therapeutic herb in several traditional recipes worldwide. This medicinal herb does not only have health benefits, but their beautiful flowers are also cultivated for floriculture purposes resulting in variety of the orchid cultivars. The appreciable of orchid for cosmetic proposes especially for skin aging protection and treatment are enclosed. Traditional uses of orchid for astringency or tonic effects are later scientifically proved by its anti-inflammatory activities

Table 3 Orchid extracts commercially supplied for cosmetics

Commercial name	INCI name	CAS number	Orchid/part	Claim/application	Supplier	Use level (%)
Orchid Water	–	–	–	Anti-inflammatories Anti-redness/anti-couperose agents Softening/texturing agents	Biogründl	–
Dermalab Orchid extract	–	–	–	Skin care (facial care, facial cleansing, body care, baby care)	Dermalab	2–4
Hyacinth orchid root extract	BLETILLA STRIATA ROOT EXTRACT		*Bletilla striata* root	Regenerating/revitalizing agents Anti-inflammatories		2–4
Oriental Orchid Extract-SG	–		Stem	Anti-inflammatories Antioxidants	The Garden of Natural Solution	–
Orchid Extract Liquid	–	–	Flower	Moisturizing agents Lightening/whitening agents Antioxidants	Specialty Natural Product	–
Herbal Extract Orchid EG	VANILLA PLANIFOLIA BEAN EXTRACT	8024-06-4	Seed pods of *Vanilla planifolia*	Skin softening. Used in moisturizers	Peter Jarvis	–
Herbal Extract Orchid EO						–
Herbal Extract Vanilla EO		–	–			
Herbal Extract Vanilla EG						
Orchid Complex WS	CYMBIDIUM GRANDIFLORUM FLOWER EXTRACT		*Cymbidium* orchids	Emollients Lubricants/slip agents Moisturizing Lubricious feeling	United-Guardian	2–10
Orchid Complex OS		65381-09-1				

Orchid Extract	ORCHIS MASCULA FLOWER EXTRACT	90082-24-9	*Orchis mascula* flower	Anti-inflammatories Antioxidants Moisturizing agents Nourishing agents Nourishing Moisturizing	Carrubba	–
Akomilk® Orchid			*Orchis macula*	Skin care (facial care, facial cleansing, body care, baby care)	Akott	–
Orchid Complex™ OS ester	CYMBIDIUM GRANDIFLORUM FLOWER EXTRACT	65381-09-1	–	Emollients Moisturizing	Ashland Specialty Chemical	–
ORCHID EXTRACT H. GL.-M.S.	ORCHIS MASCULA FLOWER EXTRACT	90082-24-9	Orchid flowers	Emollients Moisturizing agents Soothing agents Antiaging agents Antioxidants Conditioning agents Moisturizing	Provital	0.5–5
ORCHISTEM™	CALANTHE DISCOLOR EXTRACT	7732-18-5	Stem	Antiaging agents Anti-wrinkle agents Radiance promoters Shine/radiance Smoothness		

associated with prevention and treatment of skin dryness as per allergic skin as well as those of cellular inflammatory lesions that resulted from cellular oxidative stress. These benefits are accumulating in antiaging of skin capability. In addition, antioxidant activities of orchid contribute on suppression or downregulate adverse effects of oxidative stress in dermal cells surplus with diminishing overproduction of skin melanin pigments. Accordingly, cosmetic benefits of orchid would be initiated by the skin hydrating potency and delineated for antiaging of skin eventually. Some of the orchid species are already commercialized in cosmetic industry. Thus, the researchers both in academic and industrial sections are encourage to versatile those of evidence-based ones into higher value cosmetic ingredient and finished products as well. Despite some being already commercialized, the scientific-based information seems inadequate. This presenting context will be therefore available for those who are interested in cosmetic applications of orchids and for sufficient data study to ensure efficacy and safety of orchids used in health promotion products including cosmetics. Eventually, maximize benefits or orchid will be pursed and flow in the stream of the consumers' choice and preference on natural or bio-based products.

Acknowledgments Mae Fah Luang University is acknowledged for facility supports during the manuscript preparation.

References

1. Bulpitt CJ, Li Y, Bulpitt PF, Wang J (2007) The use of orchids in Chinese medicine. J R Soc Med 100:558–563
2. Singh A, Duggal S (2009) Medicinal orchids-an overview. Ethnobot Leafl 13:399–412
3. Kong JM, Goh NK, Chia LS, Chia TF (2003) Recent advances in traditional plant drugs and orchids. Acta Pharmacol Sin 24:7–21
4. Hossain MM (2011) Therapeutic orchids: traditional uses and recent advances – an overview. Fitoterapia 82:102–140
5. Ng TB, Liu J, Wong JH, Ye X, Sze SCW, Tong Y, Zhang KY (2012) Review of research on *Dendrobium*, a prized folk medicine. Appl Microbiol Biotechnol 93:1795–1803
6. Kanlayavattanakul M, Lourith N (2015) An update on cutaneous aging treatment using herbs. J Cosmet Laser Ther 17:343–352
7. Lourith N, Kanlayavattanakul M (2016) Biopolymeric agents for skin wrinkle treatment. J Cosmet Laser Ther 18:301–310
8. Kammeyer A, Luiten RM (2015) Oxidation events and skin aging. Ageing Res Rev 21:16–29
9. Pradhan S, Madke B, Kabra P, Singh AL (2016) Anti-inflammatory and immunomodulatory effects of antibiotics and their used in dermatology. Indian J Dermatol 61:469–481
10. Kanlayavattanakul M, Lourith N (2015) Biopolysaccharides for skin hydrating cosmetics. In: Ramawat KG, Mérillon JM (eds) Polysaccharides. Springer, Switzerland
11. Kanlayavattanakul M, Lourith N (2018) Plants and natural products for the treatment of skin hyperpigmentation – a review. Planta Med 84:988–1006
12. Kanlayavattanakul M, Lourith N (2018) Skin hyperpigmentation treatment using herbs: a review of clinical evidences. J Cosmet Laser Ther 20:123–131
13. Bhattacharyya P, Van Staden J (2016) *Ansellia africana* (leopard orchid): a medicinal orchid species with untapped reserves of important biomolecules – a mini review. S Afr J Bot 106:181–185

14. Chinsamy M, Finnie JF, Van Staden J (2014) Anti-inflammatory, antioxidant, anti-cholinesterase activity and mutagenicity of South African medicinal orchids. S Afr J Bot 91:88–98
15. Athipornchi A, Jullapo N (2018) Tyrosinase inhibitory and antioxidant activities of orchid (*Dendrobium* spp.). S Afr J Bot 119:188–192
16. Kanlayavattanakul M, Lourith N, Chaikul P (2018) Biological activity and phytochemical profiles of *Dendrobium*: a new source for specialty cosmetic materials. Ind Crop Prod 120:61–70
17. Wang Q, Sun P, Li G, Zhu K, Wang C, Zhao X (2014) Inhibitory effects of *Dendrobium candidum* Wall ex Lindl. on azoxymethane- and dextran sulfate sodium-induced colon carcinogenesis in C57BL/6 mice. Oncol Lett 7:493–498
18. Dong FW, Luo HR, Wan QL, Xu FQ, Fan WW, Wang KJ, Li N, Hu J-M (2012) Two new bibenzyl glucosides from *Dendrobium chrysotoxum*. Bull Kor Chem Soc 33:2247–2250
19. Lin Y, Wang F, Yang L-J, Chun Z, Bao J-K, Zhang G-I (2013) Anti-inflammatory phenanthrene derivatives from stems of *Dendrobium denneanum*. Phytochemistry 95:242–251
20. Ge J-C, Zha X-Q, Nie C-Y, Yu N-J, Li Q-M, Peng D-Y, Duan J, Pan L-H, Luo J-P (2018) Polysaccharides from *Dendrobium huoshanense* stems alleviates lungs inflammation in cigarette smoke-induced mice. Carbohydr Polym 189:289–295
21. Li Y, Li F, Gong Q, Wu Q, Shi J (2011) Inhibitory effects of *Dendrobium* alkaloids on memory impairment induced by lipopolysaccharide in rats. Planta Med 77:117–121
22. Kim JH, Oh S-Y, Han S-B, Uddin GM, Kim CY, Lee JK (2015) Anti-inflammatory effects of *Dendrobium nobile* derived phenanthrenes in LPS-stimulated murine macrophages. Arch Pharm Res 38:1117–1126
23. Zeng Q, Ko C-H, Siu W-S, Li L-F, Han X-Q, Yang L, Lau CB-S, Hu J-M, Leung P-C (2017) Polysaccharides of *Dendrobium officinale* Kimura & Migo protect gastric mucosal cell against oxidative damage-induced apoptosis *in vitro* and *in vivo*. J Ethnopharmacol 208:214–224
24. Liang J, Chen S, Chen J, Lin J, Xiong Q, Yang Y, Yuan J, Zhou L, He L, Hou S, Li S, Hong S, Lai X (2018) Therapeutic roles of polysaccharides from *Dendrobium officinale* on colitis and its underlying mechanisms. Carbohydr Polym 185:159–168
25. Wu Y-Y, Liang C-Y, Liu T-T, Liang Y-M, Li S-J, Lu Y-Y, Liang J, Yuan X, Li C-J, Hou S-Z, Lai X-P (2018) Protective roles and mechanisms of polysaccharides from *Dendrobium officinal* on natural aging-induced premature ovarian failure. Biomed Pharmacother 101:953–960
26. Zhang Z, Zhang D, Dou M, Li Z, Zhang J, Zhao X (2016) *Dendrobium officinale* Kimura et Migo attenuates diabetic cardiomyopathy through inhibiting oxidative stress, inflammation and fibrosis in steptozotocin-induced mice. Biomed Pharmacother 84:1350–1358
27. Huang X, Nie S, Cai H, Zhang G, Cui SW, Xie M, Philips GO (2015) Study on *Dendrobium officinale O*-acetyl-glucomannan (Dendronan®): part VI. Protective effect against oxidative stress in immunosuppressed mice. Food Res Int 72:168–173
28. Cai H-L, Huang H-J, Nie S-P, Xie M-Y, Philips GO, Cui SW (2015) Study on *Dendrobium officinale O*-acetyl-glucomannan (Dendronan®): part III – immunomodulatory activity *in vitro*. Bioact Carbohydr Diet Fibre 5:99–105
29. Huang X, Nie S, Cai H, Zhang C, Cui SW, Xie M, Philipes GO (2015) Study on *Dendrobium officinale O*-acetyl-glucomannan (Dendronan): part IV. Immunomodulatory activity *in vivo*. J Funct Foods 15:525–532
30. Zhang Y, Zhang L, Liu J, Liang J, Si J, Wu S (2017) *Dendrobium officinale* leaves as a new antioxidant source. J Funct Foods 37:400–415
31. Wu C-T, Huang K-S, Yang C-H, Chen Y-C, Liao J-W, Kuo C-L, Chen C-L, Lo S-F, Hsieh C-C, Tsay H-S (2014) Inhibitory effects of cultured *Dendrobium tosaense* on atopic dermatitis murine model. Int J Pharm 463:193–200
32. Schuster R, Zeindl L, Holzer W, Khumpirapang N, Okonogi S, Viernstein H, Mueller M (2017) *Eulophia macrobulbon* – an orchid with significant anti-inflammatory and antioxidant effect and anticancerogenic potential exerted by its root extract. Phytomedicine 24:157–165
33. Tatsuzawa F, Saita N, Seki H, Yokoi M, Yukawa T, Shinoda K, Honda T (2004) Acylated anthocyanins in the flowers of *Vanda* (Orchidaceae). Biochem Syst Ecol 32:651–664

34. Simmler C, Antheaume C, Lobstein A (2010) Antioxidant biomarkers from *Vanda coerulea* stems reduces irradiated HaCaT PGE-2 production as a result of COX-2 inhibition. PLoS One 5:e13713
35. Bonté F, Simmler C, Lobstein A, Pellicier F, Cauchard J-H (2011) Action of an extract of *Vanda coerulea* on the senescence of skin fibroblasts. Ann Pharm Fr 69:177–181
36. Nayak B, Suresh R, Rao AVC, Pilai GK, Davis EM, Ramkisson V, McRae A (2005) Evaluation of wound healing activity of *Vanda roxburghii* R. Br (Orchidaceae): a preclinical study in a rat model. Int J Low Extrem Wounds 4:200–204
37. Uddin MN, Afrin R, Uddin MJ, Uddin MJ, Alam AHMK, Rahman AA, Sadik G (2015) *Vanda roxburghii* chloroform extract as a potential source of polyphenols with antioxidant and cholinesterase inhibitory activities: identification of a strong phenolic antioxidant. BMC Complement Altern Med 15:195
38. Simmler C, Antheaume C, André P, Bonté F, Lobstein A (2011) Glucosyloxybenzyl eucomate derivatives from *Vanda teres* stimulate HaCaT cytochrome c oxidase. J Nat Prod 74:949–955
39. Singh DK (2001) Orchid diversity in India: an overview. In: Patak P, Sehgal RN, Shekhar N, Sharma M, Sood A (eds) Orchid: science and commerce. Bishen Singh Mahendra Pal Singh, Dehradun
40. Lin CC, Huang PC, Lin JM (2000) Antioxidant and hepatoprotective effects of *Anoectochilus formosanus* and *Gynostemma pentaphyllum*. Am J Chin Med 28:87–96
41. Shih CC, Wu YW, Lin WC (2002) Antihyperglycaemic and anti-oxidant properties of *Anoectochilus formosanus* in diabetic rats. Clin Exp Pharmacol Physiol 29:684–688
42. Kumar S (2002) The medicinal plants of North-East India. Scientific Publishers, Jodhpur
43. Maridass M, Hussain MIZ, Raju G (2008) Phytochemical survey of orchids in the Tirunelveli hills of South India. Ethnobot Leafl 12:705–712
44. Jalal JS, Kumar P, Pangtey YPS (2008) Ethnomedicinal orchids of Uttarakhand, Western Himalaya. Ethnobot Leaft 12:1227–1230

Orchids from Basilicata: The Scent 26

Maurizio D'Auria, Simonetta Fascetti, Rocco Racioppi, Vito Antonio Romano, and Leonardo Rosati

Contents

Abstract

HS-SPME-GC-MS analysis of the scent of some orchid species found in Basilicata (Southern Italy) has been performed. The analysis of *Platanthera* species showed the presence of two different behaviors. The samples from Basilicata showed the presence of lilac derivatives, while samples from Calabria or Abruzzo had as main component benzyl benzoate. The analysis of *Cephalanthera* orchids showed that, probably, scent does not have a relevant role on the pollination strategy of the plants in comparison to the role of color. The analysis of *Serapias* orchids showed the presence of alkanes and alkenes but with a lower molecular weight than those reported in previous works on these species. Furthermore, α-amorphene was found as main component in *Serapias cordigera* and *Serapias cordigera* subsp. *lucana*. The analysis of *Barlia robertiana* showed a high variability in the composition of the scent.

M. D'Auria (✉) · R. Racioppi · V. A. Romano
Dipartimento di Scienze, Università della Basilicata, Potenza, Italy
e-mail: maurizio.dauria@unibas.it; rocco.racioppi@unibas.it

S. Fascetti · L. Rosati
School of Agricultural Forestry, Food, and Environmental Science, Università della Basilicata, Potenza, Italy
e-mail: simonetta.fascetti@unibas.it; leonardo.rosati@unibas.it

J.-M. Mérillon, H. Kodja (eds.), *Orchids Phytochemistry, Biology and Horticulture*, Reference Series in Phytochemistry, https://doi.org/10.1007/978-3-030-38392-3_32

Keywords

Orchids · Platanthera · Cephalanthera · Serapias · Barlia · Volatile organic compounds · Solid phase micrextraction · Gas chromatography – mass spectrometry

1 Introduction

The determination of the volatile organic compounds emitted from a natural source is smell and taste which are the oldest of our senses. They probably developed in very primitive organisms as a means of obtaining information about chemical changes in the organism's environment. Animals use smell and taste to find food and to assess its quality. The smell of food has a powerful effect on animals.

Living organisms use the chemical sense as a means of communications. If the communication is between different parts of the same organism, the messenger is referred to as a hormone. Chemicals used to carry signals from one organism to another are known as semiochemicals. In the case of flowers, the aroma components are mainly devoted to attract pollinator insects.

The main problem in the determination of the composition of the volatile organic compounds is the correct methodology to be used in order to determine the presence of chemical compound in so low concentration. Pawliszyn at the end of the last century developed a new method known as solid phase microextraction (SPME) [1]. It consists in a gas-chromatographic syringe, where the needle contains a little section of the fiber where a stationary phase such as that present in gas-chromatographic column. The fiber was exposed to the mixture of volatile organic compounds, and then desorbed in the injection sector of the gas-chromatographic apparatus.

During this time, our research group used SPME in order to determine the composition of volatile organic compounds in a lot of natural matrices. Thus, we determined the composition of volatile compounds mixtures in olive oil [2–4], wine [5–9], saffron [10, 11], fungi and truffles [12–17], honey [18], horseradish [19], thymus [20, 21], origanum [20], lavandula [20], acinos [20], rosmarinus [21], laurus [21], salvia [21], actinidia [22], crude oil [23], onion [24], shallot [24], and beer [25]. Furthermore, recently, we used this methodology in the determination of the scent of a flower plant [26]. Based on this work, when we received the proposal for a study on the scent of the spontaneous orchids of Basilicata (Southern Italy), we were ready to accept the idea.

2 *Platanthera* Orchids

In Europe, the genus *Platanthera* shows the presence of 13 taxa; in Italy, six species have been described (*Platanthera algeriensis*, *Platanthera bifolia* subsp. *bifolia*, *Platanthera bifolia* subsp. *osca*, *Platanthera bifolia* subsp. *subalpina*, *Platanthera chlorantha*, and *Platanthera kuenkelei* var. *sardoa*). In Basilicata, only *P. bifolia* subsp. *osca* (Fig. 1a) and *P. chlorantha* (Fig. 1b) are present.

There are some differences between these two taxa. However, in *P. chlorantha*, the pollinator insects (*Sphingidae* and *Noctuidae*) can get close to the spur, while, in *P. bifolia* subsp. *osca*, this is not possible to reach nectar. Furthermore, *P. bifolia* subsp. *osca* emitted an intense scent during the night, while *P. chlorantha* emitted an aroma during all day.

Previous study on the aroma of these taxa showed some differences. In a study on *P. bifolia*, Kaiser found, as main components of the aroma, methyl benzoate (58.0%) and linalool (18.8%). A small quantity of benzyl benzoate has been determined [27]. Another study analyzed the aroma of the same orchid in different places in Pays-Bas and Sweden and Switzerland. In the site of Sandby (Pays-Bas), linalool (75%), lilac derivatives (44%), geraniol (35%), and methyl benzoate (32%) were found to be the main components of the aroma [28]. In Lenstad (Pays-Bas), they found linalool (77%) and geraniol (47%); in the site of Horn (Pays-Bas), lilac compounds (75%), linalool (63%), and methyl benzoate (47%) were found in the scent [28]. In Sweden, they found lilac compounds, linalool, and methyl benzoate [28]. Finally, in Swiss plants, the main components of the aroma were methyl benzoate and linaool [28]. Another study performed in Sweden showed the presence in the aroma of methyl benzoate, benzyl acetate, benzyl benzoate, benzyl salicylate, and lilac aldehydes

Fig. 1 (**a**) *Platanthera bifolia* subsp. *osca*; (**b**) *Platanthera chlorantha*

[29]. Finally, in a work performed in Belgium, 3,7-dimethyl-1,3,6-octatriene, and santolinatriene were detected in the aroma [30].

The authors of these studies, by using different chromatographic techniques, found different results for plants in different sites, showing an adaptive behavior. The same behavior was observed in *P. chlorantha*. In Switzerland, linalool, methyl benzoate, and geraniol were the components of the aroma [27, 28]. In Pays-Bas, lilac aldehydes and linalool were found [28]; in Sweden, methyl benzoate and linalool [28]; and in Belgium, the main components were 3,7-dimethyl-1,3,6-octatriene, 3-carene, and lilac aldehyde [30].

Considering the observed high variability of the results, we performed a study on both *P. bifolia* subsp. *osca* and P. chlorantha collected in different sites [31, 32]. The results of analyses are reported in Table 1. The structures of the compounds involved are collected in Fig. 2.

There is not a significant variation in the composition of the aroma between *P. bifolia* subsp. *osca* reported in Basilicata and P. chlorantha deriving from the same region. We observed a significant variation between the composition of the aroma (mainly formed by lilac compounds and linalool) in *P. bifolia* subsp. *osca* deriving from sites in Basilicata and those deriving from Calabria and Abruzzo where benzyl benzoate is the main component of the scent.

We can justify the observed behavior only admitting an adaptive modification of the scent as reported in the previous work on these taxa in relation with possible differences in the nature of pollinating agents. Lilac compounds are known to be attractive compounds [29]. However, benzyl benzoate is known to be an attractive compound present in some other Orchidaceae floral volatiles [33, 34].

3 *Cephalanthera* Orchids

Eight species of *Cephalanthera* orchids have been observed in Europe. Only three species have been found in Basilicata. They are *Cephalanthera longifolia* (L.), *Cephalanthera rubra* (L.), and *Cephalanthera damasonium* (Mill.) (Fig. 3). There are some differences between these three species: in particular, while *Cephalanthera damasonium* is autogamous, *Cephalanthera logifolia* and *Cephalanthera rubra* are allogamous [35]. The analysis of the scent of these species has been performed, and the results are reported in Table 2 and Fig. 4 [36].

Except for some terpenes and terpenoid compounds, *cis*-β-farnesene (found only in *C. longifolia*), α-farnesene (found in *C. rubra* and *C. damasonium*), and pristane (found in *C. damasonium*), the main components of the scent for all the species are hydrocarbons (tetra-, penta-, hexa-, and heptadecane). It is interesting to note that, considering that *C. damasonium* is a self-pollinating species, while the other two are allogamous, there is no significant difference in the scent. Furthermore, it is known that *C. rubra* mimics bellflower (*Campanula*) colors in order to receive pollination service [37, 38].

At this purpose, it is noteworthy that the components of the scent of *C. rubra* are completely different from those of *Campanula* species [39]. The mimic effect is then

Table 1 Volatile organic compounds from *Platanthera*. Only the compounds with area % higher than 3% are reported

		Platanthera bifolia subsp. *osca*									
	Compound	Rifreddo (Basilicata)	Pignola (Basilicata)				Marsico Nuovo (Basilicata)			Grisolia (Calabria)	Palena (Abruzzo)
			Sample 1	Sample 2	Sample 3	Sample 4	Sample 1	Sample 2	Sample 3		
		Area %									
1	Linalool	8.83							10.00	17.77	
2	Methyl benzoate										9.04
3	Lilac aldehyde A	6.16									
4	Lilac aldehyde B	11.80	8.61		3.01	7.80	8.64	8.02			
5	Lilac aldehyde C	3.11					1.39				
6	Lilac alcohol A		10.73								
7	Lilac alcohol B	11.48	7.41	9.90	13.08	13.52	15.10	15.86	12.99		
8	Lilac alcohol C	11.39	10.10	14.96	10.13	18.45	9.08	9.75	13.99		
9	Lilac alcohol D	11.43	7.50	21.02	32.45	34.28	32.80	25.07	34.53		
10	Geraniol										4.05
11	Benzyl benzoate									55.83	74.72
12	Galaxolide									6.66	
13	Benzyl 2-hydroxybenzoate										3.19
		Platanthera chloranta									
		Satriano (Basilicata)									
		Area %									
3	Lilac aldehyde A	15.13									
4	Lilac aldehyde B	5.39									
9	Lilac alcohol D	25.78									

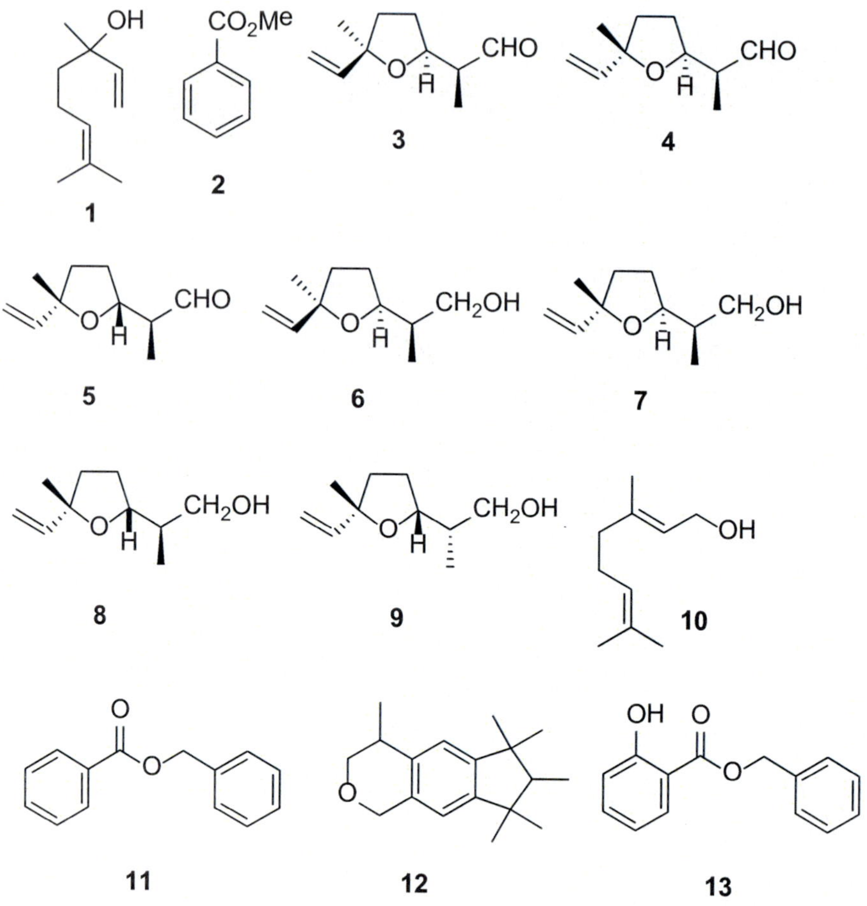

Fig. 2 Volatile compounds found in *P. bifolia* subsp. *osca* and *P. chlorantha*

due to a similar color and not to the presence of the similar attractive compounds in the aroma.

The absence of differences between different species with different pollinating mechanisms, and the absence of correlation between the scent of *C. rubra* and *Campanula* species, induces us to consider that, probably, scent has not an important role to attract insects.

4 *Serapias* Orchids

Five species of *Serapias* orchids can be observed in Basilicata. We can find allogamous diploid *Serapias cordigera* L. subsp. *cordigera*, *Serapias cordigera* subsp. *lucana* [40], *Serapias vomeracea* subsp. *longipetala* (Ten.), autogamous diploid

Fig. 3 (**a**) *Cephalanthera longifolia* (L.); (**b**) *Cephalanthera rubra* (L.); and (**c**) *Cephalanthera damasonium* (Mill)

Table 2 Volatile organic compounds of *Cephalanthera* orchids from Basilicata. Only the compounds with area % higher than 3% are reported

	Compound	*C. longifolia*	*C. rubra*	*C. damasonium*
		Area %		
14	Tetradecane			3.49
15	*cis*-β-Farnesene	5.04		
16	Pentadecane	48.94	5.52	7.41
17	α-Farnesene		3.60	4.35
18	Hexadecane		6.00	7.01
19	Ethyl dodecanoate	5.18		
20	Heptadecane	3.83	5.15	5.92
21	Pristane			3.04
22	Hexadecanal	3.08		

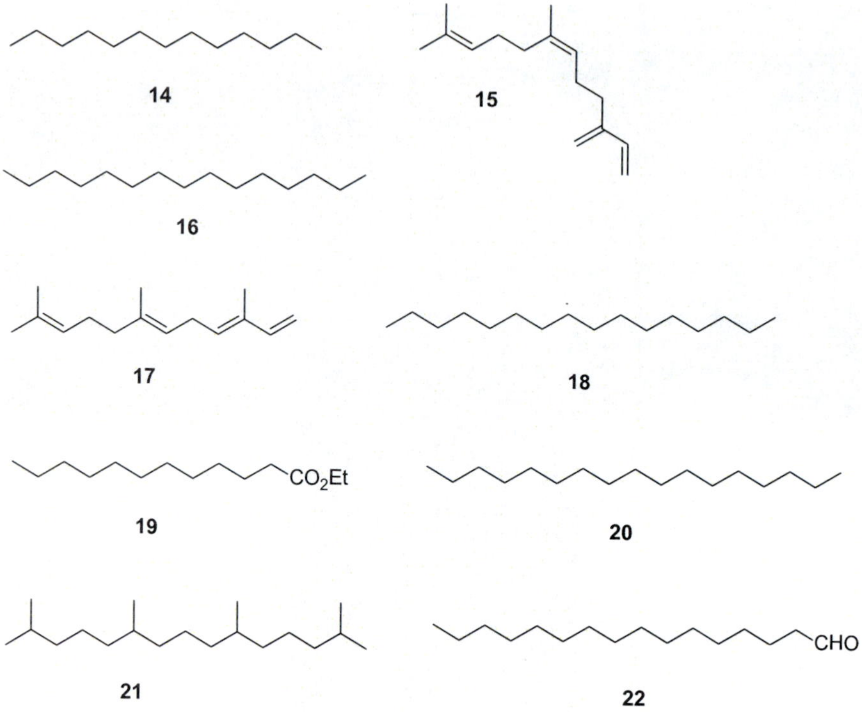

Fig. 4 Volatile organic compounds found in *Cephalanthera* orchids

Serapias parviflora Parl., and sexual deceptive tetraploid *Serapias lingua* L. (Fig. 5). We analyzed the scent of all the species recovered in Basilicata.

Previous studies performed in Calabria and in Campania on this species have been performed on the hexane extracts of the labellum [41, 42]. They found longchain (with more than 21 carbon atoms) alkanes and alkenes in different ratio. *S. lingua* showed 45% alkanes and 55% alkenes; in *S. cordigera,* they found 32% alkanes and 68% alkenes; S. parviflora had 93% alkanes and 7% alkenes. In another study, different ratios have been found [43]. In this case, in *S. lingua,* 40/60 (alkenes/alkanes) was found, while *S. cordigera* showed a 30/70 ratio. The results of our analyses are reported in Table 3 and Fig. 6 [44].

In *S. cordigera* subsp. *Cordigera,* the analysis did not find alkanes and alkenes with high molecular weights as in the previous works on this species; on the contrary, the main component was α-amorphene. The same main component was found in *S. cordigera* subsp. *lucana*. In *S. vomeracea,* the main components were an alkane (pentadecane) and some alkenes; however, in previous work [41], heptacosane and tetracosane were the main components. We found only alkanes and alkenes with lower molecular weight. The same behavior has been observed with the other species.

The different molecular weight of the compounds we determined in comparison with those found in the previous works can depend on intrinsic limits of SPME

Fig. 5 (**a**) *Serapias cordigera* L. 1753; (**b**) *Serapias cordigera* subsp. *lucana* R. Lorenz and V. A. Romano 2014; (**c**) *Serapias vomeracea* subsp. *longipetala* (Ten.) H. Baumann and Künkele 1989; (**d**) *Serapias lingua* L. 1753; and (**e**) *Serapias parviflora* Parl. 1837

determination. On the other hand, the compounds whose presence has been determined in the previous work in this field [41–43] are compounds with a high boiling point and very low vapor pressure, and probably, they cannot represent the actual components of the aroma, but only the precursors.

5 *Barlia robertiana*

Barlia robertiana (Loisel.) Greuter (Orchidaceae) (Fig. 7) is among the numerous orchids which present nonrewarding flowers, and it is obligatorily insect pollinated. It has a steno-mediterranean distribution [45] and can be found in open habitats (e.g., clearings in the scrublands and termophilous woods, dry grasslands, and roads edges), usually on silty-sandy soils. In Italy, it is present in all the regions except for some of the Alpine sector [46]. In the Basilicata region, (Southern Italy) it is rather widespread from the coast to the inner mountainous area of Apennines.

Unlike most of the mediterranean orchids, which are pollinated by specialized bees occupying narrow ecological niches, *B. robertiana* attracts many different

Table 3 Volatile organic compounds of *Serapias* orchids from Basilicata. Only the compounds with area % higher than 3% are reported

	Compound	*S. cordigera*	*S. lucana*	*S. vomeracea*	*S. lingua*	*S. parviflora*
23	2-Undecanone			3.31		
24	Pentadecene					7.05
16	Pentadecane		11.13	25.43	7.41	11.03
25	α-Amorphene	28.85	38.06			
18	Hexadecane					3.96
26	3-Heptadecene		3.02			
20	Heptadecane	3.96	4.55	13.89	11.70	6.38
11	Benzyl benzoate			4.18	3.80	
27	Octadecane			3.24	4.49	
28	Isopropyl myristate				7.80	
29	Farnesylacetaldehyde					3.15
30	5-Nonadecene			14.48		
31	Nonadecene			5.16		
32	Nonadecane				6.38	4.28
33	2-Heptadecanone	3.42				
34	Isopropyl palmitate				4.86	
35	Propyl undecanoate		5.15			
36	Heneicosane	4.14				4.31
37	Ethyl elaidate	3.39	3.47			
38	Ethyl oleate	15.33				
39	Ethyl stearate		3.33			

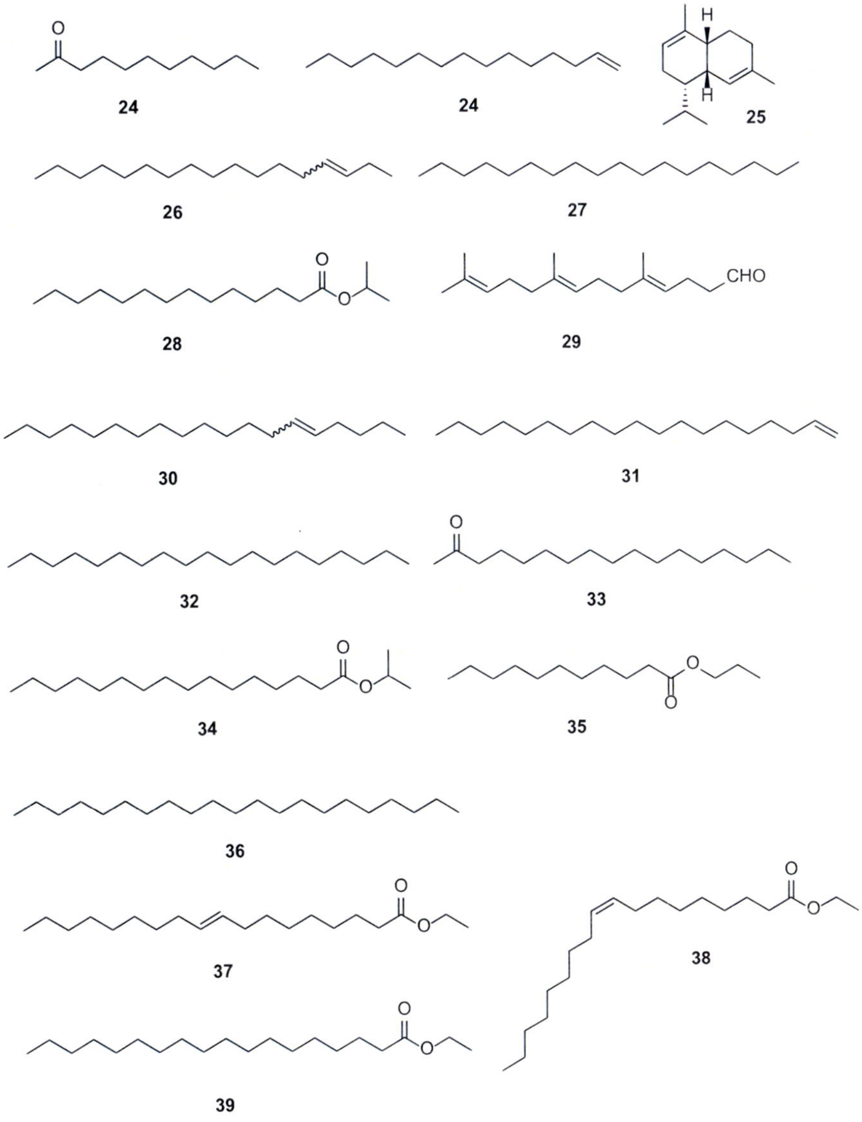

Fig. 6 Volatile organic compounds found in *Serapias* orchids

species of pollinators [47] such as Apoidea (Hymenoptera) and Cetoniidae (Coleoptera) [48, 49]. In this context, it is worth to note as an appropriate knowledge about pollination and reproduction biology is not only important for scientific purpose, but is also crucial to develop effective management and conservation actions of rare and endangered plants [26] such as terrestrial orchids.

Overall, within the 11 samples, 68 compounds were identified (min 14; max 28; and mean 19.2) with a high between samples variability both for composition and

Fig. 7 *Barlia robertiana*

abundance of VOCs (Figs. 8 and 9, Table 4). The most frequent compounds (i.e., found in >60% of plants) were hexadecane, which was present in 9 samples, β-bisabolene, β-sesquiphellandrene and ethyl dodecanoate found in 8 samples, and caryophyllene, cis-α-bergamotene, δ-selinene, and octadecane which were detected in 7 samples. As concern dominant compounds, each sample returned a different result: We alternatively found as dominant in the samples: α-zingiberene, *i*-propyl 14-methyl-pentadecanoate, farnesol, *p*-Menth-8-en-1-ol, and pristane. An exception was represented by β-sesquiphellandrene and verbenone which were found both prevailing in 2 samples. β-Sesquiphellandrene was dominant in the samples Pisticci 2 and Vietri 2, whereas verbenone was dominant in Pisticci 1 and Sant'Arcangelo 2. Remarkably, these 2 pairs of samples group together localities resulting rather distant, geographically and ecologically.

The results of VOCs compositions for each sample are reported in Table 4. Two samples from the lower course of Basento river, the closest to the Ionian coast (Pisticci municipality, 10 m a.s.l.), were found and analyzed (Table 4). The first one showed the presence of verbenone (45.22%), a monoterpene, α-zingiberene (6.88%) (Fig. 9), a sesquiterpene, and ethyl tetradecanoate (15.96%). Verbenone shows a camphor mentholic flavor and acts as a pheromone. Also α-zingiberene is a pheromone [50]. The second sample showed a completely different composition of the aroma (Table 4). The main products were caryophyllene (8.51%) (Fig. 9),

Fig. 8 *Barlia robertianai* during the determination

β-sesquiphellandrene (29.06%), and farnesal (5.81%). All the compounds are sesquiterpenes. β-Sesquiphellandrene has a herbal fruity flavor. It is interesting to note that α-zingiberene and β-sesquiphellandrene are compounds with a very similar structure (Fig. 9), both deriving from a possible cyclization of *E,Z*-farnesyl cation (Fig. 10). On the contrary, caryophyllene derives from *E,E*-farnesyl cation (Fig. 10).

Two samples were collected at Sant'Arcangelo (218 m amsl) (Table 4). The first one showed the presence of α-zingiberene (17.14%), δ-selinene (9.06%), *i*-propyl 14-methylpentadecanoate (3.74%), β-curcumene (Fig. 9) (2.43%), nonadecane (2.06%), longipinene (1.77%), ethyl dodecanoate (1.75%), and nerolidol (Fig. 9) (1.69%). The second sample has as component of the scent verbenone (31.48%), \δ-selinene (17.59%), pristane (9.61%), longipinene (4.49%), limonene (Fig. 9) (1.65%), decanal (2.12%), and β-curcumene (1.56%). Longipinene and δ-selinene are both sesquiterpenes deriving from *E,E*-farnesyl cation (Fig. 10). Pristane is present in shark liver, fish oil, ambergris, plankton, and anise fruits [51].

The samples collected at Tolve (307 m) showed the presence of the following compounds: sample 2, caryophyllene (10.11%), ethyl dodecanoate (8.29%), *i*-propyl

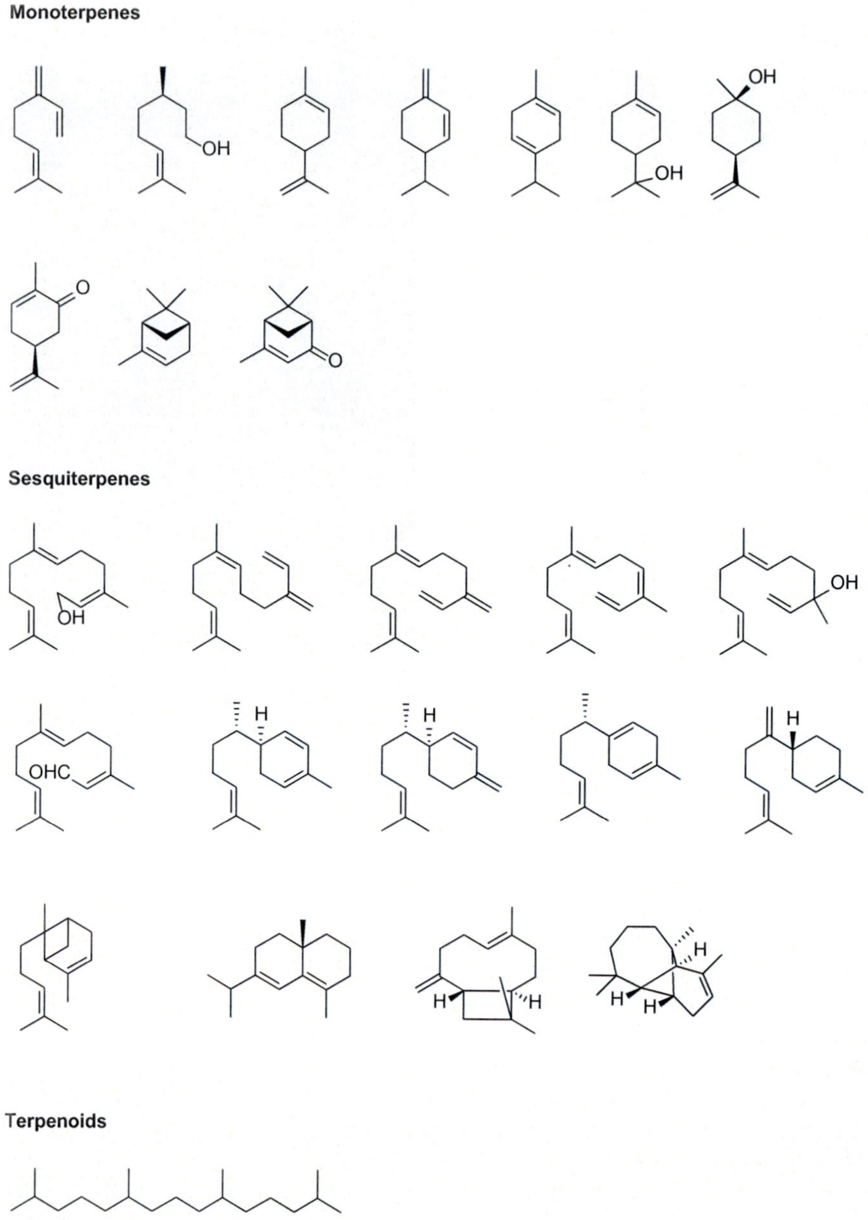

Fig. 9 Volatile tepenes and terpenoids found in *Barlia robertiana* scent

14-methyl-pentadecanoate (12.26%), tetradecane (3.59%), *Z*-β-farnesene (1.49%) (Fig. 9), β-sesquiphellandrene (2.94%), hexadecane (3.09%), and ethyl tetradecanoate (2.57%); sample 3, β-sesquiphellandrene (8.39%), farnesol (36.63%)

Table 4 Volatile organic compounds of *Barlia robertiana* from different sampled sites in Basilicata region (Italy). Only the compounds with area % higher than 3% are reported

Compound	Pisticci 1	Pisticci 2	S. Arcan.1	S. Arcan. 2	Tolve 2	Tolve 3	Pomarico	Vietri 1	Vietri 2	Potenza	Savoia
	Area %										
α-Pinene										7.87	
β-Myrcene							3.26				
β-Phellandrene							4.51				
D-limonene							3.21				
γ-Terpinene							5.24				
p-Menth-8-en-1-ol							21.68				
α-Terpineol							5.39				
Verbenone	45.22			31.48							
Citronellol									17.96		
Methyl citronellate								3.01	3.61		
Tetradecane								3.09			
α-Zingiberene	6.88		17.14								
Caryophyllene		8.51			10.11		3.68	3.07	5.21	6.75	17.96
Cis-α-bergamotene								14.04	4.15		
E-β-farnesene											7.10
Longipinene				4.40							

(continued)

Table 4 (continued)

Compound	Pisticci 1	Pisticci 2	S. Arcan.1	S. Arcan. 2	Tolve 2	Tolve 3	Pomarico	Vietri 1	Vietri 2	Potenza	Savoia
	Area %										
β-Bisabolene							4.61	8.98	3.17		5.32
δ-selinene			9.06	17.59							
β-Sesquiphellandrene		29.06				8.39	17.59	43.08	15.14	8.97	25.05
Ethyl dodecanoate					8.29			3.76			2.60
Hexadecane					3.09			4.01			3.36
Pristane				9.61						22.57	
Farnesol						36.63				19.36	
Farnesal		5.81									
Ethyl tetradecanoate	15.96										
Dihydrofarnesol						22.23					
i-Propyl 14-methyl-pentadecanoate			3.74		12.26						

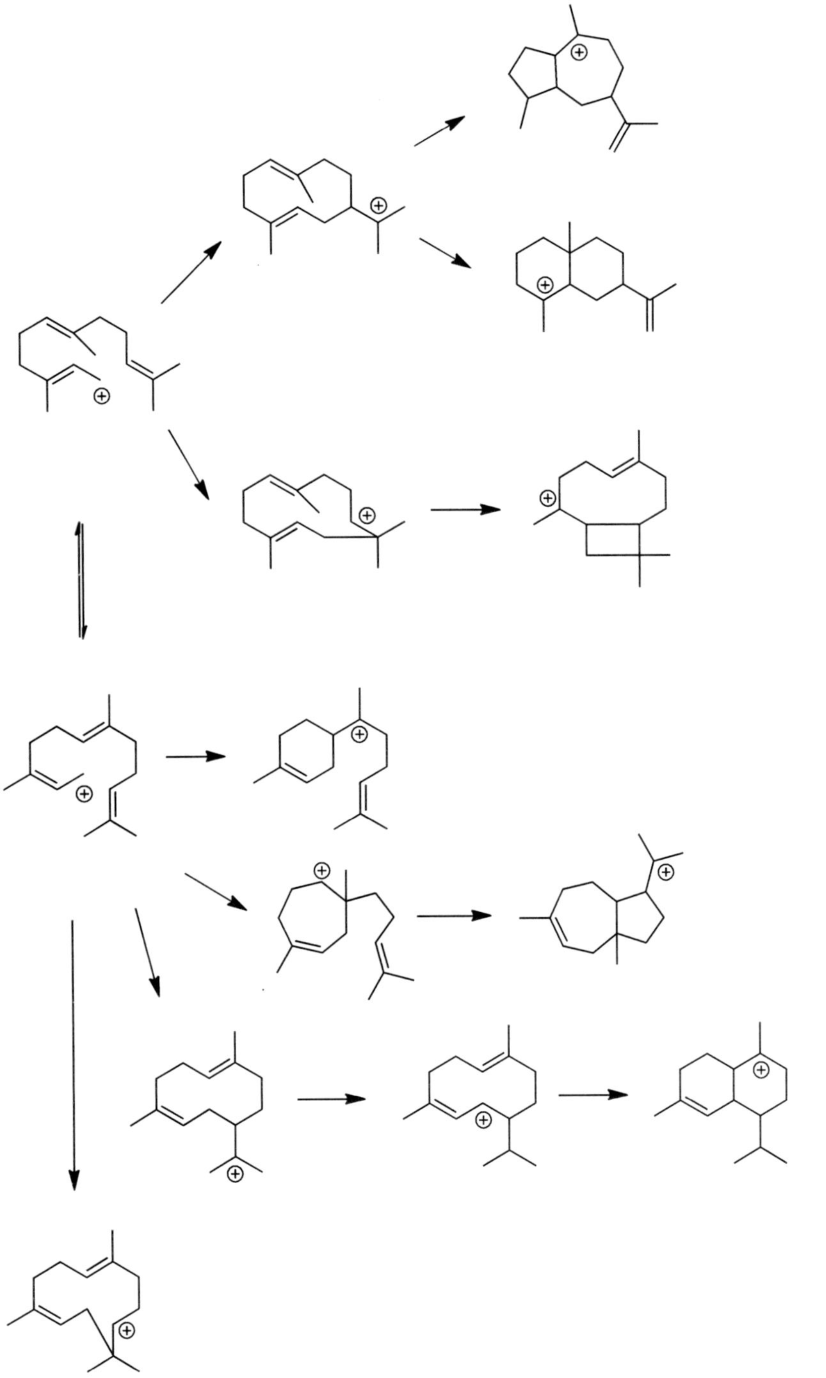

Fig. 10 Biosynthesis of some sequiterpenes

(Fig. 9), dihydrofarnesol (22.23%), *E*-β-farnesene (1.97%), and β-bisabolene (2.13%).

In the sample collected at Pomarico (348 m amsl), several monotepenes were found: β-myrcene (3.26%) (Fig. 2), β-phellandrene (4.51%), limonene (3.21%), γ-terpinene (5.24%), *p*-menth-8-en-1-ol (21.68%), and α-terpineol (5.39%). Furthermore, some sesquiterpenes were found: caryophyllene (3.68%), β-bisabolene (4.61%), and β-sesquiphellandrene (17.59%).

The site of Vietri (410 m) gave two samples. They contain almost the same compounds in different amounts: citronellol (1.77% in sample 1 and 17.96% in sample 2), methyl citronellate (3.01% and 3.61%), tridecane (2.35% and 1.29%), tetradecane (3.09% and 1.39%), caryophillene (3.07% and 5.21%), *cis*-α-bergamotene (Fig. 9) (14.04% and 4.15%), β-bisabolene (8.98% and 3.17%), δ-selinene (2.49% and 1.16%), β-sesquiphellandrene (43.08% and 15.14%), ethyl dodecanoate (3.76% and 2.63%), hexadecane (4.01% and 1.80%), and heptadecane (2.24% and 1.49%).

A sample collected at Potenza (727 m msl) showed the following compounds: α-pinene (7.87%), caryophillene (6.75%), β-bisabolene (2.00%), β-sesquiphellandrene (8.97%), pristane (22.57%), and farnesol (19.36%).

Finally, the sample collected at Savoia di Lucania (760 m) showed the presence of caryophillene (17.96%), *E*-β-farnesene (7.10%), β-bisabolene (5.32%), and β-sesquiphellandrene (25.05%).

Considering only the main components, we observed this behavior: In Pisticci, the main components were verbenone and β-sesquiphellandrene. In Sant'Arcangelo, the main components were α-zingiberene and verbenone. Tolve showed the presence of caryophyllene, *i*-propyl 14-methyl-pentadecanoate, and farnesol. The site of Pomarico showed as main components *p*-menth-8-en-1-ol and sequiphellandrene. The main components found at Vietri were β-sesquiphellandrene, and citronellol. The analyses of the sample at Potenza showed as the main component pristane. The main component of the sample collected at Savoia di Lucania was β-sesquiphellandrene.

The high variability of *Barlia robertiana* floral scent detected in this study involved both qualitative and quantitative aspects of VOCs composition. Compositions of our samples were not in agreement with the volatile organic compounds found in the scent of *Barlia robertiana* (*sub Himantoglossum robertianum* (Loisel.) P. Delforge) in a previous work [52]. For example, concerning the main components, Gallego and coworkers found as prevailing compounds: α-pinene, β-pinene, and limonene, whereas our samples were characterized by β-sesquiphellandrene, verbenone, zingiberene, *i*-propyl 14-methyl-pentadecanoate, farnesol, *p*-menth-8-en-1-ol, and pristane. However, α-pinene, β-pinene, and limonene were often detected in our samples too, but with lower abundance. The high variability of floral scent detected between sampled sites from Basilicata does not seem at all related in any way to environmental factors or to geographic locations. Therefore, these findings suggest that there is no adaptation by the species to local conditions or specific communities of pollinators. A high variation in floral signals such as color and floral scent has been highlighted in several deceptive orchids [53]. Even if some compounds detected in our study can act as pheromone (e.g., verbenone and α-zingiberene), this specific function probably is not realized in *Barlia robertiana*. The characteristic morphology and coloring of the flowers of *Barlia robertiana* seem to exclude attraction mechanisms widely used by other deceptive orchids such as shelter imitation, brood-site imitation, psudoantagonism, or sexual attraction. In fact, *B. robertiana* strategy to attract pollinators involves together scent, appearance of the flowers, and showiness of the plant during the flowering phase. This pollination system of rewardless species that does not imitate specific nectariferous plants is defined as "generalized food deception" [54]. In this framework, it has been suggested as in such species a huge variation in floral characters both within and between populations may behave the effect to disrupt the associative learning of visiting insects [55]. In fact, rewardlessness can be considered a dangerous strategy because some insects can soon learn to avoid such flowers [56], having as a consequence reduced fruit set via reduced pollinium removal and deposition. The causes of the evolution of rewardlessness have been frequently discussed but are still incompletely known [57, 58] because few experiments specifically addressed this question. A specific study involving *Barlia robertiana* [49] demonstrated probably for the first time an evolutionary advantage for rewardlessness in the Orchidaceae. In particular, in this study, the experiments comparing flowers artificially supplied with sucrose solution showed an increasing probability (approximately ten times) of pollinia removing by pollinators for flowers without nectar (i.e., the natural condition), probably leading as a consequence to an higher seed paternity. The striking results are due to the fact that bees visiting a rewardless flower were more vigorous to seek a potential reward inside the corolla; therefore, pollinia were more likely detached.

However, it must be stressed again that this (male) reproductive advantage can only occur if pollinating insects do not learn to associate the floral signals of a species with the lack of nectar inside the flowers.

6 Conclusion

HS-SPME-GC-MS analysis of the scent of some orchid species found in Basilicata (Southern Italy) has been performed. The analysis of *Platanthera* species showed the presence of two different behaviors. The samples from Basilicata showed the presence of lilac derivatives, while samples from Calabria or Abruzzo had as main component benzyl benzoate. Probably, it is the result of an adaptive behavior. The analysis of *Cephalanthera* orchids showed that, probably, scent does not have a relevant role on the pollination strategy of the plants. The scent has the same composition both for allogamous and autogamous species. Alkanes are the main components of the scent. The analysis of *Serapias* orchids showed the presence of alkanes and alkenes but with a lower molecular weight than those reported in previous works on these species. Furthermore, α-amorphene was found as a main component in *Serapias cordigera* and *Serapias cordigera* subsp. *lucana*. The analysis of *Barlia robertiana* showed a high variability in the composition of the scent.

At this time, it is not possible to understand the reason for this high variety of compounds in the scent. This situation leads us to think that we do not yet perfectly know the real reasons that lead to an effective composition of the aroma. The study of this phenomenology will have to be continued.

References

1. Pawliszyn J (1997) Solid-phase microextraction: theory and practice. Wiley, New York
2. Bentivenga G, D'Auria M, De Luca E, De Bona A, Mauriello G (2001) The use of SPME-GC-MS in the analysis of flavor of virgin olive oil. Riv Ital Sostanze Gr 78:157–162
3. Bentivenga G, D'Auria M, De Bona A, Mauriello G (2002) On the flavor of virgin olive oil. Riv Ital Sostanze Gr 79:101–105
4. Bentivenga G, D'Auria M, Mauriello G, Racioppi R, Viggiani L (2004) Characterization of virgin olive oils from Basilicata by using ^{1}H-NMR and SPME-GC-MS. Riv Ital Sostanze Gr 81:86–89
5. D'Auria M, Emanuele L, Mauriello G, Racioppi R (2003) On the origin of "Goût de Lumiere" in champagne. J Photochem Photobiol A Chem 158:21–26
6. Mauriello G, Capece A, D'Auria M, Garde-Cerdán T, Romano P (2009) SPME-GC method as a tool to differentiate VOC profiles in *Saccharomyces cerevisiae* wine yeasts. Food Microbiol 26:246–252
7. D'Auria M, Emanuele L, Racioppi R (2009) The effect of heat and light on the composition of some volatile compounds in wine. Food Chem 117:9–14
8. Cellamare L, D'Auria M, Emanuele L, Racioppi R (2009) The effect of light on the composition of some volatile compounds in wine: an HS-SPME-GC-MS study. Int J Food Sci Technol 44:2377–2384
9. Acquaviva V, D'Auria M, Racioppi R (2014) Changes in aliphatic ester composition in white wines during exposition to light. An HS-SPME-GC-MS study. J Wine Res 25:63–74
10. D'Auria M, Mauriello G, Rana GL (2004) Volatile organic compounds from saffron. Flav Fragr J 19:17–23
11. D'Auria M, Mauriello G, Racioppi R, Rana GL (2006) Use of SPME-GC-MS in the study of time evolution of the constituents of saffron aroma: modifications of the composition during storage. J Chromatogr Sci 44:18–21
12. Mauriello G, Marino R, D'Auria M, Cerone G, Rana GL (2004) Determination of volatile organic compounds from truffles via SPME-GC-MS. J Chromatogr Sci 42:299–305
13. D'Auria M, Rana GL, Racioppi R, Laurita A (2012) Studies on volatile organic compounds of *Tuber borchii* and *T. asa-foetida*. J Chromatogr Sci 50:775–778
14. D'Auria M, Racioppi R, Rana GL (2013) Volatile organic compounds of *Schenella pityophilus*. Nat Prod Res 27:41–44
15. D'Auria M, Racioppi R, Rana GL, Laurita A (2014) Studies on volatile organic compounds of some truffles and false truffles. Nat Prod Res 28:1709–1717
16. Mang SM, Racioppi R, Camele I, Rana GL, D'Auria M (2015) Use of volatile metabolite profiles to distinguish three *Monilinia* species. J Plant Pathol 97:55–59
17. Mang SM, Zotta T, Camele I, Racioppi R, D'Auria M, Rana GL (2017) Morphological, physico-chemical and molecular investigations on *Tuber bellonae* from Basilicata – Italy. J Anim Plant Sci 27:528–541
18. Bentivenga G, D'Auria M, Fedeli P, Mauriello G, Racioppi R (2004) SPME-GC-MS analysis of volatile organic compounds in honey from Basilicata. Evidence for the presence of pollutants from anthropogenic activities. Int J Food Sci Technol 39:1079–1086
19. D'Auria M, Mauriello G, Racioppi R (2004) SPME-GC-MS analysis of horseradish (*Armoracia rusticana*). Ital J Food Sci 16:501–504

20. D'Auria M, Mauriello G, Marino R, Racioppi R (2005) Composition of volatile fraction from *Thymus*, *Origanum*, *Lavandula* and *Acinos* species. J Essent Oil Bear Plants 8:36–51
21. D'Auria M, Racioppi R (2015) The effect of drying on the composition of volatile organic compounds in *Rosmarinus officinalis*, *Laurus nobilis*, *Salvia officinalis* and *Thymus serpyllum*. A HS-SPME-GC-MS study. J Essent Oil Bear Plants 18:1209–1223
22. Celano G, D'Auria M, Xiloyannis C, Mauriello G, Baldassarre M (2006) Composition and seasonal variation of soluble cuticolar waxes in *Actinidia deliciosa* leaves. Nat Prod Res 20:701–709
23. D'Auria M, Racioppi R, Velluzzi V (2008) A comparison of results obtained using liquid injection and headspace solid-phase microextraction for crude oil analysis by GC with mass spectrometer detection. J Chromatogr Sci 46:332–338
24. D'Auria M, Racioppi R (2017) HS-SPME-GC-MS analysis of onion (*Allium cepa* L.) and shallot (*Allium ascalonicum* L.). Food Res 1:161–165
25. D'Auria M, Emanuele L, Racioppi R, Stefanizzi N (2020) Determination of simple carbohydrates in beer using SPME, on fiber derivatization and gas chromatography mass spectrometry. Food Res 4:1156–1161
26. Rosati L, Romano VA, Cerone L, Fascetti S, Potenza G, Bazzato E, Cillo D, Mecca M, Racioppi R, D'Auria M, Farris E (2020) Pollination features and floral volatiles of *Gymnospermium scipetarum* (Barberidaceae). J Plant Res 132:49–56
27. Kaiser R (1993) Von Duft der Orchideen. Editiones Roche, Basel
28. Tollsten L, Bergström LG (1993) Fragrance chemotypes of *Platanthera* (Orchidaceae) – the result of adaptation to pollinating moths? Nord J Bot 13:607–613
29. Peplys D, Ibarra F, Francke W, Löfstedt C (2002) Odour-mediated nectar foraging in the silver Y moth, *Autographa gamma* (Lepidoptera: Noctuidae): behavioural and electrophysiological responses to floral volatiles. Oikos 99:75–82
30. Esposito F, Vereecken NJ, Gammella M, Rinaldi R, Laurent P, Tyteca D (2018) Characterization of sympatric *Platanthera bifolia* and *Platanthera chlorantha* (Orchidaceae) populations with intermediate plants. PeerJ 6:e4256
31. D'Auria M, Lorenz R, Racioppi R, Romano VA (2017) Fragrance components of *Platanthera bifolia* subsp. *osca*. Nat Prod Res 31:1612–1619
32. D'Auria M, Lorenz R, Mecca M, Racioppi R, Romano VA, Viggiani L (2020) Fragrance components of *Platanthera bifolia* subsp. *osca* and *Platanthera chlorantha* collected in several sites in Italy. Nat Prod Res. https://doi.org/10.1080/14786419.2019.1593166
33. Knudsen JT, Tollsten L, Bergström LG (1993) Floral scent – a checklist of volatile compounds isolated by head-space techniques. Phytochemistry 33:253–280
34. Williams NH, Written WM (1983) Orchid floral fragrance and male euglossine bees: methods and advances in the fast sesquidecade. Biol Bull 163:355–395
35. Claessens J, Kleynen J (2013) The pollination of European orchids: part 2: *Cypripedium* and *Cephalanthera*. J Hardy Orch Soc 10:114–120
36. D'Auria M, Lorenz R, Mecca M, Racioppi R, Romano VA (2020) Aroma components of *Cephalanthera* orchids. Nat Prod Res. https://doi.org/10.1080/14786419.2019.1616724
37. Nilsson LA (1983) Mimesis of bellflower (*Campanula*) by the red helleborine orchid *Cephalathera rubra*. Nature 305:799–800
38. Claessens J, Beentjes KK, Heijrman T, Miller J, Gravandeel B (2015) Beobachtungen von *Miarus campanulae* als Bestäuber von *Cephalanthera rubra*. J Eur Orch 47:77–87
39. Milet-Pinheiro P, Ayasse M, Dötterl S (2015) Visual and olfactory cues of *Campanula* (Campanulaceae) and their significance for host recognition by an oligolectic bee pollinator. PLoS One 10:e0128577
40. Lorenz R, Romano VA (2014) Beiträge über die morphologische Differenzierung in der Gattung *Serapias*: 1. *Serapias cordigera* in Süditalien. J Eur Orch 46:616–660
41. Pellegrino G, Luca A, Bellusci F, Musacchio A (2012) Comparative analysis of floral scents in four sympatric species of *Serapias* L. (Orchidaceae): clues on their pollination strategies. Plant Syst Evol 298:1837–1843

42. Schiestl FP, Cozzolino S (2008) Evolution of sexual mimicry in the orchid subtribe orchidinae: the role of preadaptations in the attraction of male bees as pollinators. BMC Evol Biol 8:27
43. Vereecken NJ, Wilson CA, Hötling S, Schulz S, Banketov SA, Mardulyn P (2012) Pre-adaptations and the evolution of pollination by sexual deception: Cope's rule of specialization revisited. Proc R Soc B 279:4786–4794
44. D'Auria M, Lorenz R, Mecca M, Racioppi R, Romano VA (2020) The composition of the aroma of *Serapias* orchids in Basilicata (southern Italy). Nat Prod Res. https://doi.org/10.1080/14786419.2020.1713127
45. Delforge P (2016) Orchidées d'Europe, d'Afrique du Nord et du Proche-Orient: la bible des orchidophiles, plus de 600 espèces et de nombreuses variétés et illustrées, 4th edn, revue et augmentée. Delachaux et Niestlé, Paris
46. Bartolucci F, Peruzzi L, Galasso G, Albano A, Alessandrini A, Ardenghi NMG, Astuti G, Bacchetta G, Ballelli S, Banfi E, Barberis G, Bernardo L, Bouvet D, Bovio M, Cecchi L, Di Pietro R, Domina G, Fascetti S, Fenu G, Festi F, Foggi B, Gallo L, Gottschlich G, Gubellini L, Iamonico D, Iberite M, Jiménez-Mejías P, Lattanzi E, Marchetti D, Martinetto E, Masin RR, Medagli P, Passalacqua NG, Peccenini S, Pennesi R, Pierini B, Poldini L, Prosser F, Raimondo FM, Roma-Marzio F, Rosati L, Santangelo A, Scoppola A, Scortegagna S, Selvaggi A, Selvi F, Soldano A, Stinca A, Wagensommer RP, Wilhalm T, Conti F (2018) An updated checklist of the vascular flora native to Italy. Plant Biosyst 152:179–303
47. Vereecken NJ, Dafni A, Cozzolino S (2010) Pollination syndromes in Mediterranean orchids – implications for speciation, taxonomy and conservation. Bot Rev 76:220–240
48. Doneddu M (2015) Osservazioni sull'impollinazione di *Barlia robertiana* ad opera dei coleotteri *Tropinota squalida* e *Oxythyrea funesta* (Cetoniidae) in Sardegna. GIROS Orch Spont Eur 58:262–265
49. Smithson A, Gigord LDB (2001) Are there fitness advantages in being a rewardless orchid? Reward supplementation experiments with *Barlia robertiana*. Proc R Soc Lond B Biol Sci 268:1435–1441
50. McBrien HL, Millar JG, Rice RE, McElfresh JS, Cullen E, Zalom FG (2002) Sex attractant pheromone of the red-shouldered stink bug *Thyanta pallidovirens*: a pheromone blend with multiple redundant components. J Chem Ecol 28:1797–1818
51. Tulp M, Bohlin L (2005) Rediscovery of known natural compounds: nuisance or goldmine? Bioorg Med Chem 13:5274–5282
52. Gallego E, Gelabert A, Roca FJ, Perales JF, Guardino X (2012) Identification of volatile organic compounds (VOC) emitted from three European orchid species with different pollination strategies: two deceptive orchids (*Himantoglossum robertianum* and *Ophrys apifera*) and a rewarding (*Gymnadenia conopsea*). J Biodivers Environ Sci 2:18–29
53. Salzmann CC, Cozzolino S, Schiestl FP (2007) Floral scent in food-deceptive orchids: species specificity and sources of variability. Plant Biol 9:720–729
54. Jersáková J, Johnson SD, Kindlmann P (2006) Mechanisms and evolution of deceptive pollination in orchids. Biol Rev 81:219–235
55. Moya S, Ackerman JD (1993) Variation in the floral fragrance of *Epidendrum ciliare* (Orchidaceae). Nord J Bot 13:41–47
56. Smithson A, Macnair MR (1997) Negative frequency-dependent selection by pollinators on artificial flowers without rewards. Evolution 51:715–723
57. Ackerman JD (1986) Mechanisms and evolution of food-deceptive pollination systems in orchids. Lindleyana 1:108–113
58. Johnson SD, Nilsson LA (1999) Pollen carryover, geitonogamy, and the evolution of deceptive pollination systems in orchids. Ecology 80:2607–2619

Index

J.-M. Mérillon, H. Kodja (eds.), *Orchids Phytochemistry, Biology and Horticulture*, Reference Series in Phytochemistry, https://doi.org/10.1007/978-3-030-38392-3

W

Y

Z